Lecture Notes in Networks and Systems

Volume 590

The series "Lecture Notes in Networks and Systems" publishes the latest developments in Networks and Systems—quickly, informally and with high quality. Original research reported in proceedings and post-proceedings represents the core of LNNS.

Volumes published in LNNS embrace all aspects and subfields of, as well as new challenges in, Networks and Systems.

The series contains proceedings and edited volumes in systems and networks, spanning the areas of Cyber-Physical Systems, Autonomous Systems, Sensor Networks, Control Systems, Energy Systems, Automotive Systems, Biological Systems, Vehicular Networking and Connected Vehicles, Aerospace Systems, Automation, Manufacturing, Smart Grids, Nonlinear Systems, Power Systems, Robotics, Social Systems, Economic Systems and other. Of particular value to both the contributors and the readership are the short publication timeframe and the world-wide distribution and exposure which enable both a wide and rapid dissemination of research output.

The series covers the theory, applications, and perspectives on the state of the art and future developments relevant to systems and networks, decision making, control, complex processes and related areas, as embedded in the fields of interdisciplinary and applied sciences, engineering, computer science, physics, economics, social, and life sciences, as well as the paradigms and methodologies behind them.

Indexed by SCOPUS, INSPEC, WTI Frankfurt eG, zbMATH, SCImago.

All books published in the series are submitted for consideration in Web of Science.

For proposals from Asia please contact Aninda Bose (aninda.bose@springer.com).

More information about this series at https://link.springer.com/bookseries/15179

Danilo Tardioli · Vicente Matellán ·
Guillermo Heredia · Manuel F. Silva ·
Lino Marques
Editors

ROBOT2022: Fifth Iberian Robotics Conference

Advances in Robotics, Volume 2

Springer

Editors
Danilo Tardioli
Centro Universitario de la Defensa (CUD)
Zaragoza, Spain

Vicente Matellán
Grupo de Robótica
Universidad de León
León, Spain

Guillermo Heredia
GRVC Robotics Lab
Escuela Técnica Superior de Ingeniería
Universidad de Sevilla
Sevilla, Spain

Manuel F. Silva
School of Engineering
Polytechnic Institute of Porto
Porto, Portugal

Lino Marques
Institute of Systems and Robotics
University of Coimbra
Coimbra, Portugal

ISSN 2367-3370 ISSN 2367-3389 (electronic)
Lecture Notes in Networks and Systems
ISBN 978-3-031-21061-7 ISBN 978-3-031-21062-4 (eBook)
https://doi.org/10.1007/978-3-031-21062-4

This Springer imprint is published by the registered company Springer Nature Switzerland AG
The registered company address is: Gewerbestrasse 11, 6330 Cham, Switzerland

Preface

This book contains a selection of papers accepted for presentation and discussion at ROBOT2022, the Fifth Iberian Robotics Conference, held in Zaragoza, Spain, during November 23–25, 2022. This edition should have taken place in 2021 but, due to the COVID-19 pandemic, and the lockdown measures imposed by governmental health authorities, was delayed to November 2022. ROBOT2022 is part of a series of conferences that are jointly organized by SEIDROB—Sociedad Española para la Investigación y Desarrollo en Robótica/Spanish Society for Research and Development in Robotics and by SPR – Sociedade Portuguesa de Robótica/Portuguese Society for Robotics. The conference organization had also the collaboration of several universities and research institutes, including University of Zaragoza, Engineering Research Institute of Aragon, Centro Universitario de la Defensa—Zaragoza, University of Seville, University of León, University of Coimbra, Instituto Superior de Engenharía do Porto, and INESC TEC.

ROBOT2022 builds up on several previous successful events, including three biennial workshops (Zaragoza—2007, Barcelona—2009, and Sevilla—2011) and the four previous editions of the Iberian Robotics Conference held in Madrid in 2013, Lisbon 2015, Seville 2017, and Porto 2019. The conference is focused on presenting research results and new developments and applications in the field of Robotics in the Iberian Peninsula, although open to contributions from all over the world. ROBOT2022 featured four plenary talks on state-of-the-art subjects on Robotics by **Estela Bicho**, Full Professor at the Department of Industrial Electronics & Algoritmi Research Centre, School of Engineering (EEUM), at the University of Minho, Portugal, where she coordinates the Control, Automation and Robotics Group (Research line Industrial Electronics) and leads the MAR Lab—Mobile and Anthropomorphic Robotics Lab and currently Vice President of EEUM in charge of interaction with society, **Wolfram Burgard**, Professor at the University of Technology Nuremberg, Head of the research lab for Autonomous Intelligent Systems, and former Head of the research lab for Autonomous Mobile Systems at the University of Bonn, **Margarita Chli**, Professor of Computer Vision for Robotics at the Cyprus University of Technology and Director of the Vision for Robotics Lab at ETH Zurich, and **Anibal Ollero**, Full Professor and Head of the

GRVC Robotics Laboratory, University of Seville, and Scientific Advisor of the Center for Aerospace Technologies (CATEC), Seville.

ROBOT2022 featured a main Robotics track plus fifteen special sessions on the topics: Autonomous Driving and Driver Assistance Systems, Best Practices and Lessons Learnt in Robotics Applied to Inspection and Maintenance, Collaborative Robots Applications, Current Challenges in Research and Development of Multi-Robot Systems, Educational Robotics, Intelligent Robotics, Localization Systems for Autonomous Mobile Robots, Machine Learning in Robotics, Mobile Robot Olfaction, Multi-Robot for Industrial Applications, Rehabilitation and Assistive Robotics, Service Robotics in Challenging Environments, Unmanned Aerial Vehicle Applications, Versatile and Heterogeneous Robots in Agriculture, and Workshop on Physical Agents.

In total, after a careful review process with, at least, three independent reviews for each paper, 98 high-quality papers were selected for publication, with a total number of 291 unique authors, from 22 countries.

We would like to thank all Special Sessions' organizers for their hard work on promoting their special session, inviting the Program Committee, organizing the Special Session review process, and helping to promote the ROBOT2022 Conference. This acknowledgment goes especially to Adrià Colomé Figueras, André Schneider de Oliveira, Angel Sappa, Angela Ribeiro, Anibal Ollero, Antidio Viguria, Antonio González, Armando Sousa, Arturo de la Escalera, Brígida Mónica Faria, Diogo Matos, Fernando Caballero, Francisco Bellas, Francisco Gómez-Donoso, Francisco Rodríguez-Lera, Gonzalo López-Nicolás, Guido Szekir, Guillem Alenyà, Javier Gonzalez-Jimenez, João Braun, João Valente, José Lima, Juan C. Moreno, Juan Jesús Roldán Gómez, Lino Marques, Luis Merino, Luís Piardi, Mar Ariza-Sentís, María Guinaldo, Mario Andrei Garzón Oviedo, Miguel Aranda, Miguel Cazorla, Miguel Oliveira, Milena Pinto, Pascual Campoy, Paulo Costa, Pedro Gomes Costa, Pedro Neto, Rosario Aragüés, Santiago Marco, Sergio Vélez, Vicente Matellán, and Vitor Santos.

We would also like to take the opportunity to thank the rest of the local organization members (Luis Riazuelo, Alejandro R. Mosteo, Luis Montano, José Luis Villarroel, Francisco Lera and Domenico Sicignano) for their hard and valuable work on the local arrangements, publicity, proceedings publication, financial issues, and website management. We also express our gratitude to the members of all the program committees and additional reviewers, as they were crucial for ensuring the high scientific quality of the event and to all the authors and delegates that, with their research work and participation, made this event a huge success. Lastly, a special thank goes to Springer, which edited this conference proceedings edition, and to Dr. Thomas Ditzinger in particular.

November 2022

Danilo Tardioli
Vicente Matellán
Guillermo Heredia
Manuel F. Silva
Lino Marques

Organization

Local Organizing Committee

Alejandro R. Mosteo	Centro Universitario de la Defensa, Zaragoza, Spain
Luis Riazuelo	Universidad de Zaragoza, Spain
Luis Montano	Universidad de Zaragoza, Spain
José Luis Villarroel	Universidad de Zaragoza, Spain
Francisco Lera	Universidad de Zaragoza, Spain
Domenico Sicignano	Centro Universitario de la Defensa, Zaragoza, Spain

Program Committee Manager

Alejandro R. Mosteo	Centro Universitario de la Defensa, Zaragoza, Spain

Publicity Chairs

Eduardo Montijano	Universidad de Zaragoza, Spain
Alejandro R. Mosteo	Centro Universitario de la Defensa, Zaragoza, Spain

Program Committee

Antonio Adan	Universidad de Castilla-La Mancha
Jon Agirre Ibarbia	Tecnalia Research & Innovation
António Pedro Aguiar	University of Porto
Teodoro Aguilera-Benítez	University of Extremadura
Eugenio Aguirre	University of Granada
Daniel Albuquerque	ESTGV
Guillem Alenyà	Institut de Robòtica i Informàtica Industrial CSIC-UPC

Luis Almeida	University of Porto
Angelos Amanatiadis	Democritus University of Thrace
Josep Amat	Universitat Politècnica de Catalunya
Rosario Aragüés	Universidad de Zaragoza
Miguel Arand	Universidad de Zaragoza
Rui Araújo	University of Coimbra
Mar Ariza-Sentís	Wageningen University & Research
Manuel Armada	CSIC
José María Armingol	Universidad Carlos III de Madrid
Guillermo Asín Prieto	Gogoa
Helio Azevedo	University of São Paulo
Jose Azevedo	University of Aveiro
Carlos Balaguer	University Carlos III of Madrid
Stephen Balakirsky	Georgia Institute of Technology
Richard Balogh	Slovak University of Technology in Bratislava
Antonio Bandera	University of Malaga
Juan Pedro Bandera Rubio	ISIS Group, University of Malaga
León Barbed	University of Zaragoza
José Barbosa	Polytechnic Institute of Bragança
Antonio Barrientos	Universidad Politécnica de Madrid
Richard Bearee	Arts et Métiers ParisTech
José Antonio Becerra Permuy	Universidade da Coruña
Marcelo Becker	USP-EESC-SEM
Francisco Bellas	UDC
José M. Bengochea-Guevara	CSIC
Luis M. Bergasa	University of Alcala
Daniel Beßler	University Bremen
Estela Bicho	University of Minho
Jose-Luis Blanco-Claraco	Universidad de Almeria
Ole Blaurock	Fachhochschule Lübeck
Francisco Bonnín	UIB—SRV
Adrian Burlacu	The Gheorghe Asachi Technical University of Iasi
Pablo Bustos	Universidad de Extremadura
Fernando Caballero Benítez	University of Seville
Itziar Cabanes Axpe	University of the Basque Country (UPV/EHU)
João Calado	ISEL - Instituto Superior de Engenharia de Lisboa
Rafael Caldeirinha	Polytechnic Institute of Leiria, Instituto de Telecomunicações
Jesus Capitan	University of Seville
Joel Luis Carbonera	UFRGS
Carlos Carreto	Instituto Politécnico da Gauarda
Alicia Casals	Universitat Politècnica de Catalunya
José Cascalho	Universidade dos Azores

Jose A. Castellanos	University of Zaragoza
Roberto Cavicchioli	Università di Modena e Reggio Emilia
Miguel Cazorla	University of Alicante
José-María Cañas-Plaza	Rey Juan Carlos University
Marco Ceccarelli	University of Rome tor Vergata
Adrià Colomé Figueras	Universitat Politècnica de Catalunya
Andre Conceicao	Federal University of Bahia
Dorin Sabin Copaci	Universidad Carlos III de Madrid
Daniel Cores Costa	Centro Singular de Investigación en Tecnoloxías da Información (CiTIUS), Universidade de Santiago de Compostela
Juan Antonio Corrales Ramon	University of Santiago de Compostela
Luís Correia	Universidade de Lisboa
Paulo Costa	University of Porto
Pedro Costa	University of Porto
Hugo Costelha	Instituto Politécnico de Leiria
Micael Couceiro	Ingeniarius, Ltd.
Vincent Creuze	LIRMM, University of Montpellier—CNRS
Nuno Cruz	INESC TEC/FEUP-DEEC
Arturo De La Escalera	Universidad Carlos III de Madrid
Félix de La Paz	Universidad Nacional de Educación a Distancia (UNED)
Fernando De La Prieta	University of Salamanca
Jorge Dias	University of Coimbra
Paulo Dias	DETI/IEETA—Universidade de Aveiro
João Fabro	UTFPR—Federal University of Technology-Parana
Brígida Mónica Faria	Polytechnic of Porto
Paulo Farias	UFBA
Vicente Feliu	Universidad de Catilla-La Mancha
Fernando Fernandez	Universidad Carlos III de Madrid
Roemi Fernandez	Centre for Automation and Robotics CSIC-UPM
Jesus Fernandez-Lozano	Universidad de Málaga
Camino Fernández	Universidad de León
Antonio Fernández-Caballero	Universidad de Castilla-La Mancha
Manuel Ferre	CAR UPM-CSIC
Bruno Ferreira	University of Porto
Hugo Ferreira	LSA-ISEP
Paulo Ferreira	ISEP
Maria Flores Tenza	Miguel Hernandez University
Pedro Fonseca	Instituto de Telecomunicações, Universidade de Aveiro
Juan M. Gandarias	Istituto Italiano di Tecnologia

Gabriel J. Garcia	University of Alicante
Nicolas Manuel Garcia Aracil	Universidad Miguel Hernández
Emilio Garcia-Fidalgo	University of the Balearic Islands
Jose Garcia-Rodriguez	University of Alicante
Miguel Garcia-Silvente	University of Granada
Ismael Garcia-Varea	University of Castilla-La Mancha
Abraham García	Universidade da Coruña
Luis Javier García Villalba	Universidad Complutense de Madrid
Antonios Gasteratos	Democritus University of Thrace
Fernando Gomez-Bravo	Huelva University
Francisco Gomez-Donoso	University of Alicante
Paulo Goncalves	Polytechnic Institute of Castelo Branco
Javier Gonzalez-Jimenez	University of Malaga
Antonio González	Universitat Politècnica de València
José Gonçalves	ESTiG-IPB
Pedro Guedes	Instituto Superior de Engenharia do Porto
Josechu Guerrero	Universidad de Zaragoza
María Guinaldo	UNED
Carlos Guindel	Universidad Carlos III de Madrid
Francisco Luis Gutiérrez Vela	University of Granada
Augusto Gómez Eguílluz	Ulster University
Tamás Haidegger	ABC Center for Intelligent Robotics, Óbuda University
Florian Haug	Åland University of Applied Sciences
Guillermo Heredia	University of Seville
Daniel Hernandez	ULPGC
Pedro Javier Herrera	UNED
Alberto Herreros	University of Valladolid
Roberto Iglesias Rodriguez	University of Santiago de Compostela
Hiroshi Ishida	Tokyo University of Agriculture and Technology
Eduardo Iáñez	Brain-Machine Interface Systems Lab, Miguel Hernández University of Elche
Antonio J. del Ama	Rey Juan Carlos University
Alberto Jardón Huete	University Carlos III of Madrid
Nuno Lau	University of Aveiro
Agapito Ledezma	Universidad Carlos III de Madrid
Paulo Leitao	Polytechnic Institute of Braganca
Francisco Lera	University of Zaragoza
Frederic Lerasle	LAAS-CNRS
Achim Lilienthal	Orebro University, AASS
José Lima	Polytechnic Institute of Braganca
Pedro U. Lima	Institute for Systems and Robotics, Instituto Superior Técnico, U. Lisboa
Alejandro Linares-Barranco	University of Seville
Cunjia Liu	Loughborough University

Ángel Llamazares	Universidad de Alcalá
Victor Lobo	CINAV—Escola Naval
Ana Lopes	University of Coimbra
Joaquin Lopez	University of Vigo
Jose Manuel Lopez-Guede	Basque Country University
Gonzalo Lopez-Nicolas	Universidad de Zaragoza
Maria Teresa Lorente	Instituto Tecnológico de Aragón (ITA)
Jesus Lozano	Universidad de Extremadura
María T. Lázaro	ITAINNOVA
Joao Macedo	University of Coimbra
Pedro Machado	Nottingham Trent University
Antonio Madrid	Universitat de Barcelona
Erik Maehle	University of Luebeck
Anthony Mandow	University of Malaga
André Luís Marcato	Federal University of Juiz de Fora
Santiago Marco	Institut for BioEngineering of Catalonia—Universitat de Barcelona
Lino Marques	University of Coimbra
David Martin Gomez	Carlos III University of Madrid
Dominique Martinez	CNRS
Humberto Martinez Barbera	Universidad de Murcia
Ramiro Martinez de Dios	University of Seville
Jorge Martins	IST, TULisbon
Francisco Martín	Universidad de Málaga
Francisco Martín Rico	Universidad Rey Juan Carlos
Pablo Marín Plaza	Universidad Carlos III de Madrid
Vicente Matellán	University of León
Qing-Hao Meng	Tianjin University
Luis Merino	Pablo de Olavide University
Jose M. Molina	Universidad Carlos III de Madrid
Javier Monroy	University of Malaga
Luis Montano	University of Zaragoza
Hector Montes Franceschi	Universidad Tecnologica de Panama
José María Martínez Montiel	Universidad de Zaragoza
Eduardo Montijano	Universidad de Zaragoza
Raul Morais	UTAD/ECT
Antonio Morales	Universitat Jaume I
Jesus Morales	Universidad de Malaga
Juan C. Moreno	CSIC
Luis Moreno	Universidad Carlos III de Madrid
Paulo Moura Oliveira	UTAD University
Manuel Mucientes	Universidade de Santiago de Compostela
Ana C. Murillo	University of Zaragoza
Vicente Mut	INAUT—Universidad Nacional de San Juan
Victor Muñoz Martinez	University of Malaga

Rafael Muñoz-Salinas	University of Cordoba
Alberto Nakano	Federal University of Technology—Paraná
Martin Naya	Integrated Group for Engineering Research
Pedro Neto	University of Coimbra
António J. R. Neves	University of Aveiro
Urbano Nunes	University of Coimbra
Pedro Núñez	University of Extremadura
Manuel Ocaña	University of Alcala
Andre Oliveira	UTFPR
Paulo Oliveira	IST/ISR
Gabriel Oliver	UIB
Joanna Isabelle Olszewska	University of West Scotland
Jorge Ortín	Centro Universitario de la Defensa, Zaragoza
Alberto Ortiz	Universitat de les Illes Balears
Jun Ota	The University of Tokyo
Javier P. Turiel	University of Valladolid
Patricia Paderewski	University of Granada
Jordi Palacín	Universitat de Lleida
Dimitrios Paraforos	University of Hohenheim
Nieves Pavon-Pulido	Universidad Politécnica de Cartagena
Luis Paya	Universidad Miguel Hernandez
Alejandro Paz-Lopez	Universidade da Coruña
Eurico Pedrosa	University of Aveiro
Artur Pereira	University of Aveiro
Marcelo Petry	INESC TEC, UFSC, INESC P&D Brasil
Tatiana M. Pinho	INESC TEC
Andry Pinto	University of Porto
Vítor Hugo Pinto	University of Porto
Jorge Pomares	University of Alicante
Mihai Pomarlan	Institute for Artificial Intelligence, University of Bremen
Cristiano Premebida	University of Coimbra
Pablo Pueyo	Universidad de Zaragoza
Noé Pérez-Higueras	Pablo de Olavide University
Joao Quintas	Instituto Pedro Nunes
Bogdan Raducanu	Computer Vision Center
Saeed Rafee Nekoo	Departamento de Ingeniería de Sistemas y Automática, Escuela Técnica Superior de Ingeniería, Universidad de Sevilla
Francisco Ramos	University of Castilla-La Mancha
Iñaki Rañó	The Maersk Mc Kinney Moller Institute
Carlos V. Regueiro	Universidade da Coruña
Giulio Reina	Politecnico di Bari
Oscar Reinoso	Universidad Miguel Hernandez
Luis Paulo Reis	University of Porto

Anthony Remazeilles	Tecnalia Research and Innovation
Ely Repiso	LAAS-CNRS
Luis Riazuelo	Universidad de Zaragoza
Lluís Ribas-Xirgo	Universitat Autònoma de Barcelona
A. Fernando Ribeiro	University of Minho
Angela Ribeiro	CAR-CSIC
Isabel Ribeiro	Instituto Superior Técnico/ISR
Miguel Armando Riem De Oliveira	University of Aveiro
Branko Ristic	RMIT University
Carlos Rizzo	Eurecat Technology Center
Luis Rocha	INESC TEC
Rui P. Rocha	University of Coimbra
Nelson Rodrigues	University of Porto
Angel Rodriguez Castaño	University of Seville
Francisco J. Rodríguez-Lera	Universidad de León
Francisco Rodríguez-Sedano	Universidad de León
Ronnier Rohrich	Universidade Tecnológica Federal do Paraná - UTFPR
Juan Jesús Roldán Gómez	Universidad Autónoma de Madrid
Roseli Romero	University of São Paulo
Juan Adrian Romero Garces	University of Málaga
Jan Rosell	Universitat Politècnica de Catalunya
Rosaldo Rossetti	University of Porto
Claudio Rossi	Universidad Politécnica de Madrid
José Rufino	Polytechnic Institute of Braganca
Diana Teresa Ruiz Bueno	Instituto Aragonés de Ciencias de la Salud
Mohammad Safeea	Coimbra University
Carlos Sagues	Universidad de Zaragoza
Jorge Sales	Universitat Jaume I
Miguel A. Salido	Universitat Politècnica de València
José Luis Sanchez Lopez	SnT - University of Luxembourg
Pedro Sanchez-Cuevas	Advanced Center for Aerospace Technologies
Alberto Sanfeliu	Universitat Politècnica de Catalunya
Cristina Manuela Peixoto Santos	Industrial Electronics Department, School of Engineering, University of Minho
Filipe Santos	University of Porto
Vitor Santos	University of Aveiro
Ricardo Sanz	Universidad Politécnica de Madrid
Rafael Sanz Dominguez	Universidad de Vigo
Angel Sappa	Computer Vision Center
Alexander Schlaefer	Hamburg University of Technology
Michael Schukat	NUI Galway
Luís Seabra Lopes	Universidade of Aveiro
Joao Sequeira	Universidade de Lisboa

Domenico Sicignano	Centro Universitario de la Defensa, Zaragoza
Bruno Siciliano	Univ. Napoli Federico II
João Silva	Altran Portugal
Manuel F. Silva	ISEP-IPP and INESC TEC CRIIS
Marco A. C. Simões	Universidade do Estado da Bahia (UNEB)
Danijel Skocaj	University of Ljubljana
Filomena Soares	University of Minho
Armando Jorge Sousa	INESC TEC + FEUP
Anne Spalanzani	Inria Rhône-Alpes/Laboratoire d'informatique de Grenoble
Alejandro Suarez	University of Seville
Juan D. Tardos	Universidad Zaragoza
Ines Tejado	University of Extremadura, Industrial Engineering School
Elisa Tosello	University of Padua
Andrés Úbeda	Human Robotics Group, University of Alicante
Pablo Urcola	Bitbrain
Jesus Urena	University of Alcala
Alberto Vale	Universidade de Lisboa
Antonio Valente	UTAD University
João Valente	Wageningen University & Research
Germano Veiga	INESC TEC
Sergio Velez Martín	Wageningen University & Research
Rodrigo Ventura	Universidade de Lisboa
Antidio Viguria	Center for Advanced Aerospace Technologies
Blas M. Vinagre	University of Extremadura
Carlos Vázquez	Universidade da Coruña
Matias Waller	Åland University of Applied Sciences
Eduardo Zalama	Universidad de Valladolid

Additional Reviewers

Armesto, Leopoldo
Baptista, Rui
Barrientos, Antonio
Barros, Tiago
Belmonte-Baeza, Alvaro
Cimarelli, Claudio
Colombo, Leonardo
Cruz, Christyan
Cuiral-Zueco, Ignacio
Dogru, Sedat
Escalona, Félix
Espinosa, Henry Paúl
Garrote, Luis
Gil, Pablo
Hani Daniel Zakaria, Mélodie
Herguedas, Rafael
Jiménez-Naharro, Raúl
Lilienthal, Achim
López de Ahumada Gutiérrez, Rafael
Magalhães, Hugo
Millan Romera, Jose Andres
Ojeda, Pepe
Oliveira, Daniel
Palacin, Jordi

Paya, Luis
Satish Srivastav, Chinmay
Shirafuji, Shouhei
Sánchez, José
Takamido, Ryota
Teruel, Enrique
Tourani, Ali
Zalama, Eduardo

Contents

Mobile Robot Olfaction

Multi-robot for Industrial Applications

Intelligent Robotics

Deep Learning for Victims Detection from Virtual and Real Search and Rescue Environments

Christyan Cruz Ulloa(✉), Miguel Garcia, Jaime del Cerro, and Antonio Barrientos

Centro de Automática y Robótica, Universidad Politécnica de Madrid - Consejo Superior de Investigaciones Científicas, 28006 Madrid, Spain
christyan.cruz.ulloa@upm.es

Abstract. Robotic interventions in post-disaster environments to carry out search and rescue explorations allow optimizing time for identifying victims and safeguarding the rescuer's integrity. The rising in the neural networks field and their application in image detection algorithms have made it possible to facilitate the early detection of victims in these first phases of exploration. This article analyses the effectiveness of applying neural network models obtained from training with different datasets of both: images captured in real environments and synthetic images from recreated virtual post-disaster environments for detecting victims. For this development, tests have been carried out in environments at the ETSII-Universidad Politécnica de Madrid, generating models from images obtained with the ARTU-R robot (A1 Rescue Task UPM Robot) and have been validated with real search and rescue exercises at the University of Malaga. The main results show that the models obtained from virtual environments apply to real ones.

Keywords: Virtual reality · Deep learning · Convolutional neural networks · ROS · Search and rescue · Unity · Quadruped robot

1 Introduction

According to historical records, post-disaster environments resulting from natural events (earthquakes, eruptions, hurricanes, floods, etc.) or attacks have resulted in the loss of thousands of lives [1]. Historically, robotic interventions have been registered that have supported emergency brigades, such as those

This work has received funding from the RoboCity2030-DIH-CM Madrid Robotics Digital Innovation Hub "Robótica aplicada a la mejora de la calidad de vida de los ciudadanos, fase IV"; S2018/NMT-4331), funded by "Programas de Actividades I+D en la Comunidad de Madrid" and cofunded by Structural Funds of the EU. and TASAR (Team of Advanced Search And Rescue Robots), funded by "Proyectos de I+D+i del Ministerio de Ciencia, Innovacion y Universidades. Financed by MCIN/AEI/10.13039/501100011033.

D. Tardioli et al. (Eds.): ROBOT 2022, LNNS 590, pp. 3–13, 2023.
https://doi.org/10.1007/978-3-031-21062-4_1

registered in the United States (Twin Towers - 2001) [2,3], Japan (Fukushima - 2011) [4], Italy (Amatrice -2016) [5], Mexico (2017) [6].

An information survey has been carried out corresponding to the different post-disaster events where robotic interventions exist, as shown in Fig. 1(a).

Figure 1(b) shows, on the other hand, the potential countries susceptible to having a post-disaster event through natural causes (volcanic eruptions, earthquakes, hurricanes or tsunamis), especially those between the pacific ring of fire and those that historically have periodically presented these events with Haiti.

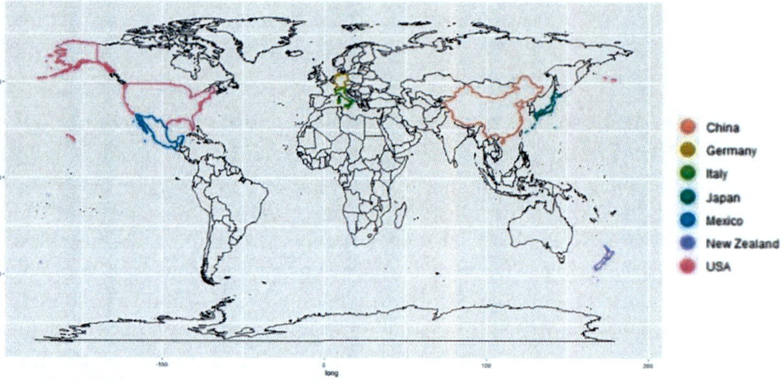

(a) Countries where robotic interventions have been carried out after a catastrophic event. Source: Authors.

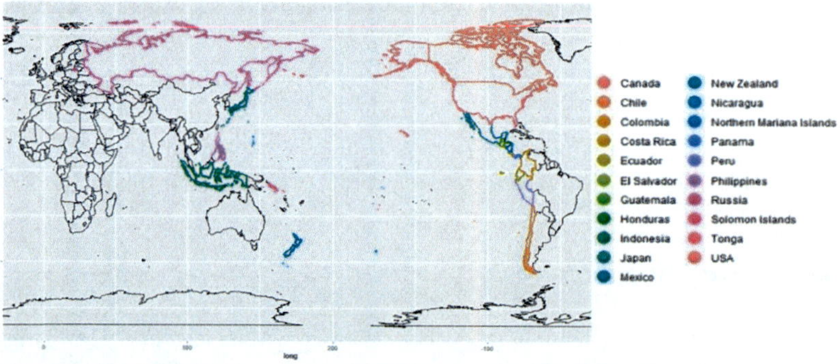

(b) Countries with a potential risk of suffering a catastrophic event, mainly those of the Pacific Ring of Fire. Source: Authors.

Fig. 1. Maps of countries that have had or are susceptible to catastrophic events.

New robotics and artificial intelligence developments have become support tools for emergency brigades in search and rescue missions. Its main advantages in this field are the optimization of time in recognizing dangerous areas through exploration with robotic platforms and the early recognition of victims through artificial intelligence techniques.

On the other hand, the constant growth of the applications developed in Virtual Reality environments has allowed preliminary execution of search strategies, training operators to tele-operate these platforms and preliminary testing of algorithms. [7–9].

YOLOv5 (You Only Look Once) has been used in this development. The authors have extensively validated this CNN model against other types of networks used for the same purpose, such as the Faster R-CNN (Region Based Convolutional Neural Networks) and the SSD (Single-Shot Detector), with the work developed in [10–14].

In state-of-the-art, previous works focused on identifying small objects in closed-circuit television (CCTV) has been found, for which they generate a dataset from images captured from CCTV and environments reconstructed in simulation through the Faster R-CNN network [15]. Other developments focus on generating synthetic data in Unity for later use in computer vision applications [16].

Other works train network models trained from synthetic images captured from virtual environments, which can recognise objects in real environments, for which tests are carried out with geometric figures and fruits, while other works do it with medical objects [17,18].

In the field of victim detection with CNN, several works have currently been carried out using images and neural networks such as [19,20], and the one described by the authors in [10].

This work's main contribution focus on generate different models (tree) of convolutional neural networks (CNN), trained from three types of datasets: real, virtual and mixed (real + virtual) to identify victims automatically. Implementing a method that allows obtaining CNN models to identify victims in places of any part of the world that have satellite information on relief and texture, which would greatly support rescue teams for primary phases of inspection.

The real datasets have been obtained in the ETSII-UPM (Escuela Técnica Superior de Ingenieros Industriales - Universidad Politécnica de Madrid). On the other hand, synthetic data has been obtained from reconstructed post-disaster scenarios.

The selected environments are the virtual reconstructions of ETSII-UPM (Madrid) and EII-UMA (Escuela de Ingeniería Industrial - Universidad de Málaga), together with areas where one day some catastrophic event occurred, such as the Three Mile Island plant [21], and other places where an event of this magnitude could exist, such as one of the countries that make up the pacific ring of fire [22].

The evaluation performed has been carried out crosswise, analyzing the detection of victims in each of the three neural network models generated, with different images obtained (real, virtual). An efficiency greater than 80.

This paper is structured as follows. Section 2 describes the materials and methods in detail-Sect. 3, containing the experiments and results. Finally, Sect. 4 presents the conclusions and main findings.

2 Materials and Methods

2.1 Materials

The experimental development has been carried out both indoors and outdoors, respectively, in the ETSII of the Polytechnic University of Madrid (Fig. 2(a)), taking as reference the scene inside the building and in the EII of Malaga University, where the scenario has been a tunnel-type area (Fig. 2(d)).

The respective reconstructions of both scenarios, Fig. 2(b)-(e) have been developed in Unity.

The robotics platform for environments exploration was the ARTU-R robot (A1 Rescue Task UPM Robot), a quadrupedal robot equipped with an RGB-D (Deph Camera Sensor) and laser sensory system for data acquisition of the environment (Fig. 2(c)-(f)).

The software versions used respectively for this work have been: for the ARTU-R robot, ROS Melodic (Ubuntu 18.04), while the Unity version has been 2018.3.6f.

2.2 Reconstructed Environments

The environments have been reconstructed in two ways. The first one is shown in Fig. 2(b), where planimetry has been used to recreate (using Unity geometry elements) in details the ETSII-UPM.

The second method for reconstructing environments captures the spatial geometry of areas through satellite images and reliefs. For this purpose, Render Doc and Blender Software have been used together with Maps. In this way, the EII-UMA test environment has been incorporated into Unity (Fig. 2(e)). The environments shown in Fig. 3 have been integrated into Unity in order to increase the dataset and incorporate areas where there has been a disaster or places where it could be carried out.

Figure 3(a)-(b) correspond to the reconstruction of the Three Mile Island nuclear plant, where a reactor failure occurred in 1979 [21]. Figure 3(d)-(e) are the reconstruction of the surroundings of the Obelisk in Argentina, a country with the potential risk of an earthquake because it is located within the Pacific Ring of Fire. Finally, Fig. 3(g)-(h), the surroundings of the"Basilica del Pilar" (Zaragoza - Spain), have been reconstructed to increase the dataset and validate the reconstruction of environments applicable to any site on the planet.

Figure 3(c)-(f)-(i) are the corresponding real scenes of each reconstruction in Unity. Within these environments, the ARTU-R robot model has been included to execute movements (in a tele-operated way) and capture images for the dataset.

(a) ETSII-UPM Frontal View. (b) ETSII-UPM Virtual recreation Frontal View. (c) Interior of the ETSII-UPM together with the robot ARTU-R.

(d) EII-UMA test environment. Tunnel post-disaster scenario. (e) Virtual test environment - global view tunnel. (f) ARTU-R robot at one end of the EII-UMA tunnel.

Fig. 2. Real and simulated environments where the tests have been carried out. Source: Authors.

2.3 Convolutional Neural Network

The convolutional neural network model implemented in this development has been YOLOv5, tested by the authors in a previous development, showing good results over other types of networks for classification and detection [10]. The hyper-parameters used in this work have been: Learning Rate Schedule: (burn_in = 1000, steps = 400,000, scales = 0.1, size = 1280 px), Initial Learning rate: 0.001, Batch size: 16, and Training Epochs: 80.

The datasets to train the three models have been divided into training (70%) and validation (30%) of each total of images. The datasets are respectively: Model 1. Images (1000) of the real environment (ETSII-UPM). Model 2. Images (1200) of virtual environments (ETSII-UPM, EII-UMA, Three Mile Island, Buenos Aires, Zaragoza). Model 3. Combination of datasets 1 and 2 (2200 images).

(a) Three mile island nuclear accident recreation in Unity. (b) Three mile island -nuclear accident recreation in Unity. (c) Three mile island -nuclear accident - real plant.

(d) Obelisk of Buenos Aires, fire scene recreation in Unity. (e) Obelisk of Buenos Aires, fire scene recreation in Unity. (f) Buenos Aires's Obelisk.

(g) Basilica of Zaragoza, recreation in Unity. (h) Basilica of Zaragoza, recreation in Unity. (i) Basilica of Zaragoza.

Fig. 3. Environments reconstructed in Unity and original versions.

Data augmentation has been applied to these datasets (mainly modification of brightness, contrast and slight rotations of less than 30 degrees) to generate greater robustness in the models. The labels used were: victim, victim-head, victim-torso, victim-arm, victim-leg, rescuer and rescuer-legs.

The performance results of the trained models will be evaluated using confusion matrices. The main diagonal of the matrices shows the percentage of correct predictions $[0 - 100\%]$ (True Positive - TP). Off-diagonal values are analyzed by: False Positives - FP (Rows) predictions assigned to a class that does not correspond and False Negatives - FN (Columns) classes assigned as negative but belonged to the class.

$$\text{Precision} = \frac{TP}{TP + FP} \tag{1a}$$

$$\text{Recall} = \frac{TP}{TP + FN} \tag{1b}$$

From these metrics, the precision and recall values are obtained, according to the Eq. 1, to evaluate the performance of the best model. In the specific case of work applied to search and rescue, it is convenient to have false negatives; that is, it is better to assume that there is a victim in an area than to go unnoticed.

3 Experiments and Results

3.1 Trained CNN Metrics

Figure 4(a)-(b)-(c) show the confusion matrices of the three trained models. The differences in the confusion matrices of the three trained models are minimal, demonstrating the effectiveness of virtual images for model training, presenting results similar to real images. In the next section, the cross-evaluation of models and data is carried out.

The most difficult classes to detect are *Victim Legs* and *Rescuer*. Instead, the class that best detects is *Victim Head* and *Victim* with a reasonable average of 80%. Figure 4(d) shows the Precision-Recall curve, where precision degrades for mean values of 0.8. In this type of Figure, the curves must tend to be as close as possible to the upper right corner, where there are better Precision-Recall values; the classes with the best model performance are victim head and victim.

3.2 Cross Evaluation of Models for Victim Detection

This test evaluates the effectiveness of the trained models compared to data captured in real scenarios that have not been considered for training. This allows to validate the hypothesis of recognizing victims from network training with synthetic data. In this first phase, the inference has been executed offline on a MSI Nvidia 1660Ti GTX, i7 10th laptop, with an average inference time of 200 ms.

Figure 5 shows the detections (bounding boxes, classes and percentage of success in the detection) of the victims with the different network models. Figures 5(a)-(b)-(c) respectively show the result of the inference of the three trained models: Model 1 (Real Dataset), Model 2 (Virtual Dataset), Model 3 (Mixed Dataset). This figure has been captured inside the test tunnel, by the robot's onboard RGB sensor, for post-processing. The image contains a victim wearing white clothes and a helmet, which has been made up to give realism to the exercise.

From the three models, Fig. 5(c) (Model 3) shows the most significant number of classes detected by the network. This is because the dataset (D-Real) used

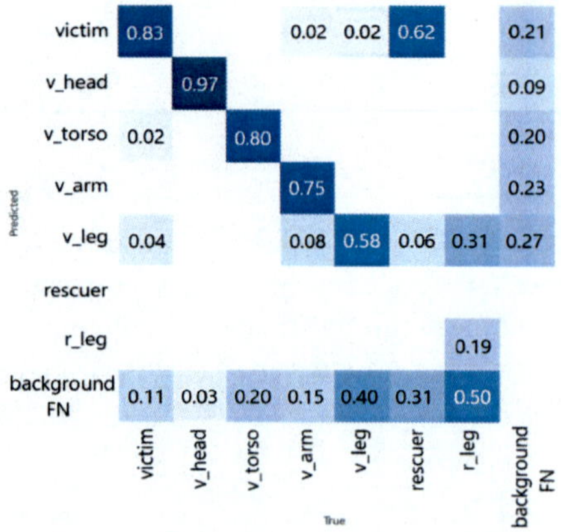

(a) Model 1: Confusion matrix of the model trained with images of real environments.

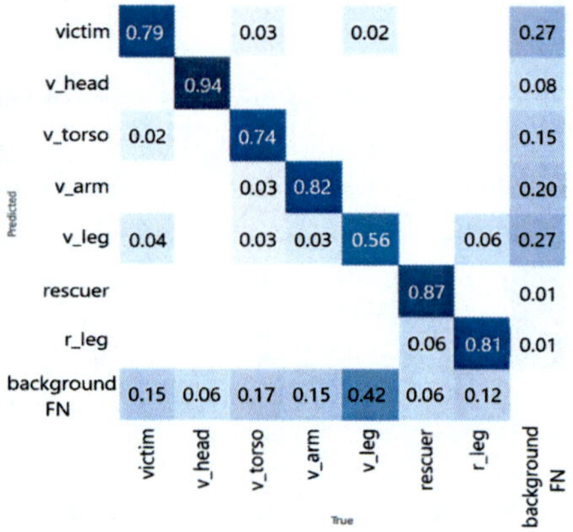

(b) Model 2: Confusion matrix of the trained model with images of simulated environment.

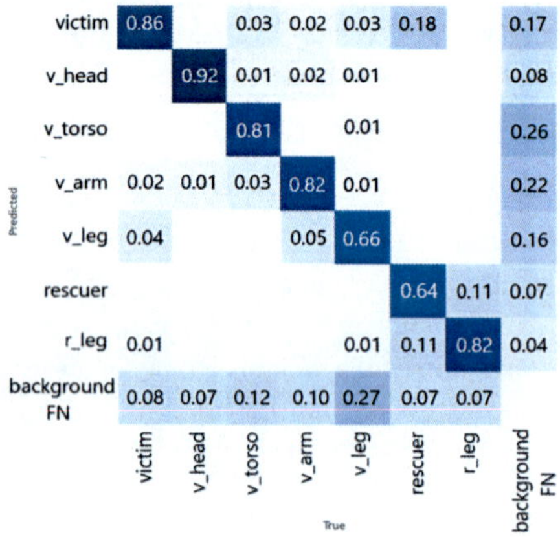

(c) Model 3: Confusion matrix of the trained model with combined images.

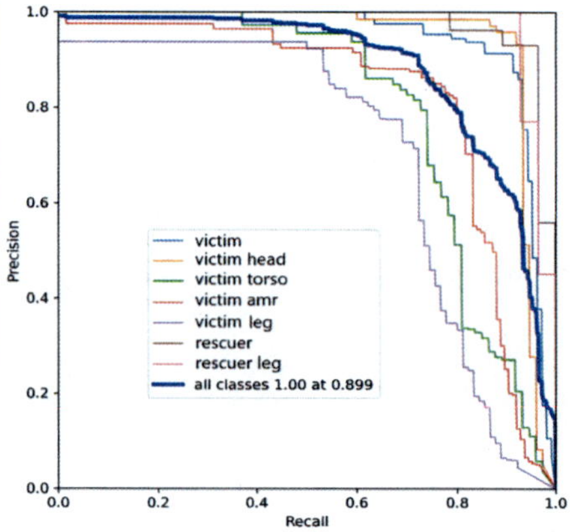

(d) Precision-Recall curve of model 2.

Fig. 4. Performance metrics of trained models.

is broader, and the generated model is more robust than the others. Model 1 and Model 2 have many similarities in the detected classes. More relevantly, the victim class is better detected with Model 2 (68%).

This scene shows another perspective of the tunnel and another victim in the scene. Figure 5(d)-(e)-(f) evaluate the three models in the same order as in the previous Fig. 5(a)-(b)-(c). The results obtained disagree with the previous case. Model 2 has had greater efficiency both in the classes detected and in the high percentage of detection; on the other hand, model 3 presents many false negatives in detecting classes, and Model 1 is pretty limited in detecting classes.

(a) Evaluation of the model trained with D-Real, post-disaster scenario / tunnel.
(b) Evaluation of the model trained with D-Virtual, post-disaster scenario / tunnel.
(c) Evaluation of the model trained with D-Real-Virtual, post-disaster scenario / tunnel.

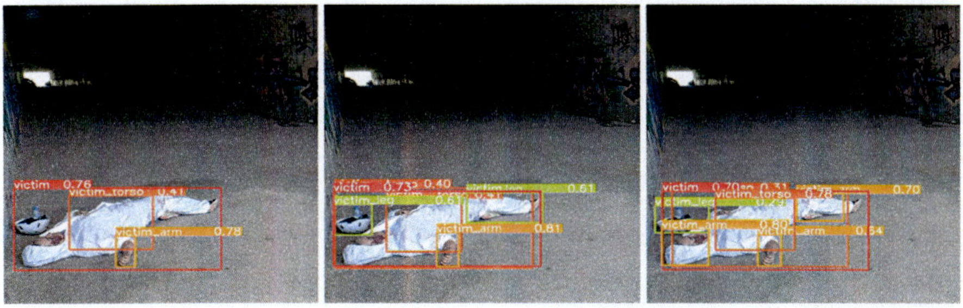

(d) Evaluation of the model trained with D-Real, post-disaster scenario / tunnel.
(e) Evaluation of the model trained with D-Virtual, post-disaster scenario / tunnel.
(f) Evaluation of the model trained with D-Real-Virtual, post-disaster scenario / tunnel.

(g) Evaluation of the model trained with D-Real, scenario / indoors ETSII.
(h) Evaluation of the model trained with D-Virtual, scenario / indoors ETSII.
(i) Evaluation of the model trained with D-Real-Virtual, scenario / indoors ETSII.

Fig. 5. Cross-evaluation of the trained models with real photographs captured from a post-disaster environment in the EII-UMA and the ETSII-UPM.

Figure 5(g)-(h)-(i) have the environment inside the ETSII, showing the evaluation of the three models and the detections obtained for both victims

(Fig. 5(g) and (i)) and rescuers (Fig. 5(h)). For Model 2, the result has been that the detection of a rescuer has been exceptional, even identifying her legs.

Although the three models have detected the corresponding classes reasonably well in these cases, it should be noted that the proposed hypothesis has been validated.

4 Conclusions

This work has presented a method to automatically identify victims through CNN as a basis for training real, synthetic and combined images obtained from real and reconstructed post-disaster environments. Among the main conclusions drawn from this work, it can be established that:

The models generated from virtual environments can be extrapolated for use with images of real environments, generating results in detecting victims in post-disaster environments. This opens up alternatives to generating CNN models in places or countries that cannot be accessed but can be applied in a hypothetical local post-disaster scenario.

The trained models can detect all classes with high efficiency. In particular, the best detected are a victim and a head victim. The three models have presented equivalent results, showing that it is possible to train a model applicable to real cases from synthetic images.

The CNN model generated through the synthetic image dataset has shown satisfactory results for inferring victims in real images, especially in the models trained with synthetic data to potentially real cases, such as the EII-UMA environment.

Acknowledgment. Special acknowledgment to the organization of the "JORNADAS INTERNACIO- NALES DE LA UNIVERSIDAD DE MÁLAGA SOBRE SEGURIDAD, EMERGEN- CIAS Y CATÁSTROFES 2022".

References

1. Wannous, C., Velasquez, G.: United nations office for disaster risk reduction (unisdr)—unisdr's contribution to science and technology for disaster risk reduction and the role of the international consortium on landslides (icl). In: Sassa, K., Mikoš, M., Yin, Y. (eds.) WLF 2017, pp. 109–115. Springer, Cham (2017). https://doi.org/10.1007/978-3-319-59469-9_6
2. Blackburn, M.R., Everett, H.R., Laird, R.T.: After action report to the jointprogram office: center for the robotic assisted search and rescue (crasar) related efforts at the world trade center. Technical report, Space and Naval Warfare Systems Center San Diego CA, August 2002
3. Murphy, R.R.: Disaster robotics. MIT press, 2014
4. Eguchi, R., Elwood, K., Lee, E.K., Greene, M.: The 2010 canterbury and 2011 christchurch new zealand earthquakes and the 2011 tohoku japan earthquake. Technical report, Earthquake Engineering Research Institute (2012)

5. Kruijff, I., et al.: Deployment of ground and aerial robots in earthquake-struck amatrice in italy (brief report). In: 2016 IEEE International Symposium on Safety, Security, and Rescue Robotics (SSRR), pp. 278–279 (2016)
6. Whitman, J., Zevallos, N., Travers, M., Choset, H.: Snake robot urban search after the 2017 mexico city earthquake. In: 2018 IEEE International Symposium on Safety, Security, and Rescue Robotics (SSRR), pp. 1–6 (2018)
7. Roldán, J.J., Peña-Tapia, E., Martín-Barrio, A., Olivares-Méndez, M.A., Del Cerro, J., Barrientos, A.: Multi-robot interfaces and operator situational awareness: study of the impact of immersion and prediction. Sensors **17**(8), 1720 (2017)
8. Craighead, J., Gutierrez, R., Burke, J., Murphy, R.: Validating the search and rescue game environment as a robot simulator by performing a simulated anomaly detection task. In: 2008 IEEE/RSJ International Conference on Intelligent Robots and Systems, pp. 2289–2295 (2008)
9. Rosen, E., Whitney, D., Phillips, E., Ullman, D., Tellex, S.: Testing robot teleoperation using a virtual reality interface with ROS reality. In: Proceedings of the 1st International Workshop on Virtual, Augmented, and Mixed Reality for HRI (VAM-HRI), pp. 1–4 (2018)
10. Ulloa, C.C., Sánchez, G.P., Barrientos, A., Del Cerro, J.: Autonomous thermal vision robotic system for victims recognition in search and rescue missions. Sensors **21**(21), 7346 (2021)
11. Li, M., Zhang, Z., Lei, L., Wang, X., Guo, X.: Agricultural greenhouses detection in high-resolution satellite images based on convolutional neural networks: Comparison of faster R-CNN, yolo v3 and SSD. Sensors **20**(17), 4938 (2020)
12. Lee, Y.-H., Kim, Y.: Comparison of CNN and yolo for object detection. J. Semicond. Display Technol. **19**(1), 85–92 (2020)
13. Cheng, R.: A survey: comparison between convolutional neural network and yolo in image identification. J. Phys. Conf. Ser. **1453**, 012139 (2020). IOP Publishing
14. Xu, W., et al.: Real-time multi-class disturbance detection for ϕ-otdr based on yolo algorithm. Sensors **22**(5), 1994 (2022)
15. González, J.L.S., Zaccaro, C., Álvarez-García, J.A., Morillo, L.M.S., Caparrini, F.S.: Real-time gun detection in CCTV: an open problem. Neural Netw. **132**, 297–308 (2020)
16. Borkman, S., et al.: Unity perception: generate synthetic data for computer vision. CoRR, abs/2107.04259 (2021)
17. Tobin, J., Fong, R., Ray, A., Schneider, J., Zaremba, W., Abbeel, P.: Domain randomization for transferring deep neural networks from simulation to the real world. In: 2017 IEEE/RSJ International Conference on Intelligent Robots and Systems (IROS), pp. 23–30 (2017)
18. Neurolabs: Using neurolabs' retail-specific synthetic dataset in production. https://neurolabs.medium.com Accessed Jun 2022
19. Sulistijono, I.A., Risnumawan, A.: From concrete to abstract: multilayer neural networks for disaster victims detection. In: 2016 International Electronics Symposium (IES), pp. 93–98 (2016)
20. Hartawan, D.R., Purboyo, T.W., Setianingsih, C.: Disaster victims detection system using convolutional neural network (CNN) method. In: 2019 IEEE International Conference on Industry 4.0, Artificial Intelligence, and Communications Technology (IAICT), pp. 105–111 (2019)
21. Miller, K.L.: The nuclear reactor accident at three mile island. Radiographics **14**(1), 215–224 (1994)
22. Paton, D.: Living on the ring of fire: Perspectives on managing natural hazard risk in pacific rim countries. J. Pac. Rim Psychol. **3**(1), 1–3 (2009)

Full-stack S-DOVS: Autonomous Navigation in Complete Real-World Dynamic Scenarios

Diego Martínez(✉), Luis Riazuelo, and Luis Montano

Engineering Research Institute of Aragon (I3A), University of Zaragoza, Zaragoza, Spain
{diegomartinez,riazuelo,montano}@unizar.es

Abstract. Autonomous navigation in dynamic environments is a nowadays unsolved challenge. Several approaches have been proposed to solve it, but they either have a low success rate, do not consider robot kinodynamic constraints or are not able to navigate through big scenarios where the known map information is needed. In this work, a previously existing planner, the Strategy-based Dynamic Object Velocity Space, S-DOVS, is modified and adapted to be included in a full navigation stack, with a localization system, an obstacle tracker and a global planner. The result is a system that is able to navigate successfully in real-world scenarios, where it may face complex challenges as dynamic obstacles or replanning. The final work is exhaustively tested in simulation and in a ground robot.

Keywords: Autonomous navigation · Dynamic environment · Obstacle avoidance

1 Introduction

Robotic autonomous navigation is a problem that has been addressed in many works. Traditional planners have proven to be able to navigate successfully and efficiently in static scenarios. They use the information received by the robot sensors to compute a path that it must follow. Nevertheless, most of that commonly used planners fail when they face dynamic obstacles, as they do not take their variables into account, as in the scenario shown in Fig. 1.

A dynamic obstacle is an object that does not have null velocity and the robot should avoid. Not only the current position of the obstacle must be taken into account, but also other variables as its velocity or its trajectory. Moreover, the velocity of the obstacle may vary. All of those aspects must be considered. Otherwise, the robot could collide with dynamic obstacles or have suboptimal behaviors.

A planner designed to navigate in dynamic scenarios is the S-DOVS, proposed in [1]. It uses the Dynamic Object Velocity Space (DOVS) to identify in which state the robot is and its relation with the environment. Based on that, it identifies a situation and applies a velocity to the robot according to a predefined strategy. The planner differs from other planners in the fact that it considers the

D. Tardioli et al. (Eds.): ROBOT 2022, LNNS 590, pp. 14–25, 2023.
https://doi.org/10.1007/978-3-031-21062-4_2

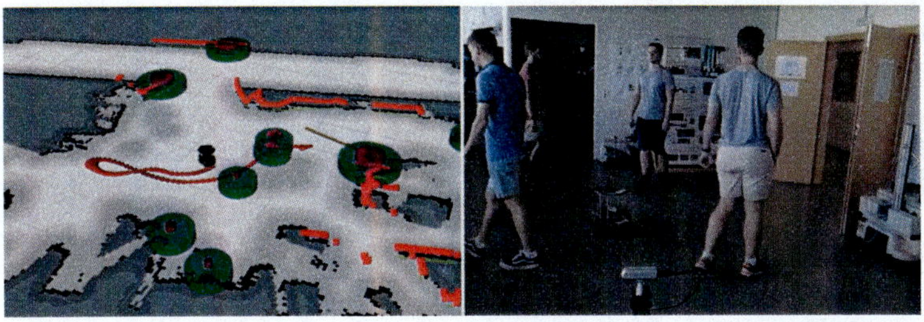

Fig. 1. An autonomous robot navigating in a dynamic scenario.

robot kinodynamic constraints and uses a deep abstraction of the environment, which is the DOVS.

The S-DOVS planner, however, has some limitations. It is able to navigate through two points that are directly connected, but it is not able in isolation to reach a goal in a big known map, where it should navigate through intermediate goals to exploit the map information and connections. It is not able to avoid convex objects or with big walls, or trap situations, where a replanning behavior would be intended. Moreover, it assumes perfect knowledge of the environment, itself and other obstacles.

In this work, we improve and complete the work presented in [1] to include a modified version of the S-DOVS planner in a full navigation stack, obtaining an autonomous robot that may navigate autonomously through real-world scenarios.

2 Background

2.1 Related Work

We can find in the literature a wide range of iterative motion planners. They compute a safe plan to reach the goal, recomputing it reactively if they find any obstacle that interferes with the path. The Dynamic Window Approach (DWA) [2] is a well established and widely used local planner for static environments and non-holonomic robots, which computes the next motion command in the robot velocity space by optimizing a criterion that weights maximum velocities, distance to obstacles, and motion towards the goal. It only contemplates the velocities reachable by the robot within its kinodynamic constraints, but does not consider future collisions with obstacle trajectories. In [3] the global dynamic window approach is also applied for computing high-velocity motions.

Other approaches include the representation of the robot movement as the result of potential forces, as the artificial potential fields presented in [4] or representing the environment as a spring-mass system, as the elastic band [5].

The previous works develop planning methods for static or no highly dynamic environments. In [6–8], planners for dynamic environments are developed. Including dynamic obstacles in the model requires a more complex representation than just using obstacles' positions as if they were static. There is a wide range of methods that use the velocity of the obstacles in the model. Two models have been widely used in the literature. One of them are the *inevitable collision states (ICS)*, introduced in [9], where the dynamics of the system and the obstacles are used to compute future states where a collision is inevitable. The other one is the *Velocity Obstacle (VO)*, introduced in [10]. The VO is defined with the trajectories the dynamic obstacles have in time, computing those that could cause a collision with the robot. The velocities that are not inside the velocity obstacles are available to be chosen by the robot. The model ensures also that the robot dynamics are respected.

In [1], a model to represent the environment dynamism is defined for non-holonomic robots, the Dynamic Object Velocity Space (DOVS), where the robot kinodynamic constraints are included. Using this model, a strategy-based planner is developed, the S-DOVS, which identifies a situation among several defined as a function of the relative robot and moving objects variables. For each situation, the method applies a different action (robot linear and angular velocities), balancing the time to goal and robot safety.

In recent years, learning approaches to solve the problem of motion planning in dynamic environments have gained importance, in works like [11–13] or [14], which also uses the DOVS. They try to use reinforcement learning to solve the limitations of model-based approaches in problems with high complexity and a big number of variables. However, the gap between the light-weight simulations used to train and the real world is huge. The simulators are not able to reproduce exactly the real-world physics and randomness, meaning that the interactions with the environment are biased. They do not consider robot kinodynamics, assuming the robots do not have acceleration constraints, may choose any velocity at any time and they may have maximum linear and angular acceleration at the same time. Moreover, they usually suppose the robot is holonomic, which is not common in real life, or has perfect knowledge of the environment.

2.2 Dynamic Object Velocity Space (DOVS)

The velocity space (VS) is defined as the set of velocities the robot may reach with the constraints of maximum and minimum velocities imposed. Some of the velocities would make the robot collide with another object, while others are safe velocities. The Dynamic Object Velocity Space (DOVS) is addressed in [1]. This model abstracts the dynamic environment, representing safe and unsafe velocities the robot can select every sampling period. The Dynamic Object Velocity (DOV) is defined as the velocities of VS that can lead to a collision in a time horizon. The methods that use this model, may choose the next velocity command in a control period among the safe velocities outside DOV.

DOVS is used to navigate through dynamic scenarios by computing safe motion commands within a time horizon. The robot size is reduced to a point, enlarging the obstacles with the robot radius. Then, for each moving obstacle, the collision band is defined as the area swept by an enlarged obstacle whilst it moves. In each possible pair of angular and linear velocities, the robot motion results in a circular trajectory while the velocities are kept. The intersection of the trajectories with the collision band define the possible collision points of the robot with the obstacle, if at the time in which the robot passes through that point, the obstacle is there too. The maximum and minimum velocities that lead to a collision with every obstacle are computed, obtaining from the sets of safe and unsafe velocities. A dynamic window is used to set the velocities the robot may take in the next step within its kinodynamic constraints.

The DOVS may be graphically represented, as it is shown in Fig. 2. In the image, safe velocities are represented in white, unsafe are dark, the dynamic window is represented with a green rhombus, the velocities that lead to the goal with a magenta line, safe velocities close to the ones that align the robot to the goal with a green line, and the big triangle represent the robot velocity limits corresponding to a differential drive robot.

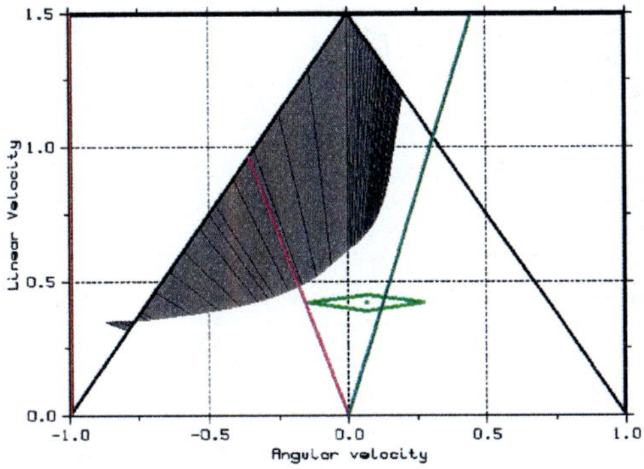

Fig. 2. A graphical representation of the DOVS.

The robot may choose a velocity among a set of only 8 possible velocities: The velocities of the extremes of the rhombus, the current velocity of the robot (center of the rhombus), the intersection of the green line with the rhombus (minimum and maximum velocities that lead to the goal) and the velocity that leads to the goal with the current linear velocity. Those last three velocities are only available if the velocities that lead to the goal are inside the dynamic window. The robot selects an action based on strategies, as explained in [1]. Depending on the situation of the agent and the objects that surround it, a state is identified as one of the predefined situations in a tree. For each recognized situation, a different behavior is designed, so that the robot avoids collisions and reaches the goal, balancing the safety and the maxima robot velocities in every sampling time. There is an extension of this work [14], where the predefined strategies are substituted by a decision-making process using a basic reinforcement learning method based on Q-tables for representing the accumulated reward for each state.

2.3 Contribution

Many of the works using learning based approaches for navigation in dynamic environments are typically trained in unrealistic simulations, while the S-DOVS model-based planner was designed regarding real-world physics and dynamics. Nevertheless, the original S-DOVS planner has some limitations, as it is a local planner to move between goals. It is not integrated with a global planner to navigate in environments with multiple rooms or convex objects due to the absence of a map, and has not a replanning capability in trap situations. It also assumes that the robot and obstacle localization is perfectly known, focusing on the navigation problem. The contributions of this work are:

- Modifications and improvements in the S-DOVS original model.
- A decentralization of the DOVS model, allowing a multi-robot execution, with each robot performing its own computations.
- An adaptation of an obstacle tracker to be used alongside the S-DOVS, getting rid of the perfect knowledge of the environment.
- An inclusion of a global planner that is able to work with the S-DOVS while enhancing its strengths.
- A ROS implementation and adaptation of the model, allowing the execution in both simulation and the real world, respecting robot kinodynamics in both of them.
- Evaluation in simulation and real-world experiments.

3 Approach

The approach taken was designing a whole robotic system to be able to navigate in complete real-world dynamic scenarios by using S-DOVS.

3.1 Decentralization and ROS

The first step of the work was adapting the program to be used directly on a real robot. The initial setup proposed in [1] was purely written in C++, and all the computations were achieved by the main program. The ground truth information was also provided by the simulator, not by any sensor. The initial step was adapting the algorithm to be used with the Robotic Operating System (ROS) [15], an open-source framework that includes a set of libraries and tools for robotic applications. A simple diagram of the ROS aspects used in this work and the way they work is shown in Fig. 3.

Stage [16], which is one of the best simulators integrated in ROS, was used. It is a lightweight simulator that is able to include multiple robots in the same scenario with few computational resources. In addition, it simulates realistically the robots' dynamics and kinematics, as well as their sensors and world physics. The simulator is modified so that collisions are detected and the agents may be teleported to start a new episode without restarting the simulation. Two types of agents were developed. Some of them, active agents, use the S-DOVS

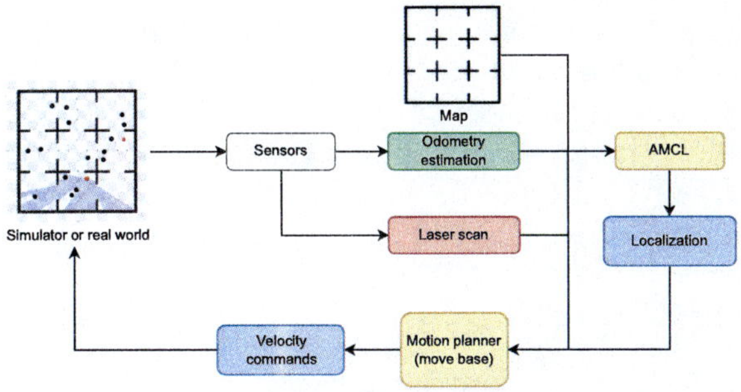

Fig. 3. A simple diagram of ROS. Sensors information is used for localization (AMCL node) and planning (typically move base node).

algorithm for navigation. The others have a preferred linear and angular velocity, which they keep until they find a wall. Then, they turn around and restore their previous velocities. Therefore, passive agents do not avoid collisions and may act as dynamic obstacles. The launching is performed with an automatic script that creates the agents, the environment and launches them. An example of simulations with different number of agents is shown in Fig. 4.

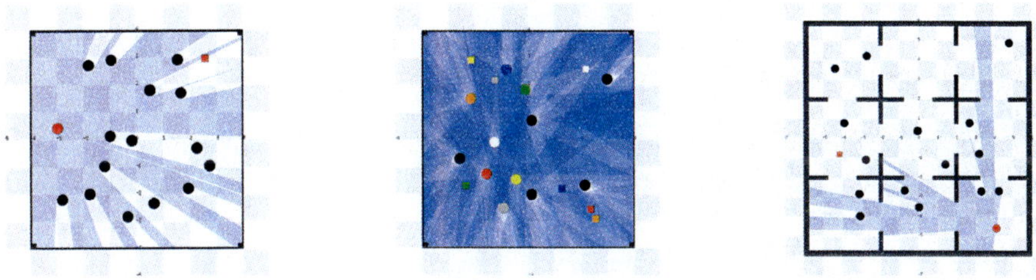

(a) 1 active, 15 passive (b) 7 active, 5 passive (c) 9 rooms scenario

Fig. 4. Example of different scenarios in Stage. Passive agents are in black and active agents (with a laser sensor) in the same color as their goals. (Color figure online)

3.2 Localization

The robot needs to know where it is to be able to navigate in a scenario. In a centralized program and in simulation, the position of the robot is known. Stage does it, and this approach is used for passive agents, which do not perform any computation. Nevertheless, active agents should behave as in the real world, and they should not use that information. As sensors, the robots only have a 2-D LIDAR sensor and the wheels encoders for estimating the odometry.

Having the static map of the scenario and the 2-D laser scans from the onboard rangefinder sensor, the approach applied for localization is using Adaptive Monte Carlo Localization (AMCL) [17], provided in the ROS navigation stack. It uses a high number of particles that are placed in possible robot poses. As more measurements are collected, the particles are updated and sampled in the places where the robot is more likely to be, converging to the robot's current position after a few iterations. Therefore, the robot is localized by using the AMCL, which takes as the input the map, the 2-D laser scan, the initial position and the encoders' measurements; and publishes the robot pose in the map frame. A representation of the AMCL in a real execution is shown in Fig. 5(b).

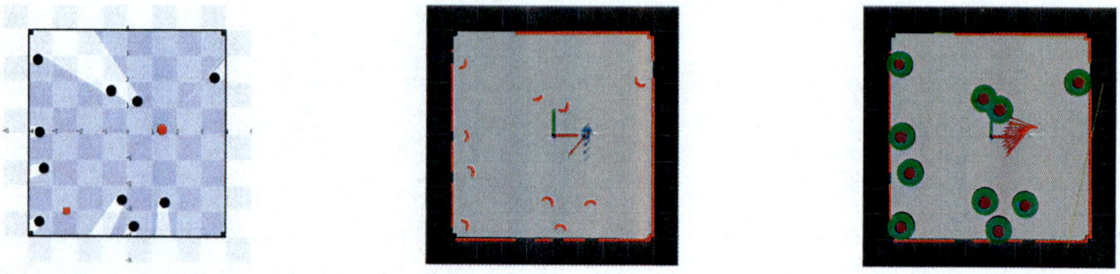

(a) Scenario in Stage

(b) RVIZ with AMCL

(c) RVIZ with obstacles.

Fig. 5. The visualization of Stage and what the robot senses in RVIZ; with AMCL pose (red), particles (blue) and laser scans in (b); and obstacles and odometry in (c). (Color figure online)

3.3 Global Planner

The S-DOVS local planner is able to navigate in dynamic environments, where the initial position and the goal are connected with a straight line, as the other planners for dynamic environments in the state-of-the-art. Nevertheless, to navigate in the real world, using the information extracted from a known map is helpful. The robot could find a large wall that should surround or could be trapped by two walls or by other obstacles temporally stopped, problems that could be sorted out by a global planner. Moreover, the original static map may be dynamically updated, so that the robot could add sensed static obstacles and take them into account in new plans. For instance, the robot may try to enter a room through a door, find out that the door is closed, and replan to find a path through a different entrance.

The typical approach in robot navigation is having a global planner computing a dense path of points and the local planner in charge of following them. However, one of the key aspects of the S-DOVS is that it takes only one point as a goal that could be far from the initial pose, and plans the trajectory in advance with the estimated velocities of the obstacles. Thus, the S-DOVS should receive points that are connected in the map and far enough from each other, so that it has the freedom to maneuver properly.

In this work, the A* implementation of the ROS standard move base has been used as a preliminary global planner, to compute an initial global path. Moreover, the map is kept updated so that the global cost map takes into account newly seen static obstacles, and the A* algorithm plans accordingly. The inflation ratio of the cost map is set to the double of the robot size, so that the S-DOVS has enough space to choose its own direction.

From the initial global plan precomputed by the A*, our global planner takes one point per square meter, ending up in a plan with separated points. Intermediate points in a straight line of the plan are undesirable, as the S-DOVS could have suboptimal performance if trying to reach all of them instead of computing the velocities to reach the last one. To remove them, for each three points, if they could be considered to be in the same line, the point of the middle is removed. To check it, we test whether the slope of the line that passes through the first and the second points is similar than the slope of the line that passes through the first and the third points. This is done with the following equation:

$$\frac{y_1 - y_2}{x_1 - x_2} \approx \frac{y_1 - y_3}{x_1 - x_3} \implies |(y_1 - y_2)(x_1 - x_3) - (y_1 - y_3)(x_1 - x_2)| < \epsilon, \tag{1}$$

where ϵ is a constant set to 0.05 and the points coordinates are (x_i, y_i).

When the robot gets close to an intermediate goal, but before it is reached, the S-DOVS is fed with the next goal of the path, so that the robot does not stop in the points of the path and moves fluently. In addition, when the robot gets too far from the plan computed by the global planner or stops because it is in a trap situation, a new plan is recomputed. With this behavior, the robot follows points that are adapted to the trajectory of the S-DOVS and is able to replan when the map is updated with new static obstacles that block its way. An example of this replanning behavior is shown in Fig. 6.

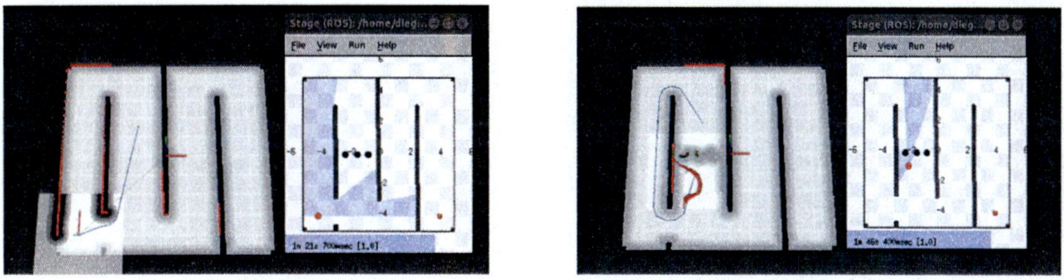

(a) The original path (b) The path after replanning

Fig. 6. A visualization of the global planner behavior. The computed path is shown in blue. The robot tries to reach the goal with the shortest way, but replans a new path after updating the map with the obstacles. (Color figure online)

3.4 Obstacle Tracker

In a realistic scenario, each active agent must gather the information of the other obstacles from the sensor measurements. The DOVS model uses the current obstacles' position, radius and velocity to predict their trajectory and their collision band; which must be computed using the LIDAR sensor.

To solve this problem, the work presented in [18] is used. It first takes the laser measurements and group them to form circles or segments. The obstacles' center and their size (radius or length, respectively) are stored. Then, an Extended Kalman Filter is used to track the measurements in time and check whether new obstacles discovered belong to previous measurement or not. The size, position and linear velocity of the obstacles is tracked through the time. Moreover, if an obstacle is no longer seen, it keeps its track a few steps in case it is seen again.

That work is extended, as the orientation and angular velocity of the obstacles are needed too. The orientation is computed by using the direction of the obstacle trajectory, while the angular velocity is computed by tracking the change in the orientation with a high frequency. This is performed to be aware of sudden changes in the obstacles' motion. Nevertheless, spurious measurements or occlusions lead to wrong estimations. Thus, a median filter is used to remove wrong angular velocity estimations. A visualization of the obstacle tracker in a real execution is shown in Fig. 5(c).

3.5 Modifications in S-DOVS

One of the main contributions to the original model is the inclusion of segment obstacles. In the original version, all the obstacles were treated as circles, defined by their position, velocity and radius, and walls would be treated as a group of circles (inaccurate and curvy). The solution designed was introducing segment obstacles detected by the obstacle tracker to the model. The way to introduce the segments in the DOVS was by saving the points of the extremes. If a trajectory intersects with a point that is between the extremes of the segment, the velocity that leads to that trajectory is forbidden (any velocity in an early implementation, changed as shown later).

Strategies that make the agent choose velocities that lead to trapping situations or situation with small margin to react have been removed, as well as some other risky situations, such as trying to accelerate when there are no free velocities to try to find any (the new behavior is braking to avoid running over pedestrians).

Another change made is a modification in the calculations of the forbidden velocities for static obstacles. An obstacle is considered static if it is a segment or its velocity is smaller than 0.1 m/s (in the original model, the minimum velocity to be static was 10^{-5} m/s, but in the real world estimation errors may be found). The original model sets that any trajectory that collides with an obstacle in any time is forbidden. However, that could be too conservative for static obstacles, which may be easily avoided. In that case, the trajectories are only considered forbidden if the distance following that trajectory is bigger than the distance necessary to make the robot brake, multiplied by a safety factor. That distance is defined with the following formula:

$$brake\_distance = \frac{v^2}{2 * \max(\Delta v)} \tag{2}$$

4 Experimental Results

The videos of all the experiments may be accessed in the following link: https://youtu.be/fGHZiMt-Yao.

4.1 Simulation Experiments

In the simulation experiments, the effectiveness of the global planner is tested. The global planner should help the S-DOVS to avoid collisions by using the previous static map information, and also by adding new static obstacles to the costmap that are avoided with the global plan.

Two different kinds of experiments are performed. The first ones use a simple 4 m x 4 m room as the scenario. In that scenario, static and dynamic obstacles are randomly placed, whose velocities are random too. The robot starts in a random position and has to navigate to a goal. A set of 200 scenarios are generated for any number of obstacles in the range 1 to 15. The robot ties to complete the same scenario both with only the S-DOVS and with the global planner too. With these experiments, we want to prove that the global planner not only does not worsen the performance, but also improves it. Theoretically, in that kind of scenario, the global planner would not be needed. Nevertheless, the addition of newly seen static obstacles to the costmap could help.

The other kind of experiments use the same structure as the previous one. The only difference is that the map changes. Instead of only one room, 9 rooms are included, as in Fig. 4. In this kind of scenario, the S-DOVS planner will face convex obstacles that could not avoid.

The results obtained in the experiments are shown in Table 1. The robot is able to reach the goal in more episodes when it uses the complete navigation stack, even on the map with only one room, due to the fact that obstacles could remain static in the way to the goal. The global planner adds them to the cost map, helping the S-DOVS. On the 9-room map, the benefits are clear. The S-DOVS needs to reactively know how to navigate through the rooms. The intermediate goals sent by the global planner makes it know how to avoid properly the walls, which could be interpreted as convex obstacles. The approaches perform similarly in terms of time in the single-room map, and comparably in the multi-room map although the whole system is clearly more safe and successful.

Table 1. Difference in success rate ($\frac{\text{number of success complete system}}{\text{number of success S-DOVS only}}$) and time rate ($\frac{\text{mean time complete system}}{\text{mean time S-DOVS only}}$), on a map with a single room and with 9 rooms, for the same 200 random scenarios for each number of agents.

Map	Metric	1	2	3	4	5	6	7	8	9	10	11	12	13	14	15
Single	Success	1.02	1.12	0.99	1.05	1.02	1.19	1.19	1.02	1.16	1.06	1.23	1.06	1.06	1.29	1.39
	Time	1.02	1.01	0.97	0.99	0.98	1.09	1.05	1.00	1.06	1.08	0.93	1.03	0.89	1.06	1.11
Multi	Success	3.09	2.89	3.41	2.84	3.09	4.16	4.11	3.08	3.67	3.21	3.73	2.86	3.93	3.52	3.53
	Time	1.09	1.32	1.11	1.07	1.21	1.08	1.13	1.07	0.95	1.11	1.25	1.24	1.10	1.10	1.11

4.2 Real-world Experiments

The whole system has been tested on a Turtlebot 2 platform with a NUC with Intel Core i5-6260U CPU and 8 GB of RAM, equipped with a Hokuyo 2D-LIDAR sensor.

The system was analyzed in two experiments. One experiment checked whether the robot could replan properly in the real world. It was sent to a goal and, when the robot tried to follow the closest path, it found a closed door; so it needed to replan and go through an alternative path, as seen in Fig. 7, when the robot has just replanned. The other experiment performed was navigating autonomously in a room with other people acting as dynamic obstacles. In both experiments, the robot was successful, navigating autonomously by using the whole integration of the system. An example of navigation may be seen in the previously shown Fig. 1.

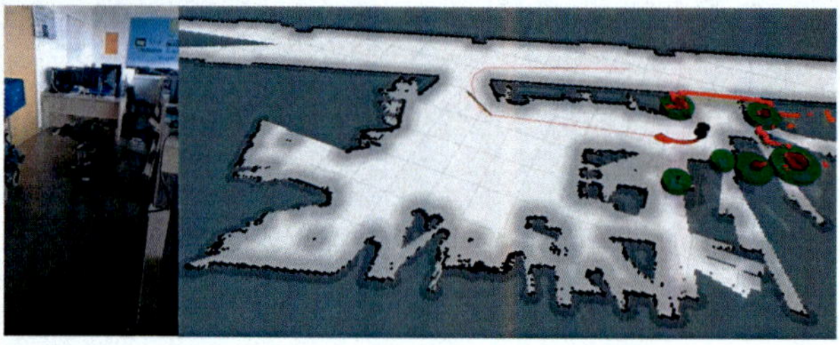

Fig. 7. An image where the robot has just replanned after detecting a closed door, with the global plan shown in red. The global costmap is also shown. (Color figure online)

5 Conclusion

In this work, we develop a complete system that is able to navigate autonomously in dynamic environments, avoiding collisions while minimizing the time to reach the goal. The S-DOVS local planner has been improved and adapted to work in the real world and in realistic scenarios, also considering the robot kinodynamic constraints, which are avoided in many works in the literature. The local planner is integrated in a ROS full navigation stack, a localization node and a global planner. In addition, the proposed navigation stack could be used in other planner intended for dynamic scenarios. The system has been tested in several simulation and real-world experiments, proving a good performance, mainly in more complex scenarios having several rooms.

References

1. Lorente, M.-T., Owen, E., Montano, L.: Model-based Robocentric planning and navigation for dynamic environments. Int. J. Robot. Res. **37**(8), 867–889 (2018)

2. Fox, D., Burgard, W., Thrun, S.: The dynamic window approach to collision avoidance. IEEE Robot. Autom. Mag. **4**(1), 23–33 (1997)
3. Brock, O., Khatib, O.: High-speed navigation using the global dynamic window approach. In: Proceedings, IEEE International Conference on Robotics and Automation (Cat. No. 99CH36288C), vol. 1, pp. 341–346. IEEE (1999)
4. Warren, C.W.: Global path planning using artificial potential fields. In: 1989 IEEE International Conference on Robotics and Automation, pp. 316–317. IEEE Computer Society (1989)
5. Quinlan, S., Khatib, O.: Elastic bands: Connecting path planning and control. In: [1993] Proceedings IEEE International Conference on Robotics and Automation, pp. 802–807. IEEE (1993)
6. Stachniss, C., Burgard, W.: An integrated approach to goal-directed obstacle avoidance under dynamic constraints for dynamic environments. In: IEEE/RSJ International Conference on Intelligent Robots and Systems, vol. 1, pp. 508–513. IEEE (2002)
7. Minguez, J., Montano, L.: Sensor-based robot motion generation in unknown, dynamic and troublesome scenarios. Robot. Auton. Syst. **52**(4), 290–311 (2005)
8. Hsu, D., Kindel, R., Latombe, J.-C., Rock, S.: Randomized kinodynamic motion planning with moving obstacles. Int. J. Robot. Res. **21**(3), 233–255 (2002)
9. Fraichard, T., Asama, H.: Inevitable collision states-a step towards safer robots? Adv. Robot. **18**(10), 1001–1024 (2004)
10. Fiorini, P., Shiller, Z.: Motion planning in dynamic environments using velocity obstacles. Int. J. Robot. Res. **17**(7), 760–772 (1998)
11. Shi, H., Shi, L., Xu, M., Hwang, K.-S.: End-to-end navigation strategy with deep reinforcement learning for mobile robots. IEEE Trans. Industr. Inf. **16**(4), 2393–2402 (2020)
12. Lei, X., Zhang, Z., Dong, P.: Dynamic path planning of unknown environment based on deep reinforcement learning. J. Robot. **2018**, 5781591 (2018)
13. Chen, C., Liu, Y., Kreiss, S., Alahi, A.: Crowd-robot interaction: crowd-aware robot navigation with attention-based deep reinforcement learning. In: 2019 International Conference on Robotics and Automation (ICRA), pp. 6015–6022. IEEE (2019)
14. Mackay, A.K., Riazuelo, L., Montano, L.: RL-DOVS: reinforcement learning for autonomous robot navigation in dynamic environments. Sensors **22**(10), 3847 (2022)
15. Stanford Artificial Intelligence Laboratory: Robotic Operating System. https://www.ros.org
16. Gerkey, B., et al.: The player/stage project: tools for multi-robot and distributed sensor systems. In: Proceedings of the 11th International Conference on Advanced Robotics, vol. 1, pp. 317–323. CiteSeer (2003)
17. Fox, D.: KLD-sampling: adaptive particle filters and mobile robot localization. Adv. Neural Inf. Process. Syst. (NIPS) **14**(1), 26–32 (2001)
18. Przybyła, M.: Detection and tracking of 2D geometric obstacles from LRF data. In: 11th International Workshop on Robot Motion and Control (RoMoCo), vol. 2017, pp. 135–141. IEEE (2017)

Artificial Stupidity in Robotics: Something Unwanted or Somehow Useful?

Juan Jesús Roldán-Gómez(✉)

Departamento de Ingeniería Informática, Escuela Politécnica Superior, Universidad Autónoma de Madrid, Francisco Tomás y Valiente, 11, 28049 Madrid, Spain
juan.roldan@uam.es

Abstract. Artificial stupidity has been reported in multiple computer science applications. This phenomenon can appear in two ways: artificial stupidity by accident is the result of artificial intelligence failures, whereas artificial stupidity by design is an intended development with a purpose. However, these concepts have not been studied in the context of robotics. This paper analyzes artificial stupidity in robotics, searching to answer the question: "Is artificial stupidity something that we must avoid or, on the contrary, something that can be useful for us?" It addresses the definition of the artificial stupidity problem and analyzes some potential methods to solve it.

Keywords: Robotics · Artificial stupidity · Artificial intelligence · Machine learning · Reinforcement learning

1 Introduction

The concept of artificial stupidity (AS) has emerged as the opposite of artificial intelligence (AI) and can refer to two opposite realities:

1. **AS by accident.** It occurs when artificial intelligence does not work properly.
2. **AS by design.** It is a behavior deliberately sought to fulfill some purpose.

Before analyzing this phenomenon, it is worth discussing its name. After reviewing the literature, we found artificial stupidity is the most common term to refer to this reality. Other synonyms are not used, such as artificial silliness, idiocy, imbecility, dumbness, ignorance, ineptitude, or nonsense.

The literature contains multiple approaches to artificial stupidity as the inability of AI to perform some tasks. This concept is discussed from a philosophical perspective in [1], which focuses on the Frankenstein syndrome, i.e., the fear that a human creation will turn against its creator. The author applies the two kinds of stupidity distinguished by Kant to this context: the stupidity of understanding and judgment. The first is when the agent misses the concepts required to make sense of a situation, whereas the second is when it has these

D. Tardioli et al. (Eds.): ROBOT 2022, LNNS 590, pp. 26–37, 2023.
https://doi.org/10.1007/978-3-031-21062-4_3

concepts but misapply them. The conclusion is that most artificial agents suffer from the stupidity of judgment.

A more technical approach to this topic is addressed in [2]. In this work, artificial stupidity is described as the result of a bad selection and training of models. To avoid this phenomenon, the authors pose some guidelines for choosing adequate neural networks and optimization algorithms.

The work in [3] addresses another interesting topic: when artificial intelligence learns from natural stupidity. Authors analyze how machine learning (ML) decision-making systems reproduce and even amplify human errors and biases. They point out incomplete datasets, learning from human decisions, and biased inference and evaluation processes as the main causes of this phenomenon.

The work in [4] proposes some potential solutions to prevent these errors and biases that shape artificial stupidity. Specifically, the authors pose two requirements to AI: explainability and verifiability. In this way, humans can understand the behavior of AI models and detect and correct their errors and biases.

However, the literature also presents artificial stupidity as the desired result in some contexts. In computer science, the Turing Test aims at building an AI that can reply to a set of questions like a human. For this purpose, the agent should not answer all the questions correctly, but make human-like mistakes and show human limits and biases [5]. There is a consensus that AIs must be as intelligent as people to enable efficient and safe interactions. Thus, there are some proposals to apply restrictions on AI, such as limiting the hardware, storage capacity, memory access, processing, clock speed, and computing capacity [6].

This approach is used in the video game industry for developing non-playable characters (NPC) [7]. These agents must be designed to lose against the players in a challenging manner, which is more difficult than either winning or losing against them. Some proposals to reach this goal in the context of first-person shooters (FPS) are being visible to the player, warning the player about what they are doing, moving before firing, having poor aim, etc.

Finally, the work in [8] proposes the development of artificial stupidity through the capture of mistakes of human behavior. The author concludes that an artificially stupid system can be as complex to develop as the best artificial intelligence.

This paper analyzes the concept of artificial stupidity in robotics, searching to answer the question: "Is artificial stupidity something that we must avoid or, on the contrary, something that can be useful for us?" As far as we know, this approach is completely original since there are no further studies on this topic.

The remainder of this paper is organized as follows: Sect. 2 reflects on artificial stupidity in the field of robotics, analyzing its causes, forms, and consequences. Section 3 presents the AS problem and compares it to the AI ones. Section 4 analyzes the methods that could solve this problem. Finally, Sect. 5 summarizes the main conclusions of this work.

2 Artificial Stupidity in Robotics

The two forms of artificial stupidity are also present in robotics. The literature contains many examples, although most works do not call it by its name. AS by accident is easier to recognize because AI is being applied in robotics and its models often fail (see Sect. 2.1). However, AS by design can have its place as long as it has utility in some types of missions (see Sect. 2.2).

2.1 AS by Accident

Artificial intelligence has been applied in robotics for decades, and its use is expected to grow even more in the near future [9]. The counterpart is the appearance of artificial stupidity in the form of model failures, which can compromise the missions and even the integrity of robots. These failures are normally due to an inadequate selection or training of the models. Some challenges for the generalization of AI in robotics, according to [9], are data quality and biases, process transparency and predictability, and human trust and adoption.

The rise of deep learning has allowed new applications in perception (scene and object recognition, human-robot interaction, and sensory integration) and motion (path planning, motion control, and grasping) [10]. It also poses some challenges related to the development and training of these models, such as using pre-trained models, sim-to-real transferring them, discovering domain-invariant features, and learning by imitation or demonstration.

Although AS by accident can appear in any use of machine learning in robotics, it is more likely to occur when the applications are complex. That is the case of vision-based grasping and manipulation [11], which can be performed by model-free and model-based methods. The current challenge is the generalization of capabilities to grasp novel objects from different locations and angles.

Two ML techniques that deserve discussion are reinforcement learning and imitation learning. In reinforcement learning, sim-to-real transfer allows training models without damaging robots but must deal with the gap between simulations and reality [12], which is a potential source of AS. For its part, imitation learning reduces this gap by demonstrating the desired behaviors in a controlled environment [13]. However, other challenges arise related to demonstration (capture devices), representation (generation of states from the acquired information), and learning (generalization from small samples to complex behaviors).

An important issue is that not all the failures in robots can be attributed to AS. A classification of robot failures is proposed in [14], which distinguishes between technical and interaction failures. Technical failures include hardware and software errors, whereas interaction failures are related to the operator or environment. In turn, according to [14], software errors can be classified into communication (e.g., incorrect data, bad timing, extra data, and missing data) and processing failures (e.g., abnormal terminations, missing events, incorrect logic, and timing and ordering). Among all these failures, only some software and, more specifically, processing errors can be considered a consequence of AS.

Figure 1 shows a classification of robot failures more focused on artificial stupidity. Failures can occur in the robot, operator, interface, human-robot interaction, and communications. Robot failures can affect hardware (e.g., sensors, actuators, control, or power system) and software (e.g., planning, perception, data processing, decision-making, or control). Artificial stupidity can appear in all these software failures, as long as there is an intelligent algorithm behind these functions.

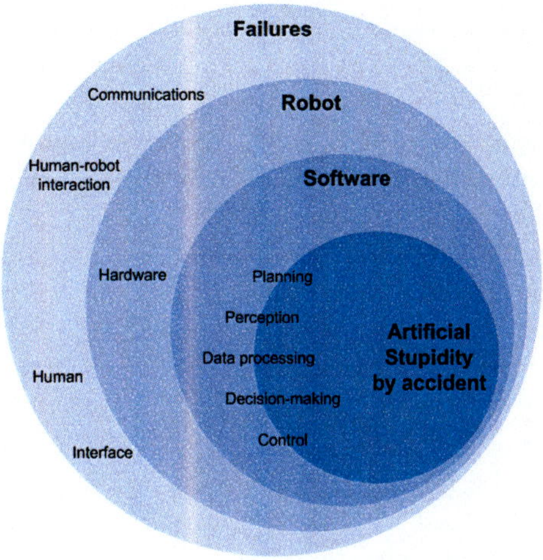

Fig. 1. Artificial stupidity by accident in the context of robot failures.

Failures happen in every robotic application: industrial, service, social, search and rescue, etc. A set of 26 interventions of ground robots in disaster scenarios between 2001 and 2017 are collected by [15]. These interventions were developed in the context of natural or human-made disasters, such as mining accidents, geological events, meteorological conditions, and terrorist attacks. Among these interventions, 6 were field tests and 18 actual missions: 11 were successful and 7 failed. The failures were due to the difficulty of scenarios, design of robots, lack of robot autonomy, human errors, and other causes.

Robot failures have an impact not only on mission performance but also on human trust [16]. Human-robot interaction errors can be performance errors, which degrade the user perception of robot capability to achieve a task, or social errors, which affect the user perception of robot socio-affective skills [17]. The impact of failures on the user perception of robot reliability and trust also depends on previous expectations [18], which complicate even more the management of these faulty situations. Finally, errors in human-robot interaction can affect learning performance in social contexts [19].

2.2 AS by Design

In Sect. 1, we saw how artificial stupidity is practical in some computer science applications, such as developing enhanced social agents or more challenging video games. For this purpose, designers and developers degrade hardware or software components to limit the performance of artificial intelligence agents. Taking this as a reference, if we want to explore AS by design in robotics, we must find a utility and then a way to implement it.

The most immediate utility of AS in robotics can be risk assessment and failure detection. In the same way that AI agents can find optimal plans, AS agents can detect potential errors. For instance, considering a robot that intervenes in a disaster scenario, AI can generate a mission plan (e.g., select a task, choose a target area, and plan a path), whereas AS can discover potential problems (e.g., blocked roads or rough terrains). Below is an analysis of the opportunities in the different fields of robotics.

Industrial robots were the first robots applied on a large scale. They could perform manipulation, welding, painting, cutting, and assembly tasks efficiently and safely. Traditionally, these robots were confined in cells and separated from people, creating controlled environments with reduced uncertainty. In this way, operators could program the robots to perform their tasks without requiring them to be intelligent (or stupid).

However, Industry 4.0 is leading to more flexible manufacturing, where environments are changeable, tasks require adaptation, and robots and humans must share spaces [20]. The paradigms are changing from robotic cells to human-robot coexistence, synchronization, cooperation, and collaboration [21]. Collaborative robots have become the most relevant trend in industrial robotics, requiring humans and robots to share spaces and execute tasks together [22]. These scenarios have become possible due to the advances in robot sensors [23] and immersive technologies [24]. And more related to this article, robots require intelligence to adapt to their environment (or stupidity to identify potential failures before executing their tasks).

Service robots were developed after industrial ones because they have to perform a wider range of missions in more complex scenarios. For these reasons, there is a greater variety of designs considering both hardware and software, given that new applications often require new models. We can find domestic, educational, healthcare, military, construction, mining, agricultural, and environmental robots [25]. Each of these applications will require a different level of autonomy and, therefore, intelligence (or stupidity).

From now on, this article focuses on search and rescue robotics since it is a demanding case, and its conclusions can be generalized. Robot autonomy is especially relevant in these scenarios [26]. If the operators have to control most of the robot actions, human factors issues like high workload and degraded situational awareness can affect their performance. If the robots can act most of the time autonomously, they require more intelligence to find the best actions (or stupidity to detect the worst ones).

In summary, this paper proposes the use of artificial stupidity to detect potential failures before robot missions. Logically, artificial stupidity must operate in simulations to avoid damaging the robots and their environment. Let a virtual-stupid agent interact with its environment, and it will detect all the potential errors of the physical-intelligent agent. Section 3 will define the artificial stupidity problem in robotics.

3 Definition of the Problem

There are multiple problems with relevance in the field of robotics, such as mission planning, task allocation, path planning, perception, control, and grasping and manipulation.

Mission planning focuses on the design of robot fleets, covering not only the selection of robots but also their communications, coordination mechanisms, and decision-making architectures [27]. Task allocation involves assigning a set of tasks to a set of robots, searching to maximize the fleet utility while meeting the defined constraints [28]. Path planning involves discovering the optimal path that goes from an origin to a destination avoiding all the obstacles in the scene [29]. Robot perception encompasses tasks like image classification, object detection, depth estimation, and semantic segmentation, which have been relevant to improving the performance of autonomous cars and field robots in recent years [30]. Robot control has evolved from classical approaches, which provided generic tools to drive robots, towards modern ones, which allow learning the tasks and adapting to the environments [31]. Finally, grasping and manipulation combine perception and control in a complex task that highly depends on the features of the target object [32].

Several of these problems can be defined as optimization problems, where a set of target variables $\mathbf{x} = \{x_1, x_2, ..., x_n\}$ must be computed to maximize a target function $f(\mathbf{x})$ while meeting a set of constraints $g_i(\mathbf{x}) \leq 0$ with $i = 1, ..., m$ (inequality) and $h_j(\mathbf{x}) = 0$ with $j = 1, ..., p$ (equality).

Artificial stupidity problems differ from optimization problems in their main objective. They do not search for a unique solution - $\mathbf{x}^*$ as the global maximum of the target function $f(\mathbf{x})$ - but multiple ones - $\mathbf{x}_k^*$ with $k = 1, ..., q$ as the local minima of that function. These solutions must meet the following criteria:

1. **They must lead to bad enough results.** This means that $\mathbf{x}$ can be a solution of the AS problem if $f(\mathbf{x}) \leq \theta$, where θ is a threshold defined by designers to ensure the solution is bad enough.
2. **They must have a sufficient probability of occurrence.** On the one hand, the solutions must meet the defined constraints, similarly to the optimization problems described above. On the other hand, AS solutions must be close to AI solutions and even derive from them. This way, we can ensure that these errors likely occur in a practical context.
3. **They must be different enough from each other.** The objective is to achieve a minimum diversity in the solutions so that they represent different

behaviors in the missions. For this purpose, we must avoid finding bunches of solutions that provide similar results and imply a computational cost.

Let's analyze the case of path planning to understand the AS problem and compare it with the AI one. Conventional path planning aims at obtaining the optimal path between current and goal positions. Therefore, this problem usually has only one solution and, in some particular cases, a few solutions. These solutions are paths that optimize some target metric, such as covered distance, elapsed time, energy cost, or a combination of them. However, AS applied to path planning looks for potentially dangerous paths that meet three conditions. First, their utility must be low enough, which means they threaten robot integrity, mission success, environment elements, or just mission performance, depending on the designer criteria. Second, their likelihood must be high enough, which means they can occur during the normal execution of the mission. This condition is satisfied when AS paths are close to AI ones: e.g., if the optimal path goes near an obstacle, one of the AS paths will collide with that obstacle. And third, these paths must be different enough to avoid unnecessary repetitions. For instance, if one AS path leads the robot to fall down a slope, the rest must not pass through the same area.

Figure 2 illustrates this phenomenon in outdoor (a) and indoor (b) scenarios. In this figure, the green lines are path planning solutions computed by an intelligent agent, whereas the red lines are the solutions discovered by a stupid agent.

In the first situation, the ground robot must descend from an elevation to a valley, avoiding falling into the lakes and hitting the trees. In that scenario, AI methods may discover two different trajectories where the robot descends through soft slopes. For its part, AS algorithms should discover the relevant deviations of these paths that threaten the robot and mission, including falling down the surrounding slopes, colliding with the closest trees, etc. Figure 2 (a) also represents two potential deviations from the solution criteria: the path that goes in the opposite direction of the optimal ones from the beginning and the multiple routes that fall down the same slopes.

In the second situation, the ground robot must traverse a room with a high density of obstacles. Like in the previous case, AI methods may find two different paths with comparable utilities, avoiding the tables and chairs by their two sides. Meanwhile, AS algorithms should discover potential failures in this task, such as colliding with the closest obstacles (doors, tables, and chairs) or exiting through the wrong door. Again, bunches of similar AS solutions are represented in Fig. 2 (b). Although this effect is not desired for AS problems, it could appear when some algorithms are applied to this context.

In both cases, we can observe that AS provides a double utility. First, we can predict unexpected failures in mission planning and prevent them during mission execution. In this circumstance, the operators that either control or supervise the robots can play a relevant role. Second, we can use this information to modify optimal paths for safety. For instance, increasing the distance between the trajectories and some obstacles can make them safer at the cost of reducing their efficiency.

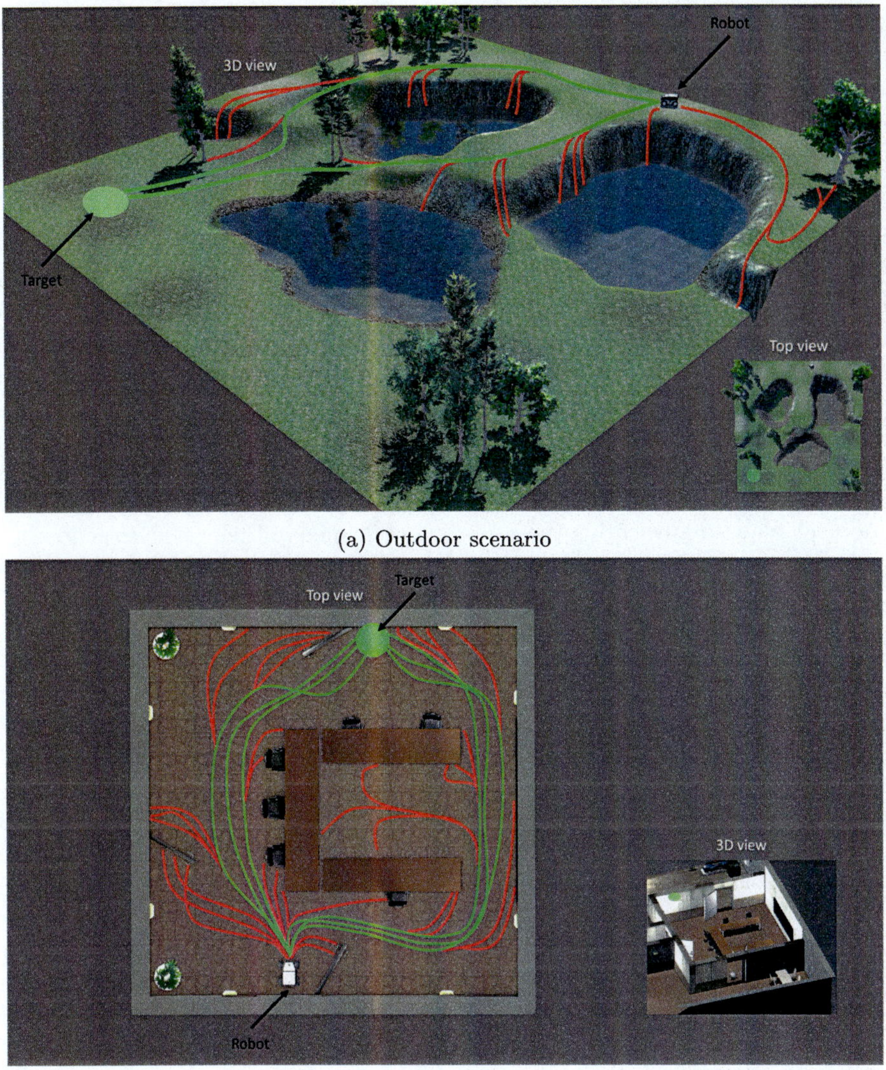

(a) Outdoor scenario

(b) Indoor scenario

Fig. 2. A comparison of AS with AI in the context of robot path planning: green paths are solutions computed by an AI, whereas red ones are potential errors found by an AS. (Color figure online)

4 Analysis of Methods

Once we have defined the artificial stupidity problem, the interest is on which methods can help solve it. Artificial intelligence is years ahead of artificial stupidity, but its algorithms may be too adapted to its problems. However, a good starting point can be analyzing current AI methods and adapting them to AS problems. Note that this paper aims at posing the AS problem and not proposing methods to solve it. This work will be performed in the future and considering the contributions of the robotics community.

In this section, we continue the analysis referring to path planning, but most results can be generalized to other contexts. Over the decades, multiple methods have been developed to solve the path planning problem [33]. These algorithms can be classified in two main groups: classic and heuristic [34].

Classic methods follow three main approaches: roadmap, cell decomposition, and potential fields [35]. Roadmap methods represent the free space on a graph and perform a graph search to find the paths. This approach is functional when the origin and destination are known but may lose utility when there are multiple unknown goals. Cell decomposition algorithms represent the free space on a grid and compute the paths cell by cell. Again, this approach is oriented to find routes between known origins and destinations, but the grid may keep information about the potential obstacles. Potential fields generate an attractive force to the goal and repulsive forces from the obstacles that influence the robot. This approach allows representing the AS problem defined in the previous section but may not be feasible in real-world scenarios. All these methods are adapted to the path planning problem and may not be transferable to other contexts.

Heuristic algorithms overcome some of the problems of classical methods, such as the high computational cost, lack of scalability, and possible convergence to local extrema. However, these algorithms do not guarantee finding a solution and, in that case, do not guarantee that solution is optimal [36]. This group usually covers probabilistic algorithms, neural networks, evolutionary algorithms, and swarm intelligence methods. Probabilistic methods include well-known widely-used algorithms such as probabilistic roadmaps (PRM) and rapidly-exploring random trees (RRT). These algorithms improve the scalability of classical ones. Nevertheless, they also focus on path planning and cannot be adapted to other problems.

Neural networks are more flexible than previous methods, given that they can encode many domains in a few architectures, receiving the available inputs and producing the required outputs. Among all the methods used for training these models, reinforcement learning is the most promising because it can work online in simulated scenarios. Reinforcement learning algorithms usually combine exploration and exploitation. The first is relevant at the beginning of training to try as many actions as possible, whereas the second is pertinent at the end to maximize the outcome. However, artificial stupidity requires continuous exploration to find all the significant failures, while it usually does not obtain value from exploitation.

Evolutionary algorithms are inspired by biological evolution and perform stochastic searches for solutions. They often encode the solutions as chromosomes, create an initial population, perform operations with that population during multiple iterations, and terminate when some conditions are satisfied. Three of these operations are especially relevant: the selection of the individuals with the highest fitness of every generation, the crossover to create new individiduals from the existing ones, and the mutation to introduce new genetic material into the population. When evolutionary algorithms like genetic algorithms are applied to optimization problems, they look for convergence to the optimal solution or a suboptimal one with enough quality. In this context, keeping population diversity is desired to avoid converging to local extrema. However, coming back to AS, keeping population diversity is even more interesting because the algorithm must find multiple and diverse solutions. There are multiple ways to obtain that diversity, such as increasing the size of the initial population, selecting not only the best individuals but also the most diverse, and increasing the mutation ratio.

Finally, swarm intelligence methods are based on biological systems such as bird flocks or insect colonies. These algorithms reproduce the collective behavior of decentralized self-organized systems. The most prominent are particle swarm optimization (PSO) and ant colony optimization (ACO). In these algorithms, a large number of simple agents interact with each other and their environment to explore the problem space and find suboptimal solutions. As evolutionary algorithms, swarm intelligence methods operate with a large number of potential solutions, providing a clear advantage for the exploration required in the AS problem. However, both approaches tend to concentrate these individuals around a unique solution instead of keeping the diversity to return multiple solutions.

5 Conclusions

This paper explores the potential utility of artificial stupidity in robotics, finding the main application of these techniques in error detection.

The AS problem is defined as a search of local minima, conversely to the AI problem that seeks the global maximum. The solutions to this problem must be sufficiently bad, probable, and different from each other.

Methods like neural networks, evolutionary algorithms, and swarm intelligence methods are the best suited to solve this problem.

In future works, we will create a simulation environment to explore the potential of artificial stupidity and try different algorithms to implement it.

Acknowledgments. The author would like to thank some interesting natural stupidities who have crossed his way and inspired this article.

References

1. Falk, M.: Artificial stupidity. Interdisc. Sci. Rev. **46**(1–2), 36–52 (2021)
2. Murphy, C.M., Koehler, G.J., Fogler, H.R.: Artificial stupidity. J. Portf. Manag. **23**(2), 24 (1997)
3. Rich, A.S., Gureckis, T.M.: Lessons for artificial intelligence from the study of natural stupidity. Nat. Mach. Intell. **1**(4), 174–180 (2019)
4. Umbrello, S., Yampolskiy, R.V.: Designing AI for explainability and verifiability: a value sensitive design approach to avoid artificial stupidity in autonomous vehicles. Int. J. Soc. Robot. **14**(2), 313–322 (2022)

5. Trazzi, M., Yampolskiy, R. V. : Building safer AGI by introducing artificial stupidity. arXiv preprint arXiv:1808.03644 (2018)
6. Trazzi, M., Yampolskiy, R.V.: Artificial stupidity: data we need to make machines our equals. Patterns **1**(2), 100021 (2020)
7. Lidén, L. : Artificial stupidity: the art of intentional mistakes. AI Game Program. wisdom **2**, 41–48 (2003)
8. Lo, A.W.: Why artificial intelligence may not be as useful or as challenging as artificial stupidity (2019)
9. Wang, W., Siau, K.: Artificial intelligence, machine learning, automation, robotics, future of work and future of humanity: a review and research agenda. J. Database Manage. (JDM) **30**(1), 61–79 (2019)
10. Károly, A.I., Galambos, P., Kuti, J., Rudas, I.J.: Deep learning in robotics: survey on model structures and training strategies. IEEE Trans. Syst. Man Cybern. Syst. **51**(1), 266–279 (2020)
11. Kleeberger, K., Bormann, R., Kraus, W., Huber, M.F.: A survey on learning-based robotic grasping. Curr. Robot. Rep. **1**(4), 239–249 (2020)
12. Zhao, W., Queralta, J. P., Westerlund, T. : Sim-to-real transfer in deep reinforcement learning for robotics: a survey. In 2020 IEEE Symposium Series on Computational Intelligence (SSCI), pp. 737–744. IEEE (2020)
13. Fang, B., Jia, S., Guo, D., Xu, M., Wen, S., Sun, F.: Survey of imitation learning for robotic manipulation. Int. J. Intell. Robot. Appl. **3**(4), 362–369 (2019). https://doi.org/10.1007/s41315-019-00103-5
14. Honig, S., Oron-Gilad, T.: Understanding and resolving failures in human-robot interaction: Literature review and model development. Front. Psychol. **9**, 861 (2018)
15. León Rivas, J. D.: Dynamic analysis of the alternating tripod gait of a hexapod robot with C-legs, Doctoral dissertation, Escuela Técnica Superior de Ingenieros Industriales, Universidad Politécnica de Madrid (2020)
16. Rossi, A., Dautenhahn, K., Koay, K. L., Walters, M.L.: How the timing and magnitude of robot errors influence peoples' trust of robots in an emergency scenario. In: Social Robotics. ICSR 2017. LNCS, vol. 10652. Springer, Cham (2017). https://doi.org/10.1007/978-3-319-70022-9_5
17. Tian, L., Oviatt, S.: A taxonomy of social errors in human-robot interaction. ACM Trans. Human-Robot Inter. (THRI) **10**(2), 1–32 (2021)
18. Washburn, A., Adeleye, A., An, T., Riek, L.D.: Robot errors in proximate HRI: how functionality framing affects perceived reliability and trust. ACM Trans. Human-Robot Inter. (THRI) **9**(3), 1–21 (2020)
19. Kim, S. K., Kirchner, E. A., Schloßmüller, L., Kirchner, F. : Errors in human-robot interactions and their effects on robot learning. Front. Robot. AI **131**, 558531 (2020)
20. Evjemo, L.D., Gjerstad, T., Grøtli, E.I., Sziebig, G.: Trends in smart manufacturing: role of humans and industrial robots in smart factories. Curr. Robot. Rep. **1**(2), 35–41 (2020)
21. Matheson, E., Minto, R., Zampieri, E.G., Faccio, M., Rosati, G.: Human-robot collaboration in manufacturing applications: a review. Robotics **8**(4), 100 (2019)
22. Sherwani, F., Asad, M.M., Ibrahim, B.S.K.K.: Collaborative robots and industrial revolution 4.0 (IR 4.0). In: 2020 International Conference on Emerging Trends in Smart Technologies (ICETST), pp. 1–5. IEEE (2020)
23. Li, P., Liu, X.: Common sensors in industrial robots: a review. J. Phys. Conf. Ser. **1267**(1), 012036 (2019). IOP Publishing (2019)

24. De Pace, F., Manuri, F., Sanna, A., Fornaro, C.: A systematic review of augmented reality interfaces for collaborative industrial robots. Comput. Ind. Eng. **149**, 106806 (2020)
25. Gonzalez-Aguirre, J.A., et al.: Service robots: trends and technology. Appl. Sci. **11**(22), 10702 (2021)
26. Roldán, J.J., Peña-Tapia, E., Garcia-Aunon, P., Del Cerro, J., Barrientos, A.: Bringing adaptive and immersive interfaces to real-world multi-robot scenarios: application to surveillance and intervention in infrastructures. IEEE Access **7**, 86319–86335 (2019)
27. Yan, Z., Jouandeau, N., Cherif, A.A.: A survey and analysis of multi-robot coordination. Int. J. Adv. Rob. Syst. **10**(12), 399 (2013)
28. Khamis, A., Hussein, A., Elmogy, A.: Multi-robot task allocation: a review of the state-of-the-art. Coop. Robots Sensor Netw. **2015**, 31–51 (2015)
29. Yang, L., Qi, J., Song, D., Xiao, J., Han, J., Xia, Y. : Survey of robot 3D path planning algorithms. J. Control Sci. Eng. **2016**, 7426913 (2016)
30. Rahman, Q.M., Corke, P., Dayoub, F.: Run-time monitoring of machine learning for robotic perception: a survey of emerging trends. IEEE Access **9**, 20067–20075 (2021)
31. Kroemer, O., Niekum, S., Konidaris, G.D.: A review of robot learning for manipulation: challenges, representations, and algorithms. J. Mach. Learn. Res. **22**(1), 1395–1476 (2021)
32. Jin, L., Li, S., Yu, J., He, J.: Robot manipulator control using neural networks: A survey. Neurocomputing **285**, 23–34 (2018)
33. Muhammad, A., Ali, M.A., Shanono, I.H.: Path planning methods for mobile robots: a systematic and bibliometric review. ELEKTRIKA-J. Electr. Eng. **19**(3), 14–34 (2020)
34. Masehian, E., Sedighizadeh, D.: Classic and heuristic approaches in robot motion planning-a chronological review. World Acad. Sci. Eng. Technol. **23**(5), 101–106 (2007)
35. Hwang, Y.K., Ahuja, N.: Gross motion planning-a survey. ACM Comput. Surv. (CSUR) **24**(3), 219–291 (1992)
36. Atyabi, A., Powers, D. : Review of classical and heuristic-based navigation and path planning approaches. Int. J. Adv. Comput. Technol. (IJACT) **5**(14), 14 (2013)

Machine Learning in Robotics

Hand Gestures Recognition for an Intelligent Wheelchair Steering Command

Patricia Almeida[1], Brigida Monica Faria[2,3(✉)], and Luis Paulo Reis[1,3]

[1] Faculty of Engineering of University of Porto (FEUP), Rua Dr. Roberto Frias, sn 4200-465 Porto, Portugal
up201605165@edu.fe.up.pt, lpreis@fe.up.pt

[2] School of Health, Polytechnic of Porto (ESS-P.PORTO), Rua Dr. António Bernardino de Almeida, 400 4200-072 Porto, Portugal
monica.faria@ess.ipp.pt

[3] Artificial Intelligence and Computer Science Laboratory (LIACC- Member of LASI LA), Rua Dr. Roberto Frias, sn 4200-465 Porto, Portugal

Abstract. The independence and autonomy of both elderly and disabled people have been a growing concern of today's society. Consequently, the increase in life expectancy combined with the ageing of the population has created the ideal conditions for the introduction of Intelligent Wheelchairs (IWs). For this purpose, several adapted sensors should be used to optimize the control of a wheelchair. During this work, the Leap Motion sensor was analyzed to convert the user's will into one of four fundamental driving commands, move forward, turn right, left, or stop. Leap Motion aims to determine the direction to follow according to the hand gesture identified. For this task, data was collected from volunteers while they were performing certain gestures. Thereby it was possible to produce a data set that after being processed and extracted some features enabled the classification of the data with an F1-Score higher than 0.97. Additionally, when tested in a real-time application, this sensor reinforced its high performance.

Keywords: Hand gestures · Leap motion · Intelligent wheelchair

1 Introduction

Independence and autonomy in mobility are two of the most important conditions for determining the quality of life of people with disabilities or with low mobility capacities [1]. Limited mobility could have origin in a broad range of situations, from accidents to disease, to the ageing process.

According to World Health Organization (WHO) [2], approximately 10% of the world's population, or around 740 million people, suffer from disabilities and between those people, almost 10% require a wheelchair. Therefore, it is estimated that about 1% of the total population needs wheelchairs, which translates into 74 million people worldwide [3].

D. Tardioli et al. (Eds.): ROBOT 2022, LNNS 590, pp. 41–52, 2023.
https://doi.org/10.1007/978-3-031-21062-4_4

The importance of providing multifaceted wheelchairs which can be adapted to the most diverse conditions of their users is thus emphasised. Different interfaces are being developed enabling to overcome existing barriers of use. In particular, special attention has been dedicated to voice control techniques, joysticks, tongue or head movements. However, hand gesture recognition systems are proving to be extremely successful methods of wheelchair control due to their accessible price and non-invasiveness. The present study intends to introduce insights into this field of intelligent wheelchairs in order to overcome this perceived problem.

2 State of the Art

Over time and with the trend towards a smart society, humans are experiencing a growing connection to electronics in the digital world, which can greatly enhance our life and productivity [4]. Communication between human beings and machines, specifically computers, has been facilitated by the newest innovations in electronics and wearable technologies. This Human Machine Interface (HMI) system will be increasingly relevant to the Internet of Things (IoT) and universal computing [5]. Commonly, communication begins when an object/machine, collects and interprets a human's desire. Therefore, for the HMI, is essential an input device, for example an adaptive sensor, which can capture the user's intention [6].

2.1 Hand Gestures

The recognition of hand gestures consists not only in following the human movement, but also in interpreting that movement as semantically significant commands [7]. Consequently, gesture recognition research intends to develop systems capable of identifying specific gestures and using them to communicate or control a device. However, gestures need to be segmented in spatial and temporal domains for the right interpretation of human's intention [8]. The static structure of the hand contains information about the hand posture, while the dynamic part corresponds to the gestures. For this gesture detection, two techniques are commonly used: Data Gloves or Vision based methods [7]. The first method determines the hand posture through sensors (mechanical or optical) attached to a glove, which allow the conversion of finger flexions into electrical signals. A method that requires several cables connected between the computer and user, hindering the spontaneity of interaction. The second one, computer vision based technique, is non-invasive and based on the way human beings understand information from their surroundings. In order to perform more human-like interfaces between humans and machines, an efficient and real time gesture recognition must be ensured [9]. According to this, Murthy et al., 2009 [7], enumerated some requirements for a successful vision based system: Robustness, Computational Efficiency, Scalability and User's Tolerance. Considering all these features, four devices for hand gesture and motion tracking could be distinguished: Microsoft Kinect, Intel RealSense Depth Camera, Microchip's MGC3X30 and Leap Motion [10].

The Microsoft Kinect is a body tracking device used for games and real-world applications as diverse as digital signage, virtual shopping, education and telehealth [11,12]. The Intel RealSense Depth Camera has the ability to calculate depth and enable devices to see, understand, interact with, and learn from their environment [13]. It allows not only face tracking and recognition capabilities, but also hand motion tracking. Microchip's MGC3X30 are controller chips based on GestIC®technology which recognize 3D gestures and track motion [14]. This technology uses an electric field (E-field) for advanced proximity sensing that can detect, track and classify a user's hand or finger motion in free space [15]. The Leap Motion is an optical hand tracking controller that captures hand movements with outstanding accuracy [16].

2.2 Leap Motion

As previously mentioned, Leap Motion Controller is an optical sensor capable of detecting and tracking hand movements. This controller simplifies the interaction with digital content, making it more authentic and effortless. Since Leap Motion is the sensor used in this work, its architecture and Application Programming Interface (API) will be analysed in detail in the next subsections.

2.2.1 Hardware

Leap Motion is a small device, as shown in Fig. 1, with an aluminium and scratch-resistant glass construction created by Ultraleap.

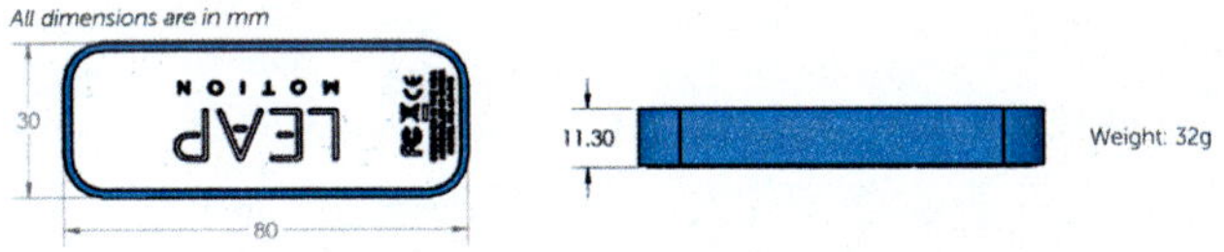

Fig. 1. Leap motion dimensions (adapted from [17]).

Regarding its components, two cameras and three LEDs can be mentioned (Fig. 2). The two 640 × 240-pixel near-infrared cameras are spaced 40 mm apart and operate in the 850-nm spectral range. Due to the large angles of the camera lenses, the sensor has a wide area of interaction. This interactive zone could be described as an upside-down pyramid with a depth between 10 cm to 60 cm preferred, up to 80 cm maximum, directly above the controller device. It should be noted that the range is limited by the propagation of LED light through space [17].

The LEDs are spaced on both sides and between the cameras to avoid overlaps. These detect infrared light with a wavelength of 850 nm, which is outside the visible light spectrum [18].

Once purchased, the package includes the Leap Motion Controller, a quick setup guide and an USB 2/3 hybrid cable.

Fig. 2. Leap motion structure (adapted from [17]).

2.2.2 API Overview

The Leap Motion system is configured assuming a right-handed Cartesian coordinate system with the origin centered on the top of the device. The configuration of the x-, y- and z-axes could be observed in Fig. 3. Just note that the positive direction of z-axes matches to the user's direction.

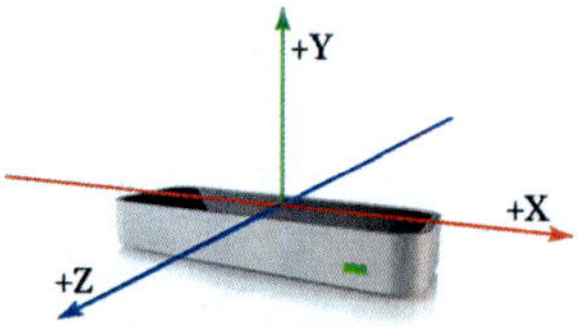

Fig. 3. Leap motion coordinate system (adapted from [18]).

Concerning the measurement units, Leap Motion API evaluates the distance in millimeters, the time in microseconds, the speed in millimeters/second and the angles in radians. The device could track hands, fingers, and tools within its field of vision. The results of the tracked entities could be found in a frame of data that contains information related to the overall motion detected. For calculating these motions, data from two frames are required. As result, different types of motions could be detected, particularly Scale (e.g., one hand moves closer to the other), Rotation (e.g., change of the orientation of the hand) and Translation (e.g., move hands to the left) [19]. Besides that, the list of fingers and/or tools associated with the detected hand could also be provided. The classification in tools or fingers depends on their shape because a tool is usually longer, thinner, and straighter than a finger. For both situations, the position and direction vectors are given to the user.

In addition, the Leap Motion Software could identify some movement patterns as gestures which are also reported in a frame. The four gestures recognized (Swipe, Circle, Key Tap, and Screen Tap) are schematized in Fig. 4.

The full information about the API Overview could be found in the Leap Motion Developer Portal - API Overview [18].

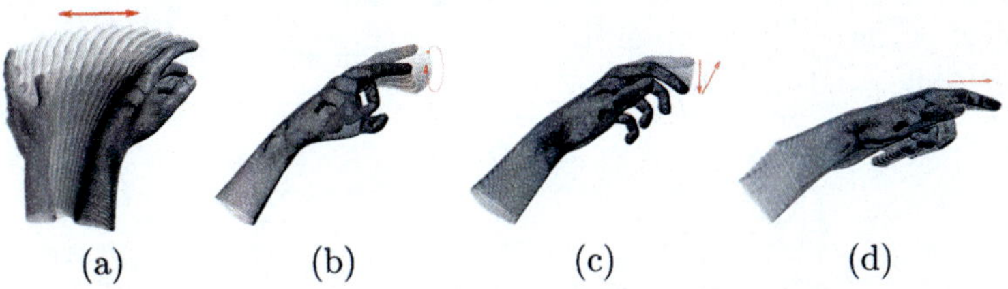

Fig. 4. Gestures recognized by the Leap Motion software (adapted from [18])): (a) Swipe; (b) Circle; (c) Key Tap; (d) Screen Tap.

3 Methodology

Driving an IW requires several commands. In this work, the four basic commands, forward, stop, go to the left and to the right will be assumed as the desired classes. Analysing the specificities of the wheelchair area and considering the properties of the sensors, Leap Motion demonstrated to be an interesting option for being incorporated in the interface of an IW. In summary, Leap Motion is an alternative to the conventional joystick wheelchair and through hand gestures will be able to interpret the user's intention. The approach followed was based on four steps, which means starting with data acquisition, then processing, feature extraction and lastly classification.

3.1 Data Set Description

Based on Boyali et al., 2015 [20], four gestures were selected to achieve a wheelchair autonomous control (Fig. 5). Therefore, this data set incorporates data from five different people, three of them aged between 20 and 24 years and the other two aged 52 years. This age difference and having three participants be female and two male greatly increased the variety and quality of data, given that the hand sizes were different.

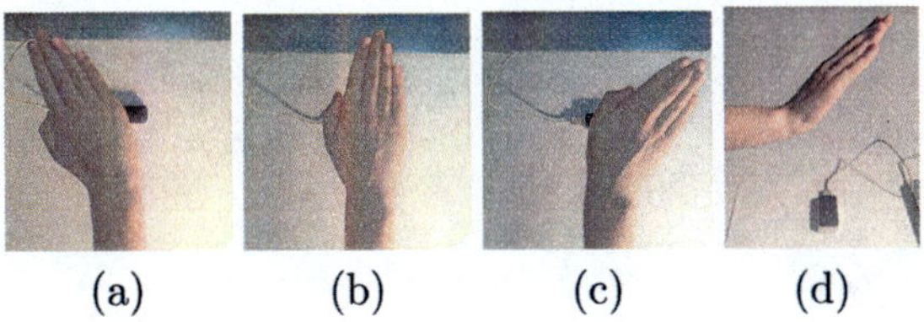

Fig. 5. Gestures representing the wheelchair steering commands: (a) Left; (b) Front; (c) Right; (d) Stop.

The acquisition protocol is explained in the diagram presented in Fig. 6. The first phase was preparation in which the volunteers were snugly seated in a chair placed in front of a computer. The next 4 s were the time to observe a visual cue, which represents the movement to be executed. After a sound was emitted, the subject performed the action, using the gesture shown in Fig. 5. At last, the data recording was concluded with a beep.

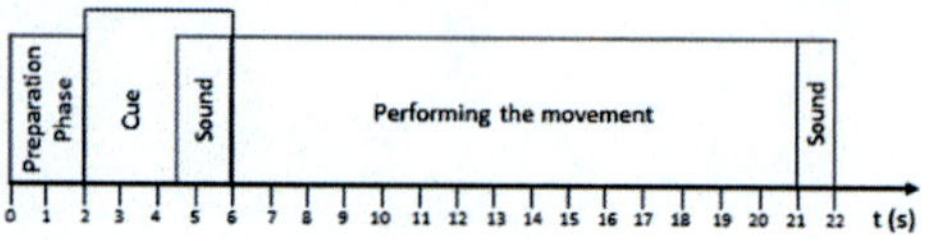

Fig. 6. Timing schedule of the acquisition protocol for the hand gesture data set.

3.2 Data Division

For each gesture, 1 500 samples were collected per participant. The data of all gestures were randomly organised and afterward partitioned into two parts, 80% for training and 20% for testing, which results in 24 000 and 6 000 respectively. When analysing the data per subject, in total this translates into 4 800 for training and 1 200 for testing.

3.3 Feature Extraction

The Leap Motion device provides an extensive API where numerous parameters are available to use as features without requiring pre-processing. Taking into consideration the purpose of this work, the angles of yaw and pitch were chosen. These two variables were adopted as well by Boyaly et al., 2015 [20] and Mohandes et al., 2015 [21] in their studies on Gesture and Posture Recognition. Furthermore, the palm velocity was also included. The detailed description of these features is given below.

1. **Yaw Angle**
   The yaw angle quantifies the rotation around y-axis, thus it is an important value to consider for horizontal hand movements, particularly for radial and ulnar deviation or, in other words, for left and right movements, respectively [22]; hence, for example, when subjects perform the movement to the right, it is expected an increase in this angle value to positive values, reaching the maximum values while the opposite behaviour should happen in movements to the left.
2. **Pitch Angle**
   The pitch angle evaluates the rotation around the x-axis, an interesting parameter for hand flexion and extension movements [22]. Looking at the gestures mentioned (Fig. 5), the stop motion is associated to an extension of the wrist, making this value play an important role.

3. **Palm Velocity**
   This parameter expresses the change rate of the palm position in millimeters/second [18] and it is subdivided into the velocities according to the x, y and z axes. Although the gestures used are static, certain people could have particular hand tremors that influence the classification. Therefore, this feature intends to capture data to avoid this problem.

3.4 Feature Selection and Dimensionality Reduction

Feature selection and Dimensionality reduction are two crucial points that contribute to the achievement of better outcomes. The first concept corresponds to the identification of the most relevant features for the classification problem. While the second, dimensionality reduction, is defined by Van Der Maaten et al., 2009 [23] as the conversion of high dimensional data into a significant representation of reduced dimensionality. Regarding feature selection, due to this being a classification task and the input data being numerical, the ANOVA F test was chosen. This method estimates the F distribution through the relationship between the group variance and within-group variance [24]. Subsequently, the significance level, p, is analysed and when lower than 5% indicates that the feature is statistically significant.

For dimensionality reduction, Principal Component Analysis (PCA) was the approach used. PCA aims to transform data by projecting it into a new orthogonal coordinate system based on an eigenanalysis of the data [25].

It should be highlighted that before dimensionality reduction, two feature scaling techniques, normalization and standardization, were tested to understand which one was more suitable.

3.5 Classification

The classification aims to discriminate four classes (Front, Stop, Left and Right) and for that several classifiers and combinations of classifiers were implemented. In order to evaluate the performance of each predictive model created, two metrics were studied - F1-Score and Confusion Matrix.

Concerning the classifiers, linear and non-linear classifiers were trained, among which: Gaussian Naïve Bayes (GNB), K-Nearest Neighbours (K-NN), Linear Discriminant Analysis (LDA), Decision Tree (DT), Random Forest (RF), Logistic Regression (LR), Support Vector Machine (SVM) and Multilayer Perceptron (MLP).

After all the explained classifiers were tested, **Voting and Boosting** techniques were also implemented. These new approaches are Ensemble methods designed to create a stronger model by combining different predictions.

Voting classifiers combine predictions of multiple models, typically of different types, and the final results come from a majority vote [26]. Consequently, the final predicted class could be either the class that the majority of classifiers predicted or the averaged probability of belonging to the class, leading to **hard and soft voting**, respectively.

AdaBoost (Adaptive Boosting) trains and implements trees in sequence. Initially, it assumes several classifiers (weak learners) and combines their individual predictions [27]. After an iterative process, the weight assigned to each sample is modified, with incorrect classifications having a higher weight than correct ones. In this manner, the classifier tries to improve the misclassified samples by focusing on the higher weights.

Furthermore, during this training process the hyper-parameters of the classifiers were optimised. For this purpose, a grid search with a 10-fold cross-validation was applied.

3.6 Application

A real-time testing represents an essential step to evaluate if the proposed method operates as expected. According to this, two approaches were considered: an implementation of the developed models in the Intellwheels project's wheelchair simulator or the creation of a simple application with the main goal of controlling a robot/object that could symbolize a wheelchair. Comparing both approaches, the choice fell on the second one. The integration of the work in the project simulator would be a long and time-consuming process since it implied the joining of several programming languages. Considering the available time, this process was not feasible and therefore, the second method was selected. In conclusion, the proof of concept involved developing an application similar to a game in which a patient was supposed to drive a robot along a defined path where the four classes studied were present.

4 Results and Discussion

Leap Motion data were acquired while the subjects performed a pre-defined movement for each direction. During this acquisition process, and even though the acquisition protocol was the same for all participants, it was possible to identify a factor that made the data differ from participant to participant - the hand tremors. Some volunteers presented noticeable hand tremors.

Succinctly, when the subject has more hand tremors, this leads to data unsteadiness. However, this variation happens essentially in the same range. For example, in the left-hand movement, the pitch angle could change, but it is still the direction with the lowest pitch angle value. In addition, the palm velocities were also considered as features to reduce this problem.

4.1 Data Set

After all classifiers were trained with the data related to each participant, the best results were recorded. Every participant has a high F1-Score value, greater than 0.97, which are excellent results given that 1 is the maximum value. Regarding the classifiers, MLP was the one that led to the best classification. However, as can be seen in Table 1 (table with the results of all the classifiers for each participant), several classifiers exhibit similar or extremely close results. In terms of the feature scaling technique used, it is evident that the standardization of features achieves the best performances.

Table 1. Results of each classifier obtained on the Leap Motion Data Set.

Classifier	GaussianNB			SVM			DTC			RF			KNN			LR			LDA			LSVM			MLP		
Feature Scaling	-	Stand	Norm	-	Stand	Norm	-	Stand	Norm	-	Stand	Norm	-	Stand	Norm	-	Stand	Norm	-	Stand	Norm	-	Stand	Norm	-	Stand	Norm
1	0,65	0,98	0,50	0,82	1,00	0,84	0,84	1,00	0,76	0,85	1,00	0,83	0,81	1,00	0,70	0,74	0,99	0,78	0,78	0,99	0,81	0,78	0,99	0,49	0,83	1,00	0,84
2	0,96	0,93	0,63	0,97	0,99	0,85	0,98	0,97	0,65	0,98	0,98	0,77	0,96	0,98	0,60	0,97	0,93	0,81	0,97	0,97	0,82	0,60	0,94	0,43	0,98	0,99	0,84
3	0,28	0,97	0,70	0,50	0,98	0,66	0,50	0,98	0,62	0,50	0,99	0,68	0,44	0,98	0,67	0,31	0,96	0,66	0,31	0,97	0,66	0,34	0,93	0,36	0,52	0,99	0,66
4	0,33	0,95	0,79	0,52	0,98	0,73	0,49	0,98	0,66	0,49	0,98	0,69	0,43	0,97	0,67	0,28	0,95	0,71	0,29	0,95	0,75	0,32	0,92	0,35	0,51	0,98	0,63
5	0,33	0,95	0,83	0,52	0,97	0,87	0,50	0,97	0,74	0,50	0,97	0,78	0,43	0,96	0,74	0,29	0,96	0,86	0,29	0,95	0,87	0,22	0,93	0,40	0,53	0,97	0,88

Analysing the confusion matrix of the best participant, in this case the number 1, it is possible to conclude that, firstly as indicated by the F1-Score value, all samples had the class correctly assigned. Second, the test set is composed of approximately the same number of samples from each class which further reinforces the success of this classification. For understanding the feasibility of developing a generic model instead of a personalised version for each patient, the data of all participants was randomly organised and the classifiers were trained. The results obtained are shown in Table 2.

Table 2. F1-score for all participants' data.

		F1 - Score	Classifier	Feature Scaling Technique
Best model		0,983	RF	Standardization
Voting	Hard	0,982	RF + MLP	Standardization
	Soft	0,984		
AdaBoost		0,982	RF	Standardization

Under this scenario, the number of samples has increased from 6000 to 30000. Nevertheless, the results are equally great. The best performance belongs to an Ensemble method, Voting Soft, where the RF and MLP models are the estimators considered. Through the confusion matrix of this classifier, it could be noted that the samples with wrong predicted classes are mainly Right and Stop, although in a low number. The cause of this misclassification could stem from a tendency for people to perform the Stop movement with a small deviation to Right.

4.2 Real Time Application

Subject 1 obtained the best classification. Hence, he was the participant selected for the real-time testing. The system consists of applying the stored MLP classifier on the extracted standardized features. During this experiment, the participant was asked to control a robot using the same gestures that he performed in the data acquisition process. Figure 7 illustrates the application developed for

(a) Application developed (b) Experimental Setup

Fig. 7. Real-time testing.

testing and the setup used. The track in grey corresponds to the path to follow having two stops indicated by the STOP sign. The object in blue is intended to simulate a wheelchair.

In the participant's first contact with the application, he demonstrated some problems in controlling the robot, since the application has some specificities. For example, when the command assigned is left/right the robot turns 30°, so if a 90° rotation is intended, the subject needs to perform 3 times the gesture for that direction. Moreover, the assigned command takes about 1 s to be processed and executed which could induce some confusion. For this reason, a training phase was required. After the adaptation phase, the subject was able to drive the robot during the whole route almost without leaving the grey section. Only 3 misclassifications were noticed, while the participant executed the forward class the model assigned the stop class. To conclude, these were extremely promising results which proved the success and feasibility of using this sensor for wheelchair driving.

5 Conclusion

Different health conditions imply the existence of multiple alternatives for the interaction between a human and a machine, replacing the traditional modes of communication. Consequently, manual wheelchairs are no longer viable solutions for everyone; therefore, with the trend towards an increasingly technological world, different options have emerged, especially that of IW. During this study, Leap Motion was carefully analyzed to decode the user's intention on one of four fundamental control commands for an IW, move forward, turn left, right or stop. Through the proposed methodologies, Leap Motion achieves an optimal result, an F1-Score greater than 0.97. Moreover, it performed really well in the real-time application, being possible to control a robot on a path. To conclude, since the results were successful, the possibility of controlling an IW using hand gesture recognition was supported. This is a promising approach that, with improvements, could be incorporated into a real IW.

Acknowledgements. This work was financially supported by Base Funding UIDB–00027–2020 of LIACC - funded by national funds through the FCT/MCTES (PIDDAC), Sono ao Volante 2.0 - Information system for predicting sleeping while driving and detecting disorders or chronic sleep deprivation (NORTE–01–0247-FEDER-039720), and Intellwheels 2.0 - IntellWheels2.0 - Intelligent Wheelchair with Flexible Multimodal Interface and Realistic Simulator (POCI–01–0247-FEDER-39898), supported by Norte Portugal Regional Operational Programme (NORTE2020), under the PORTUGAL 2020 Partnership Agreement.

References

1. Davies, A., De Souza, L.H., Frank, A.O.: Changes in the quality of life in severely disabled people following provision of powered indoor/outdoor chairs. Disabil. Rehabil. **25**(6), 286–290 (2003)
2. Guidelines on the provision of manual wheelchairs in less resourced settings. https://www.who.int/publications/i/item/guidelines_on_the_provision_of_manual_wheelchairs_in_less_resourced_settings
3. Worldwide Need - Wheelchair Foundation. https://www.wheelchairfoundation.org/fth/analysis-of-wheelchair-need/
4. Zhu, M., He, T., Lee, C.: Technologies toward next generation human machine interfaces: from machine learning enhanced tactile sensing to neuromorphic sensory systems. Appl. Phys. Rev. **7**(3), 031305 (2020)
5. Pavlovic, V., Sharma, R., Huang, T.: Visual interpretation of hand gestures for human-computer interaction: a review. IEEE Trans. Pattern Anal. Mach. Intell. **19**(7), 677–695 (1997)
6. Han, H., Yoon, S.W.: Gyroscope-based continuous human hand gesture recognition for multi-modal wearable input device for human machine interaction. Sensors **19**(11), 2562 (2019)
7. Murthy, G.R.S., Jadon, R.S.: A review of vision based hand gestures recognition. Int. J. Inf. Technol. Knowl. Manage. **2**(2), 405–410 (2009)
8. Trigueiros, P., Ribeiro, F., Reis, L.P.: Hand gesture recognition system based in computer vision and machine learning. In: Tavares, J.M.R.S., Jorge, R.N. (eds.) Developments in Medical Image Processing and Computational Vision. LNCVB, vol. 19, pp. 355–377. Springer, Cham (2015). https://doi.org/10.1007/978-3-319-13407-9_21
9. Hasanuzzaman, M.D., Ampornaramveth, V., Zhang, T., Bhuiyan, M.A., Shirai, Y., Ueno, H.: Real-time vision-based gesture recognition for human robot interaction. In 2004 IEEE International Conference on Robotics and Biomimetics, pp. 413–418. IEEE (2004)
10. Sciberras, J.: Interactive gesture controller for a motorised wheelchair, Ph. D. thesis, Murdoch University (2015)
11. How It Works: Xbox Kinect. https://www.jameco.com/Jameco/workshop/Howitworks/xboxkinect.html
12. Kinect - Windows app development. https://developer.microsoft.com/en-us/windows/kinect/
13. Intel® RealSense™ Technology. https://www.intel.com/content/www/us/en/architecture-and-technology/realsense-overview.html
14. Microchip: MGC3030/3130 3D Tracking and gesture controller data sheet. Technical report. http://ww1.microchip.com/downloads/en/DeviceDoc/MGC3030-3130-3D-Tracking-and-Gesture-Controller-40001667F.pdf

15. GestIC Technology Basics—Microchip Technology. https://www.microchip.com/design-centers/capacitive-touch-sensing/gestic-technology/gestic-technology-basics
16. Tracking—Leap Motion Controller—Ultraleap. https://www.ultraleap.com/product/leap-motion-controller/
17. Ultraleap: Leap motion controller TM. Technical report. https://www.ultraleap.com/datasheets/Leap_Motion_Controller_Datasheet.pdf
18. How Hand Tracking Works—Ultraleap. https://www.ultraleap.com/company/news/blog/how-hand-tracking-works/
19. Afonso, S.: Development of a gestural master interface for tele-surgery applications, Ph. D. thesis, Instituto Superior Técnico (2014)
20. Boyali, A., Hashimoto, N., Kavakli, M.: Continuous and simultaneous gesture and posture recognition for commanding a robotic wheelchair; towards spotting the signal patterns. arXiv:1512.00622 (2015)
21. Mohandes, M., Aliyu, S., Deriche, M.: Prototype arabic sign language recognition using multi-sensor data fusion of two leap motion controllers. In: IEEE 12th International Multi-Conference on Systems, Signals & Devices (SSD15), pp. 1–6. IEEE (2015)
22. Widodo, R.B., Quita, R.M., Setiawan, R., Wada, C.: A study of hand-movement gestures to substitute for mouse-cursor placement using an inertial sensor. J. Sens. Sens. Syst. **8**(1), 95–104 (2019)
23. Maaten, L.V.D., Postma, E., Herik, J.V.: Dimensionality reduction: a comparative. J. Mach. Learn. Res. **10**(66–71), 13 (2009)
24. Harpale, V., Bairagi, V.: An adaptive method for feature selection and extraction for classification of epileptic EEG signal in significant states. J. King Saud Univ-Comput. Inf. Sci. **33**(6), 668–676 (2018)
25. Tomaszewski, J.E., Hipp, J., Tangrea, M., Madabhushi, A.: Machine vision and machine learning in digital pathology. In: McManus, L.M., Mitchell, R.N. (eds.) Pathobiology of Human Disease, pp. 3711–3722. Academic Press, San Diego (2014). ISBN 978-0-12-386457-4
26. Ensemble machine learning algorithms in Python with scikit-learn. https://machinelearningmastery.com/ensemble-machine-learning-algorithms-python-scikit-learn/
27. AdaBoost algorithm: boosting algorithm in machine learning. https://www.mygreatlearning.com/blog/adaboost-algorithm/

Object Segmentation for Bin Picking Using Deep Learning

Artur Cordeiro[1,2(✉)], Luís F. Rocha[2], Carlos Costa[2,3], and Manuel F. Silva[1,2]

[1] ISEP/IPP - School of Engineering, Polytechnic Institute of Porto, Porto, Portugal
[2] INESC TEC - INESC Technology and Science, Porto, Portugal
{artur.j.cordeiro,luis.f.rocha,carlos.m.costa, manuel.s.silva}@inesctec.pt
[3] FEUP - Faculty of Engineering, University of Porto, Porto, Portugal

Abstract. Bin picking based on deep learning techniques is a promising approach that can solve several analytical methods problems. These systems can provide accurate solutions to bin picking in cluttered environments, where the scenario is always changing. This article proposes a robust and accurate system for segmenting bin picking objects, employing an easy configuration procedure to adjust the framework according to a specific object. The framework is implemented in Robot Operating System (ROS) and is divided into a detection and segmentation system. The detection system employs Mask R-CNN instance neural network to identify several objects from two dimensions (2D) grayscale images. The segmentation system relies on the point cloud library (PCL), manipulating 3D point cloud data according to the detection results to select particular points of the original point cloud, generating a partial point cloud result. Furthermore, to complete the bin picking system a pose estimation approach based on matching algorithms is employed, such as Iterative Closest Point (ICP). The system was evaluated for two types of objects, knee tube, and triangular wall support, in cluttered environments. It displayed an average precision of 79% for both models, an average recall of 92%, and an average IOU of 89%. As exhibited throughout the article, this system demonstrates high accuracy in cluttered environments with several occlusions for different types of objects.

Keywords: Bin picking · Deep learning · Mask R-CNN · Grasping · Artificial intelligence · RGB-D multimodal data

1 Introduction

Robotic bin picking based on deep learning techniques is a technology that applies a relatively recent concept, utilizing neural networks, intending to enhance the standard bin picking approaches. These approaches are increasingly being used in bin picking areas due to the potential of deep learning policies to further increase the intelligence and learning capabilities of robots.

D. Tardioli et al. (Eds.): ROBOT 2022, LNNS 590, pp. 53–66, 2023.
https://doi.org/10.1007/978-3-031-21062-4_5

This article provides an accurate solution to robotic bin-picking problems in highly cluttered environments for any type of object, similar to an industrial scenario, where the robot's objective is to pick several objects from different bins, each of them composed of only one type of object. To accomplish this task the framework has different models that accurately detect different types of objects, alternating them according to the determined action, achieving a robust process that detects objects in highly cluttered environments by essentially modifying only the particular Mask R-CNN trained model.

Given these ideas, and after this brief introduction, the second section details several related works to this article. In the third section is described the deep learning training applied to Mask R-CNN neural network with TensorFlow 2, and details how the dataset was generated. The fourth section provides a brief overview of the bin picking algorithm and specifies the main stages of the implemented framework. The succeeding section displays several results acquired with evaluation datasets. To conclude, the paper briefly summarizes the implementation and presents possible future developments.

2 Related Work

Over the years, several methods have been developed to solve bin picking problems, such as the one proposed by Doumanoglou *et al.* [1] in which an unsupervised feature learning from depth-invariant patches for highly cluttered environments. Pochyly*et al.* [2] implemented a revolving vision system with a specific gripper for organized industrial environments. Choi*et al.* [3] developed a voting-based pose estimation algorithm for highly cluttered environments with several occlusions. Yan*et al.* [4] implemented a pipeline for different types of objects in cluttered scenarios using an adaptive threshold segment to accelerate the pose estimation. Leão *et al.* [5] proposed a framework for entangled objects.

This article focus on solutions based on deep learning techniques, resorting to distinct approaches, namely neural networks, and picking orientations. Lenz *et al.* [6] implemented one of the first Red, Green, Blue - Depth (RGB-D) multimodal parallel gripper-oriented methods, evaluating grasp candidates for random objects (Cornell grasping dataset objects) in a scenario without occlusions; to select the final grasps pose were applied two deep networks with two inputs: RGB and depth data. Similar to Lenz *et al.*, Mahler *et al.* [7] developed a research project for generating synthetic datasets, parallel grasps, and metrics of grasps based on physics, called Dex-Net. Dex-net 2.0 [8] presented a parallel gripper-oriented method to rapidly predict the probability of success of grasps from depth images.

Diverging from gripper-oriented approaches, and focusing more on object-oriented segmentation methods, Zeng *et al.* [9] developed an RGB-D multi-view object-oriented approach for the Amazon Picking Challenge (APC). This approach estimates the 6D pose by segmenting and labeling multiple views of the scenario with a fully convolutional neural network and then fitting pre-scanned

3D object models. Le *et al.* [10] developed a work with similarities to the proposed implementation in this article, using the same neural network (Mask R-CNN); however, the solution proposed by these authors was for planar objects in cluttered environments using a vacuum gripper.

3 Deep Learning Training

Deep neural networks are mainly known and selected from their architectures, presenting different features, such as bounding-box, and classifications, among many others. As for this work, it was intended to detect and segment instances of objects in a cluttered environment. Mask R-CNN was chosen from all the possibilities considering that, in the first place, was the newest stable version of R-CNN, a neural network with good results and a lot of progress made in the segmentation field. Secondly, in several published papers, Mask R-CNN obtained results above average, surpassing a lot of other neural networks. Thirdly, as previously described, the main objective of this implementation was to find a specific known place of the model to grab the object, this way excluding most of the gripper-oriented neural networks. Lastly, it is an open-source neural network with good community support, and with exceptional results concerning the detection and segmentation of different objects with several occlusions [10–12].

Mask R-CNN consists of two stages. The first stage, called a Region Proposal Network (RPN), proposes candidate object bounding boxes. The second stage, which is, in essence, Fast R-CNN [13], extracts features using region of interest pooling (RoIPool) from each candidate box and performs classification, bounding-box regression, and outputs a parallel binary mask for each Region of Interest (RoI) [14].

Mask R-CNN was trained in a system composed of a Ryzen 5 5600x processor, GTX 1060 6 GB graphic card, and 32 GB DDR4 memory. Unlike the original neural network [15], this framework was processed with TensorFlow version 2 (2.4.1), ubuntu 20.04, CUDA toolkit 11.0, Nvidia driver 450.191.01, cuDNN 8.0, and Keras 2.4.0. This adjustment included a few changes to the neural network, assuring the compatibility between the used Mask R-CNN framework, functions, and libraries [15,16]. Nearly all the required adjustments were due to the incompatibility from TensorFlow functions, such as log_graph, set_intersection, among others, that changed between TensorFlow version 1 and TensorFlow version 2 [17]. Furthermore, the dataset class in the main python training script was adapted to load different annotations formats (JSON and COCO) from the custom dataset generated, different neural network parameters were adjusted, and lastly, Mask R-CNN neural network layers (2+,3+, heads,+4,...) training procedure was adapted according to the learning results.

3.1 Dataset Generation

The dataset generation was executed with real data, acquired by an RGB-D sensor, and manually labeled by a human operator with third-party annotator support. The RGB-D sensor, a Photoneo PhoXi 3D Scanner, is capable of acquiring depth and grayscale data, allowing different types of deep learning training inputs. The scanner has up to 3.2 million 3D points of resolution with a throughput of 16 million points per second and from 384 to 520 mm of scanning range. The resolution of the 2D images taken is up to 2060×1544, including grayscale and depth images. It can be controlled by software, called PhoXi Control, via ethernet, which includes a Graphical User Interface (GUI) and an Application Programming Interface (API) [18]. In this implementation, the datasets used for both training and evaluation were divided into two models: 90° elbow tube (Model A) and triangular wall support (Model B). These two datasets were generated in order to enhance the labelling precision and, consequently, the training quality, in such manner, for the moment the proposed neural network is limited to the two objects as proposed in the initial implementation objective.

The training was executed with 2D grayscale data, although in later stages of the process was necessary point cloud data, in order to implement pose estimation and segmentation methods. The 3D point cloud data is automatically generated by the sensor through multi-modal approaches from the initial data acquisition (RGB or grayscale and depth map).

The generated dataset for model A was divided into 4 batches. The first batch was introduced with different 3D views of the scenario, including only one tube or model. Similar to the first, the other batches, from second to fourth, presented a similar scenario with different views, differing in the number of objects presented, in this case, 2 to 4 objects, respectively. The higher the number of batches, the more complex the displayed environment.

The second resorted approach was very similar to model A: the dataset was divided into three different batches, differing in the number of objects presented. To add robustness to the model, an augmentation process was implemented, with added noise, rotating, cropping, flipping, saturation, and brightness, among other processes, enhancing the training process.

Table 1 depicts information about the dataset generation of the two models. Model A dataset is larger than model B, and with more geometrically complex objects to segment, requiring on average 21 s to annotate a single object. Taking into account that some images have 4 objects, it requires on average 84 s to complete an image annotation. Model B dataset was originally much smaller, including 59 images with 155 annotations; however, with the application of augmentation steps, were generated 179 images from the original 59.

Table 1. Dataset information

Dataset	Nr of images	Nr of annotations	Time per annotation (s)
Model A	488	737	21
Model B	59	155	11.49
Model A + Augmentation	732	–	–
Model B + Augmentation	238	–	–

4 Proposed Bin Picking System

The basis of the system developed to detect objects in a cluttered environment is depicted in Fig. 1. As perceived in this image, there are three main areas, that will be described, namely 2D detection, 3D Segmentation, and Pose Estimation.

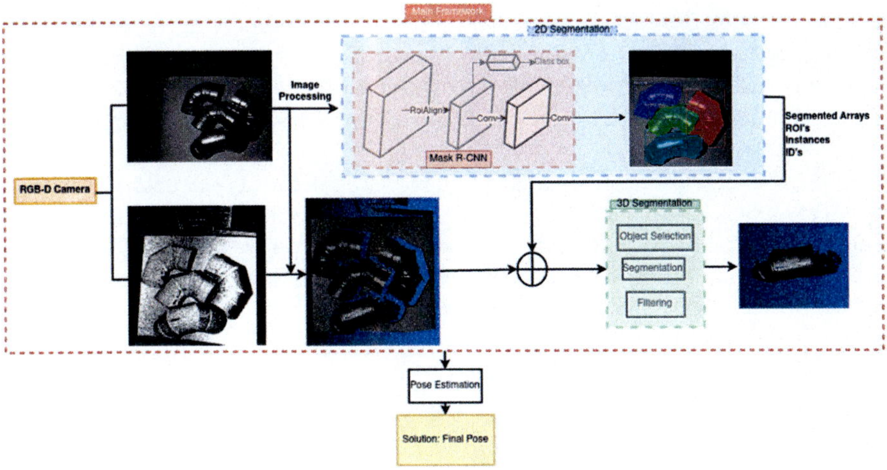

Fig. 1. Object detection pipeline overview

The idea of the proposed system was to enhance the performance of the recognition framework, and part of the segmentation approach, with techniques based on deep learning methods, whilst adopting 3D point cloud based heuristics for the final segmentation of the objects and their respective pose estimation. The primary reason for this proposed adjustment is that deep learning models when well trained, can provide a high level of robustness to the system, meaning that, when encountering novel and difficult environments (highly cluttered environments), this system can accurately segment the target objects. Another reason is the easy system modification to detect a novel object, only requiring a new neural network model with the knowledge to identify this new object, therefore not involving several complex steps.

4.1 2D Object Detection

The detection is generated in the 2D space, resorting to an object instance segmentation framework (Mask R-CNN), which can identify several segments, with a mask and bounding box, of the same types of objects.

The detection algorithm, depicted in Algorithm 1, starts by loading the specified model and selecting the correct device, among GPU or CPU (in particular, every result displayed in this article was generated using CPU). After the loading is completed, and is called the process command from the client, the algorithm waits for an image message publication to read its data.

After reading the input image, image processing techniques are performed to enhance the object's visualization. First is implemented a denoise method, with the fastNlMeansDenoising OpenCV function, followed by a Contrast Limited Adaptive Histogram Equalization (CLAHE) according to the image pixels mean values, *i.e.* the algorithm is going to generate a higher contrast if the image presents low details because of the brightness levels employing an exponential function. The functions were applied in this order to first remove noise from the image, because CLAHE causes noise in near-constant regions as the background, and to improve the contrast in the image, improving several detections with darker images.

Mask R-CNN detection, explicit in line 5 of Algorithm 1, extracts the number of objects, mask (segmentation) of each object, and bounding boxes. Afterward is generated a 2D array with all the masks discriminated by a different id for each instance, the number of objects, the center coordinates (x,y) of each bounding box, and the image resolution, and is published as an 8-bit image (2D array) topic to be received by the 3D point cloud segmentation algorithm. In the end is published a result to the client: successful if is collected an acceptable point cloud from the 3D point cloud sensor, and unsuccessful if the point cloud is not acceptable or is not received.

Algorithm 1. $mask\_img \leftarrow Object\_Detection(img, m)$

Input: $img = image, m = trained\_model$

1: $img = fastNlMeansDenoising(img)$
2: $limit = exponential\_function(image\_meanvalue)$
3: $CLAHE = createCLAHE(cliplimit = limit, tileGridSize = (3, 3))$
4: $img = CLAHE.apply(img)$
5: $detection = m.detect(img)[0]$
6: $info\_array = instances, img.size$
7: **for** i in range(0, $instances$) **do**
8: $\quad roi\_x = (roi[i, 1] + roi[i, 3])/2$
9: $\quad roi\_y = (roi[i, 0] + roi[i, 2])/2$
10: $\quad info\_array = info\_array, roi\_x, roi\_y$
11: $\quad binary\_mask = masks * (i + 1)$
12: $\quad id\_mask = (binary\_mask == (i + 1))$
13: $\quad final\_mask[id\_mask] = binary\_mask[id\_mask]$
14: **end for**
15: $mask\_img = append(final\_mask, info\_array)$
16: Publish $mask\_img$

4.2 3D Point Cloud Segmentation

The 3D point cloud segmentation algorithm expects to receive an image array containing all the detection information provided from Object_Detection, and the original point cloud acquired at the same time as the grayscale image. When the point cloud is received, it is transformed from a ROS message `sensor_msgs::PointCloud2` to PCL `pcl::PointCloud< XYZRGBA >`.

As described in the previous sub-section, the information exported through the image array is organized in specific variables, as presented in Lines 5-8 of Algorithm 2. Succeeding the data management, the original point cloud is resized (Line 9), allowing to treat this point cloud as a 2D array with the same shape as the original grayscale image acquired, and is computed the mean 8×8 square center region `z` value of each segment (characterized by the variable h), *i.e.* the height of each segment, shown in lines 10-20 (Algorithm 2), to predict the object closest to the sensor in line 21 (Algorithm 2).

After estimating the highest point cloud segment, it is extracted from the original point cloud through its ID (lines 22-28 of Algorithm 2) and are applied several filters, such as voxel grid, pass-through and statistical outlier removal (line 29 of Algorithm 2), that essentially are used to remove noise and unnecessary information. This algorithm structure improves the total time consumed because, instead of selecting and filtering every point cloud segment detected, it only selects and filters the highest segment points, which is the essential instance to pick up.

The images displayed in Fig. 2 are results from the detection and segmentation system from one random sample in each dataset of models A and B. As observed, the system receives the original image (left column of Fig. 2) and detects the distinct instances in the scene, identified by colour segmentation masks for easy visualization (Middle-Left column). Analyzing the results, even in novel and arduous environments, the system can detect and segment each object with high accuracy, as observed in the last column of Fig. 2.

4.3 Pose Estimation and Grasping

The pose estimation algorithm is implemented after the point cloud segmentation and is based on cloud matching, between the reference point cloud (displayed in Fig. 3) and the acquired point cloud. Essentially, the segmented point cloud is matched with a reference point cloud provided through the Computer-Aided Design (CAD) model of the specific part that is intended to be picked; subsequently is generated a grasp solution from the grasp evaluation system, detecting the best possible grasp (from all the pre-computed candidates) to pick the object in its current position [19].

The algorithm begins loading the pipeline configurations and the reference point cloud, acquired either from the 90° elbow tube CAD model or the triangular wall support CAD model, displayed in Fig. 3. Subsequently, the point cloud is filtered, and are estimated the surface normals and curvatures. Concluding the first step, the reference point cloud is stored, and are applied and stored the keypoints and their descriptors.

Algorithm 2. *PointCloud_Segmentation(mask_im)*

```
Input: cloud = pointcloud
while !cloud_received_flag do
    ros :: Duration(0.01).sleep();
    ros :: spinOnce();
end while
for int i = 0; i < Instances; i + + do
    cloud_center_x[i] = info_array.roi_x
    cloud_center_y[i] = info_array.roi_y
end for
Resize cloud_received(height * width)
for l = 0; l < Instance, l + + do
    for i = cloud_center_x[l] − 8; i ≤ cloud_center_x[l] + 8, l + + do
        for j = cloud_center_y[l] − 8; j ≤ cloud_center_y[l] + 8, l + + do
            if cloud_received → at(i, j).z > 0 then
                cnt[l]++
                h[l] = h[l] + cloud_received → at(i, j).z
            end if
        end for
    end for
    h[i]= h[i] / cnt[i]
end for
higher_cloud = std :: distance(h, std :: min_element(h, h + Instances))
for int i = 0; i < width; i + + do
    for int j = 0; j < height; j + + do
        if id_cloud(i,j)==higher_cloud+1 then
            partial_cloud = push_back(cloud_received− > at(i, j))
        end if
    end for
end for
filtered_cloud = Filters(partial_cloud)
Publish filtered_cloud
```

The initial pose is estimated through RAndom SAmple Consensus (RANSAC), and an initial alignment of the two point clouds is refined, using the ICP algorithm, as detailed in Fig. 4, applying a transformation from the reference point cloud, displayed at the middle right image of Fig. 4 to the detected point cloud in the middle left image of Fig. 4.

After the pose estimation, possible object collisions are checked to evaluate the best grasp candidate among all the previously taught grasp poses exhibited in Fig. 5. This decision is formed through scores that are given based on heuristics, such as euclidean distance, depth distance, roll, pitch, and yaw distance, among others. In the end, the grasp candidate with the lower cost is chosen [20].

After the full procedure (detection, segmentation, pose estimation, and grasp estimation) the robot manipulator starts the bin picking action, displayed in Fig. 6, moving from its initial static pose (Fig. 6 (a)) to the estimated grasping

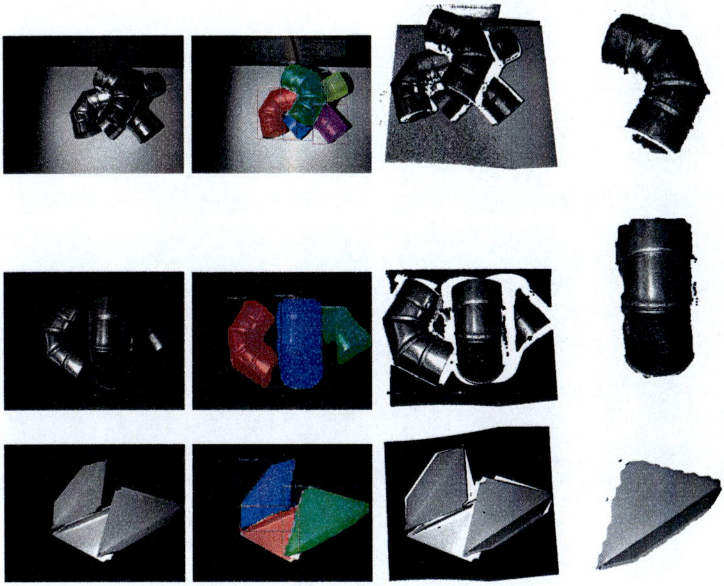

Fig. 2. Evaluation results. (Left) original grayscale image (Middle-Left) Detection mask (Middle-Right) Original point cloud (Right) Partial point cloud segmentation

Fig. 3. Reference point cloud. (a,c) CAD model (b,d) Generated point clouds

Fig. 4. Pose estimation overview

pose ((b) and (d)), and concluding with a trajectory to place the object in the bin ((c) and (e)). The same process can be repeated for any other object, modifying only some features of the main object detection framework.

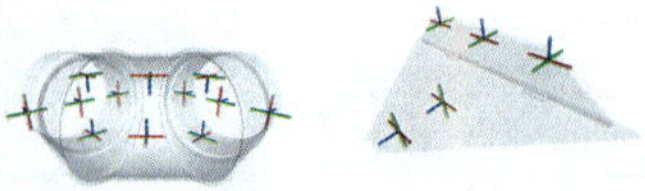

Fig. 5. Grasping candidates

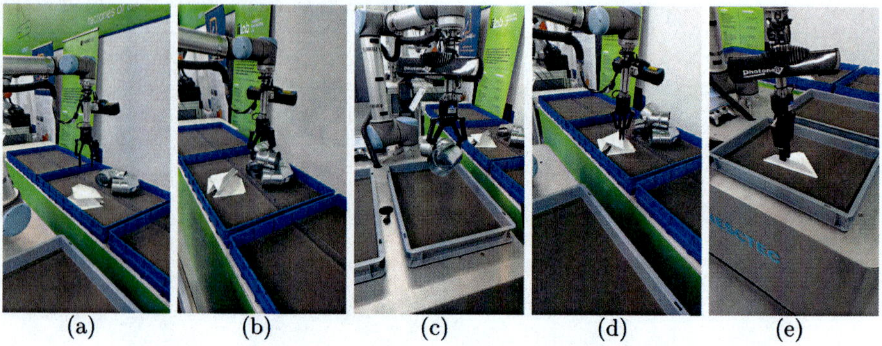

Fig. 6. Bin picking action sequence

5 System Tests and Evaluation

The object detection and segmentation framework was tested on eleven types of scenarios (datasets), detailed in Table 2, divided into six types of datasets for model A and five types of datasets for Model B. The characteristics of the computer used for this evaluation were described in Sect. 3, being only used the CPU for the inference stage, and only using the GPU for the neural network training. The evaluation datasets (depicted in Fig. 7) are divided according to the number of objects presented in the bin and the type of environment, varying between a known environment (images that trained the Neural Network (NN)) and a novel environment, with less illumination, new positions, viewpoints, and backgrounds, creating an arduous scenario to apply and test the developed framework.

Table 2. Evaluation dataset

Evaluation dataset	Number of images	Number of annotations
Tubes_eval	118	269
Triangles_eval	54	102

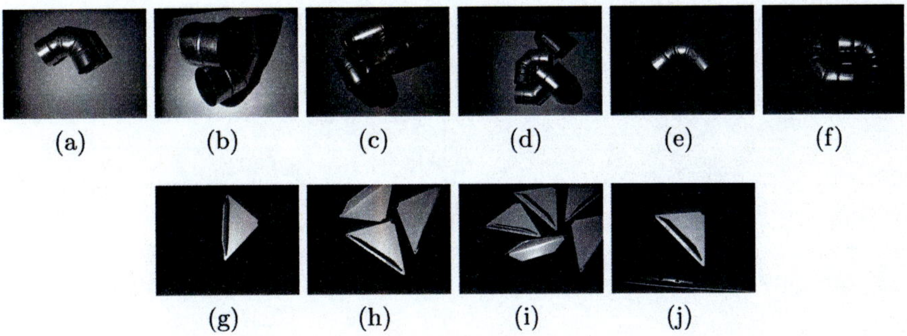

Fig. 7. Evaluation datasets. Top- Model A (knee tube), Bottom- Model B (triangle)

5.1 Object Detection and Segmentation Efficiency and Performance

The object segmentation framework average results are illustrated in Table 3, depicting all the standard deep learning metrics, such as Intersection over Union mask (IoU_mask, represented as mIOU), Average precision (Ap), Average Recall (Ar), and F1 score.

Table 3. Metrics precision

Overall	Ap	Ap_50	Ap_75	mIOU	Ar_50	F1_50
Model A	0.829	0.971	0.937	0.905	0.955	0.904
Model B	0.751	0.893	0.819	0.873	0.890	0.885

Compared to model A, model B results display a lower average score on every metric, however still exhibiting high values (around 85%) on every metric, apart from Ap (with 75%). These results were expected because model B had a lower quality training compared to model A, with less training time, fewer images, just one camera viewpoint, and fewer batches with a distinct number of objects.

Comparing these results to the ones of the original Mask R-CNN [14], it is observed that the results obtained with these trained models are better. The improvement is understandable due to the type of objects displayed in an image (classes), the simpler background due to the application in an industrial environment, and the training phase of the neural network.

The execution time of the object detection and segmentation framework, illustrated in Table 4, is divided into the detection stage, segmentation, and segmentation with filtering. The values presented are an average measurement from all datasets samples.

The detection stage, on average, requires around 0.49 s to generate the necessary masks of the input image. The segmentation without filters requires an average of 0.04 s, generating the segmented point cloud of the input mask closest to the sensor, however exhibiting all the noise and outliers produced by the

Table 4. Execution time (CPU)

Overall	Instances number	Det (s)	Seg (s)	Seg + Filters (s)
Model A	1	0.475	0.048	0.452
	2	0.481	0.070	0.431
	3	0.519	0.045	0.417
	4	0.518	0.038	0.483
Model B	1	0.464	0.033	0.299
	3	0.508	0.037	0.278
	5	0.504	0.028	0.193

depth sensor in the original point cloud. The segmentation with filtering requires around 0.36 s, applying pass-through, voxel grid, and remove outlier filters.

The average system total time is 0.53 s without filtering, and 0.85 s with filtering. Therefore, the application of filters requires on average 0.32 s, increasing or decreasing the time required according to the number of points given to the filters.

Analyzing these results, one big advantage is that the time elapsed is not dependent on the instance number, due to the segmentation algorithm structure. As displayed in Table 4, the only noticeable difference is in the segmentation and filtering stage between model A and model B, diverging from 0.44 s on average to 0.25 s, on average, respectively. This difference is due to the original point cloud generation, the model B point cloud has much less noise and outliers than the model A point cloud.

These average execution time values can fluctuate according to the model resolution, namely the values presented in this paper are for 1024×1024 image resolution, which was the resolution that the images were resized when training the model. The time can be reduced by decreasing this resolution.

6 Discussion and Future Work

This paper addressed bin picking problems in clutter environments, implementing a system based on deep learning techniques.

After analyzing different frameworks, it was proposed an approach to detect and segment instances based on instance segmentation techniques and point cloud procedures, using grayscale images and point cloud data generated with an RGB-D sensor. This implementation estimates the different instance segments in a grayscale image by applying image processing techniques and Mask R-CNN neural network. Subsequently, it is received point cloud data to estimate and extract the object (partial point cloud) closest to the sensor, to be sent to the pose and grasp estimation systems, and finally, the robot manipulator grasps the object through the desired position. The proposed system is capable of segmenting several types of objects in cluttered environments. However, in this case, it was only trained to detect two parts, a 90° elbow tube model (model A) and a triangular wall support model (model B).

The accuracy and execution time achieved by the proposed system allows to detect and segment instances with several occlusions in hard environments, being able to estimate the closest object to the sensor, which will probably be the easiest to pick up, with acceptable execution times for an industrial solution, providing a precise solution (averaging 79% Ap) to pick up specific objects with an easy configuration (pre-inference) system, allowing its application in industrial scenarios where numerous objects are placed into a bin.

The current implemented system can be further improved by estimating an initial pose of the segmented object, given the fact that the partial point cloud has the same (x, y) position as the original one, improving the efficiency of the matching algorithm in the pose estimation system, and decreasing the iterations required to match the two point clouds. In addition to initial pose estimation, the system can be implemented with multiple channels of neural network inputs, modifying Mask R-CNN to predict the output based on, for example, two types of data (such as grayscale and depth images), instead of segmenting instances based only on a 2D image. On the other hand, to enhance system performance with Mask R-CNN models with less resolution and similar precision values.

Considering that the neural network training stage is very time-consuming and exhausting to create instance segment labels for all the original images, this stage can be improved with an automated image generation and labelling system, for example, using 3D creation software, such as Blender.

Acknowledgements. The research leading to these results has received funding from the European Union's Horizon 2020 research and innovation programme under grant agreement No 101006798.

References

1. Doumanoglou, A., Kouskouridas, R., Malassiotis, S., Kim, T.-K.: Recovering 6d object pose and predicting next-best-view in the crowd. In: 2016 IEEE Conference on Computer Vision and Pattern Recognition (CVPR). IEEE, Jun 2016. https://doi.org/10.1109/cvpr.2016.390
2. Pochyly, A., Kubela, T., Singule, V., Cihak, P.: Robotic bin-picking system based on a revolving vision system. In: 2017 19th International Conference on Electrical Drives and Power Electronics (EDPE). IEEE, Oct 2017. https://doi.org/10.1109/edpe.2017.8123228
3. Choi, C., Taguchi, Y., Tuzel, O., Liu, M.-Y., Ramalingam, S.: Voting-based pose estimation for robotic assembly using a 3d sensor. In: 2012 IEEE International Conference on Robotics and Automation. IEEE, May 2012. https://doi.org/10.1109/icra.2012.6225371
4. Yan, W., Xu, Z., Zhou, X., Su, Q., Li, S., Wu, H.: Fast object pose estimation using adaptive threshold for bin-picking. IEEE Access **8**, 63 055–63 064 (2020). https://doi.org/10.1109/access.2020.2983173
5. Leão, G., Costa, C.M., Sousa, A., Veiga, G.: Detecting and solving tube entanglement in bin picking operations. Appli. Sci. **10**(7), 2264 (2020). https://doi.org/10.3390/app10072264

6. Lenz, I., Lee, H., Saxena, A.: Deep learning for detecting robotic grasps (2013). https://arxiv.org/abs/1301.3592
7. Mahler, J., et al.: Dex-net 1.0: A cloud-based network of 3d objects for robust grasp planning using a multi-armed bandit model with correlated rewards. In: 2016 IEEE International Conference on Robotics and Automation (ICRA). IEEE, May 2016. https://doi.org/10.1109/icra.2016.7487342
8. Mahler, J., et al.: Dex-net 2.0: Deep learning to plan robust grasps with synthetic point clouds and analytic grasp metrics (2017). https://arxiv.org/abs/1703.09312
9. Zeng, A., et al.: Multi-view self-supervised deep learning for 6d pose estimation in the amazon picking challenge (2016). https://arxiv.org/abs/1609.09475
10. Le, T.-T., Lin, C.-Y.: Bin-picking for planar objects based on a deep learning network: A case study of USB packs. Sensors **19**(16), 3602 (2019). https://doi.org/10.3390/s19163602
11. Höfer, T., Shamsafar, F., Benbarka, N., Zell, A.: Object detection and autoencoder-based 6d pose estimation for highly cluttered bin picking (2021). https://arxiv.org/abs/2106.08045
12. Danielczuk, M., et al.: Segmenting unknown 3d objects from real depth images using mask r-cnn trained on synthetic data (2018). https://arxiv.org/abs/1809.05825
13. Girshick, R.: Fast r-cnn (2015). https://arxiv.org/abs/1504.08083
14. He, K., Gkioxari, G., Dollár, P., Girshick, R.: Mask r-cnn (2017). https://arxiv.org/abs/1703.06870
15. Abdulla, W.: Mask r-cnn for object detection and instance segmentation on keras and tensorflow (2017). https://github.com/matterport/Mask_RCNN
16. Tested build configurations. https://www.tensorflow.org/install/source#gpu. (Accessed: 03 Mar 2022)
17. Mask r-cnn for object detection and segmentation using tensorflow 2.0. https://github.com/ahmedfgad/Mask-RCNN-TF2. (Accessed 13 Mar 2022)
18. Industrial 3d scanner: Phoxi®. https://www.photoneo.com/phoxi-3d-scanner/?gclid=EAIaIQobChMI_4y38LLp-AIVCcPVCh2E0A8eEAAYASAAEgJMwfD_BwE
19. Costa, C., Sobreira, H., Sousa, A., Veiga, G.: "Robust 3/6 dof self-localization system with selective map update for mobile robot platforms. Robot. Autonom. Syst. **76**, 113–140 (2016)
20. de Souza Carvalho, J., et al.: Reconfigurable grasp planning pipeline with grasp synthesis and selection applied to picking operations in aerospace factories. Robot. Comput.-Integ. Manuf. **67**, 07 (2020)

Robot Navigation Anticipative Strategies in Deep Reinforcement Motion Planning

Óscar Gil(✉) and Alberto Sanfeliu

Institut de Robòtica i Informàtica Industrial, CSIC-UPC, Barcelona, Spain
{ogil,sanfeliu}@iri.upc.edu

Abstract. The navigation of robots in dynamic urban environments, requires elaborated anticipative strategies for the robot to avoid collisions with dynamic objects, like bicycles or pedestrians, and to be human aware. We have developed and analyzed three anticipative strategies in motion planning taking into account the future motion of the mobile objects that can move up to 18 km/h. First, we have used our hybrid policy resulting from a Deep Deterministic Policy Gradient (DDPG) training and the Social Force Model (SFM), and we have tested it in simulation in four complex map scenarios with many pedestrians. Second, we have used these anticipative strategies in real-life experiments using the hybrid motion planning method and the ROS Navigation Stack with Dynamic Windows Approach (NS-DWA). The results in simulations and real-life experiments show very good results in open environments and also in mixed scenarios with narrow spaces.

Keywords: Deep reinforcement learning · Robot navigation · Social force model · Human-robot interaction · Anticipation

1 Introduction

Collaborative tasks [1] are essential in areas like social robotics and human-robot interaction. Usually, social-aware robot navigation [2] is involved in these tasks in different ways like, for example, assistive robots [21], approaching task, collaborative searching [6] or side-by-side navigation [18].

The most typical robot navigation approaches take into account the map knowledge to compute a global plan to a goal avoiding obstacles in combination with a local planner to react in a short range. Dynamic Windows Approach (DWA) [9] or Social Force Model (SFM) [14] are reactive local planners and they are useful in presence of unpredictable situations in dynamic environments. However, when there are moving obstacles (e.g., person, bicycles, scooters, etc.),

O. Gil—Work supported under the Spanish State Research Agency through the Maria de Maeztu Seal of Excellence to IRI (MDM-2016-0656) and ROCOTRANSP project (PID2019-106702RB-C21/AEI/10.13039/501100011033). Óscar Gil is also supported by the Spanish Ministry of Science and Innovation under an FPI-grant, BES-2017-082126.

D. Tardioli et al. (Eds.): ROBOT 2022, LNNS 590, pp. 67–78, 2023.
https://doi.org/10.1007/978-3-031-21062-4_6

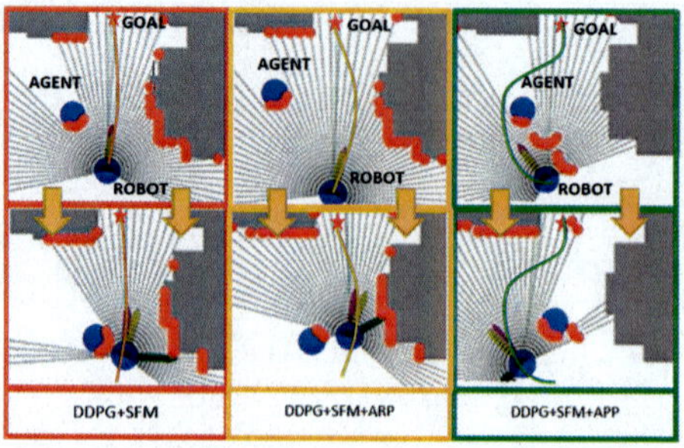

Fig. 1. Comparison between models. The image shows the trajectory changes when two anticipative strategies are applied in the same situation. The ARP model follows a more secure trajectory than the hybrid model and the APP model chooses a trajectory to avoid collisions passing from behind the agent.

reactive behaviors are not enough. It is more useful to have an anticipative behavior to deal with them [8].

Due to the noise and the incomplete environment knowledge, planners which require accurate data like Artificial Potential Fields (APF) [3] can fail in real life applications. In these cases, Deep Reinforcement Learning (DRL) can provide a feasible solution through algorithms like Deep Deterministic Policy Gradients (DDPG) [15] and Proximal Policy Optimization (PPO) [13]. These approaches yield good performance in robot navigation without complete map knowledge.

In this work, we present three different anticipative strategies that can be used in different motion planners. We have tested these strategies using our hybrid motion planning model for robot navigation that combines a DDPG policy based on [4] and SFM. The three models denominated "Anticipative Turn (AT)", "Anticipative Robot and Pedestrian's Propagation (ARP)" and "Anticipative Pedestrian's Propagation (APP)", compute the best hybrid navigation decision taking into account the prediction of the mobile objects in open and complex dynamic scenarios (refer to Fig. 1). We have validated the models through real-life experiments and using simulation in dynamic scenarios for different noise values and moving obstacle velocities.

The remainder of this paper is organized as follows. In Sect. 2, the related work is introduced with a description of the hybrid model that combines DDPG and SFM. The anticipative strategies are described in Sect. 3. In Sect. 4, a quantitative and qualitative evaluation is presented through simulations and real-life experiments. Finally, in Sect. 5, the conclusions are provided.

2 Related Work

2.1 AutoRL for Robot Navigation

Automated Reinforcement Learning (AutoRL) [4,7] is a training algorithm that uses DDPG or other DRL approaches [12,17] to train policies searching for the reward and neural network optimal hyperparameters that maximize an objective function.

In [4], the algorithm is applied to point-to-point navigation task (P2P), which consists on navigating from an initial position to a goal using as policy actions, the velocities at each navigation step. The environment information is the robot Lidar observations in goal polar coordinates. The reward function is:

$$R_{\theta_{r_{\text{P2P}}}} = \boldsymbol{\theta}^T_{r_{P2P}}[r_{\text{step}}\, r_{\text{distg}}\, r_{\text{col}}\, r_{\text{turn}}\, r_{\text{clear}}\, r_{\text{g}}] \tag{1}$$

where $\boldsymbol{\theta}^T_{r_{\text{P2P}}}$ are the reward hyperparameters for the P2P task, r_{step} is a constant penalty at each step in the episode with value 1, $r_{distgoal}$ is the negative Euclidean distance to the goal, r_{col} is 1 when there are collisions and zero otherwise. r_{turn} is the negative angular speed, r_{clear} is the distance to the closest obstacle and r_g is 1 when the agent reaches the goal and zero, in the remaining cases.

AutoRL have been implemented as a local planner that complements a global planner as PRM [10] in large indoor environments. The results in real environments show high performance and robustness to noise. However, AutorRL requires a very large training time and it has high computational cost.

2.2 Hybrid Policy with SFM

This subsection explains the hybrid model proposed in [11] resulting from a DDPG training and the SFM with collision prediction (CP) [23].

First, the model is trained using the DDPG algorithm with the same noise and optimal hyperparameters, obtained in [10] for a differential drive robot model. The training is performed with only static obstacles.

This algorithm uses feed-forward networks for both the actor and the critic. The actor network is the policy, $\pi(\mathbf{o};\theta_a)$, with network parameters θ_a, that takes the environment observations $\mathbf{o}$ and provides an action (linear and angular velocities): $\pi(\mathbf{o};\theta_a) = (a_l, a_\phi)$.

Second, in the evaluation phase, the people velocities obtained using a tracker [16,22] are used in the SFM with CP to compute velocity changes $(\boldsymbol{\Delta}\mathbf{v})_{rep} = (\Delta v_l, \Delta v_\phi)$ as is explained in [11]. These velocity changes are added to the DDPG actions to calculate the actions provided by the hybrid model:

$$(a'_l, a'_\phi) = (a_l, a_\phi) + (\Delta v_l, \Delta v_\phi) \tag{2}$$

This hybrid model takes the agent velocities as additional entries regarding the pure DDPG model. Another option is to exploit this information as an entry for training a new DDPG model. This option has several disadvantages like searching for the new optimal hyperparameters for the implementation and a longer time to learn the policies in a more complex environment.

2.3 Anticipative Planning

Anticipation is the ability to react taking into account not only the actual situation, but also a prediction in a future time window. It can be useful in human-aware motion planning to avoid dangerous situations in the future, where finding a solution could be very difficult or maybe impossible.

Some works [5,8] have implemented anticipation in path planning based on human trajectory prediction [19]. There are several methods to predict trajectories based on previous human positions or other environment cues. Despite the unpredictable and multimodal nature of the problem, the constant velocity model (CVM) [20] gives a good approximation in many cases.

3 Anticipative Models

The hybrid model described in the previous section avoids collisions but when the robot is close to the person. This section explains the three anticipative strategies proposed in this paper to avoid collisions in a human-aware way.

3.1 Problem Formulation

As in [4], this work tackles the robot P2P navigation task as a Partially Observable Markov Decision Process (POMDP). The action space, A, includes the linear and angular velocities for a unicycle robot model: $\mathbf{a} = (a_l, a_\phi)$ where $a_l \in [-0.2, 1]$ m/s and $a_\phi \in [-1, 1]$ rad/s. The robot and the agents or moving obstacles (e.g., pedestrian) are modelled in the environment as circles of radius R_r and R_p respectively. The reward function is the one described in (1).

The observations, $\mathbf{o} = (\mathbf{o}_l, \mathbf{o}_g)^{\theta_n} \in O$, are divided in two parts: $\mathbf{o}_l$, includes the 64 1-D Lidar distance measures between 0 and 5 m, which are taken during the last θ_n simulation steps and $\mathbf{o}_g$ includes the goal polar coordinates. The Lidar field of view is 220°.

3.2 Anticipative Turn (AT)

This approach propagates the robot during a time-step using the actual velocity command before moving the robot with that command, and checks the possible collisions in the next movement with the last Lidar's points.

In case of detection of a future collision, the robot does not execute the velocity command given by the hybrid model, but changes it for a command sequence of 6 steps. During the first 4 steps, the robot only turns using the angular velocity in a random way. In the next 2 steps, the robot moves only with the linear velocity. If a collision is detected when the robot movement is propagated the last 2 steps, the sequence is initialized until the robot can execute the complete sequence without collision.

3.3 Anticipative Robot and Pedestrian's Propagation (ARP)

The AT approach has been designed for static environments. For this reason, it only works well when the velocity of the moving agents is very low or with only static obstacles. For these reasons, the AT model is not activated when a possible collision with moving obstacles is detected.

To avoid static and moving obstacles the ARP model is proposed, where at each step, the robot and the moving agents are propagated during n time-steps in a local map. In this propagation, the static obstacles and moving agents of the environment are used to build the local map using the information of the last Lidar's observations. The robot motion is propagated using the hybrid policy and the moving agents are propagated using the CVM. The propagation ends before the n steps when there is a collision or the goal is reached.

After the propagation, the action $\mathbf{a}(t) = (a_l(t), a_\phi(t))$ that will be taken by the robot at the present time t is the same as the action taken when the robot was at the minimum distance of the obstacle during the propagation phase. This action increases the probability of avoiding a collision n steps before it occurs.

3.4 Anticipative Pedestrian's Propagation (APP)

The ARP model, described in the previous section, requires that the local map, with the static and dynamic objects, has to be modified taken into account the Lidar's future propagation measurements. In some cases, the future Lidar's measurements cannot detect objects, because of partial occlusions in the present time, before the robot or agent's propagation. For these reasons, the model in this subsection does not create a local map or propagate all the environment.

The APP model considers a constant velocity prediction for all moving agents detected by the Lidar in the next n time-steps. Each new predicted agent position has an associated uncertainty that can be modelled as a two-dimensional Gaussian distribution $N(\mathbf{c}, \Sigma)$, centered in the position $\mathbf{c}$ with variance Σ.

A circle centered in $\mathbf{c}$ with radius, $r = \Sigma$, represents the area with more probability to localize the agent. If we consider that the uncertainty of the agent's motions at each position is independent from other agent's motions, the propagation of the uncertainty and the radius during the n steps is the sum of the variances, $\Sigma_n = n\Sigma$.

The total radius r_n of the anticipative circle after n steps is the agent radius R_p plus the uncertainty Σ_n:

$$r_n = \Sigma_n + R_p \tag{3}$$

The APP model considers anticipative circles as virtual obstacles detected using the Lidar. At each step, the robot detects all the anticipative circles of each one of the agents for the next n time-steps. If the robot collides with a circle, it is not considered a real collision. That circle and the next circles disappear and only the circles from the present time to the circle previous to the collision remain.

A scheme of the approaches is shown in Fig. 2.

4 Simulations and Experiments

The training and evaluation in simulation are executed in maps (refer to Fig. 3), which are the same floor maps used in [11]. A new empty scenario has been

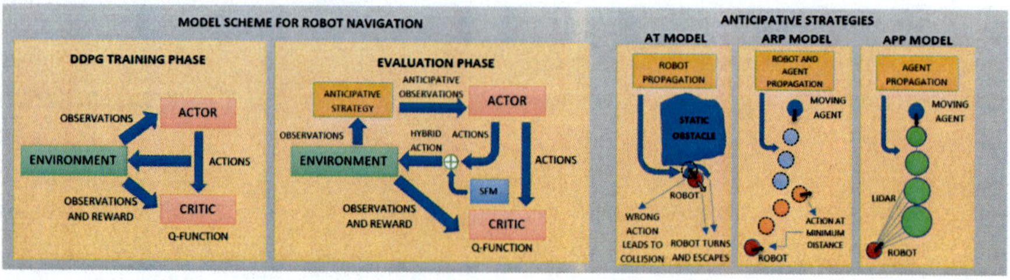

Fig. 2. Model scheme. Training is performed without SFM. In evaluation, SFM actions and anticipation can be added.

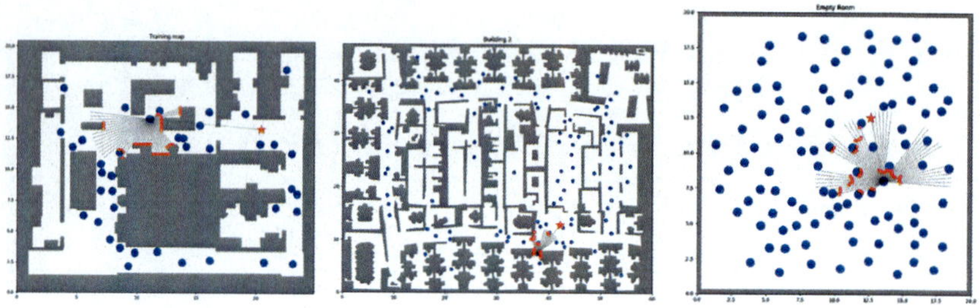

(a) Training Map 24x20 m (b) Building 2 60x47 m (c) Empty Map 20x20 m

Fig. 3. Floor maps. These are some of the floor maps with their sizes in meters. Building 1 and Building 3 maps used for this work are the ones used in [11].

added to evaluate the performance in presence of only moving obstacles. In this evaluation, the OP-DDPG model is the DDPG trained model without adding the velocity changes or anticipative strategies in the evaluation phase.

4.1 Metrics

To evaluate the models, 3 metrics have been used:

- **Success Rate (SR)**: Percent of episodes in which the robot can achieve the goal without collision. This metric shows the overall performance.
- **Collision Rate (CR)**: Percent of episodes that ends in collision.
- **Timeout Rate (TR)**: Percent of episodes where episode time limit is reached.

Other metrics, as for example, the number of steps and the goal distance, give a less global description of the performance.

4.2 Simulation Results

Several versions have been evaluated in different conditions to explore the performance and the limits of the anticipative models. In this evaluation, the trained

Table 1. Model comparison in static environments without moving obstacles for the Euclidean distance (ED) and Path distance (PD) cases.

Environment	OP-DDPG	Hybrid Model	AT
	SR-CR-TR	SR-CR-TR	SR-CR-TR
Building 1 ED	60 - 34 - 6	66 - 11 -23	78 - 0 - 22
Building 2 ED	50 - 46 - 4	74 - 15 - 11	87 - 1 - 12
Building 3 ED	65 - 35 - 0	81 - 11 - 8	92 - 0 - 8
Building 1 PD	75 - 11 - 14	88 - 3 - 9	94 - 0 - 6
Building 2 PD	87 - 13 - 0	96 - 1 - 3	97 - 0 - 3
Building 3 PD	87 - 12 - 1	97 - 1 - 2	98 - 0 - 2

OP-DDPG model is the same for all the simulations. The differences are in the evaluation phase, where the hybrid model and the different anticipative strategies are used. It is important to remark that the AT, ARP and APP models are always used in combination with the hybrid model, rather than the OP-DDPG model. All the evaluations have been performed during 100 episodes as in [4].

Two ways for evaluation have been considered: using a Euclidean Distance (ED) between the goal and the initial position of the robot in a distance between 5 and 10 m; and using the approximate path distance (PD) between 5 and 10 m of a RRT planner between the initial position and a goal.

The path distance offers episodes without very large, complex navigation for which the P2P task has not been designed, because the trained model does not use the complete map knowledge.

The results shown in Table 1 demonstrate that the hybrid model increases the success rate in all the evaluation maps. The AT model increases the performance even more avoiding most of the collisions, but with some timeouts caused by the large-scale local minima. The ARP and APP models are not included because there are not moving obstacles.

The OP-DDPG model results are worse than the results in [4], due to the AutoRL optimal hyperparameters [4] are suboptimal in a different implementation. In this work, the results are focused in the advantages of applying the hybrid model and the anticipative strategies, regardless the AutoRL optimization.

Table 2 shows that the anticipative models are robust against the number of agents. Only in environments with a very high density of moving obstacles, like Building 2 with 100 obstacles, the success rate is significantly reduced. The AT+APP model gives the best results for all the combination of models when there are static and moving obstacles.

The empty map scenario with only moving obstacles is very useful to evaluate the anticipative strategies. In this way, in Table 3, we demonstrate the positive effect of the anticipative strategies in collision avoidance with moving obstacles. In very crowded scenes, where the Euclidean distance for the goal is between 5 m and 10 m, the success rate improves a 10 percent.

Table 2. Model comparison in dynamic environments with 20 and 100 moving agents using the path distance to sample the goals.

Environment	OP-DDPG	Hybrid Model	ARP	APP	APP+AT
	SR-CR-TR	SR-CR-TR	SR-CR-TR	SR-CR-TR	SR-CR-TR
Building 1-20	74 - 11 - 15	84 - 4 - 12	84 - 6 - 10	85 - 5 - 10	**96 - 1 - 3**
Building 2-20	80 - 20 - 0	89 - 3 - 8	88 - 4 - 8	89 - 2 - 9	**93 - 4 - 3**
Building 3-20	84 - 16 - 0	88 - 2 - 10	88 - 3 - 9	87 - 3 - 11	**95 - 1 - 4**
Building 1-100	67 - 16 - 17	84 - 3 - 13	84 - 5 - 11	86 - 4 - 10	**97 - 1 - 2**
Building 2-100	61 - 39 - 0	76 - 16 - 8	75 - 16 - 9	76 - 14 - 10	**81 - 17 - 2**
Building 3-100	83 - 17 - 0	92 - 0 - 8	93 - 1 - 6	93 - 1 - 6	**94 - 3 - 3**

Table 3. Model comparison in an empty map for 35, 50 and 70 moving agents.

Environment	Hybrid Model	ARP	APP
	SR-CR-TR	SR-CR-TR	SR-CR-TR
E. Map ED-35	95 - 5 -0	95 - 5 - 0	**98 - 2 - 0**
E. Map ED-50	89 - 11 - 0	90 - 10 - 0	**96 - 4 - 0**
E. Map ED-70	81 - 19 - 0	86 - 14 - 0	**93 - 6 - 1**

The robustness to noise has been evaluated in building 2 with 20 moving agents. Like in AutoRL work, Gaussian noise is considered, $\mathcal{N}(0, \sigma_{lidar})$, for 4 different standard deviations, $\sigma_{lidar} = [0.1, 0.2, 0.4, 0.8]$. The success rate in these cases decreases less than a 3%. In these evaluations, the maximum velocity for moving agents is 1.2 m/s and the SFM is used to simulate a pedestrian as in [4].

The last evaluation is in the empty map scenario through intersections between the robot and moving agents for different agent maximum velocities. In this challenging scene, the intersections are forced to check how the robot reacts to avoid collisions. The agents do not avoid obstacles because they use the CVM.

This evaluation has been performed using 12 cases with up to 3 agent intersections in different directions regarding robot orientation. These directions are $\theta = [-60^o, -30^o, 0^o, 30^o, 60^o]$. The results in Fig. 4 show less decay for the APP model. For more than 18 km/h it is almost impossible to avoid collisions because in the Lidar's range, the robot has less than a second to react.

4.3 Real-Life Experiments

Real indoor experiments have been performed using the Helena robot. Helena is a transporter robot with a Pioneer P3-DX base and a RS-LiDAR-16. The Lidar observations have been adapted to be entries of the trained policy.

Two experiments have been performed to check the feasibility of the APP model with people in real environments:

- **Frontal Encounter (FE)**: A person and the robot cross paths walking in opposite directions.
- **Perpendicular Encounter (PE)**: A person and the robot cross paths walking in perpendicular directions.

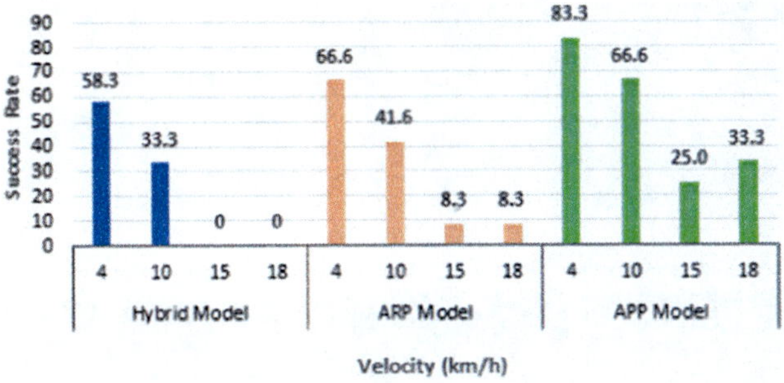

Fig. 4. Success rate variation with agent velocities in the empty map. The least rate decay occurs in the APP model.

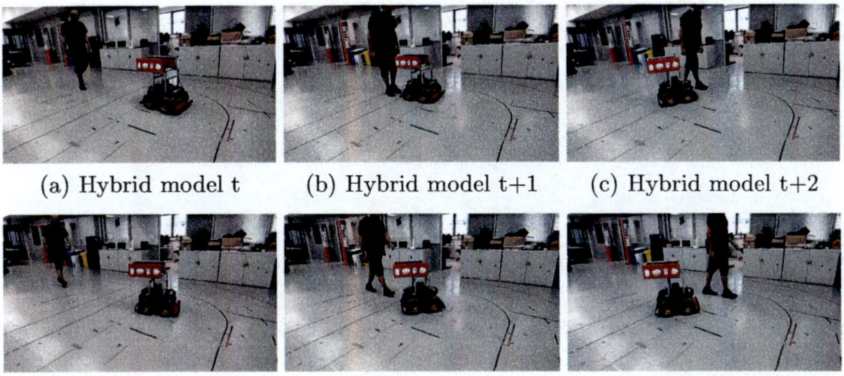

(a) Hybrid model t (b) Hybrid model t+1 (c) Hybrid model t+2

(d) APP model t (e) APP model t+1 (f) APP model t+2

Fig. 5. FE experiment. From time t to t+2 the images show the Helena robot behavior when the robot uses the hybrid model and the APP model.

In Fig. 5, the FE experiments are shown. With the hybrid model the robot only reacts close to the person. Using the APP model, the robot reacts around 2 m before the person arrives. In Fig. 6, the PE experiments show that the robot using the APP model not only reacts before, but chooses to navigate behind the person to avoid collisions which is a more efficient solution.

In Fig. 7, the PE experiment is performed using the ROS Navigation Stack with DWA (NS-DWA) and the APP model through anticipative circles created in the costmap. The NS-DWA reduces the robot velocity too close of the person but using the APP model the robot stops before reach the person. Overall, the hybrid model robot movements are more natural than the ones observed with the NS-DWA when a person is close to the robot, because the robot does not stop to replan its path, avoiding in a better way the frozen robot problem. The minimum distance to the robot increases and the navigation time is reduced (refer to Table 4).

(a) Hybrid model t (b) Hybrid model t+1 (c) Hybrid model t+2

(d) APP model t (e) APP model t+1 (f) APP model t+2

Fig. 6. PE experiment. From time t to t+2 the images show the Helena robot behavior when the robot uses the hybrid model and the APP model.

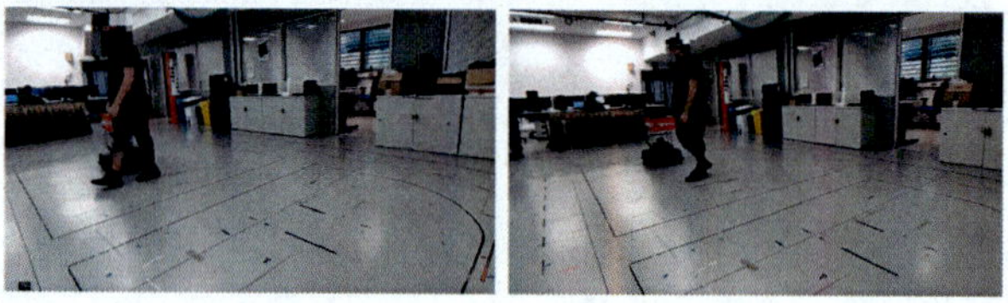

(a) NS-DWA (b) NS-DWA + APP

Fig. 7. PE experiment with NS-DWA. The APP model stops the robot before the encounter.

Table 4. Navigation time and minimum distance to the person for the PE experiment. Helena maximum velocity is 0.5 m/s and the goal distance is 6.5 m.

	NS-DWA	NS-DWA+APP	Hybrid Model	APP model
Navigation time (s)	18.8	19.6	20.0	**17.4**
M. Distance (m)	0.35	**0.57**	0.36	0.53

4.4 Implementation Details

In this subsection, the main features used in our implementation are detailed. All the hyperparameters used in OP-DDPG have been taken from [10].

For the neural network, the critic joint, critic observation and actor layer widths are $(607, 242) \times (84) \times (241, 12, 20)$ with ReLU as activation function. For the reward, the parameters $\boldsymbol{\theta}_{r_{P2P}}^{T}$ are $(-0.43, 0.38, -57.90, 0.415, 0.67, 62.0)$. At each step, the number of observations considered is $\theta_n = 1$. The noise parameters are the ones described in [10]. All the results have been obtained with the same trained model during 2 million steps and 500 steps as maximum episode length.

The implementation has been performed in PyTorch using the Adam optimizer, the Gym library and the Shapely library.

The force parameters used in SFM are $A_i = 0.7$ and $B_i = 10/17$. For the ARP and APP models the propagation number of steps is $n = 20$. The anticipative circles in APP model are created every 5 steps with $\Sigma = 0.01$ m. This value produces circles with a shorter diameter than corridors where the robot and a moving agent can navigate.

For all simulations, each simulation step is 0.2 s. The agents and the robot are circles with a radius of 0.3 m. To perform real experiments, the implementation has been transferred to the Robot Operation System (ROS) middleware and the people velocities are obtained through the Spencer tracker [16].

5 Conclusions

In this work, we have analyzed different methods to avoid collisions with people in a reactive and anticipative way, using a model trained in a static environment.

In scenarios with static obstacles, the hybrid model obtains better results than the OP-DDPG model. In scenarios with static and moving obstacles, the combination between the hybrid model and 3 different anticipative strategies to prevent collisions in a 5 m range, obtains better results than the hybrid model.

The AT model results show an important improvement avoiding collisions with static obstacles. For dynamic environments, the ARP and APP models improve the performance of the other models, because the robot anticipates the collision and reacts before the collision happens. Specifically, the APP model can avoid in some situations high-velocity agents with speeds of up to 18 km/h. Furthermore, the APP improves the robot anticipation in real experiments with people for the hybrid model and the NS-DWA. These anticipative models could be applied for velocities greater than 18 km/h with a longer Lidar's range.

References

1. Ajoudani, A., Zanchettin, A.M., Ivaldi, S., Albu-Schäffer, A., Kosuge, K., Khatib, O.: Progress and prospects of the human-robot collaboration. Auton. Robot. **42**(5), 957–975 (2018)
2. Charalampous, K., Kostavelis, I., Gasteratos, A.: Recent trends in social aware robot navigation: A survey. Robot. Auton. Syst. **93**, 85–104 (2017)
3. Chiang, H.T., Malone, N., Lesser, K., Oishi, M., Tapia, L.: Path-guided artificial potential fields with stochastic reachable sets for motion planning in highly dynamic environments. In: 2015 IEEE international conference on robotics and automation (ICRA), pp. 2347–2354. IEEE (2015)
4. Chiang, H.T.L., Faust, A., Fiser, M., Francis, A.: Learning navigation behaviors end-to-end with autorl. IEEE Robot. Autom. Lett. **4**(2), 2007–2014 (2019)
5. Cosgun, A., Sisbot, E.A., Christensen, H.I.: Anticipatory robot path planning in human environments. In: 2016 25th IEEE international symposium on robot and human interactive communication (RO-MAN), pp. 562–569. IEEE (2016)

6. Dalmasso, M., Garrell, A., Domínguez, J.E., Jiménez, P., Sanfeliu, A.: Human-robot collaborative multi-agent path planning using monte carlo tree search and social reward sources. In: 2021 IEEE International Conference on Robotics and Automation (ICRA), pp. 10133–10138 (2021)
7. Faust, A., Francis, A., Mehta, D.: Evolving rewards to automate reinforcement learning. arXiv preprint arXiv:1905.07628 (2019)
8. Ferrer, G., Sanfeliu, A.: Anticipative kinodynamic planning: multi-objective robot navigation in urban and dynamic environments. Auton. Robot. **43**(6), 1473–1488 (2019)
9. Fox, D., Burgard, W., Thrun, S.: The dynamic window approach to collision avoidance. IEEE Robot. Autom. Mag. **4**(1), 23–33 (1997)
10. Francis, A., et al.: Long-range indoor navigation with prm-rl. IEEE Trans. Robot. (2020)
11. Gil, O., Garrell, A., Sanfeliu, A.: Social robot navigation tasks: Combining machine learning techniques and social force model. Sensors **21**(21), 7087 (2021)
12. Haarnoja, T., Zhou, A., Abbeel, P., Levine, S.: Soft actor-critic: Off-policy maximum entropy deep reinforcement learning with a stochastic actor. In: International Conference on Machine Learning, pp. 1861–1870 (2018)
13. Han, Y., Zhan, I.H., Zhao, W., Pan, J., Zhang, Z., Wang, Y., Liu, Y.J.: Deep reinforcement learning for robot collision avoidance with self-state-attention and sensor fusion. IEEE Robot. Autom. Lett. **7**(3), 6886–6893 (2022)
14. Helbing, D., Molnar, P.: Social force model for pedestrian dynamics. Phys. Rev. E **51**(5), 4282 (1995)
15. Lillicrap, T.P., et al.: Continuous control with deep reinforcement learning. arXiv preprint arXiv:1509.02971 (2015)
16. Linder, T., Breuers, S., Leibe, B., Arras, K.O.: On multi-modal people tracking from mobile platforms in very crowded and dynamic environments. In: 2016 IEEE International Conference on Robotics and Automation (ICRA), pp. 5512–5519 (2016)
17. Mnih, V., et al.: Human-level control through deep reinforcement learning. Nature **518**(7540), 529–533 (2015)
18. Repiso, E., Garrell, A., Sanfeliu, A.: People's adaptive side-by-side model evolved to accompany groups of people by social robots. IEEE Robot. Automat. Lett. **5**(2), 2387–2394 (2020)
19. Rudenko, A., Palmieri, L., Herman, M., Kitani, K.M., Gavrila, D.M., Arras, K.O.: Human motion trajectory prediction: A survey. Int. J. Robot. Res. **39**(8), 895–935 (2020)
20. Schöller, C., Aravantinos, V., Lay, F., Knoll, A.: What the constant velocity model can teach us about pedestrian motion prediction. IEEE Robot. Autom. Lett. **5**(2), 1696–1703 (2020)
21. Urdiales, C., et al.: A new multi-criteria optimization strategy for shared control in wheelchair assisted navigation. Auton. Robot. **30**(2), 179–197 (2011)
22. Vaquero, V., Repiso, E., Sanfeliu, A.: Robust and real-time detection and tracking of moving objects with minimum 2d lidar information to advance autonomous cargo handling in ports. Sensors **19**(1), 107 (2019)
23. Zanlungo, F., Ikeda, T., Kanda, T.: Social force model with explicit collision prediction. EPL (Europhysics Lett.) **93**(6), 68005 (2011)

An Inductive Logic Programming Approach for Entangled Tube Modeling in Bin Picking

Gonçalo Leão[1,2(✉)], Rui Camacho[1,2], Armando Sousa[1,2], and Germano Veiga[2]

[1] Faculty of Engineering, University of Porto (FEUP), 4200-465 Porto, Portugal
{goncalo.leao,rcamacho,asousa}@fe.up.pt

[2] Institute for Systems and Computer Engineering, Technology and Science (INESC TEC), 4200-465 Porto, Portugal
germano.veiga@inesctec.pt

Abstract. Bin picking is a challenging problem that involves using a robotic manipulator to remove, one-by-one, a set of objects randomly stacked in a container. When the objects are prone to entanglement, having an estimation of their pose and shape is highly valuable for more reliable grasp and motion planning. This paper focuses on modeling entangled tubes with varying degrees of curvature. An unconventional machine learning technique, Inductive Logic Programming (ILP), is used to construct sets of rules (theories) capable of modeling multiple tubes when given the cylinders that constitute them. Datasets of entangled tubes are created via simulation in Gazebo. Experiments using Aleph and SWI-Prolog illustrate how ILP can build explainable theories with a high performance, using a relatively small dataset and low amount of time for training. Therefore, this work serves as a proof-of-concept that ILP is a valuable method to acquire knowledge and validate heuristics for pose and shape estimation in complex bin picking scenarios.

Keywords: Bin picking · Inductive logic programming · Machine learning · Pose and shape estimation · Simulation

1 Introduction

Bin picking is a classic robotics problem where the aim is to remove, one-by-one, a set of objects from an unorganized heap using a robotic manipulator. Despite the vast body of knowledge on this topic, there is very little research on how to deal with entangled objects. This sort of items poses the additional challenge of not being easily separable: if not handled properly, the robot has a high chance of picking multiple items at once, which can cause disruptions in manufacturing and production lines, such as having objects falling out of bounds. As such, in bin picking scenarios, there is a need for algorithms that can accurately model the objects within the bin using as input sensor readings from the robot. This model

D. Tardioli et al. (Eds.): ROBOT 2022, LNNS 590, pp. 79–91, 2023.
https://doi.org/10.1007/978-3-031-21062-4_7

should allow for adequate grasp and motion planning strategies that enable the robotic system to detect and overcome entanglement issues.

In previous work [1], an algorithm was developed and tested for creating a geometric model for sets of tubes placed in a bin using a point cloud captured by a depth sensor. These tubes were made of either Polyvinyl Chloride (PVC) or rubber and were highly prone to entanglement and occlusion due to the high amount of curves throughout their length. Figure 1 depicts an example of each step of the tube modeling algorithm. Before the last step of the pipeline, the point cloud is converted into a set of disjoint cylinders, which represent disjoint portions of the tubes, using Random Sample Consensus (RANSAC). The aim of the last step is to combine these cylinders in order to form complete models for the tubes. Previously, this was achieved using a greedy heuristic which prioritizes combining pairs of cylinders whose endpoints minimize a cost function that uses their distance and relative angle, as illustrated in Fig. 2. This perception algorithm was then integrated into a complete bin picking system, where it enabled a physical robotic gripper to successfully handle cases of entanglement and occlusion between tubes [2].

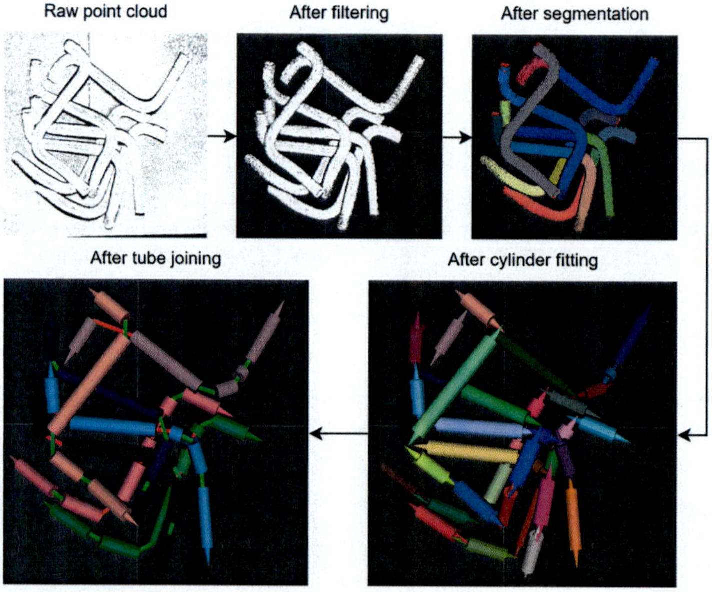

Fig. 1. Illustration of the tube modeling pipeline from previous work [2]

This paper aims to assess whether Inductive Logic Programming (ILP) can be used to derive alternative suitable heuristics to perform the final step of the tube modeling system (joining the cylinders). The ILP system was fed with labeled examples generated via simulation in Gazebo in previous work [3]. This machine learning (ML) technique allows complex relational information (such

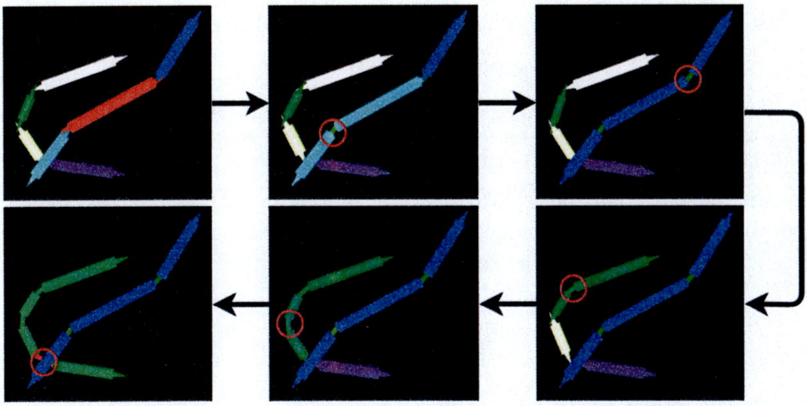

Fig. 2. Illustration of the tube joining step with two tubes [1]

as graphs) to be fed to the model, which is generally not possible with classic models, and the models that are produced are explainable. Thus, using ILP has the potential of bringing to light new strategies and conditions to combine tubes, different from those used on the original greedy algorithm.

The remainder of the manuscript is structured as follows. Section 2 presents an overview of related work on pose and shape estimation in the context of bin picking, and a brief background on ILP, with a focus on 3D data processing. Section 3 summarizes the system developed in previous work [3] to generate synthetic datasets of entangled tubes. Section 4 presents a method to build ILP-based theories for tube modeling. Section 5 provides and discusses experimental results to evaluate the time needed to learn theories using ILP and the predictive performance of the induced theories. Finally, Sect. 6 summarizes the contributions of this work, the main results and several lines of future research.

2 Literature Review

2.1 Pose and Shape Estimation in Bin Picking

Pose and shape estimation techniques in picking scenarios can be divided into those based on model-fitting and learning-based methods. Model-fitting techniques fit mathematical models to a point cloud which can either be shape primitives, typically using RANSAC [1,4], or more complex shapes, such as superquadrics [5]. When significant portions of the items are occluded, namely with entangled tubes, heuristics are employed to determine how to combine the models to form complete objects [1,6]. Machine learning solutions train using labeled point clouds, and can either use more classic methods, such as Support Vector Machines (SVM) [7], or, more recently, deep learning, using Convolutional Neural Networks (CNN) [8]. The weaknesses of this latter family of methods are that they typically require large quantities of training data and the models lack explainability, two weaknesses that ILP is able to address.

2.2 Inductive Logic Programming with 3D Data

Inductive Logic Programming (ILP) is a ML technique that manipulates logic programs, rather than data vectors as in conventional classification and regression algorithms [9]. These logic programs are often written in the Prolog programming language. The models built during training correspond to *predicates*, which express logic relations, instead of functions (as with classic ML methods). The set of rules derived during training for the predicate to be learnt correspond to *theories*. Training in ILP requires three ingredients: background knowledge (facts and rules that express information about the problem and test cases), positive examples (correct instances of the predicate to be learnt) and negative examples (incorrect instances of the predicate). The advantages of ILP include:

- the production of easily interpretable models that can enable data scientists to acquire new knowledge of the application domain.
- the ease of providing complex context-specific information (such as graphs) to the system, which is much harder to achieve with classic ML techniques.
- the capacity of learning models with high predictive capabilities even when a small dataset is used for training.

To the best of the authors' knowledge, this is the first work that proposes using ILP to reconstruct entangled tubes, or that uses this technique in a bin picking scenario. However, it shares many key similarities to other research on ILP, as several researchers have already proposed feeding 3D information to ILP systems in various areas, namely chemistry [10,11] and robotics [12]. Sternberg et al. used ILP to derive Structure Activity Relationships (SAR) in molecules, which they modeled using graphs [10]. Their system worked with the 3D coordinates of atoms and their distances. The system proposed in this work also reasons with graphs and the 3D distance between nodes to build theories. Farid et al. used point clouds captured by a 3D scanner for object recognition [12]. Each object is defined by a set of planar surfaces, which are obtained by segmenting the point clouds. This is similar to this work, where geometric primitives are fed to the ILP system, rather than using the point clouds directly, which could be computationally expensive due to clouds' sheer size.

3 Dataset Generation

The examples used to teach the ILP system were generated using simulation in Gazebo [13], since precise data of the tubes' shape and pose is required (and it would not be feasible to generate a large dataset with such requirements using physical tubes and sensors). The dataset generator was developed and presented in detail in previous work [3].

Within a virtual world created in Gazebo, a set of curved tubes are spawned inside a bin. A simulated tube is composed of a linked list of cylinders with the same radius but differing lengths that are connected by their endpoints. The usage of multiple cylinders allows for the emulation of curved tubes. Figure 3 depicts an example of a simulated bin with 7 tubes.

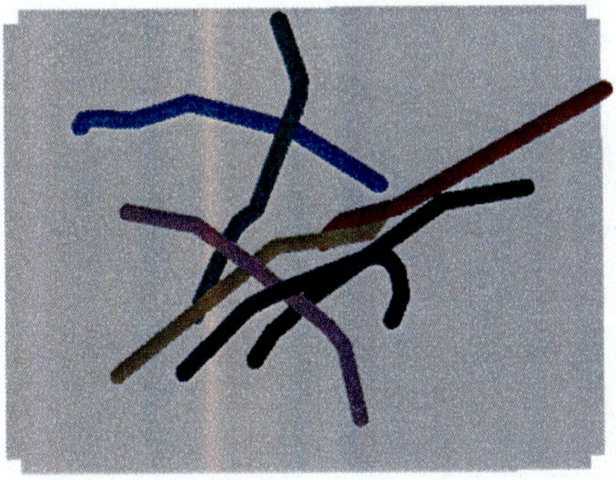

Fig. 3. Example of a bin with 7 tubes generated in Gazebo [3]

The shape of each tube is generated procedurally using a previously devised algorithm [3] that randomly defines the number of cylinders, their length and the relative angle between consecutive cylinders. The algorithm receives as input various parameters controlling the general shape of each tube, such as the minimum/maximum angle of adjacent cylinders. Thus, all the tubes have slight variations amongst them, which is realistic in most picking scenarios with tubes.

Once all the spawned tubes stop moving inside the bin, the position of all the cylinder endpoints (nodes) and their connections (edges) are read using Gazebo's API and stored in a file (in the form of a set of graphs, one for each tube). A point cloud of the bin's interior is also captured using a virtual 3D scanner. A test case corresponds to the data of a bin filled with some tubes.

4 Learning Theories Using Inductive Logic Programming

The goal of the ILP system is to build a set of rules that, given an identifier of a test case and a cylinder endpoint (node), indicate to which other endpoint the input node should be connected to. As such, to create training and test instances, for each test case, starting at one of the tubes's endpoints, the edges (cylinders) that connect two nodes (endpoints) were colored in alternating colors, between green and red, as illustrated in Fig. 4. The ML system was tasked with learning how to form the red edges, assuming that all of the green edges were known, which emulates closely the tube joining step of the original tube modeling algorithm, of combining disjoint cylinders obtained using RANSAC. Since a random angle is generated between consecutive cylinders, the upper bound of the relative angle between two green edges/cylinders is increased, which allows for a more realistic emulation of RANSAC's output, which may have some noise due to imperfect fitting of cylinders to the point cloud.

To perform ILP, Aleph [14] was used alongside SWI-Prolog [15]. The target predicate for the ILP system to build rules on is `join(+TID,+Node1,-Node2)`, where the input variables are `TID`, with the test case identifier, `Node1`, with the first cylinder endpoint identifier, and the output variable is `Node2`, with

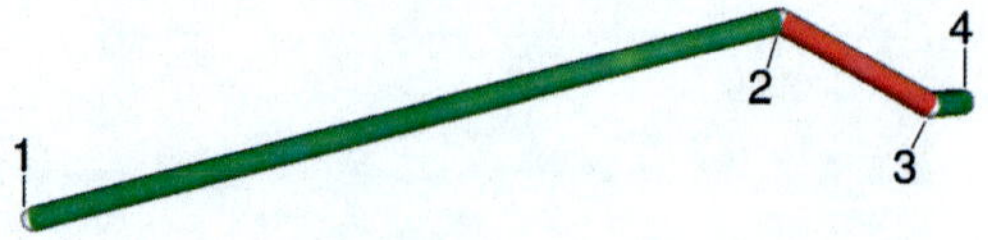

Fig. 4. Test case of a single tube, with four nodes, two green edges and a red edge (Color figure online)

the corresponding endpoint identifier (i.e. `Node1` and `Node2` correspond to the endpoints of a red edge/cylinder). The positive examples correspond to the actual red edges of the test case, whereas the negative examples correspond to random incorrect pairings of nodes. These negative examples include pairings that contain an endpoint of a tube, which should not be connected to any other cylinder. For the testing phase, the input variables are instantiated, whereas the output variable is non-instantiated and must be determined by using an induced model's set of rules.

If a test case has N nodes, then, for each node, there is a single positive example (or none if the node is a tube endpoint) and (N-1) negative examples (or N for tube endpoints). Having an overly large set of examples has a strong impact on the time needed to train with ILP. Thus, to achieve shorter training times, only 20% of the negative examples were used (they were selected randomly). For this same reason, all symmetries were removed from the positive and negative examples. In other words, only the terms `join(TID,Node1,Node2)` where `Node1` $\leq$ `Node2` were used for training. One of the key advantages of learning with ILP is that often a small training set is sufficient for the system to learn a theory with suitable predictive capabilities.

The following predicates from the background knowledge contain all of the information of a test case:

- `tube_node(?TID,?Node,?X,?Y,?Z)` - 3D position of a node (cylinder endpoint).
- `tube_edge(?TID,?N1,?N2)` - identifiers of the nodes of a green edge (cylinder).
- `n_tubes(?TID,?NTubes)` - number of tubes in a test case.
- `avg_angle(?TID,?AvgAngle)` - average angle of the axes of two consecutive cylinders of all the tubes in a test case.
- `tube_radius(?TID,?Radius)` - radius of all the tubes in a test case.
- `tube_length(?TID,?Length)` - length of all the tubes in a test case.

Having prior knowledge of the tubes' topology, such as their curvature (indicated by `avg_angle/2`), is realistic in many industrial scenarios, where there is typically a Computer-Aided Design (CAD) model of the items being manufactured and processed.

Listing 1.1 presents the Prolog code with the background knowledge, and the positive and negative example corresponding to the test case of Fig. 4.

In order for the ILP system to build theories, it must be given a set of predicates that are used to build the body of the induced rules. The body of all rules start with a special term, `get_all_node_ids/3` , that receives the identifiers

```
% Background knowledge
tube_node(test_case_1,1,0.276129,0.232478,0.013956).
tube_node(test_case_1,2,-0.099268,0.134276,0.037429).
tube_node(test_case_1,4,-0.194226,0.174318,0.012384).
tube_node(test_case_1,3,-0.173784,0.174716,0.028453).
tube_edge(test_case_1,1,2).
tube_edge(test_case_1,3,4).
n_tubes(test_case_1,1).
avg_angle(test_case_1,40.000000).
tube_radius(test_case_1,0.012500).
tube_length(test_case_1,0.500000).
```

```
% Positive examples (without symmetric pairs)
join(test_case_1,2,3).

% Negative examples (without symmetric pairs)
join(test_case_1,1,2).
join(test_case_1,1,3).
join(test_case_1,1,4).
join(test_case_1,2,4).
join(test_case_1,3,4).
```

Listing 1.1. Background knowledge, positive and negative examples for Figure 4

```
join(TID,Node1,Node2):-
        get_all_node_ids(TID,Node1,AllNs),
        <0 or more preconditions>, <0 or more filters>, <1 selector>.
```

Listing 1.2. General format of a `join/3` rule induced by the ILP system

of a test case and of a specific node, and returns the identifier of all the nodes that are different from it. The other terms of the body belong to one of three types: preconditions, filters and selectors. The general format of any rule that can be induced by the system is presented in Listing 1.2.

A custom refinement operator was defined in Aleph to enforce the induced rules to have the format of Listing 1.2 [14]. This operator alters the enumeration process of the candidate clauses during the reduction step of the induction process. User-defined refinement operators have the advantage of significantly reducing the search space during this step. This operator was defined in a way that prevents the same clause from being built from two distinct refinement paths, thus avoiding redundant computations.

Preconditions define certain conditions that the test case and/or a specific node must satisfy:

- `precond_cylinder_len_leq/3` and `precond_cylinder_len_gt/3` enforce that the length of the green cylinder connected to the input node must be less or equal to, or greater than, a specific constant (respectively).
- `precond_avg_angle_leq/3` and `precond_avg_angle_gt/3` enforce that the average angle between two consecutive cylinders of a tube must be less or equal to, or greater than, a specific constant (respectively).

Filters receive as input a list of candidate nodes and remove those that do not satisfy a condition:

- `filter_faraway_points/5` removes the nodes whose Euclidean distance to the input node is above a certain value.
- `filter_angular_distance/5` removes the nodes for which the angle between the green cylinder it is connected to and the green cylinder of the input node is above a certain value.
- `filter_long_cylinders/5` removes the nodes that are connected to a green cylinder whose length is above a certain value.

Selectors choose the node from a list of candidates that minimizes a certain cost function. The only selector used was `select_closest_node/5` , which selects the node that minimizes a distance function with respect to the input node that balances the Euclidean distance and relative angle of the respective green cylinders using the formula: $dist = euclid\_dist + coefficient * angular\_dist$.

5 Results

The experiments aimed to create ILP-based theories and evaluate their capacity of correctly modeling entangled tubes. All of the results were obtained with an Intel Core i7-8750H processor, with 2.20 GHz. Datasets of bins with varying amounts of entangled tubes were created using the generator described in Sect. 3. All of the tubes have a radius of 1.25 cm, a length of 50 cm, and varying curvatures. All of the datasets produced in this work and Prolog code used to perform ILP with Aleph were made publicly available at Github[1]. The repository also contains instructions on how to perform training and testing.

Different ILP theories were induced using the same training set with different parameter values for Aleph. The positive and negative examples of the training set were obtained from only four test cases, in order to assess how capable ILP systems can be of learning general rules from a few examples. Two of the test cases contain 1 and 10 tubes with a low curvature (having an angle between consecutive cylinders ranging from 0° to 10°). The other two cases contain 1 and 10 tubes with a high curvature (having an angle between 35° and 45°). In total, 49 positive examples and 834 negative examples were used.

A single value was used for each precondition and filter. The constants have a physical meaning and the values were chosen empirically, with most having been previously used with success in previous work [3]. The constants are indicated in Listing 1.3. For the `select_closest_node/5` selector, two constants were used in the experiments: 0 (which only considers the Euclidean distance) and 0.002.

Regarding Aleph's parameters [14], *language* was set to 1 to prevent a precondition/filter/selector from appearing twice in a clause, while *minpos* (minimum number of positive examples covered by a rule) was set to 5 to reduce the training time and prevent the system from creating ground unit clauses (which accept a specific positive example).

In all of the experiments that were conducted, the resulting set of rules (theory) was always one of the four presented in Listing 1.3.

Since ILP produces models that are easily understandable by humans, analyzing the content of the theories can enable one to discover new knowledge about entangled tubes and strategies to model them. For instance, all four theories share a common rule (Rule 1), which suggest that removing candidate endpoints with a high angular distance was often a good strategy to correctly rebuild a red edge. It is also interesting to note that Theory D built a rule (Rule 3) that joins a node with the closest one using only the Euclidean distance if it is less than 3 cm, without the angular distance (as the coefficient for

[1] https://github.com/GoncaloLeao/Entangled-Tubes-Bin-Picking-Dataset.

```
% Theory A
% [Rule 1] [Pos cover = 21 Neg cover = 0]
join(TID,Node1,Node2) :- get_all_node_ids(TID,Node1,AllNs),
        filter_angular_distance(AllNs,TID,Node1,[30.0],FilterNs),
        select_closest_node(FilterNs,TID,Node1,[0.002],Node2).
% [Rule 2] [Pos cover = 16 Neg cover = 4]
join(TID,Node1,Node2) :- get_all_node_ids(TID,Node1,AllNs),
        precond_cylinder_len_gt(TID,Node1,[0.05]),
        precond_avg_angle_gt(TID,Node1,[20.0]),
        select_closest_node(AllNs,TID,Node1,[0.002],Node2).
% [Rule 3] [Pos cover = 14 Neg cover = 1]
join(TID,Node1,Node2) :- get_all_node_ids(TID,Node1,AllNs),
        precond_cylinder_len_leq(TID,Node1,[0.05]),
        select_closest_node(AllNs,TID,Node1,[0.002],Node2).

% Theory B
% [Rule 1] [Pos cover = 21 Neg cover = 0]
join(TID,Node1,Node2) :- get_all_node_ids(TID,Node1,AllNs),
        filter_angular_distance(AllNs,TID,Node1,[30.0],FilterNs),
        select_closest_node(FilterNs,TID,Node1,[0.002],Node2).
% [Rule 2] [Pos cover = 25 Neg cover = 5]
join(TID,Node1,Node2) :- get_all_node_ids(TID,Node1,AllNs),
        precond_avg_angle_gt(TID,Node1,[20.0]),
        select_closest_node(AllNs,TID,Node1,[0.002],Node2).
```

```
% Theory C
% [Rule 1] [Pos cover = 21 Neg cover = 0]
join(TID,Node1,Node2) :- get_all_node_ids(TID,Node1,AllNs),
        filter_angular_distance(AllNs,TID,Node1,[30.0],FilterNs),
        select_closest_node(FilterNs,TID,Node1,[0.002],Node2).
% [Rule 2] [Pos cover = 14 Neg cover = 1]
join(TID,Node1,Node2) :- get_all_node_ids(TID,Node1,AllNs),
        precond_cylinder_len_leq(TID,Node1,[0.05]),
        select_closest_node(AllNs,TID,Node1,[0.002],Node2).

% Theory D
% [Rule 1] [Pos cover = 21 Neg cover = 0]
join(TID,Node1,Node2) :- get_all_node_ids(TID,Node1,AllNs),
        filter_angular_distance(AllNs,TID,Node1,[30.0],FilterNs),
        select_closest_node(FilterNs,TID,Node1,[0.002],Node2).
% [Rule 2] [Pos cover = 16 Neg cover = 4]
join(TID,Node1,Node2) :- get_all_node_ids(TID,Node1,AllNs),
        precond_cylinder_len_gt(TID,Node1,[0.05]),
        precond_avg_angle_gt(TID,Node1,[20.0]),
        select_closest_node(AllNs,TID,Node1,[0.002],Node2).
% [Rule 3] [Pos cover = 13 Neg cover = 2]
join(TID,Node1,Node2) :- get_all_node_ids(TID,Node1,AllNs),
        filter_faraway_points(AllNs,TID,Node1,[0.03],FilterNs),
        select_closest_node(FilterNs,TID,Node1,[0.0],Node2).
% [Rule 4] [Pos cover = 14 Neg cover = 1]
join(TID,Node1,Node2) :- get_all_node_ids(TID,Node1,AllNs),
        precond_cylinder_len_leq(TID,Node1,[0.05]),
        select_closest_node(AllNs,TID,Node1,[0.002],Node2).
```

Listing 1.3. Theories induced from the dataset using ILP

`select_closest_node/5` is 0). This result is coherent with the original heuristic of the modeling Algorithm [1] where the angular distance is ignored if there is a pair of points sufficiently close to each other, thus illustrating how ILP theories can also be used to validate previously devised heuristics.

According to Table 1, the resulting theory and execution time (which is the average of 3 runs) vary greatly with the *noise* (maximum number of negative examples covered by a rule) and *clauselength* (maximum number of terms in a rule) parameters. If either these parameters is set to relatively low value, then the training time increases significantly (in particular, the reduction step time) since the search space for acceptable rules is significantly reduced, and thus it is harder to find a theory for which at least 5 positive examples are covered

Table 1. Average time and output to induce theories with respect *noise* and *clause-length*, and the available values for the constant of `select_closest_node/5`

Available coefficients for `select_closest_node/5`	Noise	Clause length	Training time	Induced theory
Only 0.002	4	4	74.69 s	C
		5	23.35 s	A
		6	23.55 s	A
		7	23.81 s	A
	2	7	78.38 s	C
	3		79.62 s	C
	5		21.44 s	B
	6		21.64 s	B
0 and 0.002	4	7	236.92 s	D

(due to *minpos* being set to 5). However, if *noise* is too high, then the system may create theories that cover too many negative examples (which may result in underfitting). Decreasing any of the parameters to values lower than those in the table causes the ILP system to take much longer to perform the reduction step and prevents a theory from being generated at the end.

Using multiple coefficients for each predicate increases significantly the training time since more predicate combinations are possible. In fact, using two values for a constant of a predicate is equivalent to using two different predicates.

In order to test the predictive performance of all the theories, three disjoint test sets were created. Test Set 1 contains 36 test cases: 9 test cases were created for each of the 4 test case configurations used in the training set (where only 1 test case was used for each configuration). It aims to test the theories against other sets with the same characteristics. Test Set 2 contains 20 test cases of 15 tubes each: 10 test cases with low curvature tubes and 10 with high curvature ones. Test Set 3 contains 40 test cases, 20 for two different types of tubes: tubes with a minimum and maximum angle between consecutive cylinders of 10° and 20°, and of 25° and 35°. Both shapes have 50 cm for the length and 2.5 cm for the diameter. For both types of tubes, 10 test cases were created with bins containing 1 and 10 tubes. Test Sets 2 and 3 aim to determine whether the theories are general enough to cope with bins with more tubes/complexity (Set 2) or tubes of other shapes (Set 3).

Equation 1 presents the Matthews Correlation Coefficient (MCC), where TP, TN, FP and FN correspond to the *true positives*, *true negatives*, *false positives* and *false negatives*, respectively, of the results of a binary classification [16]. This metric was used to evaluate the performance since there is a great unbalance between the positive and negative classes, since each test case produces much more negative examples than positive examples (as discussed in Sect. 4). Unlike the accuracy metric, the MCC is robust against unbalanced datasets, and, unlike the F1 score, it also takes into account the true negatives.

$$\mathrm{MCC} = \frac{TP \cdot TN - FP \cdot FN}{\sqrt{(TP + FP) \cdot (TP + FN) \cdot (TN + FP) \cdot (TN + FN)}} \tag{1}$$

According to Table 2, theory C has a much lower score than the other theories, which suggests that the former theory is not able to accurately model the training set (underfitting). The lower score is probably due to this theory not containing a rule similar to Rule 2 from the other three theories. This result confirms the previous observation that, when using a lower value for *noise*, the system is forced to find rules in a smaller search space (since the rules must cover less negative examples), which may cause the ILP system to miss out on more general rules. Theories A, B and D have a similar performance. Their scores for test sets 1 and 2 are very close to the ones for the training set, which suggests that the rules are general enough to handle bins with various amounts of tubes and where they have different poses. These results highlight one of ILP's strengths of being able to create rather general rules with little training data. The performance of both theories on Test Set 3 was lower, which indicates that the rules generalize less well to test cases of tubes with other shapes.

Table 2. MCC score of the theories using the training set and three test sets

Theory	Training Set	Test Set 1	Test Set 2	Test Set 3
A	0.915	0.929	0.911	0.822
B	0.915	0.938	0.920	0.784
C	0.760	0.726	0.751	0.613
D	0.908	0.929	0.911	0.822

By combining the results from both tables, one can conclude that, when working with ILP systems, namely Aleph, it is essential to understand well the meaning of all their parameters and fine-tune them to each specific application. The choice of the value for each parameter has a big impact on the structure of the induced theory.

Using the knowledge gained from the four theories (namely theory A), a new heuristic was defined for the previously devised tube joining step: in test cases with low curvature tubes (average angle ≤ 20), penalize pairing cylinder endpoints which have an angular distance below 30° and where both cylinders are longer than 5 cm. Table 3 compares applying the tube modeling algorithm with the original and modified heuristic to the point clouds of the four datasets. The performance metric, micro-recall, introduced in previous work [3], provides a reliable estimate of the models' accuracy. The new heuristic resulted in a very similar performance, with Test Set 2 having a slight increase in accuracy. The authors hypothesize this is due to Test Set 2 providing more opportunities for this penalty to be (correctly) applied since there is more occlusion and cylinders produced by RANSAC, as there are more tubes present in the bin.

Table 3. Micro-recall of the tube modeling algorithm using the original heuristic [1] and a modified one based on the induced theories, for the training and test sets

Heuristic	Training Set	Test Set 1	Test Set 2	Test Set 3
Original	0.9668	0.9147	0.8987	0.8956
Modified	0.9668	0.9146	0.9025	0.8949

6 Conclusions

This paper addressed the problem of pose and shape estimation in bin picking, with a focus on entangled tubes, using an unconventional technique: Inductive Logic Programming (ILP). The manuscript detailed the steps to apply ILP to this problem, from translating an instance of the problem to Prolog, to building sound rules for joining cylinders that describe disjoint portions of the tubes. The experiments illustrated how new modeling heuristics can be discovered and how previously devised ones can be validated, using a relatively small dataset.

This work opens several lines for future research. Firstly, the induced theories had a lower predictive performance with tubes of different shapes. This issue could be mitigated and potentially solved by using a larger variety of predicates to construct rules, richer background knowledge or an expanded dataset encompassing other tube shapes. Secondly, more in-depth experiments on the learner's parameters can be conducted, given their large impact on the theories induced for this problem. Finally, the ILP system can be expanded to learn how to model objects with a larger variety of shapes, namely tubes with bifurcations.

Acknowledgments. This work was funded by National Funds through the Portuguese funding agency Fundação para a Ciência e a Tecnologia (FCT), within the PhD studentship 2020.06923.BD, and by the Environmental Funds nr 01/C12-I01/2021 and nr 02/ C12-I01/2021, under the Portuguese Recovery and Resilience Plan.

References

1. Leão, G., Costa, C.M., Sousa, A., Veiga, G.: Perception of entangled tubes for automated bin picking. In: Silva, M.F., Luís Lima, J., Reis, L.P., Sanfeliu, A., Tardioli, D. (eds.) ROBOT 2019. AISC, vol. 1092, pp. 619–631. Springer, Cham (2020). https://doi.org/10.1007/978-3-030-35990-4_50
2. Leão, G., Costa, C.M., Sousa, A., Veiga, G.: Detecting and solving tube entanglement in bin picking operations. Appli. Sci. **10**(7), 2264 (2020)
3. Leão, G., Costa, C.M., Sousa, A., Reis, L.P., Veiga, G.: Using simulation to evaluate a tube perception algorithm for bin picking. Robotics **11**(2), 46 (2022)
4. Taylor, G., Kleeman, L.: Robust range data segmentation using geometric primitives for robotic applications. In: Proceedings of the Fifth IASTED International Conference on Signal and Image Processing, pp. 467–472. Acta Press, Honolulu, HI, USA (2003)
5. Boughorbel, F., et al.: Laser ranging and video imaging for bin picking. Assembly Autom. **23**(1), 53–59 (2003)
6. Qiu, R., Zhou, Q.-Y., Neumann, U.: Pipe-run extraction and reconstruction from point clouds. In: Fleet, D., Pajdla, T., Schiele, B., Tuytelaars, T. (eds.) ECCV 2014. LNCS, vol. 8691, pp. 17–30. Springer, Cham (2014). https://doi.org/10.1007/978-3-319-10578-9_2
7. Li, Y., Chen, C.F., Allen, P.K.: Recognition of deformable object category and pose. In: IEEE International Conference on Robotics and Automation (ICRA), pp. 5558–5564. IEEE, Hong Kong, China, May 2014
8. Kleeberger, K., Huber, M.F.: Single shot 6D object pose estimation. In: IEEE International Conference on Robotics and Automation (ICRA), pp. 6239–6245. IEEE, Paris, France (2020)
9. Cropper, A., Dumančić, S., Evans, R., Muggleton, S.H.: Inductive logic programming at 30. Mach. Learn. **111**, 147–172 (2022)
10. Sternberg, M.J., Muggleton, S.H.: Structure activity relationships (SAR) and Pharmacophore discovery using inductive logic programming (ILP). QSAR Comb. Sci. **22**(5), 527–532 (2003)

11. Costa, V.S., Fonseca, N.A., Camacho, R.: LogCHEM: Interactive discriminative mining of chemical structure. In: IEEE International Conference on Bioinformatics and Biomedicine (BIBM), pp. 421–426. IEEE, Philadephia, Pennsylvania, USA (2008)
12. Farid, R., Sammut, C.: Plane-based object categorisation using relational learning. Mach. Learn. **94**(1), 3–23 (2014)
13. Koenig, N., Howard, A.: Design and use paradigms for gazebo, an open-source multi-robot simulator. In: IEEE/RSJ International Conference on Intelligent Robots and Systems (IROS), vol. 3, pp. 2149–2154. IEEE, Sendai, Japan (2004)
14. Srinivasan, A.: The Aleph Manual. Tech. rep., University of Oxford (2001)
15. Wielemaker, J., Schrijvers, T., Triska, M., Lager, T.: SWI-Prolog. Theory Pract. Logic Program. **12**(1–2), 67–96 (2012)
16. Chicco, D., Jurman, G.: The advantages of the Matthews correlation coefficient (MCC) over F1 score and accuracy in binary classification evaluation. BMC Genom. **21**(1), 6 (2020)

Mobile Robot Olfaction

Aerial Robot with Electronic Nose for Air Quality Monitoring

Patricia Arroyo, Jaime Gómez-Suárez, José Luis Herrero, and Jesús Lozano(✉)

Industrial Engineering School, University of Extremadura, 06006 Badajoz, Spain
jesuslozano@unex.es

Abstract. This paper presents an air quality measurement system onboard an Aerial robot. Specifically, the system is capable of measuring NO2, CO, O3 and NO pollution levels. Furthermore, it also includes temperature and humidity measurements. An air sampling system has been designed to avoid false readings due to the downwash effect. The air quality measurement system is based on electrochemical sensors. Therefore, it has been previously calibrated in the field using multilinear regression techniques. The data transmission is done wirelessly in real-time via ZigBee. In addition, data is stored locally in case wireless communication is not possible. Finally, a flight test has been performed to verify the correct operation of the system. A stable signal is achieved regardless of the flight of the drone. Besides, it is also possible to detect a NO2 source.

Keywords: Electronic nose · Unmanned aerial vehicle (UAV) · Air quality

1 Introduction

Air pollution is a pressing environmental concern all over the world. As industrialization and motorization increase, so does the emission of harmful pollutants into the atmosphere. These pollutants can cause a variety of respiratory and other health problems and can also damage the environment [1–3].

There are several ways to monitor air pollution. Air quality monitoring stations are set up in cities to measure levels of pollutants in the air. These stations measure a variety of pollutants, including particulate matter, ozone, nitrogen dioxide, and sulfur dioxide. Data from these stations are used to help assess air quality and to make decisions about air pollution control.

One promising method of air pollution monitoring is the use of electronic noses, which are devices that can detect and identify various chemicals in the air [4]. These devices have the potential to provide real-time data on air pollution levels, which can be used to inform air quality management decisions. There are several types of gas sensors that can be used for this type of application [5]. Specifically, electrochemical gas sensors have a very high sensitivity and can detect gases at low concentrations [6]. They are also relatively selective, meaning that they can often discriminate between different gases.

D. Tardioli et al. (Eds.): ROBOT 2022, LNNS 590, pp. 95–104, 2023.
https://doi.org/10.1007/978-3-031-21062-4_8

This makes them very useful for applications where it is important to know not only the presence of a gas, but also the identity of that gas.

UAVs equipped with e-noses can be used to map the distribution of pollution in the atmosphere. This information can be used to identify sources of pollution, assess the health risks of exposure, and develop strategies for reducing emissions. Moreover, they can also be used to monitor air quality in real-time. This information can be used to provide warnings to the public in the event of a release of any pollutant. Therefore, this technology has the potential to improve our understanding of air pollution and help us develop strategies to reduce emissions and protect public health. To this end, several studies have been published proposing technologies for the integration of different types of sensors in drones for detecting atmospheric pollutants [7–9].

It should be noted that such devices require prior and periodic calibration. Manufacturers of gas sensors usually perform a laboratory calibration before selling them. However, it has been observed that electrochemical gas sensors perform better when used in controlled laboratory conditions than when working in the field [10]. Therefore, field calibrations should be carried out in addition to the laboratory calibration. It is advisable to perform these calibrations under conditions similar to those under which the sensors will work. For this purpose, calibration algorithms based on multilinear regression or Machine Learning techniques such as Artificial Neural Networks or Support Vector Machines are used in the literature [10–12].

On the other hand, the downwash effect generated by drone propellers can generate pressure changes that can adversely affect the sensors, as well as generate a dispersion of pollutants [13, 14]. In addition, it is worth considering that flight can also interfere with the performance of the pollution measurement system [15]. Therefore, the placement of the system and the sampling method design are crucial.

This paper proposes a low-cost, low-power system for measuring pollutants in the air on board a drone. This system includes electrochemical sensors that can quantify the presence of NO_2, NO, CO and O_3 and that have been calibrated. Downwash and interference problems that may occur in flight are taken into account. Firstly, the design of the system is described, specifying the characteristics of the electronic nose and the drone used, as well as their integration. Next, the followed calibration process and the setup of the flight test carried out are detailed. The results of the calibration and flight test are discussed. Finally, the main conclusions of this work are presented.

2 Material and Methods

2.1 UAV and E-Nose Integration

The electronic nose used, because of the aim of quantifying specific pollutants, is based on electrochemical sensors for NO_2, O_3, CO and NO. Data acquisition is done through an 8-channel external 24-bit A/D converter, thus achieving more resolution than using the A/D built into the microcontroller. In addition, it incorporates a humidity and temperature sensor. The data collected from these sensors is stored on a microSD card and, in addition, is sent to a remote computer via ZigBee wireless communication. This allows real-time visualization of the pollution, temperature and humidity values achieved through acquisition software. XBee PRO modules are used for this purpose, achieving a range of 100 m in urban areas and 1200 m in the open field. The electronic nose is controlled by a 32-bit microcontroller from STM32.

With regard to the sampling system, there is an air pump that conducts the ambient air to the sensor cell during the whole process. The air inlet is located outside the turbulence zone generated by the propellers using a tube. In order to know this distance, tests have been carried out using a smoke generator. In this way it can be seen at what distance from the drone there is no influence on the air from the turbulence of the propellers. Finally, a 35 cm long tube is used. This way allows for undisturbed air sampling. In addition, the pump generates a constant flow rate which avoids false measurements on the sensors due to sudden pressure changes. It is placed downstream of the sensors according to the manufacturer's recommendations [16].

Both the electronic nose and the sampling system, incorporated in the drone, are powered by a Li-Ion battery. A block diagram is shown in Fig. 1, where the communications and models of the components used are specified.

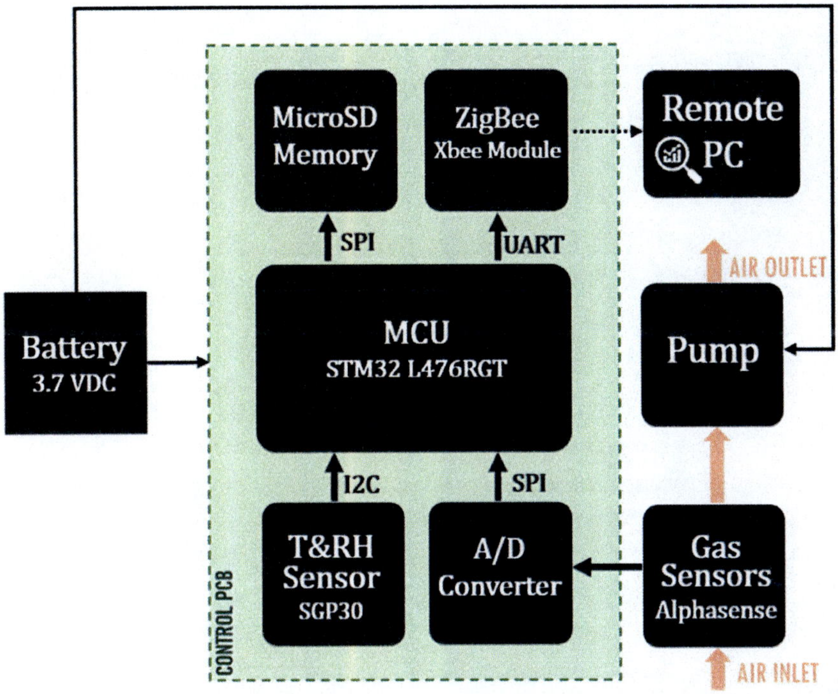

Fig. 1. Block diagram of the electronic nose and sampling system on board the drone.

An unmanned aerial vehicle (UAV), commonly known as a drone, is an aircraft without a human pilot aboard. UAVs are a component of an unmanned aircraft system (UAS); which include a UAV, a ground-based controller, and a system of communications between the two. The flight of UAVs may operate with various degrees of autonomy: either under remote control by a human operator or autonomously by onboard computers. In the present application, the DJI Phantom 4 Pro commercial drone has been used. It is a drone of 1388 g weight and 250 mm diagonal length (without propellers). It includes GPS, 3-axis stabilization and a ToF detection system. The system is controlled through the manufacturer's mobile application that communicates on a frequency of 2.4 GHz ISM with the drone.

In order to ensure a stable flight, the location of the in-air pollution measurement system on the UAV is crucial. It becomes important to keep the center of gravity stable. In addition, the position of the sampling system is also of great concern as explained above. Figure 2 shows the position of the various components of the system on the drone. The pump is located below the air sampling tube. The incorporation of all the components from the measuring system adds 256 g to the weight of the drone.

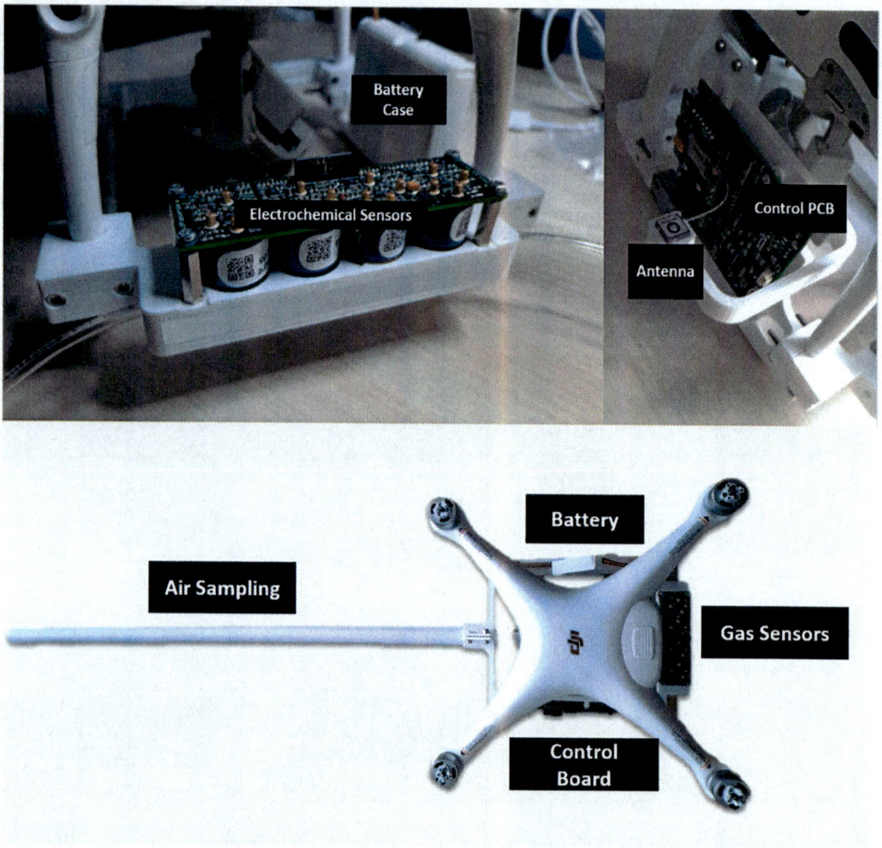

Fig. 2. Assembly of the electronic nose and sampling system on the base of the UAV or drone.

2.2 Data Processing

Before conducting any experiment in field with the sensors coupled to the UAV, the sensor array had to be calibrated. The device was collocated nearby an air quality station for two months.

Initially, our device showed a clear deviation when it was compared with the reference data. This issue was expected since this type of sensor was designed in the first place to work under controlled conditions and not in outdoor environments. Therefore, a calibration step is necessary. Since the target gas concentration is calculated from the electrical values of the electrodes (working electrode (WE) and auxiliar electrode (AE)), we decided to work with raw data from these two electrodes rather than with the concentration values. For sensor calibration, these electrical values were fed into a model, which was a multilinear regression algorithm. There are also two problems to be tackled when working with electrochemical gas sensors: cross-sensitivity and interference caused by humidity and temperature variations. To deal with these problems, for the calibration of each pollutant, the electrical data of all the remaining gas sensors were fed into the calibration algorithm as well as the temperature and relative humidity. Thus, the calibration function might be written as follows:

$$f(WE_i, AE_i, T, RH) = [Pollutant] \tag{1}$$

where WE_i and AE_i are the working and auxiliar electrode for each pollutant ($i = NO_2$, O_3, NO, CO), T is the ambient temperature and RH is the relative humidity.

The function f was a multilinear regression with ten inputs and one output which was the estimated concentration of each pollutant. This function is used because previous work using these sensors has produced enough good results with low computational power [17]. However, other more advanced techniques could be used to achieve better results. To train this algorithm, the dataset was split into 30% training and 70% testing.

To evaluate the performance of the sensors and the calibration algorithm, the data derived from (1) was compared with the data from the reference equipment. The metrics that were used in this comparison were the coefficient of determination (R^2), the Root Mean Squared Error (RMSE) and the slope and offset of the fit line.

2.3 Experimental Setup

Calibration measurements were first performed to validate the system, followed by field measurements.

The first ones allowed to calibrate the electrochemical sensors before performing the field test. In this way, interferences with temperature and humidity variation or cross-sensitivity between compounds are corrected. These measurements have been carried out in parallel with a reference station installed in a location where there is a significant range of pollution. This station includes O_3 (Absortion UV), NO_X (chemiluminescence) and CO (infrared absorption spectroscopy) measurement equipment. The calibration algorithm developed in the previous section is implemented in the embedded software of the electronic nose.

Once the calibration has been performed, a flight test has been carried out in the field. In this way it is checked that the drone has a stable flight, the electronic nose is not affected by it and, in addition, a response from the gas sensors is obtained. In particular, the ability to detect a NO_2 source has been studied. Figure 3 shows a picture of the drone in flight with the integrated electronic nose.

Fig. 3. Drone flight tests and measurement with electronic nose prototype

3 Results and Discussion

In this section, we describe, first, the results of the measurements in parallel with the reference equipment and the calibration process. Secondly, we present the results of the field test.

3.1 Calibration

The following figure (Fig. 4) shows the data of each pollutant after applying the calibration detailed in Sect. 2. From top to bottom, we show NO, NO_2, O_3 and CO. On the left panel, the scatter plot Model vs Reference and the least-squares adjustment are shown. On the right, the temporal evolution of the reference and the calibrated sensor are presented.

It should be noticed that NO, NO_2 and O_3 are expressed in $\mu g/m^3$. However, CO is in mg/m^3 because the reference station uses this unit since it is expected to have a high concentration of CO in this location due to traffic and human activity. Although it was expected to have a high concentration of this pollutant, during the period in which the measurements took place, CO did not reach high values. This means the resolution is lower compared with the rest of the pollutants and, consequently the result of the calibration for this pollutant is poorer.

As Table 1 shows, the results are particularly good for NO and NO_2, with R^2 of 0.90 and 0.94. The RMSE for these pollutants is around 4% of the maximum concentration reached in the measurement period. The slope is also very close to 1, which indicates a good adjustment to the reference data. For O_3, the results are slightly worse but still quite satisfactory. A coefficient of determination of 0.79 shows a strong correlation, the RMSE is around 10% of the maximum concentration and the slope is very close to 1. Finally, in the case of CO, the results are moderately good for the mentioned reasons.

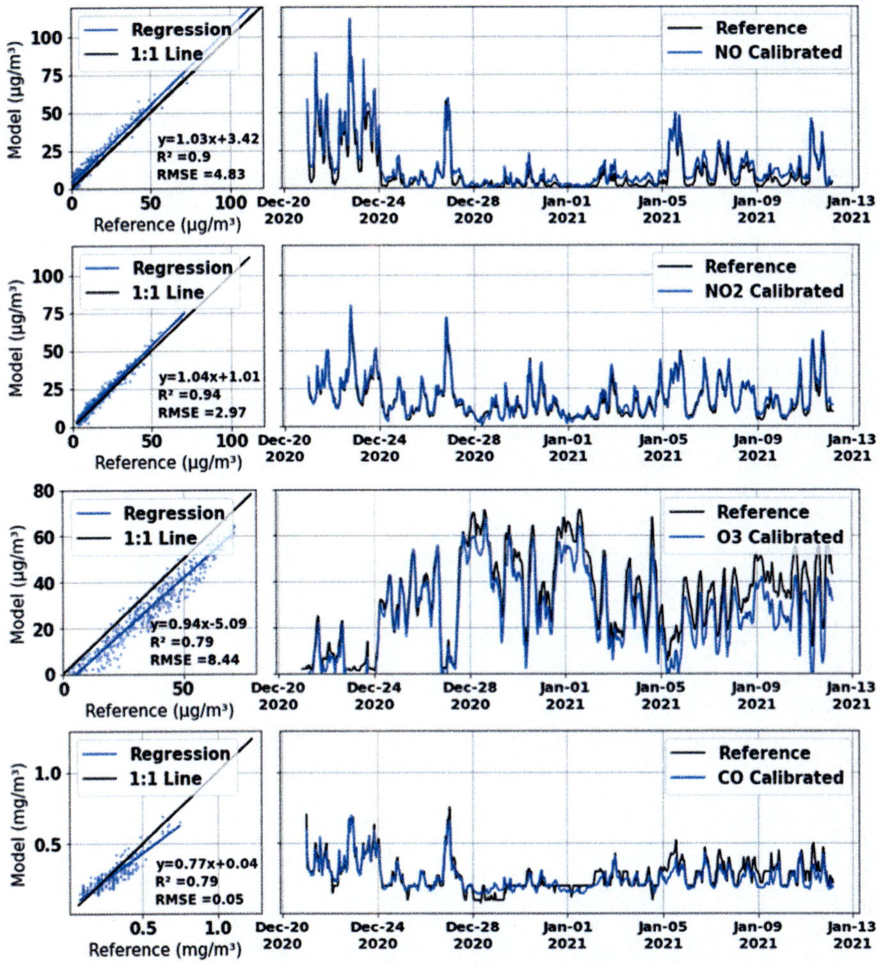

Fig. 4. Results of the calibration measurements.

Table 1. Metrics used to evaluate the calibration of the sensors.

Pollutant	R_2	RMSE	Slope	Offset
NO	0.90	4.83 μg/m^3	1.03	3.42 μg/m^3
NO_2	0.94	2.97 μg/m^3	1.04	1.01 μg/m^3
O_3	0.79	8.44 μg/m^3	0.94	5.09 μg/m^3
CO	0.79	0.05 mg/m^3	0.77	0.04 mg/m^3

3.2 Field Test

An experiment has been carried out to verify that the system operates in a stable way during flight. For this purpose, a flight around an artificially generated source of NO2 has been performed. This gas has been chosen because, when artificially generated (with nitric acid and zinc), it takes a brownish color that allows us to visualize the cloud during the whole flight.

The flight consists of approaching the drone at a constant speed to the generated gas cloud, and moving away from it until returning to the initial point of the test. For this purpose, the drone is placed 20 m away from the source (with the electronic nose and the sampling system turned on). Then, the chemical reaction is started and the flight test begins. Finally, the system is kept on after the drone has landed again.

Figure 5 shows the concentration of NO_2 (target gas in this case) measured at each moment during the experiment. It can be observed that the flight of the drone does

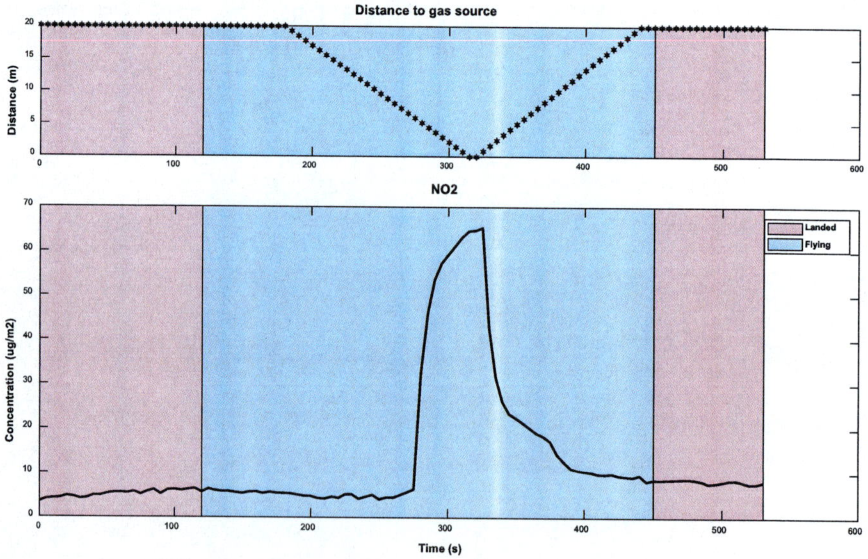

Fig. 5. Results of the field test. The graph depicts the NO2 concentration obtained during the experiment and the distance of the drone from the gas source. The red areas correspond to the time when the drone was landed, while the blue areas represent the drone in flight.

not seem to affect the sensor signal, since it remains stable in the changes from the red zone (landed) to the blue zone (flight) and vice versa. On the other hand, as the drone approaches the source (at 5 m approx.), the NO_2 concentration value shoots up. Similarly, as it moves away, the concentration value decreases. It is observed that the rate of decrease is slower, and the final value is not the same as the initial one. This could be because the gas cloud moved in the direction of the flight of the drone.

4 Conclusions

In this manuscript we have presented the design and development of an electronic nose coupled with an unmanned aerial vehicle for air monitoring, specifically, the system measures NO, NO_2, O_3 and CO. The system allows measuring these pollutants in areas with little accessibility, being a useful tool along with traditional pollutant monitoring methods. The electronic nose provides local storage as well as wireless communication with a PC onshore.

Two different experiments have been conducted. First, the device was collocated in parallel with a reference station, which allowed us to evaluate the performance of the sensors in outdoor environments. A calibration step was necessary due to the noticeable deviation from the reference data. By means of a multilinear regression algorithm the sensors were calibrated, and they achieved high accuracy (R^2 up to 0.94). This calibration was implemented in the electronic nose for the second experiment, in which we assessed that the flight of the drone did not affect the measurements. A novel sampling design has also considered the downwash effect caused by the movement of the rotors. In addition, an artificially generated source of NO2 is detected. More comprehensive studies with 3D maps of the drone's position and the gas plume are planned for future work. These will be used to study the response speed of the sensors and the offset.

References

1. Guan, W.-J., Zheng, X.-Y., Chung, K.F., Zhong, N.-S.: Impact of air pollution on the burden of chronic respiratory diseases in China: time for urgent action. Lancet **388**, 1939–1951 (2016). https://doi.org/10.1016/S0140-6736(16)31597-5
2. Boningari, T., Smirniotis, P.G.: Impact of nitrogen oxides on the environment and human health: Mn-based materials for the NOx abatement. Curr. Opin. Chem. Eng. **13**, 133–141 (2016)
3. Loomis, D., Grosse, Y., Lauby-Secretan, B., et al.: The carcinogenicity of outdoor air pollution. Lancet Oncol. **14**, 1262–1263 (2013). https://doi.org/10.1016/S1470-2045(13)70487-X
4. Liu, X., Jayaratne, R., Thai, P., et al.: Low-cost sensors as an alternative for long-term air quality monitoring. Environ. Res. **185**, 109438 (2020). https://doi.org/10.1016/j.envres.2020.109438
5. Dhall, S., Mehta, B.R., Tyagi, A.K., Sood, K.: A review on environmental gas sensors: Materials and technologies. Sens. Int. **2**, 100116 (2021)
6. Khan, M.A.H., Rao, M.V., Li, Q.: Recent advances in electrochemical sensors for detecting toxic gases: NO2, SO2 and H2S. Sens. (Basel) **19**, 905 (2019). https://doi.org/10.3390/S19040905

7. Chen, Q., Li, X.-B., Song, R., et al.: Development and utilization of hexacopter unmanned aerial vehicle platform to characterize vertical distribution of boundary layer ozone in wintertime. Atmos. Pollut. Res. **11**, 1073–1083 (2020). https://doi.org/10.1016/j.apr.2020.04.002
8. Pochwała, S., Gardecki, A., Lewandowski, P., et al.: Developing of low-cost air pollution sensor-measurements with the unmanned aerial vehicles in Poland. https://doi.org/10.3390/s20123582
9. Burgués, J., Marco, S.: Environmental chemical sensing using small drones: a review. Sci. Total Environ., 141172 (2020). https://doi.org/10.1016/j.scitotenv.2020.141172
10. Rai, A.C., Kumar, P., Pilla, F., et al.: End-user perspective of low-cost sensors for outdoor air pollution monitoring. Sci. Total Environ. **607–608**, 691–705 (2017). https://doi.org/10.1016/j.scitotenv.2017.06.266
11. Cordero, J.M., Borge, R., Narros, A.: Using statistical methods to carry out in field calibrations of low cost air quality sensors. Sens. Actuators B Chem. **267**, 245–254 (2018). https://doi.org/10.1016/j.snb.2018.04.021
12. de Vito, S., Esposito, E., Massera, E., et al.: Crowdsensing IoT architecture for pervasive air quality and exposome monitoring: design, development, calibration, and long-term validation. Sensors **21** (2021). https://doi.org/10.3390/s21155219
13. Wu, Y., Qi, L., Zhang, H., et al.: Design of UAV downwash airflow field detection system based on strain effect principle. Sensors **19**, 2630 (2019). https://doi.org/10.3390/s19112630
14. Guo, Q., Zhu, Y., Tang, Y., et al.: CFD simulation and experimental verification of the spatial and temporal distributions of the downwash airflow of a quad-rotor agricultural UAV in hover. Comput. Electron. Agric. **172**, 105343 (2020). https://doi.org/10.1016/j.compag.2020.105343
15. Shen, A., Zhou, S., Peng, S.: Atmospheric environment detection method based on multi-rotor UAV platform. Int. Arch. Photogramm. Remote Sens. Spat. Inf. Sci. **XLII-3/W5**, 67–71 (2018). https://doi.org/10.5194/isprs-archives-XLII-3-W5-67-2018
16. Alphasense Ltd.: Alphasense Application Note AAN 010 Design Considerations in Gas Detector Sampling (2013)
17. Arroyo, P., Gómez-Suárez, J., Suárez, J.I., Lozano, J.: Low-cost air quality measurement system based on electrochemical and pm sensors with cloud connection. Sensors **21**, 6228 (2021). https://doi.org/10.3390/s21186228

Simulating a Gas Source Localization Algorithm with Gas Dispersion Produced by Recorded Outdoor Wind

Takumi Haratsu[1], Motoki Sakaue[1], Haruka Matsukura[2], Patrick P. Neumann[3], and Hiroshi Ishida[1](✉)

[1] Tokyo University of Agriculture and Technology, 2-24-16 Nakacho, Koganei, Tokyo 184-8588, Japan
h_ishida@cc.tuat.ac.jp

[2] University of Electro-Communications, 1-5-1 Chofugaoka, Chofu, Tokyo 182-8585, Japan

[3] Bundesanstalt für Materialforschung und -prüfung (BAM), Unter den Eichen 87, 12205 Berlin, Germany

Abstract. This paper reports the use of the first gas dispersion simulator capable of introducing large wind fluctuations into simulations. The proposed simulator enables testing of a modification made to a gas source localization algorithm in a realistic scenario in order to study how the change affects it. Gas source localization in an outdoor environment is a challenging task mainly due to the complexity of the gas spread caused by the unpredictable nature of constantly changing wind. Therefore, a novel use of outdoor wind in developing a gas source localization system by simulation is presented in this paper. To consider the characteristic of sudden but large and unpredictable changes in wind direction, we propose to use recorded outdoor wind to simulate a realistic outdoor gas dispersion which has been done for the first time to the best of our knowledge. With the use of this simulator, we have tested a modification to a mobile robot-based gas source localization algorithm. Multiple simulations of the modified and the original particle filter-based algorithm have been done to study the effect of the tested modification. The results showed that a small difference in the algorithm can greatly impact the results. From this study, we show that the use of simulation consisting of the necessary traits to evaluate outdoor gas source localization, has the potential to accelerate the development of a reliable localization system.

Keywords: Gas dispersion simulator · Gas source localization · Mobile robot olfaction · Particle filter

1 Introduction

Robot-based gas source localization (GSL) is the task of localizing the origin of a released gas using a mobile robot equipped with sensors [1]. This technology is in demand as a means to timely and accurately determine the source of explosive, hazardous, or environmentally unhealthy gas. Negating gas terrorism against a public institution,

D. Tardioli et al. (Eds.): ROBOT 2022, LNNS 590, pp. 105–116, 2023.
https://doi.org/10.1007/978-3-031-21062-4_9

preventing gas accident at an industrial site, and monitoring the emission of greenhouse gas at a landfill is a list of a few outdoor scenarios where robot-based GSL is in need. To perform the GSL task, a reliable GSL algorithm to estimate the location of the emission source using information obtained from sensors is needed. The fluctuating nature of outdoor wind results in a constantly changing and complex gas dispersion field requiring a sophisticated algorithm to accurately localize the emission source.

To achieve reliable gas source localization in such fluctuating environments, various algorithms were proposed over the past two decades, starting from simple reactive chemotactic/anemotactic algorithms [2] to more sophisticated probabilistic ones including "infotaxis" [3, 4], Bayesian inference methods [5, 6], and particle filter-based methods [7, 8]. However, most of these algorithms were just proposed in their original papers and never underwent thorough testing under various environment. It is also very rare to see some modifications or improvements made to the original algorithms. The use of simulation as a tool to develop and validate algorithms is well known [1]. However, ordinary methods to simulate wind such as computational fluid dynamics (CFD) have limitations in producing the sudden but large and unpredictable changes in wind direction [9]. Such changes cannot be reproduced by simply adding random numbers to the mean wind direction without considering the characteristic power spectrum and histogram of the fluctuations of the natural wind [9].

Therefore, in order to utilize wind containing the appropriate property, here we propose to use actual wind data recorded in an outdoor environment to simulate the gas dispersion. In the work described in this paper, gas dispersion is simulated under wind fields given by recorded wind data to enable its fluctuating properties to take effect on the released gas. By doing so, modifications to GSL algorithms can be made in simulations and studied for improvement in its performance in outdoor environments. In the work described in this paper, a modification is made to a particle filter-based GSL algorithm proposed by Neumann, et al. [8], as a test case.

2 Gas Dispersion Simulator Using Outdoor Wind

As mentioned in the introduction, simulation is a conventional tool for developing and validating algorithms. Multiple simulators capable of generating gas plumes under restricted wind conditions using CFD have been proposed [10–12]. However, a simulator providing a gas dispersion affected by wind containing large fluctuations has not been proposed. In an outdoor environment, especially in open areas such as landfills, large fluctuations in wind direction can be observed as shown in Fig. 1. From the figure, it can be seen that even within a period of 5–10 min occasional full reversal in wind direction can be observed. Such full reversal (180° change in wind direction) cannot be reproduced by CFD which is always conducted under pre-defined inflow and outflow boundary conditions.

Since the spread of gas is highly influenced by airflow, large fluctuations in the wind need to be taken into consideration to simulate a realistic outdoor gas dispersion. Therefore, instead of using traditional CFD methods to calculate wind, we propose to use real wind measured in outdoor environments to reproduce gas dispersions. In the simulator we prepared for this work, recorded outdoor wind was applied to the filament-based atmospheric gas dispersion model proposed by Farrell et al. [13]. By using a time series of wind data with real fluctuations to computationally reproduce outdoor gas dispersion, GSL algorithms can be examined under a realistic gas dispersing environment.

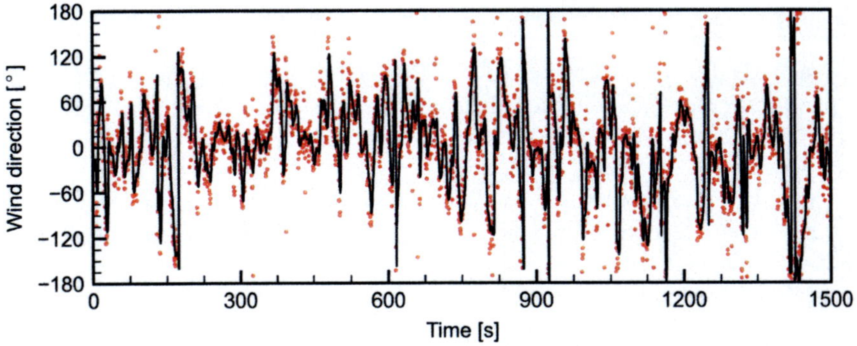

Fig. 1. Time series of wind direction recorded at an open field. The red scatter plot shows the raw wind data, and the black plot is the low pass filtered wind data.

In the filament-based atmospheric gas dispersion model proposed by Farrell et al. [13], a group of gas molecules is represented by a tracer particle called "filament." Each filament is assumed to have a Gaussian gas concentration distribution which is used in calculating a concentration at a certain point. The gas released into the atmosphere is spread by turbulent and molecular diffusion while being advected by the wind. Hence in this model, the transporting vector of the i-th filaments $\dot{\boldsymbol{p}}_i$ is represented as

$$\dot{\boldsymbol{p}}_i = \boldsymbol{v}_{\mathrm{a}} + \boldsymbol{v}_{\mathrm{m}_i}, \tag{1}$$

where $\boldsymbol{v}_{\mathbf{a}}$ is the large scale advection vector and $\boldsymbol{v}_{\mathbf{m}i}$ is the vector showing the transport of gas by mid-scale eddies, which is represented by a Gaussian white noise. In addition, the transport of gas by the small-scale eddies and molecular diffusion is modeled as the increasing radius of the filament's Gaussian distributed gas concentration as a function of time.

By giving the advection vector $\boldsymbol{v}_{\mathbf{a}}$ acting on the filament with recorded outdoor wind, wind field containing real fluctuations can be used in the simulation space. For an open outdoor field with no obstacles, we can assume uniformity of the wind in the simulating environment, at least to some extent [7]. Thus, the advection vector is given by a time series of a single wind vector recorded outdoors throughout the simulation space. By doing so, the spread of filaments is manipulated by wind with real fluctuations enabling a realistic outdoor gas concentration distribution.

3 Simulated GSL Algorithm

Many mobile robot-based GSL algorithms have been proposed by previous studies [2–8]. In this work, we have placed our focus on the particle filter based GSL algorithm proposed by Neumann et al. [8]. This algorithm is designed to localize a gas source under a turbulence-dominated outdoor environment using gas and wind measurements. Particles with weight representing the likelihood of being the gas source location are first uniformly distributed in the searching area. The particle weights are then constantly updated using gas and wind measurements throughout the GSL task. To avoid a degeneracy of particles, a resampling step takes place when most particles have negligible weights. In the resampling step, particles with large weights are duplicated and particles with small weights are randomly discarded. By iteratively updating the particle weights and performing the resampling step, the particles in the searching area will start to converge. The converged coordinate is then considered as the current estimate of the gas source location.

In this algorithm, an area called a patch path envelope (PPE) is constructed from the location of a mobile robot using wind measurements. A schematic of the PPE is shown in Fig. 2. The opening angle of the PPE $\delta_{\mathrm{l,r}}$ is determined by the stability of the wind and is calculated as follows:

$$\delta_{\mathrm{l,r}} = \theta_{\mathrm{u}} \pm 90^{\circ} \cdot S_0, \tag{2}$$

where θ_{u} is the mean wind vector and S_0 is the circular variance [14] of the wind measurements obtained from the sensors loaded on the robot. The wind measurements of the latest T_{w} s are used in this calculation. The constructed PPE describes the region in which the gas patch is estimated to have passed when the mobile robot detects gas. Consequently, it can be assumed that the chance of having the gas source being within the PPE is high during a gas detection. On the other hand, it is also assumed that during a non-gas detection the chance of having the gas source within the PPE is low. By using the wind stability as an indicator in determining the opening angle of the PPE, the turbulence acting on the spreading gas is taken into consideration which is crucial in outdoor GSL.

During the weight update step of this algorithm, the search area is divided into three regions. As shown in Fig. 2, the divided regions can be classified as (a) the region within the PPE, (b) the region outside the PPE but upwind with respect to the robot, and (c) the region downwind with respect to the robot. ω_t^i, the weight of the i-th particle at time step t, is then updated from the equation below according to the binary gas concentration value z_t and the region in which the particle is located.

$$\omega_t^i = \begin{cases} \omega_{t-1}^i & \forall \boldsymbol{x}_t^i \in (\mathrm{a}) \wedge z_t = 1 \\ \alpha \cdot S_0 \cdot \omega_{t-1}^i & \forall \boldsymbol{x}_t^i \in (\mathrm{b}) \wedge z_t = 1 \\ \alpha^2 \cdot S_0 \cdot \omega_{t-1}^i & \forall \boldsymbol{x}_t^i \in (\mathrm{c}) \wedge z_t = 1 \\ \beta^2 \cdot S_0 \cdot \omega_{t-1}^i & \forall \boldsymbol{x}_t^i \in (\mathrm{a}) \wedge z_t = 0 \\ \beta \cdot \omega_{t-1}^i & \forall \boldsymbol{x}_t^i \in (\mathrm{b}) \wedge z_t = 0 \\ \omega_{t-1}^i & \forall \boldsymbol{x}_t^i \in (\mathrm{c}) \wedge z_t = 0 \end{cases} \tag{3}$$

In Eq. (3), $\boldsymbol{x}_t^i$ is the coordinate of the i-th particle. $z_t = 1$ indicates gas detection, and $z_t = 0$ indicates non-gas detection. $\alpha \in [0.1, 1.0]$ and $\beta \in [0.1, 1.0]$ are model parameters adjusting the degree of contribution of the past particle weights. After the update of the particle weights, the weight values are normalized so that the sum of all weights equals one. During a gas detection, it is assumed that region (c) does not contain the gas source and it is also unlikely that the gas source is in region (b) as well. Therefore, the particles within regions (b) and (c) are punished. During a non-gas detection, it is assumed that region (a) does not contain the gas source and it is also unlikely to have the gas source in region (b) as well, because a small fraction of gas may reach the current position of the robot if the gas source exists in region (b). Therefore, the particles within regions (a) and (b) are punished.

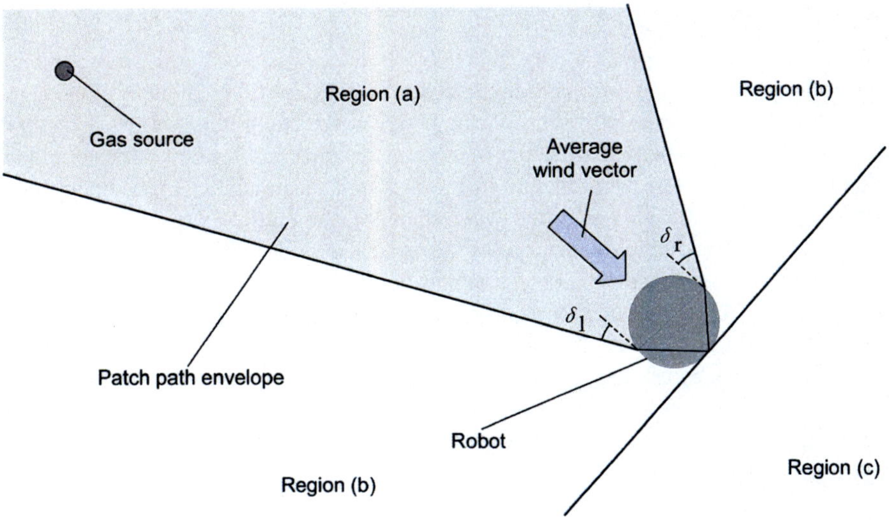

Fig. 2. Schematic of the patch path envelope. Reprinted from [15].

4 Tested Modification and Simulation Setup

One of the advantages of using simulation to study or develop an algorithm is its ability to conduct large numbers of numerical experiments. By using this advantage, a modification to an algorithm can be evaluated with ease by comparing the results with the pre-modified version. In this work, the particle weight update of the particle-filter-based GSL algorithm introduced in Sect. 3 was tweaked to see how the added difference changes the result. To observe the effect of the modification, both the original and the tested modified version of the algorithm were simulated under the same wind data. Since the gas dispersion is numerically solved using the given wind data, the two algorithms are tested under the exact same gas dispersion field. Therefore, if the results differ, the difference is purely influenced by the tested modification. It should be notified that in this work the parameters of the GSL algorithm were selected considering the work done by Neumann et al. [8] in which the parameters were optimized and validated by outdoor experiments, although the conditions are different from our work (e.g., the wind conditions).

After doing some preliminary simulations, we have modified the particle weight update procedure altering Eq. (3) to

$$\omega_t^i = \begin{cases} \omega_{t-1}^i & \forall \boldsymbol{x}_t \in (\mathrm{a}) \wedge z_t = 1 \\ \alpha \cdot S_0 \cdot \omega_{t-1}^i & \forall \boldsymbol{x}_t \in (\mathrm{b}) \wedge z_t = 1 \\ \alpha^2 \cdot S_0 \cdot \omega_{t-1}^i & \forall \boldsymbol{x}_t \in (\mathrm{c}) \wedge z_t = 1 \\ \omega_{t-1}^i & \forall \boldsymbol{x}_t \in (\mathrm{a}) \wedge z_t = 0 \\ \omega_{t-1}^i & \forall \boldsymbol{x}_t \in (\mathrm{b}) \wedge z_t = 0 \\ \omega_{t-1}^i & \forall \boldsymbol{x}_t \in (\mathrm{c}) \wedge z_t = 0 \end{cases}, \tag{4}$$

so that the particle weights are only updated during gas detections. In the original algorithm, the update of particle weights during both gas detections and non-gas detections was thought to be important since the number of assumable source locations could be quickly restricted. However, we saw in the preliminary simulations that unexpected punishment of the weights occurred when updating the particle weights during non-gas detections under large fluctuations of the wind. Therefore, in Eq. (4), the update of particle weights during non-gas detection was omitted.

To evaluate the effect of this modification, we conducted simulations using a real wind data. In the simulations the parameters for the algorithm were set considering the optimization done by Neumann et al. [8]. The model parameters α and β were set to $\alpha = 0.1$ and $\beta = 0.9$, respectively, when simulating the original version of the algorithm. For the simulation of the tested modified version, the same value was set for α and since the update of particle weights during non-gas detection was omitted in the modification, β was not a parameter in this version of the GSL algorithm. The parameter T_{w} indicating the duration of the wind measurement to determine the opening angle of the PPE was set to $T_{\mathrm{w}} = 20$ s, also referencing the work by Neumann et al. [8].

The simulated search space was a rectangular space with a width of 18 m, a length of 15 m, and a height of 3 m with the origin of the xyz coordinate system located at the bottom of the center of the simulated space. Figure 3 shows the top view of the simulated space. As shown at the bottom left part of Fig. 3, the angle of the wind velocity vector was defined counterclockwise in the range of [−180°, 180°] with the positive x-axis as origin. A gas source emitting saturated ethanol vapor at 24.85 °C with a release rate of 30 mL/s was set at coordinate (−7.0, 0.0, 0.5) m. The task of searching for this gas source by executing the GSL algorithm was given to a land-based virtual mobile robot. A predetermined path shown in Fig. 3 was prepared for the virtual robot to run at a constant speed of 0.2 m/s. It was assumed that the robot is equipped with a gas sensor and an anemometer and that those sensors can provide accurate measurements without any delay. In the simulation runs, the robot collected the sensor data at an interval of 0.5 s while running the predetermined path. The wind was assumed to be uniform in the simulation space, and therefore, the robot was given the same wind data used for the gas dispersion simulation. From the 3D gas dispersion field, the robot was given the gas concentration value at a constant height of 0.5 m at the position of the robot, accounting for the height of the equipped gas sensor. An adaptive threshold proposed by Li et al. [7] was used to determine if a gas detection is taking place, which was also used in the work by Neumann et al. [8]. The starting position of the path was set to (7.0, 8.0, 0.5) m, and the final position at (–1.0, –8.0, 0.5) m. The virtual robot executed the GSL algorithm while running the predetermined path. The path was set with the intent of using wind data consisting of large fluctuations and an mean wind direction being around 0°. By moving the virtual robot crosswind for a longer period of time, we let the robot have a better chance of intersecting with the largely meandering gas plume. It should also be said that future work must be done to optimize or validate an algorithm that is most appropriate for this GSL algorithm under open outdoor conditions.

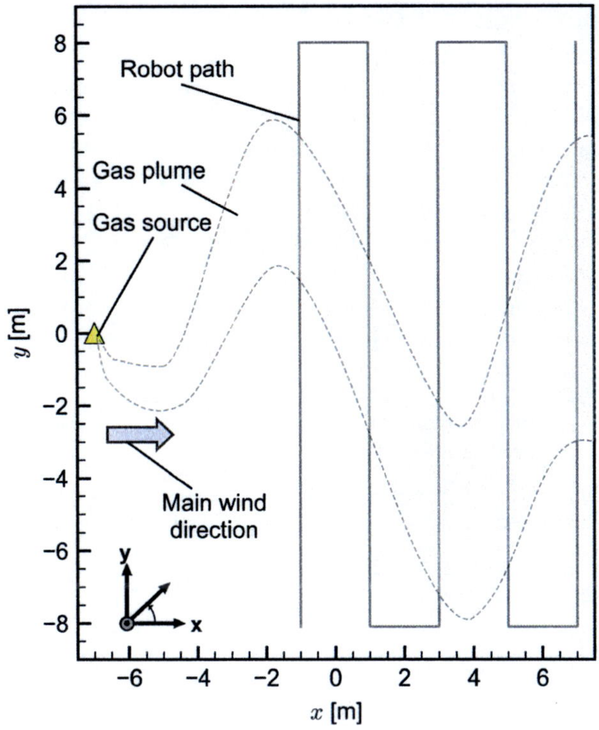

Fig. 3. 2D schematic of the simulation environment. The yellow triangle represents the simulated gas source, and the grey line represents the predetermined path along which the robot runs at a constant speed while executing the GSL algorithm.

The outdoor wind data used for this work was measured at the open athletic field of our university under the same conditions as in our previous work [16] using an ultrasonic anemometer (Model 81000, Young) with a sampling frequency of 2 Hz. The obtained wind data had an mean wind speed of 1.66 m/s with the maximum and the minimum wind speed being 7.36 m/s and 0.03 m/s respectively. The data was rotated to make the mean wind direction 0°. Frequent fluctuations were observed as shown in Fig. 1. From this 1500-s long wind data, data segments were extracted by shifting the starting time by 6 s. Each segment was used to simulate one simulation run (442.5 s) for both the original and the tested modified version of the GSL algorithm. By preparing 100 segments of wind data, the two versions of the algorithm were tested under the same conditions 100 times, where if the result differs it can be determined that the difference is the effect of the tested modification. A gas dispersion experiment was also conducted at the same athletic field with a single gas source [15]. Gas concentration measurements were taken using a photoionization detector (ppbRAE 3000, Honeywell). The measurements were used to determine the initial radius of the filament's gas concentration and the filament growth rate of the used gas dispersion model. The two parameters were set to 0.5 cm and 20 cm^2/s respectively.

5 Simulation Result and Discussion

Figure 4 shows the results of all the estimated gas source locations for both the tested modified and the original version of the algorithm. Table 1 shows the mean and the standard deviation of the localization errors of the two versions. We defined the localization error as the distance between the true and estimated source locations. The estimated source locations for the modified version had the tendency to localize near the true source compared to the original version with a mean localization error of 3.1 m. The results of the original version on the other hand had a strong tendency to estimate the gas source in the downwind area with a mean localization error of 12.8 m. From the simulation results, it can easily be concluded that the conducted change affected the results largely.

Table 1. Mean and standard deviation of the localization errors for both the tested modified and the original version of the algorithm.

Simulated algorithm	Localization error [m]	
	Mean	S. D.
Modified	3.1	2.0
Original	12.8	2.0

To investigate the cause of the difference in the two versions, the simulations were visualized. The generated video clips showed that during non-gas detections in the original version the weights of the particles within the PPE were reduced as expected but caused relative increase in the weights of the particles located elsewhere, often resulting in unexpected convergence of the particles in the regions outside the PPE. In most of these cases, the assumption that the true gas source is located in the downwind region with respect to the robot during non-gas detection failed. Instead, there were some cases in which the gas source was within the PPE even during non-gas detections. It was observed that this phenomenon was caused by the inconsistent meandering of the plume resulting in a large opening angle of the PPE. On the other hand, since in the modified version the update of the particle weights was omitted, the convergence of the particles was not misled by false assumptions and only progressed during odor detections. In gas detection cases, it was often observed that the gas source was within the PPE which allowed the particles to correctly converge near the true source location.

Figure 5 shows example snapshots of the visualized simulation runs comparing the original and the modified version of the algorithm using the same wind data. The

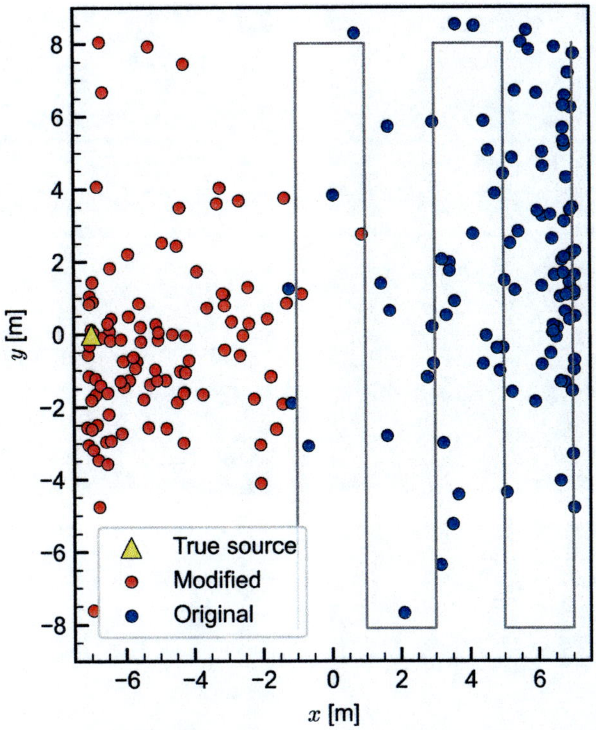

Fig. 4. Scatter plot of the estimated gas source locations and the true gas source. The yellow triangle, grey line, the blue dots, and the red dots each represents the true gas source, the robot trajectory, the estimated source locations of the original algorithm, and the estimated source locations by the tested modified algorithm.

snapshots are in chronological order within a time range of 5-time steps showing how the convergence of the particles progressed during the localization task. In the snapshots of the original algorithm, the particles converged outside the PPE. This can be seen in Fig. 5(b) and (c), where the particles outside the PPE turned red showing their increased weight values during this non-gas detection period. The estimated source location shown by the cyan square came outside the PPE at this time. However, it can be seen that the gas source was within the PPE in this non-gas detection period. It seems that the meandering of the filament plume and large opening angle of the PPE caused the region described by the PPE to contain the gas source unexpectedly during non-gas detections. Since in the modified version the update of the particle weights was omitted during non-gas detection, the convergence of the particles to an inaccurate location did not progress during the false assumption as shown in Fig. 5(B) and (C). Allowing the particles to only converge during gas detections, with the assumption that the gas source is located within the PPE, was more successful as shown in Fig. 5(A) and (a), resulting in the smaller localization error.

It should be noted that the simulation conditions such as the position of the gas source and the predetermined path, may have worked in a disadvantage to the philosophy of

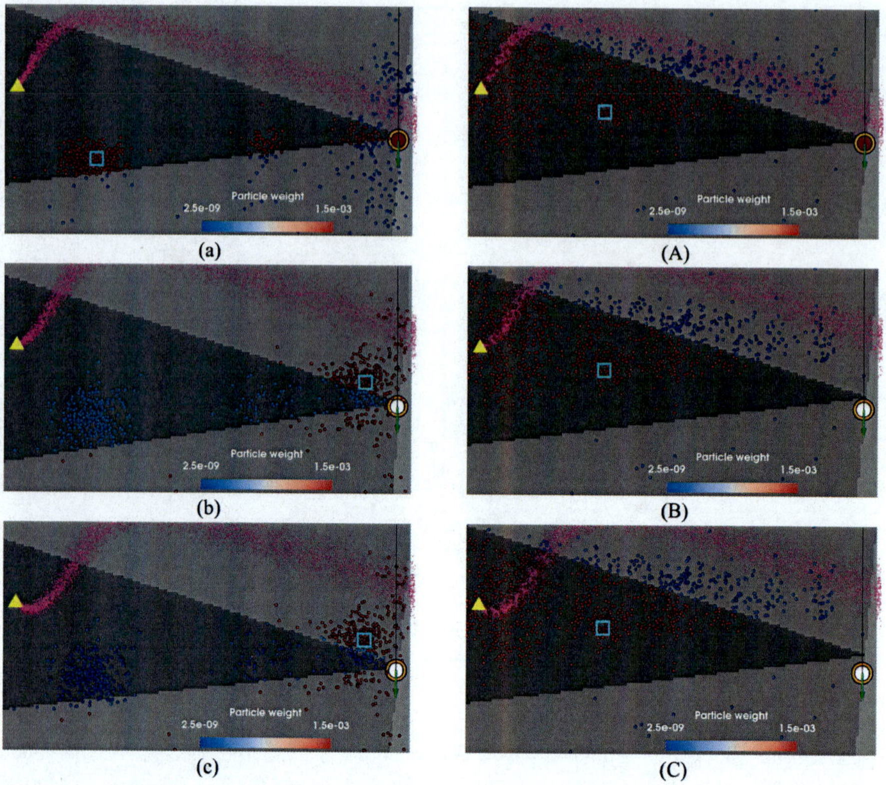

Fig. 5. Snapshots of the visualized simulation runs using the same wind data. The snapshots on the left are of the original algorithm and the snapshots on the right are of the tested modified algorithm. (a)–(c) and (A)–(C) are in chronological order with (a), (A) showing a time step during gas detection, whereas (b), (B) and (c), (C) show time steps during non-gas detection. The dark shaded area is the region described by the PPE, the light shaded area is the region outside the PPE but upwind with respect to the robot, and the unshaded area is the downwind region with respect to the robot. The circle, pink particles, colored dots, yellow triangle, and the cyan square each represent the robot, gas filaments, particles, true source, and the current estimated source location, respectively. In the snapshots, it can be seen that the particles converged outside the PPE during non-gas detections in the original algorithm while in the tested modified algorithm the convergence of the particles did not progress.

the PPE. The mean localization error in our simulation using the original algorithm was 12.8 m as shown in Table 1. On the other hand, the localization error in plume tracking experiments reported in the original paper [8] was 0.56–2.86 m. In the experiments, a drone was programmed to track a methane plume, and the sensor data collected during the tracking flight was provided to the particle filter. The deviation of the drone from the methane plume was kept minimized during the flight, and therefore, non-gas detection period must have been much shorter than in our simulation runs. In the original experiments, the drone passed the gas source as a result of plume tracking, and the source location was estimated by using all sensor data collected during that flight. In our simulations, the predetermined path for the land-based robot was terminated before reaching the gas source to see if the source location can be estimated from a partial scan. These differences may also have affected the results.

Definitely more work is needed to fully evaluate this particle filter-based GSL algorithm. Nonetheless, it can be said based on our simulations that even under outdoor wind the PPE is capable of correctly estimating the gas source location during gas detections. While during non-gas detections, the assumption that the gas source is located outside the PPE is not always correct which seemed to have led to poor results of the original algorithm. The small modification done to the algorithm impacted the results in a way that showed potential in improving the GSL algorithm. Overall, it can be concluded that the simulator consisting of real wind fluctuations showed prospects in testing modifications and validating algorithms under outdoor gas dispersion for further improvements.

6 Conclusion

This paper reported the use of the first gas dispersion simulator capable of generating gas dispersion under wind containing real fluctuations. This was achieved with a novel approach of using recorded outdoor wind as the overall wind field to simulate gas dispersions. The simulator was used to test a modification to a GSL algorithm and study how the tested modification affects the overall results. The particle filter based GSL algorithm by Neumann et al. was tested for modification in this work. Simulations of both the original and the tested modified version were done multiple times to analyze the effects of the conducted modification. From the results, it was concluded that a small change in the algorithm can impact the result greatly and a simulator consisting of the necessary traits to evaluate outdoor GSL systems has the potential in establishing a stable and precise localization system. Since the simulation conditions reported in this work were limited, future works will include simulating the algorithm using wind recorded on different days, revising the predetermined path, optimizing the parameters in the algorithm, and testing different modifications.

Acknowledgments. This work was supported in part by JSPS KAKENHI Grant Numbers 19H02103, 20H02145, and 22H04952.

References

1. Jing, T., Meng, Q.-H., Ishida, H.: Recent progress and trend of robot odor source localization. IEEJ Trans. Electr. Electron. Eng. **16**(7), 938–953 (2021). https://doi.org/10.1002/tee.23364
2. Ishida, H., Nakayama, G., Nakamoto, T., Moriizumi, T.: Controlling a gas/odor plume-tracking robot based on transient responses of gas sensors. IEEE Sens. J. **5**(3), 537–545 (2005). https://doi.org/10.1109/JSEN.2004.839597
3. Vergassola, M., Villermaux, E., Shraiman, B.I.: 'Infotaxis' as a strategy for searching without gradients. Nature **445**(7126), 406–409 (2007). https://doi.org/10.1038/nature05464
4. Voges, N., Chaffiol, A., Lucas, P., Martinez, D.: Reactive searching and infotaxis in odor source localization. PLoS Comput. Biol. **10**(10), e1003861 (2014). https://doi.org/10.1371/journal.pcbi.1003861

5. Pang, S., Farrell, J.A.: Chemical plume source localization. IEEE Trans. Syst. Man Cybern. B Cybern. **36**(5), 1068–1080 (2006). https://doi.org/10.1109/TSMCB.2006.874689
6. Ferri, G., et al.: Mapping multiple gas/odor sources in an uncontrolled indoor environment using a Bayesian occupancy grid mapping based method. Rob. Auton. Syst. **59**(11), 988–1000 (2011). https://doi.org/10.1016/j.robot.2011.06.007
7. Li, J.-G., Meng, Q.-H., Wang, Y., Zeng, M.: Odor source localization using a mobile robot in outdoor airflow environments with a particle filter algorithm. Auton. Robot. **30**(3), 281–292 (2011). https://doi.org/10.1007/s10514-011-9219-2
8. Neumann, P.P., Bennetts, V.H., Lilienthal, A.J., Bartholmai, M., Schiller, J.H.: Gas source localization with a micro-drone using bio-inspired and particle filter-based algorithms. Adv. Robot. **27**(9), 725–738 (2013). https://doi.org/10.1080/01691864.2013.779052
9. Sakaue, M., Takahashi, Y., Matsukura, H., Ishida, H.: Gas dispersion simulator with strong fluctuations for developing gas source localization systems. In: Proceedings of International Symposium on Olfaction and Electronic Nose. IEEE, New York (2022). https://doi.org/10.1109/ISOEN54820.2022.9789554
10. Awadalla, M., Lu, T.-F., Tian, Z.F., Dally, B., Liu, Z.: 3D framework combining CFD and MATLAB techniques for plume source localization research. Build. Environ. **70**, 10–19 (2013). https://doi.org/10.1016/j.buildenv.2013.07.021
11. Pashami, S., Asadi, S., Lilienthal, A.: Integration of OpenFOAM flow simulation and filament-based gas propagation models for gas dispersion simulation. In: Proceedings of the Open Source CFD International Conference, vol. 2 (2010)
12. Monroy, J., Hernandez-Bennetts, V., Fan, H., Lilienthal, A., Gonzalez-Jimenez, J.: GADEN: A 3D gas dispersion simulator for mobile robot olfaction in realistic environments. Sensors **17**(7), 1479 (2017). https://doi.org/10.3390/s17071479
13. Farrell, J.A., Murlis, J., Long, X., Li, W., Cardé, R.T.: Filament-based atmospheric dispersion model to achieve short time-scale structure of odor plumes. Environ. Fluid Mech. **2**(1–2), 143–169 (2002). https://doi.org/10.1023/A:1016283702837
14. Fisher, N.I.: Statistical Analysis of Circular Data. Cambridge University Press, Cambridge (1995)
15. Sakaue, M., Haratsu, T., Matsukura, H., Neumann, P.P., Ishida, H.: Comparison of two particle-filter-based gas source localization algorithms: use of gas dispersion simulator driven by recorded outdoor wind. Sensors (2022, submitted)
16. Bilgera, C., Yamamoto, A., Sawano, M., Matsukura, H., Ishida, H.: Application of convolutional long short-term memory neural network to signals collected from a sensor network for autonomous gas source localization in outdoor environments. Sensors **18**(12), 4484 (2018). https://doi.org/10.3390/s18124484

Potential Field Based Swarm Exploration of Processes Governed by Partial Differential Equations

Patrick Hinsen(✉), Thomas Wiedemann, and Dmitriy Shutin

Institute of Communications and Navigation, German Aerospace Center (DLR), Oberpfaffenhofen, Germany
{patrick.hinsen,thomas.wiedemann,dmitriy.shutin}@dlr.de

Abstract. In robotic exploration of gas diffusion processes, domain knowledge can improve and speed up estimation of unknown spatial gas concentrations or source localization. Here, a novel robotic swarm exploration strategy for such processes governed by a partial differential equation is presented. A probabilistic process description incorporates physical domain knowledge into the exploration strategy and allows for estimation of concentrations and source distribution, as well as for quantification of the uncertainty of estimated parameters over the spatial exploration domain. Artificial potential fields are used to guide a multi-agent system of non-holonomic robots towards regions of high uncertainty, i.e. informative sampling locations, while avoiding collisions. The developed strategy is evaluated in simulations as well as in Hardware-in-the-Loop experiments. It is shown to outperform a systematic sampling strategy for swarms of the same size.

Keywords: Swarm robotics · Robotic exploration · Uncertainty mapping · Artificial potential field control · Gas exploration

1 Introduction

This paper aims to bring domain knowledge into an exploration strategy for a robotic swarm, which will allow it to explore an environment autonomously. In particular, we consider robotic exploration of gaseous processes with the goal of mapping the gas concentration as well as potential sources—a common but challenging task in mobile robotic olfaction [3], due to the high spatial and temporal dynamics of gas propagation [12].

We take advantage of the stengths of multi-agent systems, as multiple mobile robots—a swarm—can capture spatial dynamics and space-varying gas concentrations more efficiently compared to a single robot. This does however bring with it the challenge of real-time coordination for collision avoidance and finding suitable sampling locations. The swarm should make intelligent decisions concerning its future actions based on the collected data, so new measurements, taken 'at the right place, at the right time', can capture relevant information.

Control of the swarm agents needs to account for robotic motion constraints and avoid collisions with other agents or the environment; other additional

D. Tardioli et al. (Eds.): ROBOT 2022, LNNS 590, pp. 117–129, 2023.
https://doi.org/10.1007/978-3-031-21062-4_10

constraints can also be taken into account (see e.g., [10]). Popular methods can be roughly dichotomized into optimization-based methods [7,10,11], and approaches utilizing artifical potential field based control (APFC) [6,8,14]. The shortcoming of the former is poor scalability and increasing computational complexity with a growing number of agents; methods based on APFC scale much better in this respect [6,8]. Moreover, through an appropriate adaptation of the potential field methods, other constraints can be incorporated, e.g. network connectivity constraints [2]. These aspects have been separately addressed in a number of works [10].

Nonetheless, these control schemes do not directly incorporate process exploration. The approach advocated in this work integrates physics-based, information-driven exploration of a gas propagation process and multi-agent control in a single joint framework. Specifically, we use as a mathematical model the so-called advection-diffusion (or a scalar transport) Partial Differential Equations (PDEs) to describe the gas dispersion. This allows us to reconstruct a spatial map of gas concentration, identify the parameters of the PDE [12], and quantify the spatial uncertainty distribution of the estimated gas concentration—the uncertainty map. This map is combined with APFC, which has been shown to be well suited for multi-agent control in practical settings (see e.g. [9]). The computed uncertainty map is used to construct a potential field, which attracts robots to areas of high uncertainty. Because of the process dynamics, acquiring measurements in these regions leads to changes in the uncertainty map, which in turn feed back into the distributed control scheme.

2 Fundamentals

2.1 Process Model

In this work we will rely on prior knowledge, and specifically physical modeling of gas propagation, to derive the exploration strategy. For this, we consider a gas dispersion process modelled as a linear diffusion PDE over some exploration area Ω. The corresponding PDE takes the form of Poisson's equation in 2D,

$$-\kappa\, \Delta f(\boldsymbol{x}) = q(\boldsymbol{x}), \quad \boldsymbol{x} \in \Omega \subset \mathbb{R}^2, \qquad \text{s.t.} \quad f(\boldsymbol{x}) = 0, \boldsymbol{x} \in \Gamma, \tag{1}$$

where $q(\boldsymbol{x})$ represents the spatial gas source distribution, $f(\boldsymbol{x})$ is the spatial gas concentration intensity, and κ is the diffusion coefficient, set w.l.o.g. to $\kappa = 1$. We model an open boundary Γ of the exploration domain Ω, which allows concentration to flow off (Dirichlet boundary condition).

As a first step, consider a Finite Difference Method (FDM) discretization of (1). The environment Ω is divided into N smaller subdomains (or cells), which form a regular grid with rows $i \in [1; I]$ and columns $j \in [1; J]$, so that $N = IJ$. For each cell with coordinates $\boldsymbol{x}_{i,j}$, we assume a constant concentration value. Similarly, we discretize the source distribution $q(\boldsymbol{x})$, but assume no sources in the cells at the boundary of the domain. The corresponding concentration and source strength values are aggregated into vectors,

$$\boldsymbol{f} = \left(f(\boldsymbol{x}_{1,1}) \ldots f(\boldsymbol{x}_{I,1})\, f(\boldsymbol{x}_{1,2}) \ldots f(\boldsymbol{x}_{I,2}) \ldots f(\boldsymbol{x}_{I,J})\right)^{\mathsf{T}} \in \mathbb{R}^{N}, \tag{2}$$

$$\boldsymbol{q} = \left(q(\boldsymbol{x}_{2,2}) \ldots q(\boldsymbol{x}_{I-1,2})\, q(\boldsymbol{x}_{2,3}) \ldots q(\boldsymbol{x}_{I-1,3}) \ldots q(\boldsymbol{x}_{I-1,J-1})\right)^{\mathsf{T}} \in \mathbb{R}^{Q}, \tag{3}$$

where $Q = (J-2)(I-2)$ is the number of non-boundary cells. Using the FDM approach, we discretize the Laplace operator Δ as $\tilde{\boldsymbol{A}} \in \mathbb{R}^{Q \times N}$, and define $\boldsymbol{A} = -\kappa\tilde{\boldsymbol{A}}$. This allows us to rewrite (1) as a system of linear equations

$$\boldsymbol{A}\boldsymbol{f} = \boldsymbol{q}, \qquad \text{s.t.} \quad \boldsymbol{B}\boldsymbol{f} = \boldsymbol{0}, \tag{4}$$

where $\boldsymbol{B} \in \mathbb{R}^{N \times N}$ is a selection matrix: it selects elements of $\boldsymbol{f}$ that correspond to the cells at the boundary Γ.

Finally, we model the k-th concentration measurement made by a robot a taken with an in-situ sensor at the current position of the robot $\boldsymbol{p}_{a,k}$ as

$$z_{a,k} = \boldsymbol{m}^{\mathsf{T}}(\boldsymbol{p}_{a,k})\,\boldsymbol{f} + \xi_{a,k}\,, \tag{5}$$

with selection vector $\boldsymbol{m}^{\mathsf{T}}(\boldsymbol{p}_{a,k}) \in \mathbb{R}^{1 \times N}$ selecting the concentration value from $\boldsymbol{f}$ corresponding to the currently occupied cell; the term $\xi_{a,k}$ represents an additive zero-mean Gaussian measurement perturbation. Model (5) can easily be brought into matrix notation using a vector $\boldsymbol{z} = \boldsymbol{M}\,\boldsymbol{f} + \boldsymbol{\xi}$, which aggregates all measurements made thus far.

2.2 Probabilistic Framework

Based on the process model, a spatial map—referred to as *uncertainty map*—is to be constructed over Ω, to quantify the spatial distribution of the estimated parameters' uncertainty. For this we transform our model into a probabilistic description. To this end, consider a posterior distribution $p(\boldsymbol{f}, \boldsymbol{q} \mid \boldsymbol{z}) \propto p(\boldsymbol{z} \mid \boldsymbol{f})\, p(\boldsymbol{f} \mid \boldsymbol{q})\, p(\boldsymbol{q})$ and let us consider the terms on its right-hand side in more detail.

The first term—measurement likelihood $p(\boldsymbol{z} \mid \boldsymbol{f})$—is dictated by (5). Specifically, since $\xi_{a,k}$ is assumed to be zero-mean white Gaussian, the quantity $\boldsymbol{z}|\boldsymbol{f}$ is distributed as $\boldsymbol{z}|\boldsymbol{f} \sim \mathcal{N}\left(\boldsymbol{M}\boldsymbol{f}, \lambda_{\mathrm{M}}^{-1}\boldsymbol{I}\right)$, where $\lambda_{\mathrm{M}}^{-1}$ is the variance of the additive noise. Equivalently, in log-domain we have

$$\ln p(\boldsymbol{z} \mid \boldsymbol{f}) \propto^{e} -\frac{\lambda_{\mathrm{M}}}{2}\|\boldsymbol{z} - \boldsymbol{M}\boldsymbol{f}\|_2^2\,. \tag{6}$$

where $\propto^{e}$ denotes proportionality in the log-domain, i.e. $y \propto^{e} x$ implies $\mathrm{e}^{y} \propto \mathrm{e}^{x}$.

The term $p(\boldsymbol{f} \mid \boldsymbol{q})$ describes the probability of $\boldsymbol{f}$ given some source intensity distribution $\boldsymbol{q}$ and is related to the PDE model in (1). For perfect model fit, i.e. when $\boldsymbol{A}\tilde{\boldsymbol{f}} = \boldsymbol{q}$, $p(\boldsymbol{f} \mid \boldsymbol{q}) = \delta_{\tilde{\boldsymbol{f}}}$, where $\tilde{\boldsymbol{f}}$ is the support of a Dirac delta function, such that $\boldsymbol{A}\,\tilde{\boldsymbol{f}} = \boldsymbol{q}$. In our approach, we relax this requirement by approximating

$\delta_{\tilde{f}}$ with a Gaussian pulse of width $\lambda_{\mathrm{A}}^{-1}$ (see also [13]). In a similar fashion the boundary condition (1) is relaxed, using width $\lambda_{\mathrm{B}}^{-1}$ for the Gaussian pulse:

$$\ln p\left(\boldsymbol{f} \mid \boldsymbol{q}\right) - \mathbf{0})^{\mathsf{T}} \lambda_{\mathrm{B}} \boldsymbol{I} \left(\boldsymbol{B} \boldsymbol{f} - \mathbf{0}\right). \propto^{e} -\frac{\lambda_{\mathrm{A}}}{2} \|\boldsymbol{A} \boldsymbol{f} - \boldsymbol{q}\|_2^2 - \frac{\lambda_{\mathrm{B}}}{2} \|\boldsymbol{B} \boldsymbol{f}\|_2^2 . \tag{7}$$

Naturally, for $\lambda_{\mathrm{B}} \to \infty$ and $\lambda_{\mathrm{A}} \to \infty$, Dirac distributions are recovered.

Finally, we consider the prior term $p(\boldsymbol{q})$ that incorporates a priori assumptions about the source distribution. We will assume that values of $\|\boldsymbol{q}\|_2^2$ are comparatively small, which translates to low gas inflow; specifically,

$$\ln p(\boldsymbol{q}) \propto^{e} -\frac{\lambda_{\mathrm{R}}}{2} \|\boldsymbol{q}\|_2^2 . \tag{8}$$

where λ_{R} is a tunable prior parameter. Note that in general this model can represent more complex a-priori assumptions, e.g. a sparse source distribution [13] or non-negative gas concentrations. The latter is beyond the scope of this paper.

Now we can write the logarithm of the parameter posterior as

$$\ln p\left(\boldsymbol{f}, \boldsymbol{q} \mid \boldsymbol{z}\right) = \ln p\left(\boldsymbol{z} \mid \boldsymbol{f}\right) + \ln p\left(\boldsymbol{f} \mid \boldsymbol{q}\right) + \ln p(\boldsymbol{q}) \propto^{e}$$

$$-\frac{1}{2}\left[\begin{pmatrix} \boldsymbol{M} & \mathbf{0} \\ \boldsymbol{A} & -\boldsymbol{I} \\ \boldsymbol{B} & \mathbf{0} \\ \mathbf{0} & \boldsymbol{I} \end{pmatrix}\begin{pmatrix} \boldsymbol{f} \\ \boldsymbol{q} \end{pmatrix} - \begin{pmatrix} \boldsymbol{z} \\ \mathbf{0} \\ \mathbf{0} \\ \mathbf{0} \end{pmatrix}\right]^{\mathsf{T}} \begin{pmatrix} \lambda_{\mathrm{M}} & & & \\ & \lambda_{\mathrm{A}} & & \\ & & \lambda_{\mathrm{B}} & \\ & & & \lambda_{\mathrm{R}} \end{pmatrix} \left[\begin{pmatrix} \boldsymbol{M} & \mathbf{0} \\ \boldsymbol{A} & -\boldsymbol{I} \\ \boldsymbol{B} & \mathbf{0} \\ \mathbf{0} & \boldsymbol{I} \end{pmatrix}\begin{pmatrix} \boldsymbol{f} \\ \boldsymbol{q} \end{pmatrix} - \begin{pmatrix} \boldsymbol{z} \\ \mathbf{0} \\ \mathbf{0} \\ \mathbf{0} \end{pmatrix}\right] . \tag{9}$$

Naturally, the posterior is a multi-variate Gaussian probability density function (PDF). Computing its mean lets us obtain the concentration distribution $\hat{\boldsymbol{f}}$ and source distribution $\hat{\boldsymbol{q}}$ from the measurements $\boldsymbol{z}$:

$$\begin{pmatrix} \hat{\boldsymbol{f}} \\ \hat{\boldsymbol{q}} \end{pmatrix} = \arg\max_{\boldsymbol{f}, \boldsymbol{q}} \ln p\left(\boldsymbol{f}, \boldsymbol{q} \mid \boldsymbol{z}\right) = \begin{pmatrix} \boldsymbol{M} & \mathbf{0} \\ \boldsymbol{A} & -\boldsymbol{I} \\ \boldsymbol{B} & \mathbf{0} \\ \mathbf{0} & \boldsymbol{I} \end{pmatrix}^{+} \begin{pmatrix} \boldsymbol{z} \\ \mathbf{0} \\ \mathbf{0} \\ \mathbf{0} \end{pmatrix} , \tag{10}$$

where $(\cdot)^{+}$ indicates a right pseudoinverse, making expression (10) a *maximum a posteriori* estimator. Moreover, after some straightforward manipulations, the covariance matrix of $p\left(\boldsymbol{f}, \boldsymbol{q} \mid \boldsymbol{z}\right)$ can also be computed as

$$\boldsymbol{\Sigma} = \begin{pmatrix} \lambda_{\mathrm{M}} \boldsymbol{M}^{\mathsf{T}} \boldsymbol{M} + \lambda_{\mathrm{A}} \boldsymbol{A}^{\mathsf{T}} \boldsymbol{A} + \lambda_{\mathrm{B}} \boldsymbol{B}^{\mathsf{T}} \boldsymbol{B} & -\lambda_{\mathrm{A}} \boldsymbol{A}^{\mathsf{T}} \\ -\lambda_{\mathrm{A}} \boldsymbol{A} & \lambda_{\mathrm{A}} \boldsymbol{I} + \lambda_{\mathrm{R}} \boldsymbol{I} \end{pmatrix}^{-1} = \begin{pmatrix} \boldsymbol{\Sigma}_{\mathrm{ff}} & \boldsymbol{\Sigma}_{\mathrm{fq}} \\ \boldsymbol{\Sigma}_{\mathrm{qf}} & \boldsymbol{\Sigma}_{\mathrm{qq}} \end{pmatrix} . \tag{11}$$

This covariance matrix, or more specifically it's diagonal elements, describe the variance of the marginal distributions for each element of $\boldsymbol{f}$ and $\boldsymbol{q}$. A **higher variance** in this case signifies a **higher uncertainty**. Let $\boldsymbol{\sigma}_{\boldsymbol{f}}^2 = \operatorname{diag}(\boldsymbol{\Sigma}_{\mathrm{ff}})$ aggregate the variances of the intensity distribution. By mapping the elements of $\boldsymbol{\sigma}_{\boldsymbol{f}}^2$ to the corresponding spatial cell locations, we arrive at the spatial uncertainty distribution, to which we refer as the *uncertainty map*.

The objective of exploration will thus be to reduce this uncertainty, i.e. some measure of $\boldsymbol{\sigma}_{\boldsymbol{f}}^2$, by taking measurements in strategic locations. This task we propose to solve using an APFC approach.

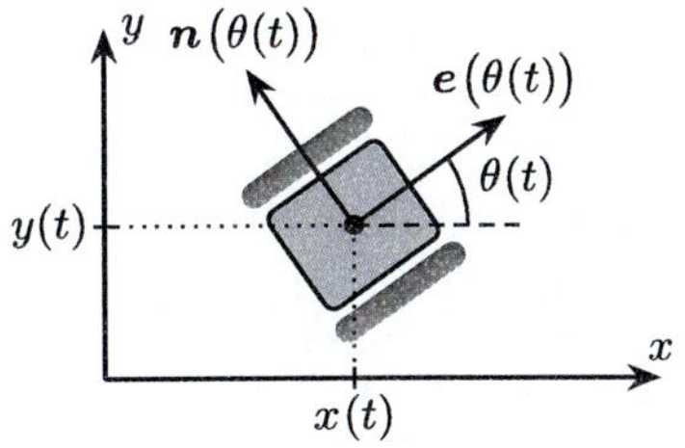

Fig. 1. Robot geometry

Fig. 2. Physical Rover

2.3 Artificial Potential Fields for Multi-robot Control

Consider a swarm $\mathcal{N} := \{1, \ldots, A\}$ of A mobile robots tasked with exploring a gas diffusion process over the exploration area of interest Ω. A simple *unicycle*-type robot model is used to describe each agent $a \in \mathcal{N}$. An agent's pose consists of the position, $\boldsymbol{p}_a(t) = \big(x_a(t)\ y_a(t)\big)^\mathsf{T} \in \mathbb{R}^2$, and the heading angle $\theta_a(t) \in [0; 2\pi)$. The heading can equivalently be expressed in terms of the forward and normal orientation vectors, $\boldsymbol{e}_a\big(\theta_a(t)\big)$ and $\boldsymbol{n}_a\big(\theta_a(t)\big)$, as depicted in Fig. 1. The dynamics of robot a are described using the state-space model

$$\mathrm{R}_a : \begin{cases} \dot{\boldsymbol{p}}_a(t) = \boldsymbol{e}_a\big(\theta_a(t)\big)\, u_{\mathrm{v},a}(t) \\ \dot{\theta}_a(t) = u_{\omega,a}(t), \end{cases} \tag{12}$$

that captures its non-holonomic constraints. In (12), $u_{\mathrm{v},a}(t)$ and $u_{\omega,a}(t)$ denote the control inputs of robot a for forward and angular velocity, respectively.

APFC [4,5] will be used to control the swarm. We will provide a brief summary of the method, as required to understand the proposed scheme. For a more in-depth explanation, the interested reader is referred to [9].

An artificial, continuously differentiable potential function $P : \Omega \to \mathbb{R}$ is constructed. Its spatial gradient $\frac{\mathrm{d}P}{\mathrm{d}\boldsymbol{x}}$, the *potential field*, generates virtual forces that act upon the robots. In our design, the potential function is the superposition $P = P_{\mathrm{rep}} + P_{\mathrm{attr}}$ of two potentials with different aims: P_{rep} is a repulsive potential ensuring that all robots maintain some separation at all times. P_{attr} is an attractive potential, designed to attract robots to regions of high uncertainty, according to $\boldsymbol{\sigma}_{\boldsymbol{f}}^2$. The latter will be discussed in Sect. 3.

To descend along the gradient of the potential function P, the robots need to align themselves with the potential field. The angular and forward velocity of robot a are set using the respective control laws,

$$u_{\omega,a}(t) = k_\omega \arctan\left(\frac{\boldsymbol{n}_a\big(\theta_a(t)\big)^\mathsf{T} \frac{\partial P}{\partial \boldsymbol{p}_a}}{\boldsymbol{e}_a\big(\theta_a(t)\big)^\mathsf{T} \frac{\partial P}{\partial \boldsymbol{p}_a}} \right) \text{ and } u_{\mathrm{v},a}(t) = -k_{\mathrm{v}} \frac{\partial P}{\partial \boldsymbol{p}_a}^\mathsf{T} \boldsymbol{e}_a\big(\theta_a(t)\big), \tag{13}$$

where $u_{\omega,a}(t)$ turns the robot until it faces in the gradient's direction with its front or rear, whichever is closest, and $u_{\mathrm{v},a}(t)$ realizes a velocity dependent on the robots angle relative to the gradient of P. The controller gains $k_\omega > 0$ and $k_\mathrm{v} > 0$ are set heuristically.

We design the repulsive potential $P_\mathrm{rep}(\boldsymbol{x})$, $\boldsymbol{x} \in \Omega$, as the sum of individual repulsive potentials $\varphi_\mathrm{rep}(\boldsymbol{x}, \boldsymbol{p}_a(t))$ around the position of each agent a:

$$\varphi_\mathrm{rep}(\boldsymbol{x}, \boldsymbol{p}_a(t)) = \begin{cases} \frac{k_\mathrm{R}}{2} \left(\|\boldsymbol{x} - \boldsymbol{p}_a(t)\| - \delta \right)^2 , & \|\boldsymbol{x} - \boldsymbol{p}_a(t)\| \leq \delta \\ 0 , & \text{otherwise} , \end{cases} \tag{14}$$

$$\text{such that } P_\mathrm{rep}(\boldsymbol{x}) = \sum_{a \in \mathcal{N}} \left(\frac{1}{2} \sum_{b \in \mathcal{N} \setminus a} \varphi_\mathrm{rep}(\boldsymbol{p}_b(t), \boldsymbol{p}_a(t)) \right) . \tag{15}$$

The fact that $\varphi_\mathrm{rep}(\boldsymbol{x}, \boldsymbol{p}_a(t))$ vanishes for distances from robot a larger than δ implies that only the local neighborhood of any robot is relevant for its collision avoidance. With the derivative of each individual $\varphi_\mathrm{rep}(\boldsymbol{x}, \boldsymbol{p}_a(t))$ (see eq. (16)), the potential field takes the form of eq. (17):

$$\frac{\partial \varphi_\mathrm{rep}(\boldsymbol{x}, \boldsymbol{p}_a(t))}{\partial \boldsymbol{x}} = \begin{cases} k_\mathrm{R} \left(1 - \frac{\delta}{\|\boldsymbol{x} - \boldsymbol{p}_a(t)\|} \right) \left(\boldsymbol{x} - \boldsymbol{p}_a(t) \right)^\mathsf{T} , & \|\boldsymbol{x} - \boldsymbol{p}_a(t)\| \leq \delta \\ 0 , & \text{otherwise} , \end{cases} \tag{16}$$

$$\text{which leads to } \left. \frac{\partial P_\mathrm{rep}(\boldsymbol{x})}{\partial \boldsymbol{x}} \right|_{\boldsymbol{x} = \boldsymbol{p}_a(t)} = \sum_{b \in \mathcal{N} \setminus a} \left(\left. \frac{\partial \varphi_\mathrm{rep}(\boldsymbol{x}, \boldsymbol{p}_a(t))}{\partial \boldsymbol{x}} \right|_{\boldsymbol{x} = \boldsymbol{p}_b(t)} \right) . \tag{17}$$

The quadratic term in (14) creates a steeply increasing potential where robots get close to each other. This leads to a linearly increasing repulsion "force" the closer two robots get, which we can scale appropriately using the parameter k_R.

3 Process Mapping via Potential Field Navigation

We will next discuss the design of the attractive part of the potential function used in APFC, $P_\mathrm{attr}(\boldsymbol{x})$, based on the computed uncertainty map $\boldsymbol{\sigma}_f^2$ (see Sect. 2.2). Our objective is to make agents gravitate towards regions of high uncertainty, where new measurements will lead to higher information gains. First, $\boldsymbol{\sigma}_f^2$ is normalized to the range $[0; 1]$, as the absolute scale of uncertainty is not of interest. To obtain from this a continuously differentiable potential function, we create an interpolation using a bivariate cubic spline,

$$\varphi(\boldsymbol{x}) = \mathrm{CubicSpline} \left(\frac{\boldsymbol{\sigma}_f^2}{\max\limits_{i,j} \boldsymbol{\sigma}_f^2} \right) , \qquad \text{s.t.} \quad \left\| \boldsymbol{\varphi} - \frac{\boldsymbol{\sigma}_f^2}{\max\limits_{i,j} \boldsymbol{\sigma}_f^2} \right\|_2^2 \leq N \sigma_\mathrm{s}^2 , \tag{18}$$

where $\boldsymbol{\varphi}$ is a vector aggregating elements $\varphi(\boldsymbol{x}_{i,j})$, in the same manner as $\boldsymbol{f}$ (see Eq. (2)). The additional constraint allows to smooth out small local extrema in $\varphi(\boldsymbol{x})$; we choose $\sigma_\mathrm{s}^2 = 0.01$. Being a polynomial representation makes $\varphi(\boldsymbol{x})$ trivial to differentiate. Notice that it is a function of space, as $\boldsymbol{\sigma}_f^2$ implicitly depends on space as well.

Differentiating $\varphi(\boldsymbol{x})$ directly is not desirable, however, as there are no bounds to its steepness. Instead, we modify the vector field one last time; specifically:

$$\frac{\mathrm{d}P_{\mathrm{attr}}(\boldsymbol{x})}{\mathrm{d}\boldsymbol{x}} = \frac{\boldsymbol{g}(\boldsymbol{x})}{\alpha + \|\boldsymbol{g}(\boldsymbol{x})\|}, \quad \text{where} \quad \boldsymbol{g}(\boldsymbol{x}) = -\nabla\varphi(\boldsymbol{x})\,. \tag{19}$$

This vector field exposes robots to a bounded virtual force, until they come close to an extremum, where they are supposed to stop. Note that Eq. (19) is akin to a sigmoid function with a saturation effect that ensures control input limits are not violated, and can be tuned with a parameter α.

It should be noted that $\frac{\mathrm{d}P_{\mathrm{attr}}(\boldsymbol{x})}{\mathrm{d}\boldsymbol{x}}$ specifies the attractive field acting on the robots. It is possible, however, that due to the saturation introduced in Eq. (19) this field can no longer be expressed as the gradient of a scalar potential function $P_{\mathrm{attr}}(\boldsymbol{x})$; in other words, it may no longer be conservative. Yet for our purposes this technicality is irrelevant, as explicit $P_{\mathrm{attr}}(\boldsymbol{x})$ is not needed for navigation.

Measurement Acquisition. As agents approach a maximum of uncertainty, a criterion is needed to decide when to stop and sample the process, i.e. take a measurement. Good results could be achieved in this work by choosing the condition

$$\left\|\frac{\mathrm{d}P(\boldsymbol{x})}{\mathrm{d}\boldsymbol{x}}\right\|_2 \overset{!}{<} s_{\min} \quad \wedge \quad t_{\mathrm{m},n+1} - t_{\mathrm{m},n} \overset{!}{\geq} \Delta t_{\mathrm{m}}\,, \tag{20}$$

which (I) establishes a cutoff for the gradient's steepness, $s_{\min}$, while also (II) giving agents time after each of their measurements to vacate a (newly formed) valley in the potential function. Here, $t_{\mathrm{m},n}$ is the time of the n-th measurement.

Note that in condition (I), the total potential P is used instead of P_{attr} alone. Consequently, robots may take measurements in slightly sub-optimal positions in terms of the uncertainty map. This will however prevent groups of agents from getting stuck in constellations of mutual repulsion, i.e. while approaching the same peak, thus never reaching a plateau. As every measurement changes $\boldsymbol{\sigma}_{\boldsymbol{f}}^2$, and thus also P, this causes agents to move elsewhere for new measurements.

4 Simulation Trials

The swarm exploration approach we presented will be evaluated by simulating it in an area Ω of $5\,\mathrm{m} \times 5\,\mathrm{m}$. In this test environment we simulate 1 to 5 sources in random locations with (unit-less) strengths uniformly drawn from the interval [50; 500]. From this source distribution, a forward solution of PDE (1) is computed, yielding a concentration distribution $\boldsymbol{f}$, which will be used as a 'ground-truth' for comparison; for this, we calculate the normalized Mean Squared Error

Table 1. Design parameters and their chosen values

$\Omega = \{\boldsymbol{x} : x, y \in [-2.5\,\mathrm{m}; 2.5\,\mathrm{m}]\}$	$\lambda_{\mathrm{A}} = 1$	$\lambda_{\mathrm{B}} = 1000$	$\lambda_{\mathrm{M}} = 25$	$\lambda_{\mathrm{R}} = 10^{-2}$
$I = 20,\ J = 20$	$k_{\mathrm{v}} = 0.5$	$k_{\omega} = 1.0$	$k_{\mathrm{R}} = 2.0$	$\delta = 1.0\,\mathrm{m}$
$\xi_a \sim \mathcal{N}(0, 0.2^2)$	$\alpha = 0.02$	$s_{\min} = 0.9$	$\Delta t_{\mathrm{m}} = 5\,\mathrm{s}$	$\sigma_{\mathrm{s}}^2 = 0.01$

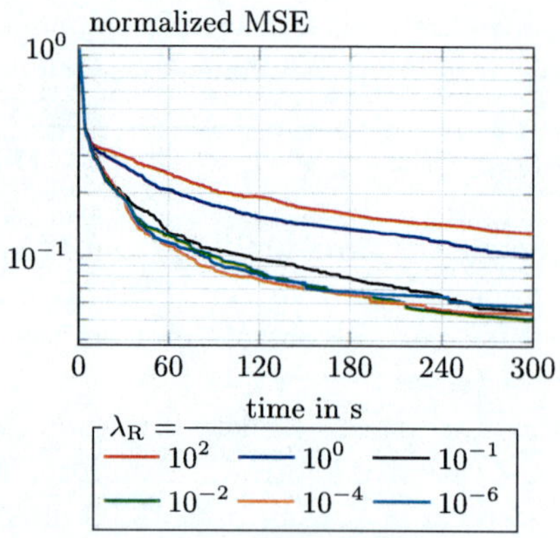

Fig. 3. Normalized MSE over time for different regularization weights λ_R

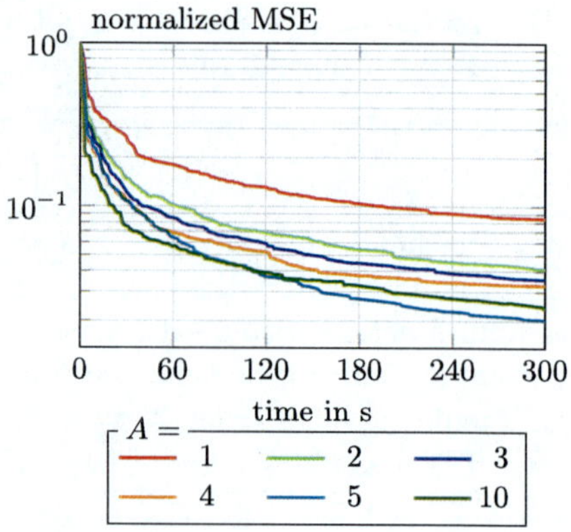

Fig. 4. Normalized MSE over time for a changing number of swarm agents

(NMSE) between $\boldsymbol{f}$ and its reconstruction $\hat{\boldsymbol{f}}$ as NMSE $= \|\boldsymbol{f} - \hat{\boldsymbol{f}}\|_2^2 \,/\, \|\boldsymbol{f}\|_2^2$. The latter is an estimation (cf. Eq. (10)) based on the measurements collected by using the proposed exploration scheme.

There are a number of design parameters to be chosen. Table 1 summarizes those used for all following simulations and experiments, unless specified otherwise. They have been tuned empirically to obtain suitable results.

Source Distribution Prior Parameter λ_R. We first want to determine the influence of the parameter λ_R on the NMSE. For that, we apply the exploration strategy in randomly generated environments, each with a randomized source distribution and a randomly drawn number of agents A from one to five. In Fig. 3, the NMSE over time is plotted for different λ_R, each line averaged over 100 such trials. We expect performance to increase for lower λ_R, which weights the prior assumption $p(\boldsymbol{q})$. Too high values of λ_R overly penalize strong sources, and will result in under-estimation of the strength of sources compared to the true values. On the other hand, for low λ_R, solving Eq. (11) will become numerically unstable, or the estimation problem will become ill-posed. This is in fact what can be observed in the simulation results. Figure 3 shows a clear trend of lower λ_R resulting in a decreased lower bound of NMSEs, while also approaching this bound faster. For $\lambda_R < 10^{-6}$, numerical issues start to occur. We observe diminishing returns for $\lambda_R < 10^{-2}$, suggesting an optimal value of $\lambda_R \approx 10^{-2}$ that offers a good trade-off between model fit and numerical stability. This value is thus chosen for all further analysis.

Swarm Size. With the newly found value of λ_R, we now want to determine the influence of the swarm size A. Figure 4 shows the NMSE over time, this time varying the number of swarm agents. As expected, a single agent performs the worst by a large margin, highlighting the benefits of robotic swarms. For the chosen environments, using more robots results in faster NMSE convergence as well as an improved lower bound—however with diminishing returns, until

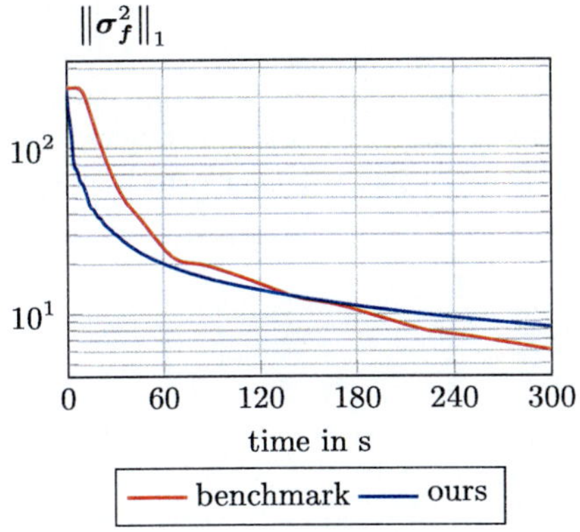

Fig. 5. Total uncertainty over time (sum of standard deviations σ_f)

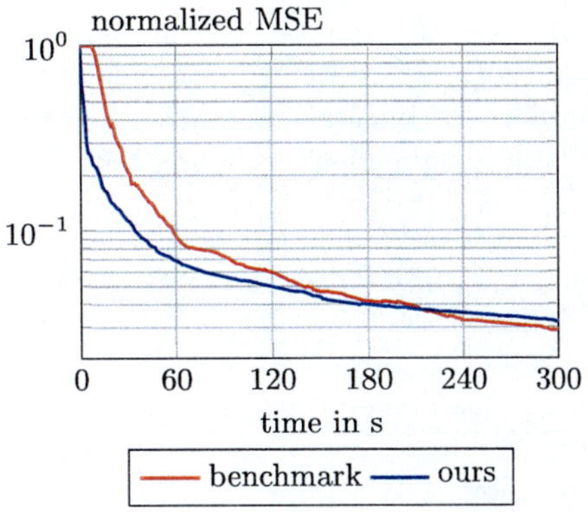

Fig. 6. Normalized MSE: best performance of our approach vs. benchmark

collision avoidance actually hinders exploration: For 10 robots, initial good performance is outweighed by navigation constraints later, as agents need to settle for sub-optimal sampling locations. The best performance thus far appears to be achievable, in the described environments, with $A = 5$ rovers.

Process Uncertainty Reduction & Benchmark. As stated in Sect. 2.2, the objective of exploration is to reduce some measure of $\boldsymbol{\sigma}_f^2$. In simulations, our strategy has been found successful in achieving this goal. To visualize this, we choose $\left\|\boldsymbol{\sigma}_f^2\right\|_1 = \operatorname{tr}\left(\boldsymbol{\Sigma}_{\mathrm{ff}}\right)$ as this metric and plot it's decrease over time (blue) in Fig. 5, averaged over 300 randomized trials with $A = 5$ and $\lambda_{\mathrm{R}} = 10^{-2}$. Particularly in the very beginning, our approach shows its strength in rapidly reducing the uncertainty, with a continued monotonic decrease being clearly visible. The corresponding NMSE is plotted in Fig. 6 (also blue), as it is the true metric of reconstruction success— that in practice could however not be gauged directly, since the ground truth is unknown. We can observe that the uncertainty is a good heuristic to achieve reduction in the NMSE, visible in how it also decreases rapidly right away, and continuously.

To judge the efficacy of the proposed exploration strategy, we compare it against a 'benchmark' strategy consisting of a swarm of equal size performing a meander trajectory, uniformly sampling the entire area of interest Ω. The results, averaged over 300 such trials, are plotted in red in Fig. 5 & 6. Comparing the evolution of $\left\|\boldsymbol{\sigma}_f^2\right\|_1$ we see that initially the benchmark strategy does not decrease uncertainty at all, since meandering tends to also cover rather uninformative locations. A similar effect is observed later (≈60 s, ≈210 s); we attribute this to sampling close to the edges of Ω. Secondly, it is important to note the Break-Even point at ≈150 s. After this time, the benchmark achieves lower uncertainties, yet at the price of higher number of collected samples (in total: ≈ 389 for benchmark *vs.* ≈ 284 for ours). This shows that if time is not a criterion, sampling every cell can be a decent strategy—which is of course only feasible if measurements are fast and cheap. When comparing the benchmark's NMSE evolution, we see that our approach reduces the reconstruction error significantly faster. Only after ≈180 s does the benchmark catch up, again at the expense of a much higher number of collected measurement samples. Nonetheless, final NMSE values are basically equivalent, bringing both approaches on a par with each other.

5 Experimental Evaluation

This section presents the results of experiments with a swarm of 4 track-driven rovers (as depicted in Fig. 2) navigating in a laboratory setting and employing the proposed exploration approach. The experiments constitute a Hardware-in-the-Loop (HIL) setup, as real robotic hardware is used in combination with synthetic sensor measurements. Based on actual robot positions obtained in real time using a motion tracking system, measurements are drawn from the ground truth intensity distribution and, just like in the simulations, superimposed with additive white Gaussian noise with $\sigma^2 = \lambda_{\mathrm{M}}^{-1}$. This way, there exists a known ground truth to compare to, which is a benefit of simulating measurements. Measured values and positions are broadcast to the other agents when required, by the use of Robotic Operating System 2 (ROS2). This way, each agent is capable of computing $\boldsymbol{\sigma}_{\boldsymbol{f}}^2$ and can navigate using APFC.

The HIL experiment is conducted in one of the environments described in Sect. 5. As can be seen from Fig. 7, which shows each robot's traveled distance, all agents participate effectively in the exploration. We observe continuous progression, with agents stopping briefly to take measurements. Notice the lack of prolonged stationary periods, which would indicate a 'deadlock' of robots in mutual repulsion. This is in part due to the fact that the uncertainty map, which the robots base their navigation upon, is ever changing, and gets updated precisely in these situations where agents would come to a halt. Also with time, robots spread out on the map due to their mutual repulsion and the fact that measurements introduce divergent regions in the potential field. Figure 8 shows

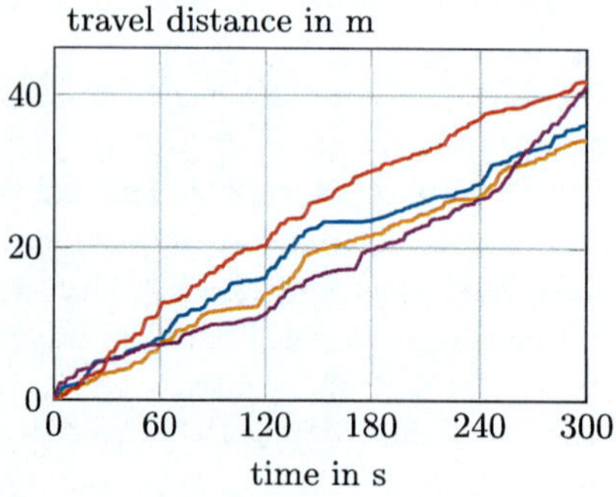

Fig. 7. Traveled distance of rovers 1, 2, 3 & 4 over time

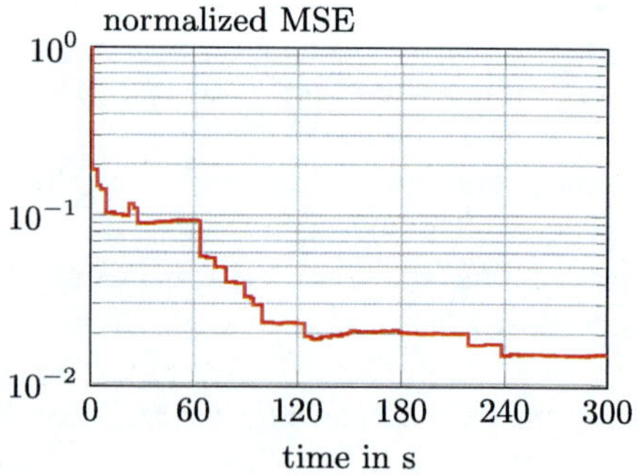

Fig. 8. Normalized MSE over time for a hardware demonstration with 4 robots

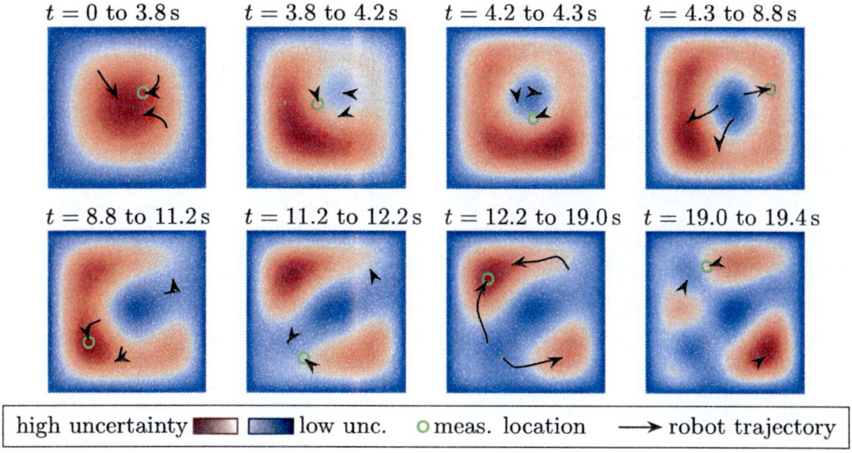

Fig. 9. Evolution of the uncertainty map with superimposed trajectories

a corresponding reduction of the NMSE as the experiment proceeds. Each measurement time $t_{\mathrm{m,k}}$ is visible in this plot as a step. Almost all measurements lead to a reduction in NMSE, and the results match well with the simulations with 4 robots shown in Fig. 4.

Figure 9 shows the evolution of the uncertainty map during an experiment with 3 rovers. Regions of high uncertainty are highlighted in red, while low uncertainty—like on the boundary—is indicated in blue. In each snapshot the colormap is normalized. Initially, because the boundary values are fixed, the center is the most informative. For this reason, all agents follow the gradient in this direction. Due to the repulsive forces between agents, the first measurement is not taken at the absolute maximum of uncertainty, but sufficiently close, while still avoiding collisions. The resulting change can clearly be seen in the next snapshot in the sequence, as a minimum forms around the measurement location. After measurements near the center, the three agents migrate outwards, following the gradient field. Each new measurement removes another maximum in uncertainty.

6 Conclusion

In this paper we have shown how to combine a potential field control scheme for a multi-robot system with an uncertainty driven, PDE-based exploration approach. We base an attractive potential field upon the uncertainty map of a gas dispersion process modelled by the diffusion PDE, which describes the physics of the explored process, forcing robots to seek regions of high uncertainty for their next measurements. During exploration, a repulsive potential field prevents inter-robot collisions.

Following the taxonomy presented in [3], our approach constitutes a probabilistic inference scheme, tightly coupled with multi-robot, information-driven navigation in the form of APFC. For an overview of other related methods, we refer the reader to [1,3]. In contrast to other APFC schemes, where robots might get stuck in local minima of the potential forever or where deadlocks might occur, our approach implicitly circumvents these issues. Whenever a robot reaches a local minimum of the potential field, the new measurement at this location influences the uncertainty map and consequently modifies the potential field. In this way the local minimum gets flattened out and the robot gets driven away from its current location.

As we have shown in simulation and Hardware-in-the-Loop experiments, the approach is able to successfully and efficiently reconstruct the gas concentration and source distribution. Further, the proposed approach can be extended easily in future to account for additional objectives. For example, additional potentials can be included to navigate around obstacles of a complex environment, or to ensure that robots in the swarm stay in communication range. Extension of the approach to convection-dominated, non-linear gas models should also be considered.

References

1. Chen, X.-x., Huang, J.: Odor source localization algorithms on mobile robots: A review and future outlook. Robot. Autonom. Syst. **112**, 123-136 (2019). https://doi.org/10.1016/j.robot.2018.11.014, ISSN 0921–8890
2. Choi, Joonwon, Song, Yeongho, Lim, Seunghan, Kwon, Cheolhyeon, Hyondong, Oh.: Decentralized multi-subgroup formation control with connectivity preservation and collision avoidance. IEEE Access **8**, 71525–71534 (2020). https://doi.org/10.1109/ACCESS.2020.2987348, ISSN 2169–3536
3. Jing, T., Meng, Q.-H., Ishida, H.: Recent progress and trend of robot odor source localization. IEEJ Trans. Elect. Electron. Eng. **16**(7), 938–953 (2021). https://doi.org/10.1002/tee.23364
4. Khatib, O.: Real-time obstacle avoidance for manipulators and mobile robots. The Intl. J. of Robot. Res. **5**(1), 90–98 (1986). https://doi.org/10.1177/027836498600500106
5. Koditschek, D.: Robot planning and control via potential functions. Robot. Rev. **1**, 349–367 (1989)
6. Liu, X., Ge, S.S., Goh, C.-H.: Formation potential field for trajectory tracking control of multi-agents in constrained space. Int. J. Control **90**, 2137–2151 (2017). https://doi.org/10.1080/00207179.2016.1237044, ISSN 0020–7179
7. Morgan, D., Chung, S.J., Hadaegh, F.Y.: Model predictive control of swarms of spacecraft using sequential convex programming, vol. 37, pp. 1725–1740. American Institute of Aeronautics and Astronautics Inc. (2014). https://doi.org/10.2514/1.G000218
8. Rezaee, Hamed, Abdollahi, Farzaneh: A decentralized cooperative control scheme with obstacle avoidance for a team of mobile robots. IEEE Trans. Indust. Electron. **61**, 347–354 (2014). https://doi.org/10.1109/TIE.2013.2245612, ISSN 0278–0046
9. Schwab, A., Littek, F., Lunze, J.: Experimental evaluation of a novel approach to cooperative control of multiple robots with artificial potential fields. In: 2021 European Control Conference (ECC). EUCA (2021). https://doi.org/10.23919/ECC54610.2021.9654958

10. Verma, J.K., Ranga, V.: Multi-robot coordination analysis, taxonomy, challenges and future scope. J. Intell. Robotic Syst. Theory Appli. **102**, 5 (2021). https://doi.org/10.1007/s10846-021-01378-2, ISSN 1573–0409
11. Wang, Peng, Ding, Baocang: Distributed rhc for tracking and formation of nonholonomic multi-vehicle systems. IEEE Trans. Autom. Control **59**, 1439001453 (2014). https://doi.org/10.1109/TAC.2014.2304175, ISSN 0018–9286
12. Wiedemann, T., Lilienthal, A.J., Shutin, D.: Analysis of model mismatch effects for a model-based gas source localization strategy incorporating advection knowledge. Sensors **19**(3), 520 (2019). https://doi.org/10.3390/s19030520
13. Wiedemann, T., Manss, C., Shutin, D.: Multi-agent exploration of spatial dynamical processes under sparsity constraints. Auton. Agent. Multi-Agent Syst. **32**(1), 134–162 (2017). https://doi.org/10.1007/s10458-017-9375-7
14. Zhou, J., Guo, Y., Li, G., Zhang, J.: Event-triggered control for nonlinear uncertain second-order multi-agent formation with collision avoidance. IEEE Access **7**, 104489–104499 (2019). https://doi.org/10.1109/ACCESS.2019.2929540, ISSN 2169–3536

Consensus-Based Distributed Source Term Estimation with Particle Filter and Gaussian Mixture Model

Yang Liu, Matthew Coombes, and Cunjia Liu(✉)

Department of Aeronautical and Automotive Engineering, Loughborough University, Loughborough LE11 3TU, UK
c.liu5@lboro.ac.uk

Abstract. Source term estimation (STE) techniques provide an effective way of understanding the key parameters of an atmospheric release in different scenarios. Following the Bayesian inference framework, this paper investigates the distributed STE problem over a sensor network based on a consensus-based particle filtering scheme. Among different consensus strategies, the posterior-based consensus method is selected, so that all the sensor nodes can reach the same belief of the source term. To effectively approximate the local posterior density functions (PDFs) and share them over the sensor network, the Gaussian mixture model (GMM) is constructed at each node by resorting to the expectation-maximization method, and the parameters of the GMMs are exchanged between the sensor nodes. The consensus between the GMMs from different nodes is realised in the sense of Kullback-Leibler average (KLA). To provide a numerical solution to this process, an importance sampling method with a novel importance density function is proposed to draw particles at each node with respect to the GMMs from the neighboring nodes. Finally, the effectiveness of the proposed distributed STE solution is demonstrated with an experimental dataset.

Keywords: Source term estimation · Sensor networks · Particle filter · Posterior consensus · Gaussian mixture model · Kullback-Leibler average

1 Introduction

The source term estimation (STE) capability has attracted considerable research attention since it can improve the situational awareness in the case of an incidental release of chemicals into the atmosphere. The common source term includes the location of the release, the release rate, the start and stop times of the release, and so on [7]. An accurate STE result provides a better understanding of possible dispersions of the hazardous material, and thus the STE methods have found many successful applications such as national security, human health, molecular biology, etc. [4,8,15].

Both sensor networks and mobile robotics have been widely used to deliver the STE function in practice. When STE is executed with a sensor network over a given

D. Tardioli et al. (Eds.): ROBOT 2022, LNNS 590, pp. 130–141, 2023.
https://doi.org/10.1007/978-3-031-21062-4_11

topology, the collected information is mostly sent to a fusion centre, where the data are processed in a centralised way. When mobile robots are introduced to search the source with the support of STE function, the solutions have been developed from a single robot (e.g., [6,11,13,15]) to multi-robots (e.g., [10,12,14]). However, most of the multi-robot solutions are still based on a centralised data processing framework. Such a strategy will unavoidably introduce heavy computation burden in the central node. Moreover, this structure demands a high communication ability of the central node since it needs to communicate with every node even when they are spatially distributed. Furthermore, the centralised mechanism is fragile to the reliability of the central node because the malfunction of the central node will make the algorithm fail completely. Summarising the discussion above, it can be seen that a distributed STE method is preferable in practice, where each node can estimate the unknown source terms with only local information and the information from its neighboring nodes. In this way, the scalability and flexibility of the STE method can be significantly improved.

Among distributed filters designed under different requirements, consensus-based filters have been extensively investigated for sensor networks. By resorting to the node-wise communication and the consensus process, an agreement in some practical sense or theoretical measure can be achieved among the filters for all the nodes in the sensor network. Because the dispersion processes are severely nonlinear and non-Gaussian with respect to the source terms, particle filter (PF), as a representative Bayesian filter, has been widely adopted in the STE problem. For PF algorithms, consensus can be realised with respect to likelihood functions [5], posterior probability density function (PDF) [9], and particle weights [16]. One of the advantages of the posterior-consensus approach is that it enables each sensor to generate particles independently but share the same belief of the source term, and no extra assumptions are needed on the basis functions of the likelihood functions or identical particles for each sensor node. It is worth mentioning that the data flooding strategy can achieve consensus by directly sharing the measurements of each node. It has been quantitatively analysed in [16] that the communication burden of the flooding-based PF is heavier than that of the distributed PF when the diameter of the network graph is large.

In posterior-consensus PFs, a key problem is how to represent the PDF at each node. The PDF for continuous random variables needs to be reconstructed based on the weighed particles generated from the importance sampling process, and the parameters of the PDF should be exchanged between the nodes to facilitate the consensus process. The Gaussian mixture model (GMM) has proven to be effective when approximating a non-Gaussian distribution sufficiently accurately [2]. Moreover, the number of parameters in a GMM is limited, which will not lead to intolerable communication loads between the sensor nodes. GMMs have been initially applied to distributed particle filtering in [9] based on a basic importance sampling process. After approximating the local posterior with the GMM, there are still some issues to be addressed: 1) how to effectively draw new particles which can support not only the local GMM but also the GMMs from the neighboring nodes? and 2) how to calculate the weight for each particle to achieve consensus? These problems will be solved for the consensus-based distributed STE issue in this paper.

In this paper, a GMM-based consensus PF is designed for the STE problem over static sensor networks, which can be easily extended to robotic mobile sensors. The sensing process is based on a chemical particle counter and a Poisson distribution is used to formulate the stochastic sensor model. The PF at each sensor node only uses the local information and the information from the neighboring nodes. GMMs are used to approximate the local posteriors and exchanged between the sensor nodes. The Kullback-Leibler divergence (KLD) is used to measure the difference between different GMMs and the consensus process based on Kullback-Leibler averaging is adopted to calculate the weights of the particles. The effectiveness of the proposed method is validated on an actual dispersion dataset. The proposed method enables all the sensors to have a similar understanding of the emission source, and thus can provide information to coordinate the robotic source searching process. The contribution of the paper is three-fold: 1) the STE problem is solved in a distributed framework, thereby improving the scalability and the flexibility of the method; 2) an effective distributed PF is designed for each node to realise the consensus of posteriors by resorting to the GMM technique, and 3) within the distributed PF, a new particle generation process is proposed to increase the support of the particles with respect to the information from all the neighboring nodes.

2 Problem Formulation

Without loss of generality, this paper considers a source with a constant release rate Q_s located at $[x_s, y_s]^T$ in a two-dimensional search area as in [15]. Thus, the unknown source term vector is set to be $\theta = [x_s, y_s, Q_s]^T$. Assuming the chemical particles emitted from the source are driven by a constant wind field along the x-axis, the encounter rate of the particles for a spherical sensor located at $[x, y]^T$ can be specified by a dispersion model as follows [18]:

$$r(\theta) = \frac{Q_s}{\ln \frac{\lambda}{a}} \exp\left[\frac{(x - x_s)V}{2D}\right] K_0\left(\frac{\sqrt{(x - x_s)^2 + (y - y_s)^2}}{\lambda}\right), \tag{1}$$

where a is the sensor size, D is the isotropic diffusivity, V is the wind speed of advection, $K_0(\cdot)$ is the zero-order modified Bessel function, and environmental model parameter $\lambda = \left(\frac{D\tau}{1+V^2\tau/(4D)}\right)^{\frac{1}{2}}$.

A Poisson distribution is employed to describe the number of the particles encountered by the sensor [17]. The probability that a detector registers $z \in \mathbb{N}$ counts in t seconds is

$$p(z; \mu) = \frac{\mu^z}{z!} e^{-\mu}, \tag{2}$$

where $\mu = r(\theta)t$ is the the mean concentration. With this sensor model, the likelihood function $p(z|\theta)$ can be constructed.

In this paper, a sensor network consisting of a set of nodes $\mathcal{N} = \{1, 2, \ldots, N\}$ is assumed to collaboratively monitor the emission source through a given topology. Every sensor node, which could be a robotic sensing platform, can collect the local measurement, process the data, and exchange certain information with its neighboring nodes. For the ith sensor node, the set of its neighboring nodes is denoted as $\mathcal{N}_i$ and the sensor measurement at time step k is denoted as $z_{i,k}$. While by resorting to a particle filtering solution, a posterior PDF of the source term $p(\theta_k|z_{i,1:k})$ can be obtained at each node via Bayesian inference, this paper seeks to make a consensus among local posteriors $p(\theta_k|z_{i,1:k})$, $i \in \mathcal{N}$ to improve the belief of source term at each sensor node. More specifically, the Kullback-Leibler average (KLA) will be exploited as the criteria to fuse local posteriors PDFs over the network, which in turn will be realised by iterative consensus steps in a distributed manner. To achieve consensus, the topology of the sensor network is assumed to be connected, i.e., there is a path between any two of sensor nodes.

It should be noted that when mobile robots are employed as the sensor network, the communication link between each pair of robot may be influenced by their relative distance and environment factors; thus, the communication topology can be time-variant and dynamic. Although this is beyond the scope of this paper, the algorithm developed in this paper is applicable to such a case with some modifications.

3 Main Results

3.1 Local Particle Filtering

At the ith sensor node, a particle filtering solution is required to calculate the local posterior distribution $p(\theta_{i,k}|z_{i,k}, \mathcal{I}_{k-1})$ at time step k, given the new sensor measurement $z_{i,k}$ and the prior distribution $p(\theta_{i,k-1}|\mathcal{I}_{k-1})$, where $\mathcal{I}_{k-1}$ denotes the information available to the ith node at $k-1$, including the local measurements $z_{1:k-1}$ and information transmitted over network via the consensus steps. Note that $\theta_{i,k} = \theta_{i,k-1}$ as the source term is assumed to be static.

Different particle filtering schemes can be used to obtain an approximation of the posterior PDF in the context of STE [6,15]. This paper adopts the particle filter in [6] to realise local Bayesian inference $p(\theta_{i,k}|z_{i,k}, \mathcal{I}_{k-1}) \propto p(z_{i,k}|\theta_{i,k})p(\theta_{i,k}|\mathcal{I}_{k-1})$, $i \in \mathcal{N}$, such that the posterior PDF at each node can be represented as follows

$$p(\theta_{i,k}|z_{i,k}, \mathcal{I}_{k-1}) = \sum_{m=1}^{M} W_{i,k}^{m} \delta\left(\theta_{i,k} - \theta_{i,k}^{m}\right), \tag{3}$$

where M is the number of the particles, $\delta(\cdot)$ denotes Dirac delta function, $\theta_{i,k}^{m}$ is the mth particle of the ith node at time step k, and $W_{i,k}^{m}$ is the weight for the particle satisfying $\sum_{m=1}^{M} W_{i,k}^{m} = 1$.

The local posterior PDFs obtained by incorporating different local measurements are inconsistent, thus representing different beliefs of the source term. To reach the consensus of the local PDFs, a measure of the difference between PDFs and the corresponding averaging process will be required, along with the numerical solution that can process the PDF representations in the form of (3).

3.2 Kullback-Leibler Average

The KLD is used in the paper to measure the distance between PDFs and calculate the average of the PDFs. For two PDFs $m(\cdot)$ and $n(\cdot)$, the KLD is defined as

$$D_{KL}(m\|n) = \int m(x)\log\frac{m(x)}{n(x)}dx. \tag{4}$$

Let

$$\mathcal{P} \triangleq \left\{p(\cdot): \mathbb{R}^n \to \mathbb{R}, \int_{\mathbb{R}^n} p(x)dx = 1, p(x) \geq 0.\right\} \tag{5}$$

represent the set of PDFs over $\mathbb{R}^n$, which includes local posterior distributions at each node i of the sensor network. Based on (5), the following definition is introduced.

Definition 1. *[1] Given N local posterior distributions $m_i(\cdot) \in \mathcal{P}$, and relative weights η_i satisfying $\eta_i \geq 0$ and $\sum_i^N \eta_i = 1$, their weighted KLA is defined as*

$$\bar{m}(x) = \arg\inf_{m\in\mathcal{P}} \sum_i^N \eta_i D_{KL}(m\|m_i). \tag{6}$$

It has been proved in [1] that $\bar{m}(x)$ in (6) can be calculated as

$$\bar{m}(x) = \frac{\prod_i^N [m_i(x)]^{\eta_i}}{\int \prod_i^N [m_i(x)]^{\eta_i}dx}. \tag{7}$$

To calculate $\bar{m}(x)$ in a recursive fashion, the following iteration can be used at each sensor node $i \in \mathcal{N}$:

$$m_i^{l+1}(x) = \bigoplus_{j\in(\{i\}\bigcup N_i)} \left(\rho_{i,j} \odot m_j^l(x)\right), \tag{8}$$

where l is the iteration number, $m_i^l(x)$ is the PDF obtained at the ith node in the lth consensus step, and $\rho_{i,j}$ are non-negative scalars satisfying $\sum_{j\in(\{i\}\bigcup N_i)} \rho_{i,j} = 1$

for any node. Operators $\oplus$ and $\odot$, standing for PDF addition and multiplication, have been defined in [1], and thus their definitions are omitted for the sake of simplicity.

At time step k, after the local posterior PDF is obtained at each node $i \in \mathcal{N}$, the consensus process will start by assigning $m_{i,k}^0(x) = p(\theta_{i,k}|z_{i,k}, \mathcal{I}_{k-1})$ and transmitting this PDF to its neighbouring nodes $j \in \mathcal{N}_i$ over the network. Meanwhile, the received PDFs $m_{j,k}^l(x)$, $j \in \mathcal{N}_i$, will be fused via (8) at each consensus iteration $l = 0, 1, \ldots, L$. However, the implementation of such a consensus process may not be straightforward, given the PDFs are approximated by Monte Carlo sampling methods as in (3). This is not only because transferring a large amount of particle vectors and weights over the network is not practical, but also due to the fact that there is no direct numerical solution to fuse different sets of particles via (8).

To tackle this problem, first, the GMM is trained to approximate the Monte Carlo representation of $m_i^l(x)$ at each consensus step for each node. In this case, the parameters of GMM are exchanged between neighboring nodes, supplying all required PDFs on the right hand side of (8). Then, an importance sampling procedure is employed to generate a Monte Carlo approximation of $m_i^{l+1}(x)$. To increase the support of the particles with respect to all the locally available GMMs, a novel importance density function is to be proposed, and the corresponding particle weights will be calculated by resorting to (7). After several consensus steps, the agreement in the sense of KLA can be seen as achieved. The final fused PDF at each node can be used as the prior distribution for local particle filter at the next time step, i.e., $p(\theta_{i,k+1}|\mathcal{I}_k) = m_{i,k}^L(x)$. The process of the proposed framework is described in the following sub-sections.

3.3 GMM-Based Approximation of PDF

After obtaining the initial particle $\theta_{i,k}^{0,m}$ and the weight $W_{i,k}^{0,m}$ at each time step using the local particle filters, the consensus process needs to be executed. At the lth consensus step, we can obtain the particle approximation for the PDF of the source term $m_{i,k}^l(\theta_k)$. A GMM is a combination of Gaussian components as follows,

$$m_{i,k}^l(\theta) \approx \sum_{c=1}^{C} \alpha_{i,k}^{l,c} \mathcal{N}(\theta; \mu_{i,k}^{l,c}, \Sigma_{i,k}^{l,c}) \triangleq \mathrm{GMM}_{i,k}^l(\theta), \tag{9}$$

where C is the number of components, $\alpha_{i,k}^{l,c}$, $\mu_{i,k}^{l,c}$, and $\Sigma_{i,k}^{l,c}$ are the weight, mean, and covariance matrix, respectively, of the cth component. With weighted particles, the GMM can be learned with the expectation-maximization (EM) algorithm [3], and the parameters in the GMM are exchanged between neighboring nodes.

It is noted that though the local posterior PDFs can be approximated with GMMs as in (9), it is intractable to directly fuse the GMMs with KLA (8) due to the mathematical complexity. Alternatively, we can generate a set of particles and then determine the corresponding weights based on (8), i.e., importance sampling is employed to generate the particles to represent the fused PDF in (8). To effectively improve the supports of the particles with respect to the GMMs from all its neighboring nodes, the importance density function in the lth consensus step needs to be developed. The particle generation process is given as follows. At the lth iteration, for each node i with ζ_i neighboring nodes, randomly choose a node j among $\{i\} \bigcup \mathcal{N}_i$ with equal probability $\frac{1}{\zeta_i+1}$. Then, a particle $\theta_{i,k}^{l+1,m}$ is drawn based on $\mathrm{GMM}_{j,k}^{l}(\theta)$.

Based on the particle generation process, the importance density is a combination of the GMMs from all the neighboring nodes. That is,

$$q\left(\theta_{i,k}^{l+1,m}|z_{i,k}\right) = \frac{1}{\zeta_i+1} \sum_{j\in(\{i\}\bigcup N_i)} \mathrm{GMM}_{j,k}^{l}\left(\theta_{i,k}^{l+1,m}\right). \tag{10}$$

After the particle generation, the next step is to calculate the corresponding weights of the particles. To achieve the consensus posterior PDF, with the GMMs calculated by (9), $m_i^l(x)$ in (8) is taken as $\mathrm{GMM}_{i,k}^{l}(\theta)$, and we choose

$$\rho_{i,j} = \begin{cases} \varpi, & \text{if } j \in N_i, \\ 1-\zeta_i\varpi, & \text{if } j = i, \end{cases} \tag{11}$$

where $\varpi = \dfrac{1}{\max\limits_i |G_i|+1}$, and $|G_i|$ is the degree of the sensor node i. It can be seen that $\sum\limits_{j\in(\{i\}\bigcup N_i)} \rho_{i,j} = 1$ is satisfied and the consensus error asymptotically converges to zero with $\rho_{i,j}$ given in (11) [19]. The iteration process can be formulated as

$$m_{i,k}^{l+1}(\theta) = \bigoplus_{j\in(\{i\}\bigcup N_i)} \left(\rho_{i,j} \odot m_{j,k}^{l}(\theta)\right). \tag{12}$$

Substituting the GMM (9) into (12) yields

$$\begin{aligned} p\left(\theta_{i,k}^{l+1,m}|z_{i,k}\right) =& m_{i,k}^{l}\left(\theta_{i,k}^{l+1,m}\right) \\ \propto& m_{i,k}^{l}\left(\theta_{i,k}^{l+1,m}\right) \prod_{j\in N_i} \left(\frac{m_{j,k}^{l}\left(\theta_{i,k}^{l+1,m}\right)}{m_{i,k}^{l}\left(\theta_{i,k}^{l+1,m}\right)}\right)^{\varpi} \\ \approx& \mathrm{GMM}_{i,k}^{l}\left(\theta_{i,k}^{l+1,m}\right) \prod_{j\in N_i} \left(\frac{\mathrm{GMM}_{j,k}^{l}\left(\theta_{i,k}^{l+1,m}\right)}{\mathrm{GMM}_{i,k}^{l}\left(\theta_{i,k}^{l+1,m}\right)}\right)^{\varpi}. \end{aligned} \tag{13}$$

The weights for the particles generated according to (10) can be determined with:

$$\hat{W}_{i,k}^{l+1,m} = \frac{p\left(\theta_{i,k}^{l+1,m}|z_{i,k}\right)}{q\left(\theta_{i,k}^{l+1,m}|z_{i,k}\right)}. \tag{14}$$

where $p\left(\theta_{i,k}^{l+1,m}|z_{i,k}\right)$ and $q\left(\theta_{i,k}^{l+1,m}|z_{i,k}\right)$ are given in (13) and (10), respectively.

Normalise $\hat{W}_{i,k}^{l+1,m}$ with

$$W_{i,k}^{l+1,m} = \frac{\hat{W}_{i,k}^{l+1,m}}{\sum_{m=1}^{M} \hat{W}_{i,k}^{l+1,m}}. \tag{15}$$

Run the consensus process (9)–(15) for L (the iteration number) times, and then the consensus of the posterior at each node is considered to be achieved.

So far, the consensus-based distributed STE problem has been solved with GMM approximation, and the detailed algorithm is summarised in Algorithm 1.

Algorithm 1: GMM-based distributed STE algorithm for the ith node at the kth time step

Set $\theta_{i,k}^{0,m} = \theta_{i,k-1}^{L,m}$ for $m = 1, 2, \cdots, M$.
Collect $z_{i,k}$ and calculate $W_{i,k}^{0,m}$ as in (3).
for $l = 1, 2, \cdots, L-1$ **do**
    Train the GMM model (9) via EM algorithm and exchange the parameters with the neighboring nodes.
    for $m = 1, 2, \cdots, M$ **do**
        Uniformly choose $j \in (\{i\} \bigcup N_i)$ and draw $\theta_{i,k}^{l+1,m} \sim \mathrm{GMM}_{j,k}^{l}(\theta)$.
        Calculate and normalise $W_{i,k}^{l+1,m}$ with (14) and (15).

4 Case Study

In this section, the proposed method is applied to a dataset collected by COANDA Research & Development Corporation based on the recirculating water channel with a fluorescein source. For more information on this dataset, please refer to Sect. 5 in [15]. In the dataset, the source is located as $[2.935, 2.935]$ mm, and the emission rate/strength of the source is 10 g/s. The flow speed is 0, the environmental model parameter $\lambda = \sqrt{1000}$, and the initial priors are $x \sim \mathcal{U}[-143.815, -2.935]$ mm, $y \sim \mathcal{U}[2.935, 143.815]$ mm, and $Q \sim \mathcal{L}(1, 1.2)$ g/s. The maximum time step of the data set is 341.

In the simulation, six sensors are considered and connected through a given topology. The positions of the sensors, the underlying topology, and the initial diffusive conditions are depicted in Fig. 1. The consensus step is set to be $L = 3$, the component number of GMM at each sensor node is $C = 5$, and the particle number of the local PF is set to $M = 1000$.

In Fig. 2, the STE results obtained by the proposed distributed PF are illustrated, where the trajectories of the mean estimate of the source location from all node are overlaid in a 2D plot. It can be seen that the consensus of the estimation result is obtained and the estimates of different nodes converge towards the ground truth.

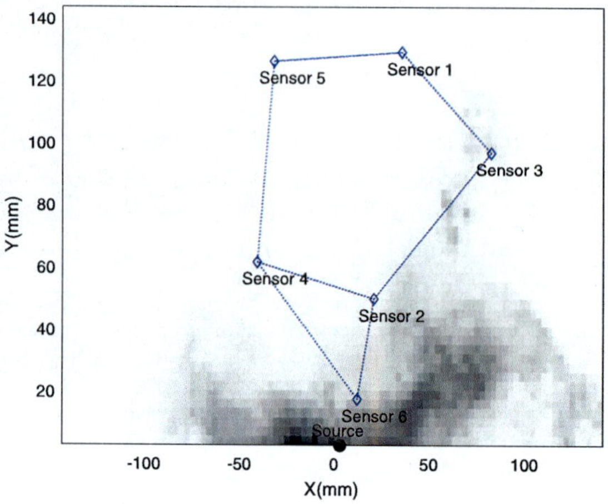

Fig. 1. Positions of the sensors, the network topology, and the initial diffusive condition

Table 1. STE Results at Every Node in 50 Monte Carlo Trials

Source term	Ground truth	Node					
		1	2	3	4	5	6
x(mm)	2.935	8.6404	8.9582	8.6459	9.0374	8.8321	9.1079
y(mm)	2.935	5.3024	5.0741	5.2141	5.0829	5.2080	5.0174
Q(g/s)	10	6.4350	6.5101	6.4581	6.5134	6.4624	6.5428

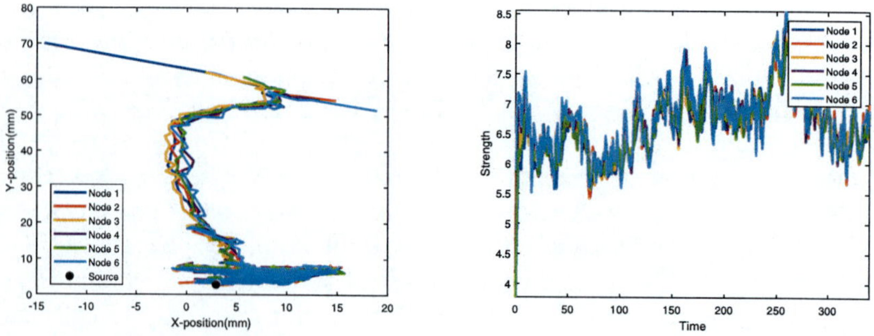

(a) Position (b) Strength

Fig. 2. STE results with 1000 particles and 5-component GMMs

Due to the convergence process, we use the mean value of the parameter estimates obtained at the last 100 time steps to characterise the estimation results of the developed method. The estimation results for all the nodes with 50 Monte Carlo trials are listed in Table 1. When the data collected via the sensor network are processed in a central node, the mean value of the source term obtained from 50 Monte Carlo trials is 4.9702 mm, 2.9264 mm, and 7.5602 g/s, respectively. It can be seen that there is still some room to improve the accuracy of the STE problem in both centralised and distributed cases. The estimation accuracy problem may result from the dispersion model and the Poisson measurement model, and further investigation is needed to improve the STE performance with respect to the modeling of the dispersion process and the measurement registration at each detector.

To further investigate the effects of the numbers of particles and GMM components, some comparative results are provided as follows. The STE performances obtained with 1000 particles (500 particles, respectively) and 3-component GMMs are depicted in Table 2 (Table 3, respectively). It can be seen that the estimation accuracy decreases when the particle number decreases from 1000 to 500 and the GMM component number decreases from 5 to 3. Therefore, sufficiently large numbers of particles and GMM components are needed to ensure the estimation accuracy in the proposed method.

Table 2. STE Results with 1000 particles and 3-component GMMs in 50 Monte Carlo Trials

Source term	Ground truth	Node					
		1	2	3	4	5	6
x(mm)	2.935	8.0815	8.4845	8.0205	8.6362	8.5244	8.9629
y(mm)	2.935	6.1844	6.1210	6.1460	6.1296	6.1726	6.0609
Q(g/s)	10	6.2322	6.2194	6.2384	6.2173	6.2184	6.2513

Table 3. STE Results with 500 particles and 3-component GMMs in 50 Monte Carlo Trials

Source term	Ground truth	Node					
		1	2	3	4	5	6
x(mm)	2.935	9.2397	9.5397	9.1359	9.7560	9.5077	9.7155
y(mm)	2.935	7.1257	6.8866	7.0361	6.8941	7.0201	6.8466
Q(g/s)	10	6.7552	6.8034	6.7651	6.8088	6.7716	6.8217

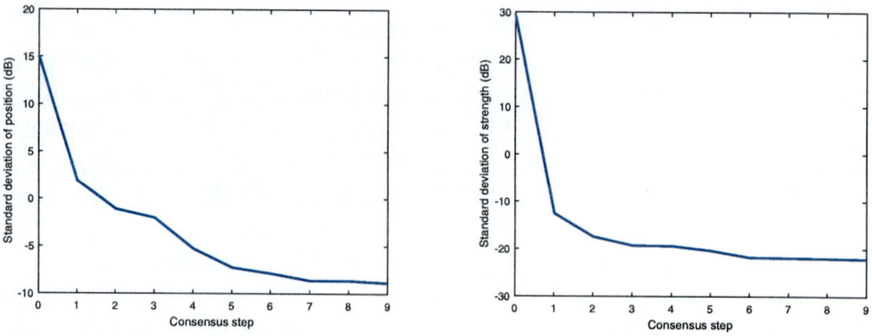

(a) Position SD (b) Strength SD

Fig. 3. SD of estimation results with different consensus steps

To analyse the relationship between the consensus error and the consensus step, the time-averaged standard deviation (SD) of the STE mean values at different sensor nodes is employed to characterise the consensus error. The SD decibels obtained with different consensus steps are depicted in Fig. 3. It is clear that more consensus steps will naturally bring in less consensus error at the cost of more computational and transmission resources.

5 Conclusion

In this paper, the consensus-based PF has been investigated for the STE problem over sensor networks. The posterior PDF at each node has been approximated by GMMs and exchanged between the sensor nodes. By resorting to KLA, a posterior-based consensus framework has been established with a novel particle generation algorithm. An actual water channel data set has been used to demonstrate the effectiveness of the proposed approach, which has shown promising results. Building on this framework, one of the future works is to apply the established consensus estimates to the source search problem for mobile robots/sensors.

References

1. Battistelli, G., Chisci, L.: Kullback-Leibler average, consensus on probability densities, and distributed state estimation with guaranteed stability. Automatica **50**(3), 707–718 (2014)
2. Bishop, C.M., Nasrabadi, N.M.: Pattern Recognition and Machine Learning. Springer, Heidelberg (2006)
3. Dempster, A.P., Laird, N.M., Rubin, D.B.: Maximum likelihood from incomplete data via the EM algorithm. J. R. Stat. Soc.: Ser. B (Methodological) **39**(1), 1–22 (1977)
4. Haley, K.: Search Theory and Applications, vol. 8. Springer, Heidelberg (2012)

5. Hlinka, O., Slučiak, O., Hlawatsch, F., Djurić, P.M., Rupp, M.: Likelihood consensus and its application to distributed particle filtering. IEEE Trans. Signal Process. **60**(8), 4334–4349 (2012)
6. Hutchinson, M., Liu, C., Chen, W.-H.: Information-based search for an atmospheric release using a mobile robot: Algorithm and experiments. IEEE Trans. Control Syst. Technol. **27**(6), 2388–2402 (2019)
7. Hutchinson, M., Hyondong, O., Chen, W.-H.: A review of source term estimation methods for atmospheric dispersion events using static or mobile sensors. Inf. Fusion **36**, 130–148 (2017)
8. Kendall, R.J., Presley, S.M., Austin, G.P., Smith, P.N.: Advances in Biological and Chemical Terrorism Countermeasures. CRC Press, Boca Raton (2008)
9. Li, J., Nehorai, A.: Distributed particle filtering via optimal fusion of gaussian mixtures. IEEE Trans. Signal Inf. Process. over Netw. **4**(2), 280–292 (2017)
10. Qiang, L., Han, Q.-L., Peng, D., Choi, Y.: Decision and event-based fixed-time consensus control for electromagnetic source localization. IEEE Trans. Cybern. **52**(4), 2186–2199 (2020)
11. Magalhães, H., Baptista, R., Macedo, J., Marques, L.: Towards fast plume source estimation with a mobile robot. Sensors **20**(24), 7025 (2020)
12. Park, M., Hyondong, O.: Cooperative information-driven source search and estimation for multiple agents. Inf. Fusion **54**, 72–84 (2020)
13. Rhodes, C., Liu, C., Chen, W.-H.: Autonomous source term estimation in unknown environments: From a dual control concept to UAV deployment. IEEE Robot. Autom. Lett. **7**(2), 2274–2281 (2022)
14. Ristic, B., Gilliam, C., Moran, W., Palmer, J.L.: Decentralised multi-platform search for a hazardous source in a turbulent flow. Inf. Fusion **58**, 13–23 (2020)
15. Ristic, B., Skvortsov, A., Gunatilaka, A.: A study of cognitive strategies for an autonomous search. Inf. Fusion **28**, 1–9 (2016)
16. Savic, V., Wymeersch, H., Zazo, S.: Belief consensus algorithms for fast distributed target tracking in wireless sensor networks. Signal Process. **95**, 149–160 (2014)
17. Tsoulfanidis, N., Landsberger, S.: Measurement & Detection of Radiation. CRC Press, Boca Raton (2021)
18. Vergassola, M., Villermaux, E., Shraiman, B.I.: Infotaxis as a strategy for searching without gradients. Nature **445**(7126), 406–409 (2007)
19. Xiao, L., Boyd, S.: Fast linear iterations for distributed averaging. Syst. Control Lett. **53**(1), 65–78 (2004)

Evolving Swarm Formations for Odour Source Localisation

João Macedo[1,2(✉)], Lino Marques[1], and Ernesto Costa[2]

[1] Institute of Systems and Robotics, University of Coimbra, 3030-290 Coimbra, Portugal
{jmacedo,lino}@isr.uc.pt
[2] Centre for Informatics and Systems, University of Coimbra, 3030-290 Coimbra, Portugal
ernesto@dei.uc.pt

Abstract. Odour source localisation is a hard problem with many applications. Over the years, researchers have drawn inspiration from Nature to devise many single-robot approaches. Swarm approaches have been growing in popularity, as they offer redundancy to the loss of agents, flexibility, scalability and enable experimenters to employ simpler robots. Many existing swarm approaches make use of robot formations. In this work, we focus on optimising the shape of a swarm formation for finding and tracking odour plumes. We do so by using a genetic algorithm, thus avoiding the cumbersome trial-and-error process that experimenters typically follow to hand-design the formations. The swarm is guided by a leader, which is controlled by a bio-inspired search strategy using the perceptions of the entire swarm. The results show that the evolved formations of three and five robots consistently outperform a single robot and that the best evolved three robot formation is more successful than the hand-designed swarms of three and five robots. As a result, one could opt by using the evolved three robot formation, minimizing the amount of robots needed. Conversely, in case there is a high risk of loss of robots, the evolved five robot formation could be preferable.

Keywords: Odour source localisation · Genetic algorithms · Swarm robotics

1 Introduction

The terms *fragance* and *odour* are typically used to refer respectively to pleasant and unpleasant scents. Nevertheless, both refer to the animals' perception of volatile chemical compounds. Through the sense of olfaction, animals often track odours to find mates or sources of food. In the literature, odour source localisation (OSL) refers to the task of locating the source that is emitting a given odour.

OSL is a hard problem with many applications. Over the years, researchers have drawn inspiration from Nature to devise many robotic approaches, mostly

D. Tardioli et al. (Eds.): ROBOT 2022, LNNS 590, pp. 142–153, 2023.
https://doi.org/10.1007/978-3-031-21062-4_12

relying on a single robot. The existing approaches can be categorised as reactive, which often attempt to mimic natural behaviours such as those of the E. coli bacteria, the dung beetle or the male silkworm moth [18], and probabilistic approaches (e.g., Infotaxis [19]), which try to estimate the location of the chemical source. Despite the success of single-robot approaches, in Nature, animals cooperate to fulfil tasks that they would otherwise be unable to accomplish or would do so with lower performance. As an example, bees and ants form swarms to forage for food, whereas birds and fish travel in flocks and schools to protect themselves [20]. Biological swarms have been the source of inspiration for various optimisation methods and, more recently, to robotic approaches. Swarm Robotics (SR) rely on highly scalable approaches employing groups of agents with limited capabilities, which would be unable or inefficient at performing the target task. Yet, through their local interactions, complex behaviours emerge enabling them to solve the task as a group [7]. The use of multiple robots provides advantages such as robustness to the loss of agents, flexibility and scalability, which are particularly interesting when working in inhospitable environments. Moreover, swarm approaches enable experiments to employ inexpensive robots, reducing the economic impact of loosing some agents. The use of multiple robots also provides the ability to sense the environment in many locations simultaneously, which is particularly helpful to locate chemical sources in environments with intermittent and meandering chemical plumes.

The existing swarm approaches can be divided based on whether the robots move in formation. The approaches that do not enforce formations often employ meta-heuristic algorithms to guide the robots. Various works have proposed Particle Swarm Optimisation (PSO) methods for guiding robotic swarms in the search for chemical sources [6,15]. Others have proposed evolutionary approaches [13,14]. Despite the differences inherent to the algorithms used, both types of approaches operate quite similarly. All of them iteratively evolve waypoints for the robots, which are evaluated through the chemical concentration sensed at each location. Thus, the methods attempt to find the location of maximum chemical concentration. In turn, formation-based approaches typically rely on attractive/repulsive forces, resulting in regular shapes [1,8]. Lochmater et al. [9] proposed the crosswind formation method, which uses a minimum of two robots moving in a crosswind line formation to perform the plume tracking stage of OSL. The robots share their wind and odour perceptions, along with relative position with each other and attempt to keep the formation centred around the chemical plume while moving to the source. Their wind-tunnel experiments showed that a formation of three robots was sufficient for tracking the chemical plume. Marjovi and Marques [12], analytically studied the optimal swarm formation for finding chemical plumes, concluding that, under a set of assumptions, the maximum probability of detection is attained by using a line formation, oriented diagonally between the crosswind and upwind directions.

Most formation-based approaches rely on experimenters to carefully hand-design the individual behaviour of each robot, often leading to cumbersome trial-and-error processes [4]. Evolutionary Algorithms (EA) are a family of stochastic search heuristics loosely inspired by the principles of evolution by natural selection and Mendel's genetics. EAs have been successfully applied to different problems from various domains [5] which either do not have analytical solutions or where such solutions would take too much effort to be found. Their use for evolving robotic morphologies [16] and controllers [2,17] yielded the research field of Evolutionary Robotics (ER). Genetic Algorithms (GA) are a family of evolutionary algorithms that iteratively improve a population of solutions for a given problem. GAs have already been used to evolve waypoints for a team of robots performing OSL [13]. In that work, no mutation operator was used, as the assignment of the locations to the robots and their navigation was considered to introduce enough randomness into the process. This method was able to perform well in a simulation environment without wind, being the odour dispersion dictated only by molecular diffusion. Later, [14] improved on it, introducing a directed mutation operator that biased the search towards upwind or crosswind depending on whether odour was sensed.

In this work, we focus on optimising the shape of a swarm formation for finding and tracking odour plumes. We do so by using a genetic algorithm, thus avoiding the cumbersome trial-and-error process that experimenters typically follow. The swarm is guided by a dynamically selected leader, which in turn is controlled by a bio-inspired strategy using the perceptions of the entire swarm.

2 Evolution of Swarm Formations

This paper proposes to locate odour sources with a robotic swarm that moves in formation to search for a chemical source. The swarm is guided by a leader, which is selected at the beginning of the trial as the one with the intended position closest to the formation centre. This approach reduces the possible behaviour instabilities caused by frequent leader changes while still being flexible, as if the leader becomes inoperative, the next closest robot may assume that role. The leader is controlled by a hand-designed reactive controller. In turn, the remaining robots are attracted to their defined positions in the formation, contributing with environmental measurements to the leader's controller.

Contrarily to the existing works, which use regular shaped hand-designed formations, this work proposes to evolve the swarm formation with a genetic algorithm. Considering a swarm of N robots, its formation is represented by an array of $2N$ real numbers, with each pair $D_{r,i}, \theta_{r,i}$ being the distance and angle of robot i to the formation's virtual centre. As an example, consider the genotype of Fig. 1, which corresponds to a formation of 5 robots.

2,00	-1,57	5,00	-0,79	1,00	0,63	3,00	2,09	5,00	3,14

Fig. 1. Example genotype for a formation of 5 robots.

The corresponding phenotype is depicted on Fig. 2, with the 5 robots (r1 to r5) shown as blue circles surrounding the virtual centre (black circle) and the large dotted circle representing the maximum admissible radius for the formation.

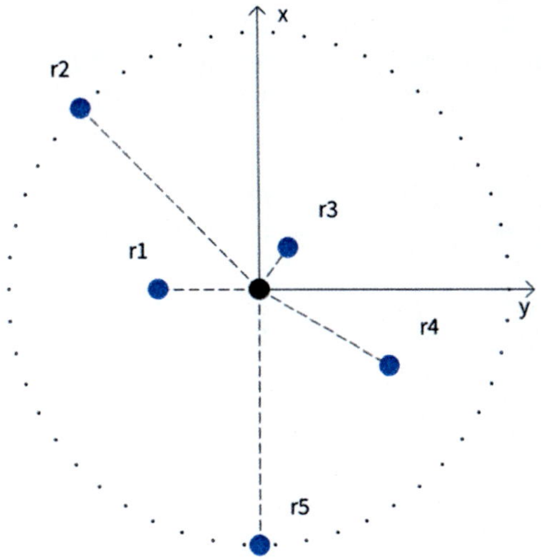

Fig. 2. Example formation of 5 robots, defined by the genotype of Fig. 1. The black circle represents the formation's virtual centre, whereas the blue circles represent the robots (r1 to r5).

The GA evolves a population of formations for a number of generations, outputting the best one found. The candidate solutions are evaluated by having the swarm perform an OSL trial, and its quality is measured by the fitness function designed in [11], which assesses how close the robots get to the source, the duration of the search and the ratio of time spent in the plume:

$$F = \sum_{i=1}^{N} \left(\beta \frac{d_i}{D} + \eta \frac{t}{T} + \zeta \frac{p_i}{t} \right) \tag{1}$$

where N is the number of robots in the swarm, d_i is the final distance of robot i to the odour source, D is the maximum distance to the odour source in this arena, t is the duration of the search, T is the maximum evaluation time and p_i is the time robot i spent without sensing odour. β and η and ζ are weight coefficients and their values are all set to 1. The goal of the GA is to minimise this function, i.e., the lower its value, the better.

3 Experimental Setup

3.1 Baseline Formations

In order to assess the performance of the evolved swarms, five baseline formations were devised: a single robot formation S1, two line formations, L3 and L5, each respectively having 3 and 5 robots spaced 2.5 m from each other; and two circular formations, C3 and C5, each with respectively 3 and 5 robots spaced equally at 5 m from the centre. Figure 3 presents these formations in the order that they were described.

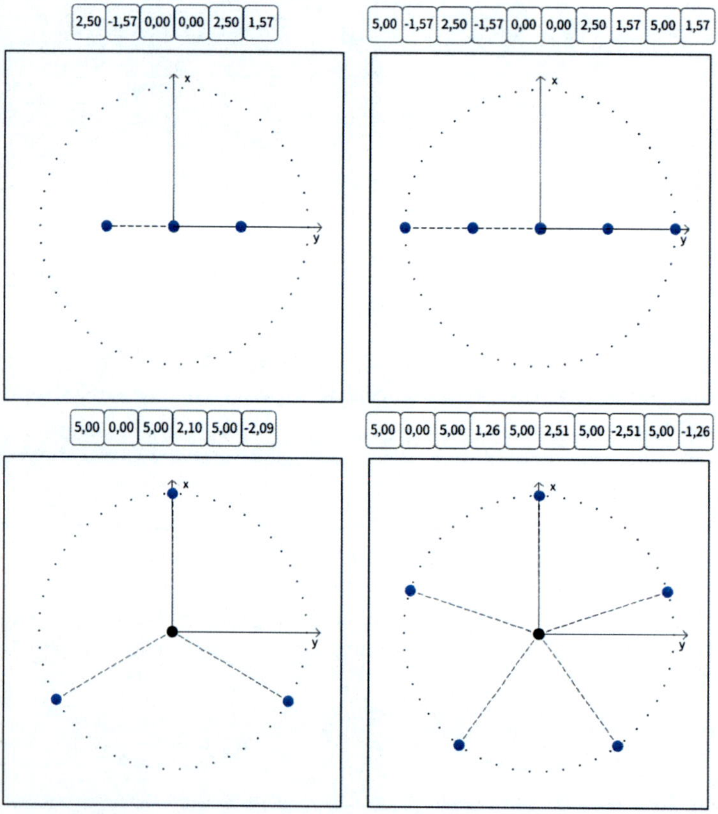

Fig. 3. Baseline formations and their genotypes.

3.2 Formation Leader Controller

The formation leader uses the perceptions of the robots in the swarm to select the actions to perform. In this work, the selection is made based on a simple reactive controller, composed by symbols that have previously been used with Genetic Programming [11]. This strategy is inspired by natural behaviours [3] moving crosswind to locate the chemical plume and upwind to track it to its source. It starts by evaluating whether odour has been sensed in the last t seconds. If so, it moves l_1 meters upwind. Otherwise, it moves l_2 meters crosswind, halting as soon as odour is sensed. In this work, t, l_1 and l_2 are respectively set to 10 s, 0.5 m and 0.5 m. Algorithm 1 presents the pseudocode for this controller.

Algorithm 1. Controller for the formation leader

```
if NPL(t) then
    moveDirection(U, l1 m, C)
else
    moveDirection(X, l2 m, SO())
end if
```

3.3 Genetic Algorithm

Table 1. Parameters of the genetic algorithm

Parameter	Value
Number of generations	100
Size of the population	100
Size of the elite	3
Crossover rate	0.7
Mutation rate	0.003
Size of the tournament	2

The overall parameters of the GA used for evolving the formations are presented on Table 1. For a swarm of N robots, the individuals are represented by a vector of $2N$ real numbers. The relative position of each robot in the formation is represented by a distance D_r and angle θ_r to the virtual centre of the formation. In this work, D_r and θ_r may respectively vary in the intervals $[0, 5]$ m and $[-\pi, \pi]$ rad. In order to maximise the diversity of the initial individuals, the initial population is created using Latin Hypercubes [5]. Moreover, to increase the local and global search abilities of the GA, on each generation, 10 random and 10 elitist immigrants are injected into the population. GAs work by selecting and recombining a set of individuals. In this work, tournament selection is used to choose the mates, which are recombined through arithmetic crossover, with $\alpha = 0.5$. Gaussian mutation may then be applied to each gene, with each σ being 10% of the domain's width, i.e., $\sigma_{Dr} = 0.5$ and $\sigma_{\theta r} = 0.2\pi$. Finally, a new population is created with the individuals chosen through elitist selection.

3.4 Evaluation Environment

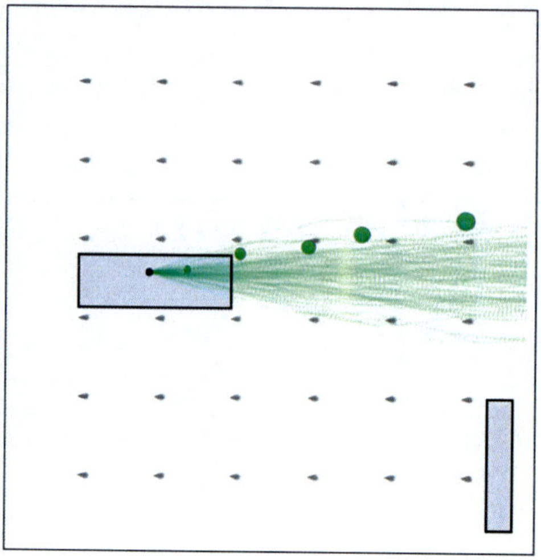

Fig. 4. Time average of air-flow (grey) and gas dispersion (green) of an instance of the evaluation environment. The green circles represent a screenshot of the odour plume at the last simulation step. The rectangle surrounding the chemical source (black circle) denotes the region from which its location is sampled, whereas the rectangle at the lower right corner represents the start region for the robots. (Color figure online)

A simulator specific for ER and odour source localisation experiments is used [10], as it showed to be much faster than real-time while adequately modelling the real world phenomena of odour dispersion and air-flow. The simulator uses a filament-based model of chemical dispersion, where each odour filament contains a given amount of molecules. An environment is devised based on a $40 \times 40\,\text{m}^2$ arena containing no obstacles and a single odour source, being its parameters presented on Table 2. The stability of the wind and chemical emission rate were set to create an intermittent meandering plume. Each robot is modelled as a circular two-wheeled differential unit, measuring 16 cm in diameter. The robots are able to measure the wind velocity and chemical concentration at their location, as well as the distance to nearby obstacles through simulated laser range finders. Each laser range finder has 10 equally spaced beams, with a field of view of π radians and a maximum range of 0.5 m. The robots move with a maximum linear speed of 0.5 m/s and a maximum angular speed of 0.3 rad. For further details, we direct the reader to [10]. Figure 4 presents one instance of the evaluation environment. Thirty instances are created with the same parameters, one for each independent trial. The position of the chemical source and the initial pose of the robot are sampled from the respective regions for each environment instance and kept fixed for all evaluations in the same trial. The robot successfully finds the source if it reaches a position closer than 0.5 m from it, within 1500 s.

Table 2. Environmental parameters

Parameter	Value
Filament emission rate (F_r)	0.1 Hz
Initial filament radius ($f_r(0)$)	0.0316 m
Filament growth rate (γ)	$0.01\,\mathrm{m}^2/\mathrm{s}$
Chemical emission rate ($\bar{Q}$)	9.86 mg/s
Detection threshold (D_t)	$1\ \mathrm{ng/cm}^3$
Initial wind speed (W_s)	0.7 m/s
Initial wind direction (W_d)	0 rad
Std. dev. of the Gaussian noise emulating turbulence (W_v)	0.075
Size of each grid cell for computing the wind velocity (W_c)	$6 \times 6\,\mathrm{m}^2$
Diffusivity constant (K_x)	$6\,\mathrm{m}^2/\mathrm{s}$
Dimensions of the arena	$40 \times 40\,\mathrm{m}^2$
Robot's initial pose	$x \sim U(36, 39)$ m,
	$y \sim U(36, 39)$ m,
	$\theta \sim N(-\pi/2, \pi/4)$ rad
Source position	$x \sim U(8, 14)$ m,
	$y \sim U(19.2, 21.2)$ m
Simulation step	0.5 s
Simulation time	1500 s

4 Experimental Results

Thirty independent trials of the evolutionary approach were conducted for evolving formations of three and five robots. Afterwards, a validation step took place, consisting on taking the best formation from each trial and re-evaluating it on the thirty environment instances. The performance of the swarms is measured through the success rates and duration of successful trials achieved in validation. As plotted in Fig. 5, using five robots seems to lead to higher success rates and lower search times. In order to draw more robust conclusions, statistical hypothesis tests are applied. The Kolmogorov-Smirnov test was applied, showing that, at a 95% confidence interval, none of the data could be considered to follow normal distributions. As a result, the Wilcoxon test was selected to perform pairwise comparisons of the success rates and duration of the searches. Its results showed that there are no statistically significant differences between the success rates of swarms of 3 and 5 robots ($p = 9.7848e{-}01$) but that there are significant differences between the duration of the successful searches ($p = 4.6796e{-}03$), implying that using 5 robots leads to locating the sources significantly faster.

To compare the performance of the evolutionary approach with the hand-designed formations, the best formation for each amount of robots must be selected. This choice is made by selecting the formation that attains the highest success rates in validation. In case of ties, the formation that takes the least

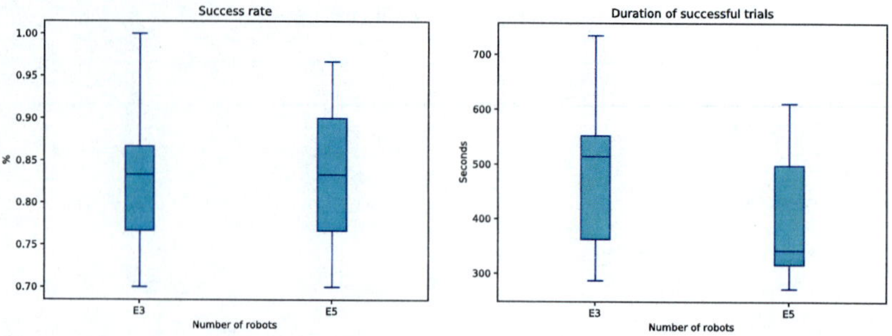

Fig. 5. Boxplots of the success rate (left) and duration of successful runs (right) of the evolved strategies in validation.

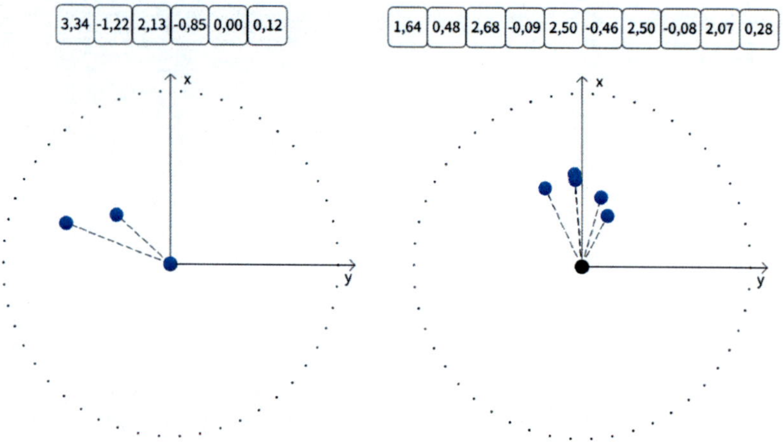

Fig. 6. Best formations evolved for three (left) and five (right) robots.

amount of time to locate the chemical source is chosen. Figure 6 presents the best formations found for three (E3) and five (E5) robots.

To better assess the performance of the evolved formations, their performance is compared to those of the baseline formations, evaluated on the thirty environmental instances (Table 3). As can be seen, using one robot leads to the overall worst success rate and slowest searches. Also, the evolved formations consistently present higher success rates than the hand designed ones. In particular, E3 presents a higher success rate than all hand-designed formations, even those using 5 robots. Conversely, E5 is slower than L5 and C5. The statistical analysis is once again performed, with the results of the Kolmogorov-Smirnov test showing that, at a 95% confidence interval, none of the data can be considered to follow normal distributions. As a result, the Wilcoxon test is selected for the remaining analysis.

Table 3. Success rates and duration of successful searches of the baseline and best evolved formations

Formation	Success rate	Mean search duration
S1	70%	685.24 ± 284.21 s
L3	76.67%	507.67 ± 287.67 s
C3	80%	521.69 ± 177.25 s
L5	93.33%	348.29 ± 127.23 s
C5	76.67%	371.91 ± 140.43 s
E3	100%	409.57 ± 130.70 s
E5	96.67%	382.97 ± 185.84 s

The first analysis consists on comparing the success and duration of the searches of S1, E3 and E5, to assess whether the amount of robots used produces substantial performance gains. The Bonferroni correction is applied, adjusting the significance value to 0.0167. The Wilxocon test shows that there are statistically significant differences between the success of S1 and E3 ($p = 2.6998e{-}03$) and of S1 and E5 ($p = 1.1412e{-}02$), but that there are no significant differences when comparing E3 and E5 ($p = 3.1731e{-}01$). Regarding the duration of the searches, the Wilcoxon test shows that there are statistically significant differences between S1 and E3 ($p = 3.1108e{-}05$) and S1 and E5 ($p = 8.1842e{-}05$), but that there are no significant differences between E3 and E5 ($p = 6.4351e{-}01$). These results show that using only one robot leads to significantly lower success rates and longer searches than using 3 or 5 robots in the formations specified by E3 and E5. Also, using E3 could be preferred to E5, as less robots are required to attain equivalent performances. However, using E5 should be more robust when there is a high risk of loosing robots.

The next step in the analysis consists on comparing the performance of hand-designed and evolved formations for the same amount of robots. Once again, the Wilcoxon test is applied with the adjusted significance value of 0.0167, showing that there are significant differences between the success of E3 and C3 ($p = 1.4306e{-}02$) and E3 and L3 ($p = 8.1510e{-}03$), but there are no significant differences between C3 and L3 ($p = 6.5472e{-}01$). Regarding the duration of the searches, the Wilcoxon test shows no significant differences between the three robot formations. Analysing the 5 robot formations, the Wilcoxon test shows no significant differences between their success rates or duration of the searches. As a result, one may conclude that the shape of the formation is more important when using less robots.

5 Conclusions and Future Work

This paper proposed an evolutionary approach for evolving the shape of a formation of a robotic swarm tasked with locating odour sources. The formation is guided by a leader, which is controlled by a simple reactive controller. The other robots are then attracted to their intended positions in the formation, which are defined as distances and angles to the formation's virtual centre. Formations of three and five robots were evolved and the best solutions were compared to each other, as well as to hand-designed formations of one, three and five robots. The results showed that there are no significant differences in the success rates of the evolved formations of three and five robots, but that using five robots leads to significantly faster searches. When comparing the overall best evolved swarms with hand-designed formations, a set of conclusions could be drawn: (1) using evolved formations of three and five robots consistently outperforms using a single robot; (2) the best evolved formation of 3 robots is significantly more successful than the hand-designed swarms, but their search times are equivalent; and (3) there are no statistically significant differences between the performances of all five robots formations. As a result, one could opt by using the evolved formation of three robots, minimizing the amount of robots needed. Conversely, in case there is high risk of loss of robots, the evolved five robots formation could be preferred, as it presents slightly better performance indicators than the hand-designed ones.

In the future, studies shall be conducted to analyse the influence of the loss of agents. Furthermore, co-evolution techniques shall be investigated for jointly evolving the shape of the formation and the controller for its leader.

Acknowledgements. This work was partially supported by the Portuguese Foundation for Science and Technology (FCT), under projects UID/EEA/00048/2020, UID/CEC/00326/2020 and Ph.D. studentship COVID/BD/152379/2022, co-funded by the European Social Fund, through the Regional Operational Program Centro 2020. It was also partially supported by the Ultrabot project, funded by the Portuguese National Innovation Agency (ANI), under reference CENTRO-01-0247-FEDER-072644.

References

1. Barnes, L., Fields, M.A., Valavanis, K.: Unmanned ground vehicle swarm formation control using potential fields. In: 2007 Mediterranean Conference on Control & Automation, pp. 1–8. IEEE (2007)
2. Bredeche, N., Fontbonne, N.: Social learning in swarm robotics. Philos. Trans. R. Soc. B **377**(1843), 20200309 (2022)
3. Chen, X., Huang, J.: Odor source localization algorithms on mobile robots: a review and future outlook. Robot. Auton. Syst. **112**, 123–136 (2019)
4. Coppola, M., McGuire, K.N., De Wagter, C., De Croon, G.C.: A survey on swarming with micro air vehicles: Fundamental challenges and constraints. Front. Robot. AI. **7**, 18 (2020)
5. Eiben, A.E., Smith, J.E.: Introduction to Evolutionary Computing. NCS, vol. 53. Springer, Heidelberg (2003). https://doi.org/10.1007/978-3-662-05094-1
6. Feng, Q., et al.: An experimental and numerical study on a multi-robot source localization method independent of airflow information in dynamic indoor environments. Sustain. Urban Areas **53**, 101897 (2020)

7. Hamann, H.: Swarm Robotics: A Formal Approach, vol. 221. Springer, Heidelberg (2018). https://doi.org/10.1007/978-3-319-74528-2
8. Hettiarachchi, S., Spears, W.M.: Moving swarm formations through obstacle fields. In: IC-AI, pp. 97–103 (2005)
9. Lochmatter, T., Aydın Göl, E., Navarro, I., Martinoli, A.: A Plume Tracking Algorithm Based on Crosswind Formations. In: Martinoli, A., et al. (eds.) Distributed Autonomous Robotic Systems. Springer Tracts in Advanced Robotics, vol. 83, pp. 91–102. Springer, Heidelberg (2013). https://doi.org/10.1007/978-3-642-32723-0_7
10. Macedo, J., Marques, L., Costa, E.: A comparative study of bio-inspired odour source localisation strategies from the state-action perspective. Sensors **19**(10), 2231 (2019)
11. Macedo, J., Marques, L., Costa, E.: Designing fitness functions for odour source localisation. In: Proceedings of the Genetic and Evolutionary Computation Conference Companion, pp. 103–104 (2021)
12. Marjovi, A., Marques, L.: Optimal swarm formation for odor plume finding. IEEE Trans. Cybern. **44**(12), 2302–2315 (2014)
13. Marques, L., Nunes, U., de Almeida, A.T.: Cooperative odour field exploration with genetic algorithms. In: Proceedings of 5th Portuguese Conference on Automatic Control (CONTROLO 2002), pp. 138–143. Citeseer (2002)
14. Marques, L., Nunes, U., De Almeida, A.T.: Odour searching with autonomous mobile robots: an evolutionary-based approach. In: Proceedings of the IEEE International Conference on Advanced Robotics, pp. 494–500 (2003)
15. Marques, L., Nunes, U., de Almeida, A.T.: Particle swarm-based olfactory guided search. Autonom. Robot. **20**(3), 277–287 (2006)
16. Murata, S., Kurokawa, H.: Self-reconfigurable robots. IEEE Robot. Automa. Mag. **14**(1), 71–78 (2007)
17. Nolfi, S., Bongard, J.C., Husbands, P., Floreano, D.: Evolutionary robotics. Commun. ACM. **56**, 74–83 (2016)
18. Andrew Russell, R., Bab-Hadiashar, A., Shepherd, R.L., Wallace, G.G.: A comparison of reactive robot chemotaxis algorithms. Robot. Autonom. Syst. **45**(2), 83–97 (2003)
19. Vergassola, M., Villermaux, E., Shraiman, B.I.: 'infotaxis' as a strategy for searching without gradients. Nature **445**(7126), 406–409 (2007)
20. El Zoghby, N., Loscri, V., Natalizio, E., Cherfaoui, V.: Chapter 8: Robot cooperation and swarm intelligence. In Wireless Sensor and Robot Networks: From Topology Control to Communication Aspects, pp. 163–201. World Scientific (2014)

Evaluating Cognitive Odour Source Localisation Strategies in Natural Water Streams

Hugo Magalhães, Rui Baptista, and Lino Marques(✉)

Institute of Systems and Robotics, University of Coimbra, Coimbra 3030-790, Portugal
{hugo.magalhaes,rui.baptista,lino}@isr.uc.pt

Abstract. Odour Source Localisation (OSL) is the task of finding the origin of a chemical emitting event by reasoning from observed measurements. Multiple cost functions have been proposed to assist in the decision-making process of cognitive strategies, but it is not yet clear which of these information metrics performs better in the OSL process. Additionally, most of these works have only been validated in simulation or in small controllable conditions such as wind tunnels with the agent starting the search close to the source position, raising additional questions about their performance in real-world scenarios. This work aims to compare the performance of cognitive search strategies between three distinct information metrics: Entropy, Kullback-Leibler (KL) divergence and Free Energy, and at the same time, evaluate their efficiency in a natural water stream with an Autonomous Surface Vehicle (ASV). All three strategies successfully locate the source in most of the trials, with Entropy and KL showing similar performance. Free Energy had a lower success rate than Entropy an KL but generated more efficient travelled paths, at the cost of a higher computational effort.

Keywords: Odour source localisation · Bayesian inference · Information-theoretic · Field robotics · Autonomous surface vehicles

1 Introduction

Rivers and lakes sustain a large bio-diversity of animals and plants and play a major role in supplying freshwater to cities and agriculture fields. A chemical spill in such environments can cause a negative impact on the balance of the ecosystem by introducing harmful substances into the water stream. The dispersion of the contaminants through the moving current originates an odour plume that can endanger humans and living organisms. Odour Source Localisation (OSL) is the task responsible for finding the origin of such polluting events which is extremely relevant to allow for fast interventions and avoid further risks to the environment. This process is usually carried out by a static sensor network or by mobile sensors that have advantage of sampling at different locations and adapting their trajectory to track a chemical plume toward its source [1].

D. Tardioli et al. (Eds.): ROBOT 2022, LNNS 590, pp. 154–165, 2023.
https://doi.org/10.1007/978-3-031-21062-4_13

The process of OSL with mobile sensors has attracted significant interest from the research community and is typically composed of three distinct tasks: (1) searching the active plume region, (2) tracking the plume towards its source after odour traces are found, and (3) confirming the source location. Bio-inspired search strategies employ different movement patterns triggered from actual perceptions of the chemical parameters or also from flow information [2]. In natural environments, turbulence originates fluctuating and intermittent odour detections, causing these reactive approaches to lose performance since the agent does not perceive the plume during significant periods of time. In situations where odour encounters are scarce, cognitive strategies guided by information-theoretic concepts, such as Infotaxis [3] are preferable to reactive or bio-inspired approaches [4]. They rely on two key components: (1) a probabilistic Bayesian inference belief of the source parameters that can be approximated with a grid or a Particle Filter, and (2) a movement decision based on the expected uncertainty reduction of the source belief at admissible positions. The expected uncertainty reduction, or by other others, the expected gain of information if the robot moved to a possible position, is quantified with a cost function employing an information metric such as Shannon Entropy (S) [3] or the Kullback-Leibler (KL) divergence [5]. A relevant topic within information-driven decision making is the development of cost functions that produce an efficient balance between exploration and exploitation. The original Infotaxis cost function is composed by an exploitation term represented by the probability of finding the source after the next movement position and an exploration term determined from the expected uncertainty reduction if the source is not found. The former term only has a significant weight when the agent is close to the expected source position, which means that the overall search is determined based on the exploration term. This conclusion led to the proposal of a modified Infotaxis cost function which only uses the former exploration term [6]. In an attempt to increase the directness of the search toward the expected source location, a form of Free Energy (FE) was proposed in [7] where a fixed "temperature" parameter adjusts the balance between the exploitation term represented through the distance between the agent position and the expected source location, and the exploration term quantified by the Entropy of the source belief. Song et al. [8] proposed an improved version of this cost function by replacing the fixed parameter with a dynamic temperature value obtained from the trace of the covariance matrix of a Particle Filter estimation process, which shifts the weight of both terms according to the uncertainty of the belief map. A different line of work seeks to guide the agent toward the position that promotes the highest divergence between the actual and the expected probability density function (PDF), quantified by KL divergence [5]. However, it is still unclear how this approach generates non-identical search patterns from Entropy or Free Energy difference minimisation. Moving to the position of maximum divergence should generate similar behaviours as to moving to the position that minimises the difference of Entropy [9,10], but to the best of the authors' knowledge, in the context of OSL, a comparison between both strategies under realistic conditions is still absent. While thoroughly evaluated under simulated environments or controllable conditions such as wind tunnels [11], field validations are still scarce. More recently, a few studies employed UAVs as the searching agents [5,12], but aquatic environments are often overlooked. Some field tests with the goal of locating chemical sources do exist but do not employ info-theoretic strategies [13,14]. The informative path planning technique proposed by Stankiewicz et al. [15] has the potential to be used in OSL applications, however, the main objective is to generate water quality models, while the information-driven search strategy from Flashpohler et al. [16] makes use of bathymetric data from a shallow coral area for validation instead of chemical readings from an odour plume.

This work aims to fill the knowledge gap about experimental validation of cognitive search strategies by comparing the performance of three different cost functions and their efficiency with a real field experiment in a natural water stream. A static surface platform acts as a chemical pollutant source, emitting a controlled amount of non-pollutant saline solution to a river, while a small ASV equipped with a conductivity probe acts as the searching agent performing all the necessary computations onboard. The main contributions of this paper are: (1) - Performance analysis of three well-known information metrics (Entropy, KL, FE) under realistic field conditions; (2) - Efficiency evaluation of cognitive search strategies with a real field experiment in a natural water stream.

2 Information-Theoretic Framework for Odour Source Localisation

Consider a mobile agent moving in known locations $\mathbf{p}_t = (x, y) \in \mathbb{R}^2$ at the water surface with the capability to sense the water at time $t \geq 0$ s for chemical measurements $c(\mathbf{p}, d, t)$ at different depths d. Furthermore, assume a bounded workspace Ω with water flowing in a dominant direction $\boldsymbol{u}$, and a source in an unknown position (x_s, y_s, d) releasing a chemical pollutant at an unknown rate Q. It is assumed that the source and measurement depth are at the same level, restricting the search to a $2D$ space. The goal is to estimate the source location (x_s, y_s) from a set of measurements across the workspace. The agent starts searching for odour traces from a random position located downstream of the suspicious source. The searching strategy consists in moving upstream with zig-zag motion patterns until detecting traces of the target substance as proposed by Marjovi et al. [17]. After a first detection (x', y'), the agent switches to one of the cognitive search strategies studied in this work. Those strategies use a Bayesian inference belief and an information-theoretic decision making (Fig. 1) to guide the searching agent. This work aims to evaluate the performance of the decision making process for different cost functions.

2.1 Dispersion Model

Dispersion models are fundamental tools to reason about the olfactory observations a robot acquires along its mission. This work adopts a Gaussian plume

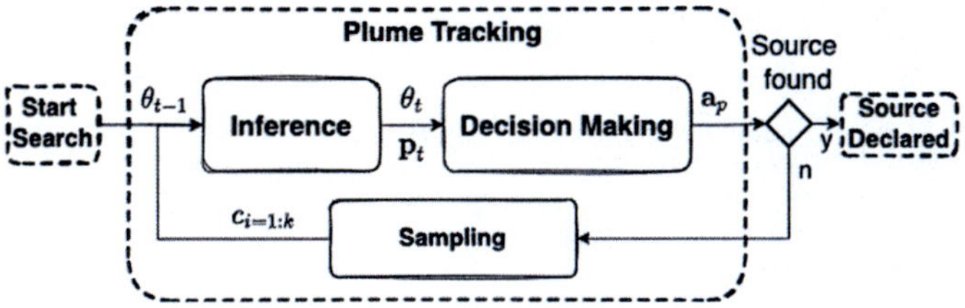

Fig. 1. OSL framework

model (1) which provides an analytical solution of the advection-diffusion equation for a point source, located at position (x_s, y_s) in a Cartesian coordinates frame, releasing at constant rate Q, with average water $\bar{u}$ flowing steadily in the direction u_d with speed u_s and assuming an infinite space.

$$\mathcal{M} = \frac{Q}{\pi \bar{u} \sigma_y \sigma_z} \exp\left(\frac{-(y - y_s)^2}{2\sigma_y^2}\right) \tag{1}$$

The lateral and vertical dispersion coefficients, σ_y and σ_z respectively, are functions of the downstream distance (x) to the source , modelled as:

$$\sigma_y^2 = 2D_y(x - x_s)/u_s \qquad \sigma_z^2 = 2D_z(x - x_s)/u_s \tag{2}$$

Assuming a laminar flow these parameters can be considered equal $(\sigma_z = \sigma_y)$. The coordinates are rotated u_d degrees (flow direction) in order to aligned the model with the actual flow.

2.2 Inference

In order to continuously update a belief of the potential source, observed data is reasoned through Bayesian inference. It is subsequently used to influence the movement decision of the searching agent, which makes this step a critical factor within the localisation process. Following the same trend of recent works [12], the process of obtaining a posterior PDF of a state vector θ_t given a set of observations $c_{1:t}$ as in:

$$P(\theta_t | c_{1:t}) \propto P(c_{1:t} | \theta_{t-1}) P(\theta_{t-1}) \tag{3}$$

is approximated with a Particle Filter [18] with a group of N weighted particles $\mathcal{O}_t := \{\theta_t^{j=1:N}, w_t^{j=1:N}\}$, including all the parameters from the dispersion model within the state vector as in $\theta_t = [x_s, y_s, Q, D_y, u_s, u_d]$.

The likelihood of the observed data given the actual hypothesis $P(c_{1:t}|\theta_{t-1})$ is evaluated with a normal distribution taking the form [19]:

$$P(c_{1:t} | \theta_{t-1}^{j=1:N}) \propto \exp\left(-\frac{1}{2} \cdot \sum_{i=1}^{k} \left(\frac{(c_i - \mathcal{M}(\mathbf{p}_i, \theta_{t-1}^j))^2}{\sigma_m^2 + \sigma_d^2}\right)\right) \tag{4}$$

where k represents the number of observed values, $\sigma_d \propto g_1 \mathcal{M}(\mathbf{p}_i, \theta_{t-1}^j)$ the standard deviation of measurement errors proportional to the predicted concentration with a constant g_1 and σ_m the standard deviation accounting for model errors.

2.3 Decision Making

The decision making process tries to optimise the search, guiding the agent towards the most informative positions, which are the ones that minimise the expected uncertainty about the source. This uncertainty is quantified with the so-called information utilities between the actual and future posterior PDF. First, a number of possible future positions N_{a_p} are generated from discrete movement steps [3]. For each of those positions a_p, a future PDF is obtained considering the actual knowledge of the source parameters and the dispersion model, allowing to compute the most informative movement position. For this step, this work follows a similar approach to [5] where, for each possible future position, N_z future concentration values are generated by drawing N_z weighted particles from the actual PDF. For each future observation, the weights are updated, and the information utility is computed with the new weights $\hat{w}$. The total expected information $\mathcal{E}$ is the average of the information utility from all future observation values as in $\mathcal{E}(a_p) \approx \frac{1}{N_z} \sum_{m=1}^{N_z} \mathcal{F}(\hat{w}^{(m,j)})$ where $\mathcal{F}$ is the utility function to be evaluated with the future observations and respective future weights. As previously stated, three information metrics are evaluated: Entropy, KL and FE. Shannon Entropy was the first information metric used in the context of OSL with a grid-based belief [3]. It quantifies the uncertainty about a given information with form:

$$S \approx -\sum_{j=1}^{N} w^j \ln(w^j) \tag{5}$$

Relative Entropy (6), also known as the KL divergence, is a measure of how much a future distribution at instance $k+1$ diverges from the actual distribution k. The lower the values, the more identical the distributions are.

$$KL = \sum_{i=1}^{N} \hat{w}_{k+1}^{(i,m)} \ln \frac{\hat{w}_{k+1}^{(i,m)}}{w_k^{(i)}} \tag{6}$$

The FE principle takes the original form of $FE = W + TS$ where W represents the potential energy, and T is the temperature that balances the relationship between the expected value W and the uncertainty S. Following the same approach as Song et al. [20], the term W is represented by the distance between the robot and the estimated source, the term T the trace of the covariance matrix $\mathbf{R}$ from the posterior PDF and S as the Entropy, given by:

$$FE = \sum_{j=1}^{N} w^j \left\| a_p - \theta_{x_s, y_s}^j \right\|^{\alpha_1} - tr(\mathbf{R})^{\alpha_2} S \tag{7}$$

with α_1 and α_2 the weights that control the balance between both terms.

Fig. 2. Testing section of Mondego River near the University of Coimbra.

3 Testing Environment

The experiments were performed in a section of the Mondego River near the Engineering Campus of the University of Coimbra (Fig. 2). It has a large variety of river features, such as shallow waters and sandy banks forming small isles in the middle of the river bed. The testing section has a length of approximately 70 m per 60 m of width and an average depth of 2 m. The measured conductivity value of clean water was 4700 $\mu S/m$ (baseline), the water temperature was at $22°C$ and the flow speed at approximately 1.5 m/s. A pollution event is simulated with a solution of salted water emitted at a constant rate of 1 l/min with a peristaltic pump from a floating platform. The solution is a mixture of 1 kg of salt for 4 l of water, with each reservoir having a total volume of 30 l, allowing 30 min of experimental time. The salted water produces a realistic non-pollutant odour plume easily distinguished from clean water with a conductivity sensor. The negative point is the higher density of the solution, which causes the plume to spread towards the riverbed as it moves further down flow. A considerable high emission rate is also required to generate a detectable plume at a significant distance from the source position. The OSL mission is performed with a small ASV composed of two floaters with a Bluerobotics T200 thruster attached at each one, having an overall dimension of 0.85×0.55 m. It can localise itself with a Global Navigation Satellite System (GNSS) module, configured in real-time kinematic positioning mode. The platform's motion is controlled by a Pixhawk 4 autopilot maintaining a moving speed of 0.5 m/s between goals positions. The high-level mission is provided by an Orange Pi 3 single board computer (SBC) running a Robot Operating System (ROS) framework. The platform uses a sonar to measure the depth of the river basin and an Atlas Scientific conductivity probe to detect the salted plume at a sample rate 2 Hz. The algorithms were coded with Python programming language and optimised with Numba JIT compiler.

Table 1. OSL parameters and Priors.

OSL framework

N	5000	N_{a_p}	8
g_1	1	length	6 m
σ_m	300 $\mu S/m$	N_z	200
α_1	1.4	α_2	1.3

Prior

x_s	$\mathcal{U}(x', x' + 150)$ m	D_y	$\mathcal{U}(0.001, 0.2)$ m^2/s
y_s	$\mathcal{U}(y', y' + 30)$ m	u_s	$\mathcal{N}(u_s', 1)$ m/s
Q	$\mathcal{U}(0.01, 200)$ g/s	u_d	$\mathcal{N}(u_d', 0.1)$ rad

3.1 Evaluation Metrics

A first experiment was performed with the agent starting downriver at approximately 80 m from the source, where the first odour encounter occurred at about 40 m. Due to the amount of time the experiment took (approximately 35 min), and considering the available volume of saline solution, the initial starting distance was fixed at 30 m from the source. By detecting traces of odour during the first crossflow movement, the plume search time is reduced, allowing more experimental time for cognitive search. A total of six reservoirs of the target solution were available, enabling four trials to be performed for each cost function, which makes the results not statistically significant. Moreover, to avoid losing potential data between movement locations and achieve a faster search strategy, at the possible cost of higher localisation error, all the observations during the last movement segment between positions are used instead of stopping the agent to collect and average data. The search ends with the uncertainty dropping below a certain threshold and the agent located within 6 m of distance from the estimated position. For a better initialisation of the water speed and direction states, an initial prior was estimated using the platform's GNSS localization by stopping it during short periods and letting it move with the water flow. The remaining parameters of the OSL framework and initial priors (u_s', u_d') are summarised in Table 1. The following metrics were evaluated for each strategy:

1. Localisation error: The error between the estimated and true source position. The state vector contains six fields, but for this work, the main interest lies in the source position;
2. Success rate: An experiment is considered successful if the located source is within 10 m of error from the true source position;
3. Exploration/exploitation ratio: A decision is considered exploitative if the chosen a_p location falls within a section of $\pm 67.5°$ in the direction of the true source position. Otherwise, the decision is considered explorative.
4. Distance efficiency: Ratio from the azimuthal distance between the starting position to the true source location and the total travelled distance;
5. Computation time: The time spent by inference and decision making processes;
6. Search efficiency: The minimum required time to travel between the starting and true source positions divided by the total extent of the experiment.

4 Experimental Results

The cognitive strategies were able to successfully locate the true source position in most of the trials. Similar to Loisy et al. [21], the cognitive search is started after the first traces of odour are found, enabling the inference process to be initialised with enhanced prior information. The particles (location states) are initialised from the agent position towards the upflow direction, considerably narrowing the search space.

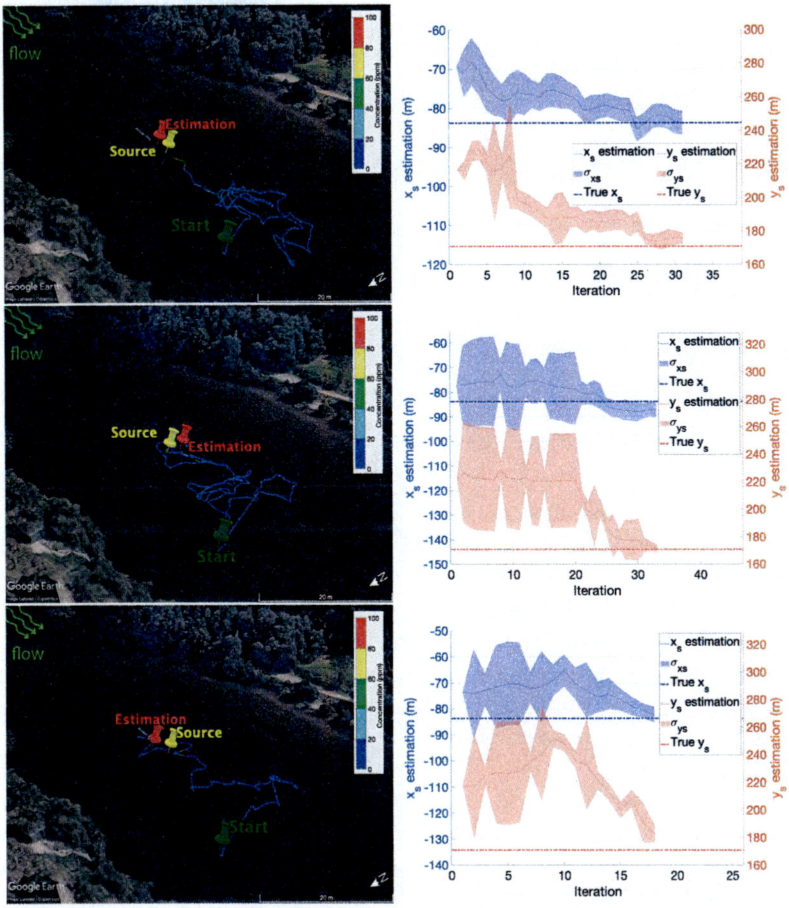

Fig. 3. Left column: Trajectory and measured concentrations for one experiment of each information metric (Entropy, KL, FE from top to bottom). Right column: Evolution of the estimation with the associated uncertainty for each of the respective experiments.

The left column from Fig. 3 show the trajectories of one experiment for each information utility. It can be seen the magnitude of the concentration values

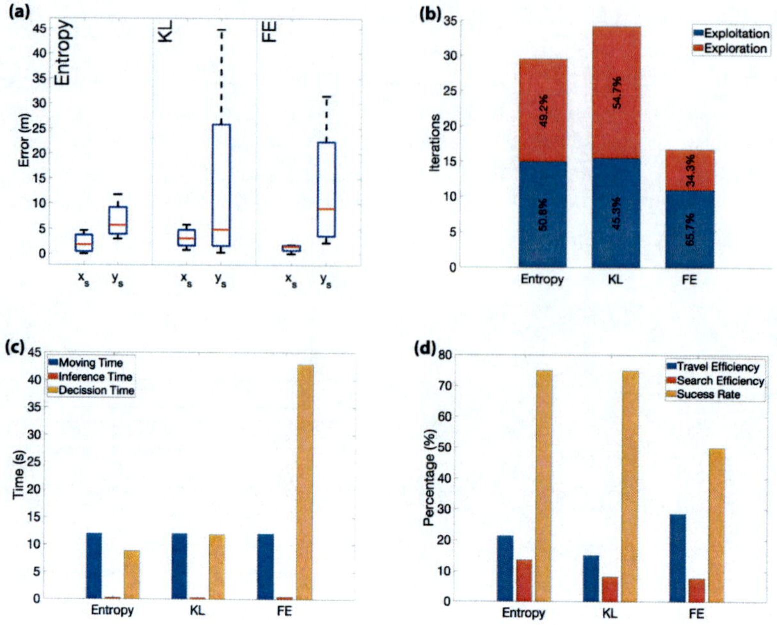

Fig. 4. Analysis of metrics' performance: (a) Localisation error, (b) Exploration/exploitation ratio, (c) Computation times, (d) Success rate, travel and search efficiency.

increasing as the agent moves closer to the source position, as well as a narrower active region. This characteristic makes the process increasingly difficult since odour encounters are restricted to a small section in the crossflow direction. Another visible feature common to all three metrics is the tendency of the agent to perform exploratory behaviours, with a significant number of crossflow movements, during the first stages of the search. It can be explained by the lower magnitude of the measured concentration values and higher intermittency, which requires additional exploration to improve the knowledge of the source parameters. The right column of the same figure demonstrates the evolution of the estimated source location and respective uncertainty regarding the true source parameters. As the process evolves, the error and the spread of the estimated states reduce until reaching a value near the true source. Only closer to the source does the uncertainty of estimation reduces considerably, a common feature of all metrics.

From Fig. 4a, it can be seen that entropy has the best results in terms of localisation accuracy with a mean error of 5 m across the experiments, with KL and FE producing higher mean localisation errors of approximately 11 m. The primary source of mean localisation error lies in the estimation of y_s state, stretching the dispersion of the particles towards the upflow direction (Fig. 5).

Fig. 5. Particles convergence. Left: Successful experiment; Right: failed experiment.

A possible explanation is the use of multiple instantaneous observations from the last movement segment in the inference process, which contributes to higher uncertainty while reasoning with the source hypothesis.

In terms of exploration/exploitation ratio (Fig. 4b), KL has the highest exploration percentage with 54.7%, Entropy is the most balanced metric with ~50% for each behaviour, and FE produces the most exploitative behaviour, as expected, with 65.7%. This additional term of FE contributed to the least travelled distance of all three metrics (approximately half of KL), but the negative point is the worst success rate compared with Entropy and KL. The higher exploitative term may lead the agent towards the source too soon, where further exploration was needed, trapping the search in regions with no odour encounters, resulting in failure. The adjustable gains in the cost function are also a drawback towards field implementations since it requires a specific tuning to balance the exploration and exploitation behaviour, which may not be the most efficient for different scenarios.

By analysing computation times (Fig. 4c), one can conclude that the decision making step is by far the most demanding operation of the search process. It is even more meaningful for FE metric with a significantly higher computational cost when compared with Entropy and KL. It is worth mentioning again that all algorithms and additional functionalities are running onboard the ASV. While significantly lower times can be achieved with more capable SBCs or processing units, the main conclusion to be drawn here is the computational weight of the decision making process towards the whole search strategy. Evaluating the results from an efficiency point of view (Fig. 4d), while FE has the best efficiency in terms of travelled distance, the lower success rate, higher computational cost, and consequently significantly higher decision time make it the least efficient method of the three. Entropy and KL have identical success rates and travel efficiency (with KL having a slightly worst travel efficiency), but Entropy's slightly lower computational cost makes it the most efficient search strategy. Analysing the experiments that were unable to locate the source within an acceptable range, two different failure scenarios were identified: (1) the filter converged towards a location that was distant from the true source position, and (2) the filter was not able to converge to a position, with the particles spread out over a large area as it can be seen in the right image from Fig. 5. The left image from the same figure shows a successful experiment with the particles converging towards a region in the vicinity of the true source location. In short, one can conclude both Entropy and KL generate similar search strategies with near-identical localisation results.

5 Conclusions

This work aimed to improve the knowledge toward OSL field experiments by performing and evaluating cognitive search strategies in a natural water stream with a small ASV. All of the three cost functions were able to locate the source within an acceptable margin of error successfully. Entropy and KL generated similar search trajectories, although KL had a slightly larger exploration ratio and, consequently, longer travel distances. FE was more efficient in terms of travelled distance, but its higher "greedy" nature due to the additional exploitation term also led to a lower success rate. A possible explanation is the weight adjustment between both terms (exploration and exploitation) that plays a major role in the performance of this cost function and is difficult to tune in natural environments where the conditions are not known in advance. As expected, the decision-making step proved to be the most computational-heavy process of the whole search strategy, with FE the most demanding cost function. This increase in computational effort did not traduce in a better performance overall, as previously stated since the success rate was lower than the other two cost functions. Efforts should be made towards researching decision strategies that require less computational effort and can still adapt to the dynamic nature of the environment, which is critical for searching agents with limited processing capabilities.

References

1. Dunbabin, M., Marques, L.: Robots for environmental monitoring: significant advancements and applications. IEEE Robot. Autom. Mag. **19**(1), 24–39 (2012)
2. Marques, L., Nunes, U., de Almeida, A.T.: Olfaction-based mobile robot navigation. Thin Solid Films. **418**(1), 51–58 (2002)
3. Vergassola, M., Villermaux, E., Shraiman, B.: 'infotaxis' as a strategy for searching without gradients. Nature. **445**, 406–409 (2007)
4. Voges, N., Chaffiol, A., Lucas, P., Martinez, D.: Reactive searching and infotaxis in odor source localization. PLoS Comput. Biol. **10**(10), e1003861 (2014)
5. Hutchinson, M., Liu, C., Thomas, P., Chen, W.-H.: Unmanned aerial vehicle-based hazardous materials response: information-theoretic hazardous source search and reconstruction. IEEE Robot. Autom. Mag. **27**(3), 108–119 (2019)
6. Ristic, B., Skvortsov, A., Gunatilaka, A.: A study of cognitive strategies for an autonomous search. Inf. Fusion **28**, 1–9 (2016)
7. Masson, J.-B.: Olfactory searches with limited space perception. Proc. Natl. Acad. Sci. **110**(28), 11261–11266 (2013)

8. Song, C., He, Y., Lei, X.: Autonomous searching for a diffusive source based on minimizing the combination of entropy and potential energy. Sensors. **19**(11), 2465 (2019)
9. Hiroshi Aoki, E., Bagchi, A., Mandal, P., Boers, Y.: A theoretical look at information-driven sensor management criteria. In 14th International Conference on Information Fusion, pp. 1–8. IEEE (2011)
10. Matthew Aughenbaugh, J., La Cour, B.R.: Metric selection for information theoretic sensor management. In: 2008 11th International Conference on Information Fusion, pp. 1–8. IEEE (2008)
11. Rahbar, F., Marjovi, A., Martinoli, A.: An algorithm for odor source localization based on source term estimation. In 2019 International Conference on Robotics and Automation (ICRA), pp. 973–979. IEEE (2019)
12. Park, M., An, S., Seo, J., Hyondong, O.: Autonomous source search for UAVs using gaussian mixture model-based infotaxis: algorithm and flight experiments. IEEE Trans. Aerosp. Electron. Syst. **57**(6), 4238–4254 (2021)
13. Farrell, J., Pang, S., Li, W.: Chemical plume tracing via an autonomous underwater vehicle. IEEE J. Ocean. Eng. **30**, 428–442 (2005)
14. Powers, C., Hanlon, R., Schmale, D.G.: Tracking of a fluorescent dye in a freshwater lake with an unmanned surface vehicle and an unmanned aircraft system. Remote Sens. **10**(1), 1–12 (2018)
15. Stankiewicz, P., Tan, Y.T., Kobilarov, M.: Adaptive sampling with an autonomous underwater vehicle in static marine environments. J. Field Robot. **38**(4), 572–597 (2021)
16. Flaspohler, G., Preston, V., Michel, A.P.M., Girdhar, Y., Roy, N.: Information-guided robotic maximum seek-and-sample in partially observable continuous environments. IEEE Robot. Autom. Lett. **4**(4), 3782–3789 (2019)
17. Marjovi, A., Marques, L.: Optimal swarm formation for odor plume finding. IEEE Trans. Cybern. **44**(12), 2302–2315 (2014)
18. Magalhães, H., Baptista, R., Macedo, J., Marques, L.: Towards fast plume source estimation with a mobile robot. Sensors **20**(24), 7025 (2020)
19. Keats, A., Yee, E., Lien, F.-S.: Bayesian inference for source determination with applications to a complex urban environment. Atmos. Environ. **41**(3), 465–479 (2007)
20. Song, C., He, Y., Ristic, B., Lei, X.: Collaborative infotaxis: searching for a signal-emitting source based on particle filter and Gaussian fitting. Robot. Auton. Syst. **125**, 103414 (2020)
21. Loisy, A., Eloy, C.: Searching for a source without gradients: how good is infotaxis and how to beat it. Proc. R. Soc. A **478**(2262), 20220118 (2022)

GadenTools: A Toolkit for Testing and Simulating Robotic Olfaction Tasks with Jupyter Notebook Support

Pepe Ojeda(✉), Jose-Raul Ruiz-Sarmiento, Javier Monroy, and Javier Gonzalez-Jimenez

Machine Perception and Intelligent Robotics Group (MAPIR-UMA), Malaga Institute for Mechatronics Engineering and Cyber-Physical Systems (IMECH.UMA), University of Malaga, Malaga, Spain
ojedamorala@uma.es

Abstract. This work presents GadenTools, a toolkit designed to ease the development and integration of mobile robotic olfaction applications by enabling a convenient and user-friendly access to Gaden's realistic gas dispersion simulations. It is based on an easy-to-use Python API, and includes an extensive tutorial developed with Jupyter Notebook and Google Colab technologies. A detailed set of examples illustrates aspects ranging from basic access to sensory data or the generation of ground-truth images, to the more advanced implementation of plume tracking algorithms, all in an online web-editor with no installation requirements. All the resources, including the source code, are made available in an online open repository.

Keywords: Robotic olfaction · Gas dispersion simulation · Python · Jupyter Notebook · Google colaborative

1 Introduction

Mobile robot olfaction (*MRO*) is the field concerned with the integration and application of the sense of smell into mobile robots. It is a widely multidisciplinary research area, involving problems such as chemical sensing and classification [4,15], dispersion modeling [5,7], optimal sampling [8] or gas source localization [10,17], among others.

There are many potential applications for autonomous, mobile robotic agents with the ability to sense the presence of gases in an environment, *e.g.* locating leaks in pipes, detecting dangerous substances, rescue missions, air quality monitoring, *etc.*. As a result of these interesting prospects, the field of robotic olfaction has steadily been gaining attention over the years, and is expected to continue to do so as the available technology improves.

A key hurdle, preventing much research from reaching the degree of maturity necessary for real-world deployment, is the lack of proper datasets and tools

D. Tardioli et al. (Eds.): ROBOT 2022, LNNS 590, pp. 166–177, 2023.
https://doi.org/10.1007/978-3-031-21062-4_14

in the field that can be used to conduct ground truth evaluations. Works dealing with these concepts are crucial, and highly popular in other fields such as computer vision [11]. In MRO, though, performing experiments in real environments is frequently not possible, and there is a serious technological limitation to obtaining ground truth information of the spatial distribution of gases in the environment [16,22].

These reasons, alongside the fact that the robotic olfaction research community is still small, make it so that not many resources (datasets, visualization tools, *etc.*) are available to facilitate the research progress. We believe that creating these assets for robotic olfaction, along with the tools to make them easily accessible to our peers, is a meaningful contribution to the field that can boost advances, and specially will help newcomers get started on a subject that has been, up to now, somewhat obscure.

In this work we present GadenTools, a toolkit to handle data from gas dispersion simulations that aims to push in that direction. It is shipped as an easy-to-use Python [6] module, and includes an extensive tutorial[1] developed with the Jupyter Notebook [9] and Google Colab technologies. This notebook-shaped tutorial was designed to illustrate both the basic interaction with the GadenTools module's API and some more advanced use cases of how to exploit its capabilities to perform powerful visualizations, data analysis, and methods' validations. For doing so, it takes advantage of the Jupyter notebooks ability to combine the expressiveness of traditional explanations from textbooks (texts, equations, figures, etc.), with the interaction capabilities of software applications and executable code with visible outputs [9,21]. As a result, GadenTools allows for a fast prototyping and development of contents related to MRO. All the resources, including the source code, are made available in an online open repository[2].

More specifically, GadenTools works with data from Gaden simulations. Gaden [16] is a gas dispersion simulator specifically designed for mobile robotics that relies on Computational Fluid Dynamics (CFD) data. Although it has been widely used in *MRO* research already [3,17,20], we believe the somewhat steep learning curve required to use it, alongside the requirement to install additional software (*i.e.* the Robotics Operating System, *ROS*), might be making researchers who would benefit from its use to decide against it. In this way, GadenTools offers a clean and accessible entry to Gaden simulations, while removing the need for other software dependencies.

2 On the Importance of Gas Dispersion Simulators and Datasets

The phenomenon of gas dispersion is very complex, and developing techniques to understand it and infer the state of the environment from sparse measurements

[1] https://colab.research.google.com/drive/1Xj7rrsmeDa_dS3Ru_UIhhzlaifGH6GS4?usp=sharing.

[2] https://github.com/MAPIRlab/GadenTools.

is a challenging task that requires extensive experimentation. This experimentation process is also non-trivial, as setting up repeatable, consistent experiments requires a way to control the airflow, temperature, source release rate, and many other factors. As a result, real-world experiments are mostly limited to simple, wind-tunnel-like environments [14,23].

Simulation is thus an essential tool for researchers, allowing them to perform perfectly repeatable experiments in environments of certain complexity. While simulations, no matter their precision, do not avoid real-world validation of the robotic techniques, they are a key tool in the early stages of development and testing.

However, because CFD simulations with state-of-the-art numerical solvers are difficult to set up and offer no integration with robotics software, many researchers end up opting for custom-made simulation tools that are more appropriate for robotic olfaction [2,13]. Since these tools are not commercial software meant to be used by the non-expert public, even when they are available online they are often difficult for new users to interact with and lack proper documentation. This ends up leading to a situation where each research group uses their own internal simulation tool, which is incompatible with those used by everyone else and hampers the fair comparison between developments. And so, getting started in the field of robotic olfaction requires learning how these complex tools work, or implementing one's own.

Monroy *et al.* attempted to solve this integration issue by implementing Gaden, a gas dispersion simulator, under the umbrella of the well-known Robot Operating System (ROS) [19]. This way, Gaden was developed as a ROS package [16], thus allowing any other ROS process to communicate with it through topics and messages. However, while this undoubtedly makes access to the simulations more convenient for those researchers who use ROS to develop their own software, it also makes the data inaccessible to those who do not. The tools presented in this work intend to help bridge that gap, making the simulation data easily accessible in multiple formats by directly querying an API through standalone Python code.

Recently, the authors of this work also released the VGR Dataset [18], a large repository of already computed simulations in complex environments, made public online[3]. Figure 1 shows some snapshots of the CFD-generated airflows for two different indoor environments (included in the dataset) that can be accessed with GadenTools. This dataset was intended to eliminate the necessity for researchers to design and create their own simulations, which is in itself a time-consuming process that requires a degree of familiarity with the simulation tools and additional knowledge of other fields (*e.g.* 3D modelling). Moreover, the data are provided in different formats, including human-readable CSV files, enabling their usage outside Gaden at the expense of developing a toolkit for managing them.

GadenTools aims to palliate this issue by reducing the barrier of entry to the field of robotic olfaction, allowing researchers to focus on developing their algorithms and techniques, rather than on implementing simulation tools and designing testing environments.

[3] https://mapir.isa.uma.es/mapirwebsite/?p=1708.

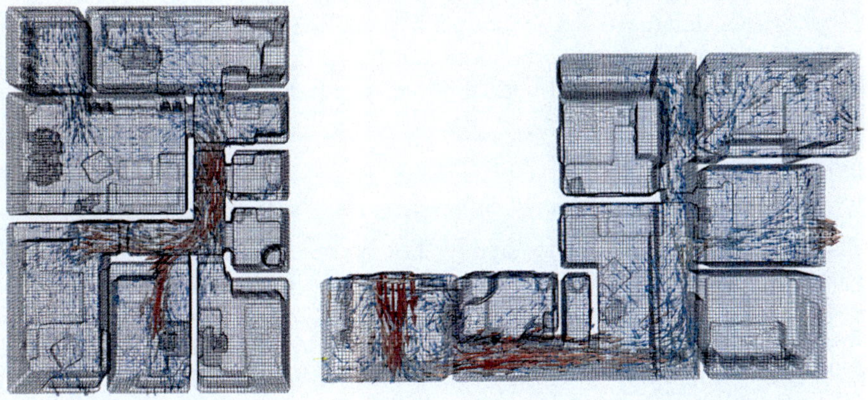

Fig. 1. Examples of airflows generated with CFD tools in two different environments. Having multiple inlets and outlets, as well as many obstacles, tends to produce flows that are more complex than would be possible in a wind tunnel.

3 Technologies Employed

In this section, we will briefly discuss the choice of technologies used for the implementation and deployment of the GadenTools Python's module and interactive tutorial.

3.1 Python

The Python programming language [6] was chosen for this project due to multiple reasons, one of the most prominent of which is that the main aim of the project is being easily accessible to everyone, regardless of their previous experience on the field or their technical skill. Python is a very popular language with an easy-to-learn syntax and a great deal of existing learning resources and useful libraries that help reduce the barrier of entry.

Another important reason why Python was a natural choice is that it is fully compatible with ROS. While GadenTools does not need to directly interact with the existing ROS implementation of Gaden (and, indeed, in its current state it is meant to be used in its stead), keeping open the possibility for their interaction makes it so that future updates to the GadenTools project may add extra functionality and create a more user-friendly way to interact with Gaden, while still leveraging the speed and stability of the existing C++ Gaden ROS nodes.

A drawback of Python is that it is not appropriate for highly performant code. While most of the operations carried out in GadenTools do not require great speed, there are some types of queries (see the visualization examples in Sect. 4.2) that may be inconveniently slow when reading simulations of big size.

For this reason, the implementation of GadenTools uses Cython [1], an extension of the Python language that is statically compiled and thus allows for greatly optimized execution. Cython modules can be used from standard Python code seamlessly, and users of the module do not need to modify their own program in any way to account for this.

3.2 Jupyter Notebooks

The Jupyter Notebook technology [9] has become the de facto choice in the scientific community for producing reproducible computational workflows. It is of special relevance in this work since it permits us to build up an interactive tutorial describing GadenTools and its usage. This type of resources are also called *computational stories* or *live documents*. This minimizes the tool's learning curve and illustrates how to perform different robotic olfaction task with it. Jupyter Notebook consists of three main components:

- **Notebook**. Text document (JSON-based) used to define the Jupyter notebook itself. It consists of a list of text or code cells. Text cells contain markdown-formatted text, which enables the inclusion of math formulas, tables, lists, rich media (images, videos, audios), etc. If needed, HTML can also be used. In its turn, code cells contain source code in the kernel language that can be executed, being their output (text, plots, etc.) printed below them.
- **Web application**. Web-based interactive development environment to create, edit and run notebooks. It allows for the hosting of the notebooks in a server in order to work with them, given that the application is also installed. Local installations are also possible using package managers such as pip or Anaconda. The classic notebook interface evolved into *JupyterLab*, which fixed different usability issues and expanded its scope.
- **Kernel**. Backend application in charge of interacting with the Jupyter application by executing code cells and making readily available their output. Although there exist kernels for more than 50 programming languages[4], here we resort to the *IPython* kernel, which executes Python code.

Instead of JupyterLab, we have relied on Google Colaboratory[5] (Colab for short) for implementing the tutorial. This application has some advantages for the design of an illustrative notebook: it provides free cloud resources so the tutorial can be executed in any browser with internet connection, it enables IDE-like features like autocompletion, function declaration inspection from any point of the notebook, a variables-watcher, etc., and other features like showing a table of contents to the user. Using Google Colab, the user can launch the tutorial in a non-writable fashion and play with it without making any local installation. Moreover, they could make a local copy (just of the tutorial) to modify it according to their needs.

---

[4] https://github.com/jupyter/jupyter/wiki/Jupyter-kernels.

[5] https://colab.research.google.com/.

4 GadenTools

GadenTools can be seen as a set of tools containing two main components: a Python module for managing datasets, and a Jupyter Notebook-based tutorial for illustrating how to operate with it, which also serves as a starting point for further developments. Next sections describe them in further detail.

4.1 The Python Module

In order to better understand the module's functionalities, it is convenient to first dive into the primitives it works with: data from Gaden simulations. The result of one of these simulations comprises a series of files in an internal binary format, which separately hold information about: i) the airflow vectors through the environment, ii) the gas distribution, as well as iii) a CSV file that describes the geometry of the 3D environment in which the simulation takes place. These binary representations are intended to speed up both reading and writing the simulation results, and to reduce their memory footprint, but regrettably they also make it difficult to manually parse the files.

This is the most immediate problem solved by GadenTools, which implements the necessary steps to decode this information and make it easily accessible through simple programmatic queries. These queries have been designed to be straightforwardly incorporated to Jupyter notebooks workflows. Moreover, as commented, the implementation has been made efficient by leveraging the advantages of Cython.

The module's core is the `Simulation` class, which permits the user to instantiate a simulation by just specifying the path to its files in the device file system. Behind this instantiation is the reading of Gaden's binary files and their parsing, so the information they encode becomes accessible through the module's API. A `Vector3` class is also provided, which in addition to basic arithmetical operations with 3-dimensional vectors also supports some extra utility functions.

This way, the API permits the MRO user to get information like the current gas concentration or the wind vector at a given location in a certain simulation iteration (expressed in *ppm* and *m/s*, respectively). The API also offers the option to query multiple positions at once. For example, one can extract the gas concentration or wind data of 2D slices of the environment, formatted as multidimensional arrays, with a single query. This functionality has multiple applications (*e.g.* obtaining the ground truth of a concentration map to compare to those generated by an algorithm that is being tested), but probably the most notable is the option to visualize the extracted data as images. The next section further explores this option.

Beyond the ability to manually query specific points of the simulation, as shown earlier, GadenTools can also replicate the traditional behavior of the ROS version of Gaden, playing back the simulation in real time. This is useful when one wants to simulate a robotic agent that takes time to move through the environment and must deal with the fact that gas dispersion is a time-dependent phenomenon. When using this mode, new snapshots of gas dispersion

```
sim = Simulation("/content/DATA/Exp C/FilamentSimulation gasType 0 sourcePosition 1.75 8.60 0.50", \
                 "/content/DATA/Exp C/OccupancyGrid3D.csv")
pointToQuery = Vector3(2.5, 4.0, 0.1)
iteration = 500
print("Gas Concentration at selected point:" + str( round(sim.getConcentration(iteration, pointToQuery), 2) ) + " ppm")
print("3D Wind Vector at selected point:" + str( vector3Round( sim.getWind(iteration, pointToQuery) ) ) + " m/s")

Gas Concentration at selected point:2.74 ppm
3D Wind Vector at selected point:[0.2, -0.25, -0.0] m/s
```

Fig. 2. Excerpt from the Jupyter Notebook tutorial that shows how to load simulation data and query for single-point measurements of the gas concentration and the airflow vector.

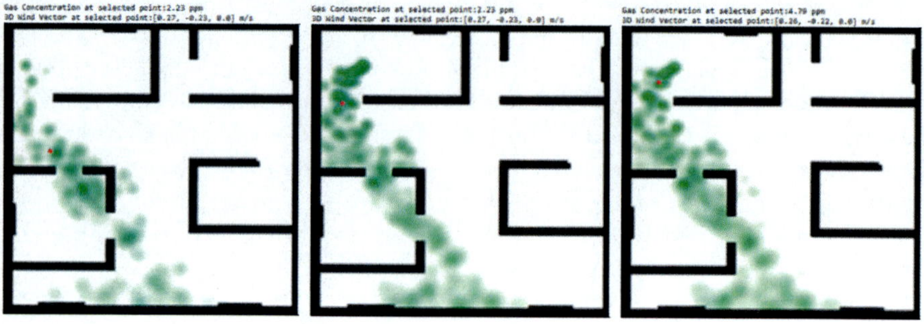

Fig. 3. Illustration of GadenTools accessing the gas concentration and 3D wind vector from a given Gaden simulation. By simply updating the robot position, the API provides the desired measurements.

and airflow information are continually loaded in the background at the specified rate without any further input from the user, and the API provides methods to access the data that pertains to the currently active iteration, rather than to a manually specified one. This functionality was designed to be Jupyter Notebook-friendly, as discussed below.

The GadenTools module also contains a "Utils" submodule, which includes several auxiliary tools which are not strictly required for the core functionality of the module. Besides the `Vector3` class, which has already been discussed, several functions are included to aid in the generation of visualizations, from simply marking the environment's obstacles in the generated images, to encoding the airflow's direction as colors. Section 4.2 explores the topic of generating visualizations in more depth.

4.2 The Notebook Tutorial

The purpose of the Jupyter notebook within GadenTools is to interactively describe its functionalities while illustrating how to use it in different MRO use cases. It starts by describing GadenTools and its API in a similar way as in Sect. 4.1, but taking advantage of code cells to exemplify its usage. For example, Fig. 2 shows a code cell that illustrates how to create a `Simulation` object, also using the API to query the ground-truth gas concentration and wind vector at a certain position and simulation iteration. We can see how the result of executing this cell code is reported just below. In the notebook, the access to existing simulation data is made available through selected downloadable examples, but one can indeed opt for any other existing resource, including the previously mentioned VGR dataset, or even newly created simulations with custom environments or source parameters.

Once the basic usage is covered, the tutorial guides the user through different GadenTools use cases. It starts by presenting different ways to extract simulation data as images. The ability to visually represent the ground truth data not only helps to gain an insight into the patterns of gas dispersion, and so help guide the development of new techniques and algorithms, but is also a convenient tool when troubleshooting methods of source localization or concentration maps creation that are not working correctly in a given scenario.

Fig. 4. (Top) excerpt from the Jupyter Notebook tutorial that shows how to generate a heatmap image of gas concentration on a 2D slice of the environment. In the notebook, a series of interactive sliders lets the user control multiple parameters regarding which information is sampled and how it is shown. (Bottom) Example of gas concentration heatmaps for different iterations (40, 100, 300 and 500).

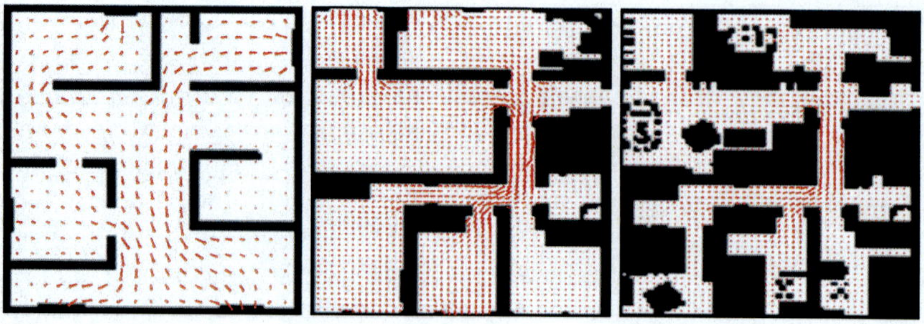

Fig. 5. Visualization of the airflow through the environment as a vector field, generated from the "2D map" method in the GadenTools API. The second and third pictures correspond to different vertical slices of the same environment, and show the variation of the geometry and airflow with height.

In the same way, Fig. 3 shows images built upon the provided API with an occupancy map of a house, and gas dispersion at different iterations. The user can simulate a mobile robot (marked as a red point in the images) to inspect said environment by sampling it to get gas concentration and wind vector measurements at different locations. The notebook also challenges the user to find the gas source by guessing its location, giving feedback on the accuracy of their guess. Figure 4 shows a notebook excerpt with a cell code illustrating how to create images with an horizontal slice of the gas concentration as a heatmap, while Fig. 5 does the same with the airflow's vectors. In both cases, a minimal GUI gives the user the possibility to set the value of relevant parameters, which helps to gain insight into the simulation behavior.

Once these basic images are obtained, the notebook also exemplifies how to apply simple image processing techniques to them in order to get additional information. Concretely, it illustrates how to retrieve the boundaries of the gas plume in order to know the area affected by the gas dispersion, for which binarization and contour finding techniques are used. A second illustrative example shows how to get concentration gradients to visualize how said concentration changes over space. Figure 6 shows the resulting images after executing these examples.

As discussed in Sect. 4.1, the API provides the required functionality to play back a simulation in real time. This is also described in the notebook, including an example of its usage to build a video of the gas dispersion at time intervals of 0.5 s. This functionality is also included in the concluding use case of the notebook: an implementation of the classic Surge-Cast source localization algorithm [12] (see Fig. 7). The Surge-Cast algorithm has the robot move through the gas plume by alternating two movement patterns, an upwind surge and a crosswind cast, according to whether it is currently sensing gas or not. By tracking the plume in this way, the robot eventually reaches the source.

This last use case is particularly illustrative of how a Jupyter notebook can be used for the fast prototyping and validation of algorithms in general, and MRO

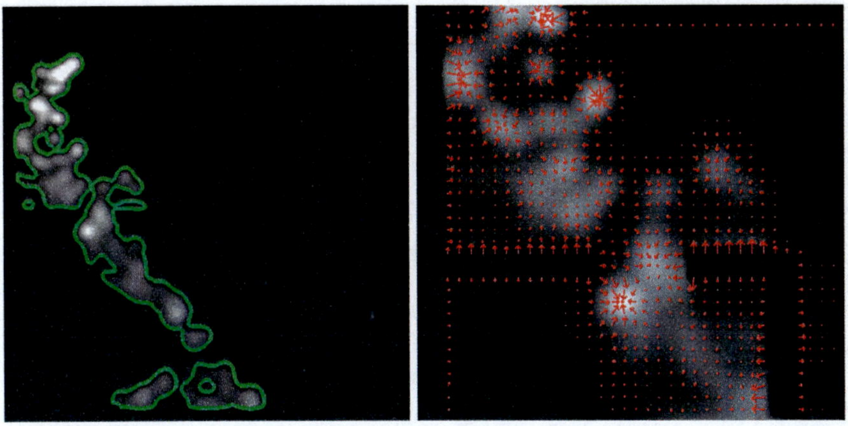

Fig. 6. Examples of using image processing techniques to extract information from the visualizations of the simulation data. The first image shows the contours of the gas plume, while the second image shows the local concentration gradients.

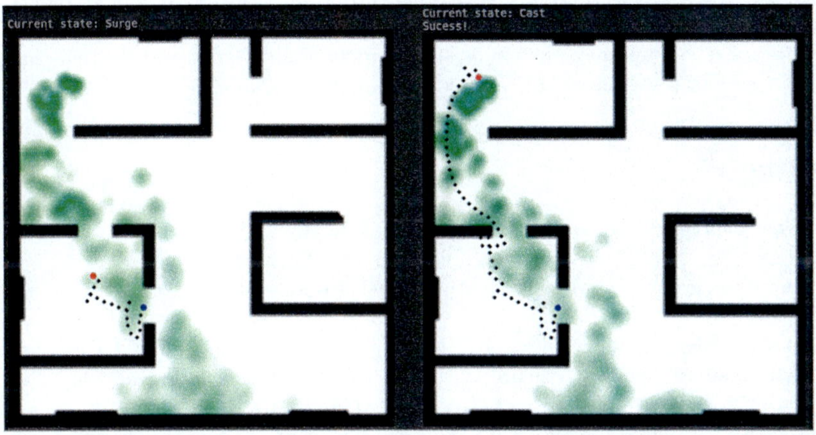

Fig. 7. An example of the proposed toolkit being used to implement an MRO algorithm. Pictured here is the Surge-Cast source localization algorithm. Both the algorithm implementation and the visualization are done fully in-notebook, and the code is presented to help users develop their own methods.

techniques in particular. GadenTools users have the chance the opportunity to continue editing and evolving the notebook according to their needs, or create new notebooks using it, since as shown, the integration of the provided Ptyhon module and Jupyter Notebook is straightforward.

5 Conclusions

This paper has presented the components of GadenTools, which pursue the boosting of research in mobile robot olfaction. For doing so, it contributes an easy-to-use Python module with a clean API that handles data from gas dispersion simulations. Specifically, the works with simulations produced by Gaden, which include data about the environments where they were carried out (*e.g.* layout and obstacles), as well as airflow and gas dispersion simulations. The paper shows how these binary data, which previously required the installation and configuration of different software dependencies for their manipulation, becomes easily accessible through the module's API. GadenTools also includes a Jupyter notebook, designed with Google Colab, that demonstrates the claimed easy-to-use and accessibility features, and also illustrates a number of use cases showing how it could be a powerful partner in the fast prototyping and validation of MRO techniques. GadenTools is an active project open to comments and contributions from the MRO community.

Acknowledgements. This work was funded by the research projects *HOUNDBOT* (P20_01302 and UMA20-FEDERJA-056), both funding by Andalusia Regional Government and the European Regional Development Fund*ERDF* and by the grant fpr the formation of pre-doctoral researchers in Andalusia (24653).

References

1. Cython. https://Cython.org/. Accessed 6 July 2022
2. Asenov, M., Rutkauskas, M., Reid, D., Subr, K., Ramamoorthy, S.: Active localization of gas leaks using fluid simulation. IEEE Robot. Autom. Lett. **4**(2), 1776–1783 (2019)
3. Bourne, J.R., Goodell, M.N., He, X., Steiner, J.A., Leang, K.K.: Decentralized multi-agent information-theoretic control for target estimation and localization: finding gas leaks. Int. J. Robot. Res. **39**(13), 1525–1548 (2020)
4. Choi, S.-I., Eom, T., Jeong, G.-M.: Gas classification using combined features based on a discriminant analysis for an electronic nose. J. Sens. **2016**, 9634387 (2016)
5. Farrell, J.A., Murlis, J., Long, X., Li, W., Cardé, R.T.: Filament-based atmospheric dispersion model to achieve short time-scale structure of odor plumes. Environ. Fluid Mech. **2**, 143–169 (2002)
6. Python Software Foundation. Python language reference, version 3.7.x. https://python.org/. Accessed 6 July 2022
7. Gongora, A., Monroy, J., Gonzalez-Jimenez, J.: Joint estimation of gas & wind maps for fast-response applications. Appl. Math. Model. **87**, 655–674 (2020)
8. Hutchinson, M., Liu, C., Chen, W.H.: information-based search for an atmospheric release using a mobile robot: algorithm and experiments. IEEE Trans. Control Syst. Technol. **27**, 1–15 (2018)
9. Kluyver, T., et al.: Jupyter notebooks: a publishing format for reproducible computational workflows. In: Positioning and Power in Academic Publishing: Players, Agents and Agendas, pp. 87–90. IOS Press (2016)

10. Lewis, T., Bhaganagar, K.: A comprehensive review of plume source detection using unmanned vehicles for environmental sensing. Sci. Total Environ. **762**, 144029–144029 (2021)
11. Lin, T.-Y., et al.: Microsoft COCO: common objects in context. In: Fleet, D., Pajdla, T., Schiele, B., Tuytelaars, T. (eds.) ECCV 2014. LNCS, vol. 8693, pp. 740–755. Springer, Cham (2014). https://doi.org/10.1007/978-3-319-10602-1_48
12. Lochmatter, T., Martinoli, A.: Tracking Odor Plumes in a Laminar Wind Field with Bio-inspired Algorithms. In: Khatib, O., Kumar, V., Pappas, G.J. (eds.) Experimental Robotics. Springer Tracts in Advanced Robotics, vol. 54, pp. 473–482. Springer, Heidelberg (2009). https://doi.org/10.1007/978-3-642-00196-3_54
13. Macedo, J., Marques, L., Costa, E.: A comparative study of bio-inspired odour source localisation strategies from the state-action perspective. Sensors. **19**(10), 2231 (2019)
14. Macedo, J., Marques, L., Costa, E.: Locating odour sources with geometric syntactic genetic programming. In: Castillo, P.A., Jiménez Laredo, J.L., Fernández de Vega, F. (eds.) EvoApplications 2020. LNCS, vol. 12104, pp. 212–227. Springer, Cham (2020). https://doi.org/10.1007/978-3-030-43722-0_14
15. Monroy, J., Gonzalez-Jimenez, J.: Gas classification in motion: an experimental analysis. Sens. Actuat. B. Chem. **240**, 1205–1215 (2017)
16. Monroy, J., Hernandez-Bennetts, V., Fan, H., Lilienthal, A., Gonzalez-Jimenez, J.: GADEN: a 3d gas dispersion simulator for mobile robot olfaction in realistic environments. MDPI Sens. **17**(7), 1–16 (2017)
17. Ojeda, P., Monroy, J., Gonzalez-Jimenez, J.: Information-driven gas source localization exploiting gas and wind local measurements for autonomous mobile robots. IEEE Robot. Autom. Lett. **6**(2), 1320–1326 (2021)
18. Ojeda,P., Monroy, J., Gonzalez-Jimenez, J.: VGR dataset: A CFD-based gas dispersion dataset for mobile robotic olfaction (2022). (to appear)
19. Quigley, M., et al.: ROS: an open-source robot operating system. In: Proceedings of the IEEE International Conference on Robotics and Automation (ICRA) Workshop on Open Source Robotics, may 2009
20. Rohrich, R.F., Piardi, L., Lima, J.L., de Oliveira, A.S.: Bio-inspired distributed sensors to autonomous search of gas leak source. In: 2020 International Conference on Manipulation, Automation and Robotics at Small Scales (MARSS), pp. 1–6 (2020)
21. Ruiz-Sarmiento, J.-R., Baltanas, S.-F., Gonzalez-Jimenez, J.: Jupyter notebooks in undergraduate mobile robotics courses: educational tool and case study. Appl. Sci. **11**(3), 917 (2021)
22. Wada, Y., Trincavelli, M., Fukazawa, Y., Ishida, H.: Collecting a database for studying gas distribution mapping and gas source localization with mobile robots. In: The Abstracts of the International Conference on Advanced Mechatronics: Toward Evolutionary Fusion of IT and Mechatronics, ICAM 2010, vol. 5, pp. 183–188 (2010)
23. Waphare, S., Gharpure, D., Shaligram, A., Botre, B.: Implementation of 3-nose strategy in odor plume-tracking algorithm. In: 2010 International Conference on Signal Acquisition and Processing, pp. 337–341 (2010)

Learning from the Past: Sequential Deep Learning for Gas Distribution Mapping

Nicolas P. Winkler[1(✉)], Oleksandr Kotlyar[2], Erik Schaffernicht[2], Han Fan[2], Haruka Matsukura[3], Hiroshi Ishida[4], Patrick P. Neumann[1], and Achim J. Lilienthal[2]

[1] Bundesanstalt für Materialforschung und -prüfung (BAM), Unter den Eichen 87, 12205 Berlin, Germany
mro@bam.de

[2] Örebro University, Fakultetsgatan 1, 702 81 Örebro, Sweden

[3] University of Electro-Communications, 1-5-1 Chofugaoka, Chofu, Tokyo 182-8585, Japan

[4] Tokyo University of Agriculture and Technology, 2-24-16 Nakacho, Koganei, Tokyo 184-8588, Japan

Abstract. To better understand the dynamics in hazardous environments, gas distribution mapping aims to map the gas concentration levels of a specified area precisely. Sampling is typically carried out in a spatially sparse manner, either with a mobile robot or a sensor network and concentration values between known data points have to be interpolated. In this paper, we investigate sequential deep learning models that are able to map the gas distribution based on a multiple time step input from a sensor network. We propose a novel hybrid convolutional LSTM - transpose convolutional structure that we train with synthetic gas distribution data. Our results show that learning the spatial and temporal correlation of gas plume patterns outperforms a non-sequential neural network model.

Keywords: Gas distribution mapping · Spatial interpolation · Sequential learning · Convolutional LSTM

1 Introduction

Gas sensor networks are an established tool for monitoring of gas concentrations within desired areas, because they provide temporally dense measurements at specific locations. To have a precise understanding of the distribution of gases, we typically want to know about the distribution also in-between individual sensors.

Inspired by the advances of machine learning for image reconstruction [1] [2], we rephrase the task of spatial interpolation between sparsely distributed sensors as an image reconstruction problem. In [3], we were able to upscale the sparse input of a sensor network to a higher-resolution representation of the gas

D. Tardioli et al. (Eds.): ROBOT 2022, LNNS 590, pp. 178–188, 2023.
https://doi.org/10.1007/978-3-031-21062-4_15

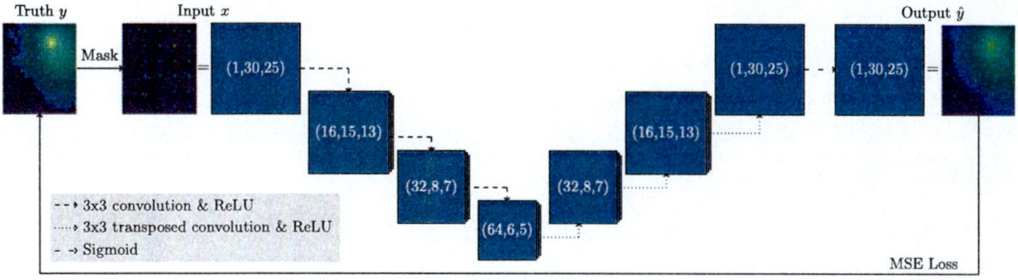

Fig. 1. Architecture of the neural network that we proposed in [3].

distribution. While that neural network model predicted the gas distribution based on measurements of a single time step, we are now investigating how a model could take advantage of multiple time steps as sequential input.

As wind is influencing the shape of a gas plume over time, a model that learns a spatio-temporal gas plume pattern should take advantage of multiple, temporally connected inputs. In this paper, we describe our first results and findings regarding a model, that is fed with a sequential input of constant sampling frequency, for gas distribution mapping.

2 Related Work

Various approaches for interpolating between sensors exist: Kernel DM+V [4] and Gaussian Markov Random Fields (GMRF) [5] are popular data-driven methods for gas distribution mapping in small to medium-sized areas. While the basic idea of both approaches is the interpolation of sparse data points, efforts are made to add more information and create more sophisticated predictions for both methods. In [5], not only a gas grid but also a wind grid is formulated as GMRF and correlated with physical laws of gas dispersal and fluid dynamics, encoded in different energy functions. Similarly, Kernel DM+V was extended with wind measurements in [6]. In [7], a time-dependent Kernel DM+V was proposed, to account for the temporal dynamics in gas distribution mapping (GDM). A different approach for the interpolation of sparse data points are Gaussian Processes [8], also known as Kriging [9]. Kriging is especially popularly used in environmental monitoring for large-scale areas.

Image inpainting [1] and *super-resolution* [2] are two popular tasks in computer vision, where machine learning is used to reconstruct a high-resolution image from a corrupted or low-resolution source image. These models are able to identify and recognize patterns in the source data as they were trained with full knowledge about the intended high-resolution output (supervised learning). In [3], we applied a super-resolution approach to GDM and successfully trained a model to reconstruct gas distributions based on the sparse input data of a sensor network. The structure of this previous encoder-decoder convolutional neural network is displayed in Fig. 1. While we could show that the encoder-decoder network outperformed basic interpolation techniques, we recognized that especially smaller plumes that are only recognized by a single sensor are difficult to predict. We also realized that our previous model is overly complicated and that network weights and therefore training time can be reduced by removing the encoder part.

3 Methodology

3.1 Data Generation

Training a neural network needs a vast amount of data, that is hard and sometimes impossible to obtain in real experiments. Therefore, we created synthetic data, based on measured wind data. We used the gas distribution simulator GADEN, which was developed for the Robotic Operating System (ROS) and enables reliable validation of gas-sensing robotic systems and algorithms [10]. Based on wind fields, GADEN calculates the gas transport with the filament-based atmospheric dispersion model proposed in [11]. This model is widely used in the research field of Mobile Robot Olfaction (MRO) because the model employs the Lagrangian approach resulting in a comparatively low computational load. A fixed number of gas molecules are treated as a single gas puff, which is called a filament. The transport and diffusion of such gas puffs with a given wind field is then calculated. Compared to calculating the gas diffusion and advection using a computational fluid solver, the computational cost is reduced by tracking only the center position and width of each filament.

To simulate realistic, synthetic gas distributions, we used real wind data measured in an outdoor environment during different seasons and years. For each season, we selected three sets of continuous wind data, each 330 s long. In total, we have twelve different wind sets across all four seasons.

To resemble the sensor network during previous experiments [12], we decided for an experiment area of $9\,\mathrm{m} \times 7.5\,\mathrm{m}$. Within this area, 30 different, evenly spaced saturated ethanol vapor positions were simulated. Only one source at a time was simulated, each source position being the size of $0.2\,\mathrm{m} \times 0.2\,\mathrm{m}$. The cell size of the simulation was set to $0.1\,\mathrm{m} \times 0.1\,\mathrm{m}$. To account for backflow of particles, the overall simulation area was $18\,\mathrm{m} \times 15\,\mathrm{m}$.

Although we have twelve different wind sets, each source was simulated with only four wind sets to save computation time. Different wind sets were used for neighboring source positions to avoid too similar plume structures. The simulation was calculated with a time step of 0.1 s. The gas concentration data was saved every 0.5 s. More information including snapshots of the simulated results can be found in [3].

3.2 Preparation of Data Sets

While each source position was simulated for 330 s, we removed the first 30 s to omit potentially undeveloped gas plumes. To ease understanding, we are going to use the term *image* for each snapshot that was saved during the simulation in a frequency of 2 Hz, i.e., every 0.5 s. As the changes between temporally neighboring images are very small, we decided to skip every second image, which leaves us with gas plumes saved at 1 Hz. With that, we obtained 36,000 individual images of gas plumes. We cropped each image down to the experiment area (which is smaller than the simulation area). This results in an image of 90 $\times$ 75 pixels. To save computation time, we sized all images down to 30 $\times$ 25 pixels using interpolation[1]. These images are the ground truth y of our gas distribution. The input x to the neural network shall mimic a spatially sparse sensor network with equally distributed $6 \times 5 = 30$ sensors. In order to obtain x, we directly extract the pixel values at each of the 30 sensor positions. With that, we ensure that x is not an averaged, downsized representation of y, but the point measurements of the sensor network.

For each source position, we allocated the first 70% for training and reserved the last 30% for validation. Because we are dealing with time-series data, few images of the validation data are connected with the temporally preceding images in the training data. The significance of this data leakage depends on the specific wind dynamics at the beginning of the validation set. Each pixel of the image contains the gas concentration in *parts per trillion* (ppt). To address problems with model training (slow or unstable training as well as exploding gradients problems), we scaled both data sets logarithmically, clamped the values between 0 and the maximum value of the training set, and min-max normalized both sets between 0 to 1.

Our training data set contains 25,200 images, generated in $4 \times 30 = 120$ simulation runs (seasons $\times$ source positions). To obtain sequential input for the neural networks, we split each of the 120 sub sets into samples containing k images, with k being the length of the network's sequential input. A sliding window of step size = 1 is used to split and group the images into the separate samples. Consequently, the total amount of training samples varies according to the sequence length k. For $k = 2$, we obtain 25,080 training samples, whereas we obtain 24,120 samples for $k = 10$.

To prevent the network from learning unwanted connections across sequences, we shuffled all samples after splitting and before training. Sequences *within* the samples remain in original order. In other words, we only look at a *stateless* case, where the network's memory shall not persist across sequences.

3.3 Neural Network Architectures

Base Model. We decided to use a *decoder* CNN as a base model for the following comparison, whose full architecture is displayed in Fig. 2. The decoder consists of multiple transposed convolution layers, through which the input image is stepwise up-scaled, i.e., interpolated. Starting from a single-channel 6 $\times$ 5 image (i.e., [1,6,5] for [channel, width, height]), we interpolate to an output image

[1] https://scikit-image.org/docs/stable/auto_examples/transform/plot_rescale.html.

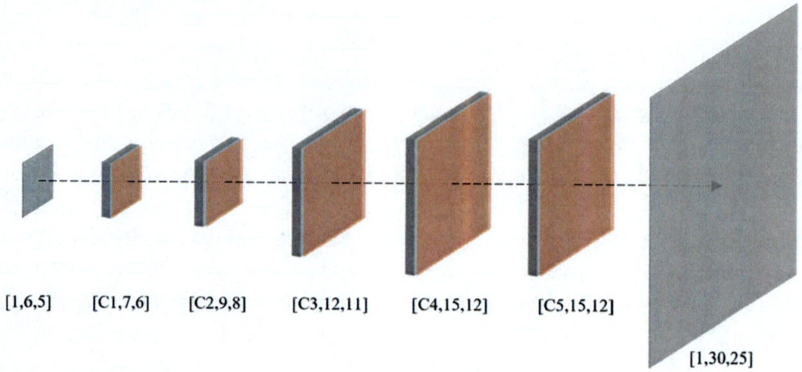

Fig. 2. Architecture of the non-sequential *decoder* CNN, which is used as base model for comparison with the sequential models. Gray layers denote the dimensions of the images, with transposed convolutions in-between them. Orange layers are rectifier activation functions (ReLU). The numbers describe the image's dimensions between the deconvolution layers: [channels, width, height]. C1 to C5 are optimized during training.

of size [1,30,25]. Between deconvolutions, rectifier functions are applied to zero negative values:

$$\mathrm{ReLU}(x) = x^+ = \max(0, x) \tag{1}$$

The height and width dimensions of the inner layers are chosen by hand and the channel dimensions are tuned by hyperparameter search during training. During prototyping experiments, we investigated different models with two to nine layers and recognized that the accuracy of the network mainly increases with additional layers. We assume that more and deeper layers enable the network to learn more abstract patterns of gas distributions. How neural networks build their understanding across layers was visualized by [13].

However, as computation cost also increases with more layers, we decided to use a six-layer network as a compromise between accuracy and cost.

Sequential Decoder. Our first sequential model is the base model, with the modification that not a single image (i.e., time step) is fed to the model, but a sequence of images (i.e., multiple time steps). The architecture is shown in Fig. 3. While CNN's are not typically used for sequential predictions, we deliberately use this architecture as a straight forward starting point from our non-sequential base model.

ConvLSTM-Decoder Architecture. We compare the Sequential Decoder with an architecture that is specifically designed for sequential predictions. In front of our decoder, we add a convolutional LSTM (ConvLSTM) block. ConvLSTM was proposed and fully described in [14]. A long short-term memory (LSTM) is a recurrent neural network, specifically designed to recognize sequence

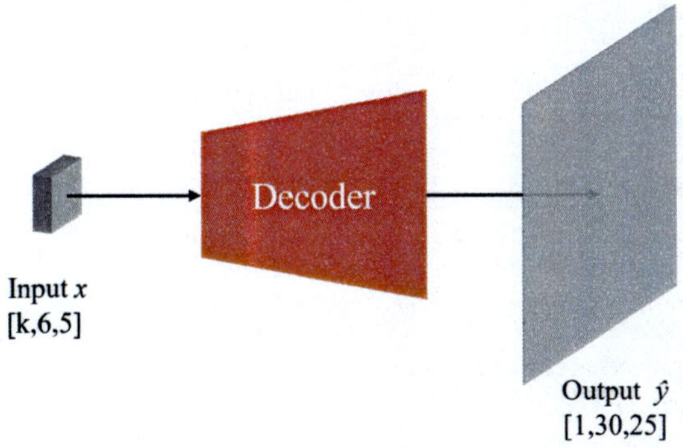

Fig. 3. Sequential decoder with an input sequence of time steps with length k.

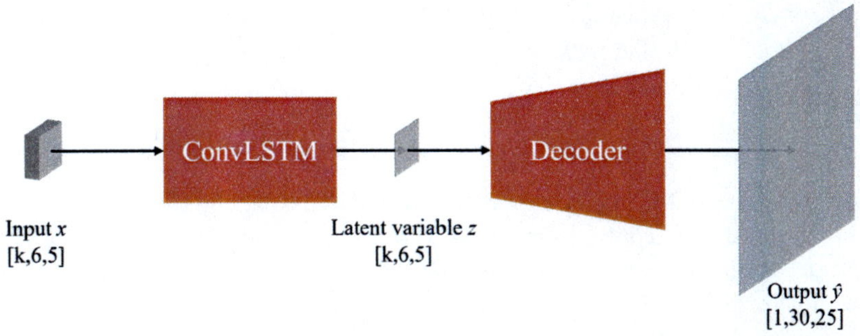

Fig. 4. ConvLSTM-Decoder architecture: The ConvLSTM is fed with the whole input sequence and its output is fed to the decoder.

patterns. In a nutshell, the ConvLSTM is a modified LSTM that adds a convolution operation and thus enables the processing of image sequences. Our ConvLSTM implementation consists of three layers, whose optimal sizes (i.e., number of hidden units) are tuned through hyperparameter search. It shall process the whole sequential input x and output a latent variable z. The decoder shall then relate the latent variable z to the true gas distribution y.

3.4 Model Implementation and Training

All models were implemented with the Python libraries PyTorch[2] and PyTorch Lightning[3]. We used PyTorch's `torch.nn.ConvTranspose2d` for transposed convolutions and `torch.nn.ReLU` for ReLU activations. As PyTorch is missing a native implementation of ConvLSTM, we relied on the implementation by Andrea Palazzi[4].

[2] https://pytorch.org/.
[3] https://www.pytorchlightning.ai/.
[4] https://github.com/ndrplz/ConvLSTM_pytorch.

Table 1. Hyperparameters that were tuned by Optuna.

Parameter	Tuning range	Description
seq_len	[2,10]	Input sequence length
inner_dims_decoder	[3,256]	Size of the decoder's layers
inner_dims_convlstm	[3,128]	Size of the ConvLSTM's layers
learning_rate	[1e-5,1e-2]	Learning rate of Adam optimizer

Table 2. Mean squared error (MSE) and sequence length k of the best model after 50 training epochs (100 trials).

Model	k (best)	MSE (best)
Base Model	1	4.64×10^{-3}
Sequential Decoder	9	4.13×10^{-3}
ConvLSTM-Decoder	6	3.99×10^{-3}
Kernel DM+V		1.46×10^{-2}

We further used Optuna[5] for automating hyperparameter search. In order to find optimal hyperparameters (see Table 1), we trained each model for 100 trials, i.e., with different sets of parameters. All models were trained with the Adam optimizer for 50 epochs on a Nvidia P4000 GPU. Data was loaded with 15 worker processes (`num_workers`) on a Intel Xeon Gold 5115 CPU.

During training, we minimized the mean squared error (MSE) between output $\hat{y}$ and true gas distribution y in batches of size $N = 30$:

$$MSE(y, \hat{y}) = \frac{1}{N} \sum_{i=0}^{N} (y - \hat{y})^2 \quad (2)$$

To compare the neural networks with the Kernel DM+V [4], we used a Python implementation of TD Kernel DM+V/W[6] with cell size $c = 1$ and kernel size $\sigma = 2.5 \times c$.

4 Results

MSEs for selected, best performing models are displayed in Table 2. The tuned, non-sequential base model achieves a MSE $= 4.64 \times 10^{-3}$. The best Sequential Decoder already performs better than this base model (MSE $= 4.13 \times 10^{-3}$). The ConvLSTM-Decoder model achieves even slightly better results (MSE $= 3.99 \times 10^{-3}$). We limited the range of sequence length k to 2 to 10. While the Sequential Decoder performs best with $k = 9$, the best ConvLSTM-Decoder we could train used an input of length $k = 6$.

5 https://optuna.org/.

6 https://gitlab.com/smueller18/TDKernelDMVW.

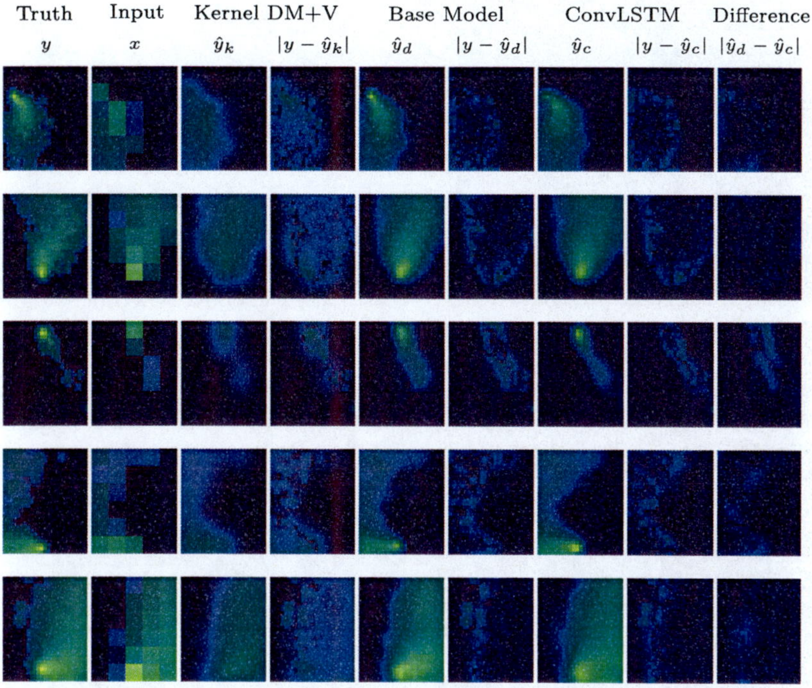

Fig. 5. Examples of gas distributions and their predictions of a base model, a ConvLSTM-Decoder model and a Kernel DM+V implementation. Differences between base model and ConvLSTM-Decoder model are highlighted in the last column.

5 Discussion

Example predictions of the ConvLSTM-Decoder model compared to the predictions of base model and Kernel DM+V are displayed in Fig. 5. Although the MSE attests a better performance, the visual differences between the base model and the sequential model are very small. As the MSE is calculated over the full image containing 750 pixels, minor differences between both models are barely distinguishable by eye. Nonetheless, we can show with our results that sequential information can enhance the predictions of gas distributions.

As we trained each model with 100 different sets of hyperparameters (see Table 1), we can understand the influence of sequence length k on the MSE. For the sequential decoder, displayed in Fig. 3, a correlation between sequence length and MSE is visible. The best models are models with sequence lengths $k >= 8$. As Optuna's search algorithms automatically focuses on most promising parameters, only few models with small k were trained.

In contrast, the ConvLSTM-Decoder model performs best with $k = 6$. A typical input sample of $k = 6$ is displayed in Fig. 7. It seems like additional input information does not improve the performance of ConvLSTM-Decoder, as displayed in Fig. 6b. Because splitting the training data yields more samples

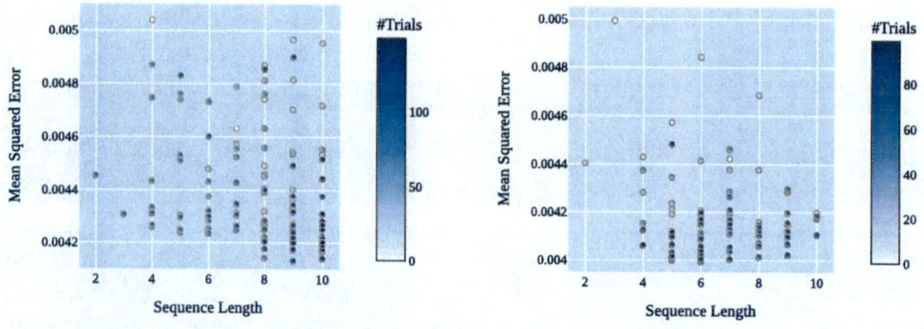

(a) Sequential Decoder

(b) ConvLSTM-Decoder

Fig. 6. MSE over sequence length k. Each circle represents one trial. As Optuna automatically selects and focuses most promising hyperparameters, the amount of models per sequence length differ. The darker the circle, the later a trial was trained. Very poorly performing trials are disregarded in the plots for better visibility.

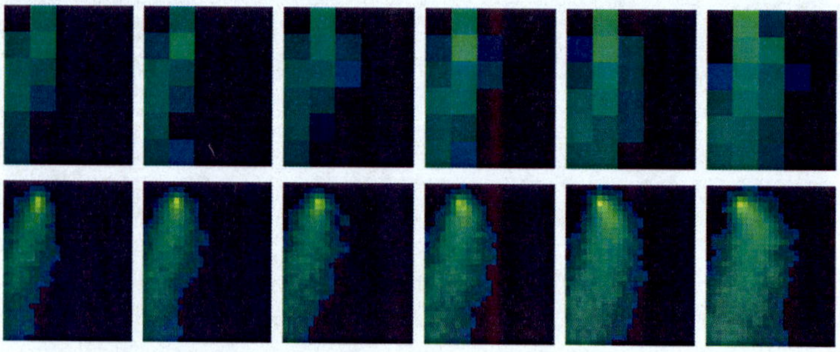

Fig. 7. Typical input sequences for a model with sequence length $k = 6$. The changes between images are often very small (with frequency 1 Hz).

for smaller k (see sect. 3.2), models with smaller k could be trained with more training samples. More training data typically leads to better performing models. However, the models with $k = 6$ were only trained with 2% more data ($n = 24,600$) compared to models with $k = 10$ ($n = 24,120$). The Sequential Decoder's results (see Fig. 6a) clearly show, that longer input sequences k contain useful information for the model. We assume that the ConvLSTM-Decoder model with $k > 6$ could not make use of this additional information, because the complexity of a LSTM grows with longer input sequences. The models with higher k are probably undertrained and therefore perform worse.

6 Conclusion and Outlook

We showed that a sequential model improve the prediction accuracy compared to a non-sequential decoder model. While the ConvLSTM-Decoder architecture performed best, the results of the Sequential Decoder are already very promising, especially when considering the smaller network complexity. Our results also highlight the importance of large data sets for training and evaluation of the model, because most of the ConvLSTM-Decoder models are probably undertrained. Our hyperparameter search emphasizes that care should be taken in optimizing models, especially when comparing different architectures with each other.

We also plan to create an independent test data set with out of domain samples, that enables more insightful evaluation than our current validation data set, that suffers from the mentioned temporal connection to the training data set. A test set, that is based on different weather conditions or untrained gas source positions, could prove the generalizability of the neural network models.

Apart from that, we want to further investigate how a neural network for gas distribution mapping could be learned with additional knowledge, for example with semantic information about the environment.

Acknowledgement. The authors would like to thank SAF€RA for funding the project RASEM, https://www.bam.de/Content/EN/Projects/RASEM/rasem.html and JSPS (KAKENHI Grant Number 22H04952) for funding part of this work.

References

1. Yu, J., Lin, Z., Yang, J., Shen, X., Lu, X., Huang, T.S.: Generative image inpainting with contextual attention. In: Proceedings of the IEEE Conference on Computer Vision and Pattern Recognition, pp. 5505–5514 (2018)
2. Dahl, R., Norouzi, M., Shlens, J.: Pixel recursive super resolution. In: Proceedings of the IEEE International Conference on Computer Vision, pp. 5439–5448 (2017)
3. Winkler, N.P., Matsukura, H., Neumann, P.P., Schaffernicht, E., Ishida, H., Lilienthal, A.J.: Super-resolution for gas distribution mapping: convolutional encoder-decoder network. In: 2022 IEEE International Symposium on Olfaction and Electronic Nose (ISOEN), pp. 1–3 (2022)
4. Lilienthal, A.J., Reggente, M., Trincavelli, M., Blanco, J.L., Gonzalez, J.: A statistical approach to gas distribution modelling with mobile robots - the Kernel DM+V algorithm. In: 2009 IEEE/RSJ International Conference on Intelligent Robots and Systems, (St. Louis, MO, USA), pp. 570–576. IEEE (2009)
5. Gongora, A., Monroy, J., Gonzalez-Jimenez, J.: Joint estimation of gas and wind maps for fast-response applications. Appl. Math. Model. **87**, 655–674 (2020)
6. Reggente, M., Lilienthal, A.J.: The 3d-kernel dm+ v/w algorithm: using wind information in three dimensional gas distribution modelling with a mobile robot. In: SENSORS, pp. 999–1004. IEEE (2010)
7. Asadi, S., Fan, H., Bennetts, V.H., Lilienthal, A.J.: Time-dependent gas distribution modelling. Robot. Auton. Syst. **96**, 157–170 (2017)
8. Stachniss, C., Plagemann, C., Lilienthal, A.J.: Learning gas distribution models using sparse Gaussian process mixtures. Auton. Robot. **26**, 187–202 (2009). https://doi.org/10.1007/s10514-009-9111-5
9. Oliver, M.A., Webster, R.: Kriging: a method of interpolation for geographical information systems. Int. J. Geogr. Inf. Syst. **4**, 313–332 (1990)
10. Monroy, J., Hernandez-Bennetts, V., Fan, H., Lilienthal, A., Gonzalez-Jimenez, J.: Gaden: a 3d gas dispersion simulator for mobile robot olfaction in realistic environments. Sensors **17**(7), 1479 (2017)

11. Farrell, J.A., Murlis, J., Long, X., Li, W., Cardé, R.T.: Filament-based atmospheric dispersion model to achieve short time-scale structure of odor plumes. Environ. Fluid Mech. **2**, 143–169 (2002). https://doi.org/10.1023/A:1016283702837
12. Bilgera, C., Yamamoto, A., Sawano, M., Matsukura, H., Ishida, H.: Application of convolutional long short-term memory neural networks to signals collected from a sensor network for autonomous gas source localization in outdoor environments. Sensors **18**, 4484 (2018)
13. Olah, C., Mordvintsev, A., Schubert, L.: Feature visualization. Distill (2017). https://distill.pub/2017/feature-visualization
14. Shi, X., Chen, Z., Wang, H., Yeung, D.Y., Wong, W.K., Woo, W.C.: Convolutional lstm network: a machine learning approach for precipitation nowcasting. In: Advances in Neural Information Processing Systems, vol. 28 (2015)

A Novel Odor Source Localization Method via a Deep Neural Network-Based Odor Compass

Zheng Yan, Tao Jing, Si-Wen Chen, Meh Jabeen, and Qing-Hao Meng(✉)

Institute of Robotics and Autonomous Systems, Tianjin Key Laboratory of Process Detection and Control, School of Electrical and Information Engineering, Tianjin University, Tianjin 300072, China
qh_meng@tju.edu.cn

Abstract. Mobile robot-based odor source localization (OSL) has broad applications in industrial and daily-life scenarios. However, subject to the limited sensing capacity of common metal oxide semiconductor (MOS) sensors, the OSL robots still lag far behind their biological counterparts. In this paper, we rethink the odor-source direction estimation paradigm of odor compass and propose a deep neural network (DNN) based method to improve both the accuracy and the generalization ability. The odor compass is composed of four wireless MOS sensors, and a DNN model, which contains a convolutional neural network (CNN) module and a long short-term memory (LSTM) module. An OSL strategy is further designed based on the proposed odor compass. Experimental results validate the feasibility of the proposed method.

Keywords: Mobile robot · Odor source localization · Odor compass · Signal processing · Deep learning

1 Introduction

As the economy progresses, hazardous gas leakages impose increasing risks on society and industry, making odor source localization (OSL) a more urgent task [1]. Mobile robot OSL, which employs robots equipped with gas sensors for autonomous source searching, has gained great attention in recent years [2–5]. However, due to the limited sensing ability of common gas sensors, mainly the metal oxide semiconductor (MOS) sensors, there is still a massive gap between robot OSL and biological olfactory searching ability. Therefore, some studies focus on the research that digs up more informative information from the gas sensor response, such as odor-source distance and odor-source direction. Among these researches, estimating odor-source direction based on in-situ sensor responses, termed as odor compass, is a vital topic for its direct benefits in OSL applications. Existing odor compass researches [6, 7] rely either on specially-designed mechanical structures or hand-crafted features, suffering from relatively limited accuracy and generalization capability. Thus, models with stronger feature representation and generalization capability are required for a more practical odor compass.

D. Tardioli et al. (Eds.): ROBOT 2022, LNNS 590, pp. 189–200, 2023.
https://doi.org/10.1007/978-3-031-21062-4_16

To address this hurdle, we rethink the paradigm of odor compass and propose a deep learning-based odor compass for mobile robot OSL. The odor compass includes two parts: a MOS sensor array to acquire odor responses, and a deep neural network (DNN) based model to estimate odor-source direction according to the responses.

The MOS sensor array is deployed on a mobile robot, which wirelessly transmits the sensor responses to the remote host computer; in return, the host computer feeds the estimated odor-source direction back to the robot for autonomous OSL. We introduce the DNN model for sensor response processing to enhance feature extraction ability, which outperforms hand-craft features in terms of accuracy and generalization. The DNN model consists of two modules, a CNN module for robot orientation (w.r.t the odor source) classification, and an LSTM module for odor-source direction regression. To train and test the DNN model, we conduct experiments to collect an odor-source direction estimation dataset.

Our main contribution includes: (1) Introducing DNN to the design of odor compass to improve accuracy and generalization; (2) Building an odor-source direction estimation dataset through experiments; (3) Putting forward an OSL strategy based on the proposed odor compass.

The rest of this paper is organized as follows. Section 2 briefly introduces related works on odor compass. The overall system design and hardware platform are stated in Sect. 3. The structure and validation of the DNN model are presented in Sect. 4. Section 5 shows the OSL strategy and OSL experiment results, followed by the conclusion in Sect. 6.

2 Related Works

During the development of OSL research for mobile robots, the estimation of odor-source direction by designing sensors and signal processing algorithms is an important element. Odor compass, the most commonly used term to describe odor-source direction prediction based on local odor sensing, refers to a device that can determine the direction from which odor patches come. The early research on odor compass [6, 7] mainly focuses on the biologically inspired design of the sensing module, which depends on intricate mechanical structures to improve performance.

Subsequently, researchers concentrate more on sensor signal analysis to extract intricate transient features for odor-source-direction estimation. Wei *et al.* [8] proposed a portable odor compass, which mainly consists of three gas sensors placed at the vertexes of a regular compact triangle. A data processing method is proposed that uses the scale-space method to obtain multi-scale information from the gas sensor response and the feature-point matching method to calculate the time difference of the odor arrivals for the odor-source-direction calculation. Luo *et al.* [9] improved the odor-source-direction inference method in [8] by applying the continuous wavelet transform to identify the arrival and departure events of odor patches, the time difference of these events is the principle for calculating the odor flow direction. However, these methods, dependent on hand-craft feature design, could only provide a rough inference of the odor flow direction, instead of accurate and detailed odor-source-direction prediction.

3 System Design

The design paradigm of odor compass generally includes signal preprocessing and extracting features to calculate the odor flow direction [10]. Odor response is a time series signal, while deep-learning models represented by recurrent neural networks have outperformed standard artificial feature approaches in time series processing, such as speech recognition [11], language modeling [12], and translation [13]. Therefore, the DNN can be used for odor-source direction estimation to improve the accuracy and generalization ability. We design a wireless MOS sensor array assisted by a remote computer that processes the signal based on deep learning algorithms to implement the DNN-based odor compass. And we present a robotic autonomous OSL system, aiming to achieve autonomous gas tracking and OSL for robots.

3.1 Overall Structure

The overall structure of the proposed odor compass and the mobile robot-based OSL system is illustrated in Fig. 1.

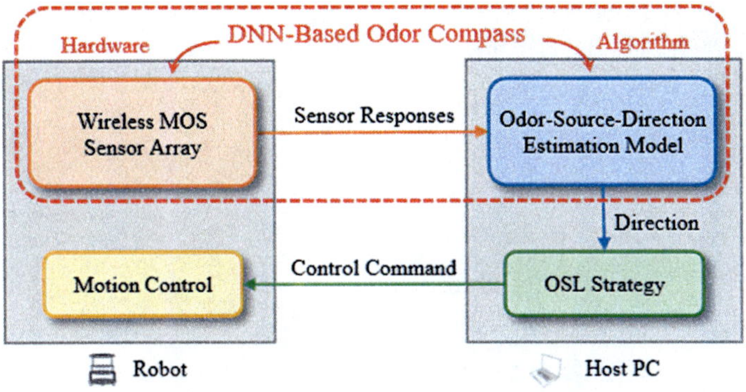

Fig. 1. Schematic diagram of the proposed OSL system.

The system is comprised of two parts: the mobile robot and the host PC. The mobile robot contains two function modules: a wireless MOS sensor array mounted on the top, which transmits gas concentration data to the host computer in real-time via wireless communication; a motion control model executes control commands from the host computer. The remote host PC leverages the DNN-based odor compass model (see Sect. 4 for details) to process the sensor array responses in real-time to estimate the odor-source direction (angle between the principal axis of the robot and the line from the odor source to the robot center, defined in Fig. 2), which is used as an input for the OSL strategy.

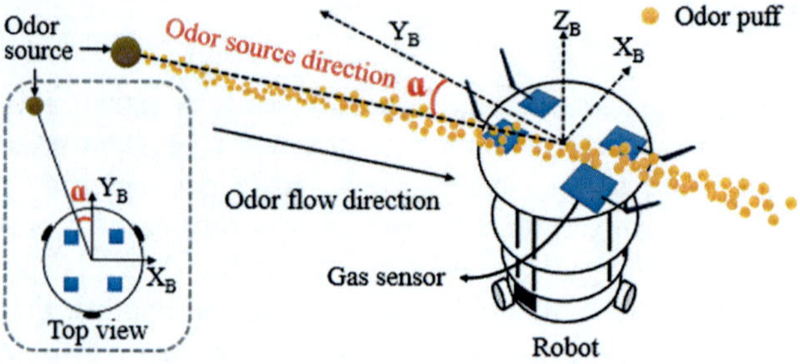

Fig. 2. The body system and odor-source-direction α, where the body frame is denoted by B.

3.2 Hardware Platform

A three-wheeled omnidirectional mobile robot shown in Fig. 3(a) is used in this research. As a vehicle for validating our approach, the robot requires no navigation system, merely a locomotion system and an odor compass. When constructing the odor compass (i.e., the wireless MOS sensor array), the number of sensors, the distance between sensors, and the overall topology are issues that need to be considered. Considering that this optimization problem is beyond the research scope of this paper, according to the previous experience of our team [8, 9] and the specific size of the robot used in this research, after many simulations and experimental tests, we utilized four MOS sensors which are arranged in a square with a side length of 0.1 m on the top platform of the robot, as illustrated in Fig. 3(b).

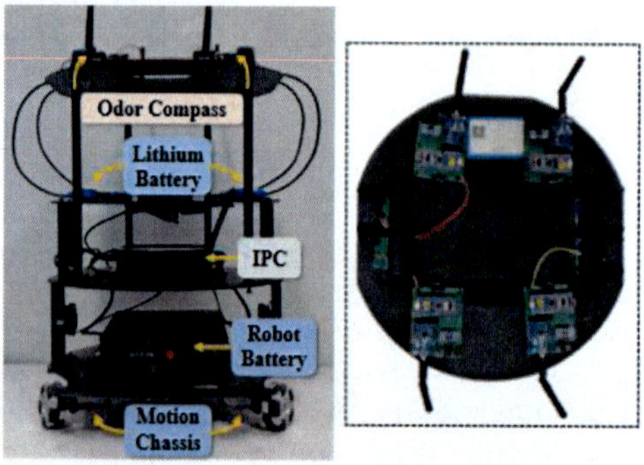

(a) Side view. (b) Top view.

Fig. 3. The robot equipped with an odor compass.

As shown in Fig. 3(b), the wireless MOS sensor array is composed of four identical modules, and each module consists of an STM32F103 microprocessor (STMicroelectronics Inc.), an SGP30 MOS sensor (Sensirion AG), a ZigBee terminal, and signal conditioning circuits, see Fig. 4(a). The microprocessor is distinctive in that it allows continuous 16-bit resolution data recording from the MOS sensor at a sampling rate of

10 Hz. The MOS sensor, the ZigBee terminal and the microprocessor are powered via the voltage regulator supported by a lithium battery.

Figure 4(b) shows the real image of one sensor module. The odor data sampled by the wireless MOS sensor array is sent to the remote host computer via a ZigBee coordinator, as shown in Fig. 4(c).

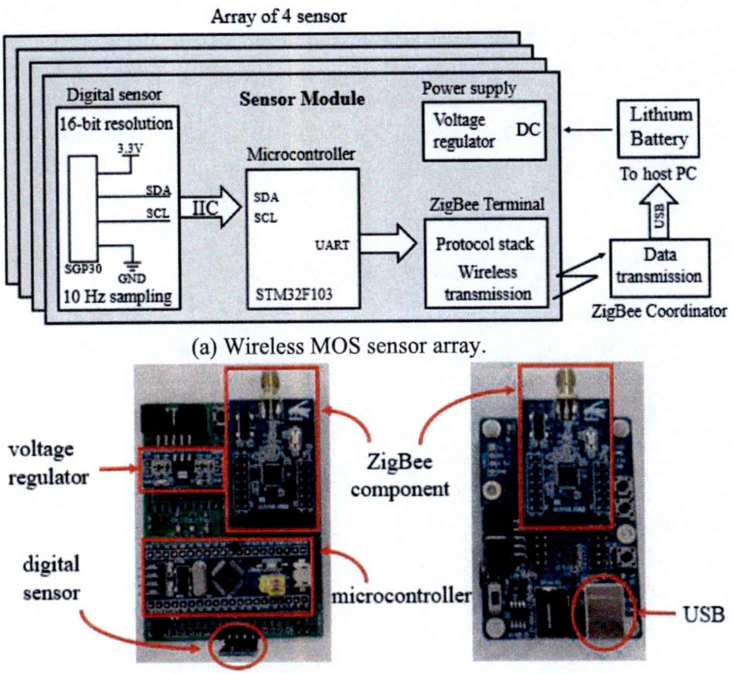

(a) Wireless MOS sensor array.

(b) Sensor module. (c) ZigBee coordinator.

Fig. 4. The block diagram and view of the custom-designed wireless MOS sensor array.

4 DNN-Based Odor Compass Model

We build the odor compass model out of common DNN bricks, including one-dimensional CNN layers and LSTM layers, making an end-to-end odor-source direction model. The model is trained and tested on experiment samples, with the four MOS sensor responses as data and the corresponding odor-source direction as the label.

4.1 Model Design

Estimating the odor-source direction is in fact regressing the angle label according to the data response. Due to the periodicity of angle, it's difficult to directly regress the odor-source direction within 0°–360°: 0° and 360° indicate the same direction but have different values. This peculiarity makes it troublesome to select a proper loss function for angle regression. Therefore, we introduce an auxiliary subtask: orientation classification before direction regression, which determines whether the robot is facing towards or away from the odor source, thus simplifying the angle regression range into −90°–90°.

As a result, the odor-source-direction estimation in this work is divided into two subtasks, orientation classification and direction regression, which are respectively implemented by CNN and LSTM, as shown in Fig. 5(a). The first subtask is to judge whether the robot is facing towards or away from the odor source based on a CNN network. If the robot is facing away from the odor source, the input sequence will be rearranged (as illustrated in Fig. 5(b)), and the robot will rotate 180° in place; otherwise, no operation is required. After the first subtask, the odor source is likely to be within ±90° in front of the robot. The second subtask is to regress the detailed odor-source direction based on an LSTM network by learning the long dependencies via the gate structure with selective passage of information [14].

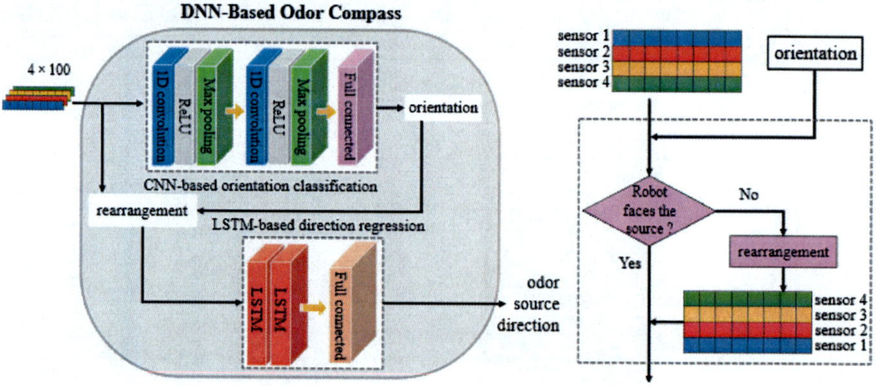

(a) Odor-source-direction estimation process (b) Rearrangement operation.

Fig. 5. Schematic of the odor-source-direction estimation model.

The CNN network consists of two one-dimensional convolutional layers and one fully connected layer, including a total of 3.8K parameters. The LSTM network uses two LSTM layers to generate hidden states followed by a fully connected layer to estimate the odor-source direction, which includes a total of 29.8K parameters.

4.2 Data Collection

Figure 6(a) depicts the indoor environment for data collection. A fan with a fixed heading was used to produce airflow, and an odor source (a humidifier with 75% alcohol) was placed on the downwind side, while the robot with the wireless MOS sensor array was on the right. The data samples were collected under four different experiment conditions, including two fan speeds (2000 RPMs and 5000 RPMs) and two release rates (0.8 L/h and 1.6 L/h). For each condition, the odor responses were collected at 6 different locations, as shown by the red circles in Fig. 6(a). At each location, the robot rotated one turn at an interval of 15° in place (as illustrated in Fig. 6(b)) to collect data, yielding 24 data samples at each location. As a result, a total of 4 × 6 × 24 = 576 data samples were acquired.

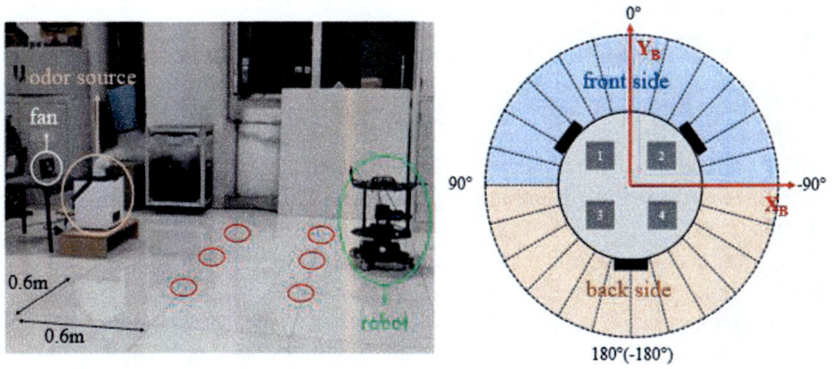

(a) Data-collection environment site. (b) Schematic diagram of robot rotation.

Fig. 6. Data-collection environment. (a) shows the red circle where the robot should be placed. (b) is a schematic of 24 directions, each covering 15°.

The steps of data acquisition are as follows. First, the robot moves to one of the six fixed positions, with one of the poses indicated in Fig. 6(b). Then, the fan is turned on and rotates at a specific speed to establish the sensor response baseline for exactly 10 s before releasing the alcohol. Next, the humidifier is turned on to release the alcohol into the environment, and the sensor array response is recorded at a sampling rate of 10 Hz. After 250 s, the humidifier is turned off and data sampling is stopped. Finally, the experimental area is ventilated for 3 min using the fan at maximum speed to completely clean the remaining alcohol gas for the next experiment. Each data acquisition cycle (i.e., baseline recording, gas exposure, and cleaning phases) lasts approximately 440 s. The resulting dataset has 576 time-series recordings—each recording contains response sequences from four sensors.

In addition, we sliced the raw long time series into fragments to meet the input length requirements (i.e., 100 sampling points for 10 s) for the odor-source-direction estimation model. We employed a sliding window to generate time fragments (4 × 100) from the original sensor response sequences (4 × 2500), where the sliding step of the window is S = 30 (corresponding to 3 s) and the window length is L = 100 (corresponding to 10 s). The number of samples in the dataset after slicing is 576 × {[(2500 − L)/S] + 1}, i.e., 46656 samples, where 576 is the number of the original recordings. Then the dataset is randomly separated into the training and test datasets by a ratio of 4:1, with 37325 samples for training and 9331 samples for testing.

The sensor response values are normalized to [0, 1] according to Eq. (1):

$$\widetilde{R} = R/Max(R) \tag{1}$$

where, R is the original sensor response value, $Max(R)$ indicates the maximum value of every original sequence and $\widetilde{R}$ stands for the normalized value. The samples are labeled with the orientation of the robot (facing towards or away from) and the odor-source direction, which are used for training the LSTM and the CNN, respectively.

4.3 Training and Validation Test

Network construction was implemented using the Paddle framework, and network training and validation were performed with NVIDIA GeForce GTX 1650 graphics cards. The odor-source-direction estimation model was trained and validated via datasets generated in Sect. 4.2. The cross-entropy and mean-square error (MSE) were respectively used as the loss functions for the CNN and LSTM networks, both of which were optimized using the Adam optimizer. The dynamic learning rate was set during the training process to accelerate the model convergence: If the model's performance does not improve after 10 epochs, the learning rate was halved to continue; After adjusting the learning rate twice and no improvement in performance, the model training would be terminated and the best model parameters were saved. The training epoch count was set to 250.

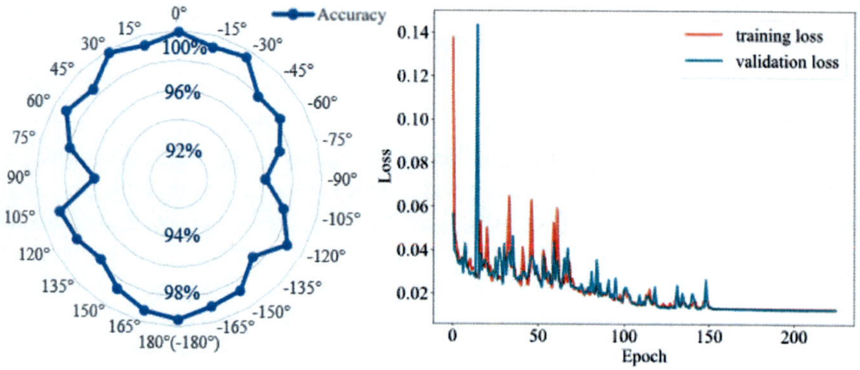

(a) CNN result. (b) LSTM result

Fig. 7. Results of odor-source-direction estimation model.

Figure 7(a) indicates the accuracy of judging whether the robot is moving towards or away from the odor source via the CNN network, the average accuracy is 98.4%, with the lowest accuracy (96%) when the odor-source direction is within the ranges of (−105°, −75°) and (75°, 105°). Figure 7(b) presents the LSTM network's training and validation loss curves, and the smallest average estimation error of the odor-source direction in the validation dataset is 4.9°. Our DNN-based odor compass model performs better compared to [8] and [9], where the deviations of the odor-source-direction estimation remain within the ranges of ±30° and ±15°, respectively.

5 OSL Strategy and Experiment

An OSL strategy was proposed based on the odor compass designed in this paper, and indoor OSL experiments were devised to verify the practical effect of the robot OSL method.

5.1 OSL Strategy

The robot follows the flowchart illustrated in Fig. 8 to locate the odor source autonomously. The tracing process begins with the robot starting to sample for 10 s, meanwhile, the robot is required to remain static to avoid fake wind interference caused by relative motion.

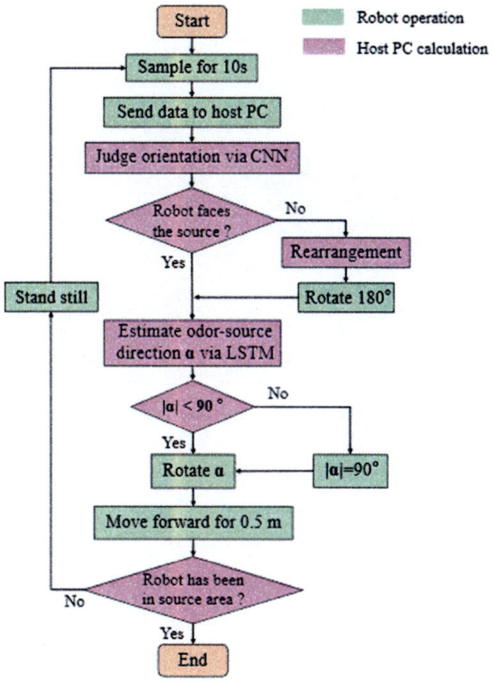

Fig. 8. Flowchart of the OSL strategy.

The operation cycle of the OSL system with the DNN-based odor compass is as follows:

1) The robot samples the gas signal for 10 s via the odor compass and sends the sensor-array data to the remote host computer for calculation.
2) After judging the robot's orientation via the CNN network, the remote host computer keeps the robot static if it is moving toward the odor source, otherwise rotates the robot 180° and rearranges the sensor array response.

3) The remote host computer rotates the robot according to the odor-source-direction (positive indicates clockwise and negative indicates counterclockwise) estimated via the LSTM network. Considering the efficiency of OSL, the robot rotates only 90° even if the absolute value of the estimated odor-source-direction is greater than 90°.
4) The robot moves forward for 0.5 m.

5.2 OSL Experiment

Figure 9 depicts the 5.0 m × 3.6 m experimental environment. The sensor response data sampled from the robot was processed in the remote host computer. When the robot reaches the circular area (named as source area) with the odor source as the center and a radius of 0.5 m, the robot will be stopped manually to avoid collision with the humidifier. The OSL is regarded success as long as the robot reaches the source area within a specified time range. Otherwise, failure.

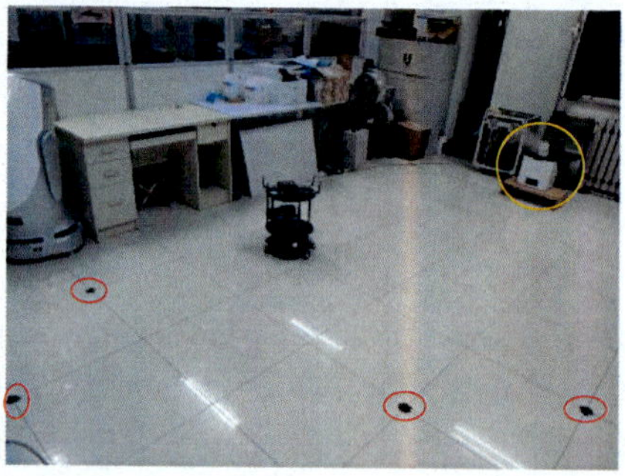

Fig. 9. Indoor OSL environment, where the size of each ceramic tile is 0.6 m × 0.6 m. The gold circle indicates the fan and odor source, and the red circles denote the starting points of the robot. The odor source is a humidifier containing 75% alcohol, with a release rate of 1.6 L/h, and the fan runs at a speed of 2000 RPMs without shaking its head.

The experimental results are presented in Fig. 10, where the ten solid circles in different colors indicate the locations at which the robot pauses during the ten OSL process, and the four hollow circles stand for the four different starting locations of the robot.

Of these 10 runs, only one failed. The ratio E of the robot's actual movement distance to the straight-line distance from the starting point to the odor source and the final distance D between the robot's final stopping point and the odor source are presented in Table 1.

The mean E value of the nine successful experiments was 1.26, with the worst-case 1.48. The OSL experiment shows that the relative position between the starting point and the odor source has almost no effect on the tracking performance. Our team has previously performed OSL experiments in the same room (similar scenarios) based on odor source direction estimation methods from the literature [8] and [9]. Respectively, the mean E value of the successful experiments was 1.28/1.53, with the worst case 1.57/2.57.

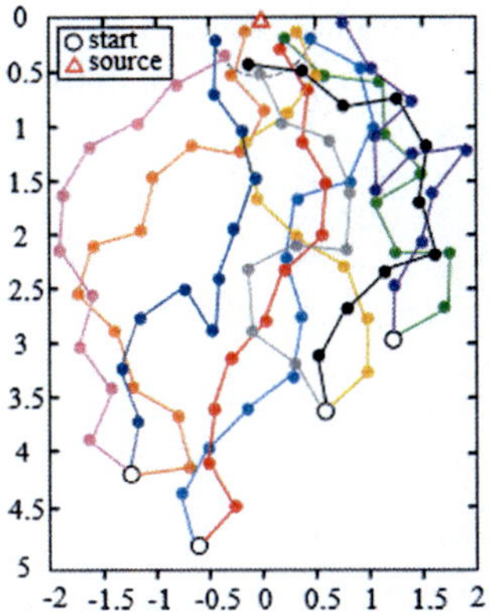

Fig. 10. Experimental results for the DNN-based OSL in an indoor environment.

Table 1. Experimental estimation for indoor OSL.

No.	E	D (m)
1	1.14	0.48
2	1.25	0.49
3	1.48	0.15
4	1.13	0.48
5	1.13	0.23
6	1.23	0.47
7	1.36	0.44
8	1.23	0.32
9	–	–
10	1.39	0.2

Overall, the results preliminarily demonstrate the feasibility of the DNN-based robotic autonomous OSL method. However, this methodology still needs to be tested for its effectiveness in large indoor scenes involving obstacles and further in outdoor windy environments, which would have been planned as a follow-up in our future work.

6 Conclusion

This work presents a robot OSL method based on a DNN-based odor compass and a derived OSL strategy. The odor compass uses a wireless gas sensor array and a deep neural network-based algorithm to estimate the direction from which the odor flows. Based on the training and validation datasets acquired in a real indoor gas diffusion environment, the DNN-based odor compass can accomplish the odor-source-direction estimation with an error of 4.9°. Indoor experiments preliminarily demonstrate the feasibility of the proposed DNN-based odor compass method for robot odor source localization. The robotic autonomous OSL method preliminarily proves effective in indoor environments, with promising results in OSL experiments. In future work, we will improve the odor-source-direction estimation model and OSL strategy, as well as test this method in outdoor windy environments.

References

1. Baetz, W., Kroll, A., Bonow G.: Mobile robots with active IR-optical sensing for remote gas detection and source localization. In: 2009 IEEE International Conference on Robotics and Automation, pp. 2773–2778. IEEE (2009)
2. Ishida, H., Suetsugu, K., Nakamoto, T., et al.: Study of autonomous mobile sensing system for localization of odor source using gas sensors and anemometric sensors. Sens. Actuators A **45**(2), 153–157 (1994)
3. Russell, R.A., Thiel, D., Deveza, R., et al.: A robotic system to locate hazardous chemical leaks. In: Proceedings of International Conference on Robotics and Automation, pp. 556–561, May 1995
4. Consi, T.R., Atema, J., Goudey, C.A., et al.: AUV guidance with chemical signals. In: Proceedings of IEEE Symposium on Autonomous Underwater Vehicle Technology (AUV 1994), pp. 450–455. IEEE (1994)
5. Sandini, G., Lucarini, G., Varoli, M.: Gradient driven self-organizing systems. In: Proceedings of 1993 IEEE/RSJ International Conference on Intelligent Robots and Systems (IROS 1993), vol. 1, pp. 429–432. IEEE (1993)
6. Nakamoto, T., Ishida, H., Moriizumi, T.: An odor compass for localizing an odor source. Sens. Actuators B Chem. **35**(1–3), 32–36 (1996)
7. Ishida, H., Kobayashi, A., Nakamoto, T., et al.: Three-dimensional odor compass. IEEE Trans. Robot. Autom. **15**(2), 251–257 (1999)
8. Wei, Y.T., Meng, Q.H., Jing, Y.Q., et al.: A portable odor-tracing instrument. IEEE Trans. Instrum. Meas. **65**(3), 631–642 (2015)
9. Luo, B., Meng, Q.H., Wang, J.Y., et al.: A flying odor compass to autonomously locate the gas source. IEEE Trans. Instrum. Meas. **67**(1), 137–149 (2017)
10. Jing, T., Meng, Q.H., Ishida, H.: Recent progress and trend of robot odor source localization. IEEJ Trans. Electr. Electron. Eng. **16**(7), 938–953 (2021)
11. Malik, M., Malik, M.K., Mehmood, K., Makhdoom, I.: Automatic speech recognition: a survey. Multimedia Tools Appl. **80**(6), 9411–9457 (2020). https://doi.org/10.1007/s11042-020-10073-7
12. Le, T.H.M., Chen, H., Babar, M.A.: Deep learning for source code modeling and generation: Models, applications, and challenges. ACM Comput. Surv. (CSUR) **53**(3), 1–38 (2020)
13. Datta, D., David, P.E., Mittal, D., et al.: Neural machine translation using recurrent neural network. Int. J. Eng. Adv. Technol. **9**(4), 1395–1400 (2020)
14. Sherstinsky, A.: Fundamentals of recurrent neural network (RNN) and long short-term memory (LSTM) network. Phys. D **404**, 132306 (2020)

Multi-robot for Industrial Applications

Modeling and Simulation of a Crossdocking System with an Integrated AS/RS

José Alves[1], Manuel F. Silva[1,2(✉)], and Fillipe Ribeiro[3]

[1] School of Engineering, Polytechnic Institute of Porto, Rua Dr. António Bernardino de Almeida, 431 Porto, Portugal
mss@isep.ipp.pt
[2] INESC TEC, Rua Dr. Roberto Frias, 4200-465 Porto, Portugal
[3] JPM Industry, Zona Industrial do Rossio, 3731-901 Vale de Cambra, Portugal
http://ave.dee.isep.ipp.pt/~mss/

Abstract. Logistics chains are being increasingly developed due to several factors, among which the exponential growth of e-commerce. Crossdocking is a logistics strategy used by several companies from varied economic sectors, applied in warehouses and distribution centres. In this context, it is the objective of the "CrossLog – Automatic Mixed-Palletizing for Crossdocking Logistics Centers" Project, to investigate and study an automated and collaborative crossdocking system, capable of moving and managing the flow of products within the warehouse in the fastest and safest way. In its scope, this paper describes the modelling and simulation of a crossdocking system, with an integrated AS/RS, to analyse possible alternatives including not only the fast movement of products, but also its storage in case needed. Different scenarios were modelled and simulated, on the FlexSim software, and the obtained results for each one were critically analysed to draw conclusions on the best storage policy.

Keywords: Crossdocking · Automated storage · Modelling · Simulation

1 Introduction

Logistics chains are increasingly developed due to several factors, among which the exponential growth of e-commerce stands out. This type of commerce drives a development in storage, distribution and logistics operations at high speed. All loads go through various collection, storage, consolidation and dispatch processes before being delivered to their final destination. Around this set of processes, competition with other companies present in the market increases, and is mainly focused on shortening delivery times, using automated state-of-the-art technology and efficient processes with regard to cargo handling.

To facilitate the management of all these processes, multiple systems have been created and adapted, as requirements and needs have developed, such as crossdocking systems (CS). The objective of CS is to facilitate products' distribution while, at the same time, reducing storage and transportation costs.

D. Tardioli et al. (Eds.): ROBOT 2022, LNNS 590, pp. 203–216, 2023.
https://doi.org/10.1007/978-3-031-21062-4_17

However, it is not always possible to send the goods to the destination without storage them first. Sometimes, it is necessary to use temporary storage, preferably less than 24 h. Although not eliminating the costs, this can reduce them substantially. Keeping the focus on current and developing technologies, one of the key storage systems in the market are the Automated Storage and Retrieval Systems (AS/RS). Both crossdocking and AS/RS systems have multiple design, operation and performance policies associated with them. To facilitate the perception of the operation and possible implementations to improve process efficiency (such as layouts, distribution methods, among others), digital 3D simulation is an extremely relevant tool in anticipating logistics operations with a view to improving efficiency. In fast cargo handling systems, such as CS, simulation is crucial, as well as in the implementation of automated warehouses such as AS/RS.

Bearing these ideas in mind, this paper presents the development of a 3D simulation model considering several policies and concepts associated with crossdocking and automated storage of goods. Following this introduction, Sect. 2 summarises the various policies and concepts addressed and applied in the model, and Sect. 3 presents the model architecture. Next, Sect. 4 discusses the implementation of the model in the FlexSim simulation software, and Sect. 5 presents the simulated scenarios, and corresponding results. Finally, Sect. 6 exposes the conclusions drawn from this simulation study.

2 Policies and Concepts

2.1 Crossdocking

Crossdocking is a logistics strategy used by companies from varied economic sectors, applied in warehouses and distribution centres [1,10]. Its objective is to receive products from the inbound trucks and send them directly to the outbound trucks, without any intermediate storage, thus increasing the efficiency of the operations. Incoming vehicles usually transport products from a certain supplier, which are placed in a CS and consolidated with products from another vehicle, and possibly another supplier. This consolidation allows products from different origins to be grouped together and distributed to different final destinations. This strategy has proven to be extremely important, and the fast movement of products allows the stock to be reduced, reducing the associated costs. Thus, crossdocking has provided distribution companies with several advantages, such as reduced storage, stock and handling costs, reduced delivery time, improved customer service, reduced storage space and reduced risk of loss. The key to success is to have the shortest possible dwell time at the product receiving and shipping facilities. The shorter this period the less temporary storage time is required. In practice, the ideal is rarely the case, so limited temporary storage is necessary and minimised as much as possible, for example by means of AS/RS.

Crossdocking characteristics can be divided into three types: physical, tactical and goods flow characteristics [6,7,10]. Physical characteristics are associated with the number of existing docks, the shape of the system, and the internal transportation of the products. The tactical characteristics are related to the tactical decisions taken about the operation of the system, and four characteristics stand out: the mode of service, pre-emption, the temporary storage capacity, and the internal resource capacity. As for temporary storage capacity, when a vehicle is not ready to pick up and transport the products to their destination, they are stored in a limited area, usually on the warehouse floor. Something common in the most evolved CS is the integration of AS/RS, allowing more efficiency and preventing possible storage inconveniences and space restrictions. The goods flow characteristics, or operational characteristics, are not always controlled by crossdocking, due to the influence of external factors to it, namely customers, suppliers and even distribution companies. The movement of goods is defined according to parameters such as arrival and departure time and the interchangeability of products.

It is also possible to distinguish CS through the number of touches or existing phases [4,7,10]. Number of touches means the times that the product is transported from one place to another, inside the system, and CS can be divided into three types: one-touch, two-touch (or one-stage) and multi-touch (or two-stage) crossdocking, and any one of these will have a direct bearing on the operation and layout of the system. One-touch crossdocking is considered "pure", because the ideal concept of this type of system is to receive the products and move them directly to the defined shipping vehicle, without any kind of intermediate storage. The products are only touched once, in the process of moving between the inbound and outbound vehicles. Two-touch or single-stage crossdocking, as the name implies, refers to two processes. In this type of system, products are received and placed in a specific location, known as "picking area". This process is considered the first touch, while the second touch refers to moving them to the outbound vehicle. Multi-touch or two-stage crossdocking is usually associated with large distribution centres. The first touch occurs when the products are received and placed in an area destined for that purpose, and then they will be conditioned, where the second touch takes place. Once this process is complete, they are moved and loaded onto the outbound vehicle, where the third touch occurs, and then shipped to their destination. The use of these three types of crossdocking varies, depending on the availability of existing space and the business objective. The "pure" (one-touch) systems are rarely used, due to the need for temporary storage of most products, which is mostly promoted by the vehicle scheduling, which has consequences, like a possible lack of coordination between inbound and outbound vehicles.

2.2 Automated Storage and Retrieval Systems

The temporary storage facility sometimes required in a CS can be just an area on the floor of a dock, for example, or a built-in warehouse into the system. To automate and improve the efficiency of processes and operations in a CS, one of the best options is to implement an AS/RS [2,8,9,11]. The characteristics presented by CS intersect perfectly with AS/RS systems, creating harmony in the process of loads movement and its eventual temporary storage.

Several factors must be considered to define methodologies and policies for storage and the implementation of AS/RS. The decision factors for the control and characterisation of AS/RS systems are the characterisation and layout of the system, assignment of storage, allotment factors, sequencing of operations and the dwell points characteristics [9].

Storage policies are a must in AS/RS systems because of the way loads and products are assigned to the available storage locations on the existing racks. Five main policies have been enumerated [5]: dedicated storage, random storage, nearest open location storage, storage based on load turnover and class-based storage. Since the topic to be addressed is an automated storage system, it is assumed that operations are performed in view of incoming requests. An important aspect in AS/RS is the sequencing of order pickings (OP). It is necessary to define selection criteria. Normally, in storage systems, requests are fulfilled according to the First-Come-First-Served principle. The same happens at the output point, where the first retrieval request will be the first to be executed. It is common for this principle to be associated with systems that work with a unit load handling method. In case of a recovery process, special attention is required since certain loads may have an associated expiration date, which leads to a more restricted approach, creating a new principle based on the priority of certain loads.

3 Crossdocking System and Model Architecture

In terms of the basic functioning of the model, items arrive in an inbound truck, are inserted into the CS, and distributed to other vehicles (outbound trucks) which will take them to their final destination. This distribution is done through previously defined dispatch orders. The vehicles which arrive to the system may only carry items of a specific type and/or supplier. These, when entering the CS, will be combined and grouped with items coming from other inbound vehicles, of different types and/or suppliers. Next, they will be packed and sent to outbound vehicles, located in the system's dispatch zone, which will transport them to their final destination. Between the arrival and the departure of the items from the CS, they may need to be temporarily stored on an automated warehouse. This way, the implementation of an AS/RS system provides the possibility of storing items, making the CS more complete. Items not used to fulfil some order will be stored and the recovery of items occurs when a dispatch order needs some of the different items to be completed and they are not in the system. This is considered the main logic to implement in this simulation model, getting as close as possible to the real context.

Before starting to implement a simulation model, decisions need to be made about the type and number of components to use, to achieve a realistic and coherent simulation. For the present simulation, two distinct models have been defined, which connect at certain points. One model is representative of the CS and the other of the AS/RS. These models are described in the sequel.

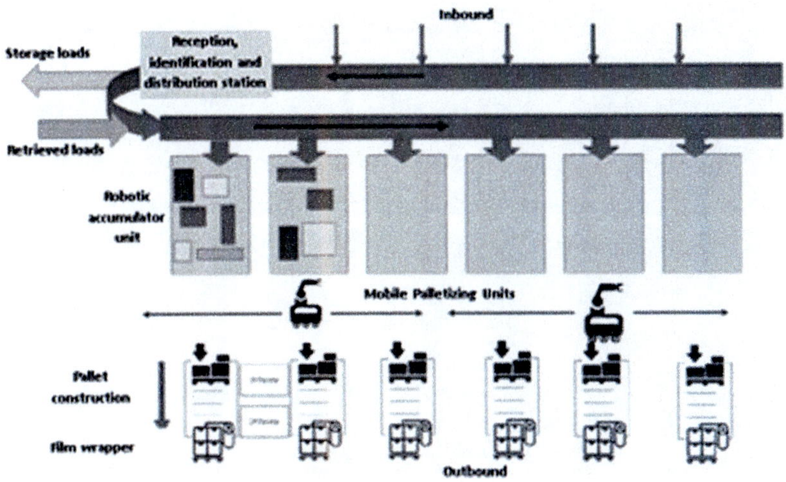

Fig. 1. CrossLog system architecture

3.1 Crossdocking System Architecture

The model of the crossdocking system follows the layout depicted in Fig. 1. It has an I-shape, an automated internal product movement, a unique service mode, and operates on multiple touches. As for the type of storage it is hybrid (some items go to storage and others directly to shipping). Regarding the number of inbound and outbound docks, it is reduced (five inbound and three outbound docks are used) to facilitate the perception of the system.

The loads enter the system through the inbound points and are transported via conveyors to a decision point, which defines whether they will be sent to the robotic accumulating units (AU), to fulfil an order being processed, or to storage (to the AS/RS). If they are sent to the first option, they will be distributed among the AU according to the orders that each one is selected to satisfy, and according to the need for that particular load. A robotic mechanical arm transfers the products for each order to a palletizer, which will put them together and then send them to a film wrapper. Once this process is complete, the orders, with their component products, are shipped to their recipients. At the mentioned decision point, if it is necessary to go to the built-in warehouse to collect products needed for an order that are not in the system, they will be entered there.

Having part of the layout implemented (illustrated in Fig. 2), with all the connections made between the source and the conveyors, six AU were created to accumulate various items to complete the dispatch orders. Since the model contains six AU, three robots and three palletisers, charges can be divided. That is, Robot1 only takes care of AU1 and AU2, Robot2 takes care of AU3 and AU4, and Robot3 takes care of AU5 and AU6.

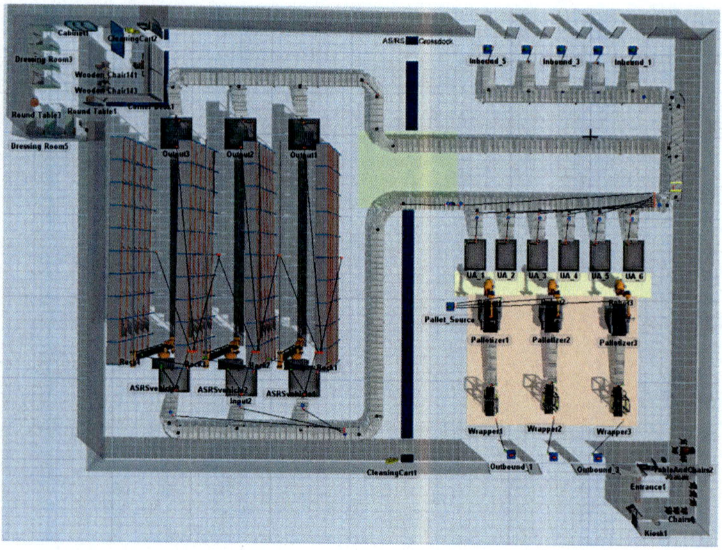

Fig. 2. Overview of the simulation model

3.2 Automated Warehouse (AS/RS) Architecture

Regarding the AS/RS model, depicted in Fig. 3a, it is placed in a lateral position to the crossdocking (as shown in Fig. 2) to create a realistic interconnection between both, taking into account the associated space restrictions. It contains three input points (the queues situated at the bottom of the system), three output points (those placed at the top), three aisles, three cranes, and four sets of racks. Each of the three inputs and three outputs is used for one aisle.

Since this system is not independent, *i.e.* it is connected to the crossdocking, its input will be one of the crossdocking outputs, and its output will be one of the inputs in the flow of items, directly to a conveyor. The loads enter through the input points, are transported to the assigned rack by the cranes, and when the crossdocking needs to retrieve any of them, they will be transported from the rack to the output point by the cranes.

The AS/RS contains seven levels in height and ten bays in length. Each bay has two storage slots with the following dimensions: (*i*) Bay Width = 2.50 m, (*ii*) Level Height = 1.40 m; and, (*iii*) Slot Width = 1.25 m.

4 Modelling Software and Model Development

FlexSim [3] was chosen for the modelling and simulation of the presented CS. It allows 3D simulating several types of systems and has some determining characteristics, such as the standard objects and elements available through an extensive library. Each object has its own processing logic, simulating them in real

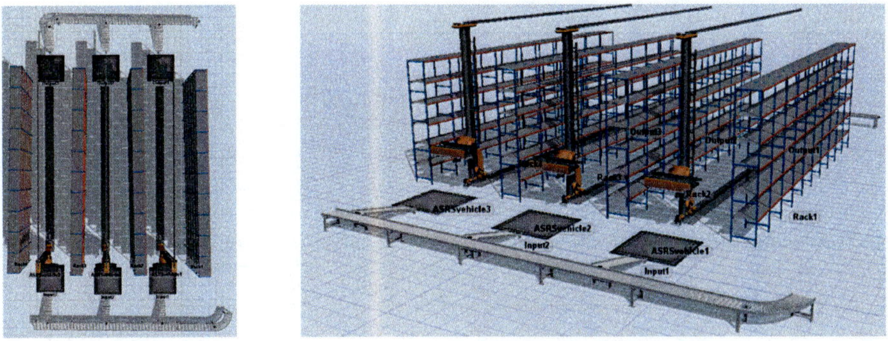

(a) 3D top view (b) 3D perspective view

Fig. 3. AS/RS system layout

context. To associate a load movement logic and simulate the system management algorithm, the software contains the Process Flow functionality, allowing to design processes, connect them and follow their performance in real time, through the use of tokens that identify each production unit [3].

After implementing the layout of both systems, several points were defined according to the functionalities offered by FlexSim. The different loads enter the system through a statistical distribution and separated by a defined time interval. Several formats of loads were also defined, each with its colour, to differentiate all loads entering the system. The system receives a job list, with the number of loads of each format corresponding to each dispatch order. A random list of dispatch orders is imported into the system through Microsoft® Excel®, and as the orders are executed they are eliminated from the list. Regarding the basic flow, as described, the loads enter the system, and if they are needed to satisfy an order they go to the AU (each one corresponding to a dispatch order), transported to the palletizer by a robotic arm, packed, and sent for dispatching. If they are not needed for any order, they are sent to storage at the AS/RS. After the items of a dispatch order are all gathered in an AU, the next step is to group these same items in a pallet. Through a resource source, enough pallets are generated to satisfy this grouping, that is, one pallet per dispatch order. Since there is a need to combine two distinct items, it was necessary to use a palletiser in the model, and a machine to wrap each order was also added. The simulation model features three palletisers, three wrapping machines, and three item transport robot, as seen in Fig. 2.

4.1 Implementation of the CS System

The system functioning logic was structured in two phases. The first phase is dedicated to the entry of items into the conveyor system and how they will be divided to satisfy the existing list of orders. The second phase is dedicated to placing each dispatch order on the pallet and sending it to dispatch.

To create a logic or flow of processes and operations in the system, the process flow (PF) [3] tool provided by FlexSim was used. Two types of interconnected PF were developed. A General Process Flow, which allows the creation of a centralised "brain" for the whole model. This type of PF is ideal for creating logic that needs to be globally accessible by all, or almost all, elements of the simulation model. In addition to this, an Order Process Flow (OPF), or fixed resource PF, in this case aimed at dispatch orders, was also used. It is intended for the creation of one or more model actors' logic, allowing to add extra logic to any element and the creation of custom logic.

With a view to greater organisation in the software's working environment and to facilitate understanding of the development phases of this flow, the OPF has been divided into three main blocks (depicted in Fig. 4), with the possibility of linking them together, as follows:

1. "Start: Get Data From Table": data collection from the dispatch orders table.
2. "Get Ordered Items Off List/Place in Queue": take the items required for an order from the list and place them in the AU.
3. "Transport Items to Combiner": transport the items to the combiner to be grouped on a pallet.

To enable the retrieval of items stored in the AS/RS, two other blocks were created:

1. "New Items to Queue": block for the activities of the items coming from the sources.
2. "Stored Items to Queue": block for the activities of existing items, which are necessary to satisfy a dispatch order, and which are stored in the warehouse.

The first block represents the beginning of the process, where the data from the dispatch order list is collected, and a Schedule source is used, with an associated uniform load generation statistical distribution. Next, a label assignment is made to each order. This assignment creates an array of data from the list and generates the number of tokens needed to enter on the flow. Then the second block contains the "Pull from List: Items" activity correlated with a "Decide" decision activity. This allows the system to check the existing items on the dispatch order list and decide whether to use items that are in circulation on the conveyor belts or whether to use the AS/RS. To make this decision functional, it was defined that existing items in circulation have higher priority (priority 1) than items in storage (priority 0). The decision activity generates two new blocks, one for each option. For the items that are already present in the system (from the sources) the "New Items to Queue" block was created, and for those that are in storage the "Stored Items to Queue" block was created. At the moment of the decision, if the required items are in circulation in the system, the token goes to the "New Items to Queue" block, where a match will be executed and sent to the AU corresponding to the dispatch order where it is needed. As for the "Stored Items to Queue" block, a Task Sequence was created so that the token can be sent to the Task Sequence, this way being possible to

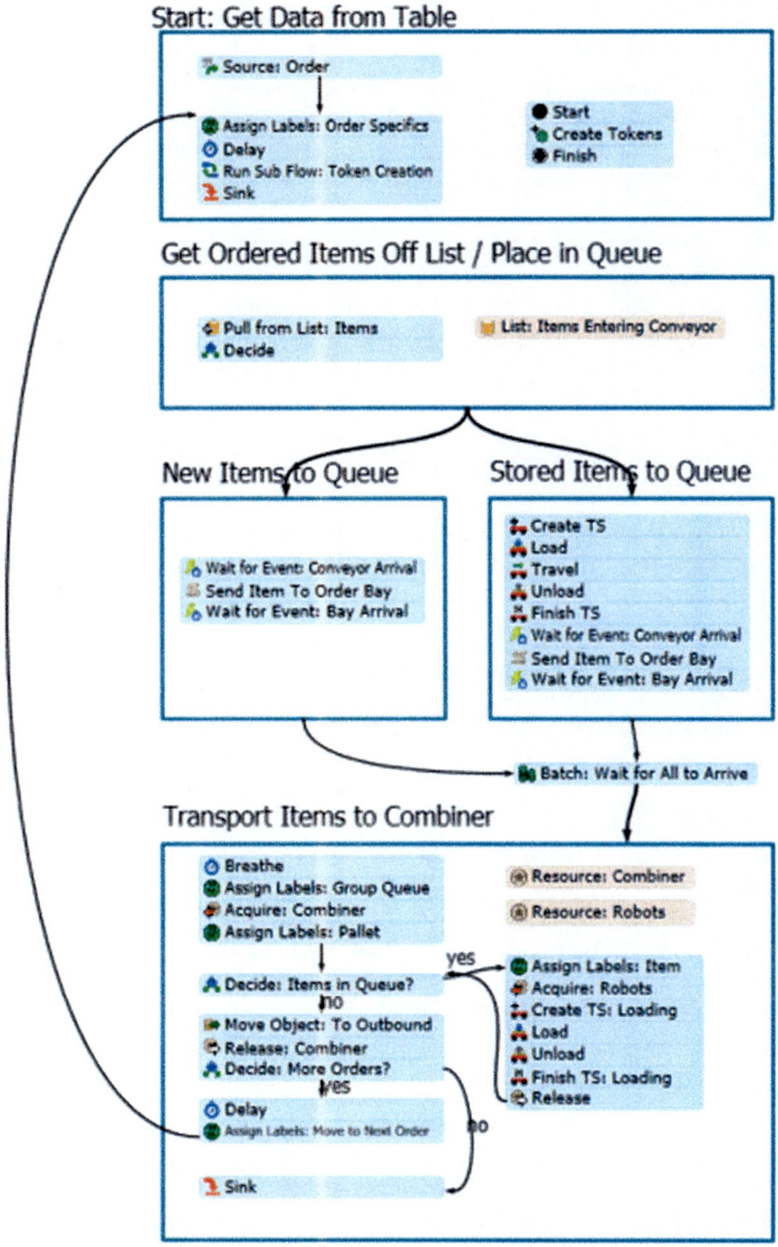

Fig. 4. Order process flow overview

retrieve the item from the warehouse and transport it to the conveyor system. The task sequence reflects how the AS/RS works. The "Create TS" activity is responsible for creating this sequence, and the "Task Executer" is the Crane, named "ASRSVehicle". Thus, the process involves taking the product from the shelves by the ASRSVehicle (Load activity), transporting it to the output point (Transport) and unloading it (Unload). At this point the sequence of tasks ends and the load enters a secondary conveyor, connected to the main one. Once it reaches a decision point strategically placed at the entrance of the main conveyor, the following activities are similar to the activities in the "New Items to Queue" block. To distinguish the item retrieved from the AS/RS, as soon as it enters the main conveyor, a trigger is triggered to change the colour of the item in question. After the execution of one of these blocks, the logic of both converges to an activity for the dispatch order being satisfied, named "Batch: Wait for All to Arrive". It aims to create a waiting period so that the items of an order are only transported to a combiner when all of them are gathered in the concerned AU. The purpose of block 3 is to transport the items to the combiners, where they will be placed on pallets and then sent for dispatch. Before being dispatched they go through a wrapping process, but this does not need to be included in the OPF as it is a purely representative process. The order is fulfilled, wrapped and sent to dispatch. The wrapping activity is predefined and its Setup Time and Process Time can be controlled. The dispatch is simulated by the sink feature.

4.2 Implementation of the AS/RS

As far as the AS/RS is concerned, besides offering several storage-related features (such as shelves, lifts, among others), FlexSim enables the use of ASRSVehicles.

Conveyors were initially created to connect the transition location of items between crossdocking and AS/RS. To improve and facilitate the storage and retrieval of items, it was defined that the items would be collected for storage at one end of the aisles and retrieved for insertion into the flow of items for crossdocking at the opposite end. In the area destined for item storage, the conveyors connect to three separate queues, named input1, input2, input3. The same is true for the item retrieval queues (output1, output2, output3), which are linked to the crossdocking return conveyor. Regarding the storage policy, dedicated storage was used, in which the items are always stored in a specific location, where their replenishment is always carried out in the same place.

5 Activity Scenarios and Results Obtained

To test the model's efficiency and whether it fits reality, taking into account the number of resources and elements in use, this section presents some possible scenarios in the activity of both the crossdocking system and the AS/RS system.

5.1 Simulated Scenarios

The scenarios considered most relevant to the functioning of the model and to its efficiency, were defined to be evaluated. They are differentiated in the

Table 1. Parameters of the different simulated scenarios

Scenarios	Scenario 1	Scenario 2	Scenario 3	Scenario 4
Nº of AU	6	6	3	3
Delay (s)	200	0	200	0

Table 2. Time table of arrival of items to the system

Source	Inbound_1	Inbound_2	Inbound_3	Inbound_4	Inbound_5
Time (s)	25	35	38	28	33

number of AU and the delay applied between dispatch orders. Four scenarios were generated, with variations in these parameters, as shown in Table 1. To evaluate the processes executed during the simulation, all scenarios were run for 3600 s.

An important factor is the input of items through the crosdocking sources, intended to simulate the unloading of products from vehicles. In the base operating scenario, the item entry times presented in Table 2 were defined for the system, through the "Inter-Arrivaltime" field existing in each source. Furthermore, in all scenarios, the number of outbounds (represented by sinks, which can be interpreted as dispatch docks) is three. Regarding the characteristics of the three robots, it was considered that they present a load capacity of one (1) item per trip AU-Palletizer, a maximum speed of 3 m/s, and an acceleration of 1 m/s^2.

To obtain credible results, it was defined that dispatch orders could have a variable number of items but not more than ten. Table 3 is an example of the order table, containing ten dispatch orders (x is the number of items per dispatch order). The existing table in FlexSim contains about three hundred dispatch orders, randomly generated and complying with the restriction that the total sum of items cannot exceed ten units.

5.2 Analysis of the Results

The results obtained are compared between them, to understand which scenario is the most advantageous and efficient in terms of dispatch orders realisation, since this is the main objective of crossdocking. The higher the number of dispatch orders fulfilled in the same period of time, the more efficient is considered the system. As the two simulated systems are integrated, the functioning of the CS influences the functioning of the AS/RS. If a smaller number of items are distributed by the AU, keeping the same input flow in the system, the AS/RS will host a larger number of them, and vice versa. Therefore, another evaluation parameter is the number of items stored in the AS/RS. The smaller the number of stored items the more efficient is considered the system, since it was not necessary to resort to the temporary storage of items, which indicates that they proceeded to the realisation of a dispatch order. It is also necessary to take

Table 3. Example of an order table used in the simulation

	Type1	Type2	Type3	Type4	Type5	x
Order1	2	2	1	3	1	9
Order2	3	4	1	1	1	10
Order3	1	0	1	1	2	5
Order4	1	4	4	0	1	10
Order5	4	0	5	0	1	10
Order6	2	1	3	2	0	8
Order7	0	1	1	4	2	8
Order8	2	2	2	0	0	6
Order9	1	3	2	1	3	10
Order10	0	1	3	3	1	8

Table 4. Summary of the results obtained in the four simulated scenarios

	Scenario 1	Scenario 2	Scenario 3	Scenario 4
No. of orders issued	44	54	24	37
Average robot utilisation (%)	32	38	17	27
Average AU use (%)	47	81	45	80
Items stored/ recovered	356/145	286/148	447/75	400/146
Average stay time (s)	690	461	1160	664
Ratio between stored and retrieved items	$\frac{356}{145} = 2.44$	$\frac{286}{148} = 1.93$	$\frac{447}{75} = 5.96$	$\frac{400}{146} = 2.74$

into account the differences between scenarios because, in a real context, there is always the probability of the existence of failures associated with the operation of the accumulator units. The robots utilisation rate is directly related to the number of dispatch orders completed, as well as the utilisation rate of the AU.

Given this, to evaluate the different scenarios, the following indicators are analysed: (*i*) number of dispatched orders; (*ii*) robot utilisation rate; (*iii*) AU utilisation rate; (*iv*) number of items stored in the AS/RS; (*v*) number of items retrieved in the AS/RS; (*vi*) content existing per rack over time; and, (*vii*) average dwell time of an item in a rack. The results are presented in Table 4.

Analysing Table 4, one can conclude that Scenario 2 is the most efficient: the number of orders dispatched, and the robots and AU usage rate present better values when compared to the others. As for the number of items stored and retrieved, the difference between both is the smallest, which indicates that there is a greater flow of items to be stored and retrieved simultaneously. This can also be concluded analysing the ratio between stored and retrieved items. Since the objective of the AS/RS is the temporary storage of products, then the smaller this ratio, the more efficient is considered the system. Scenario 2 also presents a smaller number of stored items, which indicates that the system itself, with these parameters, is actually more efficient. The dwell time associated with each scenario varies according to two factors under study: the utilisation rate of the AU and the number of items stored/retrieved. The higher the utilisation rate of the AU, the greater the flow of items in the system which leads to a higher probability of retrieving items, decreasing their dwell time in the AS/RS.

Through these analysis and comparisons, the most efficient simulated scenario, both regarding the crossdocking system and the AS/RS, is Scenario 2. In comparison with the others, this scenario presents the: (*i*) highest number of units shipped: 54; (*ii*) highest robot utilisation rate: 38%; (*iii*) highest AU utilisation rate: 81% (alongside Scenario 4, which has an AU utilisation rate of 80%); (*iv*) smallest ratio stored/retrieved items: 1.93; and, (*v*) shortest time of permanence of products in the warehouse: 461 s. After Scenario 2, the one considered most efficient through the data obtained is Scenario 1.

Being Scenario 2 considered the most efficient, it can be concluded that in a CS with an integrated AS/RS, according to the defined parameters and properties of the simulated model, the time for the attribution of a dispatch order to an AU (where the items are grouped to be dispatched) should be minimal to increase efficiency. It was proven that with the reduction of this time period (delay), more orders will be dispatched and better performance is obtained from the three robots. As for the AS/RS, its functioning data depends on the type of operation applied in the CS. This is because any item that goes to storage had a previous passage through the CS, not having been assigned to any dispatch order. When it is stored, the objective is that it is retrieved, after a few moments, to satisfy another dispatch order. The lower the number of AU in operation, the lower the number of dispatch orders dispatched, which leads to an increase in the number of stored items, also increasing their permanence time in the AS/RS. This factor can be mitigated if the entry of items in the flow decreases, thus preventing the constant entry of products in the system.

6 Conclusions and Future Developments

This work presented the modelling and simulation of a CS with an integrated AS/RS. This simulation model was developed to analyse distinct solutions for the CS and storage policies, and to check eventual logistic problems of this type of systems. The simulation model works without any limitation and adapts to new dispatch orders, new item entry flows and enables the addition of elements that are central to its operation, without the need to alter the operating logic.

Due to the range of system evaluation options available, the delay associated with the assignment of dispatch orders to AU and the number of AU in operation were considered the most relevant and with the possibility of creating a greater impact on the system results. After analysing the simulation data, the scenario containing a null delay and all AU in operation proved to be the most efficient.

Concerning future developments of this work, the simulation model will be used to analyse other parameters, such as a greater variation of incoming items, the use of more robots, palletisers and outbounds. It is also planned to check the impact on the CS of decreasing the number of AS/RS racks, their useful storage space, and the transport capacity of each ASRSVehicle. An important aspect identified during the development of the model, which was not applied, is the scheduling, both of the entry of items into the system, and their dispatch, and also the arrival of dispatch orders to them. This scheduling would allow for a more realistic operation, and could be in line with the real context of a particular logistics centre with different characteristics. The system having the entrance time of the items (arrival of vehicles for unloading) and the exit time of a dispatch order, already with the items grouped and packed (arrival of vehicles for loading) would be an excellent addition, providing a greater realism. The timing of the arrival of dispatch orders would be another interesting point, mainly in the analysis of behaviour and performance of the AS/RS system.

Acknowledgements. This work is financed by the ERDF - European Regional Development Fund, through the Operational Programme for Competitiveness and Internationalisation - COMPETE 2020 Programme under the Portugal 2020 Partnership Agreement, within project CrossLog, with reference POCI-01-0247-FEDER-039895.

References

1. Apte, U.M., Viswanathan, S.: Effective cross docking for improving distribution efficiencies. Int. J. Logistics Res. Appl. **3**(3), 291–302 (2000). https://doi.org/10.1080/713682769
2. van den Berg, J.P., Gademann, A.: Simulation study of an automated storage/retrieval system. Int. J. Prod. Res. **38**(6), 1339–1356 (2000). https://doi.org/10.1080/00207540018888
3. FlexSim Software Products Inc: 3D Simulation Modeling and Analysis Software. https://www.flexsim.com/. Access 5 May 2021
4. Fretefy: O que é Cross Docking?. https://www.fretefy.com.br/blog/o-que-e-cross-docking-tudo-que-voce-precisa-saber. Access 8 April 2021
5. Hausman, W., Schwarz, L., Graves, S.: Optimal storage assignment in automatic warehousing systems. Manag. Sci. **22**, 629–638 (1976)
6. Bartholdi, J.J., Gue, K.: The best shape for a crossdock. Transp. Sci. **38**, 235–244 (2004)
7. Ladier, A.L., Alpan, G.: Cross-docking operations: Current research versus industry practice. Omega **62**, 145–162 (2016). https://www.sciencedirect.com/science/article/pii/S0305048315001991
8. Manzini, R., Gamberi, M., Regattieri, A.: Design and control of an AS/RS. Int. J. Adv. Manuf. Technol. **28**, 766–774 (2006). https://doi.org/10.1007/s00170-004-2427-6
9. Roodbergen, K.J., Vis, I.F.: A survey of literature on automated storage and retrieval systems. Eur. J. Oper. Res. **194**(2), 343–362 (2009). https://www.sciencedirect.com/science/article/pii/S0377221708001598
10. Van Belle, J., Valckenaers, P., Cattrysse, D.: Cross-docking: state of the art. Omega **40**(6), 827–846 (2012). https://www.sciencedirect.com/science/article/pii/S0305048312000060. special Issue on Forecasting in Management Science
11. Verriet, J., Hamberg, R., Caarls, J., van Wijngaarden, B.: Warehouse simulation through model configuration. In: Proceedings - 27th European Conference on Modelling and Simulation, ECMS 2013. vol. 05, pp. 629–635 (2013)

A Novel Algorithm for Maximization of the Secondary-Task Function for Redundant Serial Manipulators

Jonathan Cerbaro(✉), Dieisson Martinelli, André Schneider de Oliveira, and João Alberto Fabro

Universidade Tecnológica Federal do Paraná - UTFPR, Av. Sete de Setembro, Curitiba, Paraná 3165, Brazil
jonathancerbaro@alunos.utfpr.edu.br

Abstract. Redundant manipulators can employ additional degrees to achieve secondary objectives when performing the trajectory tracking, such as avoiding the mechanical limits of the manipulator's joints through the Jacobian null space properties. This article presents an improvement to the classical kinematic solution for redundant serial manipulators, changing the secondary gains to a variable vector that replaces the constant scalar gain. Also, a method is suggested for the variation of those gains. The application is demonstrated in trajectories for the Movemaster RV-M2 robot to prove the effectiveness of this change. As a result, the path of the manipulator's joints remains further away from their limits along the path, maintaining the position error profile in the end-effector.

Keywords: Robot kinematics · Secondary tasks · Redundant manipulator · Jacobian null-space

1 Introduction

The contextualization of the subject treated in this paper is done in the practical industrial and mathematical academic scope. Industrial production lines that aim to handle products on conveyor belts, weld equipment, or in cargo transportation, usually automate these tasks with the use of serial manipulators. The characteristic of these manipulators of having great mobility is at the same time a solution and a problem: a structure must be able to move freely without going through singular configurations, and the robot and all its structure must avoid the limits defined for its workspace and also its physical limits like auto-collision and joint limits. These activities must be completed along with controlling the position and orientation of the manipulator tool, in a way that does not compromise the task being performed. These limitations configure a multi-task scenario, which demands from the manipulator the dexterity to perform a primary task, normally defined as the position and orientation control, and secondary tasks as many as the robot's kinematic characteristics allow to be modeled. This problem is approached in the mathematical context and such information about modeling is found in application in works in the scientific literature as here described.

D. Tardioli et al. (Eds.): ROBOT 2022, LNNS 590, pp. 217–228, 2023.
https://doi.org/10.1007/978-3-031-21062-4_18

The approach of secondary objectives using the redundant space of the Jacobian, which is the technique that underlies the algorithm proposed in this article, is also used in works such as [1,2] and [3], where the application is made for the position control of a welder robot and results in a more efficient control due to the multiplicity of solutions and greater dexterity than a non-redundant robot. Even in multi-agent systems with drones in a tethered configuration, there is room for the application of secondary objectives with the Jacobian, as in [4]. An algorithm similar to the one proposed in this paper was introduced in [5], where the secondary gain vector is adjusted in order to to maximize the manipulability, but that approach requires the computation of multiple options of joint trajectory in order to choose the best in terms of manipulability. Of course, as in all scenarios that apply a secondary task, its execution cannot compromise the performance of the primary objective [6].

In the algorithm proposed in [7], the authors do not use the Jacobian modeling to avoid the joint limits, avoiding problems inherent to the matrix calculation, but without using the algebraic advantages that its equation allows. Other secondary objective applications are done in [8], where position control is coupled to velocity and acceleration for end-effector trajectories. The degrees of redundancy of the Jacobian are explored in [9], where an algorithm is proposed to avoid joint limits while solving the inverse kinematics analytically. For collaborative robots, such techniques are applied in the works of [10]and [11] for two- and three-dimensional cases.

This paper proposes a new algorithm that aims to improve the performance of classical techniques that work with secondary tasks in serial manipulators. The formulation of differential inverse kinematics is applied and the redundancy of the transformation space of the Jacobian matrix is used to obtain solutions that allow better performance metrics of secondary tasks without compromising the primary task.

2 Problem Overview

This section contains a brief review of kinematics concepts applied to serial manipulators. These concepts are necessary for understanding the context in which the proposition of this article applies. The focus of the contribution present in this paper is on the equation for the secondary tasks of the manipulator, however, this section describes details about the kinematics of the manipulator used, to contextualize the reader regarding the experimental scenario.

The manipulator used in the validation step of the proposed algorithm is the Movemaster RV-M2. The prototype available in the laboratory was retrofitted, so its structure was modernized and adapted for academic use. The prototype modernization process aimed to collect real data instead of just simulating the techniques. The prototype is now compatible with ROS networks and has external sensoring systems. Such functionalities did not exist in the original model, whose photo before the retrofit is in the Fig. 1a. The new workstation assembled in the laboratory is shown in the Fig. 1b.

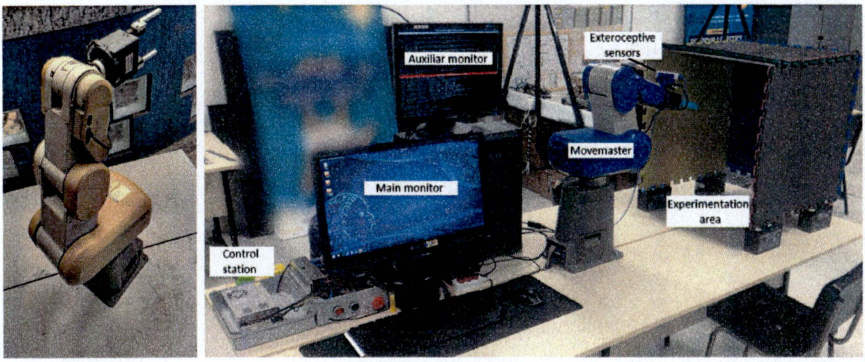

(a) Original. (b) Retrofitted workstation.

Fig. 1. The Movemaster RV-M2 robot in its original condition (a) and after the retrofitt and installed in its new workstation (b).

In addition to receiving computer vision sensors and a new controller, there was an adjustment on the limit sensors, which changed the maximum and minimum limits of each joint, which implies a change in its parameters in the experiment, different from those in the manual. Figure 2 illustrates the Movemaster manipulator highlighting its link lengths and joint limits. These parameters are used experimentally and their exposure allows the reproduction of results.

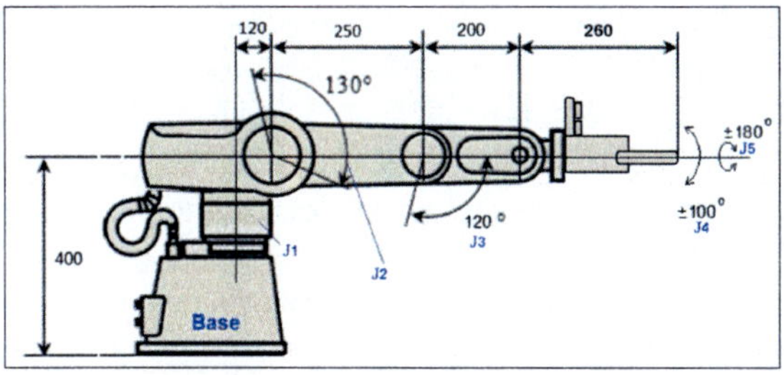

Fig. 2. Retrofitted Movemaster kinematic scheme. Adapted from [12].

For reasons of conciseness of the document and for not having a direct relationship with the proposition, the concepts of successive screws and dual quaternions, which were used for the solution of the kinematic equations, are not addressed here. These techniques were applied as part of other projects with the manipulator concomitantly with this one and its application on the RV-M2 is required to faithfully reproduce the results. In any case, the reader is encouraged to consult the references [13], and [14].

Regarding redundancy, a manipulator is redundant when the number of degrees of freedom is greater than the number of variables needed to describe the operational space [15]. From the differential kinematics point of view, one way to understand the redundancy of a manipulator is via the Jacobian mapping: The Jacobian matrix is nothing but the mapping operator between the joint space and the Cartesian space. If the manipulator has a joint space with a dimension larger than those necessary to define the operational space, then the Jacobian mapping has redundancies. The existence of redundancy implies the possibility of using the redundant movements to create secondary tasks for the manipulator using the Eq. 1 [15].

$$\dot{q} = J^{\dagger} v_e + (I_n - J^{\dagger} J)\dot{q_0} \tag{1}$$

The term $J^{\dagger} v_e$, represents the local minimization of joint velocity, which determines the primary task operation. The term, $(I_n - J^{\dagger} J)$, represents the null-space of J. In this formulation, if there is redundancy, the second term is non-zero and the value of $\dot{q_0}$ is added to the manipulator velocity space.

The secondary task is defined according to the Eq. 2, where $w(q)$ is a function representing any given information with respect to the joint variables and its maximization represents the accomplishment of the secondary task. The value of k_0 is empirically defined to weigh the influence of secondary gain in the equation. In the proposition of this paper, the gain k_0 is treated as a vector with variable values. This, along with the technique used to change said values, is the contribution of this work.

$$\dot{q_0} = k_0 \left(\frac{\partial w(q)}{\partial q} \right)^T \tag{2}$$

For the experiments exemplified in this work, the primary task is controlling the position of the end-effector in three-dimensional space — three variables being controlled in a manipulator with five degrees of freedom — and the secondary task is to maximize the angular distance from the joints to their physical limit. Such description can be done by the Eq. 3, where q_{iM}, q_{im} and $\bar{q}_i$ indicate the maximum, minimum and average values for the joint variable q_i, respectively. The value of n indicates the number of degrees of freedom that the manipulator has. With this, the secondary task $\dot{q_0}$ can be described and the secondary gain k_0 becomes the only empirical step in the process.

$$w(q) = -\frac{1}{2n} \sum_{i=1}^{n} \left(\frac{q_i - \bar{q}_i}{q_{iM} - q_{im}} \right)^2 \tag{3}$$

2.1 Proposed Algorithm

As noted in the method commonly used to insert a secondary task into the system, the weight of its influence is given by varying the gain k_0. Regardless of how this value is determined, the efficiency of the results is linked to the trajectory being executed by the manipulator. An important point to consider when the secondary gain is calibrated is the fact that the secondary task is only performed to the detriment of the primary: if the manipulator is required to increase the error of the primary objective — increase the position error, for example — to be able to avoid the limits of some joint, this will not be realized. In mathematical terms, if the robot is in a pose that does not allow this freedom, it means that there is no redundancy in the Jacobian, which causes the secondary objective to be multiplied by zero in the Eq. 2.

Observing these operating conditions, we created a simulation in the *software* Matlab, where the limits of the joints can be violated without real risk of damage to the robot, and we used examples of trajectories where the secondary task failed — in the meaning of having generated trajectories where certain joints collide or exceed their limits even when varying the gain k_0.

One approach found in the literature for scenarios of this type is to *clamping* the joint angle, that is, to assign the limit value to the joint if the inverse kinematics calculation indicates that it should exceed it. This, of course, generates a chain of errors in the kinematic chain introducing errors into the effector that must be corrected throughout iterations of a control loop, like in [16] and [17].

By observing this limitation, we theorized that it would be possible to use the moment when a given joint exceeds its limits as an indication that the gain k_0 is not satisfying the requirements of this specific joint. With this, we introduced a secondary gain vector, of dimensions $(n \times 1)$, in which the element n_i is associated with the secondary gain of the joint n, to the detriment of a scalar constant that influences all joints uniformly, thus creating the Algorithm 1.

Algorithm 1. Proposed method with variable vector gains.

for $i = 1$ to n **do**
  if $\theta_i > \theta_{max_i}$ **then**
    $\theta_i = \theta_{max_i}$
    $k_{0_i} = 1,01 k_{0_i}$
  end if
  if $\theta_i < \theta_{min_i}$ **then**
    $\theta_i = \theta_{min_i}$
    $k_{0_i} = 1,01 k_{0_i}$
  end if
  $k_0 = \text{diag}(k_0)$
end for

When the algorithm reaches the condition that the joint i is out of its limits, k_{0i} is increased by 1%, and the joint variable is set to its maximum or minimum value. Such variation is enough to generate the desired results, which are observed in the Sect. 3. The position error profile remains the same, however, the joint behavior is now compatible with what is desired in a real manipulator, where limits are not violated.

3 Experimental Scenario

To demonstrate the practical implications of the proposed algorithm, the manipulator was submitted to the generation of three arbitrarily defined trajectories:

an ellipse, a line, and a circle. The three trajectories were created using the Eq. 4 as a model, just changing its parameters to create the desired trajectories. In the model trajectory equation, p represents the instantaneous position, expressed as a function of the center c and radius r of the trajectory. The route is integrated as a function of t, which is discretized in $0.01s$ parts over a total time $T = 30s$, totaling 3000 points.

$$\begin{bmatrix} p_x \\ p_y \\ p_z \end{bmatrix} = \begin{bmatrix} c_x + r_x \cos(2\pi t/T + \pi) \\ c_y + r_y \sin(-2\pi t/T + \pi) \\ c_z + r_z \sin(-2\pi t/T + \pi) \end{bmatrix} \tag{4}$$

The path parameters were defined for the **ellipse** path as: $c = [480,\ 100,\ 180]$ and $r = [50,\ 220,\ 130]$; for the **line** path as: $c = [550,\ 0,\ 230]$ and $r = [0,\ 250,\ 50]$; and for the **circle** path as: $c = [450,\ 0,\ 200]$ and $r = [180,\ 180,\ 0]$.

In Fig. 3, the three trajectories are illustrated graphically for a better understanding of the task performed. The manipulator is represented in its corresponding position in the real model.

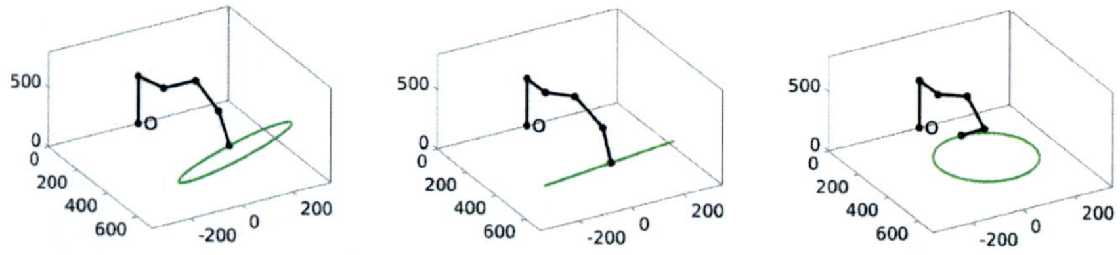

(a) Trajectory in Ellipse. (b) Trajectory in Line. (c) Trajectory in circle.

Fig. 3. Three trajectories used for experimental validation of the algorithm.

The iterative differential kinematics was performed in two scenarios for each of the trajectories: at first, contemplating the secondary task of maximizing the angular distance to the joint limits using the classical equation; in a second moment, the experiment was repeated but inserting the proposed algorithm with variable vector gain. All parameters of the differential kinematics loop are the same in all scenarios, as this ensures that the measured influences on joint behavior are due solely to the algorithm change. The secondary gain starts as $k_0 = 1$, and for the proposed algorithm this is a (5×1) vector, which has the value of each member changed along the trajectory.

Three performance metrics were used to validate the application. First, the behavior of the joints along the trajectory was plotted so that it is possible to visually compare the difference inserted by the algorithm change. To obtain quantitative information about the performance, the angular distance between each joint was measured to the nearest limit, both in the classical and in the proposed approach. Then, using the measured difference, the average angular distance of the improvement in each joint was calculated. Finally, the percentage of the trajectory in which the insertion of the algorithm caused the distances to increase was collected.

4 Results and Analysis

The experiment was created in a programming environment via Matlab. This section describes the performance metrics collected according to what was proposed in the elaboration of the experiment. This section has an expository character, where the results obtained in the alteration of joint trajectories in graphic format are observed, as well as the quantitative data of the collected performance metrics.

An important point to note in the performance graphs is the behavior of joint 1, which remained virtually unchanged when changing the secondary objective algorithm. This makes perfect mathematical sense since the secondary task can only be performed in case of redundancy: Joint 1 is the only one that allows movements around the z axis of the global coordinate system, making it impossible to have a multiplicity of solutions in this parameter.

First, the Figs. 4a–4e show the angular distance from each of the joints to the nearest limit, either clockwise or counterclockwise, along the proposed elliptic path. The classic scalar gain algorithm is represented in red and the new proposed variable vector gain algorithm is represented in blue. The angular distance is measured in degrees and the value of zero means that the joint is exactly at its physical limit. In this image, it is possible to observe the clear influence of the algorithm alteration modifying the trajectory of each joint individually without significantly altering the end-effector error profile Fig. 4f.

Another way to observe the influence of the proposed algorithm is by analyzing the improvement metric in the angular distance to physical limits. This is done as illustrated in the Fig. 5, where the graphs show the increase or decrease of the angular distance to the limits along the trajectory. If at any given moment, the influence of the new algorithm has made the joint closer to one of the limits

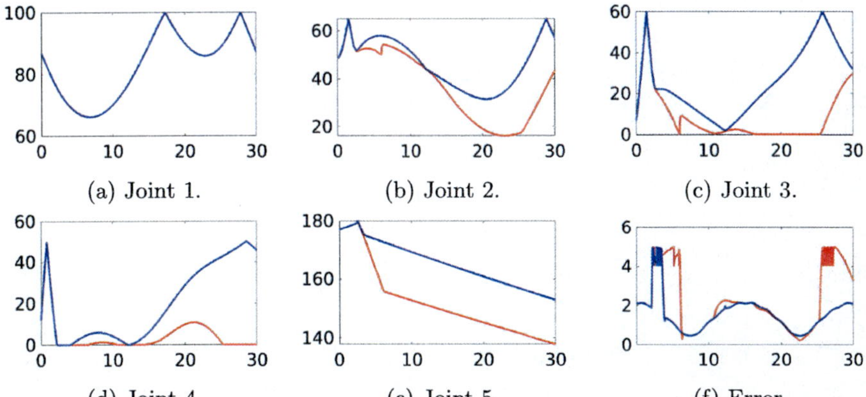

(a) Joint 1. (b) Joint 2. (c) Joint 3.

(d) Joint 4. (e) Joint 5. (f) Error.

Fig. 4. Elliptical trajectory, in seconds: each of the joint angles (a-e) in degrees and the end-effector error (f) in millimeters along the execution in both the classical (red) and proposed (blue) algorithms.

comparing the classical method, then the graph line is red. If the algorithm has improved performance at any certain point, then the line is illustrated in blue.

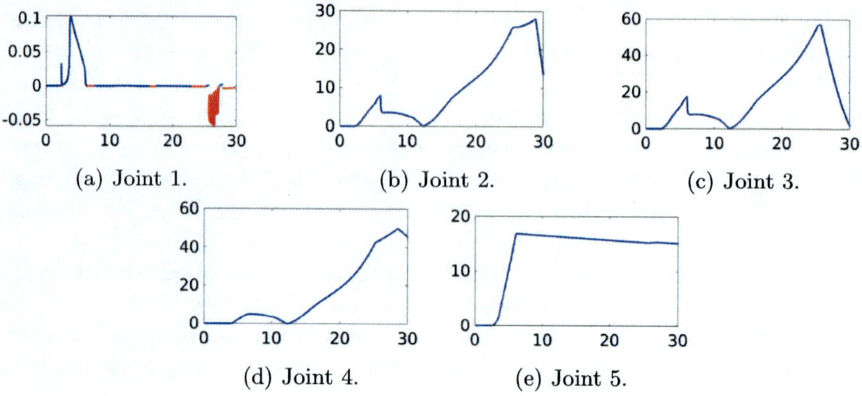

(a) Joint 1. (b) Joint 2. (c) Joint 3. (d) Joint 4. (e) Joint 5.

Fig. 5. Elliptical trajectory, in seconds: improvement (in degrees) on the secondary task. If the improvement is positive — in blue —, then the new algorithm gave better results than the classical approach. Red otherwise.

The experiment was repeated with the line trajectory. The performance measurements of the algorithms were compared in the same way as in the elliptical path and the results were similar in terms of increasing the angular distance to the physical limits. In the Fig. 6a–6e it is possible to observe the variation in the behavior of the joints along the trajectory, as well as the maintenance of the position error in the Fig. 6f.

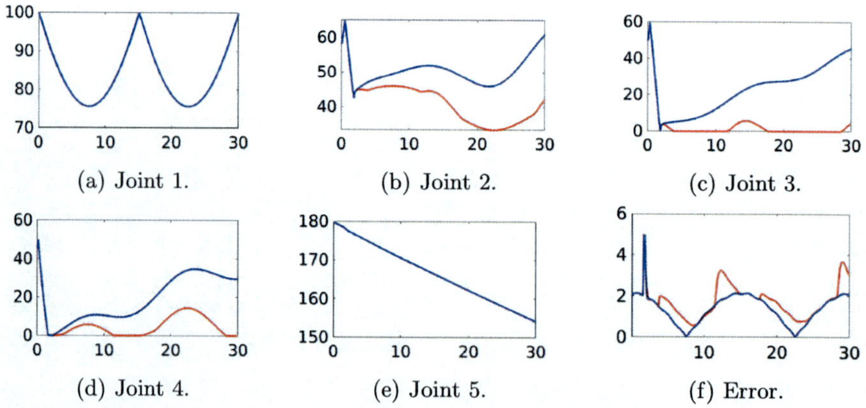

(a) Joint 1. (b) Joint 2. (c) Joint 3. (d) Joint 4. (e) Joint 5. (f) Error.

Fig. 6. Line trajectory, in seconds: each of the joint angles (a–e) in degrees and the end-effector error (f) in millimeters along the execution in both the classical (red) and proposed (blue) algorithms.

As with the elliptical path, the magnitudes of improvement along the linear path can be observed for each joint in the Fig. 7.

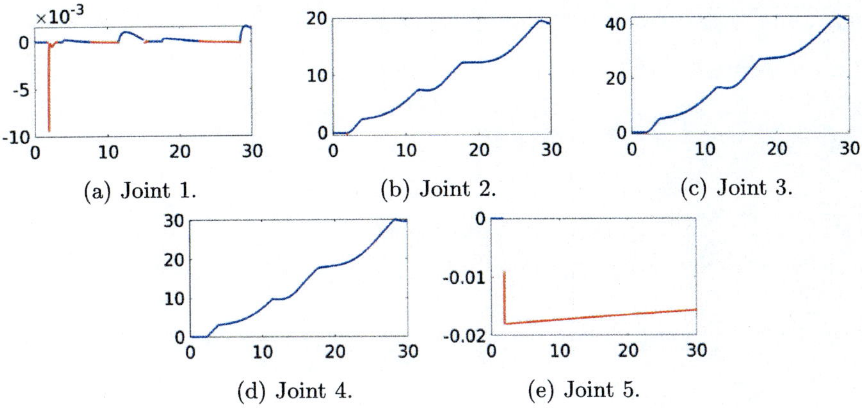

(a) Joint 1. (b) Joint 2. (c) Joint 3. (d) Joint 4. (e) Joint 5.

Fig. 7. Line trajectory, in seconds: improvement (in degrees) on the secondary task. If the improvement is positive — in blue —, then the new algorithm gave better results than the classical approach. Red otherwise.

Finally, the same data was collected on the circular path. The results are displayed similarly to the previous cases in the Figs. 8 and 9.

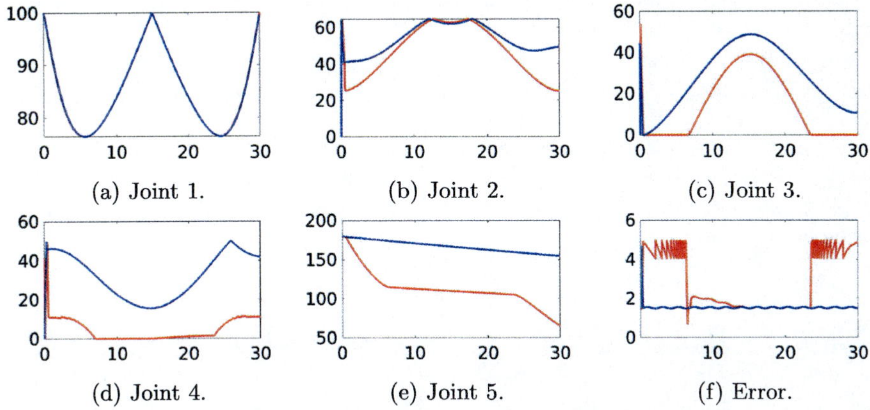

(a) Joint 1. (b) Joint 2. (c) Joint 3. (d) Joint 4. (e) Joint 5. (f) Error.

Fig. 8. Circle trajectory, in seconds: each of the joint angles (a-e) in degrees and the end-effector error (f) in millimeters along the execution in both the classical (red) and proposed (blue) algorithms.

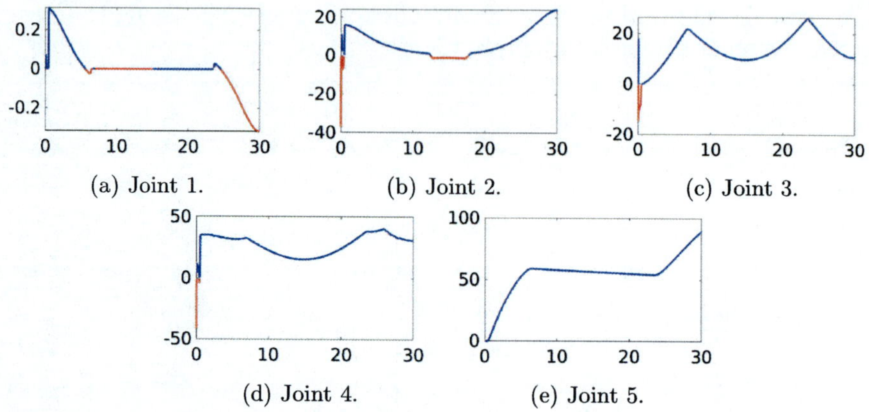

Fig. 9. Circle trajectory, in seconds: improvement (in degrees) on the secondary task. If the improvement is positive — in blue —, then the new algorithm gave better results than the classical approach. Red otherwise.

Despite being evident from the observation of the performance graphs that the application of the proposed algorithm improved the results obtained, quantitative metrics were also collected to express this improvement in numerical format as shown in Table 1. It is possible to observe that joint 1 is not, in fact, significantly altered: despite having percentage improvement results, this improvement is negligible and is due only to calculation approximations. The other joints showed an increase in the angular distance to the limits during almost the entire trajectory and in expressive angular values. Table 1 also shows the average improvement for each joint in both metrics, and taking the average of those values we get the overall performance on the three experiments with a mean improvement of 13.42° along 84.48% of the trajectory, thus proving the effectiveness of the algorithm.

Table 1. Mean improvement in the angular distance limits and percentage of the trajectory with improvement.

	Improvement (deg)					Trajectory % Improved				
Traj.	J1	J2	J3	J4	J5	J1	J2	J3	J4	J5
Ellipse	0.00	9.80	17.71	15.69	13.53	71.43%	100.0%	100.0%	100.0%	99.33%
Line	0.00	8.95	19.94	13.33	0.00	59.93%	99.67%	99.67%	100.0%	6.20%
Circle	0.00	6.54	13.65	27.51	54.67	51.00%	81.77%	98.63%	99.70%	100.0%
Mean	0.00	8.43	17.10	18.84	22.73	60.79%	93.81%	99.43%	99.90%	68.51%

5 Conclusions

In the study of serial manipulator kinematics, there is usually room for performance improvements in the algebraic and formulation characteristics of the systems. In this paper, we explored the case of redundant manipulators, which allow the execution of secondary objectives.

Given the multiplicity of solutions allowed by the kinematic characteristics of the Movemaster RV-M2 robot, as well as any other manipulator whose kinematic chain has such behavior, it was possible to change the methodology applied in the secondary gains in trajectory control. By exchanging the commonly used scalar gain for a variable vector gain, where each element is associated with one of the joints and whose value changes depending on the current state of the system, it was possible to find a solution that fulfills the secondary objective with greater quantitative performance.

To analyze the efficiency of the proposed algorithm, it was submitted to three trajectories with different characteristics, where the primary objective is to control the position of the end-effector along the path. The secondary objective exemplified is the maximization of the angular distance from each joint to its physical limits. For this, the differential kinematics equation was used and the performance of the classical solution method was compared with the proposed one. All experiments demonstrated that the new solution provides trajectories for each joint with superior secondary task performance in all aspects analyzed.

The three examples presented in this article are samples of several scenarios where it was possible to apply the proposed changes since such behavior is observed whenever there is redundancy in the solution of the kinematic equations. As it is a control system with numerical calculation and variables determined empirically, the results must be demonstrated by exhaustion and it is computationally infeasible to analyze all the possibilities of gains and trajectories. Anyway, the high efficiency in improving the results demonstrates that the algorithm is functional and should be applied in manipulator tasks containing secondary objectives.

An analysis of the multiplicity of solutions can lead future works to find the ideal solution to the problem so that it is possible to compare the best physically possible case to the results from the algorithm applied in this experiment. Changing the initial gain constant, as well as the way the vector elements vary, directly alters the solution, making it impossible—so far—to determine a combination that guarantees ideality in every scenario.

Acknowledgement. This project was partially funded by National Counsil of Technological and Scientific Development of Brazil (CNPq), and by Coordination for the Improvement of Higher Level People (CAPES).

References

1. Huo, L., Baron, L.: Kinematic inversion of functionally-redundant serial manipulators: application to arc-welding. Trans. Can. Soc. Mech. Eng. **29**(4), 679–690 (2005)

2. Huo, L., Baron, L.: The joint-limits and singularity avoidance in robotic welding. Ind. Robot: Int. J. (2008)
3. Baron, L.: Kinematic inversion of functionally-redundant serial manipulators: application to deburing tasks. In: 13th World Multi-Conference on Systemics, Cybernetics and Informatics, pp. 945–953 (2009)
4. Caruso, M., Gallina, P., Seriani, S.: On the modelling of tethered mobile robots as redundant manipulators. Robotics **10**(2), 81 (2021)
5. Huo, L., Baron, L.: The self-adaptation of weights for joint-limits and singularity avoidances of functionally redundant robotic-task. Robot. Comput.-Integr. Manuf. **27**(2), 367–376 (2011)
6. Chen, Y., McInroy, J.E., Yi, Y.: Optimal, fault-tolerant mappings to achieve secondary goals without compromising primary performance. IEEE Trans. Robot. Autom. **19**(4), 680–691 (2003)
7. Faria, C., Ferreira, F., Erlhagen, W., Monteiro, S., Bicho, E.: Position-based kinematics for 7-DoF serial manipulators with global configuration control, joint limit and singularity avoidance. Mech. Mach. Theory **121**, 317–334 (2018)
8. Faroni, M., Beschi, M., Visioli, A., Pedrocchi, N.: A real-time trajectory planning method for enhanced path-tracking performance of serial manipulators. Mech. Mach. Theory **156**, 104152 (2021)
9. Tian, X., Xu, Q., Zhan, Q.: An analytical inverse kinematics solution with joint limits avoidance of 7-DoF anthropomorphic manipulators without offset. J. Franklin Inst. **358**(2), 1252–1272 (2021)
10. Campeau-Lecours, A., Gosselin, C.: An anticipative kinematic limitation avoidance algorithm for collaborative robots: two-dimensional case. In: 2016 IEEE/RSJ International Conference on Intelligent Robots and Systems (IROS), pp. 4232–4237. IEEE (2016)
11. LeBel, P., Gosselin, C., Campeau-Lecours, A.: An anticipative kinematic limitation avoidance algorithm for collaborative robots: three-dimensional case. In: 2017 IEEE/RSJ International Conference on Intelligent Robots and Systems (IROS), pp. 3075–3080. IEEE (2017)
12. Duong, T.H., Jaksic, N.I.: Let's not throw away that big and bulky manipulator–revitalize it! In: 2018 ASEE Annual Conference & Exposition (2018)
13. de Oliveira, A.S., De Pieri, E.R., Moreno, U.F., Martins, D.: A new approach to singularity-free inverse kinematics using dual-quaternionic error chains in the davies method. Robotica **34**(4), 942–956 (2016)
14. de Oliveira, A.S., De Pieri, E.R., Moreno, U.F.: A new method of applying differential kinematics through dual quaternions. Robotica **35**(4), 907–921 (2017)
15. Sciavicco, L., Siciliano, B.: Modelling and Control of Robot Manipulators. Springer, Heidelberg (2012). https://doi.org/10.1007/978-1-4471-0449-0
16. Wan, J., Wu, H.T., Ma, R., Zhang, L.: A study on avoiding joint limits for inverse kinematics of redundant manipulators using improved clamping weighted least-norm method. J. Mech. Sci. Technol. **32**(3), 1367–1378 (2018). https://doi.org/10.1007/s12206-018-0240-7
17. Huang, S., Peng, Y., Wei, W., Xiang, J.: Clamping weighted least-norm method for the manipulator kinematic control: avoiding joint limits. In: Proceedings of the 33rd Chinese Control Conference, pp. 8309–8314. IEEE (2014)

Multi-robot Coordination for a Heterogeneous Fleet of Robots

Diogo Pereira[1,2], Diogo Matos[2,6], Paulo Rebelo[2], Fillipe Ribeiro[3], Pedro Costa[1,2], and José Lima[2,4,5](✉)

[1] Faculty of Engineering of University of Porto, Porto, Portugal
diogo.a.pereira@inesctec.pt, pedrogc@fe.up.pt
[2] INESC-TEC, Centre for Robotics in Industry and Intelligent Systems, Porto, Portugal
{diogo.m.matos,paulo.m.rebelo}@inesctec.pt
[3] JPM Industry, Zona industrial do Rossio, 3731-901 Vale de Cambra, Portugal
Fillipe.Ribeiro@jpm.pt
[4] Research Centre in Digitalization and Intelligent Robotics (CeDRI), Instituto Politécnico de Bragança, Mirandela, Portugal
[5] Laboratório para a Sustentabilidade e Tecnologia em Regiões de Montanha (SusTEC), Instituto Politécnico de Bragança, Portugal, Portugal
jllima@ipb.pt
[6] Universidade de Trás-os-Montes e Alto Douro, Vila Real, Portugal

Abstract. There is an increasing need for autonomous mobile robots (AMRs) in industrial environments. The capability of autonomous movement and transportation of items in industrial environments provides a significant increase in productivity and efficiency. This need, coupled with the possibility of controlling groups of heterogeneous robots, simultaneously addresses a wide range of tasks with different characteristics in the same environment, further increasing productivity and efficiency. This paper will present an implementation of a system capable of coordinating a fleet of heterogeneous robots with robustness. The implemented system must be able to plan a safe and efficient path for these different robots. To achieve this task, the TEA* (Time Enhanced A*) graph search algorithm will be used to coordinate the paths of the robots, along with a graph decomposition module that will be used to improve the efficiency and safety of this system. The project was implemented using the ROS framework and the Stage simulator. Results validate the proposed approach since the system was able to coordinate a fleet of robots in various different tests efficiently and safely, given the heterogeneity of the robots.

1 Introduction

With the advances in the characteristics of autonomous vehicles, such as their autonomy or weight, along with the reduction in their cost, there has been an increase in their usage in industrial environments for a wide range of tasks. Furthermore, using systems made up of a group of vehicles with differences in their brands or morphological structures is becoming more common. The ability

D. Tardioli et al. (Eds.): ROBOT 2022, LNNS 590, pp. 229–240, 2023.
https://doi.org/10.1007/978-3-031-21062-4_19

to coordinate a multi-robot system (MRS) with heterogeneous robots improves the efficiency of these vehicles in tasks where different robots are needed to perform them. An MRS is composed of multiple robots capable of executing tasks that require movement from one place to another. The coordination between various robots with the same morphology is a complex task that is still being improved today. However, a system capable of controlling a fleet of robots that are not homogeneous adds another layer of complexity to the problem and is increasingly a necessity for today's industry, whether it is controlling robots of different brands or controlling robots that perform entirely different roles in the same system. Path planning is a crucial aspect required for achieving coordinated and safe navigation in a system composed of multiple vehicles. With this, it is necessary to develop algorithms that can resolve the need for a vehicle/robot to reach a specific location without human interaction and without causing any collisions with obstacles or other robots in the MRS or deadlocks. In addition, reaching the destination is not the only goal. Finding an efficient solution is an essential aspect of these algorithms with crucial factors such as distance traveled, energy spent, or time spent. Because of this, many of the path planning methods are based on various algorithms that, on the one hand, aim to find the set of safe paths to the goal and, on the other hand, aim to link which of these possibilities lead to the most efficient solution [3]. The main goal of this work is to implement a system capable of coordinating heterogeneous robots based on the TEA* graph search algorithm. This goal is achieved by implementing a graph decomposition algorithm that will decompose the graph used by the TEA* algorithm before planning the paths and make the system more stable and robust when using a fleet of heterogeneous robots. The implemented path planning is applied to a real-time system that already exists, developed by [4] and [6] which already has the tools needed for graph based path coordination systems, such as position detection and controllers to move the robots along specific lines. Additionally, the ROS framework will be used to implement the entirety of the system. Finally, a real-time supervisor will also be implemented. This supervisor will be used to verify that all robots in the system are following the path that was planned and ensure that they do not pose safety concerns.

2 Related Works

Various heterogeneous MRS coordination methods have already been explored in the field of MRS coordination. The methods may vary a lot in terms of approaches. However, the decision-making process, or rather, whether a centralized or a decentralized group architecture is used, seems to be the most significant point of contention. For example, [2] uses a fully distributed method to coordinate the various robots in the system which communicate with each other and use a coordination protocol to coordinate with each other and a monitor connected to the communication network to supervise the behavior of the robots. A typical MRS that uses a centralized architecture is the GOFER [1] which uses a central task planning and scheduling system by viewing all the tasks that need to be resolved and what robots are available to perform those tasks. Furthermore, a hybrid approach is also possible, but the solutions they reach are often sub-optimal, as seen in the following study about MRS coordination [8]. The algorithm used in this paper's project will focus on a more centralized approach due to the nature of the TEA* algorithm.

3 Path Planning

Path planning is one of the most crucial aspects of autonomous movement systems. Path planning algorithms aim to establish paths between start and end points allowing mobile robots to navigate autonomously and safely in a working environment. They generate an efficient solution for a system to change from an initial state to a final state by avoiding static and dynamic obstacles in the environment [3]. As such, a big focus of this paper will be on the implementation of a robust path planning system capable of coordinating heterogeneous robots. To plan the paths for the robots in the system, a graph was used to define within the environment where the robots could travel or not. Using this distinction, it is possible to use graph search algorithms to plan paths for the robots. This paper will use the TEA* algorithm to plan paths within that graph and use it to coordinate paths between the robots. The main advantage of the TEA* algorithm comes from introducing a third temporal layer capable of planning a path that does not collide with any other robot in time or reach other blocking problems such as deadlocks. This feature, coupled with the capabilities of obtaining optimal paths from the A* algorithm, the TEA* accomplishes all the necessary requisites for a robust path planning algorithm.

3.1 Time Enhanced A* (TEA*)

Taking the A* graph search algorithm and adding a third temporal layer enables the neighbor node that will be explored to belong to the next temporal layer, as seen in Fig. 1. With this, it is possible to associate each link to the instant of time it will be occupied, preventing collision between robots or paths being blocked by other robots through deadlocks and creating far more efficient routes. Furthermore, in addition to exploring the neighbor nodes in the next temporal layer, the node itself is also "explored" and re-added to the list with a small increment to the cost. This addition is helpful for cases where the robot staying in place is more efficient than taking a longer route, such as having a robot wait for another robot to pass through instead of immediately attempting to take a longer route that would take longer. The run-through of the TEA* algorithm is pretty similar to the A* algorithm. Firstly, instead of a single graph and states for each node, there is an array of graphs, and each node has a specific state and stored cost at a specific time K which is known as "step". The algorithm runs for as many times as many robots that exist in the system, and, following a priority order, each robot plans its path through time and blocks that path to the lower priority robots by setting the nodes that belong to its path to the state *OBSTACLE*. The next robots plan their path, considering the new *OBSTACLE* nodes in time. It is essential to note that this algorithm is hard

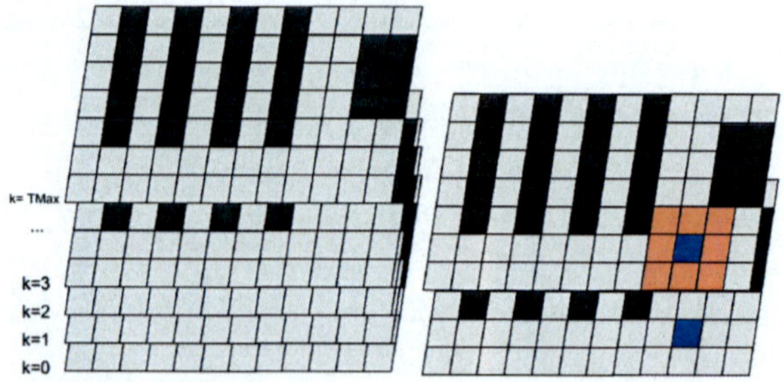

(a) Input Map over the time. (b) Analysed Neighbor Cells

Fig. 1. Input map over the various temporal layers (a) and the analyzed neighbor cells focused on the AGV position (b) [7].

to scale. This problem comes from the necessity of storing a lot of memory regarding the array of graphs used to plan the path in time [5]. On top of that, the algorithm also requires a lot of computational resources, namely read/write speed to and from memory, as the algorithm has to look for the next neighbor node to be explored constantly and update the values of the costs/states. One of the steps to reduce this potential read/write speed problem was utilizing a low-time complexity algorithm to store the graph's data. As such, in a graph, the nodes and links were stored in a C++ map container from the C++ standard library, where the keys of that map were the ID of the node/link, and the graph itself was stored in a vector of graphs. The map container was helpful because the time complexity, in Big-O notation, of searching for a member given the key is $log(n)$. Using such a container is better than using more usual containers such as C++ vectors or double-ended queues (deque), which would have at least $O(n)$ temporal complexity. Regarding using a vector to store the temporal layers of the graph, since the steps go from 0 to K_{max} sequentially, the temporal complexity of accessing a specific temporal graph in that vector is $O(1)$.

To further ensure safety, it is also common to give robots a bigger space gap between them. For example, instead of setting only the current node that the robot is, in time, as *OBSTACLE*, the neighbor nodes around that node are also set as *OBSTACLE*. Additionally, it is also common to set as *OBSTACLE* a few previous nodes in time as *OBSTACLE*. For example, if it is defined that three nodes in time are to be set as *OBSTACLE* then if a robot has a path $P_A = \{1 \rightarrow 2 \rightarrow 3 \rightarrow 4 \rightarrow 5\}$, when planning the path for that robot, in the step K that the node (4) is expected to be *OBSTACLE*, both the previous nodes (3) and (2) are also set as *OBSTACLE* to ensure further safety. These gaps are implemented in a way to depend on a safety parameter called *SAFETY_GAP*, which sets both the number of previous path nodes that are set to *OBSTACLE* but also the depth of the neighbors of the current path node to be set as *OBSTACLE*. For example, if *SAFETY_GAP* is 4 and path is $P_A = \{1 \rightarrow 2 \rightarrow 3 \rightarrow 4 \rightarrow 5\}$, in the step where node (4) is going to expected to have the robot, the nodes (4), (3), (2) and (1) are set to *OBSTACLE* and the neighbor nodes of the node (4) and the neighbors or those neighbors are also set to *OBSTACLE*. Additional examples can be seen in Fig. 2. A relevant note is that the smaller the *SAFETY_GAP* value, the more efficient the planned path is, but it also becomes less safe. With this, the TEA* algorithm can plan efficient and collision-free paths for the various robots in the system.

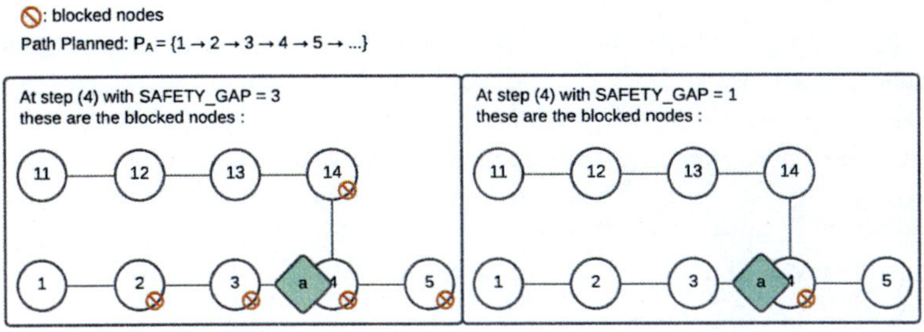

Fig. 2. Additional examples of the path planning *OBSTACLE* setup

3.2 Robot Priority Swapping

Even though the base TEA* algorithm can plan paths for all the robots in the system, extra tweaks exist to optimize it further. One of these can be the addition of a priority swap for the robots. This swap is an essential feature as sometimes the base algorithm cannot find a path for the current setup of robot priorities. For example, if there is a robot priority setup where the higher priority robot, after planning its path, blocks any possible path for the lower priority robot that means it is impossible to complete that order. But, if the priorities are switched, there is still a possibility of paths being found for all robots. Thus, having a list of possible priority setups can help eliminate this problem and find paths that the base algorithm would be unable to.

3.3 Moving Idle Robots

Another relevant feature to incorporate into the base algorithm is moving other robots when idle and in the way of another robot. This happens, for example, when there is a robot idle in a position that another robot needs to complete its path. By creating an extra order to that idle robot, if available to move, it is possible to move that robot and enable the passage of the robot that was being blocked. This solution can be achieved by first detecting if a path is not found because another lower priority robot blocks the destination by being stationary and plans the path ignoring that stationary robot. Then, by looking into the nearby stations of the stationary robot, it is possible to attempt to move the idle robot to the closest unused station, thus unblocking the way. This feature should only be applied to robots of lower priority, so combining this method with the previous process of priority swapping helps achieve more possible paths.

4 System Implementation

In order to implement and test a system capable of coordinating a group of robots, it was necessary to recreate an industrial-like environment where the conditions are similar to a real one. Furthermore, this system needs to work with the ROS framework, which is a requirement imposed. Using the *Stage* simulator, various environments were set up to host the robots that were going to be coordinated by the implemented system. With this simulation, it was possible to simulate a nearly real industrial environment by creating walls, corridors and even objects that could collide with the robots. Furthermore, these objects could be moved to create obstacles for the robot's paths and test safety features. The robots themselves were also simulated in this simulator. The robots could localize themselves using simulated LIDAR sensors on their front and back, along with collision detectors that could detect if the robots had collided with anything, forcing them to stop.

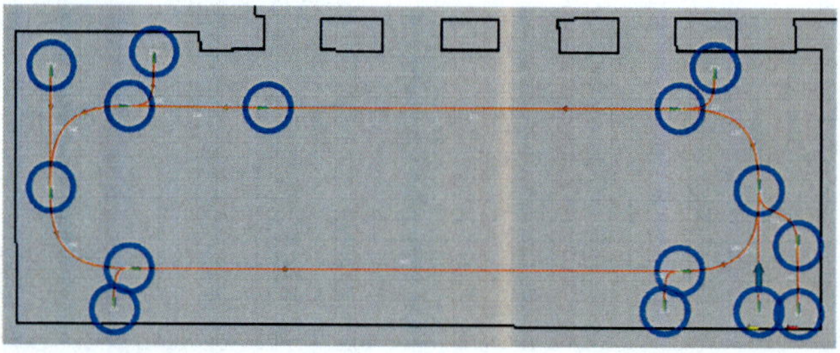

Fig. 3. An undecomposed small map and its graph used in simulations.

To view all the data being circulated, such as the current robot's estimated location and the graph points, the RVIZ visualization tool from the ROS framework is used. With this tool, it is possible to visualize the robots running through the graph and have the option to send orders directed through the RVIZ tool for robots to move from one point to another. This tool can also be used to edit and save graphs instead of manually inserting coordinates and values in the manual definition. In Fig. 3 an example of a RVIZ visualization is presented where it is possible to see the various graph links as red lines, the nodes as blue circles and the robot's position as the big blue arrow. In Fig. 4 a block diagram of the implemented system is presented.

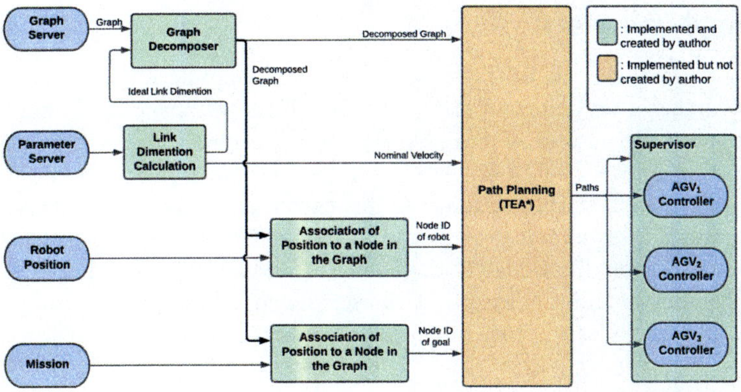

Fig. 4. Diagram of the system architecture implemented

In sum, a short description of the modules of the system implementation can be given as such:

- Firstly, a graph is created. In this project a graph is created using a list of nodes and links that describe the various characteristics of these (curves, length, ID, etc.) and storing them into a YAML file.
- Given a parameter server that holds various characteristics (size and traction type), an ideal link dimension is calculated for the decomposed graph.
- Before the system starts, these graphs can be optionally decomposed. This decomposition means that the graph links are subsequently divided so that all links have a similar length.
- Once the system is started and given an order for a robot to move, both the robot's current position and the order's destination position are associated with their position relative to a node in the graph.
- Finally, a path can be calculated by knowing the node to which the robot belongs and its destination node. This path planning also considers notable exceptions in an exception table. Once a path is calculated, it is sent to each corresponding robot.
- On top of all, a supervisor also exists to ensure that the robots are following the paths planned without the risk of problems such as collisions and deadlocks.

4.1 Graph

The graphs are defined using the YAML serialization language and specifying various fields that describe the characteristics of various nodes and links that create the graph. For nodes the fields are: ID, angle and position. For links the fields are: ID, origin and destination IDs, velocity forward/backward and other parameters for the description of a Bézier curve.

4.2 Graph Decomposition

Due to how the TEA* algorithm works, the path planning returns a sequence of edges or vertices that define a path. Therefore, if the various edges' sizes are very different, the time it takes for a robot to travel an edge may be substantially different from another robot traveling to another edge, which is even more substantial when the fact that different edges can have different velocities is taken into account. This difference can be dangerous as TEA* assumes that the time it takes to go through an edge is the same for all edges, and if that is not true, the path planned cannot ensure that there is no collision between robots or that potential deadlocks can occur. In Fig. 5 presents an example of this problem.

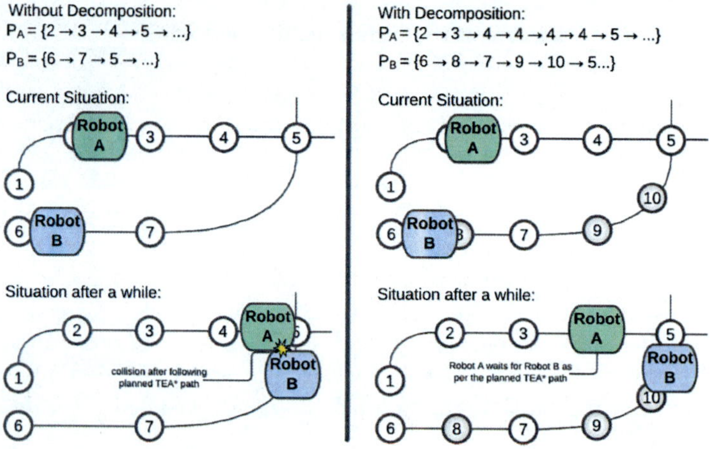

Fig. 5. Example of collision due to lack of graph decomposition

Because of these problems, to ensure that the robots are proceed safely through their path, it is necessary to decompose longer links into smaller links to minimize the differences of time that it takes to travel between the different edges. Furthermore, this decomposition also optimizes the routes for the robots, as decomposing a lengthier edge into smaller ones allows for more than one robot to travel that edge sequentially. To implement this, firstly, a decomposition algorithm was created. This algorithm's goal was to, given a specific step length, divide all links in a graph in a way that minimizes the error of the new division of the links when comparing to the specific step. The errors of this division are all taken and the average of this error is calculated. Using this error, and cycling between an interval of values, it is possible to use this algorithm to obtain the value of specific step length that minimizes the average of the error when decomposing the links. After obtaining that value the graph is decomposed using the initial algorithm.

4.3 Supervision

Because in an industrial environment, many variables exist at all times, a supervisor for such a system as a path planner is crucial to ensure safety and efficiency. As such, a supervisor was created to observe the planned path by the system and ensure that all robots within the system are following that path.

Assuming a worst-case scenario that all links are the same size as the robots (and never smaller than the robots), and two robots are following each other in a path, as long as they are not over one step ahead of the other robot regarding the route planned, they will not collide. If such a thing happens, the supervisor acts and triggers a new path planning. In Fig. 6 this is presented.

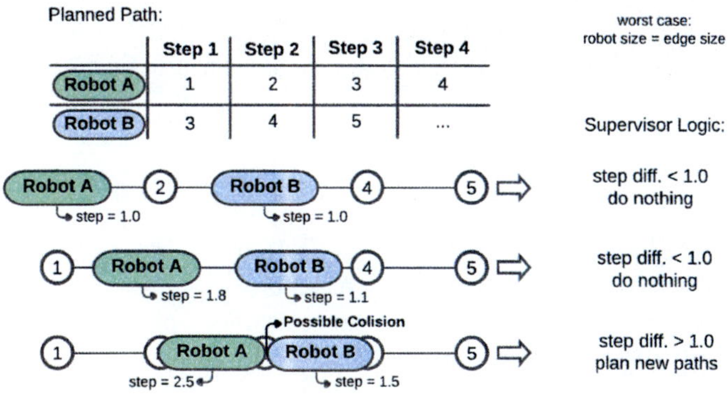

Fig. 6. Supervisor interventions in the implemented system

Because of how the system works, in the path planner, a robot belongs to a node until it reaches another. Looking at Fig. 6 when the replanning is triggered, the robot (a) that is still detected as being in node (2) will have a path that will make the robot (a) go back to node (2) and stays there for one step. In contrast, the other robot is detected as being in node (3) and plans a new path from that node, which means the robot (b) keeps advancing as he was, thus eliminating possibilities for collisions.

5 Results and Discussion

To ensure that the other tests could be done safely, the basic features of the TEA* were tested along with the extra features added to complement it[1]. Once all these tests were completed and working as intended, the supervisor was tested. To test the supervisor four robots were set on a path to follow each other in a straight line to verify the system's response when subjected to a supervisor re-planning

[1] Video of basic robot coordination tests: https://youtu.be/EEKb_r0efJY.

of the paths[2]. The robots were set in a specific order where the blue robot was at the front, followed by the red, yellow and orange robots in this order. When in movement, one of the robots was manually blocked from moving so it could trigger a new path planning and verify its effects. The test results are shown in the Table 1.

Table 1. Effects of supervisor interventions in decomposed vs undecomposed graphs

Line Test Blocking Yellow Robot				
Distance traveled by:	Blue Robot (m)	Red Robot (m)	Yellow Robot (m)	Orange Robot (m)
Decomposed	22,27	22,34	23,85	24,53
Undecomposed	22,41	22,39	24,01	26,16

Line Test Blocking Red Robot				
Distance traveled by:	Blue Robot (m)	Red Robot (m)	Yellow Robot (m)	Orange Robot (m)
Decomposed	22,29	22,91	24,22	25,09
Undecomposed	22,56	23,13	27,54	31,00

Looking at these results, it is possible to conclude that the decomposition of the graph not only makes the system safer but also faster since the distance traveled by the robots affected by the supervisor's re-planning of the paths is lower when the graph is decomposed. This is because the robots affected have to take much longer routes if their paths are re-planned when the graph is not decomposed and, as such, the orders take quite a bit longer to complete.

Some tests were also made to test the graph decomposition when attempting to coordinate a fleet of heterogeneous robots. These tests are composed of generating orders to send four heterogeneous robots simultaneously to different places in the map making them follow approximate equivalent paths with the same number of nodes[3].

In sum, the heterogeneity of the robots affects the system in two major aspects. On one hand, the size of the robots and their morphologies affect the decomposition parameters, which affects the whole decomposition. On the other hand, the traction type affects the irregularity of the velocity parameters, making them have different behaviors when moving, such as slippage or different behaviors when turning. While the first problem can be solved by re-decomposing the graph with the different parameters, the second one cannot be directly solved the same way and must depend on the supervisor to act when needed to ensure that everything is kept safe.

In the tests, the supervisor only seemed to act about 1–2 times when the system was coordinating four robots at the same time with long paths composed of about 50 nodes. These tests were an overall success as the amount of times the supervisor had to act was pretty low. Furthermore, the supervisor was acting in

[2] Video of supervisor tests: https://youtu.be/8qBK5eCUi4o.

[3] Video of heterogeneous robot coordination tests: https://youtu.be/RDjN2HEko40.

a strict mode where no matter where the robots were in the map, the difference limit for it to act was always 1.0 steps. This, in a more real scenario, would have reduced the number of interventions by the supervisor substantially, potentially to zero. Comparing to some tests done with using homogeneous robots, as seen in Table 2, the amount of times the supervisor had to act with the heterogeneous robots was a bit higher.

Table 2. The average number of actuations by the supervisor when attempting to coordinate a fleet of heterogeneous vs homogeneous robots.

	Homogeneous robots		Heterogeneous robots	
	Undecomposed	Decomposed	Undecomposed	Decomposed
Avg. Actuation of Supervisor	4.4	1	5.2	1.8

This is expected, not only due to the different traction type explained earlier but also because of the difference in sizes of the robots. The bigger the robots, the worse the decomposition of the graph. This is because the decomposition algorithm aims to have a link sizes always higher than the largest robot in the system so that the supervisor works as expected. With this, the higher the minimum size of the links, the higher the error of decomposition will be, and, as such, results are expected to be worse.

6 Conclusions and Future Work

Looking at the results obtained during the project's test phase, it is possible to conclude various aspects regarding the accomplishment of various goals. The first one is the implementation of the TEA* algorithm was done successfully allowing for the other modules to be tested without problems arising. Regarding the implemented supervisor, it was able to keep the robots from deviating too much from the expected temporal planned path by the TEA* algorithm by triggering a path re-planning for all the robots in the system to prevent extreme deviations that could lead to collisions or deadlocks. Even though the supervisor sometimes had to act four or five times to complete a full order, in no case the supervisor failed to plan new paths and complete all orders. Regarding the graph decomposition algorithm, the results proved to be quite favorable. The algorithm was able to decompose the graph's links leading to the robot movement deviating way less from the expected path generated by the path planning system. The supervisor had to act a considerably low amount of times when the graph used to plan the paths of the robots was decomposed. Considering the extreme case the supervisor was subjected to act by removing the feature that only has the supervisor work on a tight step difference when the paths of two robots intercept, these results were pretty good.

These results together make the system quite robust to the heterogeneity of the group of robots. Although the system proved quite effective in a simulation environment, it was impossible to test it in a real scenario due to time constraints. As such, it is hard to verify that the system would be able to work as expected using real robots. Furthermore, although considerably efficient and safe, the paths obtained by the implemented system are still not fully optimal. Finally, the system also suffered from some common vulnerabilities of a centralized system, such as having a single point of failure.

Acknowledgements. This work is financed by the ERDF - European Regional Development Fund, through the Operational Programme for Competitiveness and Internationalisation - COMPETE 2020 Programme under the Portugal 2020 Partnership Agreement, within project CrossLog, with reference POCI-01-0247-FEDER-039895.

References

1. Caloud, P., Choi, W., Latombe, J.C., Le Pape, C., Yim, M.: Indoor automation with many mobile robots. In: EEE International Workshop on Intelligent Robots and Systems, Towards a New Frontier of Applications, vol. 1, pp. 67–72 (1990). https://doi.org/10.1109/IROS.1990.262370
2. Iocchi, L., Nardi, D., Piaggio, M., Sgorbissa, A.: Distributed coordination in heterogeneous multi-robot systems. Auton. Robot. **15**, 155–168 (2003). https://doi.org/10.1023/A:1025589008533
3. Ángel Madridano, Al-Kaff, A., Martín, D., de la Escalera, A.: Trajectory planning for multi-robot systems: methods and applications. Expert Syst. Appl. **173**, 114660 (2021). https://doi.org/10.1016/j.eswa.2021.114660
4. Matos, D., Costa, P., Lima, J., Costa, P.: Multi AGV coordination tolerant to communication failures. Robotics **10**(2) (2021). https://doi.org/10.3390/robotics10020055, www.mdpi.com/2218-6581/10/2/55
5. Moura, P., Costa, P., Lima, J., Costa, P.: A temporal optimization applied to time enhanced a*, vol. 2116 (2019). https://doi.org/10.1063/1.5114225
6. Rocha, C., et al.: Development of an autonomous mobile towing vehicle for logistic tasks. In: 4th Iberian Robotics Conference, ROBOT 2019, pp. 669–681 (2020). https://doi.org/10.1007/978-3-030-35990-4_54
7. Santos, J., Costa, P., Rocha, L.F., Moreira, A.P., Veiga, G.: Time enhanced a: towards the development of a new approach for multi-robot coordination, vol. 2015-June (2015). https://doi.org/10.1109/ICIT.2015.7125589
8. Yan, Z., Jouandeau, N., Cherif, A.A.: A survey and analysis of multi-robot coordination. Int. J. Adv. Rob. Syst. **10**(12), 399 (2013). https://doi.org/10.5772/57313

Rehabilitation and Assistive Robotics

A Neuromechanical Model of Knee Flexion-Extension Based on Linear Regressors for Neuro-Controlled Exoskeletons

Lluis Bernat, Vicente Morell, Jose L. Ramon, Jorge Pomares, and Andres Ubeda(✉)

Human Robotics Group, University of Alicante, Alicante, Spain
andres.ubeda@ua.es
http://www.huro.ua.es

Abstract. In this paper we evaluate the use of linear regressors to reconstruct knee angles from electromyographic signals recorded from lower-limb muscles in charge of knee flexion and extension. This neuromechanical model has been tested under two different speed conditions showing a very reliable reconstruction of knee kinematics for all participants. A generalized model, including all participants, has also been tested with a small reduction of decoding accuracy, but still showing enough reliability. The application of neuromechanical modelling has important advantages in the generation of a more natural and embodied control of lower-limb exoskeletons. The proposed model is expected to have a robust behavior in a real-time control of this kind of devices.

Keywords: Neuromechanics · Machine learning · Exoskeletons · Rehabilitation

1 Introduction

During the last few years, conventional rehabilitation has seen an important development in complementary tools to increase the effectiveness of the therapy. One of the main ways of providing a faster and better recovery is the use of robotic devices. In the case of human gait, lower-limb exoskeletons have provided a way to increase the intensity and repeatability of rehabilitation protocols [1]. Most of current lower-limb exoskeletons are based on post-movement control paradigms adapting gait patterns to the patient's motion intention inferred from force/position sensors. Post-movement actuation is a robust method for lower-limb exoskeleton control. However, from the point of view of the user, the introduction of a control input based on the neuromuscular intentions of the patient, provides a better embodiment with the device, and increases ergonomy by reducing physical effort.

A possible strategy is the estimation of the needed torque to perform the movement by measuring surface electromyography (EMG) [2]. A significant

D. Tardioli et al. (Eds.): ROBOT 2022, LNNS 590, pp. 243–251, 2023.
https://doi.org/10.1007/978-3-031-21062-4_20

example of the introduction of EMG in the control loop is HAL [3], an exoskeleton that allows moving with or without electrophysiological assistance. The main drawback of neural control is that it depends on specific calibrations across participants and sessions and, thus, a precise and reliable model is necessary, which is not always possible, as electrophysiological signals have a high variability. On the other hand, EMG-based techniques do not need dynamic modelling of the limbs and of their interaction with the environment [4]. A great advantage of EMG-control is that it allows incomplete spinal cord injured patients to substitute uncomfortable wheelchair and exoskeleton control using joysticks, as they keep some voluntary muscular activation, even though it may not be totally functional [5].

In this work we propose to apply neuromechanical modelling to infer human movement from neuromuscular activity using multidimensional linear regressors. Neuromechanical modelling has been applied with great success in the implementation of neural interfaces [6]. One of the methods to infer motion kinematics from the neural system is the use of multidimensional linear regressors. First approaches to decode limb kinematics from electroencephalographic (EEG) signals show that it is feasible to extract significant information of motion from low-frequency EEG components [7,8]. However, these methods are still not very reliable due to the noisy nature of the cortical recordings. On the other hand, the application of this kind of methods to muscular electrical activity is much more robust. Examples of this are works on myoelectric control [9] or exoskeleton control [10].

Our preliminary results show that it is possible to decode knee angles during flexion and extension activities with high reliability and only using four EMG channels. We also show that the generalization of the model across different participants reduces the overall reliability but it still offers good decoding results. The neuromechanical modelling was applied at two different movement speeds showing small changes in accuracy. As a general conclusion, we expect this modelling to be feasible in a real-time control of lower-limb robotic orthoses and exoskeletons.

2 Methods

2.1 Experimental Setup

The main goal of this study is to evaluate how well knee angles can be decoded from muscular information. The ultimate goal is to apply these decoders to walking activities that involve a complex coordination of lower-limb muscles. However, to simplify the approach, exercises involving exclusively knee flexion and extension have been chosen. This simpler exercises will allow validating the proposed system in a more restricted condition by monitoring specific muscles in charge of the selected movement.

5 healthy participants (all male, age 20.6 $\pm$ 0.5) participated in the study. Informed consent was signed following the protocol of Helsinki. The exercise performed is described in Fig. 1. Each participant sat on a flat surface so that

their legs did not touch the ground. In this position, they moved their dominant leg by flexing/extending it. This movement consisted of starting with a knee angle of 90º, then extending the leg to the maximum extension, continuing to flex it to the maximum range of flexion, and finally returning to the original 90º position. Participants were asked to avoid unnecessary force to get to the extreme positions of the proposed movement. Movements were performed at two different speeds to evaluate the behavior of the neuromechanical model. For V1 condition the duration of the whole movement was fixed to 8 s (normal movement) and for V2 condition the duration was fixed to 15 s (slow movement). Participants were provided with a visual feedback to adjust movement speed. For each speed 30 repetitions were performed with short breaks of around 5 s in between.

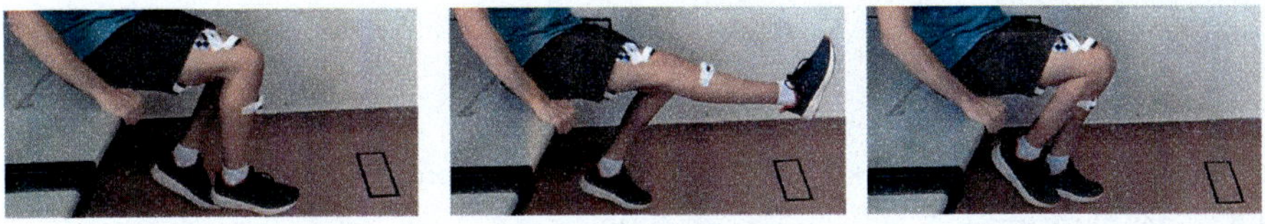

(a) *Resting position.* (b) Extension. (c) Flexion.

Fig. 1. Overview of the exercises carried out during the measurements.

2.2 EMG and Kinematic Recordings

Knee angles were obtained by using two Inertial Measurement Units (IMU) (BWT901CL) placed on the thigh and on the shin respectively. Both sensors were positioned so that their difference in orientation only varies with respect to one of the axes. To obtain a common recording of both sensors a specific Matlab acquisition code was implemented using the Bluetooth communication protocol provided with the sensors. The sampling frequency 100 Hz. To compute the knee angle, Euler angles from both sensors were transformed into quaternions and the difference between quaternions was computed to obtain the angle in the saggital plane which was the movement plane of the knee flexion and extension exercise.

EMG signals were recorded from four muscles in charge of knee flexion and extension using a Noraxon MiniDTS amplifier with bipolar wireless electrodes with a sampling frequency of 1500 Hz Hz. The muscles were selected as follows:

- Muscles responsible for extension:
  - Vastus lateralis (Fig. 2a).
  - Vastus medialis (Fig. 2b).
- Muscles responsible for flexion:
  - Biceps femoris (Fig. 2c).
  - Semitendinosus (Fig. 2d).

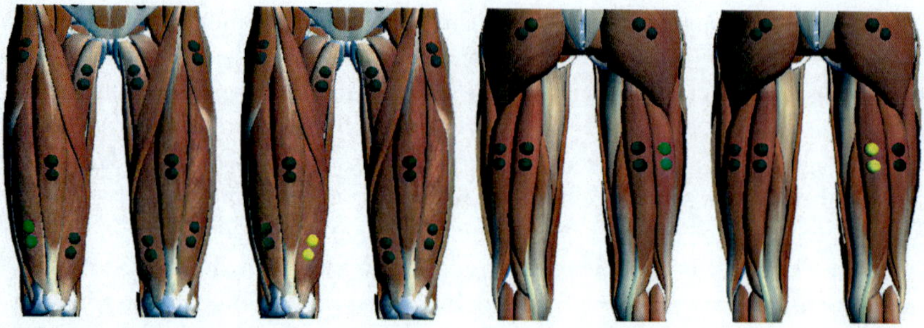

(a) Vastus lateralis. (b) Vastus medialis. (c) Biceps femoris. (d) Semitendinosus.

Fig. 2. Muscles measured during the experiments (image modified from Noraxon MR licensed software).

EMG signals were rectified and then a linear envelope was computed using a non-causal moving average filter with a 50 ms window to obtain non-negative information of muscle contractions.

Finally, both kinematic and EMG signals were synchronised using an Arduino UNO microcontroller which was in charge of sending a trigger signal to the EMG amplifier while acquiring the inertial sensors information. After the experiments, kinematic signals were resampled to 1500 Hz Hz to match EMG recordings.

2.3 Neuromechanical Model

To decode the knee angle from the EMG data, a multidimensional linear regression has been applied, following the expression defined in Eq. 1:

$$x[t] = a + \sum_{n=1}^{N} b_n S_n[t]. \tag{1}$$

were $x[t]$ are the knee angles at time t and S_n is the voltage measured at each EMG channel n at time t. N is the number of channels, and a and b are the linear regression weights. For the present work, N corresponds to 4 (number of channels introduced in the decoder, or in other words, number of muscles measured, since an individual EMG signal is obtained for each one).

According to the protocol, a total of 30 repetitions of knee flexion/extension have been obtained for each participant and for each of the two speeds. A 6-fold cross-validation has been applied to evaluate the reliability of the model. The validation metric proposed to evaluate the neuromechanical model is the correlation between the decoded and the real knee angle (decoding correlation, DC). Participants variability is very present in electrophysiological modeling due to the intrinsic and extrinsic differences in EMG signals from different participants. For this reason, the validation includes both individual and generalized models, to see if the proposed model is able to characterize the general behavior of muscles in charge of knee flexion and extension.

3 Results and Validation

Table 1 shows the set of training sessions carried out, as well as the DC and standard deviation (STD) obtained after applying the neuromechanical modelling. Training sessions (T1-T15) include individual performance of each participant (P1-P5) on each speed (V1 and V2) and combining data of both speeds (V1+V2). Generalized models are also tested in training sessions T16-T18 by combining data from all the participants. Decoding performance is also graphically presented in Fig. 3.

Table 1. Decoding performance for the different training sets and detail of the evaluated conditions.

Training	Participants	Speed	DC	STD
T1	P1	V1	0.54	0.35
T2	P1	V2	0.71	0.12
T3	P1	V1, V2	0.63	0.26
T4	P2	V1	0.71	0.24
T5	P2	V2	0.87	0.03
T6	P2	V1, V2	0.79	0.18
T7	P3	V1	0.68	0.07
T8	P3	V2	0.71	0.12
T9	P3	V1, V2	0.70	0.10
T10	P4	V1	0.70	0.35
T11	P4	V2	0.82	0.06
T12	P4	V1, V2	0.76	0.25
T13	P5	V1	0.73	0.07
T14	P5	V2	0.67	0.14
T15	P5	V1, V2	0.69	0.12
T16	P1-P5	V1	0.57	0.28
T17	P1-P5	V2	0.63	0.13
T18	P1-P5	V1, V2	0.60	0.22

As it can be seen, the DC is variable across the different training sessions, although for all of them values higher than 0.50 are obtained, and the average across all the training sessions is around 0.7.

The results are also variable across participants. For example, training sessions T1, T2 and T3 of participant P1 have a lower DC and a higher STD than training sessions T4, T5 and T6 of participant P2. A graphical example of this difference can be seen in Fig. 4. Participant P1 obtains an average of 0.63 ± 0.24, while participant P2 obtains an average of 0.79 ± 0.15.

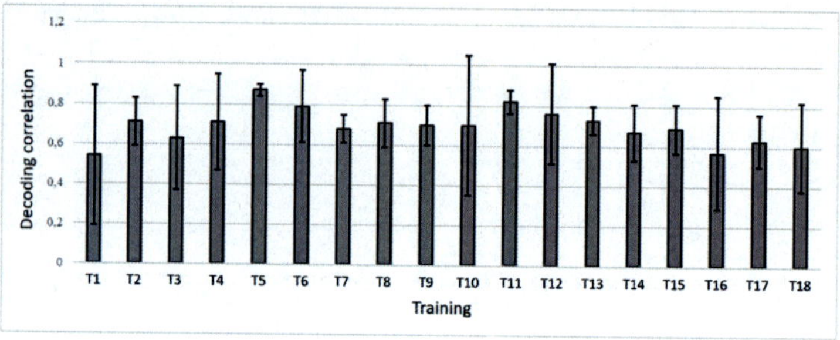

Fig. 3. Decoding performance for the different training sessions.

Regarding speed, it was found that V2 obtained a higher DC and a lower STD for most of the cases. For V1 speed, the average correlation is 0.66 ± 0.23, while for V2 speed, the average correlation is 0.74 ± 0.10. For both speeds, DC and STD values are between those obtained for each speed separately. In this case, the average correlation is 0.70 ± 0.19.

It is also relevant to study how the neuromechanical model can be generalized. T16, T17 and T18 include the whole set of participants for each different speed and the combination of both. The average correlation of combined models is 0.60 ± 0.21, while the average correlation of individual models is 0.71 ± 0.16, showing that individual models are more reliable.

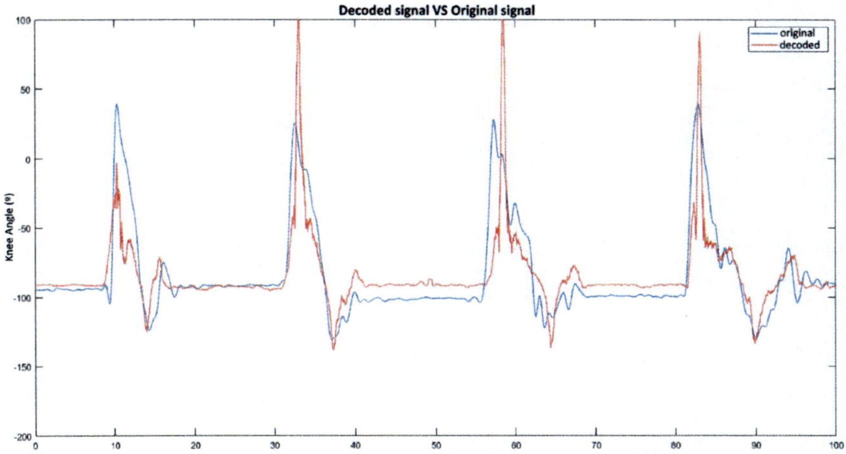

(a) T2 training.

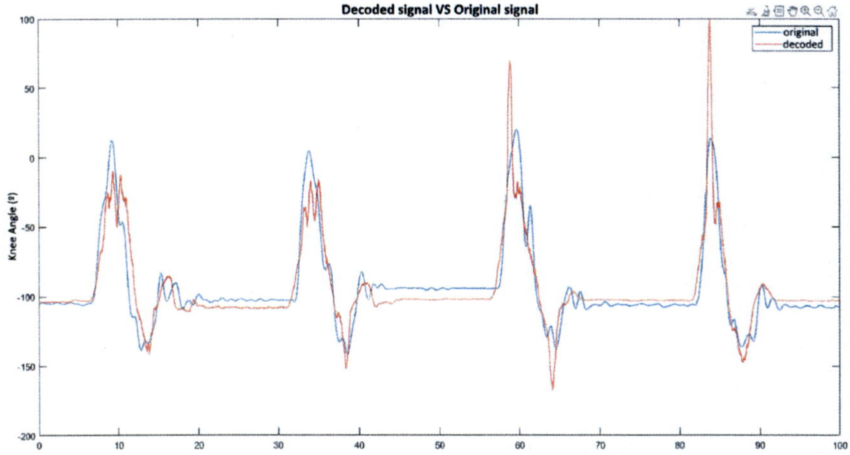

(b) T5 training.

Fig. 4. Example of the decoding of the knee angle for two significant training sets (T2 and T5).

4 Discussion

The proposed method shows robustness across folds with small standard deviations. One of the limitations of signal correlation as a metric to evaluate the reliability of the decoder is that it is very phase dependent. As a consequence, well decoded signals can obtain low correlations even if they follow accurately the actual angle. An example of this can be seen in Fig. 5. In this case, the angle is very well decoded but with a short delay, which decreases the final correlation to 0.54. To better assess the reliability of the decoder other metrics can be evaluated as proposed in [11].

Another interesting aspect is that, although individual models are more robust, a generalized model (T16-T18) can also decode knee angle from the same number of EMG channels in a quite robust way. Taking into account the limitations of the validation metric, we expect that generalized models could also be applied in a real-time basis. The same happens when analyzing different speeds. Differences in decoding performance are small, although a lower speed seems to be better decoded.

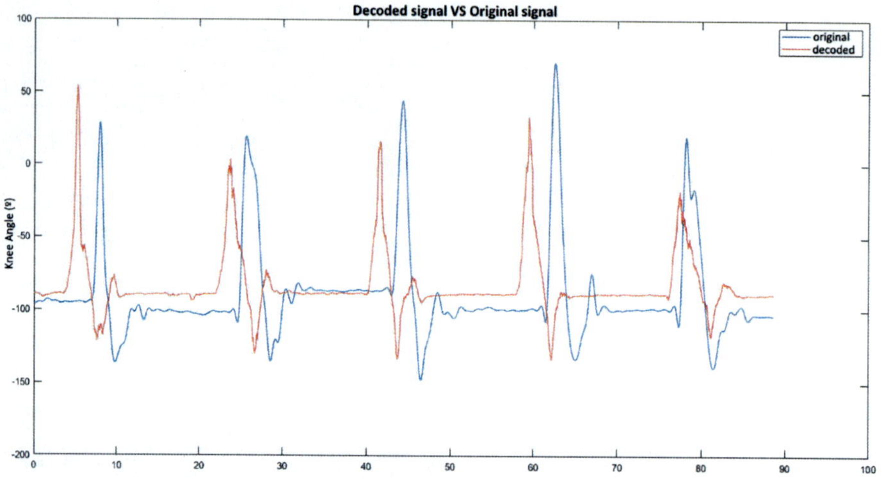

Fig. 5. Decoded angle and its original with a clear phase delay (E1 training).

5 Conclusion and Future Work

This study shows that it is possible to decode knee angles during flexion and extension activities with high reliability and only using four EMG channels. The neuromechanical modelling was applied at two different movement speeds showing small changes in accuracy. The generalization of the model across different participants reduces the overall reliability but it still offers good decoding results.

Our future work aims at optimizing the decoders to be applied in a real-time basis. For this purpose, an actuated lower-limb knee exoskeleton will be commanded by means of the proposed neuromechanical modelling during physical therapy of knee flexion and extension in motor impaired patients. EMG information from the healthy leg will replicate movement on the paralysed one with the exoskeleton using the concept of assisted-as-needed. We expect this approach will speed the recovery of motor function.

References

1. Rodríguez-Fernández, A., Lobo-Prat, J., Font-Llagunes, J.M.: Systematic review on wearable lower-limb exoskeletons for gait training in neuromuscular impairments. J. Neuroeng. Rehabil. **18**, 22 (2021)
2. Lenzi, T., De Rossi, S.M.M., Vitiello, N., Carrozza, M.C.: Intention-based EMG control for powered exoeskeletons. IEEE Trans. Biomed. Eng. **59**(8), 2180–2190 (2012)
3. Sankai, Y.: HAL: hybrid assistive limb based on cybernics. In: Kaneko, M., Nakamura, Y. (eds.) Robotics Research. Springer Tracts in Advanced Robotics, vol. 66, pp. 25–34. Springer, Berlin, Heidelberg (2011). https://doi.org/10.1007/978-3-642-14743-2_3
4. Buchanan, T.S., Lloyd, D.G., Manal, K., Besier, T.F.: Neuromusculoskeletal modeling: estimation of muscle forces and joint moments and movements from measurements of neural command. J. Appl. Biomech. **20**, 295–367 (2004)
5. Wilcox, M., Rathore, A., Ramirez, D.Z.M., Loureiro, R.C.V., Carlson, T.: Muscular activity and physical interaction forces during lower limb exoskeleton use. Healthcare Technologies Letters **3**(4), 273–270 (2016)
6. Sartori, M., Gizzi, L., Lloyd, D., Farina, D.: A musculoskeletal model of human locomotion driven by a low dimensional set of impulsive excitation primitives. Front. Comput. Neurosci. **7**, 79 (2013)
7. Úbeda, A., Azorín, J.M., Chavarriaga, R., R Millán, J.D.: Classification of upper limb center-out reaching tasks by means of EEG-based continuous decoding techniques. J. Neuroeng. Rehabil. **14**(1), 1–14 (2017)
8. Úbeda, A., Azorín, J.M., Farina, D., Sartori, M.: Estimation of neuromuscular primitives from EEG slow cortical potentials in incomplete spinal cord injury individuals for a new class of brain-machine interfaces. Front. Comput. Neurosci. **12**, 3 (2018)
9. Hahne, J.M., et al.: Linear and nonlinear regression techniques for simultaneous and proportional myoelectric control. IEEE Trans. Neural Syst. Rehabil. Eng. **22**(2), 269–279 (2014)
10. Zhang, L., Long, J., Zhao, R., Cao, H., Zhang, K.: Estimation of the continuous pronation-supination movement by using multichannel EMG signal features and Kalman filter: Application to control an exoskeleton. Front. Bioeng. Biotechnol. **9**, 771255 (2021)
11. Spuler, M., Sarasola-Sanz, A., Birbaumer, N., Rosenstiel, W., Ramos-Murguialday A.: Comparing metrics to evaluate performance of regression methods for decoding of neural signals. In: Annual International Conference IEEE Engineering in Medicine and Biology Society (2015)

A Preliminary Approach to a Wheelchair with Embedded Exoskeleton Capabilities

Alícia Casals[1(✉)], Joan Aranda[1], Manel Frigola[1], Joaquim Minguella[2], Luis Miguel Muñoz-Morgado[1], and Josep Amat[1]

[1] Center of Research in Biomedical Engineering, Technical University of Catalonia, Barcelona, Spain
alicia.casals@upc.edu

[2] Research Group in Manufacturing Technologies, Technical University of Catalonia, Barcelona, Spain

Abstract. Combining mobility with assistance for the interaction with the environment is not a common issue in designing and developing a robotized wheelchair. This work describes a robotic hybrid platform that combines mobility with assistance to the user for actuation in its environment. It is conceived for wheelchair users having limited upper limbs actuation. This hybrid platform, a chair-exoskeleton architecture, faces the challenge of providing the required degrees of freedom to achieve an acceptable performance without having a significant implication in weight, size and cost, which increase would limit its real applicability and affordability. Therefore, its design looks for a solution that combines chair performances by providing mobility, posture and manipulation accessibility, with those of an exoskeleton which allow users with some remaining arms mobility to reach and manipulate objects either on the floor, at seating level, or at high level shelters. This paper describes the preliminary design of the wheelchair and the simulation of its operativity to deal with the required user interaction with the environment and points out the need of an intuitive and adapted interface for its control.

Keywords: Robotized wheelchair · Assistive robotics · Kinematics chain

1 Introduction

The main function of a wheelchair is providing its user with the capability of moving around in seating position. Due to the limiting capabilities of such chairs, the need of facilitating other postures, from standing to lying, has led to the development of a variety of designs that provide such performances. However, wheelchairs with such posture options still fail to provide enough assistance to users with severe disability that prevents them to control their upper limbs with enough performance to interact with the environment due to their limited reaching, force or dexterous capabilities.

An increase in performance implies higher wheelchair complexity, resulting in usability drawbacks such as: larger weight and size, more demanding user's control

D. Tardioli et al. (Eds.): ROBOT 2022, LNNS 590, pp. 252–263, 2023.
https://doi.org/10.1007/978-3-031-21062-4_21

abilities and higher cost. Therefore, adding new performances for assisting the user in manipulating objects in their environment will require a careful compromise complexity-performance, as well as some level of intelligence to interpret the user's needs for its affordability and ease of use. The Posmofya project: *Orthosis-Chair Hybrid Platform to make compatible Mobility, Functionality and Acceptability for application in domestic environments*, aims to provide a wheelchair with several posture capabilities and additional assistance for manipulation. The addition of complementary links for assisting upper limbs displacements, complements the user's movements so as not only being able to move, but also to reach objects and inter-actuate with the environment at the height levels an abled person can do. To avoid the increasing complexity of adding new performances, the design should face a trade-off between the performances it can offer and its burden and cost.

Therefore, this work aims to look for a novel solution that combines mobility with manipulability for users with some remaining arm and hand actuation capabilities, but whose performance limits their upper limbs operability, mainly reaching and dexterity. In its first design, two Degrees of Freedom (DoF) are added to the articulated wheelchair, aimed to compensate the lack of user's force when trying to raise the arms. The challenges to deal with are, first offering these performances for an affordable chair and second, providing an adequate interface for its friendly and safe controllability. This paper describes the preliminary works to deal with the former, the conceptual design of the chair and its modeling for designing and evaluating the control architecture and the kinematics of the whole platform.

2 State of the Art

Most of the research in this field deals separately on mobility for displacements and body posture and on the movements and actuation of the limbs. The technical assistance is provided by wheelchairs or exoskeletons, respectively. Some chairs mobility is solved quite effectively, having some of them integrated additional performances as capabilities of moving along rough terrain. In this line, the Whill's chair is a four-wheel-drive, two of them omnidirectional allowing maneuvering with a tight turning radius, similar to the Chinese Observer all terrain [1]. The Patrafour by Toyota [2] besides being all terrain is endowed with autonomous navigation [3]. Others focus on allowing higher reachability for manipulation by elevating or bending the seat, like iLevel or Permobil chairs [4]; facilitating standing like Genie V2 [5], which is designed to reduce muscular dystrophy and bladder drainage, or iBOT [6] also able to climb stairs. A study of different mobility wheelchair architectures [7] led to the concept of the Chiba chair, in which small wheels can swivel to behave like legs when necessary. While reaching objects at a given height is important, the capability of reaching the floor is an issue when dealing with falling objects [8]. Aiming to complement mobility assistance through body rehabilitation, in [9] a specific arm support is added to the chair to assist rehabilitation of users with Duchenne muscular dystrophy while driving [10].

Transfer activities are very important to improve the users' degree of independence. However, when wheelchair users encounter height differences and gaps crossing between wheelchair and target surface, no proper transfer techniques are available. To achieve manipulation capabilities for users with a disability that restricts a normal control and actuation of their upper limbs the solution may rely on exoskeletons. Upper limb exoskeletons are electromechanical systems designed to interact alongside with the user for assistance, power amplification or substitution of the lost motor function [11]. These devices can be categorized into exoskeletons for motion amplification and exoskeletons for medical rehabilitation. Their performance depends on the degrees of freedom (number and type), type of actuation, supporting joints, and applications, as reported by different works [12, 13]. Thus, exoskeletons for motor assistance or augmentation can overcome physical weakness that affects daily life of individuals [13] and contribute to reducing the burden of manipulating loads in industrial environments. A large number of existing studies in the broader literature have identified new challenges related to research and development of upper limb exoskeletons, however here we just consider an accompanying support that helps in reaching actions by complementing the chair reconfiguration when needed.

We skip here the type of exoskeletons used in the rehabilitation of physically impaired individuals after a neuromotor disease or the complementary FES (Functional Electrical Stimulation) assistance acting directly on the activation of muscle actuation.

3 Design Trade-*Off*

Designing a solution that addresses all the users' movements is a difficult challenge. When designing such an assistive device, a compromise must be established between complexity, functionality and cost. In addition to the above points, there are other considerations that make a design acceptable to the user. There are aspects such as the adjustments of the chair to the physical characteristics of the user. Practical factors such as: the weight or the ease of folding the chair, the kind of environments in which it can be used and even aesthetic and social acceptance aspects conform a set of variables that make it very difficult to find the optimal solution.

The user's needs have been stablished based on the levels of disabilities of patients treated in the Institut Guttmann, for neurorehabilitation. Users with spine injuries, complete or incomplete, that affect vertebras from C3 to C6 that produce impairments ranging from precluding the movement of the arms (and of course the legs), to the lost of force of users with some residual mobility but unable to held loads.

In this first design phase, the main development environment of the user is considered to be a home-type environment and with the assumption that the user has some remaining arm and hand mobility that allows them to make grasping type actions with a certain voluntary control. Similar to the needs of the workers of Esclatec, an inclusive engineering participating in the project, whose employees have some disabilities and need assistance to increase their capabilities to reach and manipulate parts for wheelchairs assembling.

To start the design phase, six use cases are considered: a) the accommodation of the user's height on a desk or table, b) or in a social interaction of conversation with two people around the user, c) grab a book at different heights from a shelf, d) transfer from standing to sitting, e) horizontal rest, and f) pick up an object from the floor.

Use cases a) and b) require the height of the chair to be adjusted without changing the sitting position of the user. The difference in terms of mobility between case a) and

b) is that while in case a) the table itself is an obstacle to which the chair will have to adapt, in case b) no collision risks are expected. In case b) it would be convenient for wheelchair users to reach the height of their interlocutors. Consequently, the height to which the user must rise is considerably higher than in case a). The use case c) and d) require the chair to allow the user to stand up and keep in standing position for a while. The difference between cases c) and d) lies in the fact that for the transfer back to sitting, it is advisable for the user to be as vertical and advanced as possible. Instead, when taking an object from a shelf, for safety, the user vertical posture should prevent an advanced or prominent posture as in the case of a transfer from standing to sitting. For case of use e) that allows the user to lie horizontally, it would be convenient to be able to do so at different heights, thus allowing to make more comfortable some type of treatment by the caregiver or allow the transfer horizontally to the bed. Case f) must allow the user to lower at the same time as the backrest is tilted so that the hand reaches the zero level. The following Table 1 summarizes the actions of postural movements according to the case of use.

Table 1. Use cases and involved postural movements

Use case	Description	Involved movement (from neutral position)
a	Access to a desk or table	Seat elevation, arms reaching movements
b	Social interaction. Talking at the same height	Seat elevation, turning arms down
c	Transfer from standing to seating	Seat elevation, advancement and bending, adapting back and arm elevation
d	Take an object from a shelf	Seat elevation and bending, turning the back and arm elevation
e	Resting position or horizontal transfer	Seat elevation, turning the back and the arm
f	Pick an object from the floor	Seat descending, turning the back and the arm

The proposed design should consider these use cases and based on them try to minimize the components that allow their implementation, as well as deciding their actuation.

4 Proposed Architecture

Following the user's requirements as summarized in the previous analysis, the architecture proposed for the design of such hybrid platform, this wheelchair with added performances must be able to meet the requirements of allowing the user to (i) move, (ii) change posture, and (iii) manipulation objects in a spatial range that a non-impaired person could handle.

Concerning the movement (i), the architecture of the wheelchair will be grounded on a commercial moving platform with two drive wheels, capable of moving forward and backwards, as well as turning over its vertical axis, thanks to a couple of motors (one actuating on each drive wheel). The control of such platform is intended to be simple and capable to yield non-holonomic behaviour which has been deeply addressed in commercial systems.

Concerning the mentioned postural changes (ii), three main configurations are expected to be implementable: sitting, laying, and standing. The sitting configuration must be the standard for the wheelchair movement, and therefore must bring comfort to the user. In this configuration the wheelchair has to be able to reach two different height positions: standard sitting and low sitting. Standard sitting height is the conventional height for moving and being accommodated at a normal desk layer (armrest falls below a 700 mm level from the ground). Low sitting height is a sitting configuration in which the user bottom must reach a minimum distance from the ground, allowing the user to reach objects on the floor and/or at low height.

Secondly, the laying configuration must be a position for resting, in which the user must be able to lay almost horizontally (reaching a certain postural comfort), thus extending the legs and resting the head in a pillow.

Finally, the standing configuration must be a position in which the user must be capable of reaching objects at higher levels such as shelters, in a normal spatial range that a non-impaired person could handle.

A representation for some events made possible with this architecture is presented in Fig. 1.

4.1 Identification of Modules and Determination of the Degrees of Freedom (DoF) to Implement in the Wheelchair

Following the requirements presented in the previous section, five modules in the architecture of the wheelchair are identified; namely: a) Shoulders and head, b) Arms, c) Seat, d) Legs, and e) Wheels. These five modules are depicted in Fig. 2 (a), whilst its movement interval zones are depicted in Fig. 2 (b).

Consequent to this modular specification, the degrees of freedom (DoF) to be implemented in the wheelchair for each of them can be summarised as follows:

a) Shoulder and head: One DoF (orientation 'α_1') for achieving the laying position.
b) Arms: Two DoFs (orientations 'γ_1' and 'γ_2') for moving the arms independently, required to help the user when aiming at higher targets, as well as to provide comfort when in the laying position.
c) Seat: Three DoFs envisaged (position 'X' and 'Z', and orientation 'θ'). Actually, it is necessary to provide an ascending/descending movement (Z) as well as to be able to achieve some displacement of the seat from the relative position of the Wheels, so to balance the position of the centre of masses of the set (wheelchair plus user) in different positions (X). It is also aimed that the wheelchair can choose a Seat orientation (θ) other than the horizontal disposition (for example in the Standing position)

d) Legs: One DoF for achieving the laying position (orientation 'φ'). It will encompass a dependent movement of the feet step to provide better comfort to the user in different positions.
e) Wheels: Two DoF (orientations 'β_1' and 'β_2') for the drive and manoeuvring of the wheelchair.

The raising of the arms with the corresponding motorized elements produce a slipping of the arm with respect to the support. If the surface of the support is smooth, it does not show any dysfunction. In case a soft support is needed, which prevents slipping, we foresee an additional linear actuation produced by a rolled cable on the same joint W_A, not detailed at this stage.

Based on these actuation requirements different implementation architectures have been explored. However, although several options would be possible to produce the required actuations, when trying to minimize the number of degrees of freedom and at the same time avoid collisions between the different modules the potential classical structures fail to be acceptable. The critical module is the seat, which movements have to be compatible with the back and the legs in all different configurations, avoiding incompatibilities between them and with the user.

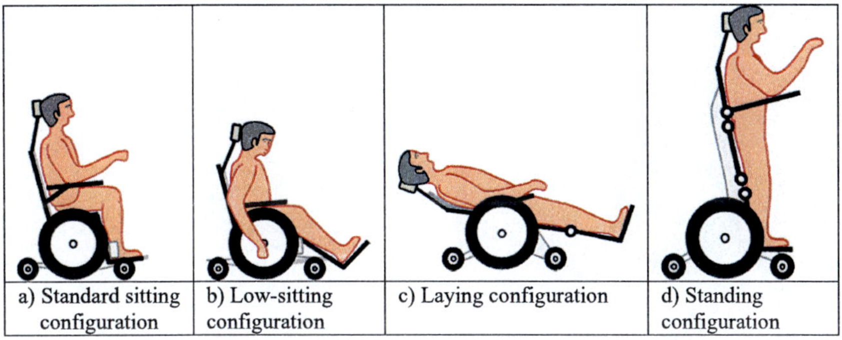

Fig. 1. Different position configurations required to the wheelchair embedded with exoskeletal capacities: (a) Standard-sitting, (b) Low-sitting, (c) Laying, and (d) Standing.

The architecture that meets these requirements has been an asymmetric parallel structure, 2-links elbow mechanism type, with two units at each side of the chair, thus producing the desired movements, the three DoF. Figure 3a shows the two modules at the right side of the chair, indicating the three actuators among the six joints of the 2-links elbow mechanism. Figure 3 b visualizes different seat poses showing its compatibility with the chair structure.

4.2 Definition of the Kinematic Pairs and Actuators for the Modular Chain

Following to the DoFs specification, the implementation of kinematic pairs and actuators for each module will adhere to this arrangement:

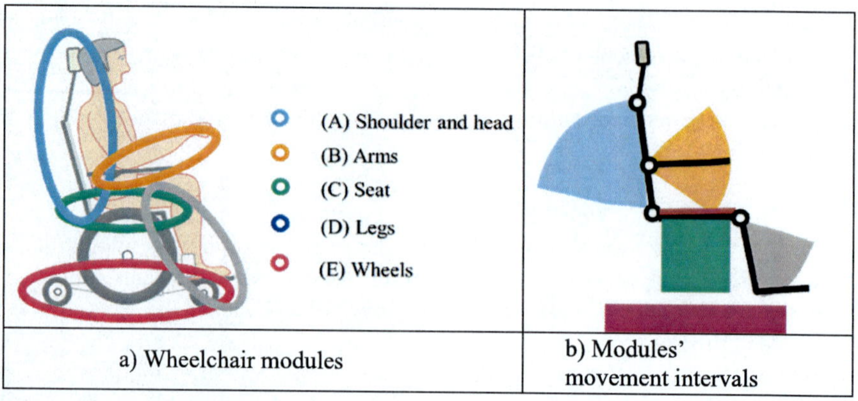

Fig. 2. (a) Five modules identified in the wheelchair with skeletal capacities, and (b) Module's movement intervals (schematic)

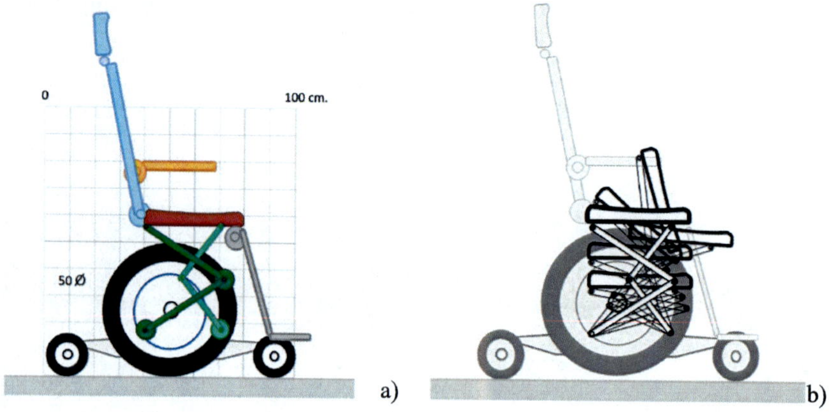

Fig. 3. Architecture for seat positioning (a) the 2-links elbow mechanism, in green. (b) A sequence of seat configurations and their corresponding actuation.

a) Shoulder and head: One motor 'W_A'.
b) Arms: Two motors 'W_{B1}' and 'W_{B2}' for moving the arms independently.
c) Seat: Three motors, two of them towards the back 'W_{CB1}' and 'W_{CB2}' and the other one towards the front of the wheelchair 'W_{CF}'. These three motors will actuate a 6-bar mechanism with the 3 DoFs required.
d) Legs: One motor 'W_D'.
e) Wheels: Two motors 'W_{E1}' and 'W_{E2}' for moving the two wheels indepen-dently, thus allowing the wheelchair to move forward, backwards and to rotate in the two senses of direction (turning left and turning right).

These motor arrangement, as well as a couple of overlapped views are depicted in Fig. 4, with indication of the actuators (a), visualization of postures in (b) and the corresponding displacements of the DoF of the seat in (c).

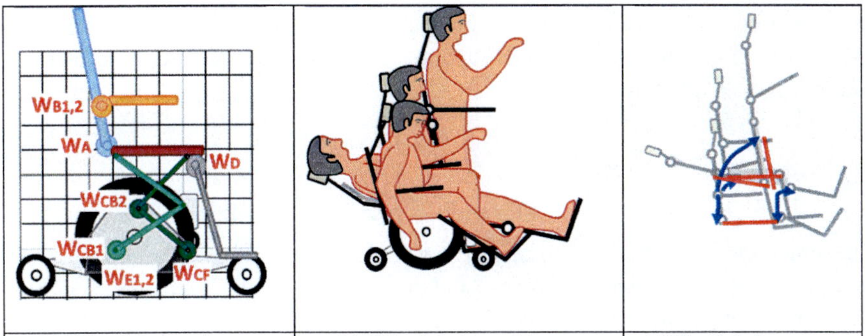

(a) Wheelchair pairs and motors | (b) Wheelchair positions overlapped view | c) Wheelchair kinematic chain overlapped view

Fig. 4. (a) Wheelchair pairs and motors, (b) Wheelchair positions overlapped view, and (c) Wheelchair kinematic chain overlapped view.

5 Simulation Features

After the first conceptual architecture design and previous to the mechanical design, the conceptual platform architecture has been simulated to evaluate the performances, detect the required adjustments and foresee any modification needed. The simulation will be the base for the control of the different modules of the hybrid platform which has to solve the kinematics of the whole architecture. Some actuations like mobility or actuation in a given position (i.e. sitting in front of a table or desk) will not present special difficulties. However, when having to reach an object at a given height, requiring some elevation of the seat and of the arm, this architecture presents multiple solutions. To solve this redundancy priority should be given to the user's comfort.

5.1 Working Environment

For the simulation of the control strategies and the study of the functionality of the proposed architecture, the use of two platforms has been considered: Gazebo and CoppeliaSim [14]. Other platforms such as Webots, ARGoS and USARSim which are more focused on mobile robots and their interactions with the environment have been discarded [15]. The main advantage of the two studied platforms is that they are software free, or at least have a free license for educational or university use, and both count with very active user communities.

It is not intended here to present an exhaustive comparison of both platforms, only to highlight the advantages of both and justify the final decision to use CoppeliaSim at this development point of our hybrid platform. From our teaching experience CoppeliaSim presents a much more intuitive and friendlier interface than Gazebo. It is an independent robotics simulation platform, which offers everything you need to work without the need to rely on other programs, as is the case with Gazebo and its de facto dependence on ROS or other packages. CoppeliaSim uses its own scripts to control the robots (in Lua language) and a hierarchy panel of all the robots, links, joints and other elements of the scene. Even so, the generated models can be also controlled from an external program, in C++ or Python, MATLAB or even ROS itself, although in the latter case Gazebo is more compatible and efficient.

5.2 Model Characterization

CoppeliaSim has been used in the first instance to test the geometry of the first prototype of the chair. Once modeled, the dimension of the different elements can be adjusted to easily satisfy the geometric constraints and determine the joint limits, the work space needed by the different mobile elements, possible collisions between them, singularities, etc.

Secondly, a kinematic study has been carried out to observe the transition between the different positions, the relationship between the linear and angular speeds, the maximum speeds reached, etc. Figure 5 presents the four main different positions of the chair, as simulated in CoppeliaSim.

Finally, dynamic data has been obtained, such as forces and torques required in the joints and union points (active and passive), as well as energy consumed during a movement, which will help to determine the required actuators' power. Figure 6 shows the evolution of the center of mass from the normal sitting position to the standing position.

a b c d

Fig. 5. Four basic positions, a) normal sitting position, b) standing, c) lying and d) low sitting

Figure 7 shows a sample obtained during a simulation from the normal sitting position to standing, seat-to-stand, in a four seconds movement. It shows the value of the controlled joints, using isochronous interpolation, of the evolution in time of velocities, torques, absolute value of power and total energy consumption.

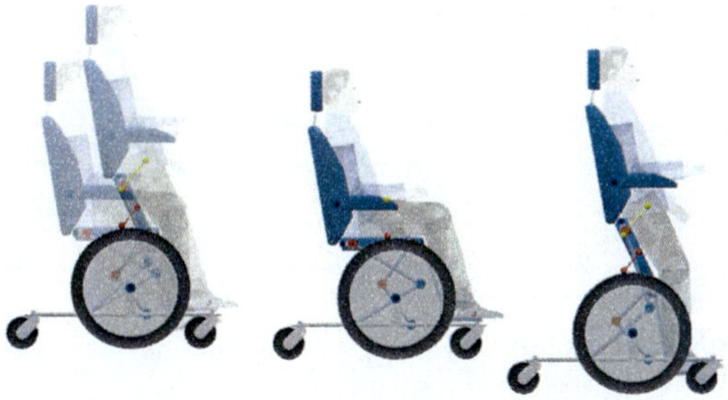

Fig. 6. Evolution of the center of mas from the normal sitting position to the standing position, of the chair-person set (in red), and the person alone (in yellow).

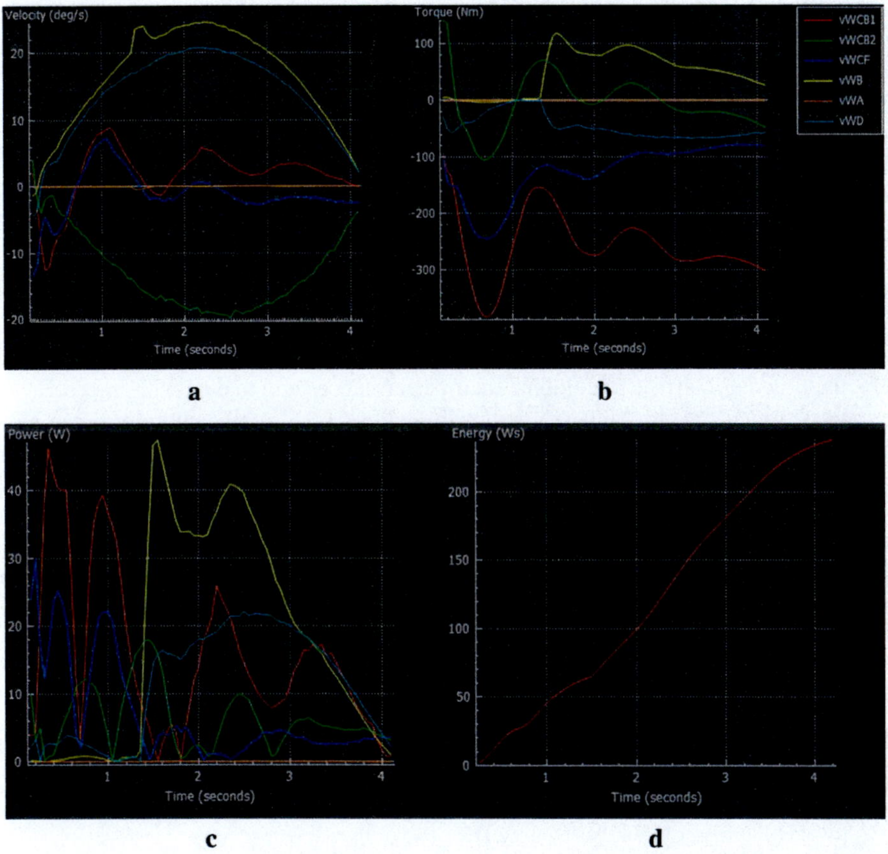

Fig. 7. Evolution in time of the joints from normal sitting position to standing position a) velocities b) torques c) absolute values of power d) total energy

6 Conclusions and Future Work

The design of a hybrid platform that combines the properties of a robotized wheelchair, together with some exoskeleton performances to assist upper limbs movements poses new challenges to its design and implementation. First to the platform design, so as to achieve an acceptable trade-off between performance, size, weight, and cost, and second to the interface and control.

Further to the described design aspects, we have considered the possibility of designing special joints that assure the adaptability to the user's anatomy when changing the posture, as studied in [16] where the design of an exoskeleton joint tries to keep an isocentric movement with respect to the user's anatomy. In addition, the minimization of cost and size of joints will consider the design of sensors integrated in the own joint, using dedicated manufacturing techniques, as applied in [17, 18].

The chair can bring the user to the working place with different poses and supports their arms in a different way than an exoskeleton would do. This is because the actuators are on the chair, instead of on the user, and their control is not direct to the arm but to position the hand. Then, it is necessary to face a coordinate transform considering chair and these additional joints. The control of this redundant structure that combines seat movements (advance/elevate) with those of the arms, implies prioritize user's comfort and collision avoidance over other parameters like energy, accelerations, or DoF involved.

Another part of the project, not the object of this paper, is to provide the user with an interface that allows an easy control of the chair and its actuation. Motor disabilities may make difficult to find the adequate device for its control and can require the use of a multimodal interface to combine their performances and to deal with imprecise or erroneous orders due to the user's motor limitations. The intelligent control can also benefit from the knowledge about the environment (i.e. position of objects) to correct them.

Acknowledgements. This work has been developed in the frame of the project: *Orthosis-Chair Hybrid Platform to make compatible Mobility, Functionality and Acceptability for application in domestic environments* (POSMOFYA). Funded by *Ministerio de Ciencia e Innovación*, Project: PID2020-116091RB-I00. The author Joaquim Minguella-Canela is a Serra Húnter Fellow.

References

1. Observer-Mobility: All-terrain wheelchair (2015). http://www.observerMobility.com
2. Toyota Motor East Japan: Patrafour (2015)
3. Grewal, H.S.. et al.: Autonomous wheelchair navigation in unmapped indoor environments. In: IEEE International Instrumentation and Measurement Technology Conference (I2MTC), Houston, TX, pp. 1–6 (2018)
4. Yukselir, M., Scarrow, K., Wilson, P., Cowan, T.: The brains behind the electric wheelchair, one of Canada's 'great artifacts'. The Globe and Mail, 27 August 2012
5. Standing wheelchair Genie V.20: EasyRise Ltd. (2019)
6. Deka-Research: Stair climbing wheelchair iBot (2015)

7. Nakajima, S.: Concept of a personal mobility vehicle for daily life. In: IEEE International Conference on Robotics and Automation (2016)
8. Hsu, P.E., Hsu, Y.L., Lu, J.M., Chang, C.-H.: Seat adjustment design of an intelligent robotic wheelchair based on the Stewart platform. Int. J. Adv. Robot. Syst. **10**, 168 (2013)
9. Lobo, J., et al.: Development and evaluation of MOVit: an exercise-enabling interface for driving a powered wheelchair. IEEE Trans. Neural Syst. Rehabil. Eng. **27**, 9 (2019)
10. Lobo-Prat, J., et al.: Evaluation of an exercise-enabling control interface for powered wheelchair users: a feasibility study with Duchenne muscular dystrophy. J. Neuroeng. Rehabil. **17**, 142 (2020)
11. Bai, S., Virk, G.S., Sugar, T.: Wearable exoskeleton systems: design, control and applications. Institution of Engineering and Technology: London, UK (2018)
12. Gull, M.A., Bai, S., Bak, T.: A review on design of upper limb exoskeletons. Robotics **9**, 16 (2020)
13. Rahman, M., Ouimet, T., Saad, M., Kenné, J., Archambault, P.: Development and control of a wearable robot for rehabilitation of elbow and shoulder joint movements. In: IEEE Conference on Industrial Electronics Society (2010)
14. Nogueira, L.: Comparative analysis between Gazebo and V-REP robotic simulators. In: Seminario de Cognicao Artificial-SICA, no. 5 (2014)
15. de Melo, M.S.P., et al.: Analysis and comparison of robotics 3D simulators. In: 21st Symposium on Virtual and Augmented Reality (SVR), pp. 242–251 (2019)
16. Amigo, L.: Design of an exocentric exoskeleton of the elbow joint with dynamic isocentricity adaptation. Ph.D. thesis, UPC Advisor A. Casals (2013)
17. Minguella, J., et al.: Comparison of production strategies and degree of postponement when incorporating additive manufacturing to product supply chains. Procedia Manuf. **13**, 754–761 (2017)
18. Muguruza, A., Minguella, J., et al.: Development of a multi-material additive manufacturing process for electronic devices. Procedia Manuf. **13**, 746–753 (2017)

Machine Learning and Inertial Sensors to Estimate Vertical Ground Reaction Force During Gait

David Martínez-Pascual(✉), Jesús Álvarez-Pastor, Clemente Irles, Jose Vicente García-Pérez, Andrea Blanco, Yolanda Vales, and Nicolás García-Aracil

Robotics and Artificial Intelligence, Bioengineering Institute, Miguel Hernández University, Elche, Spain
david.martinezp@umh.es
https://rai.umh.es/

Abstract. During gait, vertical ground reaction forces (vGRF) can reach magnitudes which could cause different lower limb injuries. Measuring and studying generated vGRF can help to prevent musculoskeletal disorders. However, in order to measure vGRF, it is required the use of expensive equipment, which may also limit the study environment. This work presents different Machine Learning models to predict vGRF employing inertial sensors.

Keywords: Vertical ground reaction force · Gait analysis · IMU · Machine learning

1 Introduction

Musculoskeletal modeling and simulation have great potential for the analysis of human biomechanics, which could allow us to prevent injuries or predict musculoskeletal disorders. Modeling tools such as OpenSim allow the simulation of the human body biomechanics introducing motion capture data and forces exerted on the subject [1]. When analyzing lower limb dynamics, one of the main components of the simulation will be the introduction of ground reaction forces.

During gait, the greatest component of the generated forces is the vertical ground reaction force (vGRF). The generated vGRF usually overcomes the body weight (BW) during walking. We can distinguish 3 main peaks within the resulting vGRF signal: first, a local maximum corresponding to the loading response (loading peak, LP), a local minimum corresponding to the mid-stance phase (mid-stance peak, MP), and a second local maximum corresponding to the terminal-stance (terminal-stance peak, TP) [2]. These peaks influence the lower limb joints and muscle loads. Therefore, the study of the peak values and their timing can be essential in order to avoid lower limb musculoskeletal injuries [3].

D. Tardioli et al. (Eds.): ROBOT 2022, LNNS 590, pp. 264–273, 2023.
https://doi.org/10.1007/978-3-031-21062-4_22

To measure vGRF, usually, force plates are employed. These plates can rest over the ground or can be embedded in treadmills. This equipment can be costly, and it might result impractical since it may limit space for gait analysis. Due to these limitations, some authors have examined the possibility of using wearable sensors to estimate the vGRF during gait. E. Shahabpoor and A. Pavic have shown that it is possible to estimate the vGRF with a single inertial measurement unit (IMU) placed over the C7 vertebra [4]. However, the method proposed is only able to estimate the complete vGRF, i.e., it cannot differ the force produced on the right and left lower extremities.

With the development and improvement of Machine Learning algorithms, it has been shown that is possible to apply these techniques to analyze gait kinematics and vGRF employing wearable sensors [6].

To estimate vGRF, one possibility could be a ridge regressor, which solves a linear regression model with L2 regularization [7]. Moreover, linear regressors are usual to incorporate polynomial therms (e.g. quadratic terms) as well as predictors interactions [8,9]. Employing this model we could examine if it is possible to estimate vGRF with a regression equation.

The vGRF could also be predicted with a random forest model [10]. This model consists of multiple decision trees, which are trained independently with a data subset. The outcome of this model will be determined by the best result given by the whole set of trees. Moreover, we can compare the performance of our trained model with X. Jiang et al. results [5].

This work aims to compare the performance of the mentioned algorithms in the task of predicting the vGRF employing lower limb kinematics data. Moreover, this work pretends to find if the algorithm's training improves when vertical acceleration data measured at the C7 vertebra is included.

2 Materials and Methods

2.1 Lower Limb Joint Measurement

To estimate the lower limbs joint angle during gait the method proposed by T. Seel et al. has been implemented [11]. This method is based on the analysis of a subject's joint movement, and it allows to estimate a joint angle with IMU sensors.

It should be noted that the authors avoided developing an algorithm that assumes a certain position and orientation of the IMU concerning the subject's limb. This method uses data from the gyroscope and accelerometer of the IMU, so the algorithm does not depend on a uniform magnetic field. Moreover, the algorithm has been shown to achieve good performance even with low acquisition rates (40 Hz) [12].

On the one hand, some methods collected in the bibliography require specific and accurate calibration movements to set up the IMU reference systems to the subject joint axis [13]. On the other hand, the algorithm proposed by T. Seel et al. makes use of arbitrary calibration movements and joints movements during gait to determine the sensors' situation with respect to the joint they measure, exploding kinematic constraints.

The method proposed by T. Seel et al. has shown a root mean squared error (RMSE) of between 1 to 3 °C in flexion and extension movements. Therefore, it can be assumed that we have enough accuracy to perform an adequate gait analysis and vGRF prediction.

2.2 Experimental Setup and Acquired Data

To obtain lower limb movement data during gait 5 XSens Dot IMU were used. These IMU were placed on the C7 vertebra, the back of the hip, the thigh, the leg, and the foot, as can be observed in Fig. 1.

XSens Dot sensors acquisition was performed at a rate 60 Hz. From the IMU placed at the lower limb and hip, acceleration and angular data were acquired. With these data, the flexion/extension angles of the hip, knee, and ankle were estimated offline. In addition, joint angles were low-pass filtered with a cutoff 6 Hz, and angular velocity was calculated for each joint. We extracted the subject's vertical acceleration from the IMU at C7 orientation and acceleration data we previously acquired [14].

The vGRF was obtained from a treadmill with a force plate h/p/cosmos 150/50. Data were acquired 100 Hz and were normalized to the subject's body weight.

Moreover, to perform a gait analysis, a common practice is to identify the beginning and the end of the gait cycle and transform data from the temporal domain to the gait cycle domain (0–100%). Hence, all data obtained has been transformed into the cycle gait domain, identifying the beginning and end of each step.

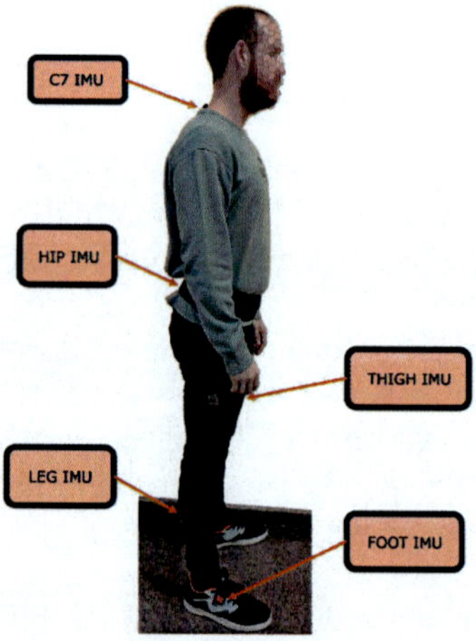

Fig. 1. IMU location to estimate vGRF. The subject wears 1 IMU placed over the C7 vertebra to acquire vertical acceleration, and 4 IMU placed over the hip, thigh, leg, and foot to estimate lower limb flexion/extension joint angles.

2.3 Experimental Protocol

Before starting the experimental session, the IMUs were positioned at the established subject locations. Once the sensors were placed, the subject proceeded to obtain the calibration parameters for the flexion/extension joint angles estimation (T. Seel et al. method). First, the lower limb joints were moved arbitrarily for 30 s on each lower limb joint. Second, the user walked for 1 min at a comfortable speed.

After the calibration process vGRF, lower limb joint angles, and the subject's vertical acceleration were obtained for different walking speeds: 1.5 km/h, 2.5 km/h, 3 km/h, 3.5 km/h, and 4 km/h. When the treadmill reached the desired walking speed, data were registered for 5 min. Between activities, 2 min were left for rest.

2.4 Machine Learning Models Training

For the training of the ridge regressor and random forest models joint angles, angular velocities, vertical acceleration, and cycle gait were introduced as inputs of the models. Moreover, since the objective is to predict the vGRF, data collected from the h/p/cosmos treadmill were introduced as the models' output. Moreover, the ridge regressor model also includes quadratic terms for all inputs, as well as all pairs of products of different inputs.

Before training, each input had been scaled from its minimum and maximum values, so we have numerical values between 0 and 1. Input scaling prevents an input with high numerical values from disproportionately affecting the cost function and dominates the optimization process. Additionally, data were split randomly into two subsets to validate the models' performance by cross-validation: 70% of the data for training and 30% as the validation set.

As mentioned before, one of this work's main objectives is to find if including vertical acceleration improves vGRF estimation. Moreover, we could also examine if we can predict vGRF with Machine Learning exclusively with vertical acceleration data, without incorporating joint angles or angular velocities as model input. To verify this possibility, we realized three different pieces of training for each model:

- **Training 1 (T1).** Joint angles, angular velocities, vertical acceleration, and cycle gait have been included as inputs.
- **Training 2 (T2).** Joint angles, angular velocities, and cycle gait have been included as inputs.
- **Training 3 (T3).** Vertical acceleration and cycle gait have been included as inputs.

2.5 Evaluation of Model Performance

The performance of the models has been evaluated as follows:

- **vGRF RMSE (BW).** A lower RMSE in the reproduction of the vGRF signal is understood as a better performance of the model.
- **LP, MP, and TP RMSE (BW).** A lower error between measured and estimated vGRF peaks is understood as a better performance of the model. To find the LP and TP peaks we searched the local maximum in the first and second periods of the foot support. To calculate the MP peak we searched the local minimum between LP and TP peaks.
- **LP, MP, and TP delay (%).** A lower delay or advance between measured and estimated vGRF peaks is understood as a better performance of the model. The delay is calculated as the difference between the estimated and measured peaks. A positive value means a delay of the estimated peak, and a negative value means an advance of the estimated peak to the measured peak.

3 Results

To validate the ridge regressor and random forest algorithms an experimental session has been carried out with a healthy subject of 26 years of age, 1.83 m tall, and with 83 kg body mass.

Both models were trained with SciKit Learn Python library [15]. Ridge regressor results have been obtained with a regularization strength α of 0.7, with a Singular Value Decomposition (SVD) solver, without a maximum of iterations and a precision of 0.001. Random forest training was performed with 170 decision trees without maximum depth, a minimum number of samples of 1, a minimum number of samples required to split an internal node of 2, and the number of inputs to consider when looking for the best split were calculated by the logarithm base 2 of the number of inputs. The hyperparameters tunned for both algorithms are the best performing based on RMSE with the training and validation dataset.

The obtained RMSE errors with the training and validation datasets are collected in Fig. 2.

The random forest models have been able to estimate the vGRF with an RMSE of less than 0.03 BW with the training and validation datasets, except for

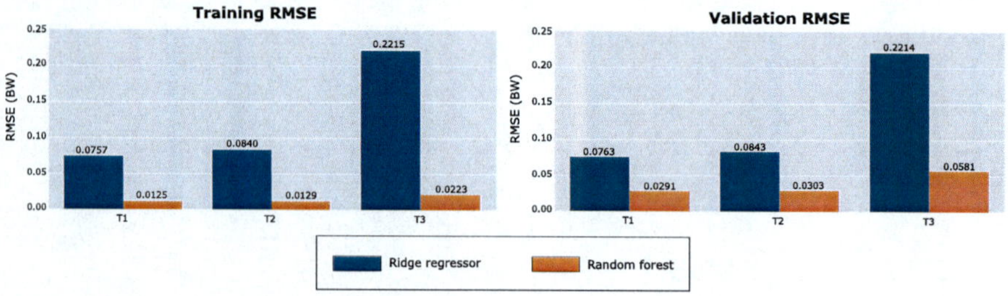

Fig. 2. Training and validation RMSE (BW) for each training of the regressor ridge and random forest models.

T3 with the validation dataset (0.0581 BW). In contrast, using a ridge regressor, a minimum RMSE of 0.0757 BW and 0.0763 BW was achieved with the training and validation sets, respectively. It should be noted that the RMSE obtained for E3 with the regression technique exceeds 0.2 BW. Since T1 has better RMSE results, the vGRF estimation with these models is analyzed in depth.

Figure 3 collects epoch graphs with measured vGRF and estimated vGRF by ridge regressor and random forest T1 models. Figure 4 shows heatmaps with RMSE errors between measured and estimated vGRF signal, the RMSE errors between each peak's magnitude, and the estimated peak's delay compared to measured peaks.

The ridge regressor model obtained an RMSE of between 0.062 to 0.092 BW for the vGRF during gait, 0.028 to 0.079 BW for the magnitude of the LP peaks, 0.027 to 0.057 BW for the TP peaks, and 0.037 to 0.150 BW for the MP peaks. With the random forest model, errors ranged from 0.017 to 0.02 BW in the estimation of the vGRF, 0.011 to 0.013 BW for the magnitude of the LP peaks, 0.009 to 0.017 BW for the TP peaks, and 0.011 to 0.016 BW for the MP peaks are obtained.

Concerning delay, the ridge regressor model estimated LP peaks with cycle gait lags between 2 and 6.4%, cycle gait lags between 0.16 and 1.7% for the TP

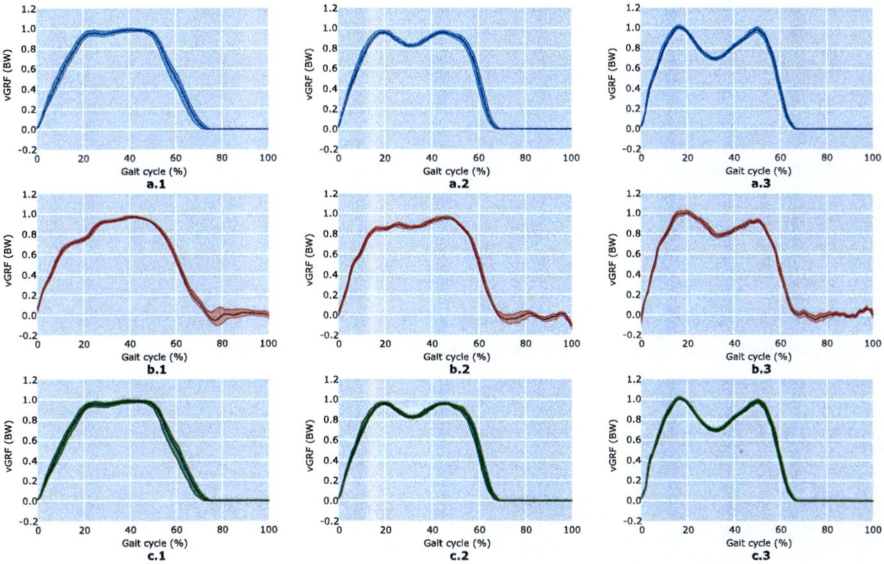

Fig. 3. Epoch graphs with vGRF (BW) according to cycle gait (%). The median data value has been represented, and values between the first and third quartile have been coloured. a) vGRF measured with h/p/cosmos treadmill for gait speeds of 1.5 km/h (1), 3 km/h (2), and 4 km/h (3); b) vGRF estimated with ridge regressor model (T1) for gait speeds of 1.5 km/h (1), 3 km/h (2), and 4 km/h (3); c) vGRF estimated with random forest model (T1) for gait speeds of 1.5 km/h (1), 3 km/h (2), and 4 km/h (3).

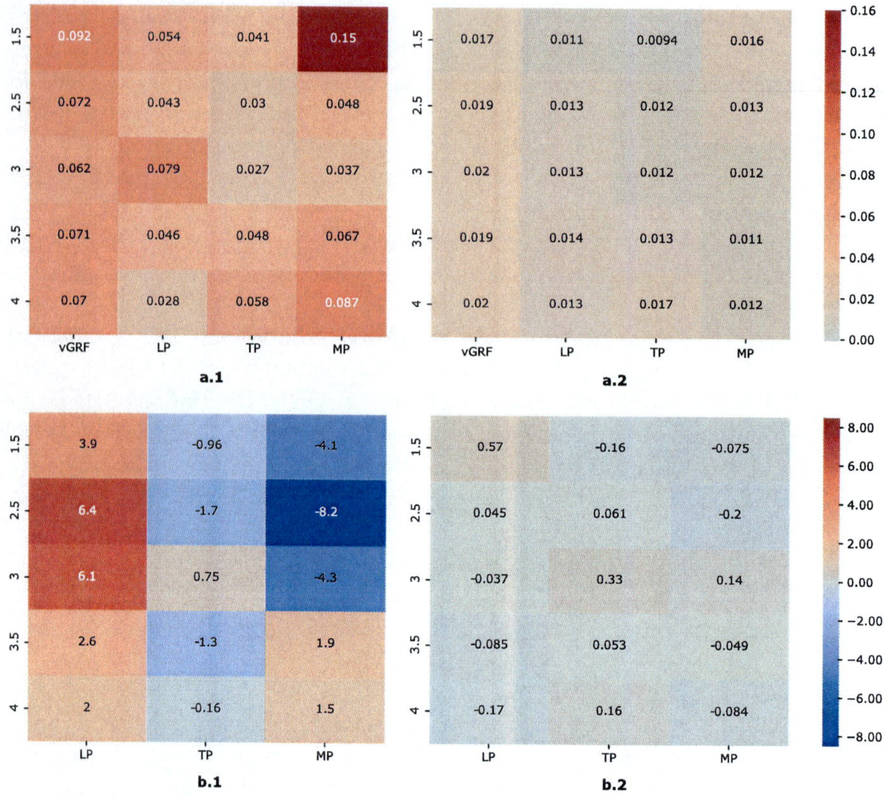

Fig. 4. Heatmaps with RMSE errors and peak delay (T1) for gait speeds of 1.5 km/h, 2.5 km/h, 3 km/h, 3.5 km/h, and 4 km/h. a) RMSE errors for vGRF signal reconstruction and the magnitude of its characteristic peaks with ridge regressor (1) and random forest model (2); b) Mean delay (%) of LP, TP y MP peaks with respect measured vGRF signal with ridge regressor (1) and random forest model (2). A positive value means a delay of estimated peak respect with measured peak, and a negative value means an advance of estimated peak respect with measured peak.

peaks, and lags between 1.5 and 8.2% for the MP peaks. The random forest model achieved a better accuracy in terms of cycle gait lags, obtaining lags between 0.037 and 0.57% for the estimation of LP peaks, 0.053 and 0.33% for the TP peaks and cycle gait lags between 0.049 and 0.2% for the MP peaks.

4 Discussion

In order to estimate the vGRF two types of Machine Learning algorithms have been used. First of all, the obtained RMSE errors with training and validation datasets suggest that random forest models are more accurate than ridge regression models in the task of estimating the vGRF magnitude.

RMSE errors get with random forest models employing lower limb kinematics data (T2) are in a similar range to results achieved by other authors for one healthy subject [5,6], so we could assume our model es accurate enough. It is worth mentioning that including vertical acceleration as input of the model, we have reduced RMSE with both regressor ridge and random forest models. Moreover, although we obtained higher RMSE values in T3, we can observe that it could be possible to estimate vGRF employing a single IMU placed over the C7 vertebra.

If we analyze further how the two models behave using lower limb kinematics and vertical acceleration data (T1), there are notable differences. First, we can observe that the random forest model obtained a lower RMSE compared to the ridge regressor model for each gait speed when performing vGRF signal reconstruction. Moreover, Fig. 2 shows that the ridge regressor model is not able to determine the swing phase. By contrast, the random forest model seems to be able to establish when the vGRF should be zero.

Attending to vGRF characteristic peaks magnitude results also suggest that the random forest model could reproduce LP, MP, and TP with higher accuracy than the ridge regressor model. It should be noted that the lowest LP and TP magnitude RMSE are obtained at a gait speed of 1.5 km/h, which might indicate that the random forest model would be especially good at estimating vGRF at low gait speeds. However, the minimum MP magnitude RMSE value has been obtained for the highest gait speeds (0.011 at 3.5 km/h, and 0.012 at 3 and 4 km/h). This may show an accuracy loss at low gait speeds.

Additionally, cycle gait lag results also show that the random forest model achieves lower error rates, unlike the ridge regressor model. While the ridge regressor model presents cycle gait delays until 6.4% (LP peak at 2.5 km/h) and cycle gait advances up to 8.2% (MP peak, at 2.5 km/h), the maximum cycle gait lag for the random forest model is a mean advance of the LP peak of 0.57% (1.5 km/h). These results agree with estimated vGRF signals shown in Fig. 3, where it can be observed that the ridge regression model barely reconstructs the vGRF measured signal.

The results achieved with the random forest model are similar to results reported by X. Jiang et al. so we could affirm that these errors are sufficient to perform a biomechanical analysis and modeling. Moreover, the results obtained suggest that the RMSE errors and the peaks' delay improve when the user's vertical acceleration is included. However, a larger dataset with diverse users is needed, as well as further analysis to confirm this hypothesis.

5 Conclusion and Future Work

This work aims to estimate vGRF employing IMU and Machine Learning techniques, comparing the ridge regression and random forest models' performance. The presented results suggest that random forest models are better than the ridge regression models at estimating the vGRF signal, as well as the magnitude and timing of the characteristic peaks.

It is worth mentioning the inclusion of vertical acceleration measured at the C7 vertebra as input of the proposed Machine Learning models. Including acceleration data has reduced RMSE with training and validation datasets, which suggests a higher accuracy when estimating vGRF.

In order to extend this study, it is intended to expand the number of users, as well as to employ new Machine Learning and Deep Learning algorithms. New methods will also be investigated to estimate as accurately as possible the vGRF with a single IMU placed over the C7 vertebra.

Acknowledgement. This work has been funded by the Ministry of Science and Innovation, belonging to the State Research Agency (AEI) through the project PID2019-108310RB-I00/AEI/10.13039/501100011033 and by the Ministry of Universities through the aid for University Teacher Training FPU20/05137.

References

1. Seth, A., Sherman, M., Reinbolt, J.A., Delp, S.L.: OpenSim: a musculoskeletal modeling and simulation framework for in silico investigations and exchange. Procedia Iutam **2**, 212–232 (2011)
2. Marasović, T., Cecić, M.O.J.M.I.L., Zanchi, V.: Analysis and interpretation of ground reaction forces in normal gait. WSEAS Trans. Syst. **8**(9), 1105–1114 (2009)
3. Shafizadegan, Z., Karimi, M.T., Shafizadegan, F., Rezaeian, Z.S.: Evaluation of ground reaction forces in patients with various severities of knee osteoarthritis. J. Mech. Med. Biol. **16**(02), 1650003 (2016)
4. Shahabpoor, E., Pavic, A.: Estimation of vertical walking ground reaction force in real-life environments using single IMU sensor. J. Biomech. **79**, 181–190 (2018)
5. Jiang, X., Napier, C., Hannigan, B., Eng, J.J., Menon, C.: Estimating vertical ground reaction force during walking using a single inertial sensor. Sensors **20**(15), 4345 (2020)
6. Wouda, F.J., et al.: Estimation of vertical ground reaction forces and sagittal knee kinematics during running using three inertial sensors. Front. Physiol. **9**, 218 (2018)
7. McDonald, G.C.: Ridge regression. Wiley Interdisc. Rev.: Comput. Stat. **1**(1), 93–100 (2009)
8. Cortina, J.M.: Interaction, nonlinearity, and multicollinearity: implications for multiple regression. J. Manag. **19**(4), 915–922 (1993)
9. Eberly, L.E.: Multiple linear regression. In: Ambrosius, W.T. (ed.) Topics in Biostatistics. Methods in Molecular Biology™, vol. 404, pp. 165–187. Springer, Heidelberg (2007). https://doi.org/10.1007/978-1-59745-530-5_9
10. Cutler, A., Cutler, D.R., Stevens, J.R.: Random forests. In: Zhang, C., Ma, Y. (eds.) Ensemble Machine Learning, pp. 157–175. Springer, Boston (2012). https://doi.org/10.1007/978-1-4419-9326-7_5
11. Seel, T., Raisch, J., Schauer, T.: IMU-based joint angle measurement for gait analysis. Sensors **14**(4), 6891–6909 (2014)
12. Seel, T., Schauer, T., Raisch, J.: Joint axis and position estimation from inertial measurement data by exploiting kinematic constraints. In: 2012 IEEE International Conference on Control Applications, pp. 45-49. IEEE (2012)
13. Cutti, A.G., Ferrari, A., Garofalo, P., Raggi, M., Cappello, A., Ferrari, A.: 'Outwalk': a protocol for clinical gait analysis based on inertial and magnetic sensors. Med. Biol. Eng. Comput. **48**(1), 17–25 (2010)

14. Esser, P., Dawes, H., Collett, J., Howells, K.: IMU: inertial sensing of vertical CoM movement. J. Biomech. **42**(10), 1578–1581 (2009)
15. Pedregosa, F., et al.: Scikit-learn: machine learning in Python. J. Mach. Learn. Res. **12**, 2825–2830 (2011)

Intelligent Wheelchairs Rolling in Pairs Using Reinforcement Learning

Nelson Rodrigues[1(✉)], Armando Sousa[1,2], Luís Paulo Reis[1,3], and António Coelho[1,2]

[1] FEUP - Faculty of Engineering of the University of Porto, Porto, Portugal
up200705576@edu.fe.up.pt, {asousa,lpreis,acoelho}@fe.up.pt
[2] INESC TEC - INESC Technology and Science, Porto, Portugal
[3] LIACC - Artificial Intelligence and Computer Science Laboratory, Porto, Portugal

Abstract. Intelligent wheelchairs aim to improve mobility limitations by providing ingenious mechanisms to control and move the chair. This paper aims to enhance the autonomy level of intelligent wheelchair navigation by applying reinforcement learning algorithms to move the chair to the desired location. Also, as a second objective, add one more chair and move both chairs in pairs to promote group social activities. The experimental setup is based on a simulated environment using gazebo and ROS where a leader chair moves towards a goal, and the follower chair should navigate near the leader chair. The collected metrics (time to complete the task and the trajectories of the chairs) demonstrated that Deep Q-Network (DQN) achieved better results than the Q-Learning algorithm by being the unique algorithm to accomplish the pair navigation behaviour between two chairs.

Keywords: Assistive robotics · Intelligent wheelchairs · Deep Q-Network · Q-Learning

1 Introduction

Intelligent robotics will be essential in improving humans' quality of life. With the increase of the elderly population, designing solutions to improve locomotion is imperative. Intelligent wheelchairs are robotics systems that need small inputs to be controlled.

This paper seeks to improve the autonomy level of intelligent wheelchairs by adding reinforcement learning to control the chair to move between two locations. Also, presents a solution to control more than one wheelchair, using a follow-the-leader strategy. The aim is to promote collaboration and improve social group behaviours and activities (e.g., visiting cultural spaces such as museums).

A simulated environment was designed using the robotics simulator gazebo with ROS to achieve the proposed objectives. Two reinforcement learning algorithms were applied: Deep Q-Network (DQN) and Q-Learning.

D. Tardioli et al. (Eds.): ROBOT 2022, LNNS 590, pp. 274–285, 2023.
https://doi.org/10.1007/978-3-031-21062-4_23

The implemented solution was tested by measuring the time that the leader chair consumes to navigate to the goal and drawing the trajectories of the chairs to verify if the following chair has a similar behaviour pattern related to the leader chair. The metrics reveal that the DQN algorithm needs fewer attempts to reach the goal and is more reliable in moving the chair to a predefined goal. The Q-Learning implementation can navigate to a goal but fails to follow a leader chair.

The developed prototype shows that it is possible to conclude that reinforcement learning is a valid technique for moving intelligent wheelchairs autonomously to a target and following another chair. The agent designed with Deep Q-Network was the more resilient solution, and it can deal with targets in movement, contrary to the agent with the Q-Learning implementation.

The rest of the paper is structured as follows. Section 2 describes the related work about the use cases used by intelligent wheelchairs and reinforcement learning. Section 3 describes the experimental design, the simulated environment, and the reinforcement learning algorithms. Section 4 presents the results, the time that the leader chair takes to get to the goal, and the trajectories of the leader and follower chairs, and Sect. 5 discusses the achieved results. The paper ends by presenting the conclusions and future work.

2 Related Work

The ageing of the human population (expected 24% by 2030 in comparison with 10% in 2000) and the lack of professional caregivers [20] raise new challenges to improve the comfort and health of the elder population in their daily lives [10]. Robotics could help to improve the quality of life by reducing the impact of social isolation by providing companion robots, help in physical or cognitive impairment with rehabilitation robots, and mobility problems with intelligent power wheelchairs [1].

This study will focus on intelligent wheelchairs. These kinds of chairs aim to improve the locomotive limitations of the disabled and elder people by providing a control mechanism that requires minimal input to control the movement of the chair [26]. The multimodal interfaces [3] are supported by methodologies from Human-computer Interface (HCI) and Human Machine Interface (HMI) to control the intelligent wheelchair smoothly. Such as Electroencephalography (EEG) records the electrical brain signals from the scalp, head orientation tracking techniques and converts the movements of the head into cursor movements on the screen, and eye-tracking techniques where a camera track the features and movement of the eye.

Nevertheless, some users will not be able to control or even define a path plan, and autonomous systems are a possible solution, where chairs move between waypoints without human intervention [12]. To overcome the multiple challenges of autonomous path navigation, machine learning algorithms have an important role in detecting and avoiding obstacles, path following [8], and dealing with localization and mapping [13].

One of the challenges is to design a system that promotes collaborative behaviour by allowing multiple wheelchairs to move together [23] in a platoon formation [21] of walking side-by-side with a caregiver [16].

Inspired by animal behaviour psychology [22], reinforcement learning enables the agents to learn from the environment according to negative and positive rewards translated to the agent's actions [11]. The agent interacts with the environment's state and originates an action in response. Then, the environment creates a new state for the agent and a reward according to the action. This process will follow in time in a sequence of steps, forming a loop of states, actions and rewards. After an intensive trial and error process, the agent is highly influenced by a Markov Decision Process (MDP) and executes actions to maximise the long-term expected cumulative reward.

The Q-Learning algorithm [25] is a value-based Reinforcement Learning algorithm that operates in discrete spaces and actions. Uses a table to store all Q-Values possible to a pair of the state-action. The Q-learning table can be updated using the Bellman equation. An extension to this approach is Deep Q-Network (DQN) [15], adding deep neural networks that can learn the best hyperparameters to accomplish a task [14]. This algorithm uses two networks: one for the prediction of the current state at each step that is constantly updating and a target network to generate a target value, that is a copy of the first network (prediction) that updates from time to time, used as "target values" checkpoints, stabilising the learning.

Reinforcement Learning has been applied to intelligent wheelchairs to predict user behaviour and help the navigation algorithms choose the best routes. Falzone et al. [4] used Reinforcement Learning in a simulated environment (CARLA simulator) on a distributed architecture to accelerate the learning process for obstacle avoidance scenarios based on sensors' data from the RGB camera, depth camera, radar, lidar, and Inertial Measurement Unit (IMU). Garrote et al. [6] designed robot-assisted navigation using Reinforcement Learning to overcome obstacles and narrow passages in virtual and real-world situations.

3 Methods and Materials

3.1 Simulated Environment

The simulated environment was designed using Gazebo [7], an open-source 3D robotics simulator that can simulate the physics of the interaction of the chairs with the world. The communication between the intelligent chairs and the simulated environment is supported by the Robot Operating System (ROS) [18] using the Noetic distribution. Additional ROS packages were added: the localization of the chairs in the virtual environment is guaranteed by the probabilistic localization system in 2D, the acml package [2]. Also, the reinforcement learning back-end is supported by the openai_ros package [17]. A reinforcement learning pipeline was designed using Keras [9] to build the model with TensorFlow [24] as the back-end.

3.2 Training the Intelligent Wheelchairs

The leader chair has the objective of navigating towards a goal (located in random positions), and the follower chair has the objective of following the leader chair. There are no obstacles between the chairs and their goals (Fig. 1). The leader-chair can reach the goal from any side of the square.

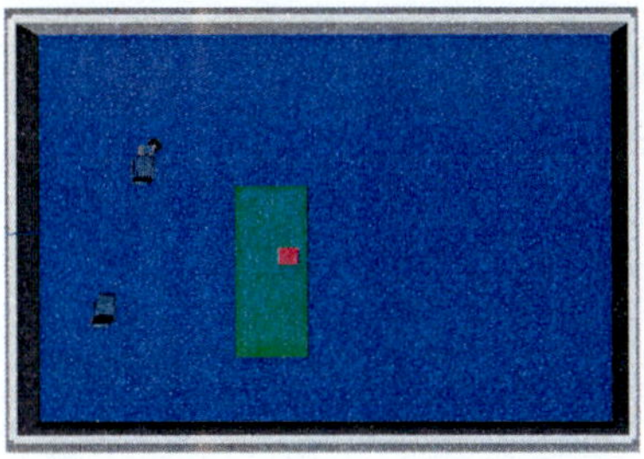

Fig. 1. The goal (red square) can be randomly locate anywhere on the green zone.

Each chair is considered an agent acting independently to the environmental constraints. Also, two different versions of each agent were designed using different reinforcement learning algorithms: Deep Q-Network (DQN) and Q-Learning.

The DQN model was trained by running 350 episodes with 1000 steps each with a learning rate of 0,00025 (Fig. 2). Then, the same neural network model was reused and applied to the follower chair. Instead of navigating towards a goal located in a random position, it should follow the leader chair that is in motion. Similar to follow a goal in random position the chair is constantly changing position.

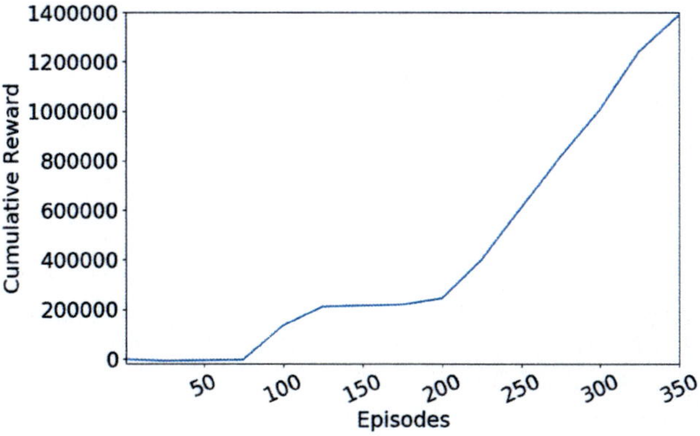

Fig. 2. Cumulative reward using DQN algorithm for the training session with a total of 350 episodes.

The Q-Learning algorithm has started the training session with the following configurations: 10000 episodes with 5000 steps by episode. It was observed that the robot achieved the goal with a certain regularity with the tuned parameters of alpha 0.1 and epsilon 0.3259. The Fig. 3 illustrated the cumulative reward using Q-Learning algorithm. A similar approach to DQN was followed for Q-Learning approach, and the parameters were reused from the navigation to a goal in a random positions training session to the follower chair scenario.

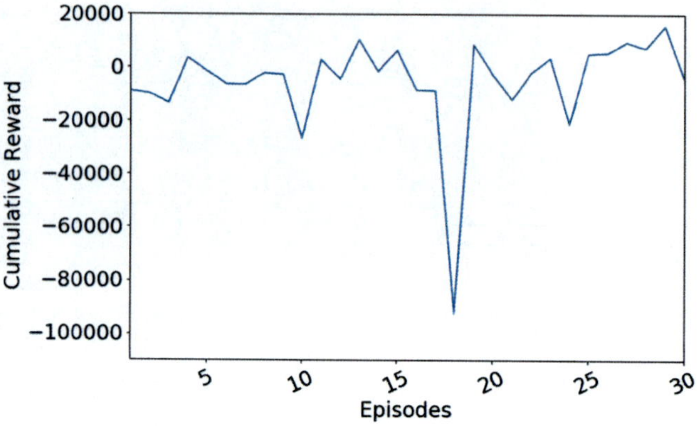

Fig. 3. Cumulative reward using Q-learning algorithm for 30 episodes using the parameters of: alpha 0.1, epsilon 0.3259.

3.3 Space and Actions

The intelligent wheelchair's Light Detection And Ranging (Lidar) can project 720 rays and collect an equal number of samples. This amount of data needs time and hardware with high specs to be processed during the training. A sampling strategy was implemented to simplify the space of the environment and feed the input neural network and Q-Learning algorithms. The sampling process starts by subscribing to the ROS topic, where the messages of the laser scan are redirected and find the ray with minimum distance to an obstacle. Then, according to a predefined threshold (10 rays) are collected the neighbour rays of that minimal hit point.

The space input is represented by the Eq. 1: 10 rays from the sampling plus the heading of the robot and the distance to the goal or the leader chair (polar coordinates). The distance is calculated using the Euclidean distance (Eq. 2), and the heading is measured by the angle of the vector between two points in 2D, from the chair and the goal (Eq. 3). The distances are measured in meters and the angles in degrees.

$$space(12) = laser\_sample(10) + heading(1) + distance(1) \tag{1}$$

$$distance = \sqrt{(x_2 - x_1)^2 + (y_2 - y_1)^2} \tag{2}$$

$$\alpha_{\text{heading}} = \operatorname{atan} 2((y_{goal} - y_{chair}), (x_{goal} - x_{chair})) \tag{3}$$

The possible chair's 'actions' are divided into five equal partitions from left to right. At each step will be assigned a different action value returned by the DQN neural network or the Q-Learning algorithm that will be related to the angular velocity of the chair.

3.4 Reward System

The chair will have a positive reward if it finds the goal position and a negative reward if the chair collides with a wall. There is a cumulative reward system according to the distance to the goal. If the chair gets close to the goal, a reward with a high value is assigned compared if the distance gets far away from the goal. If he heading of the chair is pointing to the goal, the reward is higher than if it is pointing in another direction, to reward actions that point to the goal to avoid deviations.

To accomplish the objective of navigating two chairs side by side, a relation between the leader chair and the following chair was used. The previous network was used as starting point to train the new feature. The following chair will define the position of the leader chair that is continuously changing as a goal. The position of the robots is obtained by subscribing to the odometry topic of each robot.

The reward system of the follower chair has the same basic principle as the leader chair. If the chair collides with an object, a negative reward is assigned. Similar to the previous reward system, a cumulative reward is assigned if the follower chair is getting close to the leader chair and another positive reward if the heading of the follower chair is pointing to the leader chair. Furthermore, was defined as a small positive reward if the chair is in a safe zone close to the chair. The aim is to teach the follower chair to stay in a safe zone related to the leader chair. In case of the following chair is close to a minimum safe distance, it should stop. Figure 4 resumes the reward system for the following chair.

4 Results

The comparison between the DQN and Q-Learning algorithms was performed by creating a new scenario.

4.1 Metrics Using only the Leader Chair

The metrics of the algorithms' performance were collected by launching five trials and recording the trajectories of the chairs and the time to complete the task (reach the goal). For this reason, instead of a random goal for the leader chair, it is fixed a position located in x = 1.0 and y = −5.0 coordinates. The model is learning during the trial, but when the chair reach the goal, the models are

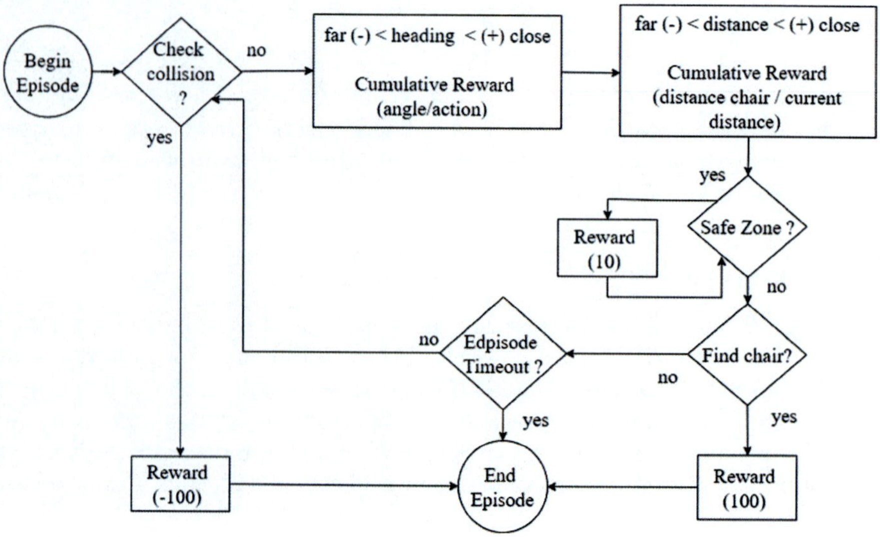

Fig. 4. Reward system for the following chair.

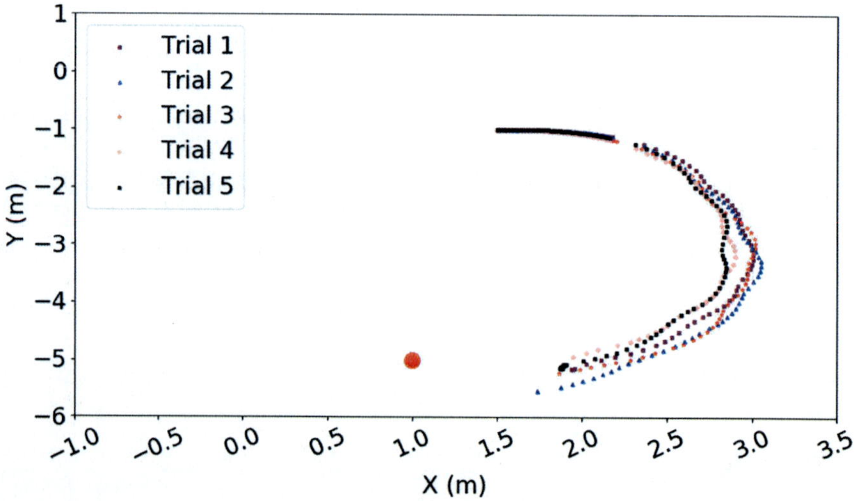

Fig. 5. Trajectories for the leader chair using DQN algorithm (the red dot represents goal's location).

restore to their initial state to ensure equal conditions at the beginning of each trial. Figure 5 presents the results of each trial for the DQN algorithm and the Fig. 6 plot the routes for Q-Learning algorithm.

The Table 1 compares the time to reach the goal and the number of attempts using DQN and Q-Learning.

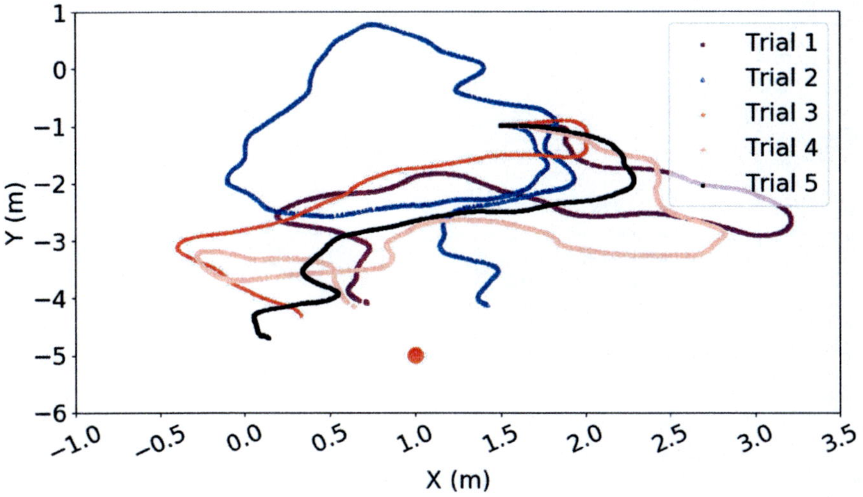

Fig. 6. Trajectories for the leader chair using Q-Learning algorithm (the red dot represents the goal's location).

Table 1. Time and number of attempts to achieve the goal using DQN and Q-Learning algorithms by using a single chair.

Trial	DQN Time (mm:ss)	DQN Attempts	Q-Learning Time (mm:ss)	Q-Leaning Attempts
1	00:46	1	01:05	7
2	00:50	1	01:37	2
3	00:46	1	00:41	1
4	00:44	1	01:14	2
5	00:43	1	00:40	2

4.2 Metrics Using Two Chairs

To measure the influence of the following chair and validate if the reinforcement learning algorithms of the "follow the leader" chair were recorded, the trajectories of the two chairs and the time to the leader chair get to the goal. The leader chair must reach the goal in less than ten attempts to validate the trial. Figure 7, illustrates the trajectories of both chairs using DQN algorithm and Table 2.

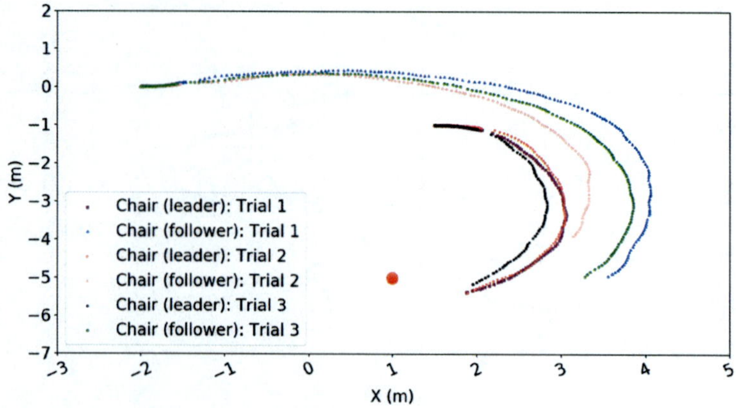

Fig. 7. Trajectories of the leader chair and the follower chair using DQN algorithm (the red dot represents the goal's location).

Table 2. Time and number of attempts to achieve the goal using DQN and Q-Learning algorithms by using follow-leader chairs.

Trial	DQN Time (mm:ss)	DQN Attempts	Q-Learning Time (mm:ss)	Q-Leaning Attempts
1	01:03	1	00:45	5
2	01:02	1	–	Failed
3	00:58	1	01:12	10
4	01:05	1	00:34	7
5	01:54	1	00:46	2

The same approach was used to measure the Q-Learning algorithm performance. The Fig. 8 and the Table 2 presents the results of the first successful trials.

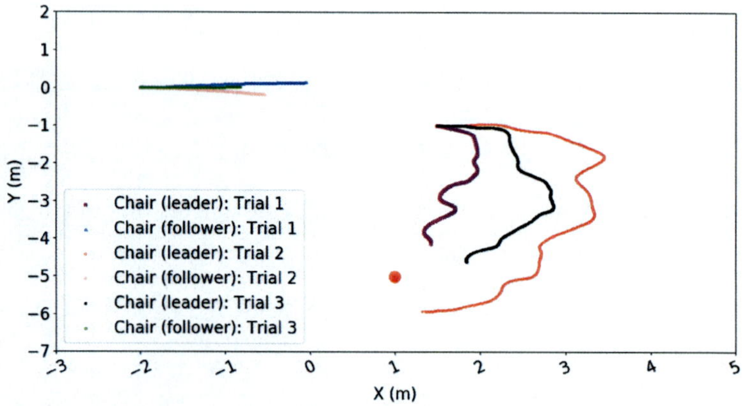

Fig. 8. Trajectories of the leader chair and the follower chair using Q-Learning algorithm (the red dot represents the goal's location).

5 Discussion

Using both algorithms is possible to get to a goal. Comparing the trajectories using only one chair (Fig. 5 and Fig. 6), the DQN algorithm's trajectories are more consistent. The trial ends when the chair collides with a wall or reaches the goal. The metrics from Table 1, show that using Q-Learning needed more attempts to get to the goal, and this algorithm is more prone to failures. The DQN trials show that was only needed one trial to get to the goal, and the chair has learned to avoid paths that will provoke a collision with a wall.

Comparing the trajectories using the two chairs is possible to observe that using DQN (Fig. 7) the follower chair follows the leader in a consistent order. Otherwise, the behaviour is not illustrated using Q-learning, and the follower chair fails to follow the leader's trajectory (Fig. 8).

Having time to complete the task, as a measure, the average time to get to the goal using DQN (average 00:46 s) is less than using Q-Learning (average 01:03 min). Adding a second chair with the objective of following a leader chair while the leader follows a goal, the average time for DQN increased to 01:12 min (Table 2), contrary to with Q-Learning the time decreases to 00:49 s. However, using Q-Learning, the algorithm failed, in the second trial, to reach the goal in less than ten attempts, so this value was discarded when calculating the average. The increase in time using DQN can be explained because of the resource consumption by running two agents (chairs) with two models at the same time in the same machine.

The automated training process of the DQN algorithm finds the best parameters for the model in order to find a goal or follow another chair. The process of trial and error to tune the parameters for the Q-Learning algorithm is hard, and it was not possible to find the best ones for the scenario of following another chair. This objective was not possible to accomplish using the Q-Learning algorithm.

6 Conclusions and Future Work

The expected ageing of the human population brings new challenges to robotics to cooperate with humans to improve their quality of life. Intelligent wheelchairs can help on augmenting the locomotion limitation of elderly people.

These chairs need little input to be controlled. However, in some cases, the user cannot provide any input, and the best solution is the autonomous navigation of the chair between waypoints.

The objective of this work was to add reinforcement learning capabilities to the chairs in order to navigate to a predefined objective and design an algorithm where the second chair follows another chair in pairs. To safely implement the solution was used a simulated environment using gazebo and ROS. Two algorithms were used to control the Q-Learning and Deep Q-Network (DQN). The collected metrics (time and trajectories) show that DQN has the best results in both tasks, finding a goal and following a leader chair. The Q-Learning implementation fails in following a leader chair.

For future work, other reinforcement learning algorithms should be compared, such as Proximal Policy Optimization [19], or Dynamic Hindsight Experience Replay (DHER) [5] optimised for dynamic goals a scenario suitable for the follower chair where the target is always changing. In this use case, the collision is simplified by assuming that if the laser scan receives a ray with less than 0.5 m is a collision but does not distinguish the type of object that collides. Furthermore, the current implementation relies only on the odometry data of the chairs. The navigation could be improved by using the information from the camera with computer vision techniques to identify the leader chair and navigate towards their direction.

References

1. Abou Allaban, A., Wang, M., Padır, T.: A systematic review of robotics research in support of in-home care for older adults. Information **11**(2), 75 (2020)
2. acml: acml - ros wiki (2022). http://wiki.ros.org/amcl
3. Coelho, F., Reis, L.P., Faria, B.M., Oliveira, A., Carvalho, V.: Multimodal intelligent wheelchair interface. In: Rocha, Á., Adeli, H., Reis, L.P., Costanzo, S., Orovic, I., Moreira, F. (eds.) WorldCIST 2020. AISC, vol. 1160, pp. 679–689. Springer, Cham (2020). https://doi.org/10.1007/978-3-030-45691-7_64
4. Falzone, G., Giuffrida, G., Panicacci, S., Donati, M., Fanucci, L.: Simulation framework to train intelligent agents towards an assisted driving power wheelchair for people with disability. In: ICAART (1), pp. 189–196 (2021)
5. Fang, M., et al.: DHER: hindsight experience replay for dynamic goals. In: International Conference on Learning Representations (2019)
6. Garrote, L., Paulo, J., Nunes, U.J.: Reinforcement learning aided robot-assisted navigation: a utility and RRT two-stage approach. Int. J. Soc. Robot. **12**(3), 689–707 (2019). https://doi.org/10.1007/s12369-019-00585-0
7. Gazebo: Gazebo, robot simulation (2022). URL http://gazebosim.org
8. Gunachandra, S. Chrisander, A. Widyotriatmo, S.: Wall following control for the application of a brain-controlled wheelchair. In: 2014 International Conference on Intelligent Autonomous Agents, Networks and Systems, pp. 36–41 (2014). https://doi.org/10.1109/INAGENTSYS.2014.7005722
9. Keras, C.F.: Keras: the python deep learning api (2022). URL https://keras.io
10. Koumpouros, Y.: A systematic review on existing measures for the subjective assessment of rehabilitation and assistive robot devices. J. Healthc. Eng. **2016** (2016)
11. Lanham, M.: Learn Unity ML-Agents-Fundamentals of Unity Machine Learning: Incorporate new powerful ML algorithms such as Deep Reinforcement Learning for games. Packt Publishing Ltd, Birmingham (2018)
12. Leaman, J., La, H.M.: A comprehensive review of smart wheelchairs: past, present, and future. IEEE Trans. Hum.-Mach. Syst. **47**(4), 486–499 (2017)
13. Mandel, C., Laue, T., Autexier, S.: Chapter 12 - Smart-wheelchairs. In: P. Diez (ed.) Smart Wheelchairs and Brain-Computer Interfaces, pp. 291–322. Academic Press (2018). https://doi.org/10.1016/B978-0-12-812892-3.00012-1
14. Mnih, V., et al.: Playing atari with deep reinforcement learning (2013)
15. Mnih, V., et al.: Human-level control through deep reinforcement learning. Nature **518**(7540), 529–533 (2015). https://doi.org/10.1038/nature14236

16. Nguyen, V.T., Jayawardena, C., Ardekani, I.: A navigation model for side-by-side robotic wheelchairs for optimizing social comfort in crossing situations. Robot. Auton. Syst. **100**, 27–40 (2018). https://doi.org/10.1016/j.robot.2017.10.008
17. openai: openai_ros - ros wiki (2022). http://wiki.ros.org/openai_ros
18. ROS: Ros - robot operating system (2022). https://www.ros.org
19. Schulman, J., Wolski, F., Dhariwal, P., Radford, A., Klimov, O.: Proximal policy optimization algorithms (2017)
20. Shishehgar, M., Kerr, D., Blake, J.: A systematic review of research into how robotic technology can help older people. Smart Health **7**, 1–18 (2018)
21. Sugano, T., Dan, Y., Okajima, H., Matsunaga, N., Hu, Z.: Indoor platoon driving of electric wheelchair with model error compensator along wheel track of preceding vehicle. In: Proceedings of the 5th International Symposium on Advanced Control of Industrial Processes (2014b), pp. 219–224 (2014)
22. Sutton, R.S., Barto, A.G.: Reinforcement Learning: An Introduction. MIT Press, Cambridge (2018)
23. Suzuki, R., Yamada, T., Arai, M., Sato, Y., Kobayashi, Y., Kuno, Y.: Multiple robotic wheelchair system considering group communication. In: Bebis, G., et al. (eds.) ISVC 2014. LNCS, vol. 8887, pp. 805–814. Springer, Cham (2014). https://doi.org/10.1007/978-3-319-14249-4_77
24. TensorFlow: Tensorflow (2022). https://www.tensorflow.org
25. Watkins, C.J.C.H., Dayan, P.: Q-learning. Mach. Learn. **8**(3), 279–292 (1992). https://doi.org/10.1007/BF00992698
26. Yadav, S., Sheoran, P.: Smart wheelchairs-a literature review. Int. J. Innovative Emerg. Res. Eng. **3**(2), 86–90 (2016)

Influence of Session Duration in Robot-Assisted Neurorehabilitation Therapies: A Pilot Study with Patients

Yolanda Vales[1], José Maria Catal[1], Luis Daniel Lled[1], Andrea Blanco[1], Jesús Álvarez-Pastor[1], David Martnez-Pascual[1], Alicia Garrote[2], Teresa Costa[2], Gemma Más[2], and Nicolás García-Aracil[1](✉)

[1] Robotics and Artificial Intelligence Group of the Bioengineering Institute, Miguel Hernández University, Avda. de la Universidad, 03202 Elche, Spain
nicolas.garcia@umh.es
[2] Hospital la Pedrera, Dénia, Spain

Abstract. Several studies have shown that robotic rehabilitation platforms are able to improve long-term outcomes of patients' neurological treatments. Furthermore, they have indicated that therapy performance improves if as many sessions as possible are performed. However, the influence of the duration of the session on the performance of the robot-assisted is unknown. This makes it difficult to adapt therapies to patients' needs. In this preliminary study the influence of session duration has been analyzed. This pilot study has involved four patients separated into two groups with different session durations: 45 min and 30 min. Results show an improvement in upper limb motor function in both groups of patients, improving their exercise performance with a lower assistance level. Differences observed between the two groups of patients are not enough to affirm that there is a difference in the improvement of the motor control of the patients due to the session duration.

Keywords: Neurorehabilitation · Rehabilitation robotics · Control

1 Introduction

In the last report from the Stroke Alliance for Europe (SAFE) [21], around 15 million people worldwide suffer from cerebrovascular accident (CVA) every year. Through the data collected by the Global Burden of Disease study in 2015 and demographic projections obtained from Eurostat (statistical office of the European Union (EU)), the number of CVAs is expected to rise a 34% between 2015 and 2035 the EU.

In most cases, CVAs causes a deterioration in the health-related quality of life (HRQoL) of patients, as it can lead to physical, cognitive, emotional and social deficits [16]. In most cases, they are partially recoverable. However, it is necessary to implement a rehabilitation program as soon as possible, and that

D. Tardioli et al. (Eds.): ROBOT 2022, LNNS 590, pp. 286–297, 2023.
https://doi.org/10.1007/978-3-031-21062-4_24

such a program comply with a set of criteria to increase the effectiveness of the therapy [1].

CVA survivors generally experience hemiparesis [19]. This implies a significant deterioration of the HRQoL [16]. An appropriate rehabilitation program and intensive therapy can help people suffering from hemiparesis regain limb movement.

The need for rehabilitation and assistance that these people present, make the care and rehabilitation exercises provided by robotic rehabilitation platforms even more important today and in the coming years [18].

Robots are presented as a useful tool during the recovery process of a patient's neurological treatment. Such systems actively participate and help the therapist to improve the performance of the rehabilitation process. Several studies have shown that robotic rehabilitation platforms have the potential to help improve long-term outcomes compared to conventional therapy [12,15].

The planning process for intervention protocols is based on different factors. Many of the intervention measures are systematized (concrete measures for specific problems), but there is no reason to believe that there is an optimal treatment that fits everyone. This is why the therapy must be adapted (level of intensity, number of repetitions and duration of exercise) to the needs and abilities of each patient [13].

In several studies, it has been shown that performance of the therapy is improved by performing as many sessions as possible and maintaining a good intensity level in each session [6,8]. However, the influence of the duration of the session on the performance of the therapies assisted by robotic rehabilitation platforms is unknown, and therefore, it is difficult to adapt the therapy to the needs and abilities of each patient.

The aim of this preliminary study is to analyze the influence of session duration on long-term therapy assisted by robotic rehabilitation platform.

2 Materials and Methods

2.1 Subjects

The study has been performed in La Pedrera Hospital (Denia, Spain) with patients suffering from neurological conditions. The rehabilitation therapists have been responsible for including patients who are receiving treatment. The experiment protocol was approved by the Ethical Committee of La Pedrera Hospital. All patients have been informed properly and they gave written consent before starting.

The inclusion criteria were: adults stroke survivors in subacute stage, with capacity of collaboration and understanding the tasks instruction and all relevant information from the study.

In addition, the Fugl-Meyer clinical scale was used to measure the functional capacity and mobility of the patient's affected arm in order to determine his or her baseline condition. This scale is a quantitative method, designed to assess motor function in upper limbs affected after stroke. The maximum motor function score is 66, being: Fugl-Meyer Assessment for upper extremity (FMA-UE) > 50, slightly impaired; FMA-UE score between 18 and 50, moderately impaired; and FMA-UE score < 18, severely impaired [9,10,20].

The inclusion of patients with the following injures was avoided: very high degree of spasticity in the upper limb (Modified Ashworth Scale (MAS)> 1+), painful shoulder, severe perceptual deficits, visual deficits (Aperceptive visual agnosia) and attention deficits.

Table 1 shows the patients that took part in this study. For this preliminary study only 4 patients have participated (2 male and 2 female) with ages between 38 and 77 years (59.25 years ± 13.97 years).

Table 1. Characteristics of impaired subjects together with the Fugl-Meyer Index.

n	Sex	Age	Diagnostic	Laterality	Sessions	Fugl-Meyer	
						Init	Final
1	M	63	Parietal Ischemic stroke	Left	Long	5	20
2	F	59	Frontal and parietal hemorrhagic stroke	Left	Long	12	44
3	F	77	Hemorrhagic stroke in the basal ganglia	Right	Short	14	36
4	M	38	Temporal and parietal ischemic stroke	Right	Short	6	14

Note. **Fugl-Meyer**, is an index to assess the sensorimotor impairment in individuals who have had stroke [9].

2.2 Experimental Setup

To carry out this experimentation, a rehabilitation platform consisting of an end-effector type arm rehabilitation robot described below and a game screen has been used, as can be seen in Fig. 1. This platform allows the height of the robot above the floor to be adjusted, especially useful when it is intended to be used with wheelchair bound patients who cannot be transferred to a conventional chair.

The screen, which was placed about 70 cm away from the study participants, displayed the rehabilitation game that the users had to interact with throughout the therapy sessions, providing them with visual feedback.

2.3 Arm Rehabilitation Robot

In this study, upper-limb rehabilitation robotic device called Rubidium has been employed (Fig. 2) [2–4,7]. The new version is called. Rubidium is commercialized and distributed by the spin-off iDRhA [11]. This rehabilitation robotic platform is a portable, desktop-type, upper limb rehabilitation robotic device with two actuated degrees of freedom. It consists of an articulated parallelogram mechanism driven by two electric motors that perform movements in a horizontal

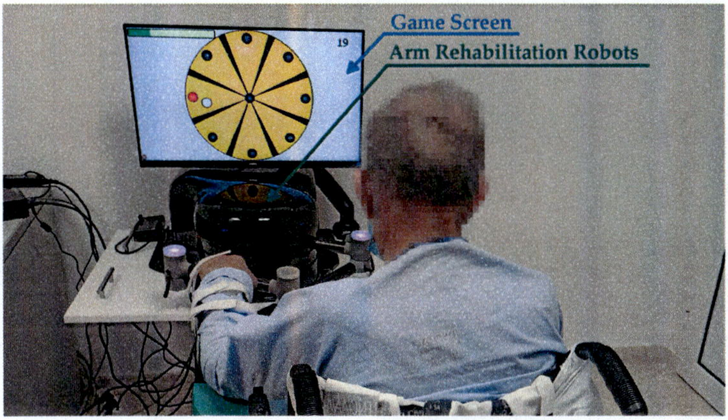

Fig. 1. Overview of the experimental setup.

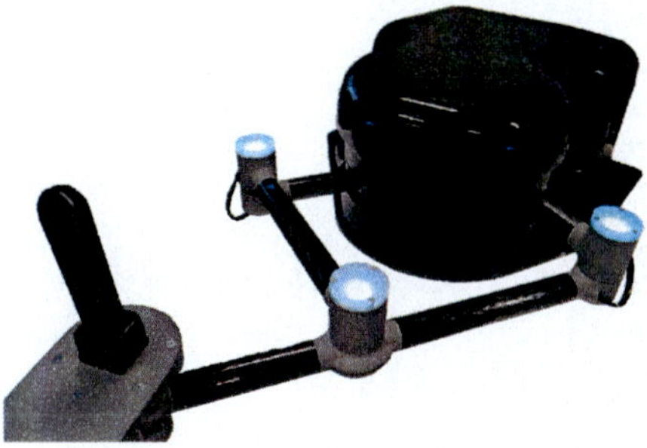

Fig. 2. Upper-limb rehabilitation robotic device called Rubidium. Rubidium is commercialized and distributed by the spin-off iDRhA [11].

plane. The patient rests the arm on an orthosis anchored to the end effector, supporting the weight of the forearm.

Rubidium incorporates an assistive mode based on force field, which have been already employed and validated in other studies with patients [3]. These force fields provide guidance and assistance to the patient during therapy.

The version of the force field employed in this study, represented in Fig. 3, help the patients to make a linear movement but also help them to reach the target if necessary. It has the shape of a tunnel with ends, with a destination point that corresponds to the position of the target and an origin point that corresponds to the initial position of the patient. The origin of this tunnel moves at constant velocity that is determined based on the among of time that the patient has to reach the target. This type of force field provide assistance only when the patient is unable to move towards the target but it allows to move freely toward the target.

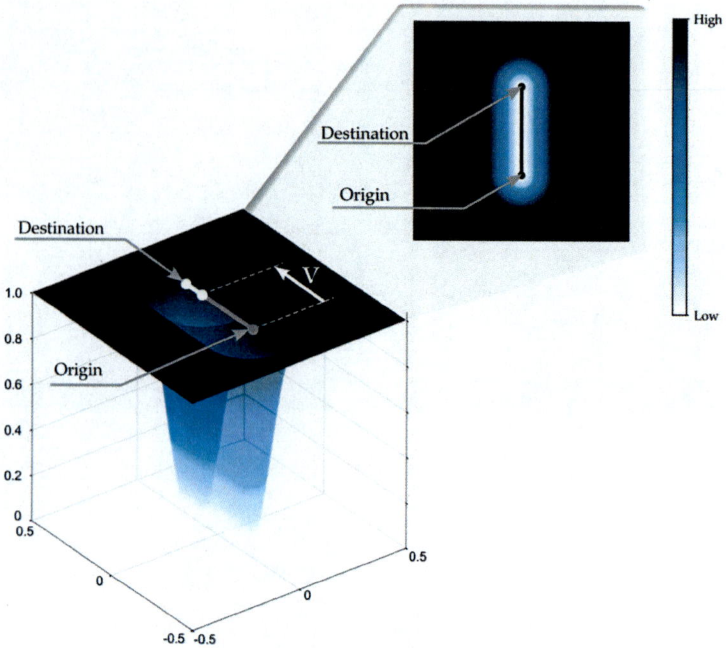

Fig. 3. Graphic representation of the force fields used in this study. The scale represents the normalized assistance force magnitude applied in each position. The maximum assistance force value is configured by means of the assistance level of the robotic rehabilitation device.

2.4 Rehabilitation Game

In this study, rehabilitation therapies were performed using a point-to-point game modality. The game consists of moving the player cursor represented by a white circle from a central position to reach one of the 8 red target points located 10 cm from the central position (Fig. 4).

In every trial, the patient has to wait in the center of the roulette until a target lights up, then he will try to reach the target before the time runs out and then return to the center. The targets to be reached are selected randomly.

The rehabilitation game used in this study, as well as the arm rehabilitation robotic platform, have been developed by the Robotics and Artificial Intelligence Group of the Bioengineering Institute, Miguel Hernández University. These have been evaluated in other studies with healthy subjects [4] and with stroke patients [2,3].

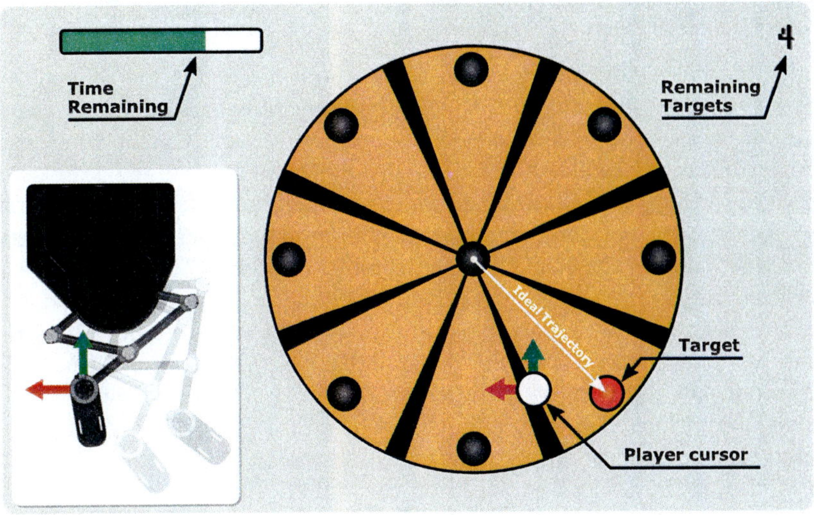

Fig. 4. Overview of the game. A screenshot of the game with an image of the robot to show the movement of the robot in correlation with the player cursor.

2.5 Analyzed Parameters

Estimation of Exercise Performance: The score of the game is a parameter widely used as a measure of exercise performance, as it allows us to evaluate when the patient is able to achieve an objective [5,17].

Other parameters related to exercise performance have also been computed. On the one hand, we have extracted maximum and mean velocity values of the hand in order to analyze how fast they are able to perform the movements. Initial movement direction error and Path length error have also been extracted as indicated in [14].

Estimation of the Assistance Level: When we talk about the assistance provided to patients, we refer to the force provided in the direction of the current target to help the patient reach it.

To estimate the assistance level, we calculate the force provided by the robotic rehabilitation device in the direction of the current target during each of the sessions. To do this, we use the Eq. 1, where F is the force provided by the robotic device, $V_{trajectory}$ is the vector defined by the ideal trajectory to follow and α is the angle formed by F_{assist} and $V_{trajectory}$. This method was evaluated in another study with patients [3].

$$F_{assist} = |F| \cdot \cos\alpha = |F| \cdot \frac{F \cdot V_{trajectory}}{|F| \cdot |V_{trajectory}|} \tag{1}$$

2.6 Study Protocol

In order to analyze the influence of session duration on robotic rehabilitation therapy with stroke patients, participants were randomly divided into two control groups: long session (LS), 45 min sessions and short session (SS), 30 min sessions. Both control groups received a total of 15 robot assisted therapy sessions over the 5 weeks of the study. All patients were evaluated in the first and last session, based on the data obtained from the Rubidium robot and the clinical scales performed by the therapist in charge of conducting the rehabilitation sessions.

Before starting the session, the patient is placed in front of the robotic rehabilitation platform in a comfortable position. The height of the robot is adjusted so that the upper limb with which the patient is to perform the task is positioned 30°C in internal rotation, with the elbow flexed to 90° C with respect to the anatomical position. Also, the patient must be at an optimal distance to be able to handle the end effector in its full range of motion.

The arm of the patient is placed on the end effector of the robot and held in place with Velcro straps. Two different types of grip were designed to better adapt to the patient's needs. On the one hand, a hemisphere was used for patients with an anatomical basal position of hand closure, or who do not have sufficient prehensile strength to grasp the end effector adequately. In this case they will perform the required movements with the hand open and forearm in pronation. For the rest of the patient, a vertical joystick type cylinder was used. In this case the user grasps the cylinder with the forearm in an intermediate position.

The game is then explained to the patient. Whenever necessary, a few minutes are spent before beginning each session to help the patient become familiar with the activity and the robotic device. Then the session begins. Both groups perform the complete session continuously.

The Fugl Meyer clinical scale was evaluated once before the first session and again after the last session.

3 Results

3.1 Exercise Performance

Table 2 shows the results of the parameters used in this study to analyze the improvement in patient performance between the first and last session. In the results obtained for the score, both groups of patients show considerable improvement. LS patients show an improvement of 24.47% between the first and last session, while SS patients improve by 33.34%. A considerable increase in average and maximum speed was also observed in the last session compared to the first session for both groups of patients. The data collected also show that patients in both groups are able to correctly perform the exercise much faster, as the average and maximum speed of both groups increases considerably in the last session with respect to the first one. As for the Route Length Error, the results suggest that this error increases with respect to the first session in both groups of patients. However, the increase in the case of patients who performed LSs (29.8

Table 2. Parameters related to exercise performance extracted from data collected by the robotic rehabilitation device.

	Group	First session	Last session
Score (%)	LS	74.34	98.81
	SS	50.0	83.34
Average speed (m/s)	LS	0.071	0.240
	SS	0.108	0.224
Maximum speed (m/s)	LS	0.333	0.671
	SS	0.414	0.631
Path length error (mm)	LS	25.2	55.0
	SS	41.7	43.1
Initial movement direction error (°)	LS	23.3	15.8
	SS	42.4	19.8

mm) is greater than the increase in SS patients (1.4 mm). On the other hand, a considerable improvement in the initial movement direction error is also shown for both groups. LS patients show an improvement of 7.5°, while SS patients improve by 22.6°.

3.2 Assistance Force

Trajectories of each patient in the first and last session with overlaying density plots indicating the location and amount of the offered assistance are shown in Fig. 5. The data for each user show, in most cases, a considerable reduction in the assistance force applied in the last session compared to the first session.

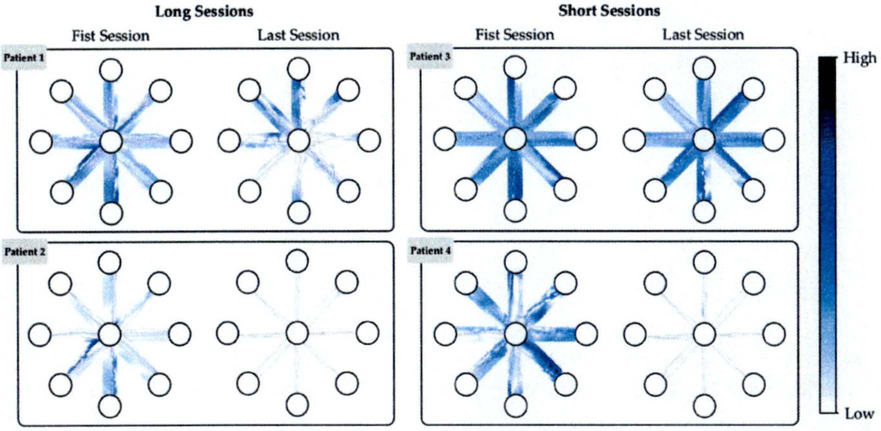

Fig. 5. Trajectories of each patient in the first and last session with overlaying density plots indicating the location and amount of the offered assistance. Darker areas indicate higher amounts of assistance.

Patient number 3 is the only one who shows no appreciable differences between the first and last session.

Mean and Maximum assistance force values in the first and last session for each of the groups were computed (Fig. 6). In the data collected from both LS and SS patients, a reduction in mean assist force was observed between the first and last session. However, the results show a reduction in the maximum assist force for patients who performed LS, but we did not observe a reduction in the case of SS patients.

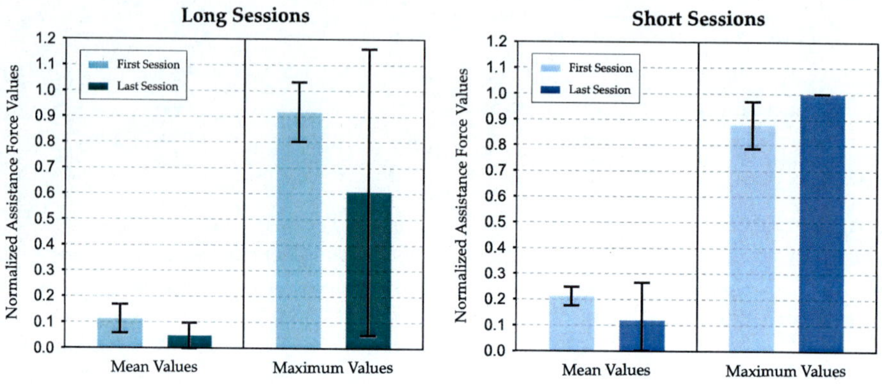

Fig. 6. Normalized assistance force values in the first and last session for each group of patients. Both bar graphs show Mean and Maximum assistance forces values in the first and last session for LS patients and SS patients. Bars indicate mean values, and error bars indicate standard deviations. In all cases, each patient's assistance force was normalized by the maximum value measured for the patient.

3.3 Assessment of Motor Function

In Fugl-Meyer clinical scale results, an increase in the scores has been observed for all users, indicating an improvement in the mobility of the affected upper limb. Patients who performed LSs have improved an average of 23.5 points, while those who performed SSs improved 15 points.

4 Discussion

The results obtained in the estimation of the assistance forces provided by the robotic rehabilitation device to patients show a considerable improvement. Figure 5 shows how the assistance provided to patients in the last session is, in general, less than in the first session. This improvement can also be seen in the results shown in Fig. 6, where a reduction in the assistance provided in the last session can also be observed.

The data collected also indicate that both groups of patients achieve a considerable improvement in score with a much higher speed (Table 2). This is a clear indication of the improvement of the patients' upper limb motor control after the 15 sessions, as both groups of patients are able to correctly perform the exercise much faster.

On the other hand, the path length error increases in the last session. However, this increase does not indicate that the trajectory is more uncontrolled. As mentioned above, there is a considerable improvement in the score. We have observed that when patients move faster with the intention of winning, they are less concerned about performing rectilinear trajectories. In addition, the assistance provided by the robotic device in the first session is significantly higher, so it is to be expected that the robot-assisted movement will be more controlled. Therefore, this increase in path length error in the last session is an expected result.

On the other hand, the Initial movement direction error improves in the last session of both groups of patients, which is another indication of the improvement in the upper limb motor control of the patients.

As we observed in the data collected by the robotic rehabilitation platform, the Fugl-Meyer clinical scale results also show an improvement in the motor function of both groups of patients. In this case, the results suggest that the improvement obtained in LS patients is better than that observed in SS patients, as they improve by an average of 8.5 points more.

Overall, the results suggest that LS patients achieve slightly better upper limb motor control compared to SS patients, as they achieve a better score, with slightly higher speeds, a lower initial movement direction error, and a slightly lower assist force, while for the remaining parameters no appreciable differences are observed. In addition, as previously mentioned, the results of the Fugl-Meyer clinical scale indicate that patients who performed LSs achieve greater improvement in motor function. These results are consistent with those observed in other studies [15].

However, it should be noted that the difference observed between the two groups is not very large. In addition, the number of patients who participated in this pilot study is very small, so that the differences observed could be due to other aspects such as the patients' recovery capacity and not only to the duration of the session.

5 Conclusion

In this pilot study we have evaluated the the influence of session duration on long term therapy assisted by robotic rehabilitation platform. To do this, we have carried out a preliminary study involving 4 patients that have performed typical point to point exercises in robot assisted therapy sessions over 5 weeks. They were separated into two groups: LS, 45 min sessions and SS, 30 min sessions. Both control groups received a total of 15 robot assisted therapy sessions.

After the last rehabilitation session, both groups of patients show an improvement in the motor function of the upper limb, improving their exercise performance with a lower level of assistance.

Although the results obtained in this preliminary study show differences between the two groups of patients, these are not enough to affirm that there is some difference in the improvement of motor control of the patients due to the duration of the session. However, these findings should be verified in a study with a larger number of patients.

Acknowledgments. This work has been supported by the Ministry of Universities and European Union, "financed by European Union - Next Generation EU" through Margarita Salas grant for the training of young doctors, the Youth Guarantee Plan through the grant EDGJID/2021/075 and by the Ministry of Science and Innovation, belonging to the State Research Agency (AEI) through the project PID2019-108310RB-I00/AEI/10.13039/501100011033.

References

1. Alessandro, L., et al.: Multidisciplinary rehabilitation for adult patients with stroke. Medicina **80**(1), 54–68 (2020)
2. Catalan, J.M., et al.: Evaluation of an upper-limb rehabilitation robotic device for home use from patient perspective. In: Masia, L., Micera, S., Akay, M., Pons, J.L. (eds.) ICNR 2018. BB, vol. 21, pp. 449–453. Springer, Cham (2019). https://doi.org/10.1007/978-3-030-01845-0_90
3. Catalán, J.M., et al.: Tele-rehabilitation versus local rehabilitation therapies assisted by robotic devices: a pilot study with patients. Appl. Sci. **11**(14) (2021). https://doi.org/10.3390/app11146259
4. Catalán, J.M., García-Pérez, J.V., Blanco, A., Martínez, D., Lledó, L.D., García-Aracil, N.: Differences in physiological reactions due to a competitive rehabilitation game modality. Sensors **21**(11), 3681 (2021). https://doi.org/10.3390/s21113681
5. Darzi, A., Novak, D.: Using physiological linkage for patient state assessment in a competitive rehabilitation game. In: 2019 IEEE 16th International Conference on Rehabilitation Robotics (ICORR), pp. 1031–1036. IEEE (2019)
6. Duret, C., Grosmaire, A.G., Krebs, H.I.: Robot-assisted therapy in upper extremity hemiparesis: overview of an evidence-based approach. Front. Neurol. **10**, 412 (2019). https://doi.org/10.3389/fneur.2019.00412
7. Díaz, I., et al.: Development of a robotic device for post-stroke home tele-rehabilitation. Adv. Mech. Eng. **10**(1), 1687814017752302 (2018). https://doi.org/10.1177/1687814017752302
8. Ferreira, F.M.R.M., Chaves, M.E.A., Oliveira, V.C., Van Petten, A.M.V.N., Vimieiro, C.B.S.: Effectiveness of robot therapy on body function and structure in people with limited upper limb function: A systematic review and meta-analysis. PLoS ONE **13**(7), 1–21 (2018). https://doi.org/10.1371/journal.pone.0200330
9. Fugl-Meyer, A.R., Jääskö, L., Leyman, I., Olsson, S., Steglind, S.: The post-stroke hemiplegic patient. 1. a method for evaluation of physical performance. Scand. J. Rehabil. Med. **7**(1), 13–31 (1975)
10. Gladstone, D.J., Danells, C.J., Black, S.E.: The Fugl-Meyer assessment of motor recovery after stroke: a critical review of its measurement properties. Neurorehabil. Neural Repair **16**(3), 232–240 (2002)

11. iDRhA: Innovative Devices for Rehabilitation & Assistance. https://idrha.es/ (2019). Accessed 24 June 2022
12. Klamroth-Marganska, V., et al.: Three-dimensional, task-specific robot therapy of the arm after stroke: a multicentre, parallel-group randomised trial. Lancet Neurol. **13**(2), 159–166 (2014)
13. Krebs, H.I., et al.: Rehabilitation robotics: performance-based progressive robot-assisted therapy. Auton. Robot. **15**(1), 7–20 (2003). https://doi.org/10.1023/A:1024494031121
14. LLinares, A., Badesa, F.J., Morales, R., Garcia-Aracil, N., Sabater, J., Fernandez, E.: Robotic assessment of the influence of age on upper-limb sensorimotor function. Clin. Interv. Aging **8**, 879 (2013)
15. Lo, A.C., et al.: Robot-assisted therapy for long-term upper-limb impairment after stroke. N. Engl. J. Med. **362**(19), 1772–1783 (2010)
16. Nichols-Larsen, D.S., Clark, P., Zeringue, A., Greenspan, A., Blanton, S.: Factors influencing stroke survivors' quality of life during subacute recovery. Stroke **36**(7), 1480–1484 (2005)
17. Norouzi-Gheidari, N., Levin, M.F., Fung, J., Archambault, P.: Interactive virtual reality game-based rehabilitation for stroke patients. In: 2013 International Conference on Virtual Rehabilitation (ICVR), pp. 220–221. IEEE (2013)
18. Oña, E., Cano-de La Cuerda, R., Sánchez-Herrera, P., Balaguer, C., Jardón, A.: A review of robotics in neurorehabilitation: towards an automated process for upper limb. J. Healthc. Eng. **2018** (2018)
19. van Peppen, R., et al.: KNGF clinical practice guideline for physical therapy in patients with stroke. review of the evidence. Nederlands Tijdschrift voor Fysiotherapie **114**(5) (2004)
20. See, J., et al.: A standardized approach to the Fugl-Meyer assessment and its implications for clinical trials. Neurorehabil. Neural Repair **27**(8), 732–741 (2013)
21. Stevens, E., Emmett, E., Wang, Y., McKevitt, C., Wolfe, C.: The Burden of Stroke in Europe: the challenge for policy makers. Stroke Alliance for Europe (2017)

Service Robotics in Challenging Environments

Data Fusion of RADAR and LIDAR for Robot Localization Under Low-Visibility Conditions in Structured Environments

David Alejo[1], Rafael Rey[1], José Antonio Cobano[1], Fernando Caballero[2], and Luis Merino[1(✉)]

[1] School of Engineering, Universidad Pablo de Olavide, Seville, Spain
{daletei,rreyarc,jacobsua,lmercab}@upo.es

[2] Department of Systems Engineering and Automation, University of Seville, Seville, Spain
fcaballero@us.es
https://robotics.upo.es

Abstract. Optical range sensors such as LiDAR and range cameras have become the most common devices for robot localization and navigation tasks. However, their performance can be degraded by meteorological hazards, such as fog, smoke, or rain. This paper proposes a new method to combine information from LiDAR sensors and low-cost RADAR sensors in structured 2D environments, in order to ensure the availability of useful information in low-visibility conditions due to smoke. Our method makes use of a novel DBScan-Line segmentation for clustering the measurements from the LiDAR sensor, and then it establishes correspondences between these clusters and the measurements from the RADAR sensors. The method has been extensively tested in field experiments with artificial smoke, and the results benchmarked against raw sensors and a state-of-the-art fusion method. Moreover, the fused measurements have been integrated into a localization method, which was able to robustly localize a ground platform in dense fog.

Keywords: Data fusion · Radar · Low-visibility · Localization

1 Introduction

Nowadays robotic systems have been successfully deployed in many scenarios for inspection purposes, including sewers, offshore oil and gas extraction platforms and medium and high voltage power lines. Currently, human operators perform these tasks, which are risky and expensive. Using autonomous robotic inspection systems aims to improve the accuracy and efficiency of inspection tasks.

For the robotic platform to be economically viable, it should operate in as many weather conditions as possible. In particular, the presence of aerosols in smoky/dusty environments leads to their detection by range sensors or to the

D. Tardioli et al. (Eds.): ROBOT 2022, LNNS 590, pp. 301–313, 2023.
https://doi.org/10.1007/978-3-031-21062-4_25

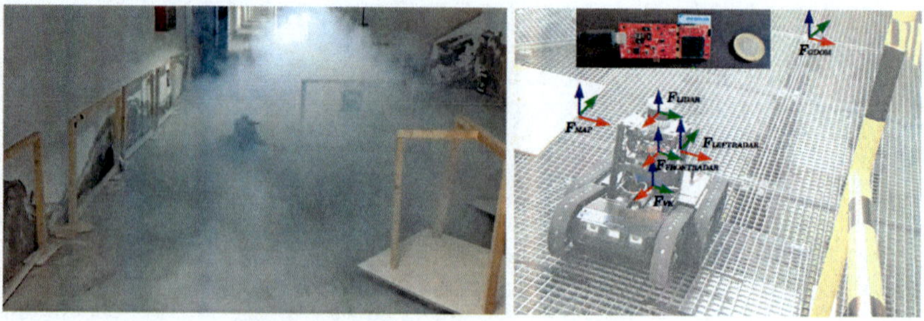

(a) (b)

Fig. 1. (a) Smoky environment considered in this paper. (b) Robot used in the experiments and its main reference frames

full absorption of the LiDAR beams. Therefore, they significantly degrade the quality of the LiDAR measurements [1]. Visual cameras are also affected by the presence of the aerosol particles in the air to a great extent. The authors of [2] evaluate the performance of several sensors, including two different IR cameras, two visual cameras, two sonar systems, a RADAR, a single-echo LiDAR, a multi-echo LiDAR, a Kinect camera, and a night vision camera in fire smoke environments. They show that RADAR systems and infrared cameras outperform the other rangefinders and cameras tested in such circumstances.

To overcome this problem, thus, it is necessary to use alternative sensors. The use of thermal infrared cameras for robot odometry and localization is subject to intense research in the last years (i.e. [3,4]). In this paper, we focus on the study of RADAR sensors. Aerosols do not affect RADAR because its wavelength is larger than most aerosol particles, penetrating through heavy dust and smoke. However, using a larger wavelength also has drawbacks, as the measurements obtained by a RADAR sensor have worse distance and angular accuracy and resolution when compared to LiDAR measurements (see [5] for a comparison of a novel RADAR sensor and a Velodyne LiDAR in localization and mapping).

In this paper, we propose a new method to fuse information from both low-cost RADAR and LiDAR sensors for robots operating in 2D structured indoor environments. The goal is to exploit as much information as possible, while discarding unreliable measurements from the LiDAR due to aerosols. For that, we leverage prior knowledge about indoor structured scenarios. The predominant presence of planar structures in such environments can be used to recognize linear patterns in LiDAR. These patterns are different than the typical scattering patterns produced by dust and/or smoke. These patterns are then validated using the low resolution RADAR, and used to increase the resolution of the point cloud provided by the latter, discarding noisy LiDAR measurements in the process. We test this fusion procedure in a set of experiments in reduced-visibility conditions (see Fig. 1-a). Furthermore, we study how this fusion of information improves the accuracy of an online localization system with a prior map acquired under normal visibility conditions. The results of our proposed method have been compared to a state-of-the-art method proposed in [6], outperforming it when using data gathered in the aforementioned experiments.

2 Related Work

Several works have been presented in the last years using a RADAR sensor for different tasks. The work in [7] reviews the recent advances in continuous wave radar systems for short-range localization and life tracking applications. A RADAR-based full graph-SLAM system was proposed in [8] for reliable localization and mapping in large-scale environments and adverse weather such as dense fog and heavy snowfall. However it uses a relatively expensive Navtech CTS350-X, a Frequency Modulated Continuous Wave (FMCW) scanning RADAR. Using a similar sensor, [9] proposes a method to fuse RADAR and LiDAR information using Gaussian Process Implicit Surfaces. Unfortunately, the computational complexity of the method makes it not adequate for real-time localization purposes. In [10], an evaluation of the potential of creating navigation maps using mmWave radar is presented, but tests do not consider low-visibility environments. On the other hand, Lu et al. [11] uses the same low-cost mmWave RADAR as us, and they implement an assisted odometry system that can estimate egomotion in combination with an IMU. The same authors use the same low-cost RADAR sensor to reconstruct a dense grid map with accuracy comparable to LiDAR, as well as providing semantic annotations of objects on the map in [12]. Even though these two last approaches are interesting, they were merely tested in a corridor scenario under reduced-visibility conditions, and more tests should be carried out in more complex scenarios. The authors of [13] combined radar odometry as well as Visual Odometry in a integrated navigation system in GNSS denied environments. Although they highlight that radar sensor is immune against rain, fog or dust, they do not consider low-visibility environments.

The work in [14] proposes a RADAR localization method in a disaster scenario in which the robot shape-matches against a dense prior map obtained with a LiDAR sensor. However, using only RADAR can lead to large localization errors that could be not adequate for autonomous navigation, as shown in their presented RMSE errors. Here we also use RADAR to leverage good LIDAR measurements, improving the precision. Castro and Peynot [15] use LiDAR and RADAR sensors to infer smoke or dust concentration by calculating the consistency between the LiDAR and RADAR measurements. However, there is no actual sensor fusion and the RADAR is used only to classify LiDAR measurements as consistent or inconsistent. In a similar approach, Fritsche and Wagner [6] detect hazards with a mobile robot in low-visibility environments. They fuse range data from a LiDAR and a RADAR sensor to map the unknown environment, being able to determine the presence of fog from the discrepancies between LiDAR and RADAR readings. In addition, they look for lines in the LiDAR readings by using the RANSAC method. However, the detected lines are included without being validated by any RADAR measurement, which can cause problems of integrating measurements from aerosol particles. Furthermore, they use a Mechanical Pivoting custom RADAR device with an estimated cost considerably higher than the mmWave RADAR used in the present work, providing a much more dense point-cloud when compared to our RADAR devices. We include results from an implementation of this method as a baseline.

3 Data Fusion Approach

This section details the approach for data fusion of LiDAR and low-cost RADAR. The method uses measurements from the two types of sensors to filter out wrong LiDAR measurements due to aerosols, and use the RADAR measurements whenever no useful measures from the LiDAR are available. The resulting scan can then be used to feed any localization or SLAM approach. Next sections describe in detail the processing steps carried out for processing the sensor data, and for fusing the information.

3.1 Enhancing the Data from Low-Cost RADARs

In this paper, we use a low-cost FMCW millimetric wavelength RADAR from Texas Instruments, the IWR6843 intelligent mmWave sensor [16]. This sensor operates in the range of 60 to 64 GHz, and can measure simultaneously the range, angle of arrival and relative radial velocity of obstacles in its field of view.

We use an array of four of such sensors in order to estimate the obstacles in the surroundings of the robot (see Fig. 1-b), each one pointing towards a different direction (front, rear, left and right). To combine the data from all sensors, we merge them in a one second wide time-window, keeping all measurements each sensor may acquire in that period of time. This way, we are able to obtain a more dense point-cloud without discarding any measures. We use the odometric frame as the common reference frame to perform the merging, taking into account the odometric estimation of the movement, which is usually accurate within this time window.

3.2 LiDAR Segmentation

The main idea is to separate LiDAR points that actually belong to a wall from the ones that might be cause by dust or fog, taking advantage of the robot being in a structured environment. To this end, we perform a clustering process on the LiDAR scan in order to search for groups of points lying in lines, with line-based segmentation, and then search for the remaining points which are closely tied together, with a variation of the DBScan method [17].

These two stages compose our proposed DBScan-Line clustering algorithm. It is a region growing method similar to the plane detector in [18], which has as inputs the points from a range sensor $\boldsymbol{x}_n$. As outputs it gives us the number of clusters $n_{clusters}$, the number of detected lines n_{lines} and a matrix C_n, which relates measurement n with the ID of the cluster it belongs, or -1 if it does not belong to any cluster.

Line Clustering. A point measurement $\boldsymbol{x}$ is part of the line L if the following equation holds:

$$\boldsymbol{v}^T \boldsymbol{x} = d, \tag{1}$$

where $\boldsymbol{v}$ is a unit vector normal to L, and d is the signed distance from L to the origin. The algorithm starts by selecting two points closer than a given ϵ distance. It uses the following incremental formulation for calculating the Mean Squared Error (MSE) from each measurement to the line, given a new measurement x_k. Note that the function *mineigenvector* returns the eigenvector associated with the minimum eigenvalue. Whenever any measurement is to be added to the line, we update the sum of the elements in the line $\boldsymbol{s}$, and matrices M and P to get the updated normal vector $\boldsymbol{v}'$ and signed distance d' as follows.

$$\boldsymbol{s}' \leftarrow \boldsymbol{s} + \boldsymbol{x}_k \tag{2}$$

$$M' \leftarrow M + \boldsymbol{x}_k \boldsymbol{x}_k^T - \frac{\boldsymbol{s}' \boldsymbol{s}'^T}{l+1} + \frac{\boldsymbol{s}\boldsymbol{s}^T}{l} \tag{3}$$

$$P' \leftarrow P + \boldsymbol{x}_k \boldsymbol{x}_k^T \tag{4}$$

$$\boldsymbol{v}' \leftarrow mineigenvector(M') \tag{5}$$

$$d' \leftarrow \frac{\boldsymbol{v}'^T \boldsymbol{s}'}{l+1} \tag{6}$$

$$MSE \leftarrow \frac{\boldsymbol{v}'^T P' \boldsymbol{v}' - 2d' \boldsymbol{v}'^T \boldsymbol{s}'}{l+1} + d'^2 \tag{7}$$

Then, a new close point is considered added to the line cluster as long as the new MSE of the line is below γ and its distance to the line is below δ. When no candidates are found, the line is kept as long as it is composed of more than N_{min} measurements. For more details please refer to [18].

DBScan Clustering. Once the search for lines finishes, the remaining points are grouped into clusters which have points at a maximum distance ϵ, again consisting of at least N_{min} points, following the DBScan method. Figure 2-c represents an example of the application of DBScan-Line on a LiDAR measurement.

3.3 Sensor Fusion

The key idea in this paper is to make correspondences between points sensed by the low-cost RADAR and clusters of LiDAR points obtained by using the proposed clustering method above to create a fused point cloud. From the clustering step we obtain two different types of LiDAR clusters: first the line clusters, which are likely to be valid measurements; and then the unstructured clusters, which are likely to be caused by perceiving fog or dust particles.

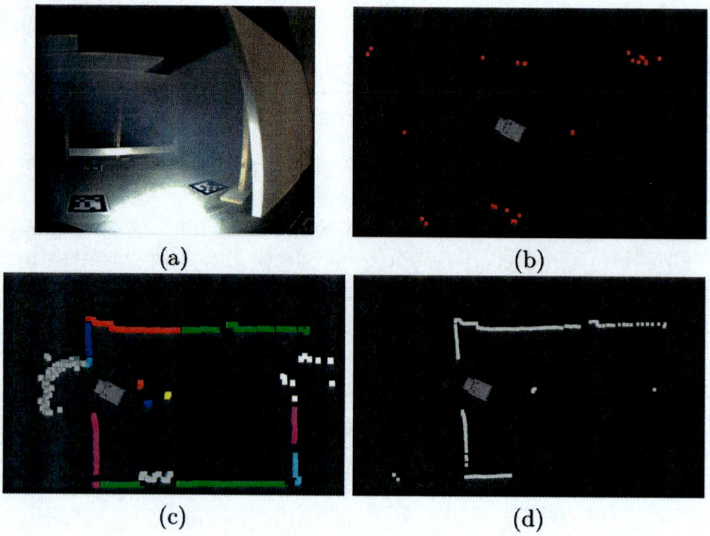

Fig. 2. Information gathered from the onboard sensors and fusion results. (a) Image from the onboard camera. (b) RADAR scan. (c) Clustered LiDAR scan. The detected lines are represented in colour, the unstructured clusters in grey and the unclassified points in white. (d) Result of the proposed fusion method.

For each RADAR measurement $\boldsymbol{x}_r$ with range ρ_r, we search for the closest point from the LiDAR measurements $\boldsymbol{x}_l$ with range measurement ρ_l that lies in its perception cone, determined by the angular resolution of the RADAR sensor. We consider that x_l corresponds with x_r as long as their associated ranges differ ρ_{min} at maximum. Then we distinguish three cases:

1. If $|\rho_r - \rho_l| < \rho_{min}$ we keep the LiDAR measurement, as the LiDAR sensor is more accurate. If x_l belongs to a line cluster, we include the whole cluster. Note that in this case we discard $\boldsymbol{x}_r$.
2. If $\rho_r - \rho_l > \rho_{min}$ and $\boldsymbol{x}_l$ belongs to an unstructured cluster, we discard the whole cluster in the fusion process, as we consider the cluster to be produced by the presence of smoke or fog. Note that whenever a cluster is discarded, it is removed from the fused scan and it cannot be included afterwards.
3. We always include $\boldsymbol{x}_r$ except in Case 1.

Note that any cluster or measurement of x_l has to be validated to be included in the fusion. However, they are only discarded if case 2 is detected, if not, the measurement or cluster could be added as long as case 1 is detected afterwards. Figure 2 represents the inputs of the fusion algorithm (RADAR and clustered LiDAR scans) and the result obtained by our fusion method. In this example, an unstructured cluster of the segmented LiDAR of the left, in gray, is discarded by applying Case 2. Moreover, several lines of the clustered scan are added by applying Case 1. The remaining samples are obtained from the RADAR scan.

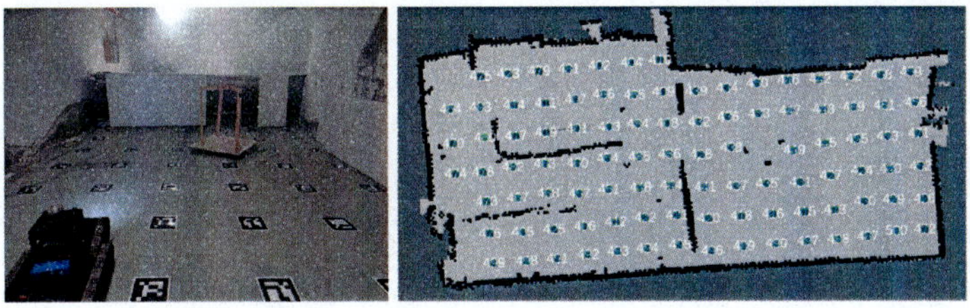

(a) (b)

Fig. 3. (a) Built low-visibility scenario. (b) 2D map of the real scenario with estimated poses of the AR markers.

4 Experimental Results

Several experiments have been performed in the scenario shown in Fig. 3, capturing data from the different sensors in a realistic environment with different corridors. We used a standard smoke machine in order to emit smoke with different densities. In particular, we performed two different long run experiments (Experiments 1 and 2) with dense smoke, whereas Experiments 3 to 7 were shorter and with a decreasing quantity of smoke in general. The dataset acquired for this paper is publicly available[1]. Moreover, the reader can reproduce the results by downloading our GitHub repository[2].

4.1 Experimental Setup

We use a tracked robotic platform Raposa from IdMind operating at a maximum speed of 0.5 m/s, equipped with a OS-1–16 LIDAR manufactured by Ouster. This sensor is able to provide roughly 10k 3D points per measurement, ranging from 0.2 to 120 m, from which we project a 2D scan with 360° FoV and 629 points. It uses 850 nm wavelenght light, outperforming LiDARs with a longer wavelength in reduced-visibility conditions [19]. We use an array of four RADAR sensors model IWR6843 intelligent mmWave sensor antenna-on-package from Texas Instruments. They provide a maximum 180° horizontal and 90° vertical FoVs, but we reduced them to 90° and 30°, respectively, as the accuracy of the sensor depends on the Angle of Arrival [16]; and to avoid the overlap of their FoVs. Again, the measurements are projected to 2D, using the same reference frame as the 2D LiDAR scan. The device has a range resolution of 3.75 cm.

4.2 Scenario

The experiments have been executed in a rectangular-shaped area of 12 × 6 meters (see Fig. 3). In order to have an external and accurate estimate of the

[1] https://robotics.upo.es/datasets/smoke.

[2] https://github.com/robotics-upo/RADAR_experiments.

position of the robot, we have installed an array of Augmented Reality (AR) markers disposed on the floor of the experimental setup, placed approximately at 70 cm from each other, as seen in Fig. 3-c. We map all markers offline as a minimum squares regression problem, implemented by using g2o library [20], with the Gauss-Newton iterative solver (implementation available at GitHub[3]). This map is then used in order to obtain a ground truth reference of the pose of the robot whenever one or more AR markers are detected with the onboard camera, obtaining a maximum error of 10 cm and 0.1 rad.

4.3 Fusion Method Statistics

We first perform an analysis of the fusion method proposed in the paper, and compare it with other baselines. To this end, we statistically analyze the number of remaining points from our whole fusion method of Sect. 3.3 - `Fusion` - and compare it to the original number points obtained by each sensor - `RADAR`, `LiDAR` -, the resultant number of points by the fusion method proposed in [6] - `Fritsche` - and the number of points identified as a line in our DBScan-Line segmentation method of Sect. 3.2 - `Cluster` -. The parameters of the DBScanLine method are $\epsilon = \gamma = 0.1\,\text{m}$, $\theta = 0.01\,\text{m}$, $N_{min} = 20$. In the fusion method, we set ρ_{min} to 0.1 m (see Sects. 3.2 and 3.3 for details).

Table 1 contains the statistics of the number of points obtained for each method in each experiment. We have configured the maximum number of points obtained for each projected 2D scan to be 629. In experiments 6 and 7, with low level of smoke or no smoke at all, the mean number of points obtained by the LiDAR sensors is close to the maximum number of points and the standard deviation is low. On the other hand, the statistics related to the RADAR sensors remain approximately constant in spite of the visibility conditions, but they are more than ten times less dense w.r.t. the LiDAR without smoke. Finally, regarding the fusion methods, the number of points obtained by our fusion method is usually lower than the one obtained by the `Cluster` method. However, the standard deviation of the `Fusion` method is lower than the one of the `Cluster` method in low-visibility conditions, which indicates that it is able to take profit of the RADAR measurements in such cases. Finally, the `Fritsche` method [6] outputs a less dense scan when compared to our method.

In order to analyze the results of one experiment in more detail, Fig. 4-a represents the number of measurements obtained with the different sensors and the proposed fusion method, in Experiment 2, as a function of time. During the first part of the movement, the LiDAR scans provides the whole range of measurements (629) because of the visibility conditions being normal. Then, from 50s to 300s the robot is immersed in an area with a dense smoke, in which the available LiDAR measurements are reduced to a great extent. However, our fusion method is able to extract the useful information from them and to discard wrong measurements that are produced by the smoke. In the last part of the experiment, the smoke is dissipated and thus the visibility is gradually recovered.

[3] http://github.com/robotics-upo/fiducials.

4.4 Localization Results

Next, we evaluate how the performance of the widely used 2D Adaptive Monte-Carlo Localization (AMCL) algorithm [21] is affected when given information from the raw sensors or from the fused information, in the hope that the AMCL

Table 1. Data fusion results: number of points obtained with each method. AMCL mean orientation and position errors. We represent the mean error of the estimated pose w.r.t. the ground truth.

Experiment	Method	Number of points stats				AMCL stats			
		Mean	Std. dev.	Max	Min	Error (m)	Error (rad)	σ_d(m)	σ_θ(rad)
1	LiDAR	243	212	629	0	1.22	0.59	0.29	0.22
Duration: 15:42	RADAR	45.6	18.3	99	0	0.49	0.11	0.21	0.029
Smoke Density:	Fritsche	58.6	15.4	79	15	0.79	0.27	0.27	0.11
Mid-High	Cluster	171	178	614	0	0.44	0.11	0.26	0.052
	Fusion	141	107	497	0	**0.29**	**0.037**	**0.17**	**0.015**
2	LiDAR	332	223	629	0	0.56	0.14	0.20	0.05
Duration: 11:48	RADAR	43.9	19.8	94	0	0.36	0.070	0.15	0.024
Smoke Density:	Fritsche	63.4	14.3	79	18	0.30	0.05	0.15	0.029
Mid-High	Cluster	220	177	610	0	0.30	0.064	0.24	0.033
	Fusion	178	119	531	0	**0.21**	**0.019**	**0.15**	**0.010**
3	LiDAR	339	189	629	0	2.32	0.75	0.41	0.22
Duration: 7:51	RADAR	42.3	21.8	100	0	0.50	0.13	0.19	0.03
Smoke Density:	Fritsche	65.0	11.0	79	34	0.37	0.09	0.17	0.03
Mid-High	Cluster	224	169	617	0	0.33	0.032	0.16	**0.01**
	Fusion	166	95	420	0	**0.25**	**0.025**	**0.15**	**0.01**
4	LiDAR	197	203	629	0	0.56	0.22	0.19	0.07
Duration: 7:20	RADAR	40.2	18.4	83	0	0.37	0.09	**0.15**	0.03
Smoke Density:	Fritsche	56.0	79	14	0	0.34	0.05	0.17	0.03
High	Cluster	124	164	615	0	0.41	0.07	0.21	0.02
	Fusion	101	82	417	0	**0.28**	**0.04**	**0.15**	**0.01**
5	LiDAR	365	176	629	0	2.70	0.20	0.72	0.09
Duration: 5:19	RADAR	42.4	17.8	89	0	0.43	0.070	0.18	0.021
Smoke Density:	Fritsche	65.5	178	79	35	0.34	0.056	0.17	0.03
Middle	Cluster	275	158	616	0	0.25	**0.021**	0.16	**0.011**
	Fusion	183	101	525	0	**0.23**	0.024	**0.14**	**0.011**
6	LiDAR	536	81.1	629	271	0.44	0.20	0.094	0.023
Duration: 3:33	RADAR	40.2	19.2	87	0	0.26	0.069	0.10	0.013
Smoke Density:	Fritsche	73.4	4.98	79	53	0.22	0.024	0.12	0.01
Low	Cluster	374	128	600	0	**0.18**	**0.013**	0.14	**0.007**
	Fusion	262	118	509	0	0.19	**0.013**	0.13	**0.007**
7	LiDAR	603	36.3	629	494	0.22	0.039	**0.089**	0.010
Duration: 4:40	RADAR	45.2	14.8	39	0	0.23	0.060	0.12	0.016
Smoke Density:	Fritsche	75.1	3.64	79	61	0.17	0.020	0.14	**0.007**
No smoke	Cluster	483	136	615	0	**0.13**	**0.012**	0.13	0.009
	Fusion	215	85.9	390	0	0.14	0.015	0.12	0.010

will perform better when using the latter. We have executed AMCL in each experiment using the resultant scan data from the different baselines described in the previous section. For each case, we run the algorithm ten times to obtain a statistical analysis of the algorithm, due to its random nature. Please note that any other localization method could be used, including Normal Distribution Transform methods [22], Iterative Closest Point methods [23] and Direct LiDAR Localization [24]. We have selected AMCL due to its capability to handle multiple hypotheses and its robustness to noisy and low-density sensors. We use the ROS implementation of the AMCL [25], with odometry parameters $\alpha_1 = \alpha_2 = 0.6$ and $\alpha_3 = \alpha_4 = 0.5$. In the laser model we set $\sigma_{hit} = 0.05\,\text{m}$, which considers the accuracy of RADAR and LiDAR sensors.

The mean localization errors obtained in each experiment with respect to the ground truth along with the mean covariance of the AMCL estimation can be also found in Table 1. Note that in good visibility conditions the uncertainty of the AMCL estimation gets lower when using information from the LiDAR sensor or the `Cluster` method. However, as the smoke makes these readings unavailable in prolonged time intervals, the filter may diverge when using these data. In contrast, this divergence is not found when using RADAR measurements at the cost of more uncertainty in the estimation due to the lesser information available. In reduced visibility scenarios (Experiments 1 to 4), AMCL performs better when using measurements from our `Fusion` method than when using clustered LiDAR measurements or measurements from the `Fristche` method.

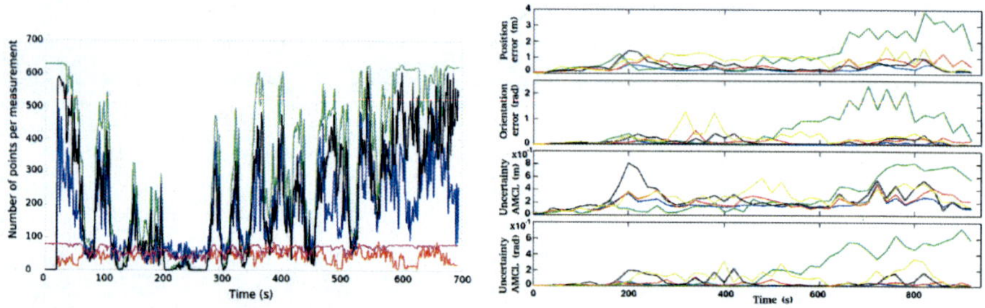

(a) (b)

Fig. 4. (a) Number of measurements obtained by the LiDAR, RADAR, Fritsche, DBScan-Lines and Fusion scans in Experiment 2, in green, red, magenta, black and blue lines, respectively. (b) RMS localization errors and related estimated uncertainties when using AMCL with data from LiDAR (green), RADAR (red), Fritsche fusion method (yellow), DBScan-Line Clustering (black) and Fusion (blue) in Experiment 1.

Figure 4-b represents the evolution of the mean angular and linear errors of the AMCL estimation when using different information obtained in Experiment 2. During the first stages of the experiment (until 200 s.) the AMCL performs better when using LiDAR information and the DBScan-Line clustering method.

In this first part of the experiment it is clear that the use of raw LiDAR data and clustered data have associated the lowest error and uncertainty estimations.

Eventually, the degradation of the LiDAR sensing data starts to make the AMCL estimation errors in position and orientation grow (from 200 to 300 s). In some executions, using raw LiDAR data results in a divergence of the estimated pose w.r.t. the ground truth. In contrast, the AMCL is able to successfully track the robot during almost the whole experiment when using the four remaining methods. On the other hand, the error results obtained when using the scans obtained from our proposed fusion method are comparable to the ones obtained with the raw LiDAR measurements in high visibility situations, while being able to keep the track of the robot in all executed runs in Experiment 2 even in the presence of dense smoke. These results indicate that the AMCL, when using our fused information, is a good choice for long-term robot localization even in low visibility conditions, clearly outperforming the results obtained when using raw RADAR data. In contrast, when using Fristche and Wagner [6] or DBScan-Line clustering methods the errors have some peaks of one meter and more than 0.5 rad in some situations, as shown in Fig. 4-b.

5 Conclusions

This paper presented a new fusion method to combine the data obtained by LiDAR and low-cost RADAR sensors in an environment with low visibility. The method is capable of validating data from LiDAR sensors according to the ones obtained by the RADAR sensors. Furthermore, by using the structural properties of the environment in which it is applied, we are able to validate or discard data clusters from the LiDAR sensors using the RADAR points. By combining the data from both sensors, we could develop a localization system able to localize a robotic platform in scenarios with reduced visibility. Future work will be devoted to extend the method to 3D, in order tu use it with an aerial robotic platform. In our view, the extension of this method to 3D will come naturally by searching for planar features instead of linear ones in the clustering algorithm.

Acknowledgment. This work is partially supported by the Spanish Ministry of Science, Innovation and Universities (COMCISE RTI2018-100847-B-C22, MCIU/AEI/FEDER, UE) and by Programa Operativo FEDER Andalucia 2014–2020 through the project DeepBot (PY20_00817). The authors would like to thank the enterprise IDMind (Lisbon, Portugal) for their help and support during the development of the experiments.

References

1. Brooker, G., Hennessy, R., Lobsey, C., Bishop, M., Widzyk-Capehart, E.: Seeing through dust and water vapor: millimeter wave radar sensors for mining applications. J. Field Robot. **24**, 527–557 (2007)
2. Starr, J., Lattimer, B.: Evaluation of navigation sensors in fire smoke environments. Fire Technol. **50**, 1459–1481 (2014). https://doi.org/10.1007/s10694-013-0356-3

3. Khattak, S., Papachristos, C., Alexis, K.: Keyframe-based thermal-inertial odometry. J. Field Robot. **37**(4), 552–579 (2020)
4. Saputra, M.R.U., et al.: Deeptio: a deep thermal-inertial odometry with visual hallucination. IEEE Robot. Autom. Lett. **5**(2), 1672–1679 (2020)
5. Mielle, M., Magnusson, M., Lilienthal, A.J.: A comparative analysis of radar and lidar sensing for localization and mapping. In: 2019 European Conference on Mobile Robots (ECMR), pp. 1–6, Prague, Czech Republic. IEEE (2019)
6. Fritsche, P., Wagner, B.: Modeling structure and aerosol concentration with fused radar and LiDAR data in environments with changing visibility. In: 2017 IEEE International Conference on Intelligent Robots and Systems (IROS), pp. 2685–26–90, Vancouver-Canada. IEEE (2017)
7. Peng, Z., Li, C.: Portable microwave radar systems for short-range localization and life tracking: a review. Sensors **19**(5), 1136 (2019)
8. Hong, Z., Petillot, Y., Wang, S.: Radarslam: Radar based large-scale slam in all weathers. In: 2020 IEEE International Conference on Intelligent Robots and Systems (IROS), pp. 5164–5170, Las Vegas-USA, 2020. IEEE (2020)
9. Gerardo-Castro, M.P., Peynot, T., Ramos, F., Fitch, R: Robust multiple-sensing-modality data fusion using gaussian process implicit surfaces. In: 17th International Conference on Information Fusion (FUSION), pp. 1–8 (2014)
10. Clarke, B., Worrall, S., Brooker, G., Nebot, E.: Towards mapping of dynamic environments with FMCW radar. In: IEEE Intelligent Vehicles Symposium (IV), pp. 147–152 (2013)
11. Lu, C. X., et al.: milliEgo: single-chip mmWave radar aided egomotion estimation via deep sensor fusion. In: Proceedings of the 18th Conference on Embedded Networked Sensor Systems (SenSys 2020), 2020, pp. 109–122. IEEE (2020)
12. Lu, C.X., et al.: See through smoke: robust indoor mapping with low-cost mmwave radar. In: ACM Conference on Mobile Systems, Applications, and Services (MobiSys), Toronto-Canada, pp. 1–14. IEEE (2020)
13. Mostafa, M., Zahran, S., Moussa, A., El-Sheimy, N., Sesay, A.: Radar and visual odometry integrated system aided navigation for UAVS in GNSS denied environment. Sensors **18**(9), 2776 (2018)
14. Park, Y. S., Kim, J., Kim, A.: Radar localization and mapping for indoor disaster environments via multi-modal registration to prior LiDAR map. In: 2019 IEEE International Conference on Intelligent Robots and Systems (IROS), Macau-China, pp. 1307–1314. IEEE (2019)
15. Castro, M., Peynot, T.: Laser-to-radar sensing redundancy for resilient perception in adverse environmental conditions. In: Australasian Conference on Robotics and Automation, Wellington-New Zealand, pp. 1–8. IEEE (2012)
16. Iovescu, C., Rao, S.: The fundamentals of millimeter wave radar sensors (2020). https://bit.ly/3nBZhaW. Accessed 14 Sept 2021
17. Ester, M., Kriegel, H. P., Sander, J., Xu, X.: A density-based algorithm for discovering clusters in large spatial databases with noise. In: Proceedings of the Second International Conference on Knowledge Discovery and Data Mining, ser. KDD 1996. AAAI Press, pp. 226-231 (1996)
18. Poppinga, J., Vaskevicius, N., Birk, A., Pathak, K.: Fast plane detection and polygonalization in noisy 3D range images. In: IEEE/RSJ International Conference on Intelligent Robots and Systems, pp. 3378–3383 (2008)
19. Wojtanowski, J., Zygmunt, M., Kaszczuk, M., Mierczyk, Z., Muzal, M.: Comparison of 905 nm and 1550 nm semiconductor laser rangefinders' performance deterioration due to adverse environmental conditions. Opto-Electron. Rev. **22**, 09 (2014). https://doi.org/10.2478/s11772-014-0190-2

20. Kümmerle, R., Grisetti, G., Strasdat, H., Konolige, K., Burgard, W.: G2o: a general framework for graph optimization. In: 2011 IEEE International Conference on Robotics and Automation (ICRA), Shanghai, pp. 3607–3613. IEEE (2011)
21. Thrun, S., Fox, D., Burgard, W., Dellaert, F.: Robust Monte Carlo localization for mobile robots. Artif. Intell. **128**(1–2), 99–141 (2001)
22. Hong, H., Lee, B.H.: Probabilistic normal distributions transform representation for accurate 3d point cloud registration. In: IEEE/RSJ International Conference on Intelligent Robots and Systems (IROS), pp. 3333–3338 (2017)
23. Chen, Y., Medioni, G.: Object modeling by registration of multiple range images. In: Proceedings 1991 IEEE International Conference on Robotics and Automation, vol. 3, pp. 2724–2729 (1991)
24. Caballero, F., Merino, L.: DLL: direct LiDAR localization. A map-based localization approach for aerial robots. In: 2021 IEEE/RSJ International Conference on Intelligent Robots and Systems (IROS), pp. 5491–5498 (2021)
25. Gerkey, B.: Amcl ROS wiki (2017). https://wiki.ros.org/amcl. Accessed 14 Sept 2021

Procedural Generation of Underground Environments for Gazebo

Lorenzo Cano[1(✉)], Danilo Tardioli[2], and Alejandro R. Mosteo[2]

[1] Instituto de Investigación en Ingeniería de Aragón, Universidad de Zaragoza, C/Pedro Cerbuna, 12, 50009 Zaragoza, Spain
lcano@unizar.es

[2] Instituto de Investigación en Ingeniería de Aragón, Centro Universitario de la Defensa, Zaragoza, Spain

Abstract. One of the challenges in the field of autonomous robotics is the thorough testing of the navigation capabilities of the different methods in a wide variety of situations. Towards this end, simulation plays an ever increasingly important role as the improvements in the technology keep closing the gap between reality and simulation. This work presents a scalable, efficient and easy-to-use framework for generating random underground environments. The system is based on the procedural placement of discrete modules in such a way that the end result are sprawling and complex networks of tunnels that can be used for testing robotic applications for underground environments.

Keywords: Procedural generation · Simulation · Testing · Navigation

1 Introduction

When developing a new autonomous robotic application, it is vital to test the new system in as many different situations as possible, to ensure its safety, reliability and robustness to the best possible degree. The difficulty of this task is mainly dependent on two factors. First, the complexity and size of the expected operational space of the robot, or in other words, how many different situations are expected to be faced by it. For example, an autonomous driving vehicle is expected to work in day and night conditions, in crowded and empty city streets, in both the city and the highway, and so on. The second main factor that drives the difficulty of testing is the accessibility to the expected operational space of the robot. For example, in space robotics, the size of the operational space is usually quite small (the Curiosity Mars Rover has only traveled 28.5 km in over 10 years[1]), but testing in the actual operational environment is not possible, so the validation process becomes a very complex, expensive and time consuming task.

One of the main technological contributions that has helped in managing these challenges has been computer simulation. Modern simulation software is

[1] http://mars.nasa.gov/msl/mission/where-is-the-rover.

D. Tardioli et al. (Eds.): ROBOT 2022, LNNS 590, pp. 314–324, 2023.
https://doi.org/10.1007/978-3-031-21062-4_26

capable of accurately recreating the behavior of a robot in a virtually endless variety of environments, which greatly decreases the need for testing in real-world scenarios (although it does not eliminate it). Computer simulation is a huge field, with many different fronts where advancements are being made: fluid dynamics, friction behaviors, visual fidelity, etc. However, among all of these fronts, there is one that has found great use in the robotic community: the procedural generation.

Procedural generation is an umbrella concept that covers any method of data generation performed by a computer, be it sound, image or any other kind of data, and it includes a wide variety of methods, many of which have been developed by the video-game industry. The reason why the video-game industry has been the one that has pushed procedural generation the most, is that the manual creation of virtual environments is a very time-consuming task. However, with the use of procedural generation techniques, it is now possible to generate new plants, creatures, terrains, or even entire planets, without any human intervention. These techniques have found use outside the video-game industry to generate realistic test scenarios for testing of robotic applications, as they can quickly generate a wide variety of situations.

Among the different types of environments for which robotic applications are being developed, underground environments rank among the most hostile to people. Usually, these tend to be difficult to get to, and in many regions of the world, it is not possible to take a robot to an underground mine or cave without traveling for hours. Also, these environments tend to be very oppressive and uncomfortable, so staying for long periods of time is not possible. Finally, they can be potentially hazardous, with danger of loose rocks falling, breathing air filled with dust particles or tripping because of the irregular floors and bad lighting.

For all these reasons, it makes sense to try to reduce the testing in real environments to a minimum. With this in mind, in this work, we propose the use of a procedural generation techniques to generate big, complex and sprawling tunnel networks for the testing of underground robotics applications.

Procedural generation is an umbrella term used to refer to the set of algorithmic techniques used for synthetic generation of data, which can take many different forms, like images, video, sound, or in the case of this work, the layout of a tunnel network.

The rest of the paper is organized as follows. In the next section, related works are introduced and discussed. Section 3 presents the proposed method discussing the modeling and the environment generation. Section 4 briefly introduces the results obtained and Sect. 5 present the conclusions.

The code described in this work has been made publicly available[2] for open use by anyone.

[2] http://github.com/LorenzoCanoAn/gzb-subt-env-proc-gen.

2 Related Work

The use of procedural generation techniques for testing robotic applications has been specially motivated by the field of autonomous driving, as researchers need to test a robotic application in a realistic environment but without risking any passenger or road user life. Works that tackle the problem of procedurally generating road networks can be found in [1–4]. In [5], a procedural generator for roads is also implemented, where the use of stochastic data from sample environments is combined with parametric control, allowing for an easier to use method for non-expert users than other alternatives. Other procedural generation techniques can focus on more concrete parts of the traffic network, as is the case of [6], where the signals of a road network are procedurally generated, or [7], where given the road network, the street appearance and other street objects are procedurally generated.

Other procedural generation systems focus on the creation of complete urban environments, as is the case with [8] and [9]. Some other work has focused on the generation of more village-like environments, like [10]. Procedural generation can also be used to generate dynamic elements inside a simulation, as is the case in [11], where the crowds of a city are generated based on its structure.

There is also work in the procedural generation of more general environments, [12], that allows the generation of room-like environments with a variety of random obstacles placed inside, that allows the testing of more general navigation approaches.

In the field of robotics, procedural generation can also be used for purposes other than testing and validation as it is the case with [13], where the authors collect a dataset exclusively from simulations of procedurally generated environments, use it to train a model, and successfully use it on the real world.

The procedural generation of underground environments has been mainly focused in the generation of cave-like scenarios for use in video-games, [14–16] and [17], which cannot be conveniently used by the robotics community.

Finally, works in underground robotics [18–20] show the recent interest form the robotics community in this field, and motivate the development of a tool suited for these kinds of environments.

3 Procedural Environments Generation

As seen in Sect. 2, the field of procedural generation is a very rich and continually evolving one. There exits both procedural generation methods centered around robotics, and procedural generation of underground environments. However, there is an unfilled niche in the use of procedural generation of underground environments focused on robotic applications, which is what motivated this work.

The proposed method generates arbitrarily complex tunnel networks, ideal to test underground robotic applications in. The method is based on the procedural placement of *tiles*, and so, it can be very easily expanded by defining new types of tiles.

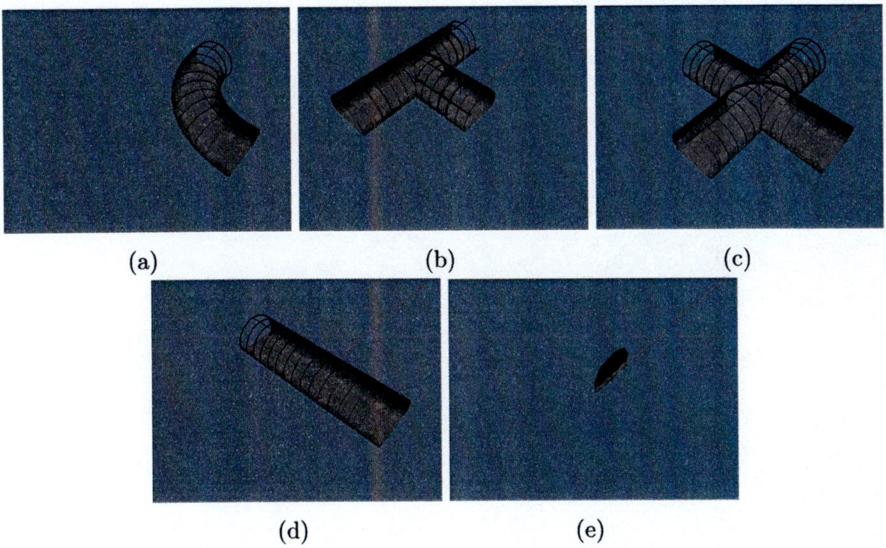

Fig. 1. Subset of the DARPA Subterranean Challenge tiles used for procedural world generation. (a) *Curve* tile. (b) *T intersection* tile. (c) *Cross* tile. (d) *Straight* tile. (e) *Hatch* tile.

This work has been designed to create environments for the Gazebo[3] simulation software. Gazebo is one of the most widely used simulators in the robotics community, as it is very well integrated with the Robot Operating System (ROS)[4], the most popular framework in the field. This integration allows the development of software in simulation that has a very direct transfer to a real robot. The Gazebo simulator loads the environments from `.xml` files, where the assets to be loaded are specified. For this reason, the output of this work is one of these `.xml` files, where the placement of the tiles that will compose the environment is specified.

3.1 Tile Modeling

The objective of this procedural generation is to quickly create complex and random tunnel networks by combining a set of *tiles* provided by DARPA for their Subterranean Challenge[5]. Even though more than 50 tiles where provided by DARPA, however only a small subset has been used in this procedural generation (Fig. 1).

To be able to assemble the tiles in an efficient and coherent way (no overlaps), it became necessary to create simplified models with the minimal information needed. The model of each tile contains: a 2D *outer shape* that allowed to check

[3] http://gazebosim.org/home.
[4] http://www.ros.org.
[5] http://www.subtchallenge.com.

for collisions with other tiles and the set of *connection points* at which it can connect with other tiles. Both the outer shape and the connection points are defined with respect to the *local reference frame* of the tile, so that by moving the tile, both the connection points and outer shapes move too. For the sake of simplicity, only 5 tiles have been used for the following explanation, however, there is no inconvenience to using more with this method, as shown in 4. The models for each of the selected tiles are shown in Fig. 2.

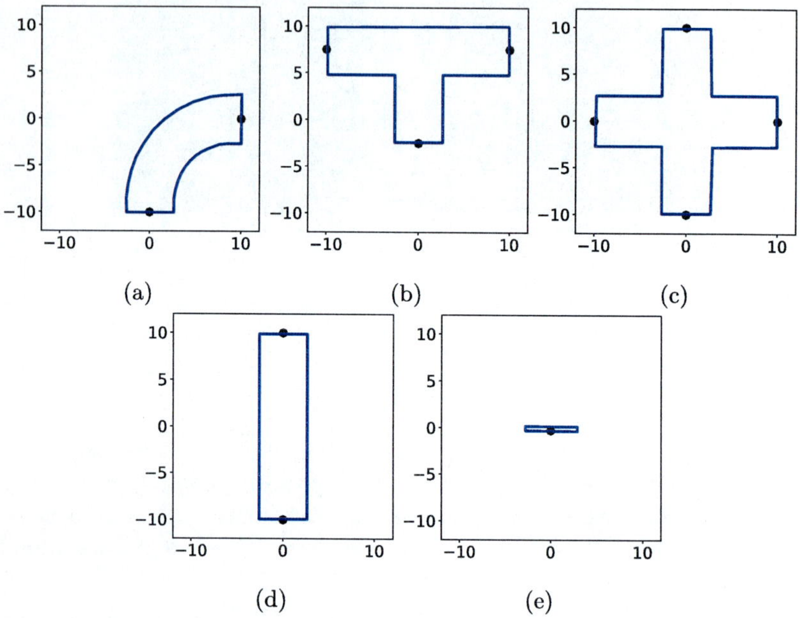

Fig. 2. Models of the tiles with outer shape in blue, connection points in black and local frame with position (0, 0) and orientation of 0°. All units in axes are in Meters. (a) Curve tile model. (b) T intersection tile model. (c) Cross tile model. (d) Straight tile model. (e) Hatch tile model.

The tiles where categorized into three classes, *intersection* tiles (T-intersection and cross), *corridor* tiles (straight and curve) and *blocking* tiles (hatch).

3.2 Tile-Based Procedural Generation

The environment generation is based on a graph-like structure, where each node represents a tile, and has a number of edges equal to the number of connection points that the tile it represents has. This means that, for example, a node corresponding to a T-intersection will have 3 edges, while a node corresponding to a hatch will have only one. These connections can be either *open* or *closed*.

Open connections are the ones available to connect to other tiles, while closed connections are the ones that already have a different tile associated to them. For example, if a cross tile is added to an empty graph, its 4 connections are *open.* If a straight tile is then added to the graph, it has to connect with one of the 4 connections of the cross tile, which goes from *open* to *closed.* Equally, the connection from the straight tile that connects to the cross tile is also *closed,* while the other connection remains *open.* Finally, these nodes also contain the 2D *position* and *orientation* of the tile with respect to the *world frame.*

The first step in the world generation is to randomly choose the type of the first tile, which can either be an intersection or corridor tile. It is then introduced as the first node in the graph with a position of $(0, 0)$ and orientation of $0°$.

After the first tile is placed in the graph, further tiles are placed iteratively until the desired number of tiles have been added. Each iteration of the tile placement is divided into two steps: *random tile placement* and *closure of possible loops*, both described in the next sections.

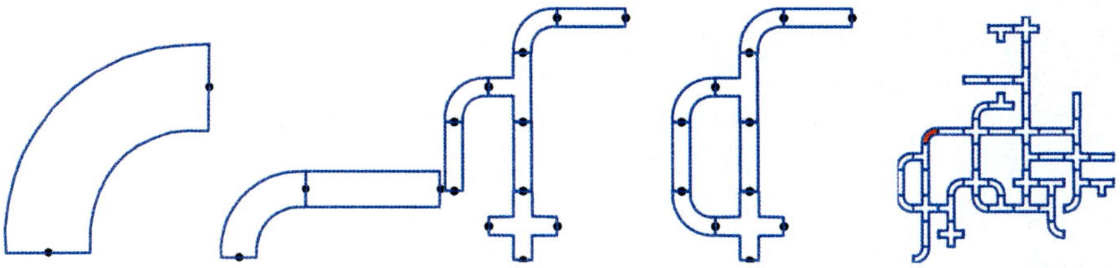

(a) (b) (c) (d) (e)

Fig. 3. Map generation process. (a) Initial tile. (b) After first random tile placement. (c) Presence of a group of two open connections near each other where a loop closure is possible. (d) After loop closure. (e) Final map with first tile in red (50 tiles).

3.3 Random Tile Placement

This step is the main reason of randomness in the final map. It starts by choosing one of the open connections of the node (referred to as *base connection*) and by selecting a random non-blocking tile as a *candidate tile*; from it, a random connection is chosen (*candidate connection*). The position and orientation of the candidate tile are adjusted so that the candidate connection would connect with the base connection.

Finally, a collision check is made between the candidate tile and the nearby tiles. In this implementation, the collision check between two tiles is performed by calculating the overlapping area between their bounding boxes, using the *Shapely*[6] library, if said area is above a threshold, it's considered a collision. If there is no collision, the candidate tile is added to the graph. Otherwise, there are two possibilities, depending on if the candidate tile is symmetric or not.

[6] http://pypi.org/project/Shapely/.

- If it is symmetric (cross and straight) it means that it does not matter which candidate connection is used, there will be a collision. Therefore the candidate tile is discarded.
- If it is not symmetric, all the other connection(s) will be used as candidates one by one. If any of these candidate connections results in no collision, the candidate tile is added to the graph. If all of them result in collision, the candidate tile is discarded.

At this point, two scenarios are possible: either a tile has been added to the graph or not. In the first case, it means that both the candidate connection and the base connection will be closed. In the second case, a new non-blocking tile is chosen as candidate tile, and the process is repeated until all non-blocking tiles have been tested. If after testing all non-blocking tiles no tile has been added to the graph, the hatch tile is used to close the connection.

3.4 Loop Closure

The **loop closure** step is executed every time the previous step adds a non-blocking tile to the graph. This step begins by gathering all the groups of open connections that are near each other (Fig. 3.c). Then, for each of these groups of open connections, the system tries to place a non-blocking tile (candidate tile) so that two or more of the connections in the group connect with it. This process gives preference to candidate tiles that can connect with as many connections as possible. For example, if a group has 4 connections that are near each other, this step will try to first use a cross tile; if it collides, a T-intersection, and if it collides, either a straight segment or a curve. If every non-blocking tile fails, no tile is added to the graph.

If a tile is placed during this process, the whole step is repeated from the beginning with the updated open connections. This process continues until a complete loop over all the connection groups finishes without adding a new tile.

Once the desired number of non-blocking tiles has been added to the graph, the iteration stops. Most often this means that there are many open connections left. So in the last step, all the open connections that are left are closed with a hatch tile, completing the graph and finishing the generation. In Fig. 3, some intermediate steps of a generation process are shown, specifically, instances of tile-adding and loop-closure steps.

Once the world generation is finished, the information of the graph is exported into a `.xml` file, that can be loaded into Gazebo, and used to test robotic applications for underground environments.

4 Results

The resulting system can generate tunnel networks of a complexity that can be specified by the user in terms of the number of tiles to be used. Figure 4 shows three examples of environments generated using the procedural generation described, from 10 to 500 tiles. The generation time scales linearly with the desired number of tiles, as can be seen in Fig. 5.

(a) (b) (c)

Fig. 4. Results of the map generation process in the Gazebo simulator with different number of tiles. (a) 10 tiles. (b) 80 tiles. (c) 500 tiles.

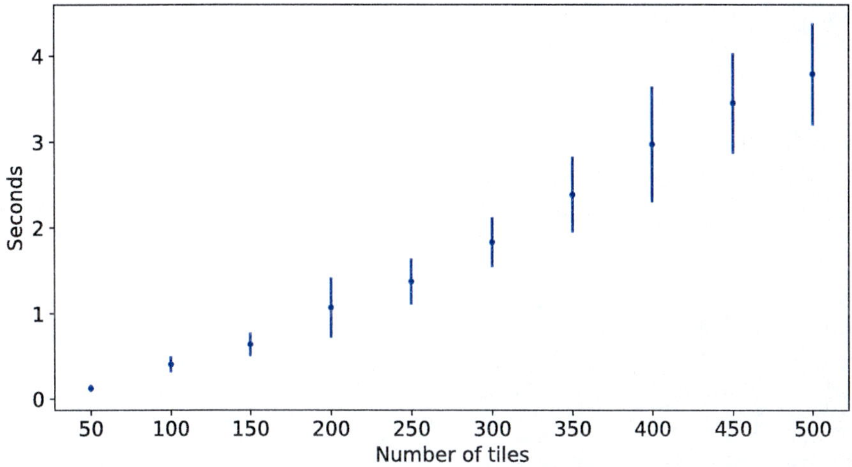

Fig. 5. Average generation time depending on the specified number of tiles. The average and standard deviation have been obtained from 10 map generations for each number of tiles.

Additionally, the method can easily be augmented by introducing new tiles. For example, in Fig. 6, new types of tile are added to the pool of intersection tiles. The only effort required is to create the model of the tile, containing the outer shape and connection points. To use the tiles provided by DARPA but at different sizes, it's enough to scale the models (bounding box and connection points) of the tile before calculating its pose and also specify the scale factor (in the `.xml` file) that is loaded to Gazebo.

This method has also been used for generating environments to test the framework presented in [21] in which a method is proposed to navigate underground environments using simple topological representations. In this work, 180 simulated environments with a different number of tiles were generated, in order to guarantee a specific number of intersection tiles (T-intersection, cross tiles). Figure 7 shows the map obtained using a GMapping SLAM by a robot navigating in one of the cited environments.

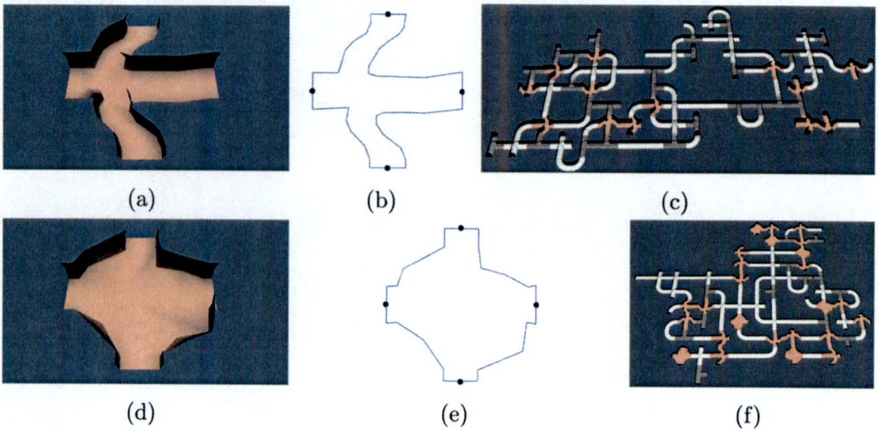

Fig. 6. Examples of adding new tiles to the procedural generator. (a) 4-way cave intersection tile. (b) Model of the 4-way cave intersection tile. (c) Sample environment with the 4-way cave intersection tile. (d) Diaphanous cave intersection tile. (e) Model of the diaphanous cave intersection tile. (f) Environment with the 4-way and diaphanous cave intersection tiles.

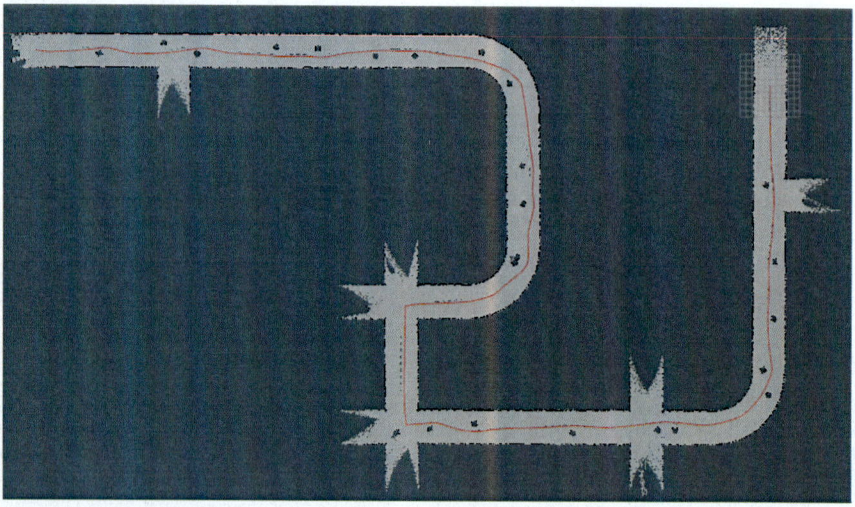

Fig. 7. Occupancy grid generated with gmapping [22] during an experiment executed in a procedurally generated environment. The red line is the path followed by the robot during the experiment.

5 Conclusions

The testing of robotic applications is a crucial part of their development, but in the case of underground environments, this can be a very inconvenient and potentially dangerous task, specially in challenging or hostile environments. However, in order to decrease the necessity of real-world testing, simulations can be used. With the objective of facilitating this task, a procedural generation system of underground environments has been implemented, that can generate arbitrarily complex tunnel networks for the Gazebo simulator using tiles provided by DARPA through its DARPA SubT Challenge.

The proposed system is based on combining a set of tiles in a random but coherent way. This tile-based approach makes it easy to expand with new types of tiles. Furthermore, the time it takes to generate any given environment depends linearly on the desired number of tiles, which allows the generation of really complex environments in a matter of seconds.

The system has been successfully used in [21] to test the proposed navigation method in hundreds of different environments in a completely automatic fashion.

In future work, the system will be further developed to handle $2.5D$ environments. These environments will be composed of different layers at different (but discrete) heights, using some of the vertical shaft and sloped tiles also provided by in the DARPA SubT Challenge set of tiles.

Acknowledgements. This work was supported by the project PID2019-105390RB-I00 founded by Spanish MICINN.

References

1. Muktadir, G., Jawad, A., Paranjape, I., Whitehead, J., et al.: Procedural generation of high-definition road networks for autonomous vehicle testing and traffic simulations. SAE Int. J. Connect. Autom. Veh. **6**(1), 2023 (2022)
2. Campos, C., Leitão, J.M., Pereira, J.P., Ribas, A., Coelho, A.F.: Procedural generation of topologic road networks for driving simulation. In: 2015 10th Iberian Conference on Information Systems and Technologies (CISTI), pp. 1–6. IEEE (2015)
3. Galin, E., Peytavie, A., Maréchal, N., Guérin, E.: Procedural generation of roads. In: Computer Graphics Forum, vol. 29, pp. 429–438. Wiley Online Library (2010)
4. Kelvin, L.Z., Anand, B.: Procedural generation of roads with conditional generative adversarial networks. In: 2020 IEEE Sixth International Conference on Multimedia Big Data (BigMM), pp. 277–281. IEEE (2020)
5. Teng, E., Bidarra, R.: A semantic approach to patch-based procedural generation of urban road networks. In: Proceedings of the 12th International Conference on the Foundations of Digital Games, pp. 1–10 (2017)
6. Taal, F,C., Bidarra, R.: Procedural generation of traffic signs. In: Eurographics Workshop on Urban Data Modelling and Visualisation, pp. 17–23 (2016)
7. Cura, R., Perret, J., Paparoditis, N.: Streetgen: In base city scale procedural generation of streets: road network, road surface and street objects. arXiv preprint arXiv:1801.05741 (2018)

8. González-Medina, D., Rodríguez-Ruiz, L., García-Varea, I.: Procedural city generation for robotic simulation. In: Robot 2015: Second Iberian Robotics Conference. AISC, vol. 418, pp. 707–719. Springer, Cham (2016). https://doi.org/10.1007/978-3-319-27149-1_55
9. Paranjape, I., Jawad, A., Xu, Y., Song, A., Whitehead, J.: A modular architecture for procedural generation of towns, intersections and scenarios for testing autonomous vehicles. In: 2020 IEEE Intelligent Vehicles Symposium (IV), pp. 162–168. IEEE (2020)
10. Emilien, A., Bernhardt, A., Peytavie, A., Cani, M.-P., Galin, E.: Procedural generation of villages on arbitrary terrains. Vis. Comput. **28**(6), 809–818 (2012)
11. Pujalt, O.R.: Procedural modeling of cities with semantic information for crowd simulation. Master's thesis, Universitat Politècnica de Catalunya (2016)
12. Bosch research. Procedural generation for gazebo. https://boschresearch.github.io/pcg_gazebo_pkgs/
13. Loquercio, A., Kaufmann, E., Ranftl, R., Dosovitskiy, A., Koltun, V., Scaramuzza, D.: Deep drone racing: From simulation to reality with domain randomization. IEEE Trans. Rob. **36**(1), 1–14 (2019)
14. Mark, B., Berechet, T., Mahlmann, T., Togelius, J.: Procedural generation of 3D caves for games on the gpu. In: Foundations of Digital Games (2015)
15. Antoniuk, I., Rokita, P.: Procedural generation of underground systems with terrain features using schematic maps and l-systems. Challeng. Mod. Technol. **7**(3), 8–15 (2016)
16. Van Der Linden, R., Lopes, R., Bidarra, R.: Procedural generation of dungeons. IEEE Trans. Comput. Intell. AI in Games **6**(1), 78–89 (2013)
17. Cui, J., Chow, Y.-W., Zhang, M.: Procedural generation of 3d cave models with stalactites and stalagmites. Int. J. Comput. Sci. Netw. Sec. (2011)
18. Rizzo, C., Tardioli, D., Sicignano, D., Riazuelo, J.L., Villarroel, L., Montano, L.: Signal-based deployment planning for robot teams in tunnel-like fading environments. Int. J. Robot. Res. **32**(12), 1381–1397 (2013)
19. Li, M., Zhu, H., Tang, C., You, S., Li, Y.: Coal mine rescue robots: Development, applications and lessons learned. In: Wu, M., Niu, Y., Gu, M., Cheng, J. (eds.) ICAUS 2021. LNEE, vol. 861, pp. 2127–2138. Springer, Singapore (2022). https://doi.org/10.1007/978-981-16-9492-9_210
20. Ebadi, K., et al.: Lamp: Large-scale autonomous mapping and positioning for exploration of perceptually-degraded subterranean environments. In: 2020 IEEE International Conference on Robotics and Automation (ICRA), pp. 80–86. IEEE (2020)
21. Cano, L., Tardioli, D., Mosteo, A.: Navigating underground environments using simple topological representations. In: IEEE/RSJ International Conference on Intelligent Robots and Systems (to appear) (2022)
22. OpenSLAM. Gmapping. http://openslam.org/gmapping.html. (Accessed 07 Jul 2022)

ROSE: Robot for Automatic Spacer Installation in Overhead Power Lines

Paul Espinosa Peralta[1], Verónica Palomino-Díaz[1], Pablo Rodríguez Herrerías[2], Carlos Rodríguez del Castillo[3], Carmen María Pedraza Cubero[2], Juan Pablo Yuste[2], Paulino Aparicio Cillán[2], Rafael Mesía López[4], Guillermo Águila Rodríguez[1], Víctor Pablo Fernández Franco[1], Miguel Ángel Sánchez-Urán[1](✉), and Manuel Ferre[1](✉)

[1] Centre for Automation and Robotics UPM-CSIC, Universidad Politécnica de Madrid, Madrid, Spain
{miguelangel.sanchezuran,m.ferre}@upm.es
[2] Red Eléctrica, Alcobendas, Madrid, Spain
[3] Elewit, Tres Cantos, Madrid, Spain
[4] Redeia, Alcobendas, Madrid, Spain

Abstract. This paper describes the prototype of a robotic system for spacer installation in duplex aerial lines called (Robot for automatic spacer installation in Overhead Power lines) ROSE. This system has been tested and evaluated on a test bench at (Universidad Politécnica Madrid) UPM and actual overhead lines at Red Eléctrica facilities. The ROSE system is designed to autonomously navigate in a span and perform the spacer installation at a configurable distance in meters, according to line configuration. The system is supervised by an operator from the ground who follows the proper execution of each operation step. The operator receives process environmental data from four different video signals, and various sensor; all this data is transmitted via WIFI. Two robotic arms do spacer installation, the first for rod manipulation and the second focusing on the screw operations. ROSE also includes a vehicle that transports all components among pylons.

Keywords: Robotics · Spacer installation · Overhead power line · Robot for construction

1 Introduction

The construction of an overhead power line is a process that offers several possibilities of automatization. In this paper, we focus on the spacer installation that implies displacement over the span to fix the spacers to the cables regularly. Line spacers are required for multi-conductor bundles, and they have been used for decades in overhead lines. In this configuration, conductor spacers are key pieces of hardware aimed at maintaining the geometry of the bundles to meet electrical and mechanical features.

Few studies have been done until now for this kind of activity [1]; however, there are more works of robotic applications in overhead for inspection and maintenance [2–4], and some examples are explained in [5–10].

D. Tardioli et al. (Eds.): ROBOT 2022, LNNS 590, pp. 325–337, 2023.
https://doi.org/10.1007/978-3-031-21062-4_27

ROSE is a prototype that successfully achieved a technology readiness levels 6 (TRL6) in installing spacers for duplex lines. Their main components are two manipulator arms for executing the corresponding spacer manipulation procedure, a wheeled vehicle for displacement over the line, a spacer store, and other auxiliary components required for these works. An operator supervises the system's operation from the ground, he/she receives all required information from video cameras and sensors that show the evolution of the installation operations.

Overhead line construction involves an important number of tasks carried out by works at height procedures. This is the case for spacer installation in bundle configuration lines, where the presence of line crews working at height is nearly permanent during working days. Even though Personal Protective Equipment (PPE) specifically designed for these tasks has evolved and improved over the past years, the use of robots for the riskiest tasks enables a significant increase in safety as the time exposure of work at heights is reduced for line workers. Red Eléctrica, the transmission system operator (TSO) is committed to ensuring people's safety. At the same time, they perform the duties associated with their jobs through training in the correct use of working equipment, the application of individual and group safety measures, and by promoting innovation, as one of the main action lines to answer the strategic target of being a "Zero accidents" group. The following section describes the main components of the ROSE prototype. Then, in Sect. 3, it is shown the procedure to install a line spacer, and Sect. 4 discusses several trials and tests performed with the prototype. Finally, the conclusion is summarised in Sect. 5.

2 ROSE Equipment Description

The ROSE prototype is shown in Fig. 1, which includes two subsystems, the robotics arms for spacer installation and a vehicle for moving along the line, which also transports several auxiliary equipment. All these components and some details about its behaviour are described in the following subsections.

2.1 Robotic Arms for Spacer Installation

Initially, the complete spacer installation was simulated, considering different configurations and tools for the spacer installation. As a result, a setup with two specialised robotic arms was decided as the best configuration. The most efficient results are using a robotic arm specialised in spacer rod manipulation and a second arm specialised in the screw/unscrew tasks. Each robot has its own tooling, which is fixed permanently, and tool exchange is not required for the spacer installation.

Robot for Spacer Rod Manipulation (SM)
The robot selected for this purpose is the UR3e with a 3 kg payload. A specific tool has been designed to pick up and move the spacer. The procedure implemented by this robot is as follows:

- Pick up the spacer from the depot.

– Transport the spacer to the manipulation area.
– Locate the spacer under the cables at the installation position.

Figure 2 shows components of this subsystem. The spacer depot (label 3) is on the right side, which can store 15 spacers. The working area is in the centre of the vehicle (label 4), where the spacer is manipulated before its final installation.

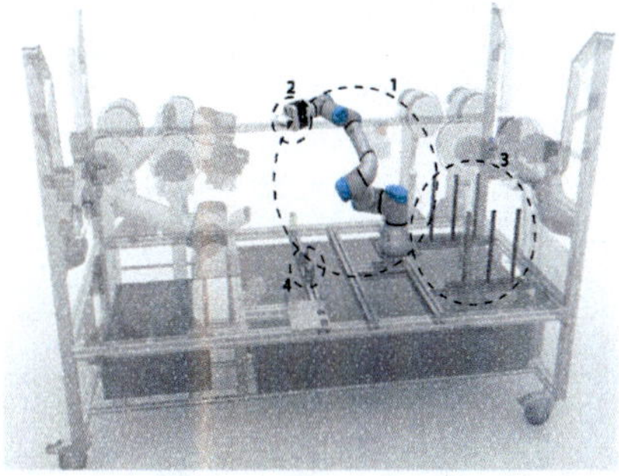

Fig. 2. Robot for spacer manipulation. Robot UR3 (1), customise tool (2), spacer depot (3), and working area previous to spacer installation (4).

Robot for Clamp Manipulation (CM)

The robot used in this task is the UR5e with a 5 kg payload. It transports a customised tool designed explicitly for screwing and unscrewing the spacer clamps. For this robot, the programmed tasks are as follows:

– Unscrew the right clamp.
– Store the right clamp at the buffer area.
– Unscrew the left clamp.
– Wait until the spacer rod is at the installation position.
– Screw the left clamp.
– Get the right clamp from the buffer.
– Screw the right clamp.

The clamp installation finalises when both clamps are properly screwed into the spacer. Afterwards, the robots move to the corresponding home positions. Both robots communicate during the spacer installation to synchronise their movements; a state machine implements the robot synchronisation according to the information provided by robot controllers and other sensors. Figure 3 shows the subsystem for clamp manipulation inside the vehicle. This robot transports a tool for clamp manipulation (label 2), and a specific location is used to temporary store the right clamp (label 5). Additionally, two video cameras supervise the clamps' operations (labels 3 and 4).

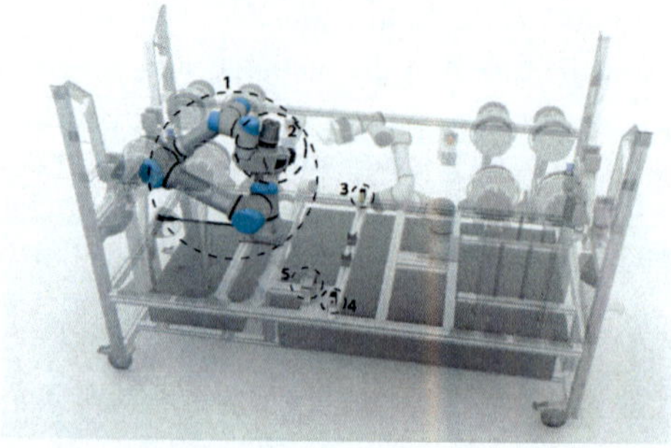

Fig. 3. Robot for clamp manipulation. Robot UR5 (1), tool for clamp manipulation (2), cameras (3–4), clamp buffer (5).

Fig. 1. ROSE prototype during the spacer installation.

2.2 Vehicle for Displacement in the Overhead Line

A vehicle for displacement in the overhead line during the construction phase of the line has been specifically designed. Figure 4 shows the main components of this vehicle: wheels, wheel support, linear actuator for wheel placement, breaks, vehicle bumpers and auxiliary components.

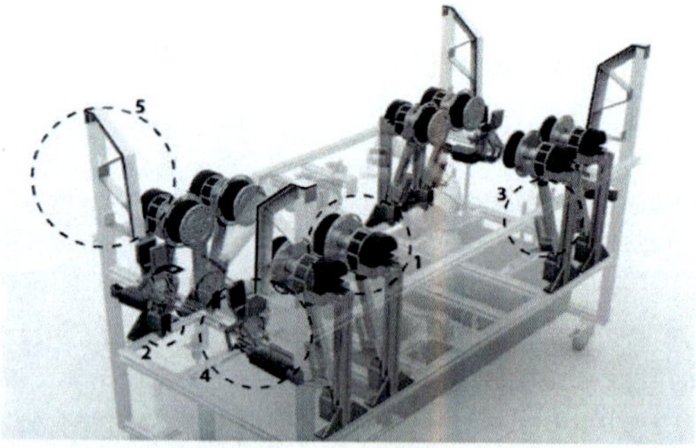

Fig. 4. Vehicle for ROSE displacement, the main components are wheels (1), wheel support (2), linear actuator for wheel placement (3), breaks (4) and vehicle bumpers (5).

Locomotion Subsystem
The locomotion is based on eight wheels driven by their corresponding actuators. Each conductor has four wheels, and their behaviour is the same in both conductors. Therefore, it implies that wheels act as four pairs.

The wheels have two predefined positions: raised and placed on the cable, controlled by a linear actuator, as shown in Fig. 5. The raised position corresponds to the maximum extension of the linear actuator; in this case, the wheels are the farthest location from the cable. The other position, called placed on the cable, corresponds to the wheels located in the line ready to turn for moving the vehicle; it implies that the linear actuator is in its shortened configuration.

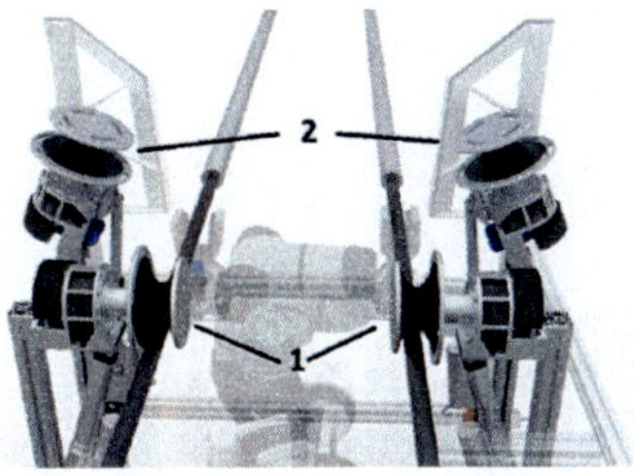

Fig. 5. Wheels positions: placed in the cable (1) and raised (2).

Vehicle Breaks
For several reasons, it is required to fix the vehicle to the conductors firmly. Hence, a break subsystem was developed, as shown in label 4 of Fig. 4. Four breaks are activated by linear actuators using a scissor mechanism. When the actuator extends the rod, the scissor is open; on the contrary, when the rod is retracted, the scissor is closed, and the vehicle holds the line.

Auxiliary Components
The vehicle also includes additional components, such as bumpers, cable detectors, video cameras, WIFI antenna, and battery. The bumpers guide the vehicle during its installation on the lines. Cable detectors are capacitive sensors of presence that inform the proper position of the vehicle when holding the line. The video cameras show different scenes of the robots during the spacer installation. The WIFI antenna transmits all communications between the robots/vehicle and the operator interface. Finally, the battery provide power to all components. In addition, several plastic containers store the electronic and electrical components necessary for the execution of the tasks. These containers protect the equipment against water and dust and facilitate the connection between the subsystems described above.

2.3 The Operator Interfaces

An operator interface has been developed to facilitate a simplified and visual control of the system by an operator at all production states (initial testing, hoist, manual and automatic production). Moreover, this GUI also enables the configuration of the manipulators, vehicle parameters and the visualisation of the cameras installed on the vehicle. These cameras show the evolution of the task and provide enough information to the state machine and the operator to decide the proper completion of tasks. Figure 6 shows the most important screens.

The initial testing interface, shown in Fig. 6.a is launched after the powering of the vehicle. All vehicle subsystems check the functionality to assure a safe and correct production state. Once all checks end, results are presented to the operator through a colour code green correct, red fail.

The raising/lowering tasks are performance manually by the operator while the interface shown in Fig. 6.b, operator can control linear actuators and brakes in function of the auxiliary sensors information, also the visualisation cameras allow to check the correct procedure and control.

The automatic mode interface, Fig. 6.c, allows the operator to establish the installation parameters, such as the number of spacers to install, the distance between them, check the installation through the video cameras, and some feedback information provided by the system.

Manual mode interface, shown in Fig. 6.d, setting individual task of linear actuators, brakes, wheels, SM, CM. This manual mode is intended to be mainly used for recovering purposes for unexpected failures and has a complete control of manipulators.

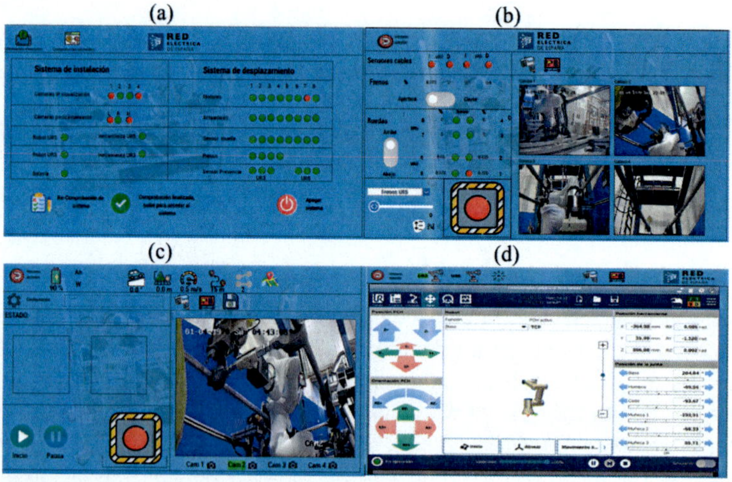

Fig. 6. Operator interface main screens. Initial testing (a), raising/lowering procedure (b), automatic mode (c), and manual mode (d).

3 Installation Procedures for the Vehicle and the Spacer

Before starting the spacer installation, the first step is to place the vehicle over the conductors. A specific method has been developed for raising and lowering the vehicle. Afterwards, checking the vehicle is in the right status, then is started the spacer installation procedure.

3.1 Procedure for Raising/Lowering the Vehicle

A crane has been used for raising/lowering the vehicle over the conductors. Figure 7 shows the method for raising the vehicle, this is performed in three steps. First, a crane hangs the vehicle suspended in the middle of the conductors. When the vehicle is a bit higher than the conductors, the wheels are retracted to be placed over the cables. Finally, the crane is retired, and breaks are activated to maintain the vehicle in a fixed position. When the spacer procedure completes, the reverse operation is made to lower the vehicle.

Fig. 7. Procedure for rising the vehicle. First, the vehicle is hung by a crane, and the wheels are in the raised position (left); second, when the vehicle is over the conductors then, the wheels are retracted to place them over the cables (centre); finally, the vehicle is fully suspended in the line, and the crane is retired (right).

3.2 Procedure for Automatic Installation of the Spacer in a Duplex Line

The installation procedure has been defined according to the line features, starting when the vehicle is held on the lines. First, breaks firmly grasp the cables and the white SM transports the spacer from the depot to the working area, located between the two robots. Then, the yellow CM removes both clamps. The right clamp is temporarily stored in a buffer, and the CM holds the left clamp. When both clamps are removed, the SM raises the spacer under the conductors in its final location. Afterwards, the CM screws the left clamp and the right clamp. This procedure finishes when the operator checks the proper spacer installation, and the robots go to their home positions.

3.3 Procedure for Jumping the Spacer Line

The spacer is installed between the vehicle wheels; the control and synchronization of wheels and linear actuators enables to jump the installed separator. The wheels perform a previously defined displacement, and linear actuators raise the corresponding wheels to avoid collision with the just installed spacer. This process guarantees that the vehicle's centre of gravity is always between the wheels supporting the vehicle.

Table 1 summarises the states, Fig. 8 shows the implemented finite state machine for spacer manipulation procedure and jump the spacer, and Fig. 9 shows the pictures of states executed.

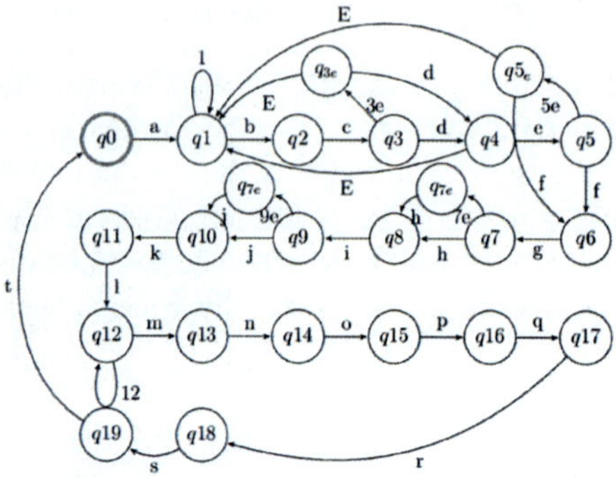

Fig. 8. Finite state machine for spacer installation and jump.

4 Tests and Performance of the ROSE System

The system has been under exhaustive tests at the laboratory test bench and Red Eléctrica overhead line. The operator controls the robots and vehicle actions from the ground using the corresponding interface; in some cases, operator inputs are required to follow the operations.

4.1 Torque Applied During Screwing

A critical task is to unscrew and screw the clamps. Figure 10 shows the torque and velocity evolution of the clamp screw when it is installed at its final position in the line. Label 1 represents the instant in which is started the screwing process. The screwing has the first step from label 1 to label 2, in which the torque and velocity are constants; the torque applied is to overcome the friction of the screw when screwing into the nut. The next step is between labels 2 and 3; in this case, the torque is mainly applied to firmly fix the clamp to the spacer since the velocity significantly reduces.

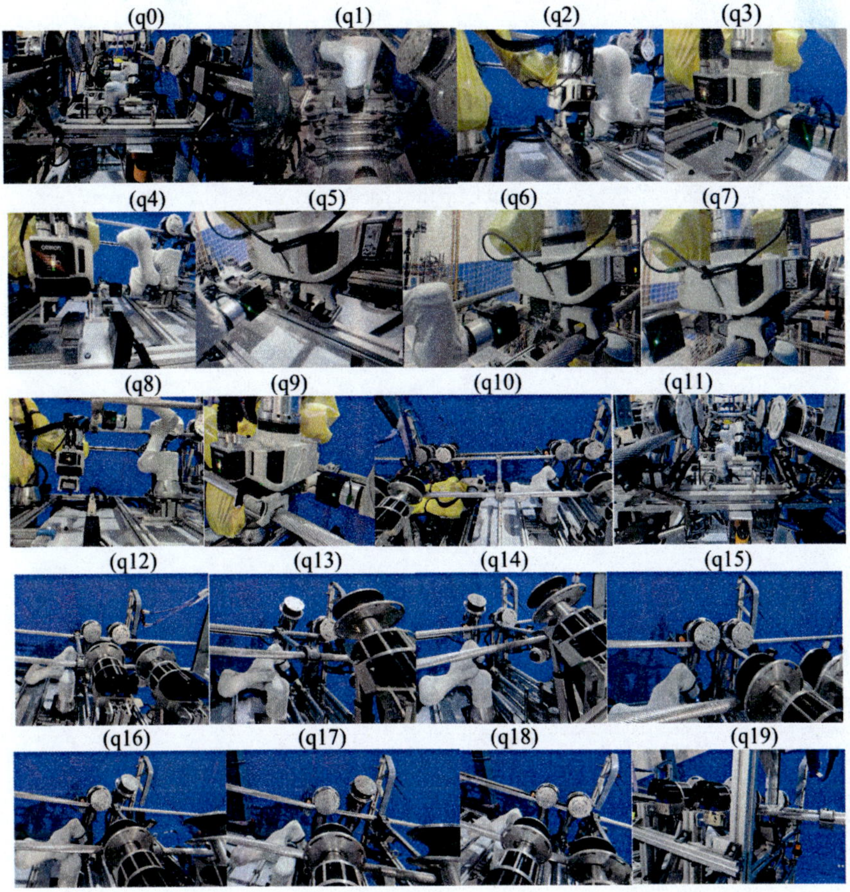

Fig. 9. Pictures of the spacer installation and jump according to Table 1.

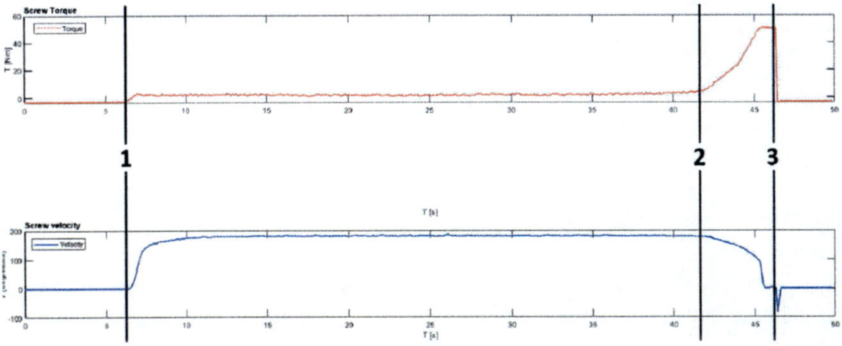

Fig. 10. Screwing process. The first step is to insert the screw into the nut (the period between 1 and 2); the second step is to fix the clamp to the spacer (the period between 2 and 3).

Table 1. States for spacer installation by using two specialized robotics arms (Fig. 9), SM, CM, and jump the spacer using displacement subsystem (DS).

State	System	Description
q0	DS	Brakes grasp the cables
q1	SM	Pick up the spacer from the depot
q2	SM	Transport the spacer to the manipulation area
q3	CM	Unscrew the right clamp
q3e	CM	Correct q3 state
q4	CM	Store the right clamp at the buffer
q5	CM	Unscrew the left clamp
q5e	CM	Correct q5 state
q6	SM	Locate the spacer under the cables
q7	CM	Screw the left clamp
q7e	CM	Correct q7 state
q8	CM	Get the right clamp from the buffer
q9	CM	Screw the right clamp
q10	CM, DS	Move robots to home
q11	DS	Brakes release the cables
q12	DS	Move vehicle until detect spacer
q13	DS	Raise the first pair of wheels
q14	DS	Move vehicle until next wheels
q15	DS	Placed the first pair of wheels
q16	DS	Raise the second pair of wheels
q17	DS	Move vehicle until pass the spacer
q18	DS	Placed the second pair of wheels
q19	DS	Pass the spacer

4.2 Battery Consumption

The battery consumption was measured to obtain an approximate value in the execution of different implemented states. Values obtained to allow for optimizing the process in the future and dimensions of a battery pack according to the number of times to perform the task. The Fig. 11 shows the measurements of voltage, amperage, consumption, and the remaining battery percentage in the execution of the states described in Table 1.

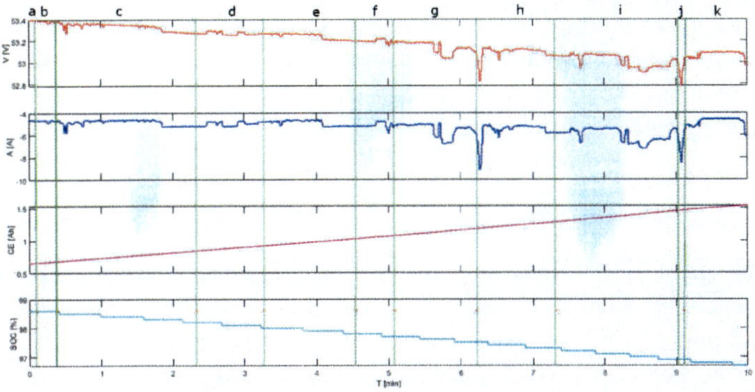

Fig. 11. Battery consumption during the installation and jump of the spacer. (red) voltage (V). (blue) amperage (A). (magenta) consumed amp hours (CE). (cyan) a State-of-charge (SOC). Each mark (green) represents the final of the sates executed according to Table 1. (a) q0; (b) q1, q2; (c) q3; (d) q4; (e) q5; (f) q6; (g) q7; (h) q8; (i) q9; (j) q10; (k) q11, q12, q13, q14, q15, q16, q17, q18, q19.

4.3 Experiments at the Testing Bench

The vehicle was tested at a testing bench simulating the power lines where the spacers will be installed (Fig. 12). These experiments affected all the vehicle subsystems and included the following functionalities.

Fig. 12. Vehicle on the testing bench at the CAR Robotics Arena.

The vehicle traction was tested at different slopes (0 and 15 degrees) and variable speeds. Several emergency situations simulating less functional traction systems available were also tested. Results showed high robustness and an overall good performance of the vehicle traction system. Wheels positioning are also tested in different configurations, such as the jump of the installed spacer and the avoidance of the cable junction. Various combinations of wheels were tested with simultaneous lifting and descending. The spacer installation has been carried out under different slopes and vehicle loads. Experiments were also repeated at different strains of the test power lines to increase the process's robustness.

Several experiments were conducted to position the brakes regarding the braking subsystem. Once positioned, the system was tested under different conditions, including slopes and emergency situations with fewer brakes available. Results showed outstanding performance and robustness of the braking system.

4.4 Trials in Overhead Line

ROSE prototype has been also tested and evaluated in an actual overhead line. Several trials have been executed in Red Eléctrica facilities. Figure 13 shows the vehicle on the overhead line. During these sessions, the vehicle was exposed to work under actual conditions. This exposure allowed the validation of the design of the vehicle subsystems.

The first experimental session focused on the structure, traction system and raising/lowering validation. The raising process was tested on the line with a prototype and then validated with the vehicle. The vehicle showed excellent mobility capabilities under restricted conditions, moving at high speed with just two active traction systems. In addition, the mechanical structure showed acceptable robustness.

The second experimental session focused on robotic and electronic systems validation. The vehicle successfully achieved the installation of a spacer in the overhead line. Both WIFI communications and user interface performed great at these experimental sessions. The vehicle antenna was able to transmit all the necessary data without interruptions, and the interface facilitated the control of the vehicle.

Fig. 13. ROSE at the overhead line during trial tests.

5 Conclusion

This work demonstrates the possibility to automatise the installation of line spacers, which is a task of great interest in constructing aerial lines. Moreover, the technologies developed also open new opportunities for inspection and maintenance since new automatic high-torque tools and autonomous vehicles that displace over conductors have been successfully demonstrated.

References

1. Jinqing, S., et al.: Research and application of spacer installation robot. E3S Web Conf. **245**, 01026 (2021). https://doi.org/10.1051/e3sconf/202124501026

2. Toussaint, K., Pouliot, N., Montambault, S.: Transmission line maintenance robots capable of crossing obstacles: state-of-the-art review and challenges ahead. J. Field Robot. **26**(5), 477–499 (2009)
3. Gonçalves, R.S., Carvalho, J.C.M.: Review and latest trends in mobile robots used on power transmission lines. Int. J. Adv. Robot. Syst. (2013). https://doi.org/10.5772/56791
4. Seok, K.-H., Kim, Y.S.: A state of the art of power transmission line maintenance robots. J. Electr. Eng. Technol. **11**(5), 1412–1422 (2016)
5. Richard, P.-L., et al.: LineRanger: analysis and field testing of an innovative robot for efficient assessment of bundled high-voltage powerlines. In: 2019 International Conference on Robotics and Automation (ICRA), pp. 9130–9136 (2019). https://doi.org/10.1109/ICRA.2019.8794397
6. Peñín, L.F., Aracil, R., Ferre, M., Pinto, E., Hernando, M., Barrientos, A.: Telerobotic system for live power lines maintenance: ROBTET. In: 1998 International Conference on Robotics and Automation (ICRA), pp. 2110–2115 (1998). https://doi.org/10.1109/ROBOT.1998.680632
7. Sawada, J., Kusumoto, K., Maikawa, Y., Munakata, T., Ishikawa, Y.: A mobile robot for inspection of power transmission lines. IEEE Trans. Power Deliv. **6**(1), 309–315 (1991). https://doi.org/10.1109/61.103753
8. Debenest, P., et al.: Expliner - robot for inspection of transmission lines. In: 2008 IEEE International Conference on Robotics and Automation, pp. 3978–3984 (2008). https://doi.org/10.1109/ROBOT.2008.4543822
9. Lorimer, T., Boje, E.: A simple robot manipulator able to negotiate power line hardware. In: 2012 2nd International Conference on Applied Robotics for the Power Industry (CARPI), pp. 120–125 (2012). https://doi.org/10.1109/CARPI.2012.6473376
10. Fan, F., Wu, G., Wang, M., Cao, Q., Yang, S.: Multi-robot cyber physical system for sensing environmental variables of transmission line. Sensors. **18**(9), 3146 (2018). https://doi.org/10.3390/s18093146

Unmanned Aerial Vehicle Applications

Benchmark on Real-Time Long-Range Aircraft Detection for Safe RPAS Operations

Víctor Alarcón[1(✉)], Pablo Santana[1], Francisco Ramos[1], Francisco Javier Pérez-Grau[1], Antidio Viguria[1], and Aníbal Ollero[2]

[1] Advanced Center for Aerospace Technologies (CATEC), Seville, Spain
vmalarcon@catec.aero

[2] Robotics, Vision and Control Group (GRVC), Universidad de Sevilla, Seville, Spain

Abstract. The growing market in Remotely Piloted Aircraft Systems (RPAS) and the need for cost-effective "Detect and Avoid (DAA)" systems are critical issues up to date towards enabling safe beyond visual line of sight (BVLOS) operations. In hopes of promoting earlier threat detection on DAA systems, we benchmark several object detection algorithms on multiple graphical processing units for the concrete DAA use case. Two state-of-the-art "real-time object detection" and "object detection" model sets are trained using our CENTINELA dataset, and their performances are compared for a wide range of configurations. Results demonstrate that one-stage architecture YOLO variants outperform ViT on all tested hardware in terms of mean average precision and inference speed despite their architecture complexity gap. Additional resources are available to the reader at https://github.com/fada-catec/detection-for-safe-rpas-operation.

Keywords: Real-time object detection · Convolutional neural networks · Visual transformers · Unmanned aerial systems · Detect and avoid

1 Introduction

1.1 Motivation

Over the last years, the Remotely Piloted Aircraft Systems (RPAS) business has grown significantly. The European RPAS market witnessed a growth of 16.6% according to the compound annual growth rate from 2017 to 2021, surpassing USD 3 billion in 2020 [1]. In addition, the RPAS market is projected to expand rapidly at 21.9% through 2032 in the European region [2].

The changes in regulations made by the European Union Aviation Safety Agency (EASA) support the deployment of RPAS in various sectors, such as infrastructure, agriculture, transport, entertainment, and security [3]. Nevertheless, one of the critical points to reaching the full potential of the RPAS market

D. Tardioli et al. (Eds.): ROBOT 2022, LNNS 590, pp. 341–352, 2023.
https://doi.org/10.1007/978-3-031-21062-4_28

is the transition from operations within Visual Line Of Sight (VLOS) to Beyond VLOS (BVLOS), which offer a higher added value. The Single European Sky ATM Research (SESAR) Joint Undertaking estimates that approximately 50% of the professional market will focus on BVLOS operations in rural environments for applications such as linear infrastructure inspection and monitoring, precision agriculture, and surveillance [4]. However, factors such as strict regulations, security and safety concerns, and lack of trained pilots, are anticipated to hamper the market growth during the forecast period.

Thus, the need for a Detect And Avoid (DAA) system for RPAS is introduced, which refers to integrating sensors to recognize the environment and navigate safely regarding unforeseen encounters and potential collisions. The problem is divided into two tasks: detecting objects close to the aerial vehicle, and managing the maneuver to avoid a collision. DAA systems' architecture can vary, and multiple types of sensors, including acoustic, visual or radars, can be used [5]. When performing BVLOS flights, the main roadblock is the absence of cost-effective systems approved by the aeronautical authorities. This is a subject of research activities, and different approaches have been tested up to date, including complex sensors and powerful but heavy computer systems. The difference in terms of cost-effective systems between visual sensors and others is a compelling argument for focusing the work on object detection on images. Given the application, an RGB camera providing high-resolution images (i.e. around 20 megapixels) is needed to acquire images with the maximum availability in terms of field of view and number of pixels for object detection, which tends to be a challenge for objects at distances larger than 1 km. In terms of computational needs, this application demands a computer with enough graphical processing capacity and advanced algorithms to perform real-time object detection.

The key enabler of this work is the lack of references in the state of the art for algorithms and their performances to be embedded on small, cost-effective, portable and energy-efficient single-board computers (SBC). The focus is laid on effectiveness in terms of speed and accuracy for the early detection of manned aircraft using RGB cameras and embedded processes over single board computers. Results allow going deeper into the design of DAA systems to detect distant aircraft in real-time, hence fostering BVLOS operations for RPAS.

1.2 Related Work

Compared to traditional Computer Vision (CV) techniques, Deep Learning (DL) achieves greater accuracies in tasks such as image classification, semantic segmentation and object detection. Since DL approaches are trained rather than deterministically programmed, applications using this approach can exploit the tremendous amount of data available in today's systems, and often require less expert analysis and hand-tuning. DL also provides superior flexibility because the models can be re-trained using custom datasets for other use cases, contrary to CV algorithms, which tend to be more domain-specific [6]. That is the main reason to go deeper into DL than classic CV approaches.

Real-time small object detection using high-resolution RGB images and DL algorithms is an open issue in DAA systems, as well as in other applications such as autonomous driving [7] or video surveillance [8]. In terms of accuracy, recent research on object detection has already achieved significant progress, but it is still challenging when the target appears as a small percentage of the entire image. It is common to find works that face the problem of detecting objects using Convolutional Neural Networks (CNN), such as Faster R-CNN [9] or Single Shot Detector [10]; however, input images are downsampled using pooling layers, hence targets could easily be filtered out in the final feature map. Other approaches try to solve this issue by using innovative architectures based on Transformers with Attention [11,12] or Feature Pyramid Networks (FPN) [13]. Although these methods address this problem and improve state-of-the-art performance in small object detection, their time complexity is higher for accurate real-time detection.

It is necessary to prepare detectors that address our specific scenario to ensure a reasonable accuracy in our application. An extensive aerial object detection dataset with objects at various scales is presented at [14]. Still, its great amount of largely-sized items deters the detection of smaller ones, making it impossible to evaluate the performance of detectors on this issue. Other aerospace-related works aim to compare different algorithms in real-time small object detection, as presented in [15] and [16], but both are focused on a different scenario.

The absence of an extensive collection of images with aircraft flying far from the visual sensor makes it difficult to focus the research activities, as there are no results from previous works. As a scientific novelty, this work compares supervised neural network architectures for object detection by training models using a proprietary dataset, where targets are small aircraft flying at least 6000 ft away. In addition, multiple computing boards with different processing capabilities are used to test the performance of the trained models.

The rest of the paper is structured as follows. In Sect. 2, the selected DL object detection algorithms are presented. Section 3 describes the software tools, the hardware platforms, the dataset and the performance metrics used in the experiments. In Sect. 4, the results of each model evaluation are presented, and the implications of the results are discussed. Finally, Sect. 5 presents the study's main conclusions and future lines of work.

2 Object Detection Algorithms

In order to solve the DAA problem using DL algorithms, one of the critical aspects is choosing a network architecture that optimally balances detection accuracy and computational cost. While the nature of the DAA problem favors lightweight DL models to approach real-time performance, detecting small-sized aeroplanes at lengthy distances demands high-resolution imagery in which targets might cover only a few pixels.

Taking into account [17], some architectures have been selected regarding two state-of-the-art benchmarks called "Object Detection" and "Real-Time Object

Detection". In the first one, a general ranking prioritizing inference accuracy is presented, while on the other one, a fast time inference while maintaining a base level of accuracy is given. Despite the absence of inference times in the "Object Detection" benchmark, our work is set to provide enough inference speed due to fast in-time hardware evolution and cost decrease of processing components. These rankings have been tested and validated using the MS COCO dataset [18], one of the most widespread free dataset used for new DL strategies.

The selected architectures are based on the You Only Look Once (YOLO) family, initially defined in [19], and the Visual Transformers (ViT) neural network types [20]. YOLOv architectures are considered the fastest DL object detection methods due to their architecture which prioritizes inference speed by producing the output without intermediary region proposals. ViT architectures are focused on a different paradigm since they avoid recurrence and rely entirely on an attention mechanism to establish global dependencies between input and output, without using sequence-aligned convolutions, prioritizing accuracy over inference speed. For each type of architecture, a simple and a more complex neural network have been selected, with the idea of testing them on multiple board computers with different Graphics Processing Units (GPU).

In terms of CNNs, YOLOv4 [21] variants are considered the fastest real-time neural network architectures for object detection with a base level of accuracy. YOLOv5 [22] architecture builds on YOLOv4 including some improvements to run better off-the-shelf; though no research paper has been made available, its promising experimental results have motivated its inclusion in our study. YOLOv4-tiny and YOLOv5N are the light versions considered, while YOLOV4-P6 and YOLOv5S are the implementations for more complex computers.

Concerning ViT architectures, DETR [23], Swin-transformer [24] and DyHead [25] have high scores in the benchmark. DAB-DETR and Swin-s are selected as light architectures from DETR and Swin-Transformers, respectively, and deformable DETR, Swin-t and DyHead as the complex architectures. In particular, the proposed ViT arquitectures are the following: a DAB-DETR, its multiscale version DAB-Deformable and the Dynamic Head (DyHead) architectures using the same ResNet50 backbone. In addition, three configurations of the Swin-transformer architecture including the two-stage detector Mask RCNN combined with Swin-T and Swin-S backbones, and a combination of the one-stage detector RetinaNet with Swin-T backbone.

The nature of the proposed DAA system demands a trade-off between accuracy and inference speed, so we consider it essential to compare both available options in the state-of-the-art, which there is no doubt about their relevance to the scientific community regarding the object detection challenge.

3 Experimental Setup

This section discusses technical details towards software and hardware selection, dataset construction and model configuration. Related material is available at https://github.com/fada-catec/detection-for-safe-rpas-operation.

3.1 Hardware Platforms

Our experiments have been carried out on different hardware platforms to compare the performance of our models in a varied range of specifications. Since the detection algorithms should optimally be run in real-time, it is interesting to obtain performance metrics using various devices, as such processing can be performed either *in situ* or remotely. In the former case, light single board computers with limited GPUs are deployed in the operating area. In the latter, more powerful computer equipment has been considered taking into account the recent advances in cloud computing and high-bandwidth communications with low latencies, enabling high-end servers for real-time data processing.

Regarding light computing, we target an array of NVIDIA SBCs, which provide great computing capabilities on compact and energy-efficient equipment [26]. These are delivered through different boards with a broad performance range, from which we select the increasingly powerful set consisting of the Jetson Nano, Jetson TX2 and Xavier NX.

In the middle range, we include a laptop with an NVIDIA RTX A2000 GPU. While such card is neither suitable to be deployed *in situ*, nor powerful enough to be considered a cloud computing solution, its study becomes relevant towards early development stages of DL projects, which usually take place on average machines with mid-end GPUs.

Finally, we assess two high-end GPU machines geared with an NVIDIA GeForce RTX 2080Ti and an RTX 3080 GPU, which stand for remote cloud computers. Though commercial servers that process information using DL algorithms typically provide higher capabilities, these still possess a decent amount of power and represent a common self-hosted solution for research teams to have as a shared resource. These two machines will also be used to train our models.

3.2 Software Tools

Regarding software, it is useful to select a common DL framework to execute training and model evaluations. PyTorch [27] has been selected, since it covers all our target models, which are run through intermediary tools that use it as a backend.

One-stage CNN architectures have been supported by three different open-source repositories: ScaledYOLOv4 [28] for YOLOv4-P6, PyTorch_YOLOv4 [29] for YOLOv4-tiny, and Yolov5 [22] from Ultralytics for YOLOv5N and YOLOv5S. On the other hand, the open-source toolbox MMDetection [30] has been used to work with the ViT-NN models: DAB-DETR, DAB-Deformable, DyHead, Mask RCNN Swin-T, Mask RCNN Swin-S and RetinaNet Swin-T.

To test our models within a common set of performance metrics, these are evaluated with the official MS COCO validation API [31], by first using the aforementioned tools to perform inference on a test split of the data, and then using MS COCO's to compare those against its ground-truth labels. Besides overall precision results, MS COCO also computes metrics by dividing the target data in three size groups: small for an area of $32 \times 32\,\text{px}^2$, large for an area over $96 \times 96\,\text{px}^2$, and medium elsewhere. Such size segregation refers to the original images, without regard to the resizing they undergo at inference time.

3.3 Datasets

In order to benchmark the models defined in Sect. 2, we have crafted a proprietary dataset composed of images of small aircraft taken at different field tests, namely the *CENTINELA* dataset. Such field tests took place in five different airfields in which a few distinctive landscapes were captured during flight tests. Natively, these images have a resolution of $5472\times3648\,\text{px}^2$. They were manually labelled using [32], being $125\times80\,\text{px}^2$ the average size of aircraft labels.

In Fig. 1, the left column shows random samples of these images, and each row belongs to a different scenario. The main pitfall is the similarity among samples within the same field test, as each location contains only a different set of landscapes over which the aircraft appear. Since many images are visually similar (except for the small aircraft within the image itself), performing a train-validation-test split by randomly shuffling all our images would lead to all three splits being very similar. Hence, this would yield overly-optimistic yet unrealistic detection performance metrics. On the other hand, splitting the images by field test makes our training set very monotonous, so that trained models can generalize to neither test nor validation sets.

To overcome these two issues, we featurized the images of each field test using a pre-trained low-resolution VGG-16 network, and then used those features to cluster the images using the K-means algorithm. We then shuffled the clusters of all field tests to achieve a 70:15:15 split for train, validation and test subsets respectively. An overview of this splitting strategy can be seen in the rest of Fig. 1, which shows a random preview of samples from different clusters and the three final splits. According to the MS COCO standard sizes, our test split is composed of 189 small, 881 medium and 533 large instances.

3.4 Model Configuration

Our detectors must process image sizes of 20 megapixels in real-time. Large image sizes would not only hinder inference speed when deployed, but also increase training costs. Besides being time-wise expensive, training on very big images can also limit the choice of batch size.

Since GPUs have a limited amount of memory, increasing the image size would restrict to training on smaller batches. Small batch sizes detriments training efficiency, as the gradient is not so well averaged across samples. Besides slower processing, this would lead to longer training times, since loss convergence is harder to reach. On the other hand, a big batch size performs less gradient updates per epoch, requiring larger learning rates that may lead to instabilities. The latter is, however, not usually an issue within object detection, where larger batch sizes are advisable [22].

Real-time models were trained on a square resolution of 1280 px. Such image size was deemed enough for the labelled aircraft not to become unrecognisable pixel blobs, while allowing the batch size to be higher than one picture in our training computer for our heaviest target model. After this resize, average label

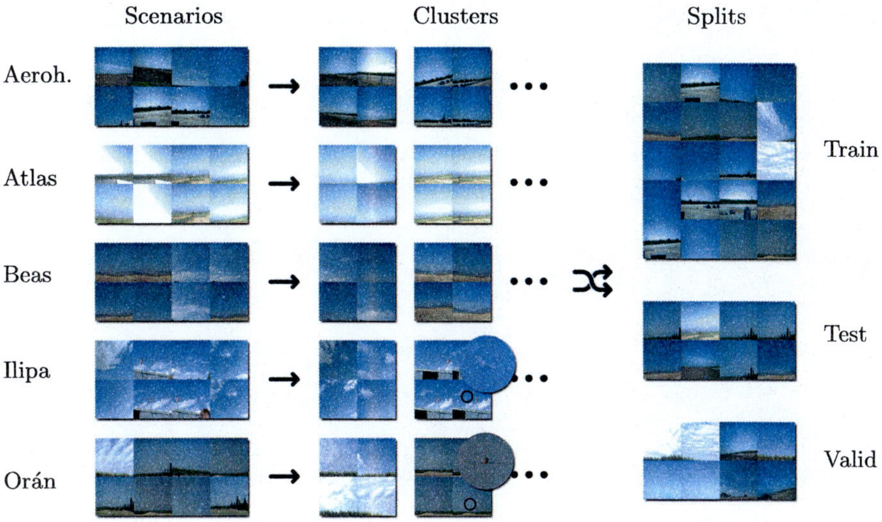

Fig. 1. *CENTINELA* dataset construction overview. Images gathered from different scenarios (left) were clustered by similarity (middle). Those clusters were finally shuffled to assemble three landscape-rich yet heterogeneous splits (right).

size approached $30 \times 30\,\text{px}^2$. Following best practices [22], the batch size was set as big as possible after the image size was established.

On the other hand, non-real-time models inherit their input resolution from the FPN standard [33], which comes by default in [34]. Because our images are taller than wide, they get resized to a height of 800 px while keeping their 3 : 2 aspect ratio, thus leaving a network input image resolution of $1200 \times 800\,\text{px}^2$. Such resolution is on par with the one used for our real-time models, still allowing small aircraft to be identifiable with an average label size of $27 \times 17\,\text{px}^2$. Again, the batch size was set at the highest value allowed by the training hardware.

Table 1 holds a summary of the training setup for all target models.

4 Results and Discussion

4.1 Performance Metrics

To obtain an evaluation of the accuracy and the inference speed, we must select the metrics on which to base our conclusions. Object detection challenges typically use their own metrics to evaluate the proposed task. Researchers who want to evaluate their work using different datasets need to either implement their version of the metrics or adhere to an existing standard.

When measuring accuracy, [17] proposes the use of MS COCO as the most popular dataset to establish their ranking, and then evaluates Mean Average Precision (mAP) metrics, a metric calculated with the help of several other metrics such as Intersect over Union (IoU), confusion matrix, precision and recall.

Table 1. Input image size (px^2), batch size and training GPU for all target models.

Model	Input-size	Batch-size	Machine
yolov4-p6	1280 × 1280	2	RTX 2080Ti
yolov5l	1280 × 1280	4	RTX 2080Ti
yolov4-tiny	1280 × 1280	24	RTX 2080Ti
yolov5n	1280 × 1280	28	RTX 2080Ti
swin-t	1200 × 800	2	RTX 3080
swin-s	1200 × 800	2	RTX 3080
dyhead	1200 × 800	3	RTX 3080
retina-swin	1200 × 800	3	RTX 3080
detr	1200 × 800	4	RTX 3080
defor	1200 × 800	1	RTX 3080

Despite using a proprietary dataset, our work is based on the assumption of results published by this source, so we have considered using the same metrics.

The case of measuring inference speed is different. The most common scenario is to provide the results in completely processed frames per second (FPS), and the inference time per frame in milliseconds (ms). Therefore, FPS and MS are the metrics selected in our work.

4.2 Inference Speed Comparison

Table 2 shows the results of testing each model on each hardware platform, where the higher speed is highlighted in bold. The results prove that one-stage detectors perform better than two-stage detectors due to their architecture type. YOLOv models are more appropriate for real-time tasks since their inference time is generally lower than those using ViT. Specifically, YOLOv tiny models are faster than others due to their lightweight architecture. By contrast, large YOLOv architectures and ViT models perform poorly on single board computers, or cannot be used due to a lack of memory (e.g. Jetson Nano). Still, executing real-time operations for the studied use case is feasible using medium and high-processing graphic boards.

4.3 Inference Accuracy Comparison

The results for all models are shown in Table 3, and the highest values are highlighted in bold. Real-time networks lead the accuracy scores, with the most accurate model being YOLOv4-P6. If we focus on comparing the real-time architectures, some behaviours worthy of study are found:

- The size accuracies indicate that smaller objects are harder to detect, and that YOLOv4 models have more difficulty dealing with them versus YOLOv5 variants. An extreme example is YOLOv4-tiny, which gets 0% of mAP on

Table 2. Inference speed per image of all different models on the target hardware.

Name	Height px	Nano ms/fps	TX2 ms/fps	Xavier ms/fps	A2000 ms/fps	2080Ti ms/fps	3080 ms/fps
yolov4-P6	1280	–	1600/0.63	670/1.5	87/12	43/23	33/30
yolov5L	1280	–	740/1.4	310/3.2	40/25	23/43	18/56
yolov4-tiny	1280	560/1.8	140/7.1	64/16	8.8/110	**6.0/167**	**4.5/220**
yolov5n	1280	**470/2.1**	**90/11**	**60/17**	**7.8/130**	7.3/140	6.9/150
dyhead	800	–	1700/0.59	1600/0.63	170/5.9	82/12	63/16
swin-s	800	–	2700/0.37	1300/0.77	120/8.3	62/16	46/22
swin-t	800	–	1800/0.56	1400/0.71	120/8.3	68/15	47/21
retina-swin	800	–	1600/0.63	1300/0.77	130/7.7	66/15	45/22
defor	800	–	1800/0.56	1500/0.67	150/6.7	77/13	51/20
detr	800	–	**710/1.4**	**640/1.6**	**74/14**	**38/26**	**28/36**

small objects. On the contrary, the results reflect that YOLOv4 architectures perform faster on large objects.

- Comparing YOLOv5 variants, YOLOv5-L has slightly lower mAP and AP50 scores than YOLOv5-N, which is interesting since the former has a more complex architecture, which should allow it to learn more features and significantly outperform the latter. This result can be explained by a slight overfitting, to which larger networks are more vulnerable, although YOLOv4 models do not share such pattern. In any case, YOLOv5-L is more accurate for higher IoU thresholds.
- Across different box scales, YOLOv5-N seems to perform better with large objects, while YOLOv5-L shows better mAPs on small and medium objects. The best mAP is for the medium size group, the most abundant in our test set.
- Comparing YOLOV4 on large objects, although YOLOv4-P6 is still ahead, the 3 points difference in mAP shows that simpler architectures like YOLOv4-tiny are close in performance to more complex architectures.
- The fact that the accuracies of YOLOv5-N are so close to those of YOLOv4-P6 suggests that small models can keep up with large ones when the number of classes is low.
- YOLOv5-L and YOLOv5-N perform better than YOLOv4 variants on smaller objects, keeping a similar mAP, thus generating a case worth studying.

Table 3. Average precision metrics (%) for all models on the dataset. Respectively: image height, mean average precision, average precision at IoU 50%, average precision at IoU 75%, plus mean average precisions for small, medium and large bounding boxes.

Name	Height px	mAP all	AP50 all	AP75 all	mAP small	mAP medium	mAP large
yolov4-P6	1280	30.2	**76.9**	15.1	18.5	30.3	35.4
yolov5L	1280	26.2	68.3	14.7	**23.7**	28.4	27.3
yolov4-tiny	1280	22.2	57.4	13.7	0.00	19.4	32.4
yolov5n	1280	26.7	70.2	14.2	22.7	26.4	30.3
dyhead	800	21.7	55.0	15.7	13.9	21.1	22.6
swin-s	800	**32.4**	60.3	**32.3**	0.00	**33.0**	**39.8**
swin-t	800	23.3	50.7	19.1	0.00	24.4	28.1
retina-swin	800	23.1	58.6	14.0	13.5	23.7	28.5
defor	800	14.2	42.5	7.1	8.4	11.9	20.5
detr	800	16.3	63.1	3.4	9.7	14.3	23.3

5 Conclusions and Future Work

This paper compares the top-ranked neural network architectures according to [17] for detecting small aircraft in high-resolution images in real-time. We present the outcomes of typical real-time-oriented architectures (YOLOv family) and good-accuracy performance-oriented architectures (ViT). The keys of the result are based on two standard performance metrics, mAP for inference accuracy and time metrics for inference speed. The results clearly show that real-time algorithms perform better than ViT concerning the use-case related to this work, represented by our proprietary dataset.

Proposing the use of ViT architectures was motivated by the good results of other scientific researchers according to their practical use cases. However, the poor results presented compared to YOLOv architectures will be subject of future work, as some possibilities are open to study. Increasing the quality and extension of the dataset, updating the input data configuration of architectures in the training phase, or combining backbones with more precise features extractors are good examples of the open paths to continue this work. These proposed items are also valid for real-time object detection networks, as some discussed behaviours are not entirely comprised.

Finally, as DL techniques are subject to continuous update, the emergence of new architectures has to be considered to be included in the benchmark, as they could improve the results of real-time object detection metrics in terms of accuracy and inference speed, offering better performances in the use case proposed in this work.

Acknowledgements. This work has been partially supported by the OMICRON project, funded by the EU H2020 programme under grant agreement 955269, and

CEL.IA, a Cervera Network for applied artificial intelligence, funded by the Spanish government through CDTI (CER-20211022).

References

1. Graphical Research. Europe Commercial Drone Market Forecast 2027. Accessed 30 Aug 2022. https://www.graphicalresearch.com/industry-insights/1016/europe-commercial-drone-unmanned-aerial-vehicle-UAV-market
2. Fact.MR. Europe Drone Market Outlook (2022–2032). Accessed 30 Aug 2022. https://www.factmr.com/report/europe-drones-market
3. EASA. Civil drones (unmanned aircraft). Accessed 30 Aug 2022. https://www.easa.europa.eu/domains/civil-drones
4. Single European Sky ATM Research. European Drones Outlook Study. Accessed 30 Aug 2022. https://www.sesarju.eu/sites/default/files/documents/reports/European_Drones_Outlook_Study2016.pdf
5. Mariscal-Harana, J., et al.: Audio-based aircraft detection system for safe rpas bvlos operation. Electronics **9**(12), 2076 (2020)
6. Arai, K., Kapoor, S.: Advances in computer vision. In: Conference Proceedings CVC, p. 104. Springer, Heidelberg (2019)
7. Cai, Y., et al.: YOLOv4-5D: an effective and efficient object detector for autonomous driving. IEEE Trans. Instrum. Meas. **70**, 1–13 (2021)
8. Jha, S., et al.: Real time object detection and tracking system for video surveillance system. Multimedia Tools Appl. **80**(3), 3981–3996 (2021)
9. Ren, S., et al.: Faster R-CNN: towards real-time object detection with region proposal networks. Adv. Neural Inf. Process. Syst. **28** (2015)
10. Liu, W., et al.: SSD: single shot multibox detector. In: Leibe, B., Matas, J., Sebe, N., Welling, M. (eds.) ECCV 2016. LNCS, vol. 9905, pp. 21–37. Springer, Cham (2016). https://doi.org/10.1007/978-3-319-46448-0_2
11. Zhu, X., et al.: TPH-YOLOv5: improved YOLOv5 based on transformer prediction head for object detection on drone-captured scenarios. In: 2021 IEEE/CVF International Conference on Computer Vision Workshops (ICCVW), pp. 2778–2788 (2021)
12. Lian, J., et al.: Small object detection in traffic scenes based on attention feature fusion. Sensors **21**(9), 3031 (2021)
13. Liu, Z., et al.: HRDNet: high-resolution detection network for small objects. In: 2021 IEEE International Conference on Multimedia and Expo (ICME), pp. 1–6. IEEE (2021)
14. Xia, G.S., et al.: DOTA: a large-scale dataset for object detection in aerial images. In: Proceedings of the IEEE Conference on Computer Vision and Pattern Recognition, pp. 3974–3983 (2018)
15. Sun, X., et al.: FAIR1M: a benchmark dataset for fine-grained object recognition in high-resolution remote sensing imagery. ISPRS J. Photogram. Remote Sens. **184**, 116–130 (2022)
16. Wang, Y., et al.: Remote sensing image super-resolution and object detection: Benchmark and state of the art. Expert Syst. Appl., 116793 (2022)
17. Papers with Code. A free and open resource with Machine Learning papers, code, datasets, methods and evaluation tables. Accessed 30 Aug 2022. https://paperswithcode.com

18. Lin, T.-Y., et al.: Microsoft COCO: common objects in context. In: Fleet, D., Pajdla, T., Schiele, B., Tuytelaars, T. (eds.) ECCV 2014. LNCS, vol. 8693, pp. 740–755. Springer, Cham (2014). https://doi.org/10.1007/978-3-319-10602-1_48
19. Redmon, J., et al.: You only look once: unified, real-time object detection. In: Proceedings of the IEEE Conference on Computer Vision and Pattern Recognition, pp. 779–788 (2016)
20. Dosovitskiy, A., et al.: An image is worth 16×16 words: transformers for image recognition at scale. ArXiv preprint arXiv:2010.11929 (2020)
21. Wang, C-Y., Bochkovskiy, A., Liao, H.-Y.M.: Scaled yolov4: scaling cross stage partial network. In: Proceedings of the IEEE/CVF Conference on Computer Vision and Pattern Recognition, pp. 13029–13038 (2021)
22. Ultralytics. YOLOv5. Accessed 30 Aug 2022. https://github.com/ultralytics/yolov5
23. Carion, N., Massa, F., Synnaeve, G., Usunier, N., Kirillov, A., Zagoruyko, S.: End-to-end object detection with transformers. In: Vedaldi, A., Bischof, H., Brox, T., Frahm, J.-M. (eds.) ECCV 2020. LNCS, vol. 12346, pp. 213–229. Springer, Cham (2020). https://doi.org/10.1007/978-3-030-58452-8_13
24. Liu, Z., et al.: Swin transformer: hierarchical vision transformer using shifted windows. In: Proceedings of the IEEE/CVF International Conference on Computer Vision, pp. 10012–10022 (2021)
25. Dai, X., et al.: Dynamic head: unifying object detection heads with attentions. In: Proceedings of the IEEE/CVF Conference on Computer Vision and Pattern Recognition, pp. 7373–7382 (2021)
26. NVIDIA. Sistemas integrados NVIDIA para las máquinas autónomas de la próxima generación. Accessed 30 Aug 2022. https://www.nvidia.com/es-es/autonomous-machines/embedded-systems
27. Paszke, A., et al.: Pytorch: an imperative style, high-performance deep learning library. Adv. Neural Inf. Process. Syst. **32** (2019)
28. Wong, K.Y.: ScaledYOLOv4. Accessed 30 Aug 2022. https://github.com/WongKinYiu/ScaledYOLOv4
29. Wong, K.Y.: PyTorch_DYOLOv4. Accessed 30 Aug 2022. https://github.com/WongKinYiu/PyTorchYOLOv4
30. Chen, K., et al.: MMDetection: open mmlab detection toolbox and benchmark. arXiv preprint arXiv:1906.07155 (2019)
31. cocodataset. COCO API. Accessed 30 Aug 2022. https://github.com/cocodataset/cocoapi
32. Roblox. ImageLabel. Accessed 30 Aug 2022. https://developer.roblox.com/en-us/api-reference/class/ImageLabel
33. Lin, T.Y., et al.: Feature pyramid networks for object detection. In: Proceedings of the IEEE Conference on Computer Vision and Pattern Recognition (CVPR) (2017)
34. OpenMMLab. mmdetection. Accessed 30 Aug 2022. https://github.com/open-mmlab/mmdetection

Sensorial Testbed for High-Voltage Tower Inspection with UAVs

Guido S. Berger[1,2,3(✉)], Alexandre Oliveira Júnior[1,2], João Braun[1,2,4], José Lima[1,2,4], Milena F. Pinto[7], António Valente[3,4], Ana I. Pereira[1,2], Álvaro R. Cantieri[6], and Marco A. Wehrmeister[5,8]

[1] Research Centre in Digitalization and Intelligent Robotics (CeDRI), Instituto Politécnico de Braganca, Campus de Santa Apolonia, Braganca, Portugal
{guido.berger,alexandrejunior,jbneto,jllima,apereira}@ipb.pt
[2] Lab. para a Sustentabilidade e Tecnologia em Regiões de Montanha (SusTEC), Braganca, Portugal
[3] Universidade de Trás-os-Montes e Alto Douro, Vila Real, Portugal
avalente@utad.pt
[4] INESC TEC - INESC Technology and Science, Porto, Portugal
[5] Federal University of Technology – Paraná, Curitiba, Brazil
wehrmeister@utfpr.edu.br
[6] Federal Institute of Paraná, Curitiba, Brazil
alvaro.cantieri@ifpr.edu.br
[7] Centro Federal Tecnológica Celso Suckow da Fonseca, Rio de Janeiro, Brazil
milena.pinto@cefet-rj.br
[8] University of Münster, Münster, Germany
wehrmeister@uni-muenster.de

Abstract. This work presents a methodology for characterizing ultrasonic and LASER sensors aimed at detecting obstacles within the context of electrical inspections by multirotor Unmanned Aerial Vehicles (UAVs). A set of four ultrasonic and LASER sensor models is evaluated against eight target components, typically found in high-voltage towers. The results show that ultrasonic sensor arrays displaced 25° apart reduce the chances of problems related to crosstalk and angular uncertainty. Within the LASER sensor suite, solar exposure directly affects the detection behavior among lower power sensors. Based on the results obtained, a set of sensors capable of detecting multiple obstacles belonging to a high-voltage tower was identified. In this reasoning, it becomes possible to model sensor architectures for multirotor UAVs to detect multiple obstacles and advance in the state of the art in obstacle avoidance systems by UAVs in inspections of high-voltage towers.

Keywords: Sensor characterizing · Unmanned Aerial Vehicles · Electrical inspections · Obstacle avoidance

1 Introduction

Unmanned Aerial Vehicles (UAVs) are currently a popular choice for robotic applications that aim to monitor infrastructure in different domains. Within the

D. Tardioli et al. (Eds.): ROBOT 2022, LNNS 590, pp. 353–364, 2023.
https://doi.org/10.1007/978-3-031-21062-4_29

context of electrical inspections, UAVs have advantages over traditional methods of electrical inspections due to their ability to access hard-to-reach places without endangering human operators, ease of implementation of sensory systems, high mobility capacity, ability to hover close to the environment to be inspected, and low cost compared to manned inspection methods [1]. When UAVs are used for electrical inspections, it is often necessary to get close to the components to assess the state of the general condition of electrical assets. However, as the UAV approaches the object, the risk of collision progressively increases. Most conventional UAVs are manually controlled. A ground operator observes the image provided by the UAV's camera and estimates safe zones, positioning the aircraft in regions far from obstacles to conduct the inspection process [2].

Even if the pilot has good operational skills, there are still risks of collisions through blind spots or maneuver errors during aircraft control. In this sense, when applied in complex environments, such as electrical inspections, using obstacle detection and prevention systems in UAVs should be seen as an essential factor in mitigating the chances of accidents. Implementing obstacle detection sensors in UAVs is a challenging task, and different issues involving the choice of sensory technology, the type of inspection environment and positioning of sensors on the aircraft must be discussed and evaluated when seeking to develop autonomous navigation systems by UAVs [3].

This paper presents a study on the detection characteristics of ultrasonic and LASER sensors in electrical inspections. The impacts caused by the positioning of sensors on the body of multirotor UAVs are raised and discussed. The main objective of this research is to contribute to developing future obstacle detection and prevention systems by multirotor UAVs when applied in the electrical inspection environment. This research is a continuation of the work presented by [4], adding new results that advance the previously proposed theme linked to electrical inspections by multirotor UAVs.

The remaining of this work is organized as follows. Section 2 presents the related works and the background of this work. The materials and methodology for characterizing the sensors are presented in Sect. 3. In Sect. 4, the results observed by the experiments are presented, and finally, in Sect. 5, the conclusion and future directions for the work presented are addressed.

2 Overview of Ultrasonic and LASER Sensor Applied in Obstacle Detection

In order to achieve the safety of UAVs when applied in electrical infrastructure inspection in urban or rural perimeter, it is essential to adopt efficient measures for obstacle detection. In this sense, different sensory technologies can be implemented in multirotor UAVs, categorized into active and passive sensors [5].

Active sensors such as ultrasonic, and LASER sensors operate by emitting and receiving a reflected signal from one or more targets. This signal, following the examples above, can be in the form of, e.g., an acoustic signal, an electrical signal, or a light wave, respectively. Passive sensors, mostly cameras such as

RGB, thermal or IR, can operate under different wavelengths and detect the energy radiated by obstacles [6].

Ultrasonic sensors (also known as sonar) are commonly used in multirotor UAVs to assist in-flight stability and perform autonomous navigation tasks in a controlled environment [7]. Ultrasonic sensors are light and inexpensive, easy to implement, have low power consumption, and can be applied in indoor and outdoor environments for obstacle detection. However, these sensors are susceptible to reading errors when exposed to environments with constant vibration, usually in regions close to the aircraft's engines or propellers [7]. In addition, these sensors cannot correctly detect spongy surfaces, and their detection accuracy can be affected by temperature variation in the environment [8].

Another critical aspect being raised is that the ultrasonic sensors, when grouped in proximity to each other, can cause the crosstalk effect, making it impossible to detect targets correctly. This effect occurs when the signal emitted by one of the sonars is received by adjacent sonars, causing errors in the target detection reading. This problem is usually encountered in projects that use many ultrasonic sensors around multirotor UAVs [9].

The detection directionality of ultrasonic sensors can be a factor that compromises the correct detection of targets. When the beam pattern of ultrasonic sensor reflects on a given object, this obstacle can be present in any area of the region belonging to the sonar range, represented by a cone-shaped beam pattern. However, the object's exact position cannot be determined if the object lies within the limits of the cone region. This problem is known as angular uncertainty and objects that are not perpendicular to the acoustic axis may remain undetected by the sensor [10].

Other approaches use LASER sensors based on Time of Flight (ToF) technology with multirotor UAVs for autonomous navigation and stability systems [11]. This technology presents advantages in directionality and data acquisition speed. Also, some LASER sensors models based on LiDAR (Light Detection And Ranging) technology rotate and detect obstacles at long distances up to 360° of range. They are applied to 2D/3D maps reconstructions [12].

However, most of the works found in the literature use LASER sensors in a controlled environment [13], that is, where there is little or no interference to the sensor, since, in conditions of high exposure to sunlight, snow or rain, this type of sensor can present unreliable distance measurements [5]. In addition, these sensors fail to detect objects from surfaces that reflect light, such as windows and chrome objects. Therefore, because they have different advantages, sensors with LASER and ultrasonic technology are often used to increase the chances of detection between different targets [14]. To fulfill this goal, the placement of ultrasonic and LASER sensors on the multirotor UAV structure must be carefully analyzed when designing multiple obstacle detection systems.

3 Materials and Methods

The components chosen for this research, shown in Fig. 1, are accessories with complex geometry and different sizes, serving as a basis for analyzing the sensor's detection response. Unlike the other components, the metal plate (P8) was used to indicate the maximum capacity of perception of the sensors due to its geometry for easy identification.

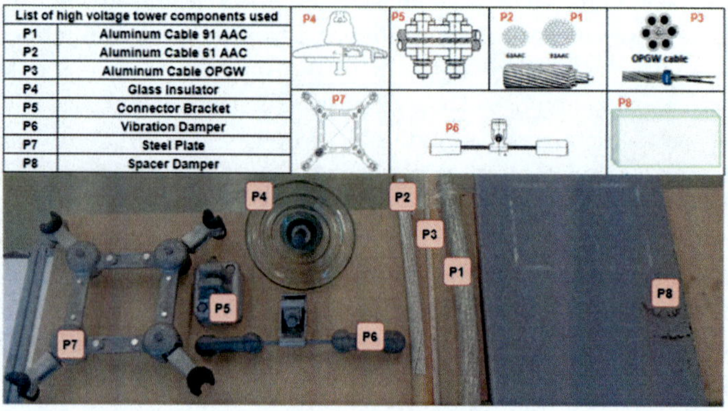

Fig. 1. Components of a high-voltage tower used in the research.

The proposed method evaluates the maximum and minimum detection capabilities of the sensors highlighted in Table 1. The objective is to identify the sensors that have capabilities to be integrated into multirotor UAVs to detect multiple obstacles commonly found in high-voltage towers. The tests are conducted in indoor and outdoor environments, evaluating different responses between the sensors when exposed to environments with high sun exposure. In addition, the detection angles of ultrasonic sensors on specific targets are evaluated, seeking to identify problems related to the crosstalk effect and angular uncertainty. In this way, the presented method helps choose sensors that can be integrated with multirotor UAVs when applied in multiple obstacle avoidance systems of electrical inspections.

Table 1. Maximum and minimum detection capabilities of the sensors.

Sensor	Range [cm]	Resolution [mm]	Current [mA]	Price [$]	Weight [g]
HC-SR04	2-400 cm	3 mm	15 mA	1$	20 g
RCW-0001	1-450 cm	1 mm	2.8 mA	1$	15 g
US-15	2-400 cm	0,5 mm	2.2 mA	2$	20 g
US-16	2-300 cm	3 mm	3.8 mA	2$	20 g
VL53L0X	0-200 cm	1 mm	19 mA	5$	0.5 g
VL53L1X	0-400 cm	1 mm	16 mA	15$	0.5 g
YDLiDAR X4	0-1000 cm	1 mm	380 mA	99$	189 g
LiDARLite V3	0-4000 cm	10 mm	135 mA	130$	22 g

3.1 Sensors Characterization in a Controlled Environment

The sensors used in this work were selected due to their low cost in the market in comparison to image sensors based on object detection, low energy consumption, and small size. The detection range of the selected sensors is at least 2 m, a distance considered safe when using multirotor UAVs in electrical inspection processes. The tests conducted in this subsection took place in a controlled environment, with little disturbance from external factors. The Fig. 2 illustrate the experimental procedures adopted in a controlled environment. In this first experimental stage, the following criteria were adopted:

- All objects used as targets have their orientation always in front of the sensors. The positioning of objects is done manually and guided by a tape measure with a resolution of 1 mm.
- All ultrasonic sensors used a temperature and humidity sensor (DHT11) to correct variations of the environmental conditions.
- All objects used were manually positioned in front of the sensors every 10 cm until they reached a distance of 450 cm. For every 10 cm of distance between the object and the sensor, 45 detection samples are collected.
- The sweep angle of the YDLiDAR X4 sensor was set to 10°, allowing the sensor to detect only the target object.
- Data from all sensors, except for YDLiDAR X4, were obtained from a data acquisition system implemented in an Arduino Mega 2560 controller board. YDLiDAR data were obtained from a data acquisition system implemented in Raspberry Pi 3B+.

With the datasets extracted in this first evaluation, the average and variance are derived, aiming to identify the detection accuracy and precision of the sensors on the different targets selected in this study. A second evaluation sought to analyze each sensor's detection reading errors on each target component, identifying the safety of the sensory perception on the target. A script was added to the sensor's code that triggers a flag when the sensor cannot detect a distance measurement over a given target obstacle.

A set of tests was established to identify potential detection failures due to the crosstalk effect and angular uncertainty to evaluate the detection performance of ultrasonic sensors at different angles. So, each ultrasonic sensor was tested individually at 0° and subsequently at angles of 15°, 25° and 35° over the P8 object, storing an amount of 45 samples for every 10 cm traveled until reaching the maximum detection distance of the sensors related to the target object.

In this evaluation, the P8 object was used because it presents a geometry that is easy to identify among all the used sensors. With this, the ultrasonic sensors could reach their maximum target detection capacity, evaluating with greater precision the responses that indicate the problems related to crosstalk and Angular Uncertainty. Thus, angular detection evaluations were performed using ultrasonic sensors in pairs.

Following the same procedures adopted in the previous experiment, a 28BYJ-48 stepper motor was coupled to each ultrasonic sensor model, precisely maintaining the positioning of the sensors under angles of 0°, 15°, 25° and 35° to

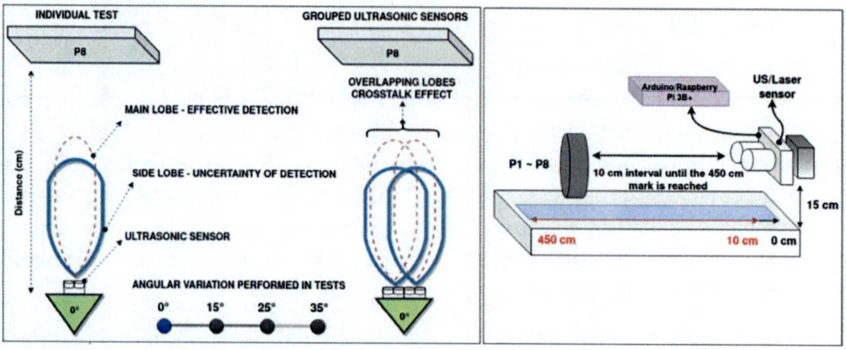

Fig. 2. Experimental procedures adopted in a controlled environment.

detect the P8 object. Considering that the crosstalk effect occurs when there is proximity between the ultrasonic sensors, this evaluation helped to identify the angular distance between groups of ultrasonic sensors without the incidence of the crosstalk effect.

3.2 Characterization of Sensors in an External Environment

In this experiment, all LASER sensors were evaluated outdoors with high solar incidence to identify which sensors could work accurately in this environment. This evaluation used a wall as a target because it is easily detectable. The YDLiDAR X4 and LiDAR Lite V3 sensors were positioned 10 m from the wall. A measuring tape was used to guide the sensors at each displacement of one meter in front of the target, obtaining 400 detection samples per meter travelled.

In a second evaluation, and following the same methodological procedures of the previous experiment, sensors with a shorter detection range were used to analyze the detection performance when used in constant movement, aiming to identify the scanning detection capabilities and if problems related to vibration interfere with the acquisition of sensor data. For this, a servo motor was attached to the sensors used in the experiment, providing a scan detection range of 180°.

4 Results and Discussion

This section presents the results and analyzes the characterization of sensors in indoor and outdoor environments.

4.1 Sensory Results in Controlled Environment

Among all the objects used, the OPGW cable (P3) presented more significant difficulties in being detected by most sensors since this component has a thin and non-uniform geometry. When compared with the metal plate (P8), an object with a uniform geometry that is easy to detect, it was possible to perceive the difficulty of most sensors in detecting the OPGW cable, as shown in Fig. 3. Among all objects used as targets in this experiment, cables P1, P2 and P3 were identified with the most significant detection difficulties by most sensors. Objects that present hollow geometry, such as the vibration damper (P7) or objects made of different materials such as the glass insulator (P4), could be identified by most sensors, presented linear detection results and were more significant than 2 m away. The LASER VL53L0X sensor, on the other hand, presented detection results of less than 1 m of distance between most objects used as a target.

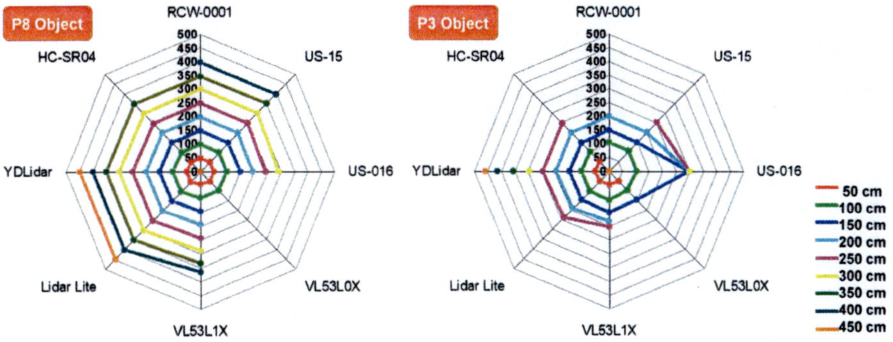

Fig. 3. Preliminary results obtained during tests in a controlled environment.

The analysis of the detection results by the ultrasonic sensors on the electrical cables (P1, P2, and P3) indicates that the HC-SR04 sonar presented detection responses superior to two meters of distance over the set of cables of different thicknesses. Among all the ultrasonic sensors used in this experiment, the HC-SR04 was the sonar that presented the best distance and accuracy in detecting targets, as shown in Fig. 4.

The results generated by sensor detection failures by objects P1 to P8 helped identify regions susceptible to distance reading errors. In this way, it was possible to extract information that preliminarily indicates which types of sensors present detection uncertainties on specific components of the infrastructure of an electrical tower. Figure 5 illustrates the numerical amount of detection errors (Out of range) among all sensors used in this experiment in an object of easy detection (P8) and difficult to detect due to its non-uniform geometry (P4).

The analysis of the results indicated that the VL53L0X, RCW0001, and US015 sensors are incompatible for detection tasks with precision on the selected components, as they presented the highest amounts of detection errors in the experiment. Therefore, they are not sufficiently safe in tasks that require a robust sensory system in obstacle avoidance systems.

Results generated from the comparison between the individual detection and the pairs of ultrasonic sensors on the P8 object showed that in the face of the

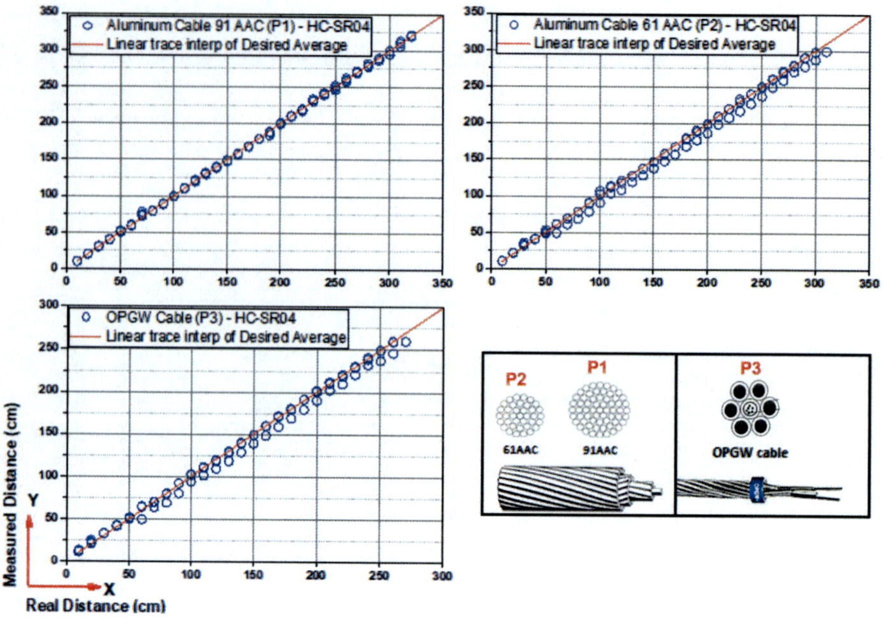

Fig. 4. Detection results of the HC-SR04 sensor on the selected electrical cables.

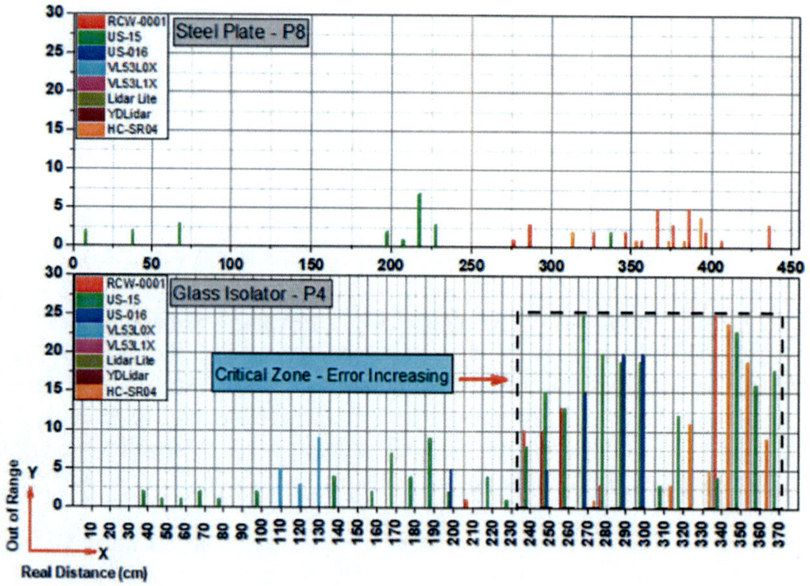

Fig. 5. Error detection ratio (Out of Range) on different targets.

positioning between 0° and 15°, problems related to the crosstalk effect introduced errors in the P8 object detection. The crosstalk effect intensifies when positioned at 0°, where the total overlap of the lobes of the ultrasonic sensors occurs. Figure 6 shows the individual and paired detection responses of the HC-SR04 and US016 sensors in front of the P8 object. When used individually, the sensors presented detection responses similar to the actual distance, containing only some variance errors, represented by the vertical lines under the response curves of the HC-SR04 sensor. When used in pairs at 0°, the ultrasonic sensors obtained detection inaccuracies on the target, informing measurements of unrealistic distances.

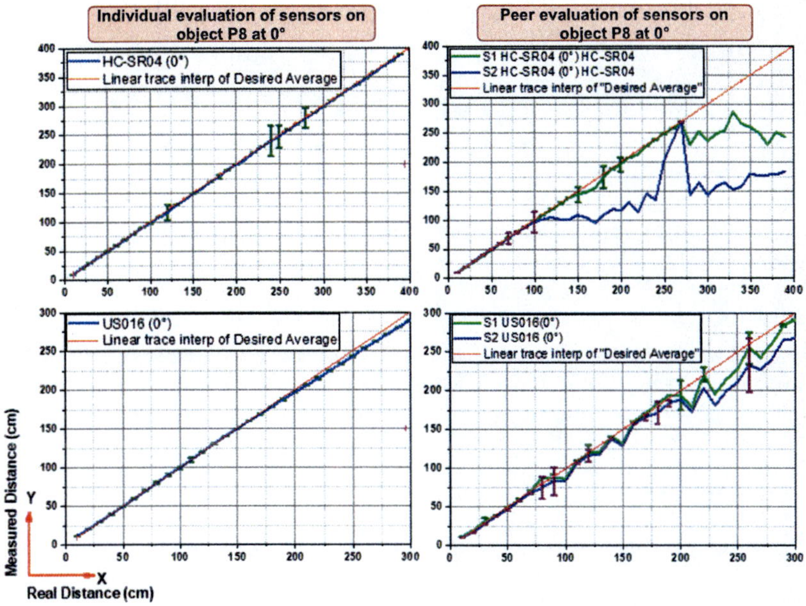

Fig. 6. Comparison of individual and peer detection responses by ultrasonic sensors.

In this experiment, it was identified that with the positioning of 25°, detection results were produced between US sensors with greater proximity to the desired average distance. Overall, the HC-SR04 sensor presented the best results, identifying obstacles at a distance range of more than two meters. When observing the HC-SR04 sensor detection behaviour in front of the P8 object, it is seen that the positioning of 35° also presents results with high precision. However, when comparing the detection performance between smaller objects such as the glass insulator (P4), using the HC-SR04 sensors at an angle of 35° does not produce reliable detection results, as shown in Fig. 7.

Since the ultrasonic sensors are not centred in front of the target obstacle, generated by the positioning of 35°, small objects cannot be adequately detected,

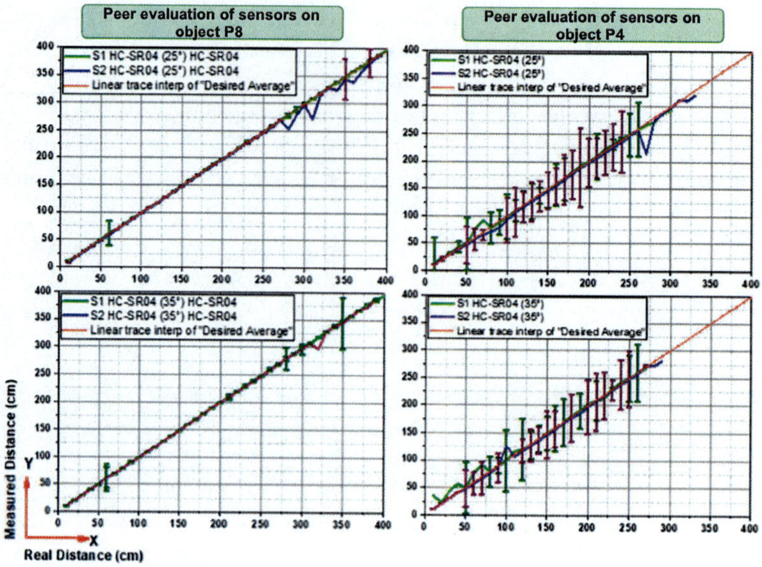

Fig. 7. Identification of problems related to angular uncertainties by objects of different geometries.

causing problems related to angular uncertainties. In this way, it was identified that the positioning of 25° between multiple ultrasonic sensors generates conditions that make it possible to identify different types of obstacles accurately.

4.2 Sensory Results in External Environment

In front of the target, represented by a wall in an open environment and with high sun exposure, the YDLiDAR X4, LiDARLite V3, HC-SR04, US-16, and VL53L1X sensors detection performance were tested. The graphs presented in Fig. 8 show the results obtained by detecting long-range sensors: the YDLiDAR X4 and LiDARLite V3 (top image), and short-range, represented by the HC-SR04, US-016, and VL53L1X sensors (image bottom).

The results presented by the upper graph show that the LiDARLite V3 sensor can detect even in sunlight, and the YDLiDAR X4 sensor showed a null detection result. Only in an environment with shaded regions does the YDLiDAR sensor show detection linearity. As for the shorter range sensors, in the lower graph, it can be seen that only the HC-SR04 presented results within the expected detection range, that is, greater than two meters away.

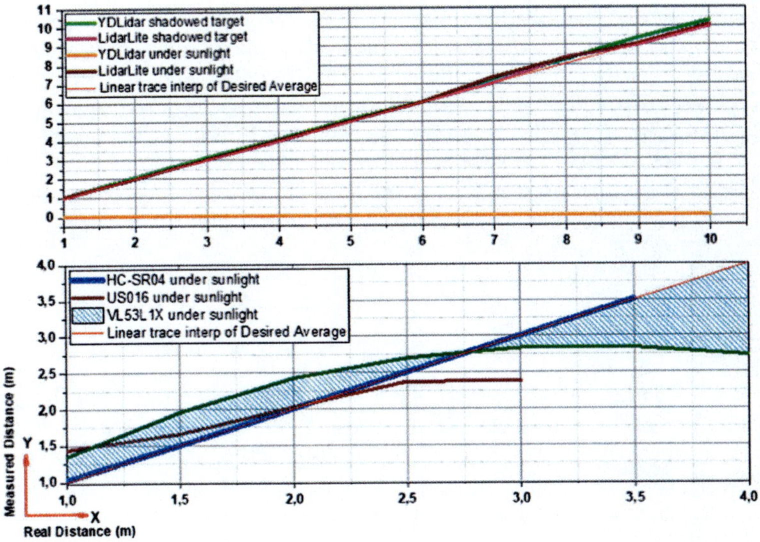

Fig. 8. Results of sensory detection obtained in the external environment.

5 Conclusion and Future Work

With the results obtained in this work, it was possible to identify low cost obstacle detection sensors capable of detecting different components commonly used in electrical towers. Objects P1, P2 and P3 presented greater detection difficulties in relation to the other objects used in the experiments. The results indicate that the HC-SR04 and LiDAR Lite V3 sensors presented the best detection performance conditions for the target application of this research work. The adopted methodology allowed identifying the best displacement between multiple ultrasonic sensors by positioning them between 25° apart to avoid problems related to the crosstalk effect and angular uncertainty. The results presented by the other sensors did not present the minimum characteristics necessary to meet the nature of the operation, that is, to correctly detect the set of components commonly used in high-voltage towers at a minimum range of two meters.

As future work, we intend to develop models of sensor architectures for multirotor UAVs. The geometric representations of HC-SR04 and LiDARLite V3 sensors' actuation beam in robotic simulators assists in analyzing, distributing, and displacing different types of sensors on multirotor UAVs. In this method, the geometric representations are created based on the patterns of actuation beams of different types of sensory technology, seeking to reliably reproduce the performance of the sensors in a scenario of inspection of electrical towers. Given these perspectives, the objective is to advance in developing autonomous navigation systems by multirotor UAVs within the context of electrical inspections.

Acknowledgment. The authors are grateful to the Foundation for Science and Technology (FCT, Portugal) for financial support through national funds FCT/MCTES (PIDDAC) to CeDRI (UIDB/05757/2020 and UIDP/05757/2020), SusTEC (LA/P/0007/2021), Oleachain "Skills for sustainability and innovation in the value chain of traditional olive groves in the Northern Interior of Portugal" (Norte06-3559-FSE-000188) and Centro Federal de Educação Tecnológica Celso Suckow da Fonseca (CEFET/RJ). The project that gave rise to these results received the support of a fellowship from "la Caixa" Foundation (ID 100010434). The fellowship code is LCF/BQ/DI20/11780028.

References

1. Tsuji, S., Kohama, T.: Omnidirectional proximity sensor system for drones using optical time-of-flight sensors. IEE J. Trans. Electr. Electron. Eng. **17**(1), 19–25 (2022)
2. Máthé, K., Buşoniu, L.: Vision and control for uavs: A survey of general methods and of inexpensive platforms for infrastructure inspection. Sensors **15**(7), 14887–14916 (2015)
3. Aswini, N., Krishna Kumar, E., Uma, S.V.: Uav and obstacle sensing techniques - a perspective (2018)
4. Berger, G.S., Wehrmeister, M.A., Ferraz, M.F., Cantieri, A.R.: Analysis of low cost sensors applied to the detection of obstacles in high voltage towers. In: Proceedings of International Embedded Systems Symposium (IESS 2019) (2019)
5. Yasin et al.: Unmanned aerial vehicles (uavs): Collision avoidance systems and approaches. IEEE Access **8** (2020)
6. Jordan, S., et al.: State-of-the-art technologies for uav inspections. IET Radar Sonar Navig. **12**(2), 151–164 (2018)
7. Suherman, S., et al.: Ultrasonic sensor assessment for obstacle avoidance in quadcopter-based drone system. In: International Conference on Mechanical, Electronics, Computer, and Industrial Technology, (2020)
8. Papa, U., Ponte, S.: Preliminary design of an unmanned aircraft system for aircraft general visual inspection. Electronics **7**, 435 (2018)
9. Shoval, S., Borenstein, J.: Using coded signals to benefit from ultrasonic sensor crosstalk in mobile robot obstacle avoidance. In: IEEE International Conference on Robotics and Automation, vol. 3, pp. 2879–2884 (2001)
10. Chandrashekar, L.: Characteristics of sonar sensors for short range measurement. Int. J. Eng. Res. Technol. (IJERT) (2014)
11. Devos, A., Ebeid, E., Manoonpong, P.: Development of autonomous drones for adaptive obstacle avoidance in real world environments. In: 2018 21st Euromicro Conference on Digital System Design (DSD) (2018)
12. Viña, C., Morin, P.: Feature-based pose estimation on-board mavs equipped with 2d laser scanners for the automatic inspection of electric towers. In: International Workshop on Robot Motion and Control (2017)
13. Kodai, et al.: Development of drone system embedded with multiple distance sensors for inspection task of social infrastructure (2018)
14. Nieuwenhuisen, M., Schadler, M., Behnke, S.: Predictive potential field-based collision avoidance for multicopters. In: International Archives of the Photogrammetry, Remote Sensing and Spatial Information Sciences, vol. 08 (2013)

Proposal of a Visual Positioning Architecture for Master-Slave Autonomous UAV Applications

Lucas Coradin Rech[1(✉)], Luciano Bonzatto Junior[1], Guido S. Berger[3,4], José Lima[3,4], Alvaro Rogério Cantieri[2], and Marco Aurélio Wehrmeister[1,5]

[1] Computer Systems Engineering Laboratory (LESC), Federal University of Technology-Campus Curitiba, Curitiba, Paraná 80230-901, Brazil
{lucasrech,lucjun,}@alunos.utfpr.edu.br, wehrmeister@utfpr.edu.br

[2] Applied Robotics and Computation Laboratory - LaRCA- Federal Institute of Paraná, Paraná, Pinhais, Brazil
alvaro.cantieri@ifpr.edu.br

[3] Research Centre in Digitalization and Intelligent Robotics (CeDRI), Instituto Politécnico de Bragança, Campus de Santa Apolónia, 5300-253 Bragança, Portugal
{guido.berger,jllima}@ipb.pt

[4] Laboratório para a Sustentabilidade e Tecnologia em Regiões de Montanha (SusTEC), Instituto Politécnico de Bragança, Campus de Santa Apolónia, 5300-253 Bragança, Portugal

[5] University of Münster, Münster, Germany
wehrmeister@uni-muenster.de

Abstract. Autonomous UAVs offer advantages in industrial, agriculture, environment inspection, and logistics applications. Sometimes the use of cooperative UAVs is important to solve specific demands or achieve productivity gain in these applications. An important technical challenge is the precise positioning between two or more UAVs in a cooperative task flight. Some techniques provide solutions, like the GNSS positioning, visual and LIDAR slam, and computer vision intelligent algorithms, but all these techniques present limitations that must be solved to work properly in specific environments. The proposal of new cooperative position methods is important to face these challenges. The present work proposes an evaluation of a visual relative positioning architecture between two small UAV multi-rotor aircraft working in a master-slave operation, based on an Augmented Reality tag tool. The simulation results obtained absolute error measurements lower than 0.2 cm mean and 0.01 standard deviation for X, Y and Z directions. Yaw measurements presented an absolute error lower than 0.5 °C with a 0.02–5 °C standard deviation. The real-world experiments executing autonomous flight with the slave UAV commanded by the master UAV achieved success in 8 of 10 experiment rounds, proving that the proposed architecture is a good approach to building cooperative master-slave UAV applications.

Keywords: Autonomous cooperative UAV · Visual UAV positioning · AR Tag UAV positioning · Master-slave UAV architecture

D. Tardioli et al. (Eds.): ROBOT 2022, LNNS 590, pp. 365–375, 2023.
https://doi.org/10.1007/978-3-031-21062-4_30

1 Introduction

Small size autonomous Unmanned Aerial Vehicles (UAVs) had an expressive increase in number and types of practical applications nowadays due to the decrease of this aircraft costs and the offer of new small size hardware and sensors proper to embed in these vehicles. The autonomous control of small UAVs presents technical challenges, like secure flight and collision avoidance, path planning, precise 3D positioning and robust control.

Collaborative arrangements between multiple UAVs are an excellent alternative for some applications. The division of tasks between several aerial vehicles reduces the task duration. Consequently, it increases the efficiency of the process, allowing to solve technical issues such as the reduced flight time of multi-rotor vehicles, for example. The literature presents works in agriculture [1], inspection of energy power plants [2], 3D mapping of an external area [3], among other.

Precise positioning is vital to assure the security of the operation in autonomous multi-UAV tasks, to ensure the correct path following. Nowadays, the most common techniques to achieve precise positioning in UAV flights is Differential Global Navigation Satellite System (GNSS). This technique is an optimum approach to solve the positioning problem, but GNSS systems demand a proper satellite signal receiving to work properly, which is sometimes hard to achieve[4].

Visual positioning techniques provide position data to UAV flights and operations. The standard approaches use monocular, binocular and omni-directional RGB cameras and stereoscopic cameras [5]. The application computer vision techniques to calculate the UAV coordinates is demanding, especially in not-structured environments. This approach presents challenges, like the sensibility of environment light variations, visual occlusion of the target, image capture noise and distortion due to the camera vibration and twist, among others.

A possible approach to decrease the computational demanding is to create auxiliary image marks to provide specific visual data for the correct 3D position estimation. Augmented reality tag positioning, proposed originally for Augmented Reality applications, has been used for this kind of application. April Tags [6] and AR Track Alvar [7] are example of these tools. The present work proposes a relative positioning architecture between two UAVs working in a master-slave cooperative arrangement. An Augmented Reality Tag tool, among a regular RGB camera embedded in the master UAV, is used to provide the relative position between the aircraft. This tag provides the quick and effective identification of each vehicle and calculates its position in the experimental environment.

The experiments are conducted in an indoor laboratory using two small aircraft. This investigation aims to provide a first proof-of-concept of the proposed architecture, allowing the proposal of future practical cooperative applications based on this master-slave architecture. The Sect. 2 reviews the use of visual tags in cooperative autonomous vehicle applications. Section 3 describes the proposed architecture and components. Section 4 presents the experimental set and obtained results. Finally, the Sects. 5 and 6 sections discusses the work results and describe conclusions and future works.

2 Related Works

The computational vision process used to calculate the relative position between two or more cooperative autonomous vehicles is a complex task which commonly demands a lot of hardware capacity. The application of printed tags to provide high visual quality information to the vision algorithms is an excellent choice to decrease the computational demand. The literature presents some investigations that use this kind of tag to provide position data to control a UAV flight.

The paper [8] proposes a UAV vision-based target positioning solution. The work proposes a position control algorithm to follow a ground target represented by a tag. The work presents a 92% reliability in real-world experiments, showing the viability of using this kind of proposal for UAV outdoor positioning.

An indoor AprilTag markers UAV localization is proposed in [9]. A set of markers are fixed in the experimental area walls providing visual clues captured by the UAV frontal camera. An off-board algorithm extracts the relative position between the aircraft and each visible tag and sends it to the aircraft control algorithm. Besides the small size of the flight area, this confirms the reliability of using the tags to provide 3D position data to a multi-rotor UAV flight.

Another similar approach is presented in the work [10]. The proposition uses an AR.Drone 2.0 autonomous controlled by a base station that processes the images of an AprilTag captured by the UAV camera. The experiments present stable aircraft control using the position data processed from the tags.

The work [11] presents an AR-Drone autonomous search algorithm. In this work, the UAV captures the images of a printed tag held in front of it by a human. These images are off-board processed to provide the relative position between the UAV and the tag, allowing the aircraft to following it using a PID flight control.

Using visual tags to provide relative position between autonomous cooperative vehicles is a possible application of this technique. Works present in the literature propose using these tools to allow autonomous landing and cooperative navigation between UGVs and UAVs in most cases.

The work [12] proposes an outdoor UAV navigation system where the aircraft follows a moving automobile and executes an autonomous landing. A group of visual tags are placed, allowing the correct positioning of the landing even when only using the visual position information provided by the tag.

A navigation and route generation system for land vehicles based on the vision obtained by a drone in cooperative work is presented in [13]. The UAV serves as a leader in the displacement. It sends position information related to the ground robot through a TAG fixed to the top of this vehicle. The tag images are captured by a down pointing camera installed in the UAV. This arrangement allows the ground vehicle to follow an obstacle-free path previously determined by the ground vehicle through the vision system.

This short review provides a basis for the present work proposition: the use of a computer vision algorithm based on AR-Tag to allow cooperative navigation between two UAVs. To the best of the authors' knowledge, no similar proposal is present in the literature. The following sections describe the architecture proposal, and experimental results are presented.

3 System Architecture and Components

The positioning architecture is based on visual feedback by capturing images of the slave UAV obtained by the master UAV camera. The experiments use two small drones, a DJI Tello model as a slave and a Bebop Drone model as a master. Drones connect to a base station composed of an Intel Core i7 processor with 16 GB RAM and an Intel® HD Graphics 520 (Skylake GT2) board computer running Ubuntu 16.4 LTS and ROS Kinect. The base station runs the ROS packages to provide communication and control interfaces for both aircraft.

All the architecture components use the Robot Operating System (ROS) to exchange information. [14]. The Bebop drone is connected through the Bebop Autonomy package [15], which allows the base station to read the UAV camera and publish *cmd_vel* commands to control the UAV flight. Bebop drone is an excellent choice to implement this architecture because it is a small size drone with stabilized flight performance and an embedded Full-HD resolution camera. The camera gimbal stabilization minimizes the image shifts during the data acquisition. DJI Tello drone offers embedded flight stabilization control. The small size of this aircraft allows the flight in reduced spaces. A *tello_driver* ROS package implements the Tello drone communication with the ground station. The Ar-Track Alvar Augmented Reality SDK [7] is used to implement the visual position reference system. The Ar Track Alvar package developed by Scott Niekum [15] provides ROS compatibility. The Images of the AR-TAG are captured by the Bebop drone camera and processed by this package that publishes position and orientation data on *ar_pose_marker* a Ros Node.

Both aircraft work in a hot-spot Wi-Fi connection architecture. The connection between the base station and the Bebop Drone works at a frequency of 5.8 GHz using a USB Wi-Fi adapter, while the connection with the Tello Drone works at 2.4 GHz using the PC onboard Wi-Fi. The base station concentrates all the ROS messages and sends commands to the agents to execute their activities.

The Bebop camera is set to point to the ground all the time to keep the perfect visual capture of the tag. Bebop can point it's camera inside a limit of a 75-degree angle so that it will provide a capture in its frontal area. The slave UAV must keep its flight position inside this area during all operation to allow the perfect control. Figure 1 a) shows the architecture coordinate reference system. A C++ code implements a PID control of the Tello drone flight's vertical, horizontal and yaw positions, publishing in two ROS nodes called */position* and */orientation*. The experiments using the Bebop drone as master are performed using manual piloting implemented by C++ code running in the base station. The code captures the commands from the computer keyboard and sends *cmd_vel* messages to the aircraft through the Bebop Autonomy ROS driver. Manual control is chosen in this case to assure the security flight of the aircraft in the reduced space present in the laboratory. Figure 1 b) shows the logical diagram of the proposed architecture and its components.

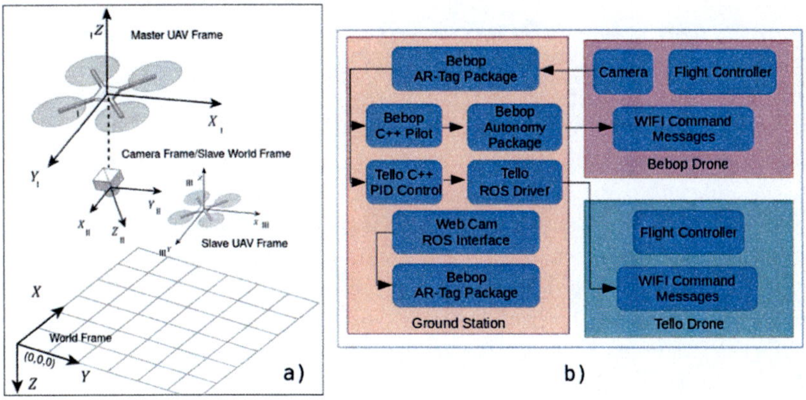

Fig. 1. a) Coordinate system representation; b) Architecture block diagram

4 Experiments Description and Results

The experimental planning aims to provide the first proof-of-concept for the master-slave following the leader's proposed architecture in a laboratory environment. This step is essential to evaluate the reliability of the technique and allow complex cooperative applications using the UAVs in a real-world environment. Three-step experimental planning is described in this section. First, a simulation of the master-slave operation using the visual control based on the AR-Tags is presented.

After that, two different real-world experimental arrangements are tested. First, a fixed position Web Cam captures the Tello Drone tag images during a flight and commands the aircraft using the tag position data and a C++ PID control code. After that, a cooperative master-slave flight where the Tello Drone autonomous follows the Bebop Drone is executed.

4.1 Simulation Experiments

The simulated environment uses the Copellia Sim [16] simulation software to evaluate the proposed UAV autonomous cooperative performance. Two multi-rotor models running a stabilization PID control LUA script are used in the experiments, allowing to the UAV's keep hovering at a static position and displace from one point to the other in the space without a crash. This stabilization code works as a flight controller hardware of a real-world UAV.

An image sensor is fixed at the bottom of the master UAV and captures the AR Tag images sent to the package, which processes the tag position calculation. Looking to avoid the camera shift during the UAV displacements, a virtual gimbal is programmed to control the camera orientation and keep it constantly pointing to the ground. The AR-Tag Alvar package publishes 6 DOF position and orientation information in */visualization_marker* ROS topic. A ROS master provides the node communication between the simulation software, the AR-Tag package and the external C++ position control code.

A programmed path point are sent to the slave UAV, that displace using the master UAV to provide the 3D position data. The absolute positioning error of the tag measurements and the UAV 3D position are evaluated. Ten experimental rounds are executed for three different paths, a square, a triangle and a cross. The absolute position error for each point is recorded for statistical calculations. Figures 2 a) and b) shows the measurements of the tracking error calculated from the AR-Tag data and the UAV world-frame position for a flight. Figure 3 shows the path followed by each UAV in space and the plot of the absolute error and AR-Tag error for a flight. Table 1 presents the statistic calculations of the absolute error for all the experiment rounds.

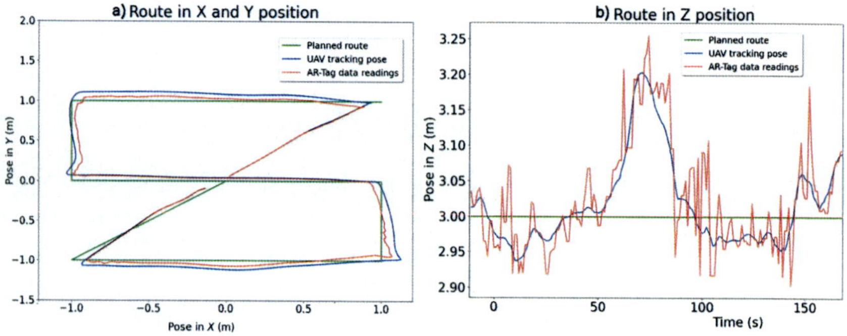

Fig. 2. a) Tracking of UAV position in X and Y axis of a flight master-slave simulation; b) Tracking of UAV height of a flight master-slave simulation

4.2 Real World Experiments

The second step is to evaluate the capacity of the real-world architecture to provide autonomous flight capability in the master-slave arrangement. The laboratory where the experiments are conducted is not equipped with a precise positioning reference system to provide ground truth for the position error measurements. An indirect error measurement is used to evaluate the experimental data based on the AR-Tag position measurements, using the (0,0,0) point of the camera frame as coordinate origins. The relative error between the AR-Tag

Table 1. Mean and Stardard Deviation for X;Y;Z ;Yaw tag measurements. These points refer to the camera frame reference system.

Set Point cm [X,Y,Z,Yaw]	X_Mean	Y_Mean	Z_Mean	Yaw Mean	X_Std	Y_Std	Z_Std	Yaw Std
0;0;230;0	0.45	0.32	0,29	2.45	0.043	0.051	0.058	1.25
60;35;230;0	0.65	0.39	0.28	3.12	0.033	0.040	0.055	1.44
60;-35;230;0	0.60	0.41	0.28	3.10	0.051	0.049	0.067	0.99
35;60;230;0	0.68	0.38	0.33	2.88	0.048	0.054	0.052	2.11
-35;60;230;0	0.71	0.35	0.38	2.75	0.049	0.066	0.050	1.31

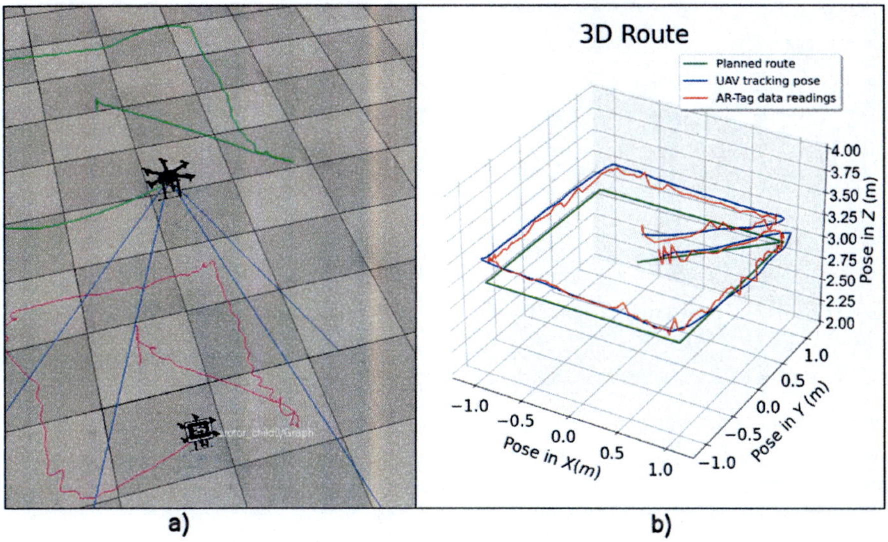

Fig. 3. 3D UAVs route and plot of tracking of UAV position in a follow-the leader flight. a) Image of the simulated environment. b) Plot of the track error.

(X,Y,Z) readings and the center of the camera frame is presented in the experiment graphs. A video of the real-world experiments is available at https://youtu.be/jDQYng34abY.

Autonomous Slave UAV Flight with an Fix Camera Reference. The first experiment evaluates if the tag-based positioning system can provide reliable data to the UAV displacement control algorithm in a controlled environment. A Webcam HD Logitech c920–1080p resolution fixed in the laboratory ceiling provides the AR-Tag capture and sends the data to the base station computer. The camera is set at an approximately 3.2 m height and provides a visual area of 2,5 × 2,0 m. These readings provide the UAV (X,Y,X,Yaw) data.

The Tello UAV with a 6.0 × 6.0 size tag fixed on its roof is commanded by the PID control code running in the base station, using the tello_driver package to command the aircraft to follow a square route in the camera vision area. The flight is executed at a 1.0 m height in 10 rounds, and the AR-Tag position data is collected and stored to allow the position graph plot. Figure 4 shows the plot of planned route *versus* the AR-Tag position collected form three flight rounds.

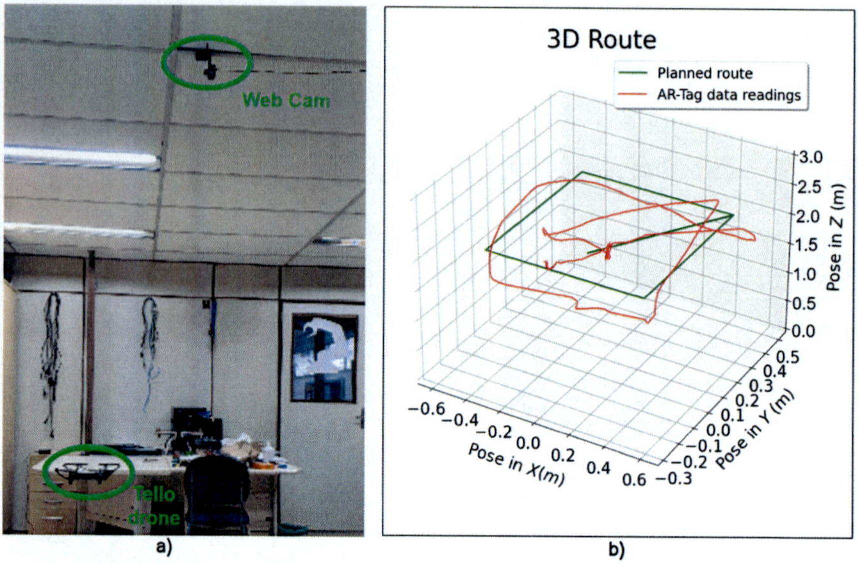

Fig. 4. The Tello UAV autonomous flight plot with the fixed WebCam generating position data. a) Image of the experiment arrangement. b) Plot of the track error.

This experiment indicates that the control algorithm based in the AR-Tag position measurements works properly, so the next step is to evaluate the master/slave UAV arrangement.

Autonomous Master-Slave UAV Follow-the-Leader Flight. The follow-the-leader experiment is based on the Bebop Drone working as the master and capturing images from the tag fixed in the Tello Drone, the slave. The control algorithm is programmed to lead the slave UAV to reach the (0.00, 0.15, –1.00) meter from the center of the image frame captured by the Bebop Drone. Bebop drone is manually commanded to displace in the flight area, and the Tello drone follows it continuously. Experiments is executed in an approximately 2.0-minute flight, at a maxim horizontal velocity of approximately 0.1 m/s. Figure 5 presents a plot of the UAV (X,Y,X) measurements for a flight round.

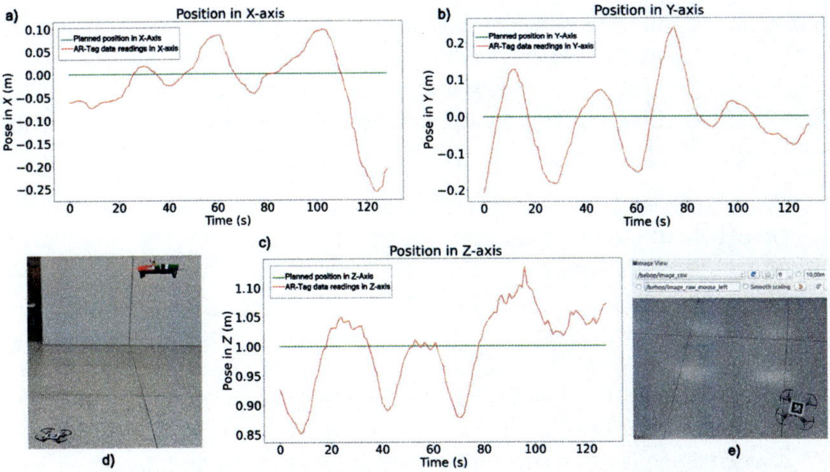

Fig. 5. Plots of the slave UAV position in a following autonomous operation and images of the experiment arrangement. a), b) and c) Plot of the actual *versus* planned pose readings. d) and e) Images of the Bebop drone and Tello drone cooperative flight.

5 Discussion

The experiments indicate that the proposed architecture is a reliable way to create a visual master-slave collaborative UAV architecture. The results presented small absolute error measurements in simulation. In real-world experiments, environment and hardware conditions impact the control stability, mainly in the master-slave flights, due to the Bebop drone image capture limitations.

The camera quality captures are a limitation for this schema. In the first case, the Web Cam offers constraints like slow brightness correction time, low data transfer to the PC due to the 2.0 USB connection and no auto-focus mechanism. These limitations impact the frame captures and the AR-Tag processing, so it is possible to evaluate the use of professional cameras to increase the data position quality and the robustness of the positioning system in indoor environments. Besides the Bebop drone camera acquiring Full HD images, the Bebop Autonomy driver only transmits 640×480 resolution frames to the base station, which decreases the AR-Tag position error calculation. Besides, the executed follow flights presented a proper operation when the tag capture provides good position data to the PID control algorithm.

The camera shift caused by the master UAV frame inclination during the route changes drives the tag position calculation error. Still, the Bebop gimbal was able to compensate for this, preventing the aircraft loss of control. Also, the small tag size demands keeping the slave UAV in a close flight with the master one. Because of that, the camera capture area is reduced, so the master UAV sometimes loses visibility of the slave, and the position control is lost. This difficulty could be minimized if a slave UAV capable of supporting a bigger tag size is used.

Some image frames sent by Bebop drone to the base station fail, which is possible to see in the experiment video. These fail to affect the position data generation of the AR-Tag package and the good work of the PID algorithm. Besides that, as shown in the video, this problem was handled by the control algorithm.

6 Conclusions and Future Works

The experiments confirm the reliability of the master-slave position control proposed architecture for real-world applications under controlled laboratory conditions. The AR-Tag tool's measurements provide adequate quality data for the UAV autonomous flight algorithm in the proposed requirements.

The use of a better resolution and image quality camera, increased capability tag capture arrangements, and multi-sensor fusion data is an interesting way to achieve better results for the proposed architecture. Practical problems must be solved to allow the application of this method in real robot operations, mainly in unstructured outdoor environments. Still, it is possible to consider that a first proof-of-concept of the proposed collaborative architecture is presented in this work.

Future works are planned to investigate new visual positioning data systems and arrangements that could provide better data to the applications, like active light tags, for example. Also, real-world practical problems that could be solved using the cooperative master-slave proposed architecture will be investigated.

Acknowledgment. The authors are grateful to the FCT - Foundation for Science and Technology, Portugal for financial support through national funds FCT/MCTES (PIDDAC) to CeDRI - Research Centre in Digitalization and Intelligent Robotics (UIDB/05757/2020 and UIDP/05757/2020) and SusTEC (LA/P/0007/2021).

References

1. Lee, H.-S., Shin, B.-S., Thomasson, J.A., Wang, T., Zhang, Z., Han, X.: Development of multiple uav collaborative driving systems for improving field phenotyping. Sensors **22**(4), 1423 (2022). https://doi.org/10.3390/s22041423, https://www.mdpi.com/1424-8220/22/4/1423, ISSN 1424–8220
2. Lu, J., Fu, J., Zhang, J., Zhang, K.: Multi-machines and Multi-tasks Scheduling for UAV Power Inspection in Smart Grid. In: Liu, Q., Liu, X., Chen, B., Zhang, Y., Peng, J. (eds.) Proceedings of the 11th International Conference on Computer Engineering and Networks. LNEE, vol. 808, pp. 1579–1587. Springer, Singapore (2022). https://doi.org/10.1007/978-981-16-6554-7_174
3. Kanellakis, C., Mansouri, S.S., Fresk, E., Kominiak, D., Nikolakopoulos, G.: Cooperative uavs as a tool for aerial inspection of large scale aging infrastructure. In: 2018 IEEE/RSJ International Conference on Intelligent Robots and Systems (IROS), pp. 5040–5040 (2018). https://doi.org/10.1109/IROS.2018.8593996

4. Zimmermann, F., Eling, C., Klingbeil, L., Kuhlmann, H.: Precise positioning of uavs - dealing with challenging rtk-gps measurement conditions during automated uav flights. ISPRS Annal. Photog. Remote Sens. Spatial Inf. Sci. **4**(2W3), 95–102 (2017). https://doi.org/10.5194/isprs-annals-IV-2-W3-95-2017, ISSN 21949050
5. Al-Kaff, A., Martín, D., García, F., de la Escalera, A., Armingol, J.M.: Survey of computer vision algorithms and applications for unmanned aerial vehicles. Expert Syst. Appli. **92**, 447–463 (2018). https://doi.org/10.1016/j.eswa.2017.09.033, ISSN 09574174
6. APRIL Robotics Laboratory. AprilTags Visual Fiducial System (2018). https://april.eecs.umich.edu/software/apriltag
7. VTT Technical Research Centre of Finland. Augmented Reality/3D Tracking (2017). http://virtual.vtt.fi/virtual/proj2/multimedia/
8. Hinas, A., Roberts, J.M., Gonzalez, F.: Vision-based target finding and inspection of a ground target using a multirotor UAV system. Sensors (Switzerland) **17**(12), 2929 (2017). https://doi.org/10.3390/s17122929, ISSN 14248220
9. Kayhani, N., Heins, A., Zhao, W., Nahangi, M., McCabe, B., Schoellig, A.: Improved tag-based indoor localization of uavs using extended kalman filter. In Al-Hussein, M. (ed.) Proceedings of the 36th International Symposium on Automation and Robotics in Construction (ISARC), Banff, Canada, pp. 624–631. International Association for Automation and Robotics in Construction (IAARC), May 2019. https://doi.org/10.22260/ISARC2019/0083, ISBN 978-952-69524-0-6
10. Wang, S., Hu, T.: Ros-gazebo supported platform for tag-in-loop indoor localization of quadrocopter. In: Chen, W., Hosoda, K., Menegatti, E., Shimizu, M., Wang, H. (eds.) IAS 2016. AISC, vol. 531, pp. 185–197. Springer, Cham (2017). https://doi.org/10.1007/978-3-319-48036-7_14
11. Boudjit, K., Larbes, C.: Detection and implementation autonomous target tracking with a quadrotor ar.drone. In: 2015 12th International Conference on Informatics in Control, Automation and Robotics (ICINCO), vol. 02, pp. 223–230 (2015)
12. Muskardin, T.: A novel landing system to increase payload capacity and operational availability of high altitude long endurance uav. In: 2016 International Conference on Unmanned Aircraft Systems (ICUAS), pp. 495–504 (2016). https://doi.org/10.1109/ICUAS.2016.7502668
13. Hood, S., Benson, K., Hamod, P., Madison, D., O'Kane, J.M., Rekleitis, I.: Bird's eye view: Cooperative exploration by UGV and UAV. In: 2017 International Conference on Unmanned Aircraft Systems, ICUAS 2017, pp. 247–255 (2017). https://doi.org/10.1109/ICUAS.2017.7991513
14. ROS.org. ROS (2019). http://www.ros.org
15. Monajjemi, M.: Bebop Autonomy - ROS Driver for Parrot Bebop Drone (2018). https://bebop-autonomy.readthedocs.io/en/latest/
16. Copellia Robotics. CoppeliaSim (2019). https://www.coppeliarobotics.com/

Modeling and Under-actuated Control of Stabilization Before Take-off Phase for Flapping-wing Robots

Daniel Feliu-Talegon(✉), Saeed Rafee Nekoo, Alejandro Suarez, Jose Angel Acosta, and Anibal Ollero

The GRVC Robotics Lab., Departamento de Ingeniería de Sistemas y Automática Escuela Técnica Superior de Ingeniería, Universidad de Sevilla, Seville, Spain
danielfeliu@us.es

Abstract. This work studies a stabilization problem of flapping-wing flying robots (FWFRs) before a take-off phase while a robot is on a branch. The claw of the FWFR grasps the branch with enough friction to hold the system steady in a stationary condition. Before the take-off, the claw opens itself and the friction between the claw and branch vanishes. At that moment, the mechanical model turns into an under-actuated multi-link (serial configuration) robotic system where the first joint can rotate freely without any friction as opposed to rotation. The stabilization and balancing are the crucial tasks before take-off. This work explores a new methodology to control an under-actuated lightweight manipulator for its future adaptation to FWFR to improve the stabilization performance before take-off. The setup tries to mimic the birds with two-link legs, a body link, and 2-DoF (degrees of freedom) arms, being all active links except the first passive one. In contrast to common arms, the lightweight-design restriction limits the frame size and requires micromotors. With all of these constraints, control design is a challenge, hence, the system is categorized: a) the leg subsystem (under-actuated), including the two first links, and b) the body and arm subsystem (fully actuated) with the rest of links. The fully-actuated links are controlled by feedback linearization and the under-actuated part with active disturbance rejection control (ADRC) for estimation and rejection of the coupling between both subsystems. The mechanical design, modeling, and control of the proposed system are reported in this work. Experimental results have been also proposed to present a proof of concept for this modeling and control approach.

Keywords: Flapping-wing robot · Under-actuated · Robust · Take-off · Aerial robot

D. Tardioli et al. (Eds.): ROBOT 2022, LNNS 590, pp. 376–388, 2023.
https://doi.org/10.1007/978-3-031-21062-4_31

1 Introduction

The operation of ornithopters, or flapping-wing flying robots (FWFRs), with manipulation capabilities, includes several phases such as launching, flapping flight, possible gliding, perching/landing, stabilization after perching, sampling/manipulation, and take-off again. The initial launching phase is usually done by hand for small to medium scale birds [1–3]; and in some cases by a special external launcher system [4], or self-launching mechanisms (take-off system) [5,6]. The robot birds for launching must be posed in a proper position to facilitate the take-off maneuver which defines the initial condition of the robot. Hoff and Kim used a launching mechanism to gain similar initial conditions for finding a data-driven model [4]. Afakh et al. developed a flapping robot that could fly from the ground without assisting mechanism [5]. The robot was placed in a high pitch angle pose to enable a successful take-off from the ground. The problem of the take-off from a branch after manipulation or completing a task requires the same condition. The FWFR must be in a proper pose with a high pitch angle. Here in this work, the stabilization of the robot bird before the take-off is investigated to put the robot in a desired position/orientation. Considering the take-off after perching on a branch using a mechanical claw with enough friction, it is assumed that the bird does not rotate or slide on the branch when the claw is closed, and it can hold its body steadily with a fixed pose. Reopening the claw removes the friction, and the FWFR will act on the branch like an inverted pendulum (an under-actuated model) for a split of a second before take-off. Here in this work, an underactuated mechanical system is modeled to analyze the problem and control the trajectory of the body before flying again. This is important since the perching might not pose the body in the desired configuration for take-off. Then the leg moves the body based on the designed trajectory.

Inverted pendulums are challenging models because of under-actuation conditions. Under-actuation imposes two difficulties to the system: controllability problem and reduced range of convergence in comparison with fully actuated mechanisms [7,8]. Previous works [9,10] study the nonlinear control problem of underactuated systems considering static friction at its first passive joint to explore the possibility to perform manipulation with FWFR once the system is perched. This work focuses on a new methodology to control an under-actuated lightweight manipulator that emulates the behavior of a bird on the branch just before the take-off phase. The main difference with the mentioned works is that we consider that the claw is opened and the friction at this first passive joint has vanished which means that the first passive joint can rotate freely without any friction as opposed to the rotation. The weight limits the configuration, and geometry and leads us to use micromotors and very lightweight materials. For example, in [11] a new sensor system is developed that emulates the beak's bird to allow FWFR to interact with the environment. Then, the mentioned work is devoted to two tasks: a) to maintain the equilibrium of the body while perching and b) to control the force interaction with the environment. Although, there are various methods and available results for under-actuated systems, the hard constraints imposed by nature to mimic the anatomy of birds, make necessary the exploration of the most appropriate control techniques.

The actual system behaves similarly to that of the acrobot [12]. The setup is a two-DoF under-actuated robot that replicates the human acrobat who hangs from a rope and tries to balance. The control of n-link under-actuated manipulators with a passive first link has been investigated in the last decades as in e.g. [13], and [14]. Unlike [9,10], there is no static friction at the base of the manipulator because just before the take-off, the claw opens itself and the friction between the claw and branch vanishes. However, manipulation with under-actuated systems has been rarely investigated due to its difficulty, and therefore it is still an open challenge. Our prototype is a multi-link planar under-actuated robot with a passive first link. A first approximation to control is the decoupling of the actuated and under-actuated systems. The first subsystem is under-actuated and composed of a passive first link and an active second one (similar to an acrobot). The second subsystem is composed of the remaining fully-actuated links. On the one hand, for the well-known fully-actuated subsystem state-feedback linearization control technique has been employed [15–17]. On the other hand, for the acrobot-like subsystem, due to the high uncertainty coming from the coupling term and the limited implementation, robust control has been used, the method is so-called active disturbance rejection control (ADRC).

The ADRC inherits the conventional linear controllers, enhanced by nonlinear feedback and robustness obtained by the addition of additional fictitious state variable [18,19]. The active disturbance rejection control is a powerful tool for under-actuated mechanical systems especially in experimentation [20,21]. Uncertainty in parameters, unknown friction values in the model, time-varying control, and unknown torque load motivated the application of the ADRC [21]. Ma et al. used a data-driven type of ADRC for a crane pendulum model using flat output [22]. Sparse regression was employed to find the interaction between the states and the flat output. Moreover, a recent theoretical breakthrough in [23] allows using the Lyapunov theory to analyze stability in cascade nonlinear systems and hence, is very suitable for the work presented here. More details are available in [24] for flatness and the generalized proportional integral (GPI) observers; [25] for academic and industrial engineering applications; and in particular works in [23,26] demonstrated that this method suits well for under-actuated systems.

The main contribution of this work is to present a model for the under-actuated state of a flapping wing robot, right before take-off from a branch, and control the trajectory of the system using feedback linearization and an active disturbance rejection approach. The rest of the paper is organized as follows. Section 2 presents the description and model of the system. Section 3 presents the strategy of the proposed control. Section 4 describes the actual prototype (IV-A), simulation results (IV-B), and experimental data (IV-C). Concluding remarks are presented in Sect. 5.

2 Description and Modeling of the System

As a first approach, we consider the study of a system composed of an actuated leg, a body, and an actuated arm in a two-dimensional space, considering that the real 3D system will be made with a pair of them. Figure 1 left depicts a scheme of the system case of study. The system is equipped with n-links ($i = 1, ..., n$) of length l_i, and masses concentrated at their tips m_i. The first link has a passive joint at its base whereas the other joints are active. The torques provided by the motors in the active joints are Γ_i. Figure 1a depicts a schematic of the system where θ_i are the angular positions of the joints. The control objective is to follow a trajectory with the tip in order to achieve a desired posture with the system while maintaining the equilibrium of the system. This is important to perform lately the take-off maneuver. As it was aforementioned, we split the system into two subsystems: the first subsystem $\theta_1 - \theta_2$ (similar to an acrobot), and the second subsystem defined through the remaining fully-actuated links. Let us now derive the dynamic model of the underactuated manipulator with a single passive joint at its base. To simplify the approach n-link fully-actuated mechanism is treated as the only one whose dynamics are all concentrated at the center of mass of those three links. Thus, let $q = [\theta_1, \theta_2, ..., \theta_n]^T \in \mathbb{R}^n$ denotes the the joint angles in generalized coordinates, and $U = [0, \Gamma_2, \Gamma_3, ..., \Gamma_n]^T \in \mathbb{R}^n$ the control input vector. Then, the Lagrange dynamics of the n-link underactuated robot manipulator can be expressed as

$$M(q)\ddot{q} + C(q, \dot{q})\dot{q} + G(q) = U, \tag{1}$$

where $M(q) \in \mathbb{R}^{n \times n}$ is a symmetric positive definite inertia matrix; $C(q, \dot{q}) \in \mathbb{R}^{n \times n}$ is the Coriolis and centrifugal forces matrix; and $G(q) \in \mathbb{R}^n$ is the gravity force vector. Following the subsystem approach described above, the Eq. (1) can be rewritten in two blocks in an obvious way as follows

$$\begin{bmatrix} M_{12} & M_c \\ M_c^T & M_{3n} \end{bmatrix} \begin{bmatrix} \ddot{q}_{12} \\ \ddot{q}_{3n} \end{bmatrix} + \begin{bmatrix} C_{12} & C_{c1} \\ C_{c2} & C_{3n} \end{bmatrix} \begin{bmatrix} \dot{q}_{12} \\ \dot{q}_{3n} \end{bmatrix} \begin{bmatrix} g_{12} \\ g_{3n} \end{bmatrix} = \tau,$$

where $q_{12} = [\theta_1, \theta_2]^T$, $q_{3n} = [\theta_3, ..., \theta_n]^T$, $\tau := [\tau_{12}, \tau_{3n}]^T$ with $\tau_{12} = [0, \Gamma_2]^T$, $\tau_{3n} = [\Gamma_3, ..., \Gamma_n]^T$.

3 Control Strategy

In this section, a cascade control scheme is derived from each subsystem separately. The proposed control scheme divides the system into two parts: 1) control of the under-actuated part of the system (first subsystem $\theta_1 - \theta_2$) using an active disturbance rejection control technique and a GPI Observer to estimate the coupling between the two subsystems and the unmodelled dynamics and 2) control the trajectory of the remaining fully actuated part using feedback linearization. Figure 1b shows a simplified scheme of the proposed methodology.

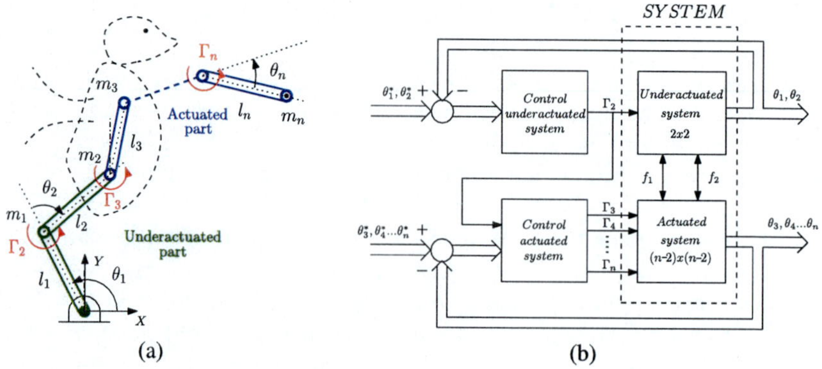

Fig. 1. a) Schematic of the system and b) general control scheme.

3.1 Underactuated Subsystem

The underactuated subsystem consists of a multivariable non-linear time-invariant system, which has two outputs θ_1, θ_2 and one input Γ_2. Recall that the proposed control technique is the ADRC. The idea is to consider the influence of the other subsystem as an external disturbance that can be estimated and approximately canceled in the feedback loop. Notice that this can be done because the other subsystem is fully actuated. The dynamics of this subsystem becomes

$$M_{12}\ddot{q}_{12} + C_{12}\dot{q}_{12} + g_{12} + F(q_k, \dot{q}_k, \ddot{q}_k) = 0, \quad k = 1, ..., 3, \tag{2}$$

where $F(\cdot)$ is the coupling term between both subsystems.

Remark 1. It is important to highlight that the disturbance $F(\cdot)$ depends on all the angular coordinates, but only on the first and second derivatives of the angular coordinates of the other subsystem, which is crucial for its implementation.

We follow the standard approach to derive the ADRC controller. To this end, we first linearize the system around an equilibrium point $\theta_1 = \theta_{10}$, $\theta_2 = \theta_{20}$, $\dot{\theta}_1 = 0$ and $\dot{\theta}_2 = 0$. The incremental variables are defined as $\theta_{1\delta} = \theta_1 - \theta_{10}$, $\theta_{2\delta} = \theta_2 - \theta_{20}$, $\dot{\theta}_{1\delta} = \dot{\theta}_1$ and $\dot{\theta}_{2\delta} = \dot{\theta}_2$.

$$a_{11}\ddot{\theta}_{1\delta} + a_{12}\ddot{\theta}_{2\delta} + c_1\theta_{1\delta} + c_2\theta_{2\delta} + f_1 = 0, \tag{3}$$
$$a_{21}\ddot{\theta}_{1\delta} + a_{22}\ddot{\theta}_{2\delta} + c_3\theta_{1\delta} + c_4\theta_{2\delta} + f_2 = \Gamma_{2\delta}. \tag{4}$$

The functions f_1 and f_2 are considered as disturbances prompted for the other subsystem that can be eliminated. Thus, for the linearized system it is straightforward to see that the output defined as $y := a_{11}\theta_{1\delta} + a_{12}\theta_{2\delta}$ is flat. From Eq. (3) and the flat output, it can be obtained

$$\begin{pmatrix} \theta_{1\delta} \\ \theta_{2\delta} \end{pmatrix} = \begin{pmatrix} a_{11} & a_{12} \\ -c_1 & -c_2 \end{pmatrix}^{-1} \begin{pmatrix} y \\ \ddot{y} \end{pmatrix}, \tag{5}$$

and therefore, the input-to-flat output relation yields

$$\Gamma_{2\delta} = \underbrace{\frac{a_{21} \cdot a_{12} - a_{22} \cdot a_{11}}{a_{11} \cdot c_2 - a_{12} \cdot c_1}}_{K_0} y^{(4)} + \underline{\xi}(t), \tag{6}$$

with K_0 a control gain and $\underline{\xi}(t)$ depending on the flat output, y, and its first and second derivatives. Thus, with the ADRC methodology (e.g. [26] and [27]) we obtain the following simplified input-output model of the system

$$y^{(4)} = \frac{1}{K_0}\Gamma_{2\delta} + \xi(t), \tag{7}$$

where $\xi(t)$ represents some neglected internal dynamics and disturbances.

GPI Observer. The GPI observer is a breakthrough of state observers that can simultaneously estimate the states and their time derivatives and the total disturbance of the system. The consecutive time derivatives of the flat output are computed using the following GPI observer

$$\begin{aligned}
\dot{\hat{y}}_0 &= \hat{y}_1 + l_4(y - \hat{y}_0), \\
\dot{\hat{y}}_1 &= \hat{y}_2 + l_3(y - \hat{y}_0), \\
\dot{\hat{y}}_2 &= \hat{y}_3 + l_2(y - \hat{y}_0), \\
\dot{\hat{y}}_3 &= \frac{\Gamma_{2\delta}}{K_0} + l_1(y - \hat{y}_0) + z, \\
\dot{z} &= l_0(y - \hat{y}_0),
\end{aligned}$$

allowing to estimate the perturbation terms $\xi(t)$ algebraically via the state of the observer $z(t)$. Additionally, from (7) and defining the observation error as $e_0 := y - \hat{y}_0$, the coefficients of the GPI observer, l_i, are chosen so that the reconstruction error dynamics becomes

$$e_0^{(5)} + l_4 e_0^{(4)} + l_3 e_0^{(3)} + l_2 \ddot{e}_0 + l_1 \dot{e}_0 + l_0 e_0 = \dot{\xi}(t), \tag{8}$$

and such that its associated polynomial $p_o(s) = s^5 + l_4 s^4 + l_3 s^3 + l_2 s^2 + l_1 s + l_0$ is Hurwitz, rendering an exponentially decreasing estimation error for any bounded $\dot{\xi}(t)$. To ease its tuning the polynomial can be defined in normal form as $p_o(s) = (s^2 + 2\zeta_o\omega_o s + \omega_o^2)^2(s + T_o)$.

Active Disturbance Rejection Controller. Given the flat output, its derivatives, and the estimation of the disturbance term $z(t)$, the GPI-ADRC scheme to track the output reference trajectory $y^*(t)$ is defined as

$$\frac{\Gamma_{2\delta}(t)}{K_0} = y^{*(4)}(t) - \gamma_3(\hat{y}^{(3)} - y^{*(3)}(t)) - \gamma_2(\ddot{\hat{y}} - \ddot{y}^*(t)) - \gamma_1(\dot{\hat{y}} - \dot{y}^*(t)) - \gamma_0(\hat{y} - y^*(t)) - z(t),$$

that upon defining the tracking error as $e_c := \hat{y} - y^*$, it forces a tracking error dynamics given by

$$e_c^{(4)} + \gamma_3 e_c^{(3)} + \gamma_2 \ddot{e}_c + \gamma_1 \dot{e}_c + \gamma_0 e_c = \xi(t) - z(t), \tag{9}$$

where the linear controller gains, γ_i, are set by using a fourth-order Hurwitz polynomial $p_c(s) = s^4 + \gamma_3 s^3 + \gamma_2 s^2 + \gamma_1 s + \gamma_0$, or in normal form $p_c(s) = (s^2 + 2\zeta_c \omega_c s + \omega_c^2)^2$, such that the closed-loop tracking error e_c converges exponentially to zero provided that $\xi(t) - z(t)$ tends to zero. The analysis of the stability and tracking of ADRC with extended observers for nonlinear time-varying uncertain plants has been studied in [28] and some recent developments in [23,29].

3.2 Actuated Part of the System

The control of the fully-actuated subsystem consists of a multivariable nonlinear time-invariant system, which has one output θ_3 and one input Γ_3. Feedback linearization is a very well-known nonlinear control technique. The core of the technique is to cancel out all the nonlinearities and impose the desired linear dynamics. In essence, it transforms the original system into an equivalent linear form by change of coordinates and feedback. More information about this technique can be found in [15–17]. The application of this technique to the fully-actuated subsystem is straightforward just defining the torques for the inner loop of the articulated joints to cancel out all the nonlinearities as follows

$$\begin{aligned}\tau_{3n} &:= [M_{3n} - M_c^T M_{12}^{-1} M_c]\ddot{q}_{3n} + [C_{c2} - M_c^T M_{12}^{-1} C_{12}]\dot{q}_{12} \\ &+ [C_{3n} - M_c^T M_{12}^{-1} C_{c1}]\dot{q}_{3n} - M_c^T M_{12}^{-1} g_{12} + g_3 + M_c^T M_{12}^{-1} \tau_{12}.\end{aligned} \tag{10}$$

The subsystem (10) can be expressed as

$$\tau_{3n} = [M_{3n} - M_c^T M_{12}^{-1} M_c] \cdot v_3 + f(\underline{X}, \tau_{12}), \tag{11}$$

where $\underline{X} = [\theta_1, \theta_2, ..., \theta_n, \dot{\theta}_1, \dot{\theta}_2, ..., \dot{\theta}_n]$ and the fictitious control signals $v_3 = \ddot{\theta}_3$ acting as an outer loop. Upon considering that the whole state vector is measurable, then $\underline{X}$ and τ_{12} are known from the underactuated subsystem. Thus, the trajectory tracking controller that renders $q_{3n}(t) \to q_{3n}^*(t)$, is easily completed with an outer-loop linear controller, e.g. with double poles located at P as

$$v_k := \ddot{q}_k^* - 2p_k(\dot{q}_k - \dot{q}_k^*) - p_k^2(q_k - q_k^*), \quad k = 3, 4, ..., n. \tag{12}$$

4 Simulation and Experimental Verification

Although the control strategy proposed in the paper is evaluated in a prototype of a 3-DOF robotic manipulator, the control strategy is developed in a general form and it can be used for an n-link underactuated manipulator.

4.1 The Manipulator Prototype

In order to validate experimentally the proposed control strategy, a 3-DOF manipulator implementing the ankle, knee and hip joints of the robotic system has been developed. Figure 2 shows a picture of the prototype and the testbed, indicating the main components along with the link lengths. The prototype, built with customized actuators and an aluminum frame structure, has been designed to be as lightweight and slim as possible, inspired by the biomechanical features of birds. The passive joint at the base emulating the claw is built with a pair of igus ESTM-06-SL polymer bearings crossed by a 6 mm section shaft used as a support point (the branch of the bird), whose friction is practically negligible. The rotation angle θ_1 is measured 500 Hz with an AMS 5047 magnetic encoder connected to the STM32 Nucleo microcontroller board through SPI (Serial Peripheral Interface). Two different actuators are employed in the knee and hip, according to the performance requirements for each joint. On the one hand, a zero-backlash actuator, weighting 70 g and built with a Maxon EC20 flat brushless motor and a Harmonic Drive CSF5-100 gearbox, is used in the knee joint, providing a rated torque of 0.6 Nm and torque/current control 200 Hz through the external ESCON 36/3 controller. The joint angle θ_2 is also measured with an AMS 5047 magnetic encoder through the same microcontroller board, that generates the PWM signal corresponding to the torque reference of the motor.

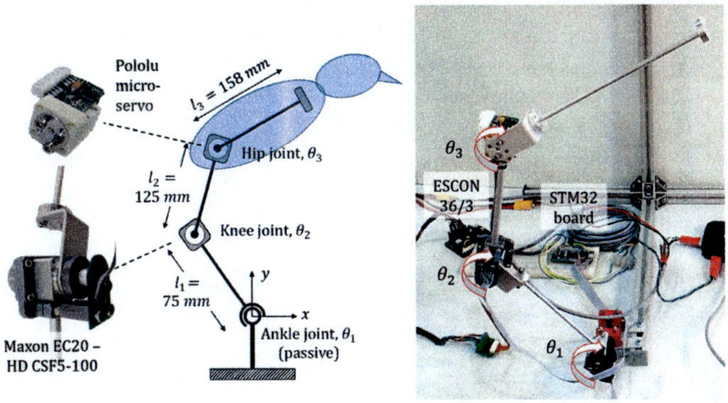

Fig. 2. The developed prototype.

On the other hand, the hip joint is built with a micro servo actuator (25 g weight) based on a Pololu micro metal gear motor, integrating a Murata SV01 potentiometer to measure the angle θ_3, along with the low-level control electronics that communicate with the microcontroller board through a serial interface. This actuator, although more compact and lightweight than the other one, presents a lower performance in terms of rated torque (0.2 Nm), clearance (around 2 °C), and positioning accuracy. Taking into account that the proposed control scheme applied in this particular case relies more on the knee that on the hip actuator to stabilize the passive joint, it is, therefore, more convenient to place the higher-performance actuator on the knee. Finally, the robotic system is controlled through the STM32 Nucleo board that interfaces with the computer where the control program is implemented through the serial interface, sending the torque references and feedback position measurements at 230400 bps. The firmware of the microcontroller was developed in C/C++ using the CubeMx IDE.

4.2 Simulated Results

Simulations have been carried out with the parameters of the prototype described in Sect. 4.1 to demonstrate the efficiency and robustness of the proposed technique. It has been considered that the mass of the links are concentrated at the joints and with values of $m_1 = 0.158$, $m_2 = 0.025$, and $m_3 = 0.0285$ kg, and the length of the links are $l_1 = 0.075$, $l_2 = 0.125$, and $l_3 = 0.158$ m. For the control of the fully-actuated subsystem, Bezier curves, which are smooth curves that can be differentiated indefinitely, have been chosen as reference trajectories. The dynamic model of the system is described by expression (1), making $n = 3$. The details about the mathematical model used in this work can be found in [30]. The results of the simulations are shown in Fig. 3. The poles of the closed-loop system using the feedback control linearization in the actuated part are located at $p_3 = -15$. The GPI observer gain parameters used in the ADRC were set to: $\zeta_o = 1$, $\omega_o = 40$, $T_o = 40$. The controller design parameters of the ADRC were set to: $\zeta_c = 1$, $\omega_c = 25$. The initial point of the simulations and the equilibrium point used for implementing the ADRC control technique are $\theta_{10} = 110°$, $\theta_{20} = -32.5°$ and $\theta_{30} = -37.8°$.

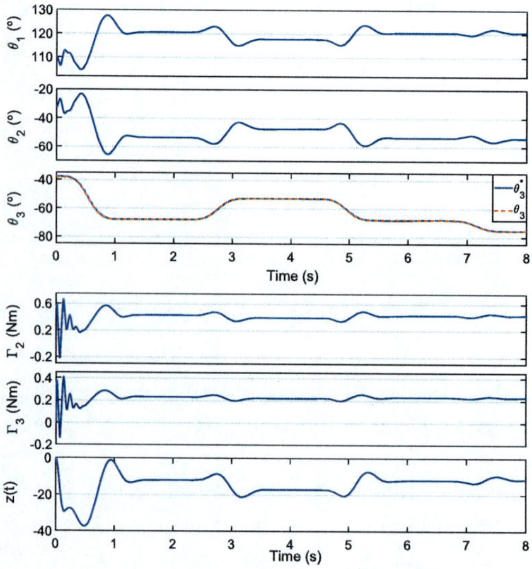

Fig. 3. Simulation results: angular coordinates and torques and estimation.

Figure 3 shows the trajectory tracking of the third joint, θ_3, that corresponds to the actuated part of the system. The results show that we achieve almost perfect trajectory tracking with the actuated part. Moreover, the position of the underactuated part of the system, θ_1 and θ_2 are also shown. The figure shows the positions that are needed to maintain the equilibrium of the system (around the unstable equilibrium point) while the third joint follows a preset trajectory. Figure 3 also shows the torques needed in joint 2 and joint 3 to control the system. Also, this figure shows the estimation of the perturbation that the GPI observer is estimating during the maneuver.

4.3 Experimental Results

Experiments have been performed with the robot manipulator described in Sect. 4.1 to show the effectiveness and robustness of the proposed control scheme in real applications. The poles of the closed-loop system using the feedback control linearization have been located at $p_k = -15$. The proposed ADRC controller is implemented and compared with the linear quadratic regulator (LQR). The GPI observer gain parameters used in the ADRC were set to: $\zeta_o = 1$, $\omega_o = 40$, $T_o = 40$. The controller design parameters of the ADRC were set to: $\zeta_c = 1$, $\omega_c = 5$. On the other hand, the LQR has been designed using the linear approximate model (3) and (4) at the equilibrium point ($\theta_{10} = -90°$, $\theta_{20} = 0°$), with weighting matrices $Q = 0.5 \cdot I$ and $R = 5 \cdot I$ (I is the identity matrix), yielding the controller gains $K = [-0.5463\ -0.0298\ -0.007\ 0.0755]$. For the experiments with the real prototype, the control technique is implemented for the case that the system is hanging and we want to control the system around this stable equilibrium point. Figure 4 shows the angular coordinates of the robot manipulator for the cases: (a) controlling only the fully-actuated subsystem, i.e. no control on the underactuated one, (b) controlling the underactuated subsystem with the LQR and the fully-actuated one with the feedback linearization and (c) controlling the underactuated subsystem with the ADRC and fully-actuated one with the feedback linearization. Figure 4 also shows the angular coordinate of the third joint, the preset trajectory is a sinusoidal signal during the first two seconds to exert a continuous disturbance in the underactuated part and validate the robustness of the ADRC. This angular coordinate is practically the same for the three cases showing the excellent performance of the feedback linearization for the three cases. Figure 4d shows the trajectory tracking error in the three aforementioned cases. These experiments show good results in cases (b) and (c) because the controller of the fully-actuated subsystem was able to track the prescribed trajectories, while the underactuated system can stabilize the system around an equilibrium point. For that reason, the trajectory tracking error is reduced in these cases in comparison with the case of not using a control technique for the underactuated subsystem. Finally, the results also show that the proposed ADRC is much more efficient in reducing the trajectory tracking error than the LQR because the former estimates the coupling between the two subsystems and it can approximately cancel it.

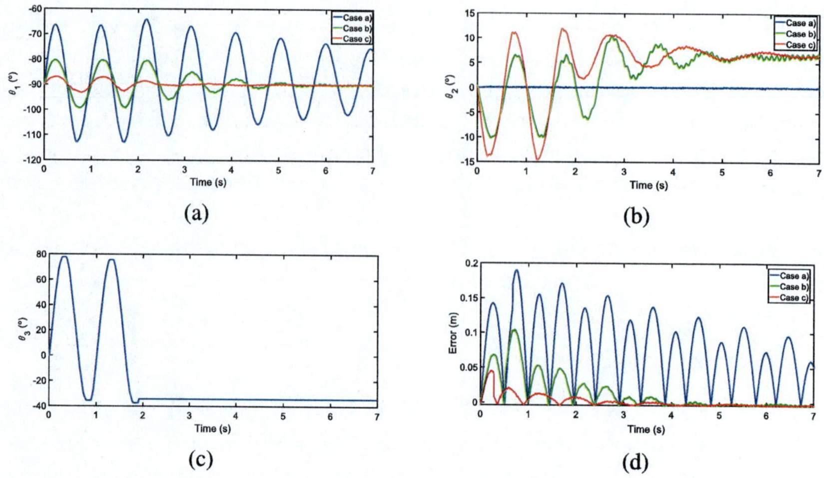

Fig. 4. Angular coordinate θ_1, θ_2, θ_3 and trajectory tracking errors.

5 Conclusions

This work presents an under-actuation model of a mechanical arm that imitates the birds having two-link legs, a body link, and two-link arms, being all active links except the first passive one. This model presents the behavior of a robotic bird, right before the take-off from a branch. The strategy of control approach includes two subsystems: a) the leg counterpart (under-actuated), which includes the two first links, and b) the body and arm subsystem (fully actuated) with the rest of the links. The control part of the fully-actuated links uses feedback linearization, and the under-actuated part, which is the difficult one, employs a linear high gain extended observer-based active disturbance rejection control approach which performs state/disturbances estimation. Although the setup is a three-DoF platform, the proposed methodology could be generalized for arms with more DoFs. The simulation results offer a more complicated under-actuated mechanism because it is controlled around an unstable equilibrium point. Experimental results demonstrate that the proposed control technique exhibits excellent behavior and it can estimate and reject the coupling between both subsystems. This technique is compared in experimentation with the well-known LQR method showing that the introduced control works efficiently for this kind of system and has a strong disturbance rejection ability in comparison with other methods.

Acknowledgment. The authors acknowledge support from the European Project GRIFFIN ERC Advanced Grant 2017, Action 788247; and also partially support from "Sistema robótico híbrido aéreo-acuático para muestreo, monitorización e intervención (HAERA)" funded by the Spanish Ministry of Science and Innovation PID2020-119027RB-I00.

References

1. Zufferey, R., et al.: Design of the high-payload flapping wing robot e-flap. IEEE Rob. Autom. Lett. **6**(2), 3097–3104 (2021)
2. Sun, W., Yu, J., He, G., Cai, Y.: Study on transmission mechanism and flexible flapping wings of an underactuated flapping wing robot. J. Intell. Robot Syst. **104**(2), 1–13 (2022)
3. Zufferey, R., Barbero, J.T., Talegon, D.F., Nekoo, S.R., Acosta, J.A., Ollero, A.: How ornithopters can perch autonomously on a branch. arXiv preprint arXiv:2207.07489 (2022)
4. Hoff, J., Kim, J.: Two-stage trajectory optimization for flapping flight with data-driven models. In: 2021 IEEE International Conference on Robots and Automation (ICRA), pp. 7688–7692. IEEE (2021)
5. Afakh, M.L., Sato, T., Sato, H., Takesue, N.: Development of flapping robot with self-takeoff from the ground capability. In: 2021 IEEE International Conference on Robots and Automation (ICRA), pp. 321–327. IEEE (2021)
6. Hudson, O.A., Fanni, M., Ahmed, S.M., Sameh, A.: Bio-inspired jumping maneuver for launching flapping wing micro air vehicles. In: 2018 IEEE International Conference on Autonomous Robot Systems and Competitions (ICARSC), pp. 104–109. IEEE (2018)
7. Zhang, Z., Suh, C.S.: Underactuated mechanical systems-a review of control design. J. Vibr. Test. Syst. Dyn. **6**(01), 21–51 (2022)
8. Rego, R.C.B., de Araújo, F.M.U.: Lyapunov-based continuous-time nonlinear control using deep neural network applied to underactuated systems. Eng. Appl. Artif. Intell. **107**, 104519 (2022)
9. Feliu-Talegon, D., Acosta, J.A., Ollero, A.: Control aware of limitations of manipulators with claw for aerial robots imitating bird's skeleton. IEEE Rob. Autom. Lett. **6**(4), 6426–6433 (2021)
10. Feliu-Talegon, D., Acosta, J.A., Suarez, A., Ollero, A.: A bio-inspired manipulator with claw prototype for winged aerial robots: benchmark for design and control. Appl. Sci. **10**(18), 6516 (2020)
11. Feliu, D., Acosta, J.A., Feliu, V., Ollero, A.: A lightweight beak-like sensing system for grasping tasks of flapping aerial robots. IEEE Rob. Autom. Lett. **7**(2), 2313–2320 (2022)
12. Spong, M.W.: The swing up control problem for the acrobot. IEEE Control Syst. Maga. **15**(1), 49–55 (1995)
13. Xin, X., She, J.-H., Yamasaki, T., Liu, Y.: Swing-up control based on virtual composite links for n-link underactuated robot with passive first joint. Automatica **45**(9), 1986–1994 (2009)
14. Zhang, A., Qiu, J., Yang, C., He, H.: Stabilization of underactuated four-link gymnast robot using torque-coupled method. Int. J. Non-linear Mech. **77**, 299–306 (2015)
15. Slotine, J.J.E., Li, W., et al.: Applied Nonlinear Control. Prentice hall, Englewood Cliffs (1991)
16. Isidori, A.: Nonlinear Control Systems, 3rd edn. Springer, Berlin (1995). https://doi.org/10.1007/BFb0006373
17. Sastry, S.: Nonlinear Systems: Analysis, Stability and Control. Springer, New York (1999). https://doi.org/10.1007/978-1-4757-3108-8
18. Defeng, W., Yuan, K., Huang, Y., Yuan, Z.-M., Hua, L.: Design and test of an improved active disturbance rejection control system for water sampling unmanned surface vehicle. Ocean Eng. **245**, 110367 (2022)

19. Hou, G., Gong, L., Wang, M., Xiaodong, Yu., Yang, Z., Mou, X.: A novel linear active disturbance rejection controller for main steam temperature control based on the simultaneous heat transfer search. ISA Trans. **122**, 357–370 (2022)
20. Ramirez-Neria, M., Gao, Z., Sira-Ramirez, H., Garrido-Moctezuma, R., Luviano-Juarez, A.: Trajectory tracking for an inverted pendulum on a cart: an active disturbance rejection control approach. In: 2018 Annual American Control Conference (ACC), pp. 4881–4886. IEEE (2018)
21. Curiel-Olivares, G., Linares-Flores, J., Guerrero-Castellanos, J.F., Hernández-Méndez, A.: Self-balancing based on active disturbance rejection controller for the two-in-wheeled electric vehicle, experimental results. Mechatronics **76**, 102552 (2021)
22. Ma, S.F., Leylaz, G., Sun, J.Q.: Data-driven robust tracking control of underactuated mechanical systems using identified flat output and active disturbance rejection control. Int. J. Control, 1–17 (2021)
23. Aguilar-Ibañez, C., Sira-Ramirez, H., Acosta, J.A.: Stability of active disturbance rejection control for uncertain systems: a lyapunov perspective. Int. J. Robust Nonlinear Control **27**(18), 4541–4553 (2017)
24. Fliess, M., Lévine, J., Martin, P., Rouchon, P.: Flatness and defect of non-linear systems: introductory theory and examples. Int. J. Control **61**(6), 1327–1361 (1995)
25. Zhao, S., Gao, Z.: An active disturbance rejection based approach to vibration suppression in two-inertia systems. Asian J. Control **15**(2), 350–362 (2013)
26. Ramírez-Neria, M., Sira-Ramírez, H., Garrido-Moctezuma, R., Luviano-Juarez, A.: Linear active disturbance rejection control of underactuated systems: the case of the furuta pendulum. ISA Trans. **53**(4), 920–928 (2014)
27. Ramírez-Neria, M., Sira-Ramírez, H., Garrido-Moctezuma, R., Luviano-Juárez, A.: Active disturbance rejection control of singular differentially flat systems. In: 2015 54th Annual Conference of the Society of Instrument and Control Engineers of Japan (SICE), pp. 554–559. IEEE (2015)
28. Zheng, Q., Gao, L., Gao, Z.: On stability analysis of active disturbance rejection control for nonlinear time-varying plants with unknown dynamics. In: Proceedings of the IEEE Conference on Decision and Control, New Orleans, LA, December 2007, pp. 12–14 (2007)
29. Xue, W., Huang, Y.: Performance analysis of active disturbance rejection tracking control for a class of uncertain lti systems. ISA Trans. **58**, 133–154 (2015)
30. Xin, X., Kaneda, M.: Swing-up control for a 3-dof gymnastic robot with passive first joint: design and analysis. IEEE Trans. Rob. **23**(6), 1277–1285 (2007)

Autonomous Landing Strategy for Vertiports: A Preliminary Case Study

A. Martinez, P. J. Sanchez-Cuevas(✉), M. A. Trujillo, and A. Viguria

Advanced Centre for Aerospace Technologies, Seville, Spain
{amartinez,psanchez}@catec.aero

Abstract. Urban air mobility is raising a lot of attention from both research and industrial communities. In near future UAM vehicles will be manned aircrafts with vertical take-off and landing capabilities, being this use case one of the priorities of current UAM regulations. Yet, the complete potential of UAS will arrive when autonomous UAS could be adapted to be used under addressing the standards. This papers aims to contribute to this activity proposing a way to execute a new autonomous landing in vertiports maneuver that follows the recently published standard presented by EASA. The landing maneuver has been studied and tested in simulation, analysing its complete casuistry. Its implementation is also described and could be deployed in the most common open-source autopilots available today.

Keywords: UAV · UAS · UAM · VTOL · Vertiports

1 Introduction

Urban air mobility (UAM) is an emerging topic in recent years [1–3]. In fact, this field is attracting a lot of attention not only from a research point of view but also from industry side. UAM technologies aim to revolutionize logistics and urban transport industries in the near future [4,5]. The growing interest in UAM is specially motivated due to the vast market and the amount of possibilities this technology will bring to the current society.

Thanks to the progress in power electronics, communications, sensors and the recent developments in data science, nowadays it is possible to build a sustainable UAM environment mostly composed of electric vehicles and green energies [6]. UAM is a very large concept that includes applications ranging from the use of autonomous UAVs for package delivery or cities monitoring to intercity and intracity people transport [7]. However, it will only be at the moment in which the UAM systems adapt to the regulations when the disruptive change in the market occurs.

Europe is at the forefront of a worldwide effort to enable this new mobility concept promoting the development of new VTOL aircraft and investing in the necessary infrastructure to transform the current urban mobility [8]. UAM fields involve not only the development of new vehicles and regulations but also ground

D. Tardioli et al. (Eds.): ROBOT 2022, LNNS 590, pp. 389–400, 2023.
https://doi.org/10.1007/978-3-031-21062-4_32

and UAM traffic management infrastructure, operations and procedures, market actors and a long list of technologies and stakeholders.

One of the core components in UAM are the so-called vertiports. Vertiports represent the ground infrastructure that is necessary for take-off and landing operations of UAM vehicles. Only four years ago, [9,10] discussed about different layout of vertiports inspired by the existing heliports and their organization. Today, EASA has recently published the technical specifications for the design of VFR vertiports specifically oriented to UAM vehicles [11]. This document offers guidance to safely design this UAM infrastructures. This regulation evidences that although UAM is a completely new field of aviation, it is a disruptive technology that will change cities in a short term. Another evidence is the amount of UAM aircraft manufacturers such as Lilium, Volocopter, or Ehang among others, which are presenting aerial vehicles that are quite close to be commercial products.

Although this regulation [11] is specifically focused on manned aviation, it is clear that in a few years UASs will be a relevant actor in the UAM environment. Companies like Amazon, DHL and Correos are actively investing in autonomous solution to increase the agility and the efficiency of their logistic systems not only between their warehouses but also in the urban package delivery. Being this an isolate case of a vast amount of applications that will involve UASs as a part of the UAM environment.

This paper proposes an autonomous landing procedure to be used in UAS when landing in vertiports. The operation has been defined according to the actual constraints considered in vertiports by EASA in the recent regulation. The landing maneuver has been implemented as finite state machine and integrated with two of the most widespread open-source autopilots (PX4 [12] and APM).

The rest of the paper is structured as follows: Sect. 2 analyses the regulation published by EASA and its different casuistry. Section 3 presents the state machine proposed to make an autonomous UAV capable of accomplishing the landing operation according to EASA's standard. Section 4 describes the implementation detail and the methodology used to validate the proposed maneuver. Section 5 shows simulated results that illustrate the performance of the state machine in different situations. Last, Sect. 6 closes the paper presenting conclusions and future works.

2 Landing in Vertiports: Regulation Analysis

2.1 Obstacle Free Volume (OFV) Description

Choosing an adequate safety landing volume that complies with EASA regulations is the first fundamental step to ensure efficient landing and takeoff maneuvers in congested areas. More precisely, according to EASA landing VTOL-capable aircrafts require of an specific Obstacle Free Volume for safe maneuvering during takeoff and landing in densely populated environments, where vertiports aim to be located. To tackle this particular task, EASA proposes a selection of reference Obstacle-Free Volumes (OFVs) which are geometrically constrained by

Table 1. List of abbreviations

Abbreviation	Definition
UAM	Urban Air Mobility
UAV	Unmanned Aerial Vehicle
UAS	Unmanned Aerial System
VTOL	Vertical Take-Off and Landing
GNSS	Global Navigation Satellite System
FOV	Field Of View
EASA	European Aviation Safety Agency
FSM	Finite State Machine
OFV	Obstacle-Free Volume
FATO	Final Approach and Take-Off
TO	Take-Off

a set of standardized reference values. The design and location of these safety landing regions are subject to various constraints, including minimum shape and size, elevation, visual aids, downwash or influence of the local environment (such as turbulence or temperature), and many others. Among the main parameters proposed by EASA to facilitate the design of the OFV in [11], there are the dimensions of the FATO (Final Approach and Takeoff Area) within the vertiport landing platform, as well as the dimensions of the upper section of the OFV (defined as TO, or Takeoff Area), and slope values for approach and departure surfaces during initial approximation and final takeoff maneuvers, respectively. At least two hover distances above the FATO must be defined so as to ensure accuracy of positioning during approach.

Three distinct approaches to the geometric characterization of the OFV can be studied according to EASA regulation metrics. First of all, a generic OFV differentiates between takeoff and landing maneuvers. This means different approach and departure surfaces must be distinguished, with different slope values (θ_{app}, θ_{dep}), which define the restrictions for the initial approach and final departure stages, respectively. Secondly, a simplified bidirectional volume can be derived, using the minimum slope value ($min(\theta_{app}, \theta_{dep})$) for both takeoff and landing. Both FATO and TO dimensions are selected based on the maximum between front and back distances, in order to leave enough space for both maneuvers from any of the two possible approaching or departure directions. The resulting OFV shape is shown in Fig. 1a. Lastly, a more versatile omnidirectional approach with a conic funnel-shaped OFV is considered (see Fig. 1c). In this case, both departure and landing can be approached from any angle, with FATO and TO sections being defined by their diameter values, as shown in Eqs. (1) and (2).

$$\varnothing TO = \sqrt{4 * max(TO_{front}, TO_{back})^2 + TO_{width}^2} \quad (1)$$

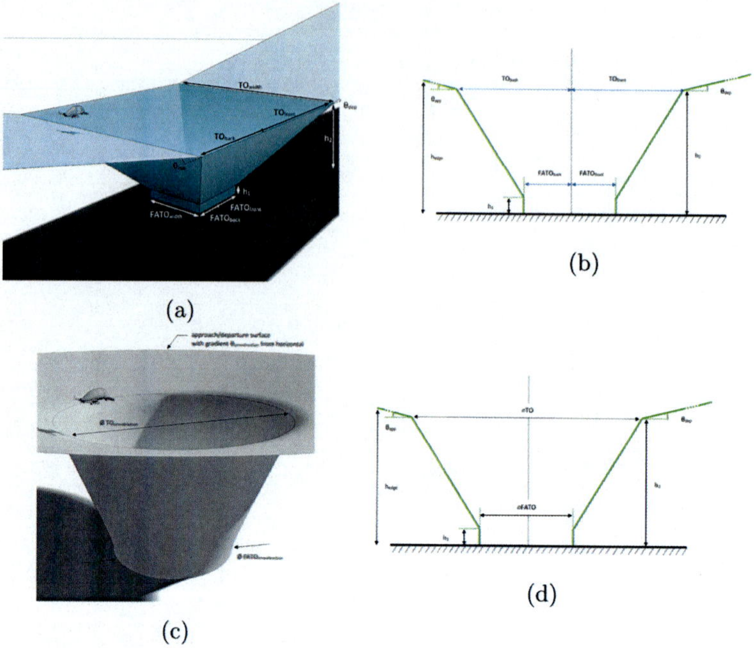

Fig. 1. 3D and 2D projections of EASA reference volumes for safe landing and takeoff maneuvers in vertiports: bilateral design approach, shown in (a) and (b) and omnidirectional approach with a *funnel-like* volume, shown in (c) and (d). Extracted from [11]

$$\varnothing FATO = \sqrt{4 * max(FATO_{front}, FATO_{back})^2 + FATO_{width}^2} \quad (2)$$

2.2 OFV Characterization

The geometric constraints for the OFV are defined as a set of equations which depend on a common standardized value to ensure compliance with EASA regulations. The base reference parameter for OFV design is the diameter D (in metres) of the smallest circle enclosing the aircraft projection on a horizontal plane while landing or takeoff maneuvers are taking place. Table 2 summarizes parameter definitions and constraints for VTOL landing maneuvers in vertiports, as defined in [11].

3 Autonomous Landing Strategy for Vertiports

3.1 Maneuver Requirements

This section presents the proposed algorithm to adapt autonomous landing operation of UAS according to EASA standard.To ensure safe and precise navigation

Table 2. Standardized parameter definitions and constraints for vertical take-off and landing. Extracted from [11]

Parameter	Description (as defined by EASA)	Standarized constraint
h_1	Lower hover height	- -
h_2	Higher hover height	$\geq h_1$
TO_{width}	Width at high hover distance h_2	$\leq 5D$
TO_{front}	Front distance at high hover distance h_2	$\leq 5D$
TO_{back}	Back distance at high hover distance h_2	$\leq 5D$
$FATO_{width}$	Width of FATO	$\geq 1.5D$
$FATO_{front}$	Front distance of FATO	$\geq 0.75D$
$FATO_{back}$	Back distance on FATO	$\geq 0.75D$
θ_{app}	Safe approach surface slope angle	$\geq 4.5\%$
θ_{dep}	Safe departure surface slope angle	$\geq 4.5\%$

within the established boundaries of the OFV, the control algorithm must comply with the following requirements:

1. The landing task must be subdivided into different stages (states) of approach to ensure precise landing within the bounds of the OFV.
2. Relative location from the aircraft to landing platform (FATO) must be known for the control algorithm to minimize the relative error in the approaching stages. As a result, the aircraft must keep the landing platform in its FOV during approach (i.e. GNSS based location, vision-based detection algorithms or U-Space services).
3. If the system loses the relative distance with the target, the approaching maneuvers must halt and a recovery strategy must ensure safety conditions during the operation.
4. Initial approach maneuvers and subsequent re-attempts must consider the influence of gusts of wind during aircraft insertion in OFV, with the possibility of it consistently getting expelled out of the volume. Thus, a safety threshold inside the OFV, defined as half the radius of the horizontal section of the volume at the current altitude of the UAS in case of an omnidirectional OFV (or half the base of the pyramid if a bidirectional volume is employed), is introduced. The approximation task must ensure that the UAS has surpassed this threshold before the airship can start descending.
5. The system must be robust against a noisy and uncertain input signal in the relative location in order to execute state transitions and therefore update control references.

3.2 Proposed Landing State Machine

Table 3. FSM transitions and associated stated with their associated codified values (left) and FSM Schematic (right).

Transition	To State
Approaching(1)	Approximation(01)
Centering(2)	Centering(02)
Not finished(3)	Descent(03)
Still enabled(4)	Completed(04)
Approached(5)	Centering(02)
Centered(6)	Descent(03)
Not centered(7)	Centering(02)
Exited OFV(8)	Recovery(05)
Disconnected(9)	Blind mode(06)
Finished(10)	Completed(04)
Entered OFV(11)	Approximation(01)
Reconnected(12)	Approximation(01)
No reception(13)	Blind Mode(06)
Outside OFV(14)	Recovery(05)

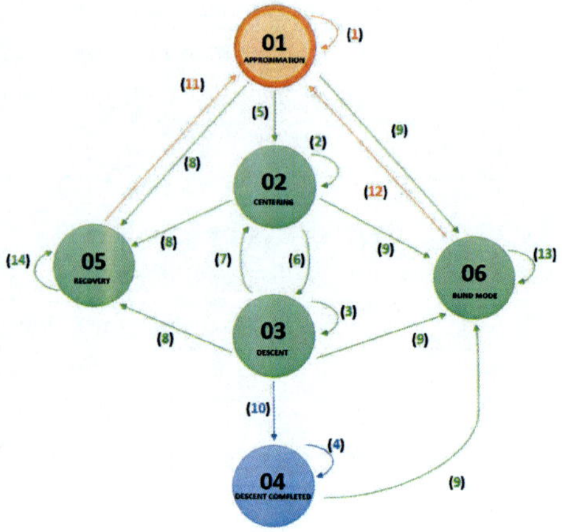

Concerning the autonomous landing procedure, a simple Finite State Machine (FSM) which appropriately manages transitions between a set of basic control modes can be integrated with a linear velocity control. Each iteration, the algorithm manages transitions based on the distance from the aircraft to the center of the FATO in the horizontal plane, while switching the altitude reference accordingly to ensure a smooth descent. This paper proposes a total of 6 states and 14 possible transitions as depicted in the left side of Table 3. A visual schematic of the FSM is also illustrated in the right side of Table 3.

- *01.Approximation:* the UAS approaches the designed landing spot by entering a safety threshold inside the OFV. First approximation must use a height reference $z = h_2$, while subsequent approximations, assuming the UAS may be expelled out of the OFV, can be tackled in the horizontal plane as long as the aircraft is above a minimum safety distance of $z = h_1$.
- *02.Centering:* when the UAS is contained within this safety threshold inside the OFV, it proceeds with a centering maneuver above the FATO at $z = h1$ before initiating a landing attempt.
- *03.Descent:* once the aircraft is contained within the FATO area at a given height, the landing maneuver can start.
- *04.Descent completed:* the aircraft has reached the designed landing target and must either hold position or give control authority back to other tasks.
- *05.Recovery:* if the vertiport is not within the UAS's FOV it will ascend until the FATO is again detected, then proceed with re-insertion into the OFV.

In the simulation experiments, we make the assumption that this situation happens whenever the UAS exits the OFV. However, the possibility that the UAS is outside the OFV while still within detection area must be considered in a real case scenario. In any case, only once there is valid detection data and the UAS has entered the OFV, approximation can follow.

- *06.Blind mode:* the UAS must interrupt any ongoing landing state and hold position. Transition to this state is required whenever the landing process is under risk of failing, usually due to a fault in data reception which could make input information unreliable.

4 Implementation

4.1 Simulation Framework

The functionalities of the custom autonomous landing strategy has been tested with a simulation setup consisting of a ROS Melodic distribution running on Ubuntu 18 and GRVC-UAL (UAV abstraction layer designed by University of Seville researchers [13,14]) simulation tools for Gazebo, using a MAVROS back-end configuration and PX4 software in the loop simulations [12]. In this way, we are able to reproduce the full flight stack and a realistic dynamic in simulation. This facilitates the development at an algorithmic level and reduces the risk while developing.

4.2 Software Description

The safe landing system has been implemented with two custom-built ROS [15] packages, the detection package and the autonomous landing controller: the first package is in charge of feeding the second one with the relative location between the aircraft and the vertiport. The second package implements the autonomous landing control strategy, which is the main focus of this paper. In practice, the first package would be replaced for a vision based algorithm to feed the controller with this relative location, with a specific node that calculates this relative location based on GNSS or a U-Space service. However, simulating this component is needed to successfully test if the control strategy proposed in this paper is correctly performing.

Therefore, in this case the first package integrates a *simulated detection* module which launches a set of nodes that allow for safety area computation and representation relative to a hypothetical landing platform located at a user-specified position. As it was aforementioned, the main purpose of this node is to provide an input flow of data to the FSM, specifically, the relative position of the vertiport in the UAS frame. The second package consists of a single *land control* node that implements the FSM for autonomous landing based on relative pose information and geometric constraints of the OFV generated around the landing platform by the previously mentioned simulated detection module. The current state is codified to its corresponding integer value, as depicted in Table 3, and published to a custom topic for monitoring.

Autonomous landing can be requested through a custom service that receives a landing mode specification, including a selection between full landing on platform or a partial landing at a positive hover altitude z_{thr} above the platform, which could be useful for quick cargo delivery.

4.3 Finite State Machine

The designed FSM has been implemented in a custom ROS node which receives the relative pose to landing target in the UAS base frame. The main control loop, implemented with 20 Hz timer, manages FSM transitions and publishes high level velocity control signals to the UAL stack. In each iteration relative position of the aircraft to the OFV and geometric constraints are evaluated, using a set of standardized parameters read and computed from a custom parameter file. Both FATO and TO dimensions are calculated with respect to the specified dimension D to comply with the regulations. Omnidirectional approach to the vertiport landing platform implies that only references in altitude must be switched so that the aircraft approaches the target from a safe region inside the OFV. In the case of a bilateral approach, x and y references may require to be switched accordingly to approach the FATO from a specific side of the OFV.

Once the state transition has been computed on a given iteration, linear velocity control signals can be calculated using a basic PI control structure with a saturation of 1 m/s on each linear component of the velocity vector to ensure a smooth and precise trajectory within the bounds of the safety volume.

5 Results

A set of experiments have been carried out in order to test the proposed landing control strategy, where a partial landing at x = 10 m, y = 10 m at a positive height of z = 0.25 m has been solicited. The two reference OFVs presented in Fig. 1, bilateral and omnidirectional, have been tested, with hover distances set at $h_2 = 30$ m and $h_1 = 3$ m, and a chosen diameter value of D = 1 m. FATO and TO dimensions correspond to the most restrictive values as shown in Table 2.

5.1 Bilateral OFV

This section presents the autonomous landing strategy performance assuming a bilateral OFV in Fig. 2, which includes the 3D trajetory followed by the UAS from different perspectives in Figs. 2a, 2b and 2c. Figures 2d and 2e show the relative error to target with respect to time in both X, Y as well as Z axis; and position data with respect to time in the X and Y axis, respectively. The approach surfaces of the OFV are defined from the sides of the FATO that are perpendicular to the X-axis, and then extended up to h_2.

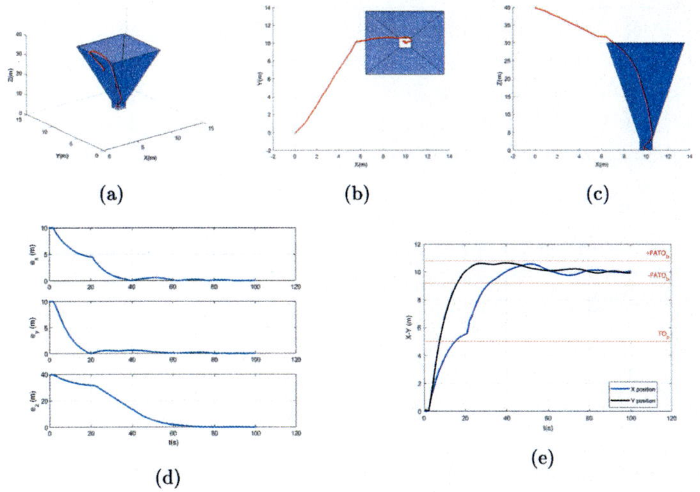

Fig. 2. (a) 3D trajectory inside the bounds of the bilateral OFV during landing, with XY (b) and ZY (b) sections; (d) shows relative error with respect to time in x, y and z axis; and (e) shows position data in the X and Y axis during approach.

5.2 Omnidirectional OFV

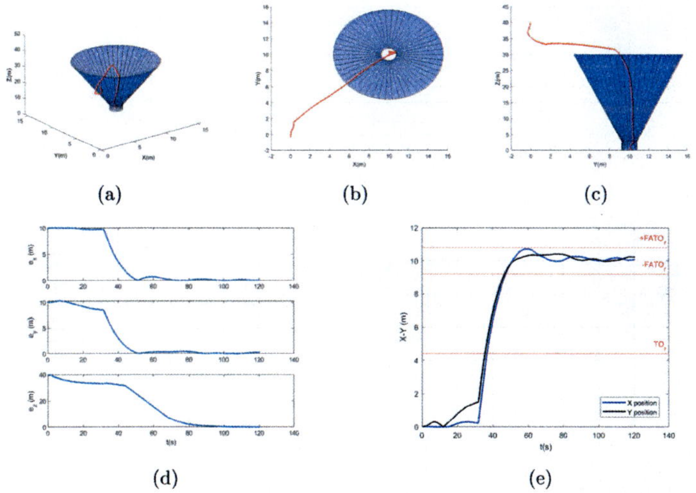

Fig. 3. (a) 3D trajectory inside the bounds of the omnidirectional OFV during landing from z = 40 m, with XY (b) and ZY (c) sections. (d) shows the evolution of the relative error in the X, Y and Z axis with respect to time, and (e) shows position data in the X and Y axis

Omnidirectional approach to landing is tested by constructing a funnel-shaped OFV with a safe approach surface of $\theta = 15^{\circ}$, and the landing maneuver is tested in the same conditions as the previous experiments. Figures 3a, 3b and 3c show the resulting 3D trajectory and 2D projections of the landing process, while relative errors and position evolution in the X and Y axis can be seen in Figs. 3d and 3e. In addition, Figs. 4a, 4b and 4c show an experiment where the UAS is solicited to start the maneuver from z = 5 m and, therefore, under the coverage of the OFV. Another experiment where the UAS initial position is in close proximity to the vertiport can be seen in Figs. 4d, 4e and 4f. These results evidence that the proposed landing maneuver is able to comply with EASA rules even in these situations.

Finally, in order to assess the behaviour of the FSM when the aircraft loses the relative location of the vertiport, the blind mode has been induced in the state machine by interrupting reception of such information (red area in Fig. 5a). The results show the aircraft keeps hovering conditions until new messages are received. As seen in Fig. 5a, when a transition to this mode (code 06) is induced around $t \approx 40s$, the UAS immediately stops descending, as reflected in Fig. 5b, then resumes with the previous mode once communication with the simulated detection module is re-established at around $t \approx 60$ s.

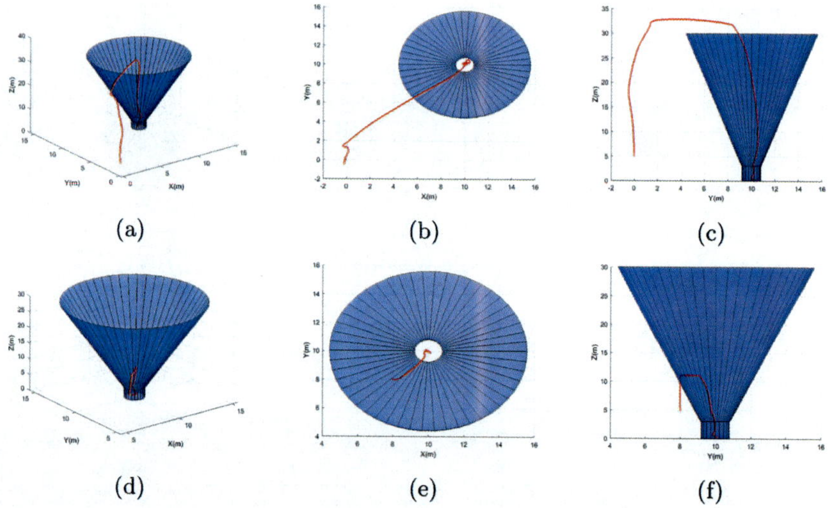

Fig. 4. (a) 3D trajectory inside the bounds of the omnidirectional OFV during landing from z = 5 m, with XY (b) and ZY (c) sections. (d), (e) and (f) show a similar experiment starting from x = 8 m and y = 8 m.

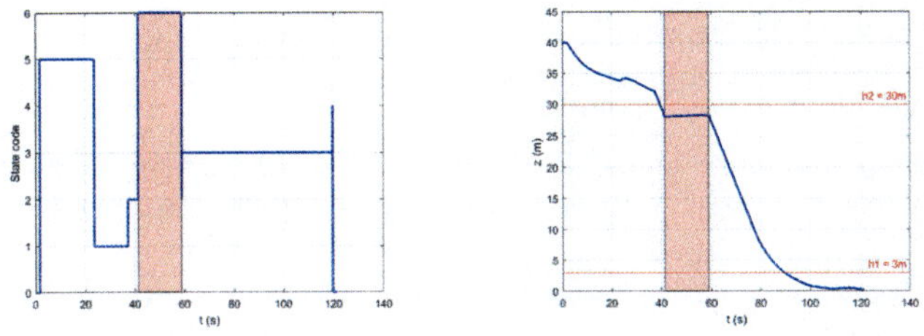

(a) (b)

Fig. 5. (a) State information during omnidirectional landing approach with a simulated loss of input data during the interval marked in red, triggering a transition to blind mode (06), which is reflected in z position evolution (b). (Color figure online)

6 Conclusions

This paper has presented an autonomous landing strategy based on a Finite State Machine to tackle safe landing maneuvers in vertiports. According to EASA regulations and standardized parameters, a geometrically constrained Obstacle Free Volume (OFV) has been determined using two distinct design approaches: a bilateral, pyramid-shaped OFV and an omnidirectional, funnel-shaped OFV. Autonomous landing has been tested in a simulated environment and trajectory and relative error data has been collected, proving the UAS manages to land safely inside the bounds of the OFV. Additional experiments with different initial approach heights have been carried out on the two different OFVs, proving the FSM ensures a sufficiently accurate and smooth trajectory towards the vertiport landing platform. We have noted that landing strategies are highly dependant on the quality and robustness of the input data flow from the perception module, which is outside the scope of this paper and has only been simulated with a Gaussian noise injection to input data. Further experiments may require more complex filtering and data fusion techniques on real input data provided for instance from a vision based detection system. Additionally, a coordinated multi-UAS landing strategy can be tested, managing schedule constraints in order to reproduce real-life situations, where many aircrafts may be requested to land on a given spot at a given time and collision must be avoided.

Acknowledgements. This work has been supported by the DELOREAN Project, funded by the European Commission (H2020-SPACE-EGNSS-2019-870251), and by the Joint Innovation Unit AGORA (602C2000003) of the IDEA Agency.

References

1. Straubinger, A., et al.: An overview of current research and developments in urban air mobility-Setting the scene for UAM introduction. J. Air Transp. Manag. **87**, 101852 (2020)

2. Silva, C., et al.: VTOL urban air mobility concept vehicles for technology development'. In: 2018 Aviation Technology, Integration, and Operations Conference, p. 3847 (2018)
3. Thipphavong, D.P., et al.: Urban air mobility airspace integration concepts and considerations. In: 2018 Aviation Technology, Integration, and Operations Conference, p. 3676 (2018)
4. Narkus-Kramer, M.P.: Future demand and benefits for small unmanned aerial systems (UAS) package delivery. In: 17th AIAA Aviation Technology, Integration, and Operations Conference, p. 4103 (2017)
5. Vempati, L., et al.: Linear program and simulation model for aerial package delivery: a case study of Amazon Prime Air in Phoenix, AZ. In: 17th AIAA Aviation Technology, Integration, and Operations Conference, p. 3936 (2017)
6. Shamiyeh, M., Bijewitz, J., Hornung, M.: A review of recent personal air vehicle concepts. In: Aerospace Europe 6th CEAS Conference, vol. 913, pp. 1–18 (2017)
7. Jensen, O.B.: Drone city-power, design and aerial mobility in the age of "smart cities". Geographica Helvetica **71**(2), 67–75 (2016)
8. Gillis, D., Petri, M., Pratelli, A., Semanjski, I., Semanjski, S.: Urban air mobility: a state of art analysis. In: Gervasi, O., et al. (eds.) ICCSA 2021. LNCS, vol. 12950, pp. 411–425. Springer, Cham (2021). https://doi.org/10.1007/978-3-030-86960-1_29
9. Vascik, P.D., Hansman, R.J.: Evaluation of key operational constraints affecting on-demand mobility for aviation in the Los Angeles basin: ground infrastructure, air traffic control and noise. In: 17th AIAA Aviation Technology, Integration, and Operations Conference, p. 3084 (2017)
10. Vascik, P.D., Hansman, R.J.: Development of vertiport capacity envelopes and analysis of their sensitivity to topological and operational factors. In: AIAA Scitech 2019 Forum, p. 0526 (2019)
11. European Union Aviation Safety Agency. Prototype Technical Specifications for the Design of VFR Vertiports for Operation with Manned VTOLCapable Aircraft Certified in the Enhanced Category. Vertiports, PTSVPT- DSN, Cologne Germany, March 2022. https://www.easa.europa.eu/downloads/136259/en. Accessed08 July 2022
12. Meier, L., Honegger, D., Pollefeys, M.: PX4: a node-based multithreaded open source robotics framework for deeply embedded platforms. In: 2015 IEEE international conference on robotics and automation (ICRA), pp. 6235–6240. IEEE (2015)
13. Real, F., et al.: UAL: an abstraction layer for unmanned aerial vehicles. In: 2nd International Symposium on Aerial Robotics (2018)
14. Real, F., et al.: Unmanned aerial vehicle abstraction layer: an abstraction layer to operate unmanned aerial vehicles. Int. J. Adv. Rob. Syst. **17**(4), 1729881420925011 (2020)
15. Quigley, M., et al.: ROS: an open-source robot operating system. In: ICRA Workshop on Open Source Software, Kobe, Japan, vol. 3, no. 3.2, p. 5 (2009)

Blueberry Row Detection Based on UAV Images for Inferring the Allowed UGV Path in the Field

Dimitrije Stefanović(✉), Aleksandar Antić, Marko Otlokan, Bojana Ivošević, Oskar Marko, Vladimir Crnojević, and Marko Panić

BioSense Institute, University of Novi Sad, Novi Sad, Serbia
dimitrije.stefanovic@biosense.rs

Abstract. Blueberries (Vaccinium corymbosum) have become one of the most popular fruits in Europe, being ranked as the top-berry fruit species. Due to the high cost and poor availability of the workforce, agricultural production of blueberries is becoming heavily reliant on robots. Within the EU-supported Flexigrobots project, unmanned ground vehicle (UGV) solutions are being developed for weeding, soil analysis, and spraying. However, to detect the problematic regions (regions of interest) and plan the UGV trajectories, fields are first scanned using unmanned areal vehicles (UAVs). In this paper, we propose a procedure for processing UAV images for inferring the allowed UGV trajectories in blueberry fields. First, the image dataset was acquired using a UAV equipped with a multispectral camera. Afterward, we utilized U-Net-based deep neural network (DNN) models for semantic segmentation of blueberry rows and enhanced the results with Radon transform and additional image post-processing steps. The quality of the obtained segmentation mask is analyzed through the ranking of trained models in terms of the intersection over union (IoU) and F1-score metrics. Lastly, based on the detected rows and the inter-row regions, we generated the trajectory along which the UGV is allowed to move to perform the weeding, soil analysis, or spraying tasks.

Keywords: Precision agriculture · Row detection · Unmanned aerial vehicles · Deep learning · Image processing

1 Introduction

In recent years, the utilization of aerial images acquired by UAVs in the most diverse areas of human interest has increased significantly, with a particular application in agriculture [1–4]. Another significant cause of this phenomenon was the technological development of various types of UAVs [5–7], as well as state-of-the-art imagery sensors (e.g., RGB, multispectral, hyperspectral, thermal, and radar) [8–10]. Before UAVs became more affordable, the primary source of aerial imagery was mostly satellite imagery and for remote sensing tasks, such

D. Tardioli et al. (Eds.): ROBOT 2022, LNNS 590, pp. 401–411, 2023.
https://doi.org/10.1007/978-3-031-21062-4_33

as crop classification on large-scale agricultural fields, these types of images were more than useful [11–13]. However, the low spatial resolution of images obtained from non-commercial satellites and their unchangeable imaging orbit limited their wide application in precision agriculture. Most such satellites have a spatial resolution greater than 10 m × 10 m per pixel, which is not accurate enough for tasks such as weed detection or orchards row detection [14]. Alternatively, UAVs allow us to collect images with high spatial resolution at different flying altitudes and also avoid the problem of occurring clouds during image acquisition, which is often in satellite imagery [15]. The high spatial resolution of UAV images is of great importance when developing machine or deep learning algorithms for weed detection [16], yield prediction [17], anomaly detection [18], and many more applications that are further being used in monitoring systems within the agricultural domain.

By combining remote sensing, deep learning, and image processing, the basis of a monitoring system for analyzing conditions of agricultural plots can be created. Multispectral and hyperspectral imaging provides additional information, as they acquire images within specific wavelength ranges across the electromagnetic spectrum, besides the visible range. This type of data can emphasize various feature engineering, which is limited when only the visible part of the spectrum is present, thus leading to the creation of the vast majority of vegetation indices. These indices are obtained as linear combinations of individual spectral bands and these combinations often express specific vegetation properties that can be used for various applications such as vegetation detection [19,20]. These kinds of problems are often referred to as image segmentation tasks and both machine learning (ML) and DNN-based algorithms are frequently used for solving them, particularly for image data obtained from UAVs. A study described in [21] utilized object-based multi-resolution segmentation of UAV imagery for the detection of agricultural fields. Besides the object-based approach, studies on the pixel-based approach have been proposed by authors in [22], where a support vector machine (SVM) classifier for agricultural mapping and reef monitoring within UAV images is used. Possible applications of random forest (RF), SVM, and gradient boosting decision tree (GBDT) for the detection of trees, grasses, bare gravel, sand bed, and water surface within UAV images have been described in [23]. One of the studies where standard ML algorithms were compared with DNN-based one for the problem of land surface classification was mentioned in [24] where the best performances are achieved with the DNN models. The authors in [25] used modified U-Net architecture for the detection of rice logging in RGB UAV images.

The application of unmanned vehicles in blueberry orchards is currently one of the more interesting use cases. Due to their profitable production, high antioxidant capacity, and nutritional value, blueberries are ranked as the top-berry fruit species [26,27]. However, there is also a need for more efficient use of resources during the growing season to make the production even more profitable as well as to enhance fruit quality. Various parts of orchards are in different microclimatic conditions and weed affects those regions in different quantities. Sometimes they do not receive the same amount of sunlight or precipitation. All this affects the health of the plants and therefore the final yield. Therefore, the reasons for the growth of interest in the utilization of UGV [28,29] can be found in man's constant search for solutions that will simplify the hard labor in the field and lead to the optimization of resources, especially in blueberry production.

Within this paper, we present a procedure for monitoring blueberry production that utilized the advantages of both types of unmanned vehicles UAV and UGV. We proposed an approach for the generation of navigation maps for UGVs that will enable them to autonomously drive through blueberry orchard rows and carry out tasks in the appropriate parts of the orchard such as weeding, soil analysis, spraying, etc., based on high-resolution multispectral images acquired by UAVs. We involve analysis of two DNN architectures for semantic segmentation, U-Net [30] and U-Net++ [31], and employ them for the blueberry rows segmentation task on multispectral UAV images. Then a procedure for inferring the allowed UGV path within the blueberry field is proposed based on the application of Radon transform [32] and medial axis transformation [33] on rows segmentation mask obtained from DNN. As a result, we get final detected rows that are then used for generating lines along which the robot is allowed to move and operate different tasks.

2 Methodology

In this section, we present in detail the image acquisition procedure as well as dataset creation. Then the selection of the DNN model based on U-Net [30,31] architecture for semantic segmentation of blueberry rows is performed. Afterward, a procedure for inferring the allowed UGV path within the blueberry field, which involves the application of the Radon transform [32,34,35] on U-Net generated binary output is presented.

2.1 Dataset Creation

The blueberry fields used for UAV mission planning and dataset creation were set at one field in Babe ($44°31'32.1''\mathbf{N}, 20°31'26.4''\mathbf{E}$), Serbia. The UAV missions and image acquisition procedure were conducted in July 2021 with DJI Phantom 4 Multispectral UAV (DJI P4). A multispectral camera array on DJI P4 captures Blue (450 nm ± 16 nm), Green (560 nm ± 16 nm), Red (650 nm ± 16 nm), Red-Edge(730 nm ± 16 nm), and Near-Infrared(840 nm ± 26 nm) spectral bands with 2.12MP spatial resolution. Across two missions, 549 georeferenced images per spectral band were captured resulting in a total of 2745 multispectral images of dimensions 1600 × 1300. Spectral bands are further preprocessed in two steps to minimize the possible appearance of artifacts in created orthomosaics. Within the first step, a registration of spectral bands is done according to the sensor offset written in image metadata and additionally by employing the Enhanced Correlation Coefficient (ECC) [36] based co-registration algorithm. These variabilities are often caused by UAV's movements when tracing predefined routes or

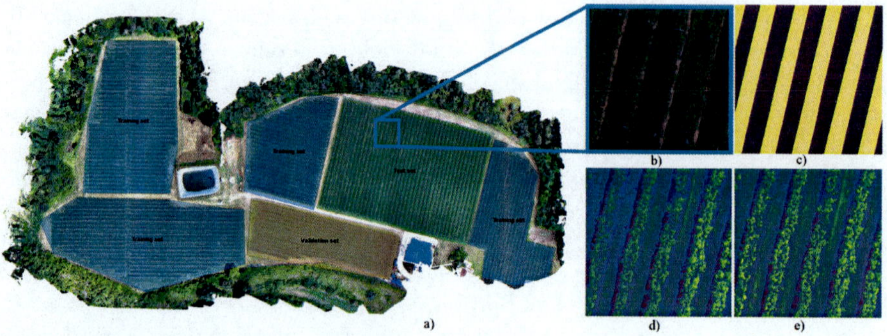

Fig. 1. Raw data: (a) RGB orthomosaic of a blueberry orchard in Babe, Serbia. Orthomosaic was split into training (blue), validation (yellow), and test (green) regions, (b) 512×512 RGB image within test region with corresponding (c) label image, and (d) NIR and (e) Red-Edge spectral bands.

Table 1. Dataset details

Dataset	Raw data	Training	Validation	Test
Images	2745	1192	272	1065
Size	1600×1300	512×512		

adjusting to the winds. After co-registration, we applied a gamma correction on those images per spectral band, which has global brightness outside the first two quartiles of brightness distribution estimated from all acquired images. These variations in images are caused by cloud shadows resulting from the passing of clouds during data acquisition.

Then, multispectral images were stitched in an orthomosaic using licensed Pix4DMapper software (see Fig. 1) and with QGis software [37] a shape file for each blueberry orchard within the field is created. With created shape files the regions of interest (ROI) for training (blue region), validation (yellow region), and testing (green region) are set. We then labeled each blueberry row in the orchard using QGis to get a ground truth segmentation mask. The created shape files are further being used for the division of ROIs into image patches of spectral bands and binary labels with dimensions of 512 × 512 (see Fig. 1). By assigning the whole blueberry orchards to only one of the sets (train, validation or test), we avoid data leakage. During the data splitting into patches, the ones that do not contain any label data are discarded. As a result, the number of images has decreased significantly. Therefore, we employed the data augmentation step by flipping and rotating patches by 90°, 180° and 270°, on training and validation sets. As for the test set, it is divided into patches with 75% overlapping, which will be explained in detail in the results section. In this way, we obtain our final dataset, which consists of 1192 training, 272 validation, and 1065 test multispectral images (see Table 1).

2.2 Segmentation of Blueberry Rows in UAV Images

Within this work, we opt for the usage of U-Net-based DNN architecture for the problem of blueberry rows detection. We tested two different models in our experiments: a U-Net from [30] and its improved version with dense and skip connections in upsampling blocks named U-Net++ proposed in [31]. These two models were slightly modified by reducing the feature maps number and the downsampling and upsampling steps in the encoder and decoder parts of both networks. As compared to the original U-Net architecture, our network had three downsampling steps and in every convolutional block, the mirroring padding was applied so that the dimension of feature maps on the output of every convolutional block was not reduced by the stride usage. The feature maps of the three downsampling steps have 16, 32, and 64 feature maps. Further, the softmax at the last output layer was omitted. In the case of the U-Net++, the number of downsampling steps was also reduced to 3 and the original number of feature maps are used. The softmax activation in the last output layer was also omitted in this case. All of the activations (ReLu) in convolutional layers in both architectures are kept as in their original implementations. Output segmentation maps generated by these two models were used as the first step of row detection in our experiments. In every experiment, both modified U-Net-based models are trained using Adam optimizer [38], and binary cross-entropy (BCE) loss function. The batch size was 4, training lasted 40 epochs and an initial learning rate is set to 1e–5 and used without schedulers throughout the whole training.

2.3 Inferring a UGV Path Based on Segmented Blueberry Rows

The final segmentation mask indicating blueberry rows is obtained by thresholding the output of the selected DNN architecture, thus converting it to a binary image $f(x, y) \in \{0, 1\}^{R \times C}$, where $x \in \mathbb{Z}_R^+$ or $y \in \mathbb{Z}_C^+$ with a finite set of positive integers $\mathbb{Z}_n^+ = \{0, 1, \ldots, n-1\}$ and R, C as rows and column dimensions respectively. Created binary segmentation mask $f(x, y)$ may have some irregularities in the shapes of the estimated rows of blueberries. Irregularities appear in the form of noisy borders of segmented rows, a possible break of the rows, or even a join of two or more rows. However, knowing the constraints in setting up the blueberry fields (i.e. rows have to be parallel to each other and it is impossible to have overlapping between blueberries of different rows) a post-processing procedure is developed for removing these irregularities in $f(x, y)$.

During a UAV mission planning procedure, a selected trajectory of the UAV for image acquisition defines an orientation of the blueberry field in the later created orthomosaic. Such information reveals possible directions of the blueberry rows in relation to the field boundary. Within this work, we propose a method for determination of the rows direction through estimation of $\theta \in \mathbb{R}$ angle between rows direction and one of the image axis x or y based on Radon transformation of $f(x, y)$ [32]. After each rotation, a summation along the y image axis is conducted resulting in the slices of sinogram $\mathcal{R}_\theta[f(x, y)] \in \mathbb{R}^K$ of $f(x, y)$ where $\mathcal{R}$ denotes the Radon transform operator and K dimension of the

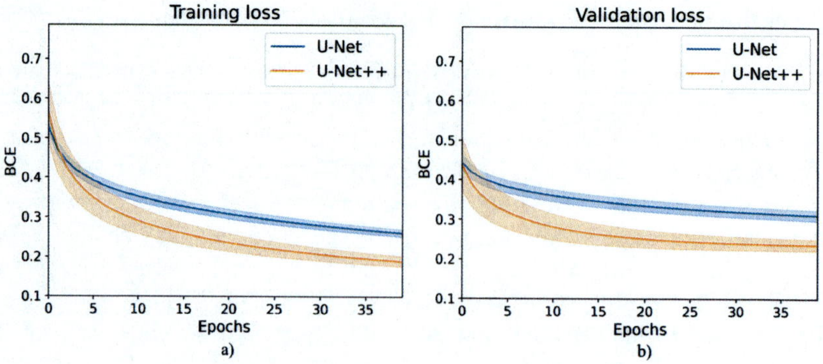

Fig. 2. Values of BCE losses during a) training and b) validation phase for modified U-Net and U-Net++ architecture.

slice. The smallest area under the 1D slices is obtained when the direction of the segmented rows is parallel to the axis of summation. Therefore, the estimation of the angle of rotation $\hat{\theta}$ is conducted by solving the following optimization problem:

$$\hat{\theta} = \underset{\theta}{\text{argmin}} \left\{ \sum_{y \vee x} \mathcal{R}_{\theta}[f(x,y)] \right\} \quad (1)$$

where notation $y \vee x$ denotes along which image axis the summation is conducted and this is determined based on prior information about the orientation of the blueberry field. Once the $\hat{\theta}$ is estimated according to the formulated criterion in (1), the normalized sinogram slice $\mathcal{R}_{\hat{\theta}}[f(x,y)]/||\mathcal{R}_{\hat{\theta}}[f(x,y)]||_1$, which corresponds to the estimated $\hat{\theta}$, is further used for determination of rows position and their width. By thresholding $\mathcal{R}_{\hat{\theta}}[f(x,y)]/||\mathcal{R}_{\hat{\theta}}[f(x,y)]||_1$ a 1D binary signal is obtained from which pulse width corresponds to the row width. Utilizing this information, all irregularities in shapes of estimated rows regions in $f(x,y)$ are corrected resulting in corrected binary segmentation mask $f_c(x,y)$. Due to the installation of the anti-hail network within the blueberry field, additional buffer regions at the end of estimated rows regions must be added to ensure that the robot trajectory circumvents it. Finally, one of the possibly allowed trajectories for the robot routing within the blueberries field is estimated by applying medial axis transformation [33] on the binary image obtained from exclusive or operation between $f_c(x,y)$ and mask, which denotes the parcel region.

3 Results

The proposed methods are evaluated in several steps. To examine the influence of stochastic data shuffling during the training, we ran 50 experiments for the training of each model using the splitting of the dataset from Table 1. In Fig. 2 are shown averaged training and validation loss curves with their corresponding confidence intervals after 50 experiments. U-Net++ achieved a steeper loss curve

Table 2. Validation loss and IoU metric from cross-validation leave-one-out parcel experiment

	Leave-one-out parcel	
Model	U-Net	U-Net++
Validation loss	0.2891 ± 0.014	**0.23356 ± 0.0114**
IoU	0.80155 ± 0.03	**0.82803 ± 0.04**

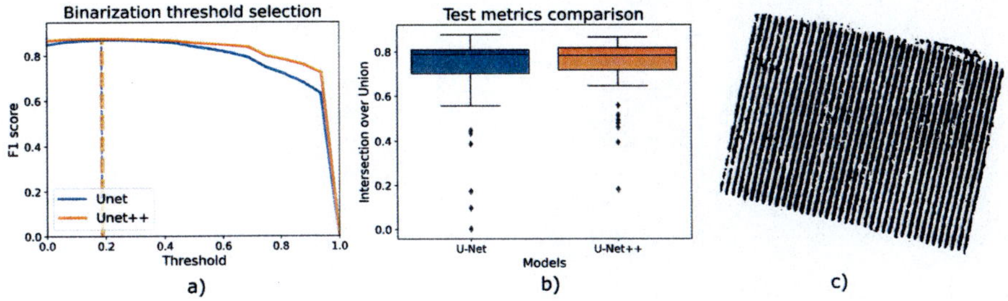

Fig. 3. Creation of a binary map of detected rows for the whole orthomosaic of test parcel from Fig. 1. a) Selection of the threshold for mask binarization, b) IoU patch-wise distribution of predicted maps for both models, and c) binary map of detected blueberry rows (black) using U-Net++ trained model with a threshold equal to 0.1875.

than U-Net in both training and validation, which lead to lower loss values, as well as higher mean IoU validation metric.

Further, a quantitative analysis of the segmentation performance is conducted through a leave-one-out parcel cross-validation experiment. The validation parcel is kept the same while the test parcel is changed (see Fig. 1), keeping always four parcels in the training set. Overall IoU metric on the test parcel is obtained by averaging the calculated IoU metrics across patches. In Table 2 are reported results of leave-one-out parcel cross-validation for both models. The presented metric with ± standard deviation in Table 2 showed that U-Net++ architecture achieves slightly better performances compared to U-Net. In further analysis, we use models that reached the lowest validation loss, reported after the 50 experiments with different data shuffling and the same splitting of a dataset depicted in Fig. 1 and described in Table 1.

Over the test region in Fig. 1, two selected models for row detection are applied using the sliding window technique with an overlap of 75%. Prediction results on patches (512 × 512) are aggregated into a test region dimension. In this way, we want to emphasize those pixels for which the model is sure to be blueberry rows, while for those pixels for which it is not completely sure, it will have the opportunity to make a decision multiple times and thereby increase the probability of an accurate prediction. After assembling prediction images, we perform min-max normalization. The next step is to determine the

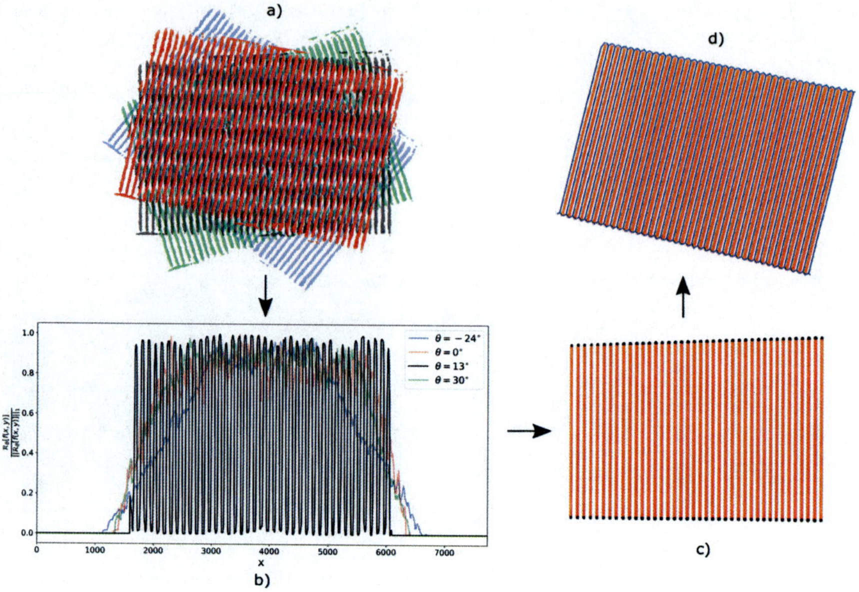

Fig. 4. Inferring robot trajectory a) rotations of the segmented rows regions $f(x, y)$ obtained from DNN output, b) sinogram slices for different values of θ, obtained by summing the $f(x, y)$ along the y-axis and normalizing them with ℓ_1 norm c) corrected $f_c(x, y)$ with row regions denoted (orange) and added buffer at their ends (black), based on normalized sinogram slice under b) which corresponds to $\theta = 13°$, d) estimated robot trajectory (blue) around corrected rows regions (red).

binarization threshold, which gives us the best results in terms of the F1 score. In Fig. 3, the dependence of the F1 score on the threshold value is shown, and the best results for both models are obtained with a threshold equal to 0.1875. Prediction maps obtained on test images using this threshold are then being evaluated in terms of IoU (see Fig. 3). We have 69 non-overlapping 512×512 test images with final predictions. Although performances of both models are expressed with similar distributions of IoU metric on test images, U-Net++ has a slightly higher IoU median value and tighter interquartile range as well as tighter whiskers. Also when we calculate IoU on the whole test parcel, we get 0.7377 and 0.7697 for U-Net and U-Net++, respectively. Considering previous analysis, we are encouraged to choose U-Net++ as our final model for generating blueberry rows segmentation mask (see Fig. 3) for further inference method.

The Radon transform is then applied to obtained row detection binary mask $f(x, y)$ to estimate the θ angle. In Fig. 4 under a) are presented several rotations of $f(x, y)$ around its center for different values of rotation angle θ. The optimal θ is obtained by solving the optimization problem in (1) by searching the sinogram slice with the minimum area under the curve. Sinogram slices for different values of θ are depicted in Fig. 4 under b). Once the optimal θ is found and its corresponding sinogram slice, the correction of rows regions is conducted. The final two steps for inferring the allowed UGV path within the blueberry field: extension of rows and medial axis transformation are presented in Fig. 4 under c) and d). As a result, the inclusion of this post-processing step increased IoU metrics from 0.7697 to 0.9021 on the test parcel from Fig. 1, which is a significant improvement.

4 Conclusions and Future Work

In this paper, we present an approach for row detection in blueberry fields using multispectral UAV images and for inferring the allowed path for UGV based on segmented row regions. The semantic segmentation of blueberry rows is conducted using U-Net and U-Net++ DNN architectures. From the obtained segmentation mask additional corrections in the shape of rows are performed using Radon transform. A final allowed path for UGV is obtained from the corrected mask and parcel mask using exclusive or logical operation and medial axis transformation. In future work, we will address the routing of UGV along the estimated allowed path for the soil sampling task.

Acknowledgement. This work was supported by FLEXIGROBOTS (Flexible robots for intelligent automation of precision agriculture operations) project funded from the European Union's Horizon 2020 research and innovation programme under grant agreement No. 101017111.

References

1. Bah, M.D., Hafiane, A., Canals, R.: Crownet: deep network for crop row detection in UAV images. IEEE Access **8**, 5189–5200 (2020)
2. Bhatnagar, S., Gill, L., Ghosh, B.: Drone image segmentation using machine and deep learning for mapping raised bog vegetation communities. Remote Sens. **12**, 2602 (2020)
3. Adão, T., et al.: Hyperspectral imaging: a review on UAV-based sensors, data processing and applications for agriculture and forestry. Remote Sens. **9**, 1110 (2017)
4. Messina, G., Modica, G.: Applications of UAV thermal imagery in precision agriculture: state of the art and future research outlook. Remote Sen. **12**, 1491 (2020)
5. Tsouros, D.C., Bibi, S., Sarigiannidis, P.G.: A review on UAV-based applications for precision agriculture. Information **10**, 349 (2019)
6. Kim, J., Kim, S., Ju, C., Son, H.I.: Unmanned aerial vehicles in agriculture: a review of perspective of platform, control, and applications. IEEE Access **7**, 1051000–105115 (2019)
7. Boursianis, A.D., et al.: Internet of things (IoT) and agricultural unmanned aerial vehicles (UAVs) in smart farming: a comprehensive review. Internet Things **18**, 100187 (2022)
8. Latif, M.A.: An agricultural perspective on flying sensors: state of the art, challenges, and future directions. IEEE Geosci. Remote Sens. Mag. **6**, 10–22 (2018)
9. Lapray, P.J., Wang, X., Thomas, J.B., Gouton, P.: Multispectral filter arrays: recent advances and practical implementation. Sensors **14**, 21626–21659 (2014)

10. Mahlein, A.K., Kuska, M., Behmann, J., Polder, G., Walter, A.: Hyperspectral sensors and imaging technologies in phytopathology: state of the art. Ann. Rev. Phytopathol. **56**, 535–558 (2018)
11. Jia, K., Wu, B., Li, Q.: Crop classification using HJ satellite multispectral data in the North China Plain. J. Appl. Remote Sens. **7**, 073576 (2013)
12. Kussul, N., Lavreniuk, M., Skakun, S., Shelestov, A.: Deep learning classification of land cover and crop types using remote sensing data. IEEE Geosci. Remote Sens. Lett. **14**, 778–782 (2017)
13. Palanisamy, S., Selvaraj, R., Ramesh, T., Ponnusamy, J.: Applications of remote sensing in agriculture - a review. Int. J. Curr. Microbiol. Appl. Sci. **8**, 2270–2283 (2019)
14. Lopez-Granados, F.: Weed detection for site-specific weed management: mapping and real-time approaches. Weed Res. **51**, 1–11 (2011)
15. Meraner, A., Ebel, P., Zhu, X.X., Schmitt, M.: Cloud removal in sentinel-2 imagery using a deep residual neural network and sar-optical data fusion. ISPRS J. Photogram. Remote Sens. **166**, 333–346 (2020)
16. Bah, M.D., Hafiane, A., Canals, R.: Deep learning with unsupervised data labeling for weed detection in line crops in uav images. Remote Sens. **10**, 1690 (2018)
17. Maimaitijiang, M., Sagan, V., Sidike, P., Hartling, S., Esposito, F., Fritschi, F.B.: Soybean yield prediction from UAV using multimodal data fusion and deep learning. Remote Sens. Environ. **237**, 111599 (2020)
18. Liu, Q., Kampffmeyer, M.C., Jenssen, R., Salberg, A.B.: Multi-view self-constructing graph convolutional networks with adaptive class weighting loss for semantic segmentation. In: Proceedings of the IEEE/CVF Conference on Computer Vision and Pattern Recognition (CVPR) Workshops (2020)
19. Hunt, E.R., Jr., Daughtry, C.S.T., Eitel, J.U.H., Long, D.S.: Remote sensing leaf chlorophyll content using a visible band index. Agron. J. **103**, 1090–1099 (2011)
20. Ponti, M.: Segmentation of low-cost remote sensing images combining vegetation indices and mean shift. IEEE Geosci. Remote Sens. Lett. **10**, 67–70 (2013)
21. Sari, N., Kushardono, D.: Object segmentation on UAV photo data to support the provision of rural area spatial information. In: Forum Geografi, vol. 29 (2015)
22. Parsons, M., Bratanov, D., Gaston, K.J., Gonzalez, F.: UAVs, hyperspectral remote sensing, and machine learning revolutionizing reef monitoring. Sensors **18**, 2026 (2018)
23. Miyamoto, H., Momose, A., Iwami, S.: UAV image classification of a riverine landscape by using machine learning techniques. In: EGU General Assembly Conference Abstracts, p. 5919 (2018)
24. Pádua, L., Adão, T., Hruška, J., Guimarães, N., Marques, P., Peres, E., Sousa, J.J.: Vineyard classification using machine learning techniques applied to rgb-uav imagery. In: IGARSS 2020–2020 IEEE International Geoscience and Remote Sensing Symposium, pp. 6309–6312 (2020)
25. Zhao, X., et al.: Use of unmanned aerial vehicle imagery and deep learning unet to extract rice lodging. Sensors **19**, 3859 (2019)
26. Li, D., Meng, X., Li, B.: Profiling of anthocyanins from blueberries produced in China using HPLC-DAD-MS and exploratory analysis by principal component analysis. J. Food Compos. Anal. **47**, 1–7 (2016)
27. Medeiros, J.G.S., Biasi, L.A., de Bona, C.M., Cuquel, F.L.: Phenology, production and quality of blueberry produced in humid subtropical climate. Revista Brasileira de Fruticultura **40** (2018)
28. Aguiar, A.S., dos Santos, F.N., Cunha, J.B., Sobreira, H., Sousa, A.J.: Localization and mapping for robots in agriculture and forestry: a survey. Robotics **9**, 97 (2020)

29. Blok, P.M., van Boheemen, K., van Evert, F.K., IJsselmuiden, J., Kim, G.H.: Robot navigation in orchards with localization based on particle filter and kalman filter. Comput. Electron. Agric. **157**, 261–269 (2019)
30. Ronneberger, O., Fischer, P., Brox, T.: U-Net: convolutional networks for biomedical image segmentation. In: Navab, N., Hornegger, J., Wells, W.M., Frangi, A.F. (eds.) MICCAI 2015. LNCS, vol. 9351, pp. 234–241. Springer, Cham (2015). https://doi.org/10.1007/978-3-319-24574-4_28
31. Zhou, Z., Siddiquee, M.M.R., Tajbakhsh, N., Liang, J.: Unet++: a nested u-net architecture for medical image segmentation. In: Deep Learning in Medical Image Analysis and Multimodal Learning for Clinical Decision Support, pp. 3–11 (2018)
32. Radon, J.: On the determination of functions from their integral values along certain manifolds. IEEE Trans. Med. Imaging **5**, 170–176 (1986)
33. Blum, H.: A transformation for extracting new descriptions of shape. In: Models for the Perception of Speech and Visual Form (1967)
34. Götz, W., Druckmüller, H.: A fast digital radon transform-an efficient means for evaluating the hough transform. Pattern Recogn. **29**, 711–718 (1996)
35. Zhang, Q., Couloigner, I.: Accurate centerline detection and line width estimation of thick lines using the radon transform. IEEE Trans. Image Process. **16**, 310–316 (2007)
36. Evangelidis, G.D., Psarakis, E.Z.: Parametric image alignment using enhanced correlation coefficient maximization. IEEE Trans. Pattern Anal. Mach. Intell. **30**, 1858–1865 (2008)
37. QGIS Development Team: QGIS Geographic Information System. QGIS Association (2022). https://www.qgis.org
38. Kingma, D.P., Ba, J.: Adam: a method for stochastic optimization (2014)

Versatile and Heterogeneous Robots in Agriculture

Collaborative Harvest Robot

Maria Nuria Conejero[1], Hector Montes[1,2], Dionisio Andujar[1], Jose Maria Bengochea-Guevara[1], and Angela Ribeiro[1](✉)

[1] Centre for Automation and Robotics (CAR), CSIC-UPM, Arganda del Rey, 28500 Madrid, Spain
angela.ribeiro@csic.es
[2] Centro de Investigación e Innovación Eléctrica, Mecánica y de la Industria (CINEMI), Universidad Tecnológica de Panamá, Panama City, Panama

Abstract. Agriculture is a field of activity characterized by physical labor, which is why many efforts are being made to introduce robotics on farms. In the specific case of vineyards, and harvesting, the difficulty lies in approximating the harvesting times of the grape harvester without damaging the product and consequently offering a cost-effective solution. This paper presents an operator tracking strategy for a prototype mobile, collaborative robotic platform to address the harvesting task with a human-robot collaboration that combines the manual dexterity of the harvester with the carrying capacity of a robot. The system designed and developed is applicable in the field of manual fruit picking. A commercial mobile robot platform has been used and a structure adapted to the transport and weighing of the contents of the box has been built. In addition, an RGB-D camera and UWB sensors have been integrated. A system for tracking a specific operator while maintaining a safety distance has been integrated into the robot. The system runs on ROS. The results show that the robot correctly performs the operation of tracking the grape harvester and is robust in the event of major changes in lighting.

Keywords: Agricultural robotics · Collaborative robotics · Harvesting · Mobile robot · Person-follower · UWB · RGB-D

1 Introduction

Vineyards' manual harvesting, like any other manual fruit or vegetable harvesting process, requires great physical effort during long working days due to the location of the bunches of grapes on the vines, irregularities in the terrain, variable weather conditions, humidity, and dust [1]. In addition, the operators are also responsible for transporting the boxes with the bunches of grapes from row to row, which can weigh up to 20 kg when the boxes are full, and they must place them at the harvesting points, where sometimes are not collected until the end of the day. In this context, the advantages of integrating robots to reduce the physical effort required by the harvester while maintaining product quality and harvesting speed are obvious.

Robots that are capable of autonomously harvesting and picking all kinds of products such as greenhouse tomatoes [2], peppers [3], strawberries [4], kiwis [5] and lettuces

D. Tardioli et al. (Eds.): ROBOT 2022, LNNS 590, pp. 415–426, 2023.
https://doi.org/10.1007/978-3-031-21062-4_34

[6] have been designed and developed. Grape harvesting is not far behind, and several projects have aimed to automate this process as well [7]. However, they do not reach the picking speed of an operator and the bunches are always damaged. For the latter reason, some appellations guarantee manual harvesting. In the field of manual fruit picking, which includes harvesting, some platforms assist autonomous transport once the box is full and after a warning from the operator [8]. Also, there is currently a diverse offer of devices based on different technologies to build robust people detection and identification methods. Some examples are Laser Imaging Detection and Ranging (LIDAR) [9], depth cameras [10], Radio Frequency Identification (RFID) [11] and Ultra WideBand (UWB) [12]. Commercial alternatives that advocate human-robot collaboration to assist grape and other manual harvesting processes are found, but only with autonomous navigation between the end of the line and the storage location, without following up the operator [13].

The aim of this work is the design and development of an economically cost-effective collaborative robotic mobile platform to assist in harvesting or manual fruit picking and to combine the manual dexterity of the operator and avoid the physical effort of transport utilizing a load-carrying robot. The development and testing of this robot on Terras Gauda (https://www.terrasgauda.com/index_en.php), a farm in Pontevedra (Spain), are one of the objectives of the European FlexiGroBots project (https://flexigrobots-h2020.eu). The current paper addresses specifically the pilot 1, which is being developed in Spain, with the aim of carrying out various tasks in vineyards autonomously. The robotic platform must be able to move following an operator at a certain and safe distance, with the appropriate speed to maintain that distance within a reactive behaviour, while the operator deposits the bunches of grapes collected in the box that the robot transports. The platform will also weigh the contents of the box continuously, so that it will be possible to report at all times to a control center and, once the harvesting task is completed, to generate a harvest map. The overall objective is to increase or maintain the rate of grapes harvested per unit of time, reducing operator fatigue and possible injuries when transporting the grapes.

2 Materials and Methods

The starting point was the RB-Vogui commercial mobile platform from *Robotnik*. A structure was designed and built to contain the sensory equipment and the box. The sensory devices include an RGB-D camera, namely the Intel RealSense D415, two UWB modules from *Decawave's* MDEK1001 kit placed at a similar height to the camera, a GRAM K3-F industrial scale on which the harvesting box is placed, and a *Septentrio* MOSAIC-H GNSS receiver (Fig. 1(a)). The box is placed on top of the scale with a robust but easy to handle anchoring system. The 4-point anchoring ensures that the box is solidly attached to the platform and is easy to undock.

On the other hand, the operator must wear an orange reflective vest and a "label" (UWB module) hanging from his/her neck (Fig. 1(b)), which allows the corresponding operator to be identified. It should be noted that in a real situation there will be more than one robot working, each one assigned to a grape harvester.

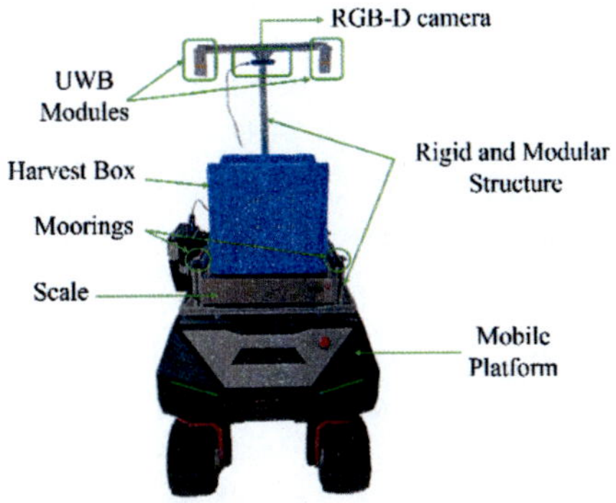

(a) Platform configuration.

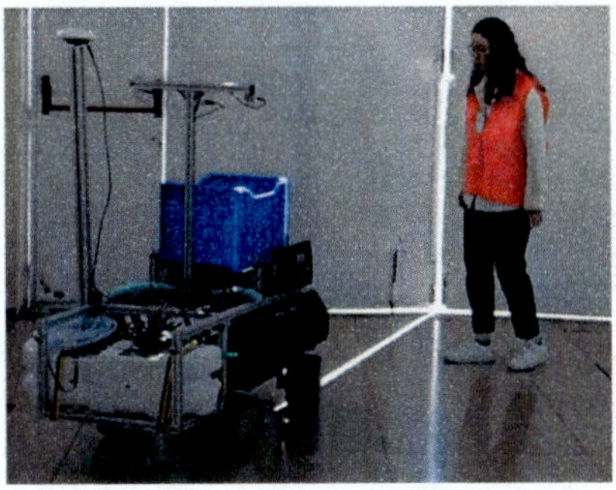

(b) Operator in relation to the robot.

Fig. 1. Elements and disposition of the robotic platform

2.1 Operator Identification System

To achieve greater robustness in tracking, the information provided by the RGB-D sensor, which allows the location of the vest worn by the operator, has been merged with the information provided by the 3 UWB devices (two onboard the robot named tags and one on the operator named anchor). The developed method allows knowing the distance and orientation from the robot to the correspondent operator.

Segmentation of the Vest and Distance Acquisition with the RGB-D Camera The identification of the vest is carried out by processing the color image (RGB), the aim of which is to isolate the pixels in the image belonging to the vest by color. To do this, the image processing is done in the HSV (Hue Saturation Value) color space, and a mask is applied to each plane of the space that selects the values between maximum and minimum coefficients, giving them a value of 1, while the rest have a value of 0. Pixels with a 1 in all three planes are then classified as part of the vest (Fig. 2). The coefficients for the segmentation have been obtained utilizing a genetic algorithm with the configuration shown in Table 1 and using a dataset of 150 RGB images with a resolution of 1280 $\times$ 720 pixels taken with the RGB-D camera both indoors and outdoors, on sunny and cloudy days.

Fitness is defined as;

$$Fitness(x) = \frac{vab}{vvb + vfb}, \tag{1}$$

where *vab* are the hits on the vest pixels, *vvb* are the true white values (vest pixels) and *vfb* are the false white values (not part of the vest). Once the coefficients for the segmentation are available, the developed algorithm obtains the distance for each image by segmenting the vest pixels, applying morphological operators: a first erosion, dilation, and a second erosion, and then searching for the vest outline in the binary image. With the associated depth image (640 $\times$ 480 resolution) the mean of the distance to each white pixel inside the vest is calculated, estimating the distance from the robot to the operator.

Table 1. Genetic algorithm setup.

Parameter	Value	Parameter	Value
Individual	H, S, V (min and max)	Crossover fraction	0.8
Lower limit	(0, 0, 0, 0, 0, 0)	Mutation function	Uniform
Upper limit	(1, 1, 1, 1, 1, 1)	Mutation probability	0.3
Individuals	40	Elitism	1 individual
Selection function	Uniform stochastic	Completion criteria	0.99 fitness or 200 generations without improvement or 1000 iterations achieved
Crossover function	Scattered		

Fig. 2. Original and segmented image.

Obtaining Position, Orientation and Distance of the Operator to the UWB Device. With the tags and the anchor, the objective is to obtain the position and orientation of the operator in order to calculate the distance from the robot to the operator. The tags are active elements that emit a Bluetooth type signal that is received by the anchor. The distance between the operator and the robot is estimated from the time of flight (ToF) and by using the information provided by the two tags integrated within the robot. In the presence of several operators, the distance received will be that of the devices on the same network, which also allows the operator to be identified.

The update frequency of these measurements has been set at 10 Hz for steady-state and during movement so that new measurements will be received every 100 ms. The range of variation of the measurements, even in static mode, is 10 cm, which makes it impossible to obtain the same measurement continuously. Consequently, a digital low-pass filter (2) has been implemented together with a moving average filter (3) in order to reduce the existing noise.

$$y[n] = b_0x[n] + b_1x[n-1] + b_2x[n-2] + \ldots + + b_Mx[n-M] - a_1y[n-1] \\ -a_2y[n-2] - \ldots - -a_Ny[n-N] \quad (2)$$

$$y[n] = \frac{1}{M}\sum_{k=0}^{M-1} x[n-k] \tag{3}$$

With the low-pass filter, a certain number of previous measurements are considered when obtaining an output value of the current measurements. The filter has been configured with an order of 2 and a cut-off frequency of 0.5 rad/s so that every 5 measurements received (M) the same measurements are returned attenuated, and the average is obtained with the moving average filter. In this way, the new measurements are updated every 0.5 s and the position and orientation of the operator are estimated with respect to the geometric center (center of rotation and advance) of the robot. Finally, the distance between the robot and the operator is calculated from the rear of the robot, where the camera and the tags are placed at a known distance d_x from the geometric center (in this case 0.308 m) and a distance d_T between them (0.47 m) in order to compare both estimations (RGB-D and UWB). The operator will always move between the first two quadrants of the Cartesian plane (Fig. 3(a)) because if he is outside the camera the robot stops. The determination of the case type is given by the distance d_{01} and distance d_{02}. If d_{01} is equal or bigger than d_{02}, the A_{01} (operator) is within the first quadrant. In any other cases A_{01} is within the second quadrant.

The estimation from the tags varies depending on the quadrant in which the operator is located. For the first quadrant the estimates of the distances d_{01} and d_{02} form the triangle shown in Fig. 3(b), where the angle θ_1 is obtained from expression (4), and the distances Y_A and X_A to anchor A_{01} from expressions (5) and (6).

$$\theta_1 = \arccos\left[\frac{(d_{01})^2 + (d_T)^2 - (d_{02})^2}{2(d_{01})(d_T)}\right] \tag{4}$$

$$X_A = (d_{01})\sin(\theta_1) - d_x \tag{5}$$

$$Y_A = (d_{01})\cos(\theta_1) - \frac{d_T}{2} \tag{6}$$

In turn, the orientation θ_R and the distance d_R (from the rear of the robot) to anchor A_{01} can be calculated from expressions (7) and (8).

$$\theta_R = \tan^{-1}\frac{X_A}{Y_A} \tag{7}$$

$$d_R = \sqrt{(X_A + d_x)^2 + (Y_A)^2} \tag{8}$$

Once the last transformation has been performed, a Kalman filter is applied for the correction and estimation of the x and y position coordinates in order to smooth the measurements and further reduce their variation, especially in a static position, reducing the yaw and pitch errors that could appear in the robot. For this specific case and taking as a reference the filter applied for the UWB devices in the work [14], a constant velocity model will be used as an approximation of the high reactive behavior of the operator since it is impossible to know its displacement velocity. Therefore, the velocity of x and y variables (dt) is set as the refresh rate of the measurements, 0.5 s.

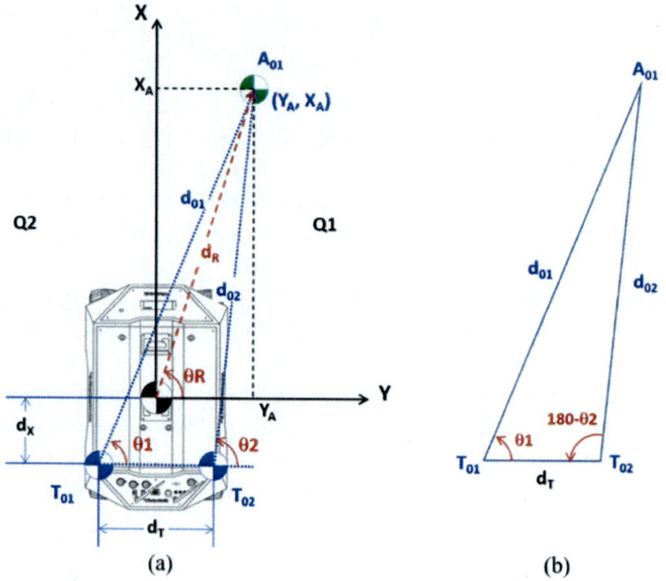

Fig. 3. Proposed human-robot collaboration system. (a) Position of the operator (A_{01}) in the first quadrant. (b) Operator's orientation obtention in the first quadrant.

2.2 Control of Robot Movement

To control the movement of the robot, there is always an acting on the linear velocity for the x-axis and the angular velocity on the z-axis assigned.

Angular Velocity Control. To assign both the module and the direction of the angular velocity corresponding to the robot, the following parameters are considered: the angle of rotation about the z-axis of the robot itself, the angle θ of the operator with respect to the robot, obtained from the UWB equipment, and an allowed orientation error tolerance of 8.6 degrees/0.15 radians. For the angular velocity, a case-by-case control is carried out, distinguishing between 3 possible situations (9).

$$\begin{cases} if\ |\theta_{user}| \leqslant \ \ tolerance \quad \rightarrow |\omega| = 0\,rad/s \\ if\ -30^\circ \leq \theta_{user} < -tolerance \cup tolerance < \theta_{user} \leq 30^\circ \rightarrow |\omega| = 0.6^\circ rad/s \\ if\ |\theta_{user}| > 30^\circ \qquad \rightarrow |\omega| = \ 0.7 rad/s \end{cases} \tag{9}$$

The operator is always in front of the robot (first or second quadrant), so fast and short movements are required. Therefore, the strategy is to apply a slightly higher speed when a larger turning angle is required, without being too high to avoid an excessive turning angle. The direction of rotation is decided between clockwise and counterclockwise, considering the robot's reference system. A clockwise rotation will have a negative ω and θ sign and a positive sign in the opposite case.

Finally, through a comparison between the θ_{robot} value given by its odometry, and the $\theta_{operator}$ value with respect to it (summed to θ_{robot}), the different situations that occur and the direction of the corresponding velocity are shown in expressions (10), when the robot is in the second quadrant ($\theta_{robot} > 0$), (11), when the robot is in the first quadrant ($\theta_{robot} < 0$) and (12), when the robot is in the center ($\theta_{robot} = 0$).

$$\begin{cases} if\ \theta_{operator} \leq 0 \rightarrow \omega\, cw. \\ if\ \theta_{operator} > 0\, and\, \theta_{robot} > \theta_{operator} \rightarrow \omega\, cw. \\ if\ \theta_{operator} > 0\, and\, \theta_{robot} < \theta_{operator} \rightarrow \omega\, ccw. \end{cases} \tag{10}$$

$$\begin{cases} if\, \theta_{operator} \geq 0 \rightarrow \omega\, ccw. \\ if\, \theta_{operator} < 0\, and\, \theta_{robot} < \theta_{operator} \rightarrow \omega\, ccw. \\ if\, \theta_{operator} < 0\, and\, \theta_{robot} < \theta_{operator} \rightarrow \omega\, cw. \end{cases} \tag{11}$$

$$\begin{cases} if\, \theta_{operator} > 0 \rightarrow \omega\, ccw. \\ if\, \theta_{operator} < 0 \rightarrow cw. \end{cases} \tag{12}$$

Linear Velocity Control. The value of the linear velocity is given by a proportional controller;

$$v\_\text{linear} = 0.5 * distance\_operator, \tag{13}$$

where the constant value 0.5 has been experimentally fixed.

2.3 Integration of the Different Systems on the Robotic Platform

The logic or operation of the platform and its follower behavior can be divided into 3 large blocks operating independently and in parallel: the weighing block, the block for collecting and synchronizing the sensory information and the block for decision-making and actuation. The latest two blocks are also interconnected. A diagram of the information exchange between them can be seen in Fig. 4.

Weighing Block. This first block synchronizes the weigh data and the GPS position and registers it for the execution of a harvest map. This functionality does not directly influence the mechanisms that generate the robot's reactive behavior and is not within the scope of the article mainly due to limitations in the length of the article itself.

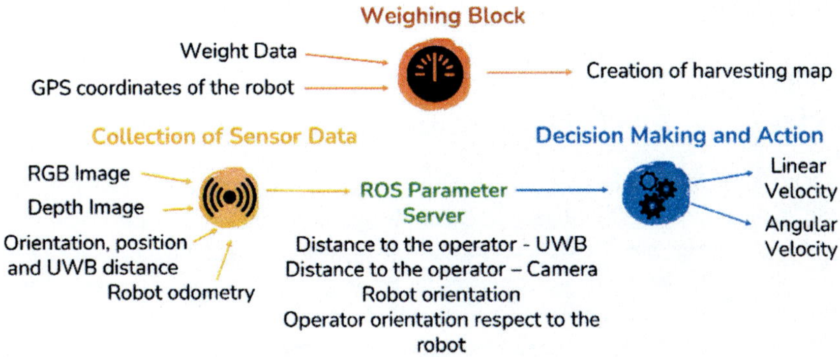

Fig. 4. Flow of information between the three operating blocks of the system: the weighing block, the collection of sensor data block and the decision making and action block.

Collection of Sensor Data. In this block, the RGB-D, UWB and odometrical information of the robot are received synchronously. With all the information, the first step is to obtain the distance from the robot to the anchor. The process is then repeated with the orientation data of the anchor and that of the robot. This data, added to the orientation given by the anchor, allows the angle of the operator to be known according to the robot's reference system when the robot is moving. Next, the RGB-D images are used to obtain the distance to the operator provided by the camera. At this point, the UWB information is used to disambiguate and use the correct distance provided by the camera. The UWB operator distance, camera operator distance, robot orientation and operator orientation are managed through the ROS parameter server.

Decision Making and Action. This block is responsible for generating the control commands for the speeds within three options.

Situation 1: No Distance Given by the Camera. If the value of the distance to the operator given by the camera is null/empty, it means that no operator has been detected or the small orange region detected is not considered relevant, and the robot stops. In this way, the robot remains stopped when the operator is depositing the crop.

Situation 2: The Distances to the Operator Given by the Camera and the Anchor are Similar. This situation occurs when the difference between them is less than 2 m. Then, if that difference is greater than 0.3 m, the value of the anchor is taken as the reference distance to the operator. In other cases, the distance to the operator is given by the camera. Once the best value has been chosen, it is checked whether it is less than or equal to 0.5 m. If it is, the robot is sufficiently close to the operator and stops. If not, the velocities are calculated to generate the tracking behavior. For the case of angular velocity, it is first checked whether the orientation of the operator is within the tolerance angle set at $\pm$ 0.15 radians (about the robot's forward axis) and the robot is sufficiently centered to move only linearly. If not, angular, and linear velocity are calculated.

Situation 3: The Distances Given by the Camera and the Anchor are Different. This last situation is considered to avoid confusion with the operator to be followed and increase safety. Therefore, when the difference between the distances obtained with the sensors is greater than the established 2 m threshold, the robot stops.

3 Results and Discussion

The calculation of the operator localization accuracy was carried out experimentally. For the evaluation of the low-pass filter with the greatest reduction of the UWB signals error, filters of order 2 or 3 with an intermediate or low cut-off frequency were selected for the experiment. With one person wearing the anchor and the vest at 1.4 m from the robot, measurements were taken at a position of 0°/0 rad to the robot, 45°/−0.7854 rad to the left of the robot and 45°/−0.7854 rad to the right of the robot in order to calculate the orientation and position error (Table 2). The filter of order 2 and cut off frequency 0.5 presents lower error in orientation, while the filter of order 2 and cut off frequency 0.3

Table 2. Selected low-pass filter errors.

Position	Order	Cut-off (rad/s)	Angle error (rad)	Distance error (m)
Center	3	0.5	0.0724	0.1239
Center	2	0.5	0.0002	0.1143
Center	2	0.3	0.0392	0.0479
Left	3	0.5	0.1740	0.0925
Left	2	0.5	0.0513	0.0703
Left	2	0.3	0.0405	0.6110
Right	3	0.5	0.4510	0.0341
Right	2	0.5	0.3671	0.0042
Right	2	0.3	0.6154	0.0738

achieves higher accuracy in the distance. However, the former is chosen as the distance most used by the robot is from the camera, while the orientation reference depends only on the UWB devices.

To quantify the impact of the addition of the Kalman filter and the stability of the position data, a second experiment was performed. The robot was positioned with its centroid at a (0.0) position, and on a matrix with equidistant markers varying by 0.4 m in one of the two plane axes, on the ground. The operator moved over each marker and

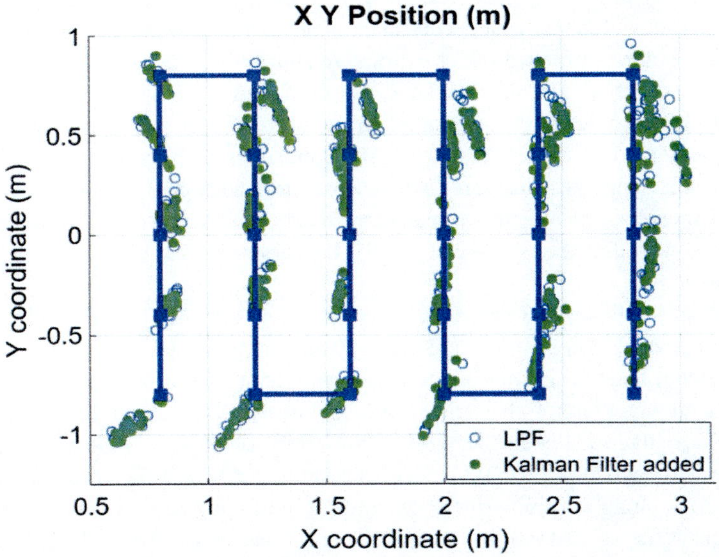

Fig. 5. Results of the Kalman filter experiment for static positions. In blue the theoretical path and measurement points, in light blue the position results using only the low-pass filter and in green the position results with the addition of the Kalman filter.

remained static on them for 10 measurements, approximately. Figure 5 shows the results, comparing the data obtained with the one without the Kalman filter.

The Root Mean Squared Error (RMSE) was used as a metric to calculate the error of the positions (x, y) and the distance between the operator and the robot with and without applying the filter. The Kalman filter correction is reduced to values in the order of millimeters, having an error of 70.8 mm and 71.6 mm with and without the Kalman filter respectively in the X position, an error of 166.6 mm and 169.7 mm in the Y position, and a distance error of 44.2 mm and 44.7 mm also with and without the Kalman filter respectively. This is considering that the calculated errors, even using only the low-pass filter, are of very small values, except for the y-axis which is approximately 17 cm. However, the observation of a smoothing of the dispersion of the static measurements indicates that it works well enough to apply to situations where both the robot and the person carrying the anchor are in motion.

Finally, to check the overall performance of the platform, two types of experiments have been carried out: 1) evaluation of indoor tracking to check the behavior of the system in the presence of objects that could lead to confusion, and 2) evaluation of outdoor tracking to check the response in the presence of the sun, the associated changes in lighting, and the presence of several operators.

Indoors, when the operator encounters the anchor in front of the robot, the tracking is correctly performed with sufficiently stable orientation data to allow forward movement without unnecessary turning. However, in the case of extremely sharp turns or movements and despite the Kalman filter, the robot moves forward with a small turning radius until the stability of the operator allows it to move forward. Furthermore, it is corroborated that, in the presence of other objects of the same color or other operators wearing vests, the robot identifies the assigned operator without problem. After carrying out the same experiment in an outdoor environment and on an uneven surface, a difference is observed. While the vibrations produced on the structure supporting the camera and the tags do not affect the color segmentation of the vest or the distance measurement by the camera, they do affect the calculation of orientation by the UWB sensors, especially when the operator is oriented laterally. In these cases, the robot does not move in a frontal direction, a situation that must be considered when working in very narrow crops. Concerning the direct incidence of the sun, when the distance data is not obtained correctly, the robot's reasoning uses the distance of the UWB equipment as a reference without major inconvenience.

An initial test under real harvesting conditions was also performed in Terras Gauda in September 2021. The results were very satisfactory, considering that the terrain had almost no slope and the distance between rows was 3.5 m. The robot was able to track the operator while maintaining the safety distance. Further experiments with the improved prototype are planned for the September 2022 harvest season. In general, the robot's reasoning and decision-making are adequate and cover all the possibilities with a fast response time along with that of the operator. In addition, safety is maximized as the robot stops at the set distance of 0.5 m and remains stationary if the distance does not increase.

4 Conclusions and Future Lines of Action

A prototype collaborative mobile robotic platform has been designed, developed, and tested in a real harvesting environment, which is also applicable to different types of manual fruit picking. The platform is robust in agricultural environments, safe for the operator, and initial tests show that it is a plausible and usable alternative in the short term in real applications. Furthermore, the identification of the operator utilizing the color segmentation of a reflective vest and combined with UWB technology allows the operator's position and orientation to be obtained. This generates a fast and robust method that also shows stable behavior in the robot when the low pass, moving average and Kalman filters are integrated. It performs well in the face of abrupt operator behavior, despite the inconvenience of direct sunlight on the camera. This indicates that the best option for the application of these characteristics is to equip the robot with robust sensors to cover the maximum possible number of cases and combine the information provided by the sensory equipment. In this way, the most reliable information is used to make decisions according to the situation detected, which allows it to be used in dynamic agricultural environments with many conditioning factors such as weather conditions, the morphology of the terrain or the presence of obstacles.

In future lines of the project, it is planned to incorporate in the robotic platform a level maintenance system for the scales on sloping terrain, where the load always remains stable and unaffected. In addition, the incorporation of two ultrasound sensors on the sides of the robot is proposed to measure the distances with respect to the crop rows and avoid damaging the hardware or the crop itself when making certain turns. Furthermore, the functionality of the platform will be extended by using the camera and the ultrasound sensors to detect and avoid external and dynamic obstacles, achieving the second desired behavior consisting of safe and autonomous navigation along a planned route. Finally, it is also possible to consider an operator detection strategy that does not depend solely on the color of the vest and its segmentation.

Acknowledgements. This work was carried out within the framework of the FlexiGroBots project funded by the European Union under the H2020 Programme with Grant Agreement Id. 101017111.

References

1. Ravankar, A., et al.: Autonomous and safe navigation of mobile robots in vineyard with smooth collision avoidance. Agriculture **11**(10), 954 (2021)
2. Feng, Q., et al.: Design and test of robotic harvesting system for cherry tomato. Int. J. Agric. Biol. Eng. **11**(1), 96–100 (2018)
3. Arad, B., et al.: Development of a sweet pepper harvesting robot. J. Field Robot. **37**(6), 1027–1039 (2020)
4. Xiong, Y., et al.: An autonomous strawberry-harvesting robot: design, development, integration, and field evaluation. J. Field Robot. **37**(2), 202–224 (2020)
5. Williams, H., et al.: Improvements to and large-scale evaluation of a robotic kiwifruit harvester. J. Field Robot. **37**(2), 187–201 (2020)
6. Birrell, S., et al.: A field-tested robotic harvesting system for iceberg lettuce. J. Field Robot. **37**(2), 225–245 (2020)

7. Vrochidou, E., et al.: An autonomous grape-harvester robot: integrated system architecture. Electronics **10**(9), 1056 (2021)
8. Peng, C., et al.: A strawberry harvest-aiding system with crop-transport co-robots: Design, development, and field evaluation. arXiv preprint arXiv:2107.13063 (2021)
9. Zhao, Y., et al.: A real-time low-computation cost human-following framework in outdoor environment for legged robots. Robot. Auton. Syst. **146**, 103899 (2021)
10. Priyandoko, G., Wei, C.K., Achmad, M.S.H.: Human following on ros framework a mobile robot. Sinergi **22**(2), 77–82 (2018)
11. Wu, C., et al.: A UHF RFID-based dynamic object following method for a mobile robot using phase difference information. IEEE Trans. Instrum. Meas. **70**, 1–11 (2021)
12. Rykała, Ł, Typiak, A., Typiak, R.: Research on developing an outdoor location system based on the ultra-wideband technology. Sensors **20**(21), 6171 (2020)
13. Burro. https://burro.ai/. Accessed on 22 Nov 2021
14. Sang, C.L., et al.: A bidirectional object tracking and navigation system using a true-range multilateration method. In: 2019 International Conference on Indoor Positioning and Indoor Navigation (IPIN), pp. 1–8. IEEE (2019)

Automatized Rapeseed Pest Detection and Management with Drones

Jere Kaivosoja[1(✉)], Ari Ronkainen[1], Juho Hautsalo[1], Juha Backman[1], Raimo Linkolehto[1], Miguel San Emeterio[2], and Juha-Pekka Soininen[3]

[1] Natural Resources Institute Finland (LUKE), Korkeakoulunkatu 7, 33720 Tampere, Finland
jere.kaivosoja@luke.fi

[2] ATOS IT Solutions and Services Iberia (ATOS), SL Calle Albarracin, 25, 28037 Madrid, Spain

[3] VTT Technical Research Centre of Finland, Kaitoväylä 1, 1100, 90571 Oulu, PL, Finland

Abstract. Efficient oil seed pest management can be challenging. Invasive insect pests can destroy the whole yield, but the reduction of applied pesticides is well encouraged. Recent drone technologies provide new tools for the pest management. In this work, we studied possibilities for implementing drones for pest invasion scouting and for precision pesticide spraying for the rapeseed fields. We verified individual components for the pest imaging, pest identification, spraying application construction and for the spraying mission and made an implementation plan for the system automatization. In terms of custom automatization, the implementation of the spraying drone remains challenging.

Keywords: Spraying · Image · UAV · ISOBUS · KML · FMIS · Regulations

1 Introduction

Pest management of oilseed crops such as rapeseed (Brassica napus subsp. Napus) and turnip rape (Brassica rape subsp. Oleifera) can be very laborious and challenging. The verification of existing pests or a real threat is the key in Integrated Pest Management (IPM). Sufficient pest management may need daily field visits at critical growth stage. Manual observations or data from IoT pest traps can provide up-to-date pest invasion information. However, developing UAV (Unmanned aerial vehicle) also called drone technologies may offer interesting new tools. UAVs have shown their great potential in agriculture. The UAV applications combine information and communication technologies, robotics, artificial intelligence, big data, and the Internet of Things [1].

Several different reviews of UAV applications in agriculture have been published recently. The study by Kim et al. [1] reviewed different UAV platforms, controls and applications, while the works done by Mogili et al. [2] and Tsouros et al. [3] focused on the applications and updated an earlier study by Zhang et al. [4]. The review [5] studied related reference measurements, and OECD report [6] presented the current state of the art of UAV spraying systems in agriculture. The use of drones for pesticide applications has the potential to provide benefits such as the reduction of applicator exposure, access scenarios and the enablement of precise zone or spot application and contribute to the

D. Tardioli et al. (Eds.): ROBOT 2022, LNNS 590, pp. 427–437, 2023.
https://doi.org/10.1007/978-3-031-21062-4_35

more sustainable use of pesticides [6]. Berner et al. [7] tested a drone-based spraying for rapeseed, and Zhang et al. [4] studied flowering of rapeseeds by applying drone imaging.

With rapeseed pest management, UAV systems may provide two interesting possibilities: 1) detecting pest invasion, 2) pest control against observed insect invasion. However, the first phase needs new tools since the rapeseed pests such as the pollen beetle (*Meligethes aeneus)* are the size of 2.5 mm. For this, a close-range UAV imaging such as 1 m [8] or 2 m [9] could provide a solution. These close-range images could point out the existence of pests within different locations in fields. For the second phase, flexible spraying tools based on UAV technologies can give better treatment when selective spraying of only invaded areas is needed. In addition, a spraying drone and a tractor-sprayer combination offer two different ways to spread pesticides. The drone can spray on-demand to the hotspots or specific areas without stamping the crops while the tractor can spray the whole field efficiently.

Our scope in this paper is to study state-of-the-art technologies that can be adapted for specifying, designing, and implementing pilot system components to provide actual ICT solutions for the automatization of the rapeseed pest management connected to an existing Farm Management Information Systems (FMIS) and Data Space ambitions while acknowledging safety aspects and related standardization. From that perspective, our main research question is: what kind of system design would be needed for the rapeseed pest management pilot automatization? To access this, we tested separate system components beginning from the summer 2021 and constructed sequence diagrams from the missions to make an implementation plan and to determine needed planning services. As the limited spraying resources of the spraying drones are well-known, we plan to develop the system in a way that the selection of the spraying equipment is flexible: if the spraying need is larger than the realistic capacity of the UAV, tractor sprayer -combination will perform the application task.

2 Methods

First, we identified four main steps for the UAV based pest management process: 1) pest imaging, 2) pest identification, 3) spraying application construction, 4) spraying mission and 5) operational requirements. These steps are presented next.

2.1 Pest Imaging

Our goal for the imaging was to be able detect individual rapeseed pests from separate images. The imaging campaigns were carried out just before the rapeseed flowering time in Finland in the early July 2021 and in the late June 2022. We tried different drone setups and cameras. As a result, Mavic 2 Zoom provided most comprehensive data. It was a compromise of the image resolution, exposure and the draft made by propels.

Fig. 1. Mavic 2 zoom drone taking rapeseed images.

We observed the following parameters to be adequate in average conditions (wind, sun light): flying altitude 1.2–1.5 m (Fig. 1), manual camera focus to flower-level, white balance according to cloud cover (we used –0.3 on average) and everything else in automatic mode. The images used were still images. In our case, the results of video streams were too blurry to be able to detect small insects. At this stage, the imaging was done manually around the field, taking about 100 images with a stop-and-go method.

According to our earlier review [5], majority of pest imaging studies utilized subjective visual observations on UAV images, so we did not collect any additional reference data. However, the pest invasions were verified at the field.

As the flying mission operation at this first stage was done manually, an automatic option was planned and verified with QGroundcontrol software (http://qgroundcontrol.com/) (QGC) and Parrot Anafi drone (Fig. 2). QGC can operate directly the Anafi drone, however the imaging quality of the Anafi drone system was not sufficient at our preliminary tests, but it demonstrated the mission execution.

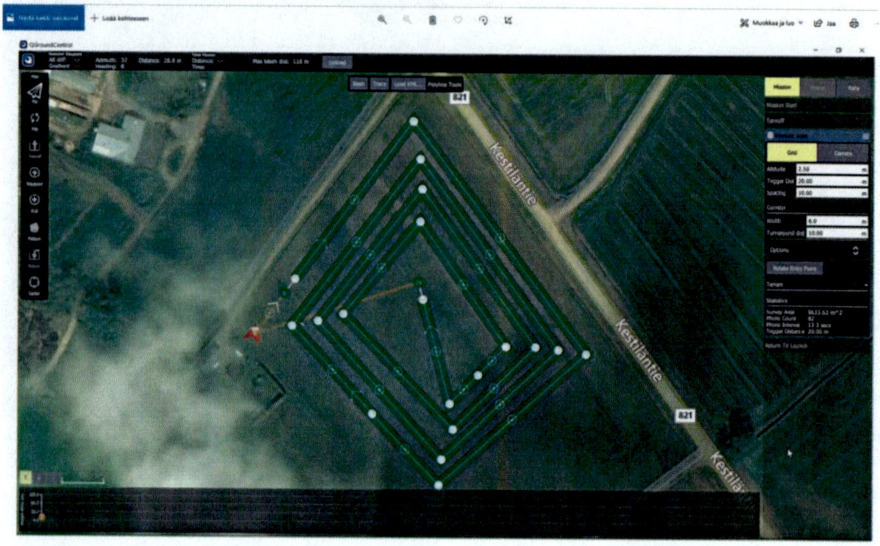

Fig. 2. Mission planning with QGroundControl.

QGC provides full flight control and mission planning with MAVLink (Micro air vehicle communication protocol) enabled drones with open-source code. This makes it possible for exploiting agriculture based standardized mission and route planning solutions with the UAV such as EFDI standard (ISO/DIS 5231) and ISOBUS-standard (ISO 11783).

2.2 Pest Identification

At this first stage, we identified pests manually from the UAV images. We collected about 500 images during the summer 2021, but at this first stage, pests were observed from about 50 images. Figure 3 shows an example of manual pest detection of pollen beetles on a part of a captured image. Pesticide control for *Phylloreta*-species was done in the early summer after the imaging campaign. Data collection was successfully repeated during the summer 2022 including about 200 images taken in a good conditions.

Fig. 3. Part of a captured image showing manually identified pest.

As a next step with the collected images, an open source CVAT Computer Vision Annotation Tool https://cvat.org/ software was used for the pest annotation.

After that, an AI model will be trained for the automatic detection of pests in rapeseed images. This computer vision algorithm will be based in deep neural networks for use supervised machine learning. The model will be trained for object detection using an annotated dataset featuring samples of adult insects of the pollen beetle species (*meligethes aeneus*). The algorithm will detect bounding boxes around individual insects, allowing for classification, precise location within the image and counting applications for pest detection and subsequent spraying activities.

The solution will need GPU computing capabilities, that could be achieved using edge computing devices or web services if connection is provided. There are several bibliographic references featuring the use of deep learning technologies for precision farming such as review by Li et al. [10] presents. The output of this tool is the number of pests in an individual image (coordinates). The detailed AI model training was out of the scope of this paper.

2.3 Spraying Application Construction

The pointwise data about the pest invasion is then interpolated to cover the whole fields in FMIS. At this development stage, we used a GIS software Microimages TNT GIS (https://www.microimages.com/) to interpolate and extrapolate simulated pest occurrence points by applying a suitable threshold. The result of this process is a zone map presenting the spraying need: no spraying, buffer area, full spraying. The areas where pests are identified are fully sprayed while close by areas are determined as buffer zones where the spraying will also be executed. At this stage, the zone map is exported as KML (Keyhole markup language) format, and further transferred to the SHP format suitable for the spraying drone.

Based on the zone map, the goal is that the FMIS creates ISOBUS-application task for the drone or for the tractor. The tractor can use the ISOBUS task as it is, but for the drone the ISOBUS task will be converted to suitable format. The criteria for the spraying tool selection include the spraying UAV (DJI Agras T-16) capacity (4*15 L), total infected area, features of active substance, weather conditions, detected pests and estimated time criticality.

2.4 Spraying Mission

As the pesticide spraying with UAV's is prohibited in Finland, we operated with tap water. The first spraying trials were made in the late June 2021 (Fig. 4) and in the June 2022. At this stage, the actual mission planning was done with commercially available tools, the flight plan was made at the field automatically with the DJI software for the KML boundaries. Aviation regulations with customized UAV controlling are becoming challenging, we used off-the-shelf tools including DJI Terra flight plan software. No adjustments or optimization as such [11] were done for the spraying unit either.

Fig. 4. Spraying UAV DJI Agras T-16 on a rapeseed field.

2.5 Operational Requirements

The professional use of pesticides or plant protection products is governed in Europe by the so-called Sustainable use of pesticide directive or SUD-directive 2009/128/EC [12]. The SUD-directive calls for implementing integrated pest management principles

(IPM). The IPM calls for first identifying the plant protection needs and then planning for the plant protection means accordingly. The objective of the IPM is that also other pest management practices, than application of pesticides, are used, and if pesticides are used, then they are used for a need.

The SUD directive article 9 prohibits aerial spraying in general [12]. As the SUD is a directive, this requirement has been put into force by the Member States by their national legislation. The European Commission has taken a position, that spraying from an UAV is considered aerial spraying. However, article 9 also allows national derogations, under which aerial spraying may be allowed under certain conditions.

For this operation, the legal acceptability of the operation was considered from the Finnish legislative point of view. The question to be answered was, what needs to be taken into consideration in the planning of the system, so that it could be allowed to be executed. The reference point is the Finnish Statue on plant protection 2011/1563, which implements the SUD directive in Finland [13].

What comes to the liquid components, there are no low volume pesticide formulations available, so there is an urgent need to develop innovative formulations to improve the efficacy of drone technology [14].

3 Results and Discussion

The main findings of the system design are presented in the following Figs. 5 and 6 in terms of sequence diagrams.

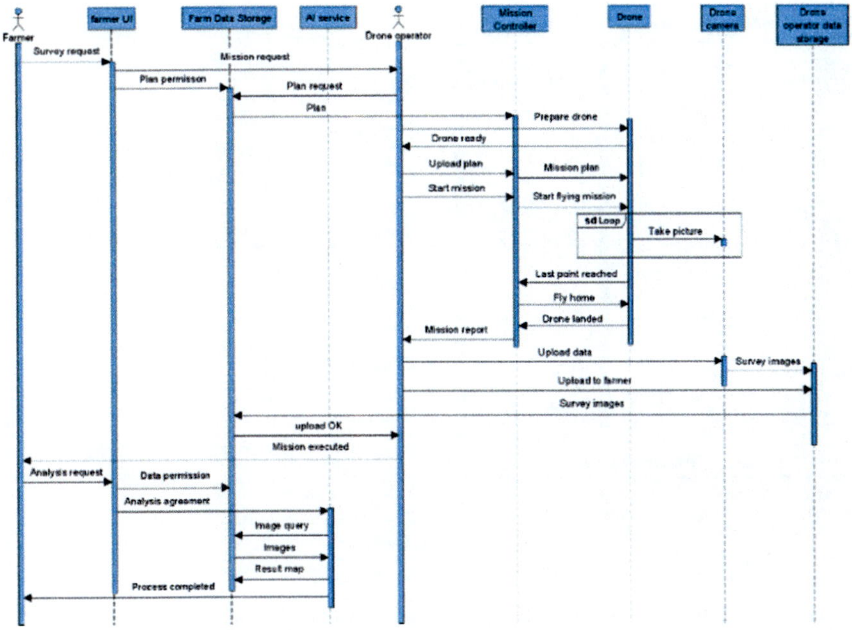

Fig. 5. Drone survey mission execution process

As presented in Fig. 5, the pest management use case starts with a creation of a request to analyse whether there are rapeseed pests in a specific field. The request includes a farm task description that contains the details of the field. The description is shared with specific "AgriDrone" company that creates a detailed drone mission task and executes the autonomous imaging mission. For this mission, QGroundControl (QGC) open-source flight control and mission planning will be used. The result of this imaging mission is a set of nadir photos of the rapeseeds. These photos are shared with "AI service" provider that analyses the images with pest detection algorithms and creates a pest map using maps of the locations of detected pests. These maps are shared with the "farmer" which confirms that spraying missions are needed. Depending on the situation, the farmer or the FMIS decides to do the UAV or tractor based sprayings or to do nothing. In case of spraying, the farmer either shares the pest map with "AgriDrone" or "AgriRobot" company, who will make respective detailed plans, execute the missions and report the results. At the end, the "farmer" accepts the tasks, and the use case is performed.

Figure 6 describes the initial drone spraying mission execution process. The process starts when the farmer gives a request for "Drone operator" to do the mission and giving a permission to access the spraying mission task description. The drone mission is sketched like any other autonomous drone mission. In practice, the spraying of pesticides is currently prohibited by legislation in rapeseed pilot location.

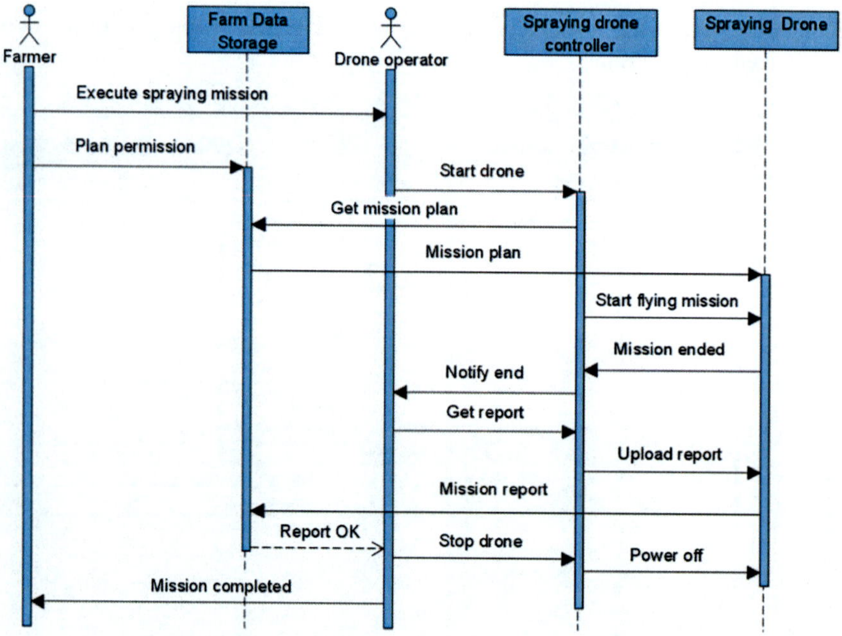

Fig. 6. Spraying mission execution process

The main objective for executing this operation in Finland is the fact that only derogation allowed for aerial spraying is an acute pest outbreak declared by the Finnish Food Authority or by the Ministry of Agriculture and Forestry from proposal of Forest

Centre [13]. So, the first requirement would be that the statue needs more possibilities for aerial spraying.

The operation could be possible by the decree of Ministry of Agriculture and Forestry on aerial spraying 8/2012 [15], if it states that a permit needs to be applied and for the permit to be accepted following conditions need to be met:

- Operator having necessary certifications
- Sufficient buffer zones to sensitive areas, such as habitation or water bodies
- Plant protection product to be approved for aerial use.

Fist two requirements are trivial and can be handled by a professional drone pilot. As the spraying operation limited category operation, the operator will need the necessary training and operation permits, in which the buffer zone restrictions can be included. The third requirement is the most difficult.

For the Plant Protection Product (PPP) to be approved its safety and biological effectiveness needs to be evaluated. The evaluation of the risks is based on risk evaluation tools, which take into consideration risks of point contamination and drift, risks for the operator and bystanders and risks of residues [16]. It turns out that the risk evaluation tools are not fit for evaluation unmanned aerial spray systems as the work cycles and background data are very different from land-based systems.

The OECD Working Party on Pesticides performed a literature review on Unmanned Aerial Spray Systems (UASS). One conclusion from this review is that the of testing done with UASS has a lot of variation in their methods and this makes the results difficult to use in risk assessment. Also, deposition studies and results vary, making the evaluation of effectiveness of the PPP difficult [6].

It turns out that the most challenging issue is the risk assessment of PPP. From the operations point, there is only limited aspects that the drone operator or the system can do to assist in risk estimation. One aspect that the system could do, is to evaluate the needed spraying area. This could be used in estimating risk per operation as some risks are related to the size and location of the application area.

Other way around if the risk assessment and approval of PPP could define safe operation conditions envelope for the spraying operation, this could be used in the automated planning of the spraying operation. Staying inside the safe operations envelope could also be ensured by the automation.

The European Commission published a new proposal for regulation on the sustainable use of plant protection products (SUR) on June 22nd, 2022 [17]. This proposal if adopted would replace the SUD. The proposal recognizes spraying from drones as separate from aerial spraying and would make it possible to use UASSs. The decision on the criteria is left for Member States to specify. How this will affect the acceptability of the concept presented, still requires additional analysis, but the requirement for risk assessment for PPP and for the operation remain.

4 Conclusions

This work presented an implementation plan for the automatization of a drone-based pest management of rapeseeds. This document provided the overview of the systems to be developed within our ongoing research. At this stage, we managed to verify individual components for the pest imaging, pest identification, spraying application construction and a spraying mission. As a result, mission execution processes are presented. At this stage, all the major steps of the automatized rapeseed pest management with drones are successfully operated separate components.

The bottleneck of the pilot automatization implementation seems to be the interface for the spraying drone in relation to aviation regulations even with the research purposes as the safety issues differ from ground vehicle robotization research, where the emergency stop is always available.

The key performance indicators of the possible end product are the reduction of the pesticide usage and the reduction of the labor needs. In addition, the sufficient product can increase the yield value and lower production risks.

This presented pest management concept is rather general, the use case dependent factors will have the greatest impacts on the rationality. Rapeseed pests require constant observations and pesticide reduction is encouraged. The pest invasion typically starts from the outer bounds of the field making relevant to focus observations mainly to certain areas.

Acknowledgements. This project has received funding from the European Union's Horizon 2020 research and innovation programme under grant agreement No 101017111. https://flexigrobots-h2020.eu/.

References

1. Kim, J., Kim, S., Ju, C., Son, H.: Unmanned aerial vehicles in agriculture: a review of perspective of platform, control, and applications. IEEE Access **2019**(7), 105100–105115 (2019). https://doi.org/10.1109/ACCESS.2019.2932119
2. Mogili, U., Deepak, B.: Review on application of drone systems in precision agriculture. Procedia Comput. Sci. **133**, 502–509 (2018)
3. Tsouros, D., Bibi, S., Sarigiannidis, P.: Review on UAV-based applications for precision agriculture. Information **2019**, 10 (2019). https://doi.org/10.3390/info10110349
4. Zhang, T., Vail, S., Duddu, H., Parkin, I., Guo, X., Johnson, E.N., Shirtliffe, S.J.: Phenotyping flowering in canola (*Brassica napus* L.) and estimating seed yield using an unmanned aerial vehicle-based imagery. *Front. Plant Sci.* ***12***, 686332. (2021).https://doi.org/10.3389/fpls.2021.686332
5. Kaivosoja, J., et al.: Reference measurements in developing UAV systems for detecting pests, weeds, and diseases. Remote Sens. **2021**(13), 1238 (2021). https://doi.org/10.3390/rs13071238
6. OECD Report on the State of the Knowledge – Literature Review on Unmanned Aerial Spray Systems in Agriculture, OECD Series on Pesticides, No. 105, OECD Publishing, Paris (2021)
7. Berner, B., Moitzi, G., Pachuta, A., Maheswari, N., Wagentristl, H., Jerzy, C.: Oil seed rape plant spraying using multirotor drone. In: Farm Machinery and Processes Management in Sustainable Agriculture 2019 | Book chapter (2019)
8. Pflanz, M., Nordmeyer, H., Schirrmann, M.: Weed mapping with uas imagery and a bag of visual words based image classifier. Remote Sensing **2018**, 10 (2018). https://doi.org/10.3390/rs10101530

9. Tetila, E., et al.: (2020), Automatic recognition of soybean leaf diseases using UAV images and deep convolutional neural networks. IEEE Geosci. Remote Sens. Lett. **17**, 903–907 (2020). https://doi.org/10.1109/lgrs.2019.2932385
10. Li, W., heng, T., Yang, Z., Li, M., Sun, C., Yang, X.: Classification and detection of insects from field images using deep learning for smart pest management: a systematic review. Ecol. Inf. **66**(2021), 101460 (2021), ISSN 1574-9541, https://doi.org/10.1016/j.ecoinf.2021.101460
11. Wen, S., Zhang, Q., Yin, X., Lan, Y., Zhang, J., Ge, Y.: Design of plant protection UAV variable spray system based on neural networks. Sensors, **19**(5), 1112 (2019). https://doi.org/10.3390/s19051112
12. The European commission. DIRECTIVE 2009/128/EC OF THE EUROPEAN PARLIAMENT AND OF THE COUNCIL of 21 October 2009 establishing a framework for community action to achieve the sustainable use of pesticides. official Journal of the European union, L 309/71 (2009)
13. Valtioneuvosto. Laki kasvinsuojeluaineista 2011/1563 (2011). https://www.finlex.fi/fi/laki/ajantasa/2011/20111563
14. Subramanian, K., Pazhanivelan, S., Srinivasan, G., Santhi, R., Sathiah, N.: Drones in insect pest management. Front. Agron. **3**, 640885 (2021). https://doi.org/10.3389/fagro.2021.640885
15. Ministry of agriculture and forestry, Maa- ja metsätalousministeriön asetus kasvinsuojeluaineen lentolevityksestä (2012). https://mmm.fi/documents/1410837/1501869/MMMa_8_2012.pdf/2f591099-6482-4973-bf36-90c8bf4c35e1/MMMa_8_2012.pdf?t=1446121954000
16. The European Commission, REGULATION (EC) No 1107/2009 OF THE EUROPEAN PARLIAMENT AND OF THE COUNCIL of 21 October 2009 concerning the placing of plant protection products on the market and repealing Council Directives 79/117/EEC and 91/414/EEC, Official Journal of the European Union, L 309/1 (2009)
17. The European commission proposal for a REGULATION OF THE EUROPEAN PARLIAMENT AND OF THE COUNCIL on the sustainable use of plant protection products and amending regulation (EU) 2021/2115 (2022). https://food.ec.europa.eu/system/files/2022-06/pesticides_sud_eval_2022_reg_2022-305_en.pdf

LiDAR-Based Topological Mapping of Orchard Environments

André Teixeira(✉), Sedat Dogru(✉), and Lino Marques(✉)

Institute of Systems and Robotics, Department of Electrical and Computer Engineering, University of Coimbra, 3030-290 Coimbra, Portugal
{a.teixeira,sedat,lino}@isr.uc.pt

Abstract. Autonomous agricultural mobile robots have seen increasingly more use for multiple tasks in recent years. The autonomous behaviour can be better accomplished if the robot is able to localize itself and build a reliable map of its surrounding environment. The large size of agricultural environments requires a more compact map representation to reduce memory and processing requirements, and using topological maps in that environments is a good option in that direction. This paper proposes a novel application of topological maps with a focus on orchards, where the trees with their distinctive features serve as the nodes, and the neighborhoods of the trees serve as the edges for the topological maps. We showcase this new topological map approach using 3D LiDAR data collected in an orchard, and we also present preliminary results showing that Iterative Closest Point (ICP) approach can be effectively used to recognize the nodes of the topological map of the orchard.

Keywords: Topological mapping · Orchard · ICP

1 Introduction

In recent years, automation in agricultural environments has become more common with the use of autonomous robots that can perform several agricultural tasks such as weed detection and removal, crop collection, seeding, mowing, among others. [7,14]. This arises from the need to save labour cost, prevent people from performing heavy and risky operations, and provide the farmer with precise information that will help increase productivity and efficiency [12]. One of the main requirements for automation is the ability of a mobile robot to navigate entire crop fields or orchards in an autonomous and reliable way. To do so, the robot should be capable of localizing itself and build a map of its surrounding environment on-line.

There are two main strategies for robot localization: absolute and relative. The most common approaches for absolute localization are supported by the

This work was supported by FCT - Fundação para a Ciência e a Tecnologia funded projects UID/EEA/00048/2019, and GreenBotics: Intelligent Robotic System for Digital Agriculture, under reference PTDC/EEI-ROB/2459/2021.

D. Tardioli et al. (Eds.): ROBOT 2022, LNNS 590, pp. 438–450, 2023.
https://doi.org/10.1007/978-3-031-21062-4_36

use of Real-Time Kinematic Global Navigation Satellite System (RTK-GNSS) and although this technology is becoming increasingly available and easier to incorporate in agricultural robotic systems [17], it still has some drawbacks. Not only it adds a higher cost to the overall system both in terms of hardware (extra equipment such as a base station and antennas) and software fees (subscriptions to RTK networks), but it is also susceptible to interference from atmospheric conditions and it can be blocked by the tree canopy particularly in dense forests and dense orchards [4]. Degradation of the GNSS signals can quickly result in inaccuracies in localization, which in turn affect the mapping process negatively. Relative localization, which relies on local sensing using LiDARs or cameras, can complement GNSS-based localization, and these sensors are also often used to facilitate sensor-based navigation, and also mapping [16]. This type of localization has proven useful when GNSS signals are of low quality or unavailable, which may happen often in the field.

As mentioned above, another important aspect for automation in agricultural environments is the ability for the robot to build a map of its environment. There are several types of maps in robotics including metric maps and topological maps [1]. Metric maps represent the geometry of objects in a continuous or discrete metric space, and they have the potential for high accuracy. However, they also have a potential for high computational costs in terms of both memory and time. The computational cost is more significant in agricultural environments, which are typically large in size and lack distinctive features for a landmark or feature-based map. Topological maps can attenuate some of these issues by avoiding direct measurements of geometric features, while only considering specific places and the relations between them. The map can then be formally represented as a graph with nodes and edges. Although octrees address some of the issues by reducing memory usage while still properly mapping large regions, the use of topological maps is still preferred due to the lower overall usage of computational resources and improved scalability, which is an important consideration in field robotics.

Topological maps have seen a lot of applications in indoor or structured environments, and they have been constructed using LiDARs or cameras, sensors that also serve for navigation. Vision systems can be used for object detection and room categorization [5] as well as to perform visual place recognition [11], both of which can be used in the mapping process. LiDARs can be used to obtain geometric information for semantic labeling of places [15]. They can also be fused with vision sensors to generate topological maps [2]. In agricultural environments, LiDARs and radars provide several advantages compared to cameras due to their relative robustness to different weather conditions and luminosity changes [3]. The use of topological mapping in field applications and more specifically in agricultural environments, such as crop fields and orchards, is recent. Unmanned Aerial Vehicles (UAVs) or satellite images were used to extract topological maps [8] to be later utilized by Unmanned Ground Vehicles (UGVs) for autonomous navigation. There has been an implementation of topological mapping strategies using only UGVs [13] as well. All these topological maps focus on the free space, and represent the traversable space for the robot with nodes which are then connected according to their neighborhood in a graph. Then these maps are used for path planning for the robot.

This paper proposes a novel type of topological map to represent orchards, where each tree is represented as a node, with neighbouring trees being connected by edges. Each node contains metadata about the tree it represents, such as its position in a fixed cartesian frame, its position in GNSS coordinates and a descriptor that uniquely identifies that particular tree. The edges also contain metadata, like the distance between nodes as well as the traversability of the edge. This approach favours simplicity and expandability, as the map construction becomes straightforward and additional metadata can easily be added to the nodes and edges.

This paper is organized as follows. In Sect. 2 we explain how the topological map is structured, how the map is constructed from LiDAR point cloud data and how the node descriptor is obtained. In Sect. 3 we detail the experimental setup used to collect data and validate our strategy and present the obtained results. Finally, in Sect. 4, we conclude the work and outline future work.

2 Methodology

2.1 Map Structure

A topological map M is given by

$$M = \{N,\ E\} \tag{1}$$

where N and E represent the set of nodes and edges, respectively, which can be described by

$$N = \{n_i\}, \tag{2}$$

$$E = \{e_{ij}\}, \quad i, j = 1,\ 2,\ \ldots,\ m \tag{3}$$

with m being the total number of nodes. In this representation, both nodes and edges can be seen as data structures that hold information about each tree and its relation with the neighbouring trees. A node representing tree i is given by

$$n_i = \{p_i,\ g_i,\ c_i\} \tag{4}$$

where $p_i = \{x_i,\ y_i,\ z_i\}$ represents the tree's position in a local fixed cartesian coordinate frame, $g_i = \{\phi_i,\ \lambda_i,\ h_i\}$ is the tree's position in GNSS coordinates (latitude, longitude and altitude) and $c_i = PC_{trunk}$, i.e., the point cloud of the corresponding tree trunk, represents the descriptor associated with tree i. An edge between tree i and tree j is given by

$$e_{ij} = \{d_{ij}\} \tag{5}$$

where d_{ij} is the edge's length, representing the distance between tree i and tree j. The set of edges is specified by an adjacency matrix A, where $A_{ij} = e_{ij}$. Nodes are connected to one another based on a distance threshold to their neighbors. If the distance between tree i and tree j is less than a specific threshold, an edge e_{ij} is created. An example of the topological map structure is shown in Fig. 1.

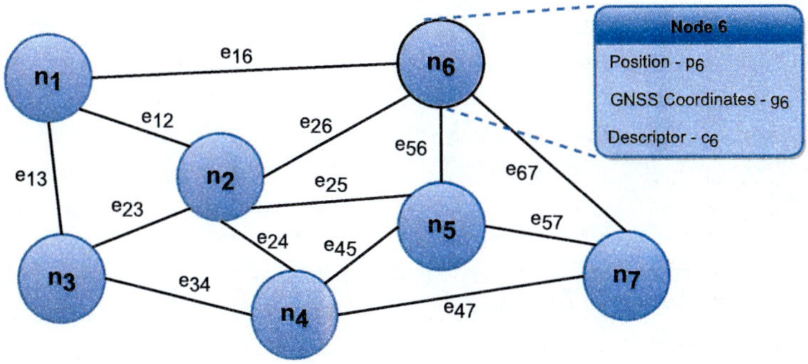

Fig. 1. Example of the topological map structure

2.2 Map Construction

A general overview of the proposed strategy for the map construction process is presented in Fig. 2. The system takes point cloud data from the 3D LiDAR as input. It then constructs the topological map as the robot moves and outputs some markers for visualization. The strategy can be divided into three major stages: filtering, segmentation and map building with each stage being sub-divided into multiple steps.

Filtering Stage. This stage receives as input raw point cloud data from the 3D LiDAR, PC_{raw} consisting of points with x, y, z and *intensity* information. It then applies a set of filters to the point cloud, in order to extract the tree trunks, while also accumulating and merging multiple filtered point clouds over time to obtain a denser and more complete point cloud for the segmentation stage. It outputs a filtered point cloud, $PC_{filtered}$. The set of filtering steps are the following: ground plane removal, low intensity and high z value points removal and outlier removal using both statistical and radius based approaches.

As outlined in Algorithm 1, the objective is to extract a point cloud containing only the tree trunks. For the first step, after copying PC_{raw} into $PC_{filtered}$, the ground plane is removed by extracting the points that support a plane model, using the RANdom SAmple Consensus (RANSAC) method [6]. For the second step, we traverse all points in $PC_{filtered}$ and remove every point with z and *intensity* values below specific thresholds, z_th and $intensity\_th$, respectively. These steps are essential to significantly reduce the number of points in the point cloud, decreasing the computation times for real time processing. In the third

Fig. 2. General system overview

step, we apply a radius outlier removal filter, meaning we traverse all points in $PC_{filtered}$ and for each point p_i compute the number of neighbouring points nn inside a specific radius r. If the number of neighbours is less than a certain threshold nn_th the corresponding point is removed from the point cloud. Finally, for the fourth step, we apply a statistical outlier removal filter, i.e., we loop through every point in $PC_{filtered}$ and compute the mean distance from point p_i to its k nearest neighbours, d_i. By assuming this distribution is Gaussian, we then compute the mean μ_d and standard deviation σ_d of these distances. All points whose mean distance d_i is greater than a threshold $th = \mu_d + \alpha \cdot \sigma_d$ are removed from the point cloud, where α is a standard deviation multiplier and a tunable parameter.

Although not specified in Algorithm 1, the resultant filtered point clouds are accumulated and merged over a moving time window, meaning at each time instant t the system also outputs a merged point cloud PC_{merged} given by

$$PC_{merged}(t) = PC_{filtered}(t) \cup PC_{filtered}(t-T) \cup \cdots \cup PC_{filtered}(t-nT) \quad (6)$$

where T is the period at which point cloud data arrives and n is a tunable parameter. This merging process is preceded by registering successive point clouds using the Iterative Closest Point algorithm (ICP) [9] to reduce noise in the final merged point cloud.

Segmentation Stage. This stage takes as input the merged point cloud PC_{merged} from the previous stage and segments it into clusters to obtain distinct tree trunks, while also computing the clusters' centroids and descriptors. It outputs an array of custom data structures containing information about the detected clusters given by

$$C_{detected} = \{C_i\}_{i=1\ldots n_c} \quad (7)$$

where n_c is the number of detected clusters. Each cluster, represented by a custom data structure, is given by

$$C_i = \{id,\ num\_points,\ centroid,\ descriptor\} \quad (8)$$

which holds the cluster's ID number, number of points, centroid and descriptor. Similar to the map's data structure, it can be easily expanded to include more information about the cluster. The segmentation stage can be sub-divided into the following steps: downsampling, normal component calculation, segmentation into clusters and centroid and descriptor computations.

An overview of this stage is outlined in Algorithm 2. Since the segmentation stage is computationally the most expensive stage, the first step, which reduces the number of points in the cloud, is crucial to reduce computation times. Point reduction is achieved by overlaying a 3D voxel grid over the input point cloud PC_{merged} and approximating the points in each voxel with their centroid. The voxel size is tunable and given by $leaf\_size$. The second step is required for the segmentation stage and it approximates the problem of determining the normal

Algorithm 1: Tree trunk extraction from 3D point clouds.

Input: A 3D point cloud $PC_{raw} = \{p_i\}_{i=1...n_p}$
Output: A filtered 3D point cloud $PC_{filtered} = \{p_i\}_{i=1...n_f}$

```
PC_filtered ← PC_raw;
// Ground plane and low intensity points removal
PC_filtered = removeGroundPlane(PC_filtered);
for p_i ∈ PC_filtered do
    if p_i.z ≤ z_th or p_i.intensity ≤ intensity_th then
        Remove p_i from PC_filtered
    end
end
// Radius outlier removal filter
for p_i ∈ PC_filtered do
    nn = numNeighbours(p_i, r);
    if nn < nn_th then
        Remove p_i from PC_filtered
    end
end
// Statistical outlier removal filter
for p_i ∈ PC_filtered do
    d_i = meanDistanceK(p_i, k);
    D ← d_i
end
μ_d = mean(D);
σ_d = std(D);
th = μ_d + α · σ_d;
for p_i ∈ PC_filtered do
    d_i = meanDistanceK(p_i, k);
    if d_i > th then
        Remove p_i from PC_filtered
    end
end
```

to a point on a surface by the problem of estimating the normal of a plane tangent to the surface, by using Principal Component Analysis (PCA) [10]. The estimated normal components are then concatenated to the downsampled point cloud, generating $PC_{normals}$ which will be fed to the segmentation step. To segment the point cloud into clusters, an Euclidean Clustering technique is used together with a custom clustering condition that needs to hold for a point to be added to the cluster. This condition requires a point to be similar in both surface normal direction and intensity value to be added to a cluster. Clusters that are considered too small or too large are also removed. In the fourth and fifth steps, we traverse all clusters in $C_{detected}$ and for each cluster compute its centroid and descriptor, which will be used as the location and identification for the map's nodes, respectively.

Algorithm 2: Segmentation of tree trunks into clusters.

Input: A 3D point cloud $PC_{merged} = \{p_i\}_{i=1...n_m}$
Output: An array of detected clusters $C_{detected} = \{C_i\}_{i=1...n_c}$
PC_{merged} = DownsampleCloud($PC_{merged}, leaf\_size$);
$PC_{normals}$ = ComputeNormals(PC_{merged});
$Clusters$ = ConditionalEuclideanClustering($PC_{normals}$);
for $cluster_i \in Clusters$ **do**
  $\bar{p} = \text{mean}_{p \in cluster_i}(p)$;
  h_i = computeDescriptor($cluster_i$);
  $C_i.id \leftarrow i$;
  $C_i.num\_points \leftarrow cluster.points.size()$;
  $C_i.centroid \leftarrow \bar{p}$;
  $C_i.descriptor \leftarrow h_i$;
end

Map Building Stage. This stage receives as input the cluster information from the previous stage, $C_{detected}$, and builds the topological map. As seen in Algorithm 3, the goal is to match newly received clusters to known nodes. As such, we traverse through the detected clusters $C_{detected}$ and for each one, we match it against the set of known nodes N. This matching process is done based on both position and descriptor similarity, i.e., if the distance between a cluster's centroid $C_i.centroid$ and a node's position $n_j.p$ is less than a certain threshold d_th and the difference between a cluster's descriptor $C_i.descriptor$ and a node's descriptor $n_j.c$ is below a specific threshold $score_th$, we can assume the identified cluster already exists as a node in the map. The node's position is then updated by computing the mean between the cluster's centroid and the current node's position while the node's descriptor is updated by replacing its current point cloud with the cloud extracted from the detected cluster. Otherwise, the received cluster is added to the map structure as a new node and the edges between neighbouring nodes within a certain radius are updated. The geographic coordinates of the trunk is computed by converting the respective Cartesian coordinates in an East-North-Up (ENU) frame to geodetic coordinates. With this strategy, the map can also be saved to a file to be analysed offline and loaded from a file when the robot's operation resumes, allowing the robot to start from a previously known map.

Algorithm 3: Topological map update.

```
Input: An array of detected clusters C_detected = {C_i}_{i=1...n_c}
Output: A 3D topological map M = {N, E}
for C_i ∈ C_detected do
    for n_j ∈ N do
        d = distance(C_i.centroid, n_j.p);
        score = descriptorDiff(C_i.descriptor, n_j.c);
        if d < d_th and score < score_th then
            n_j.p = mean(n_j.p, C_i.centroid);
            node_exists ← true ;
            break ;
        end
    end
    if node_exists then
        continue ;
    else
        N.addNode(C_i.centroid);
        E.updateEdges();
    end
end
```

3 Experimental Work

3.1 Experimental Setup

The mobile platform used in this work is a Clearpath Husky UGV equipped with a dual antenna RTK-GPS unit, a 3D LiDAR and an XSens MTi-300 IMU (Fig. 3a). The on-board LiDAR is a Velodyne Puck, a 16-channel LiDAR with 100 m range, 360° horizontal field of view, 30° vertical field of view, 3 cm range accuracy and 5–20 Hz rotation rate. The GNSS unit consists of a dual antenna Septentrio GNSS receiver operating in RTK-GNSS mode, supported by a local GNSS base station, achieving sub-centimeter position accuracy. The method described in Sect. 2 was implemented using Robot Operating System (ROS) in C++ with existing packages as well as a custom developed one.

The experiments were conducted in an olive orchard (Fig. 3b). The terrain was more like a ramp with an uneven surface, and a significant height difference between the lowest and highest point. Additionally, most olive trees did not have straight upright trunks. For practicality reasons and to speed up the development process, the data was collected in a ROS `bag` file which contained odometry, IMU and RTK-GNSS readings, apart from the 3D point clouds.

To reduce odometry errors and obtain cleaner 3D point clouds, the robot was driven between rows and stopped roughly every 2 m to collect the point clouds. The robot's position was estimated fusing data provided by wheel odometry, IMU and RTK-GNSS readings using the `robot_localization` package from ROS. The obtained map is shown in Fig. 5.

(a) (b)

Fig. 3. **(a)** Husky and **(b)** the olive orchard used in the experiments.

3.2 Results

In this section, we analyze some preliminary results on trunk extraction, topological map generation, and the matching performance in the map building stage. The algorithm was able to detect and isolate the tree trunks well, as seen in Fig. 4. Figure 4a shows an example of an original raw point cloud. Figure 4b shows the filtered point cloud with the extracted trunks. Figure 4c shows the final point cloud with the identified clusters. At the end of the mapping process, a complete topological map of the area of interest was constructed (Fig. 5), where we can see that the created topological map approximately matches the satellite view of the orchard. For the matching performance validation, we present some results of a partial run, corresponding to approximately 10 m of distance travelled by the robot, with four detected trees of the full topological map shown in Fig. 5. In the first step, the topological map was built and saved to a file. For that purpose, the mapping process used data only from the times when the robot was stopped, which helped obtain cleaner point clouds and reduce errors in the node descriptors. In this step, the node matching process was carried out using only the distance between detected clusters and existing nodes. In the second step, the previously built topological map was loaded and the algorithm was executed only when the robot was moving, to test the distinctiveness of the proposed node descriptors. Here, the matching process already included the fitness score provided by the ICP algorithm which was implemented using the Point Cloud Library (PCL), for which lower score is better fitness. Samples of the obtained fitness scores were taken at three different positions along the robot's path (Fig. 5). In each position (A, B and C), the number of detected clusters, their distance to each known node as well as their ICP fitness score were recorded

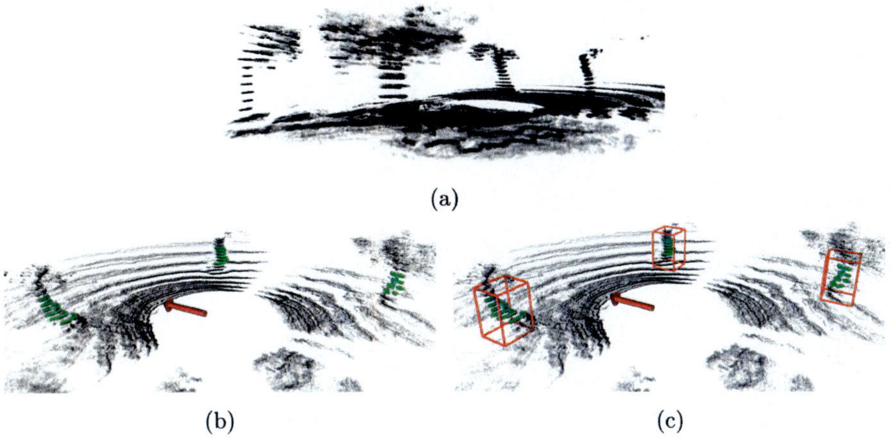

Fig. 4. Filtering and segmentation results. **(a)** Example of an original point cloud. **(b)** Filtered point cloud with the extracted trunks. **(c)** Detected clusters.

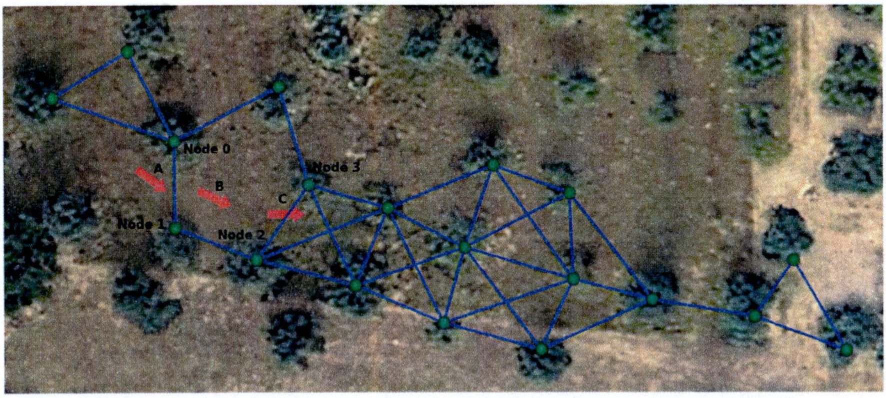

Fig. 5. Topological map of the orchard overlaid on satellite image

and are shown on Table 1. On the table we can see that in position C, three clusters were detected. For each cluster, the distance to a known node is minimum when the cluster corresponds to that respective node. For example, cluster 0 corresponds to node 3. Here we can see the fitness score is also minimum when the cluster corresponds to a particular node and that a cluster can be correctly matched with a known node, based on their ICP fitness score.

Table 1. Results obtained from the partial run - number of clusters detected in each position, their distance and fitness scores to known nodes

Position	No. of clusters	Cluster ID	Distance	Fitness score
A	2	0	**0.0288**	**0.000174**
			7.4422	0.019689
			11.4797	0.018049
			11.2053	0.023249
		1	7.4714	0.025533
			0.0462	**0.000260**
			6.2203	0.023864
			10.5716	0.030887
B	2	0	11.2200	0.055035
			10.5237	0.042715
			6.9464	0.031216
			0.0406	**0.000121**
		1	11.5227	0.029697
			6.1852	0.033277
			0.0505	**0.000123**
			7.0317	0.036585
C	3	0	11.2332	0.036298
			10.5526	0.047129
			6.97823	0.032180
			0.0248	**0.000129**
		1	7.2898	0.015534
			0.4457	**0.008789**
			6.6403	0.024667
			10.8857	0.027446
		2	11.4950	0.030312
			6.1589	0.034993
			0.0543	**0.000371**
			7.0283	0.032097

4 Conclusion and Future Work

In this paper, we proposed a new type of topological map to represent orchards and forests, and showed how LiDAR can be used to construct such a map. The mapping process used a node matching strategy based on ICP, which was shown to be effective in recognizing the trees in follow up runs as well. The proposed method favours modularity and scalability, allowing for additional information about each node and edge to easily be added to the map structure, increasing the overall robustness and utility of this strategy. Future work will include the addition of more information to both the nodes and edges. Extra node metadata will include tree signatures, trunk descriptors from multiple views, height descriptors, among others and additional edge metadata will incorporate its traversability. The whole approach will eventually be validated in a full simultaneous localization and mapping scenario in an orchard.

References

1. Aguiar, A.S., dos Santos, F.N., Cunha, J.B., Sobreira, H., Sousa, A.J.: Localization and mapping for robots in agriculture and forestry: a survey. Robotics **9**(4), 1–23 (2020)
2. Blöchliger, F., Fehr, M., Dymczyk, M., Schneider, T., Siegwart, R.Y.: Topomap: Topological mapping and navigation based on visual SLAM maps. In: 2018 IEEE International Conference on Robotics and Automation (ICRA), pp. 1–9 (2018)
3. Dogru, S., Marques, L.: Evaluation of an automotive short range radar sensor for mapping in orchards. In: 2018 IEEE International Conference on Autonomous Robot Systems and Competitions (ICARSC), pp. 78–83 (2018)
4. Falco, G., et al.: Investigation of performance of GNSS-based devices for precise positioning in harsh agriculture environments. In: 2019 IEEE International Workshop on Metrology for Agriculture and Forestry (MetroAgriFor), pp. 1–6 (2019)
5. Fernandez-Chaves, D., Ruiz-Sarmiento, J.R., Petkov, N., Gonzalez-Jimenez, J.: From object detection to room categorization in robotics. In: Proceedings of the 3rd International Conference on Applications of Intelligent Systems, APPIS 2020. Association for Computing Machinery, New York, NY, USA (2020)
6. Fischler, M.A., Bolles, R.C.: Random sample consensus: a paradigm for model fitting with applications to image analysis and automated cartography. Commun. ACM **24**(6), 381–395 (1981)
7. Fountas, S., Mylonas, N., Malounas, I., Rodias, E., Hellmann Santos, C., Pekkeriet, E.: Agricultural robotics for field operations. Sensors **20**(9), 2672 (2020)
8. Heiwolt, K., Mandil, W., Cielniak, G., Hanheide, M.: Automated topological mapping for agricultural robots. In: UKRAS20 Conference: Robots into the Real World Proceedings, pp. 27–29 (2020)
9. Holz, D., Ichim, A., Tombari, F., Rusu, R., Behnke, S.: Registration with the point cloud library - a modular framework for aligning in 3-D. IEEE Robot. Autom. Mag **22**, 110–124 (2015)
10. Jolliffe, I.T., Cadima, J.: Principal component analysis: a review and recent developments. Philos. Trans. R. Soc. A: Math. Phys. Eng. Sci. **374**(2065), 20150202 (2016)
11. Lowry, S., et al.: Visual place recognition: a survey. IEEE Trans. Rob. **32**(1), 1–19 (2016)
12. Puranik, V., Sharmila, Ranjan, A., Kumari, A.: Automation in agriculture and IoT. In: 2019 4th International Conference on Internet of Things: Smart Innovation and Usages (IoT-SIU), pp. 1–6 (2019)
13. Santos, L., Santos, F.N., Magalhães, S., Costa, P., Reis, R.: Path planning approach with the extraction of topological maps from occupancy grid maps in steep slope vineyards. In: 2019 IEEE International Conference on Autonomous Robot Systems and Competitions (ICARSC), pp. 1–7 (2019)

14. Shamshiri, R., et al.: Research and development in agricultural robotics: a perspective of digital farming. Int. J. Agric. Biol. Eng. **11**, 1–14 (2018)
15. Stachniss, C., Mozos, O.M., Rottmann, A., Burgard, W.: Semantic labeling of places. In: International Symposium of Robotics Research. San Francisco, CA, USA (2005)
16. Weiss, U., Biber, P.: Plant detection and mapping for agricultural robots using a 3D LiDAR sensor. Robot. Auton. Syst. **59**(5), 265–273 (2011)
17. Winterhalter, W., Fleckenstein, F.V., Dornhege, C., Burgard, W.: Localization for precision navigation in agricultural fields-beyond crop row following. J. Field Robot. **38**, 429–451 (2021)

Towards Automated UAV Image Processing Workflows in Precision Viticulture: Challenges and Benefits

Sergio Vélez(✉), Mar Ariza-Sentís, and João Valente

Information Technology Group, Wageningen University and Research, 6708 PB Wageningen, Netherlands
{sergio.velezmartin,mar.arizasentis,joao.valente}@wur.nl
https://saidlab.wur.nl/

Abstract. Remote sensing is a tool based on acquiring images of objects at a distance. The distance to the object is a determining factor, conditioning the strength of the analysis, and it can be reduced with lower flights and longer flight missions. Still, there is a physical limit depending on the type of platform used. In this way, Unmanned Aerial Vehicles (UAVs) allow images to be obtained with a higher spatial resolution, as they can fly only a few metres above the ground. Therefore, they have become a very useful platform as they have democratised field imaging technology, offering greater versatility in obtaining spectral information due to their small size and significantly reduced costs compared to traditional aerial platforms. In this way, several authors have conducted comparative studies on the suitability of different platforms for remote sensing data collection in viticulture; however, there is a lack of information about the image processing workflows. In the framework of the FlexiGroBots project, methodologies for analysing the spatial variability of different factors affecting the crop are being explored, and the automation of processes is a key factor that has proven their applicability.

In this article, the suitability of carrying out two different workflows for image processing is analysed: i) a manual workflow, which involves specific steps with human intervention and considers software traditionally used by pilots and remote sensing technicians, and ii) an automated workflow in Python, which allows image processing to be carried out with little or no human intervention.

Keywords: FlexiGroBots · Precision agriculture · Unmanned aerial vehicles

1 Introduction

Remote sensing is a tool based on acquiring images of objects at a distance. It has been used for over 100 years, and pictures of the Earth and the landscape have been taken from above, using cameras and other sensors mounted on several platforms, such as satellites and aircraft [1].

Historically, airborne platforms have faced various issues, e.g. weather, that interfere with proper mission performance and adequate image acquisition. In addition, the

D. Tardioli et al. (Eds.): ROBOT 2022, LNNS 590, pp. 451–462, 2023.
https://doi.org/10.1007/978-3-031-21062-4_37

distance to the object is a determining factor affecting the final image resolution. For example, satellite images usually have poorer resolution than those acquired using aircraft, and even lower resolution than ground platforms, mainly due to the greater distance to the object. This fact implies that the quality and type of sensor can help to obtain images with a higher spatial resolution. However, the distance to the object is still a determining factor, conditioning the strength of the analysis [2]. This distance can be reduced with lower flights and longer flight missions, but there is a physical limit depending on the type of platform used. Thus, satellites cannot modify their orbit easily, and aircraft, due to their size, are limited in their flight altitude. As a result, aerial photography has focused mainly on improving platforms and sensors to capture images with a higher resolution and allow better image analysis.

More recently, Unmanned Aerial Vehicles (UAVs) have solved these problems, allowing images to be obtained with a higher spatial resolution, as they can fly only a few metres above the ground. Thus, the proliferation of UAVs in both the professional and private sectors has meant that ultra-high resolution earth observation data have become increasingly common in precision agriculture and environmental research. In addition, UAVs have become a very useful platform as they have democratised field imaging technology, offering greater versatility in obtaining spectral information due to their small size and significantly reduced costs compared to traditional aerial platforms. In this way, comparative studies have been carried out on the suitability of different platforms for remote sensing data collection, showing that in the case of hedgerow crops, such as vineyards and other woody crops, where the inter-row areas involve a relevant portion of the soil, the spectral information captured by decametric resolution satellite platforms compared to UAV has difficulties in assessing the spatial variability of the crop and its status [3]. Moreover, considering plot sizes up to 5 ha, the overall cost of UAV imaging is lower compared to other aerial platforms, such as satellites and aircraft, including acquisition costs, georeferencing and orthorectifying costs and image processing costs [4].

These comparative studies are not limited to cross-platform comparisons. Different vegetation indices have also been analysed since they were initially developed for traditional airborne and spaceborne sensors and not for UAV systems. [5]. Other comparative studies have also been conducted using different drones to evaluate the performance in tests based on image processing, demonstrating that different UAVs can perform well for specific applications such as agricultural monitoring [6]. It is important to note that all these authors have already indicated the importance of analysing different factors that affect the variability of the UAV imaging process. However, it is also essential to study the time and economic cost of image processing.

Certainly, UAVs are especially useful in agriculture, as the correct management of spatial variability is a critical factor that farm managers must control. Being unaware of the development, status or evolution of the various variables affecting the crop can lead to severe problems or even complete harvest loss. In this way, precision farming is a key tool for site-specific management of crop heterogeneity, both in spatial and temporal dimensions. This fact is critical since drone imaging in the field is often approached from the perspective of an isolated mission, a single image acquisition at a specific moment in time. However, several flight missions are generally required throughout the year to manage the spatial variability of the field.

Some authors have considered optimising routes and times to carry out the UAV mission to obtain images [7]. However, after the mission, a huge amount of spectral information is generated and needs to be analysed. The quantity of information depends, among other factors, on the number of images taken and their resolution, ranging from a few GB of information, in the case of standard RGB images, to several hundred GB in the case of multispectral or hyperspectral images. Furthermore, all this data must be processed on computers with a certain capacity, and the workflow for processing all this data is relatively complex, and there is no unique method to do it. Thus, depending on the technical skills of the users processing the information, the workflow can be more or less automated.

In the framework of the FlexiGroBots project (FlexiGroBots Website, 2022), methodologies for analysing the spatial variability of different factors affecting the crop are being explored. FlexiGrobots is an Innovation Action aiming to build a platform for flexible heterogeneous multi-robot systems for intelligent automation of precision agriculture operations, providing multiple benefits to farmers worldwide. The project envisions the development of an innovative platform enabling the efficient use of novel robotic applications in the agriculture domain. The platform will be versatile for using a specific robotic system to perform different tasks, allowing heterogeneous robot cooperation to perform complex tasks. Moreover, it considers the autonomy of robotic systems with advanced capabilities to adapt missions, and efficient data management, processing, accessibility and utilisation in supporting robot operations.

The automation of processes is a key factor that has proven their applicability. However, very few studies have compared the differences between a manual and an automatic orthomosaic generation process in precision viticulture. Liba and Berg-Jürgens [8] analysed the accuracy obtained with a semi-manual process, introducing GCPs (Ground Control Points), and an automatic procedure, without GCPs. Most authors focus on the changes in accuracy gathered when modifying variables in the automatic process, such as the flying height or the algorithm used [9, 10]. Nonetheless, there are no studies comparing manual and automated methods within a precision agriculture context. Considering all those mentioned above and the importance of the analysis stage of the images obtained with the drone, the lack of information about the time required for the pre-processing, processing and post-processing of all this data is surprising. Therefore, in this paper, the suitability of carrying out two different workflows for image processing is analysed: i) a manual workflow, which involves specific steps with human intervention and considers software traditionally used by pilots and remote sensing technicians, and ii) an automated workflow in Python, which allows image processing to be carried out with little or no human intervention.

2 Material and Methods

2.1 Experimental Site

A flight mission was developed during the 2021 campaign over a 1.06 ha and 8.1% slope commercial vineyard. The experimental site was located within the 'Rias Baixas

AOP' (Appellation of Origin) in 'Tomiño, Pontevedra', Galicia, Spain (X: 516989.02, Y: 4644806.53; ETRS89/UTM zone 29N), property of 'Bodegas Terras Gauda, S.A.' (Fig. 1). The plants were *Vitis vinifera* cv. Loureiro, trained in vertical shoot positioning (VSP), and planted in 1990 with a NE-SW orientation and a 2.5 × 3 m distance between plants and rows, respectively. Spontaneous vegetation species grew as cover crops.

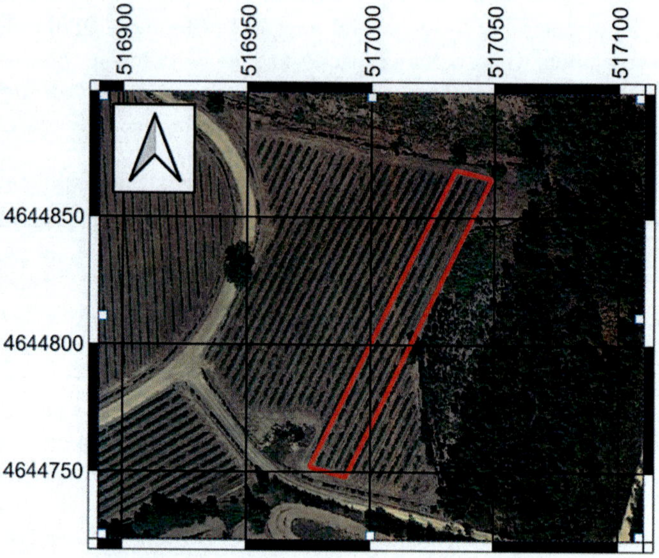

Fig. 1. Vineyard location. Coordinates in ETRS89/UTM zone 29N.

2.2 UAV Process

A typical complete workflow in commercial UAV remote sensing includes around five steps (Fig. 2). First, the UAV flight is defined and executed. Then, after the flight, images are copied from the SD card for processing. The images can also be sent in real-time/streaming from the drone, but bandwidth is often a limiting problem in the case of multispectral or hyperspectral images. The next step is to take the raw images into the software for the orthomosaic process. In this process, the orthomosaic is created with up to as many bands as the raw images have. Finally, data processing is carried out on the different layers to analyse what is happening and obtain agronomic information. Vegetation indices are often used in this last step.

The images were processed using a high-performance computer with Linux 64-bit (Ubuntu 20.04.4 LTS). The computing platform was equipped with 64 Gb RAM, a Samsung SSD 860 EVO 1TB and an Intel(R) Core(TM) i9-10940X CPU processor with 14 cores (28 threads) and a base frequency of 3.30 GHz (4.80 GHz in Turbo Boost Max mode). Moreover, the computer includes two Nvidia Titan RTX, each with a Clock of 1770 MHz, 576 Tensor Cores, 4608 CUDA cores, and 24 GB GDDR6.

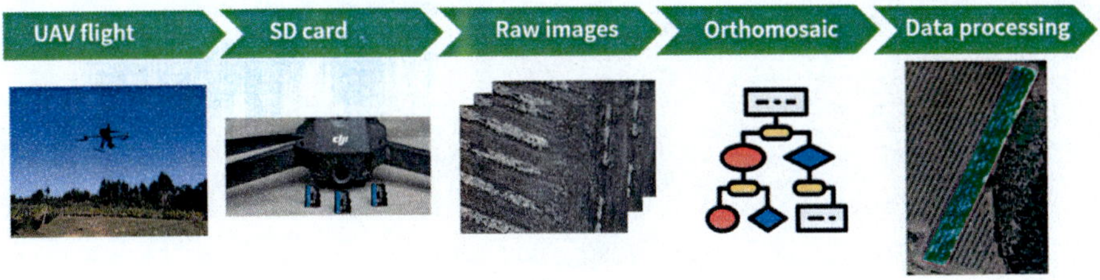

Fig. 2. UAV process.

2.3 UAV Imagery Acquisition

Seven ground control points prior to UAV flight (GCPs) were located in the vineyard and georeferenced using the Trimble R2 Integrated GNSS system to improve the geometric accuracy of the image mosaicking process. The equipment employed in this work is a DJI M210 multi-rotor platform (DJI Sciences and Technologies Ltd., Shenzhen, Guangdong, China, Fig. 3) and a Micasense RedEdge 3 multispectral camera (AgEagle Sensor Systems Inc., Wichita, Kansas, USA, Table 1).

The mission was performed on 16 September 2021, at midday. The multispectral sensor was oriented towards the ground with a nadir angle. The UAV's horizontal speed was 2 m/s, and the flying height was 30 m above ground level (AGL) with an 80% overlap.

Fig. 3. Left: DJI M210 Unmanned Aerial Vehicle (UAV) flying over the vineyard. Right: detail of the Flexigrobots banner in the vineyard.

Table 1. Characteristics of Micasense RedEdge3 sensor.

Feature	Characteristics
Sensor size	4.8 mm × 3.6 mm
Pixel size	3.75 µm
Band resolution	1280 × 960
Number of imagers	5
Blue band	Centre: 475 nm, Bandwidth: 20 nm
Green band	Centre: 560 nm, Bandwidth: 20 nm
Red band	Centre: 668 nm, Bandwidth: 10 nm
Red edge band	Centre: 717 nm, Bandwidth: 10 nm
Near-infrared band	Centre: 840 nm, Bandwidth: 40 nm
Aspect ratio	4:3
Focal length	5.5 mm
Focal aperture	f/2.8
Field of view (H)	47.2°
Field of view (V)	35.4°
Exposure time	1/523 s
Storage format	16-bit tiff raw

2.4 UAV Imagery Processing Workflows

Figure 4 shows the workflows for the two methodologies employed to obtain the final product: an NDVI (Normalized Difference Vegetation Index) map of the Region of Interest (ROI). Both share the same input: the raw multispectral imagery acquired with the UAV. However, the first methodology (Fig. 4, left) requires several steps of user intervention, such as copying the files stored in the SD card and loading the information to Agisoft Metashape, which is based on the usual operations carried out by the technicians. For orthomosaic creation, two of the most commonly used commercial software packages are Agisoft Metashape and PIX4D. In addition, some of the most widely GIS software employed are QGIS and ArcGIS. Therefore, in the present work, QGIS (QGIS developer team 2022) and Agisoft Metashape Professional v1.7.6 (Agisoft LLC, St. Petersburg, Russia) were employed, following the software developer guidelines: 1) the images were introduced into the software to process the set of images, selecting the parameters according to the 'low', 'medium', 'high', or 'ultra' workflows, 2) the Align photos step was performed, and the GCPs were located in the aligned images, 3) build dense cloud step, 4) build mesh step, 5) build texture step, 6) the DSM (Digital Surface Model) is generated using the dense point cloud, and finally the orthomosaic was generated for each band and, finally, 7) the orthomosaic is made using DEM. After the orthomosaic is generated, user intervention is needed again to copy the orthomosaic files and load them to QGIS, where the NDVI map is computed and masked using ROI.

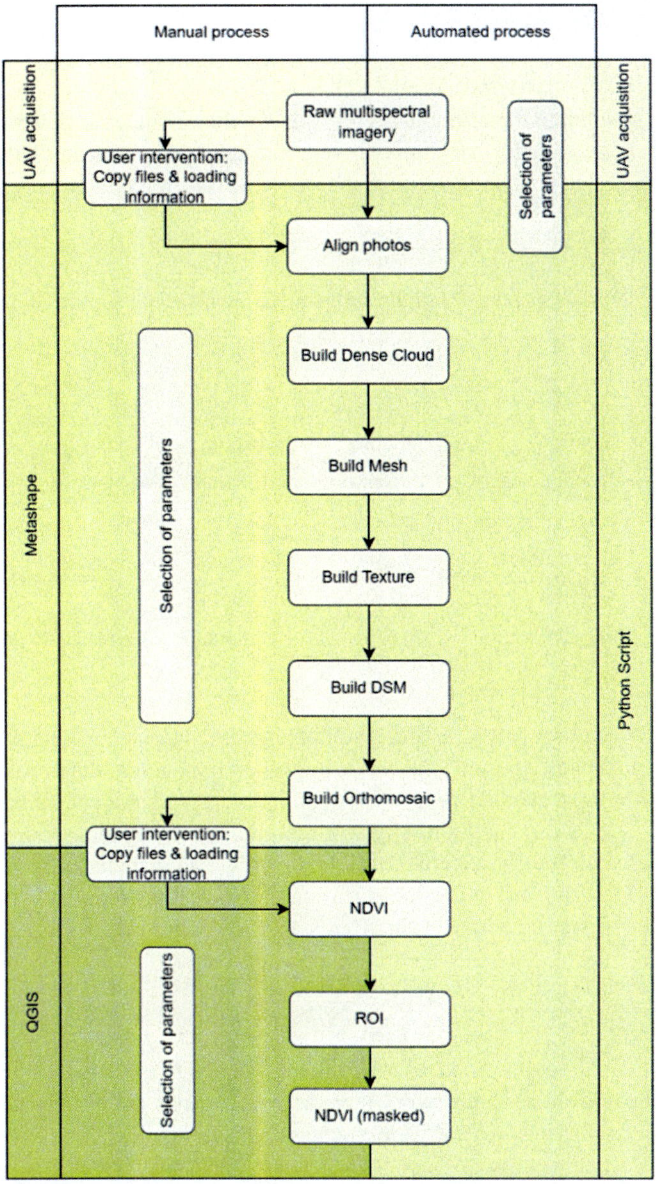

Fig. 4. Manual and automated workflows. Raw multispectral imagery is the input for the process. UAV: Unmanned Air Vehicle. DSM: Digital Surface Model. ROI: Region of Interest. NDVI: Normalized Difference Vegetation Index.

Nevertheless, the second methodology (Fig. 4, right) automates the whole procedure with a main Python script. The computer detects the SD card and copies the files to a folder. The Python script provides access to the folder to Open Drone Map (ODM) [11], and the orthomosaic is computed thanks to the command line toolkit by ODM's GitHub

[12]. ODM was selected because it is an open-source tool that eases the automation of the process. After the orthomosaic is obtained, the Python script continues running to compute the NDVI map cropped to the ROI.

In order to generate the orthomosaic in Agisoft Metashape with varying quality, all the quality parameters shown in Table 2 were modified. However, in ODM, a general parameter called 'feature quality' adjusts the whole workflow quality to reach the desired quality.

Table 2. Selected parameters for the orthomosaic process with Agisoft Metashape.

		Low	Medium	High	Ultra
Align photos	Accuracy	Medium	Medium	High	Highest
Dense Cloud	Quality	Low	Medium	High	Ultra-high
	Depth filtering	Mild	Moderate	Aggressive	Aggressive
Mesh	Depth maps quality	Low	Medium	High	Ultra-high
	Depth filtering	Mild	Moderate	Aggressive	Aggressive
DEM	Quality	Low	Medium	High	Ultra-high
	Depth filtering	Mild	Moderate	Aggressive	Aggressive

Key points were set to 10000 and resolution to 0.05 m/pixel as default parameters for both Agisoft Metashape and ODM. It is important to note that the 'lowest' option to align photos has been avoided in all combinations, as the accuracy must be adjusted at least to 'low' (ODM) or 'medium' (Agisoft Metashape) to obtain a minimum number of points that match the characteristics of the images. The displacement of the ground control points was assessed to test the performance of each procedure.

Finally, NDVI was calculated using Red and NIR bands:

$$NDVI = \frac{(NIR - Red)}{(NIR + Red)} \tag{2}$$

And as the final output product of the process, the NDVI was masked using a ROI within the vineyard. In a manual workflow, these operations are usually performed by the technician employing GIS software, such as QGIS or ArcGIS.

3 Results

Table 3 shows the time comparison of computational time between Agisoft Metashape and ODM for performing the orthomosaic process, including all the different quality options. It can be observed that ODM takes more time than Agisoft Metashape for every quality selected.

Table 4 shows the time comparison between manual and automatic methods for the whole orthomosaic procedure. The option selected is ultra.

Table 3. Comparison of computational time between Agisoft Metashape and ODM in the orthomosaic process varying the quality of the final product. Time in hh:mm:ss.

	Agisoft Metashape				ODM			
	Ultra	High	Medium	Low	Ultra	High	Medium	Low
Align photos	00:47:24	00:39:53	00:30:57	00:12:26	02:22:03	01:37:10	00:54:34	00:38:04
Build Dense Cloud	00:45:35	00:13:48	00:13:56	00:01:54	00:18:43	00:04:32	00:06:53	00:04:34
Build Mesh	00:06:09	00:01:37	00:01:38	00:00:18	00:02:30	00:02:15	00:02:35	00:00:47
Build Texture	00:55:05	00:15:07	00:15:09	00:02:54	00:04:36	00:05:25	00:05:54	00:04:15
Build DEM	00:01:59	00:01:14	00:01:15	00:00:31	00:01:02	00:00:32	00:00:52	00:00:58
Build Orthomosaic	00:03:49	00:02:01	00:02:00	00:01:46	00:02:08	00:00:54	00:01:21	00:01:44
TOTAL	02:40:01	01:13:40	01:04:55	00:19:49	2:51:02	01:50:48	1:12:09	0:50:22

Table 4. Time comparison between manual and automatic methods for the whole orthomosaic procedure. Option showed: ultra. Time in hh:mm:ss.

Task	Manual	Automatic
Copy files from UAV and load information	0:12:00	0:00:00
Orthomosaic	2:40:01	2:51:02
Load orthomosaic into QGIS	0:04:00	0:00:00
NDVI map and ROI mask	0:03:30	0:00:20
TOTAL	2:59:31	2:51:22

Figure 5 displays the eight orthomosaics generated following the two mentioned methodologies. The maps are pretty similar, and no main differences are visually observed at the employed spatial resolution (0.05 m/pix). All the orthomosaics are correctly georeferenced, however, in the case of Agisoft Metashape, all Ground Control Points (GCPs) remained perfectly in the predefined location, but in ODM, they suffered a minor displacement (Table 5). Therefore, it can be noted that Agisoft Metashape obtained better results (lower values) than ODM.

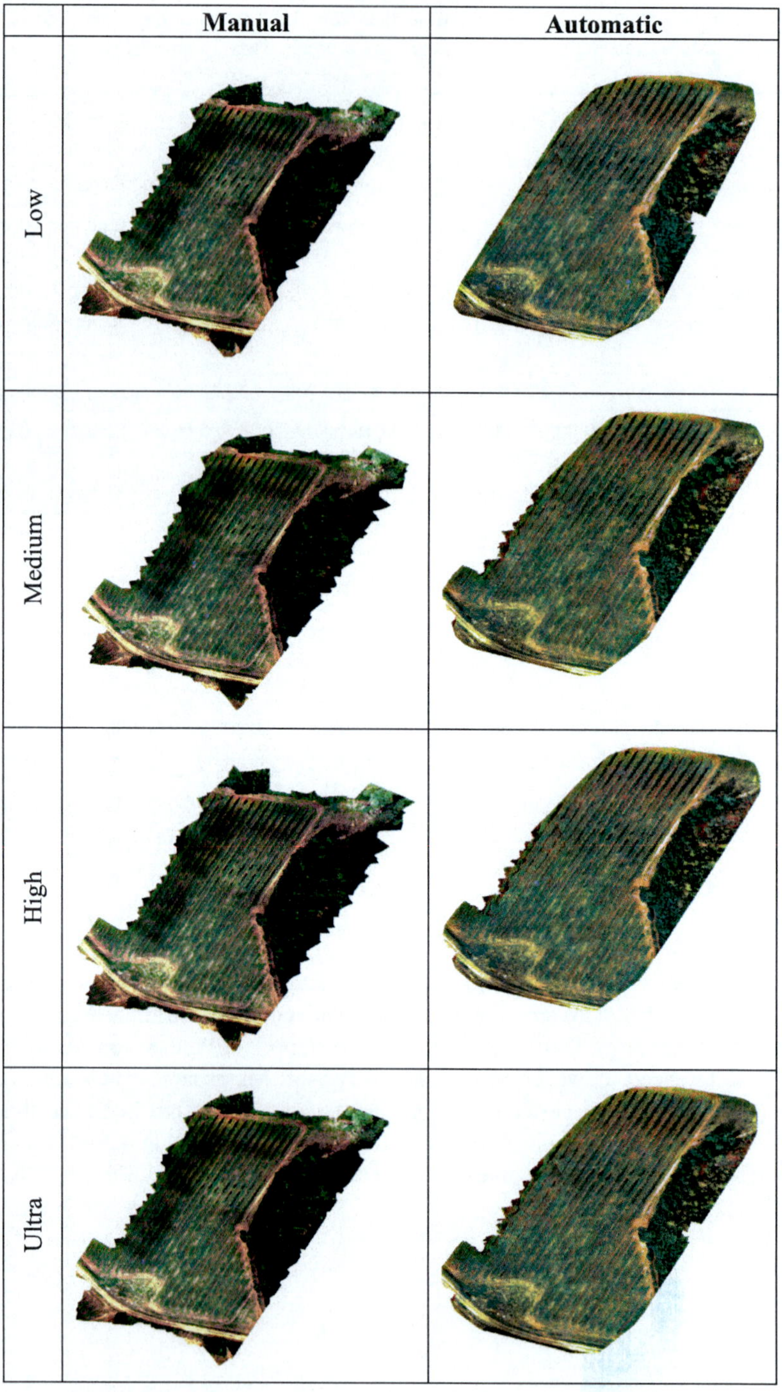

Fig. 5. Comparison of the orthomosaics generated with varying quality between Agisoft Metashape and ODM.

Table 5. Displacement of the GCP (ground-truth) and the GCP in the orthomosaic.

		Agisoft Metashape				ODM			
		Ultra	High	Medium	Low	Ultra	High	Medium	Low
Point location	RMSE X:	0.010	0.009	0.010	0.010	0.340	0.341	0.369	0.303
	RMSE Y:	0.026	0.025	0.020	0.031	0.113	0.088	0.199	0.090
Euclidean distance	Mean	0.022	0.023	0.019	0.029	0.343	0.343	0.401	0.276
	St. Dev.	0.019	0.016	0.013	0.015	0.114	0.089	0.135	0.166

4 Discussion and Conclusions

This article tries to solve the lack of knowledge about the time needed to pre-process, process and post-process an orthomosaic within a precision viticulture context, describing two methodologies commonly used during the analysis stage of the images obtained with the drone. There are many variants to these methodologies, i.e. using Pix4D instead of Agisoft Metashape or languages other than Python; however, in this work, we have tried to synthesise them into two specific workflows.

Regarding the orthomosaic process: i) the accuracy of GCP location does not rely on the quality of the process (ultra, high, medium, low), as similar results were obtained, and ii) the automated methodology takes more time than the manual procedure. However, it is not due to the workflow but to performance differences between the ODM software and the Agisoft Metashape software. In addition, the accuracy of the outputs generated by ODM was lower than those generated by Agisoft Metashape, as the GCPs had a higher displacement. This performance will probably be improved in successive versions of ODM. Nevertheless, because of the amount of time required by human intervention (Table 4), it can be concluded that the automated process is faster than the manual methodology. Furthermore, a fully automated procedure enables the user to obtain the NDVI map more conveniently and is prone to fewer human errors since the unique step to be carried out is introducing the SD card into the computer. Finally, the user can use the framework built to generate NDVI maps of the same plot. If the plot is changed, only the vector file of the Region of Interest should be modified to reuse the previously developed framework. Other factors have not been taken into account, such as the fact that the technician must be trained to use the software or that the automation task entails a longer initial setup time, as all the code has to be written. However, these factors are outside the scope of the paper and only have to be carried out once at an initial stage.

Therefore, automatic processing could be more convenient than manual processing for remote sensing workflows. Automation allows several advantages, such as i) reduction of the time needed, ii) less human error, iii) the technician only needs to input the data and can focus on other tasks while the data is being processed. However, the technician has to perform several tasks in the manual workflow process.

Acknowledgements. This work is part of the FlexiGroBots project, funded by the European Commission - H2020 programme (contract 101017111, https://flexigrobots-h2020.eu/). The authors acknowledge 'Bodegas Terras Gauda, S.A.' for its valuable contribution.

References

1. Eltner, A., et al. (eds.) UAVs for the Environmental Sciences: Methods and Applications. wbg Academic, Darmstadt (2022). ISBN 978-3-534-40588-6
2. Pierce, F.J., Clay, D. (eds.) GIS Applications in Agriculture. GIS Applications in Agriculture Series. CRC Press, Boca Raton (2007). ISBN 978-0-8493-7526-2
3. Khaliq, A., Comba, L., Biglia, A., Ricauda Aimonino, D., Chiaberge, M., Gay, P.: Comparison of satellite and UAV-based multispectral imagery for vineyard variability assessment. Remote Sens. **11**, 436 (2019). https://doi.org/10.3390/rs11040436
4. Matese, A., et al.: Intercomparison of UAV, aircraft and satellite remote sensing platforms for precision viticulture. Remote Sens. **7**, 2971–2990 (2015). https://doi.org/10.3390/rs70302971
5. Yeom, J., et al.: Comparison of vegetation indices derived from UAV data for differentiation of tillage effects in agriculture. Remote Sens. **11**, 1548 (2019). https://doi.org/10.3390/rs11131548
6. Cano, E., Horton, R., Liljegren, C., Bulanon, D.: Comparison of small unmanned aerial vehicles performance using image processing. J. Imaging **3**, 4 (2017). https://doi.org/10.3390/jimaging3010004
7. Tian, R., Cao, M., Ma, F., Ji, P.: Agricultural UAV path planning based on improved A* and gravity search mixed algorithm. J. Phys. Conf. Ser. **1631**, 012082 (2020). https://doi.org/10.1088/1742-6596/1631/1/012082
8. Liba, N., Berg-Jürgens, J.: Accuracy of orthomosaic generated by different methods in example of UAV platform MUST Q. Mater. Sci. Eng. **2015**, 9 (2015)
9. Küng, O., et al.: The accuracy of automatic photogrammetric techniques on ultra-light UAV imagery. Int. Arch. Photogram. Remote Sens. Spat. Inform. Sci. **XXXVIII-1/C22**, 125–130 (2012). https://doi.org/10.5194/isprsarchives-XXXVIII-1-C22-125-2011
10. Puniach, E., Gruszczyński, W., Ćwiąkała, P., Matwij, W.: Application of UAV-based orthomosaics for determination of horizontal displacement caused by underground mining. ISPRS J. Photogramm. Remote Sens. **174**, 282–303 (2021). https://doi.org/10.1016/j.isprsjprs.2021.02.006
11. Drone Mapping Software - OpenDroneMap Available online. https://www.opendronemap.org/. Accessed 7 July 2022
12. OpenDroneMap Authors ODM - A Command Line Toolkit to Generate Maps, Point Clouds, 3D Models and DEMs from Drone, Balloon or Kite Images. OpenDroneMap/ODM GitHub Page (2020). https://Github.Com/OpenDroneMap/ODM

Workshop on Physical Agents

Including Transfer Learning and Synthetic Data in a Training Process of a 2D Object Detector for Autonomous Driving

Miguel Antunes(✉), Luis M. Bergasa, Javier Araluce, Rodrigo Gutiérrez, J. Felipe Arango, and Manuel Ocaña

RobeSafe Research Group, Electronics Department, University of Alcalá, Madrid, Spain
{miguel.antunes,luism.bergasa,javier.araluce,rodrigo.gutierrez, juanfelipe.arango,manuel.ocanna}@uah.es

Abstract. Nowadays the use of deep learning (DL) based systems is widely extended in several areas such as facial recognition, voice and audio processing or perception systems. The training process that must be performed for proper functionality requires a large amount of data with the required characteristics needed for the task to be executed. The process of obtaining new adequate training data is complex and tedious, therefore multiple techniques such as data augmentation or transfer learning have been developed in order to have a greater amount of knowledge in the network without the need to search for new data sources. The aim of this paper is to study the effect of the inclusion of knowledge from transfer learning in a 2D image detector trained with real world and synthetic data from multiple sources. The detector that is going to be trained will be focused on autonomous driving tasks, therefore we decide to use KITTI as the real world data source and our AD PerDevkit (based on CARLA) and Virtual-KITTI as the synthetic sources.

Keywords: Deep learning · Transfer learning · 2D object detection · Simulation · KITTI · CARLA

1 Introduction

In recent years, Deep Learning (DL) based systems have become a fundamental tool for computer vision tasks, especially thanks to their remarkable performance improvement and the evolution of the hardware on which they are executed. The training from scratch of DL based detectors requires a very large amount of data rich in diverse information, therefore many techniques exist to accelerate and improve this process. The number of training data can be increased by using techniques such as traditional data augmentation, which applies geometric transformations such as rotation or compression to the images, or even using

D. Tardioli et al. (Eds.): ROBOT 2022, LNNS 590, pp. 465–478, 2023.
https://doi.org/10.1007/978-3-031-21062-4_38

synthetic data from some source to complement the real data. Another common approach is the use of transfer learning, in which weights trained on an external database are used to provide initial knowledge to the system to later specialize in the task to be performed. In this paper we study the effect of using and mixing several of these methods on the training process of a 2D object detector focused on autonomous driving tasks (Fig. 1). The images must come from specialized and open-source databases, for this reason the real data used for training, test and validation comes from two of the most used datasets for this type of tasks: KITTI [1] and Waymo [2]. On the other hand, one of the most widely used tools to test the performance of driving architectures in simulation is CARLA [3], which, using Unreal Engine and ROS, provides realistic looking environments with accurate information about the vehicle's sensors and the rest of the elements of the simulation. Taking advantage of the data provided by CARLA, we have developed our own ground truth generation tool: AD PerDevkit, which will be used to obtain part of the synthetic data. The rest of the synthetic images are obtained from the Virtual Kitti 2 dataset [4,5] which applies a recreation of the Kitti data in a virtual environment based on Unity [6] from which more synthetic sequences are generated with different conditions such as weather or time. The main difference between the two methods is that CARLA is a simulator, which means that a multitude of tests can be performed on its maps and a virtually unlimited amount of data can be generated. VKitti, on the other hand, tries to bring the knowledge provided by a real database to a virtual domain to provide a better generalization of object characteristics while maintaining similarity with the original dataset.

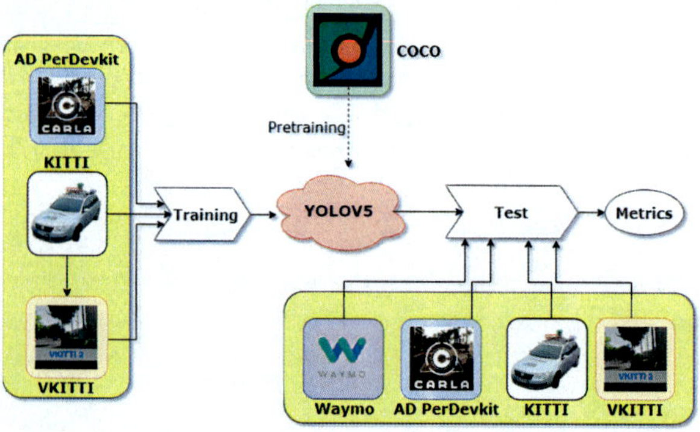

Fig. 1. Training, test and validation images mix for 2D object detection.

2 Related work

2.1 Data Augmentation

Data augmentation is one of the most common techniques in DL based system training. In [7] several traditional methods such as rotation, compression, filtering or color changes are presented along with other state of the art DL based methods such as Generative Adversarial Network (GAN) or Adversarial Training. These methods can contribute to the network with generalized knowledge that helps to improve the results and avoid overfitting with training data. In this paper we will combine synthetic data from various sources with various classic data augmentation methods such as blur, transformation to grayscale, image compression or image contrast modification.

2.2 Artificial Data from Virtual Environments

The CARLA autonomous driving simulator [3] is an open source tool that aims to provide a realistic environment in which to test the performance of complete autonomous driving architectures. In the case of perception tasks the simulator provides data from various sensors such as LiDAR, radar or camera with easily configurable parameters that mimic the real hardware with which you want to work afterwards. In the case of vehicle camera images CARLA provides different formats such as RGB or Depth. The strength of using a simulator as a data source is the possibility of obtaining new images in a practically unlimited way from user defined scenarios with parameters such as weather, time, environment and even the route and speed of the car. There are several tools that help us to obtain the data we need from the simulator. One of them is AIODrive[8] which provides a dataset with multiple environments with labeled objects based on direct information from the simulator. The main problem with this dataset is that it labels objects out of range as it does not take into account the visibility of the objects relative to the car. The other tool examined is our AD PerDevkit dataset [9] which extracts information in a similar way to AIODrive but removes objects that are not of interest for training tasks or are out of range, improving weak points of the former dataset. The data in our tool is in a similar format to KITTI with some extra information such as the frame identifier to facilitate the association with the raw images. The main problem for camera-based detection from simulation is the visual difference between objects in the synthetic world and in the real world, although with the advances in virtual environments in recent years it will become more and more realistic. Another way to obtain synthetic data is presented by Virtual-KITTI [4,5], in which the images provided have been generated by a semi-automatic procedure that aims to create virtual environments called *'proxy'* that are clones of a real-world dataset. In this case, *'proxy'* worlds are generated in Unity [6] from several KITTI scenes and transformations are applied to obtain those same environments with different weather or time of day conditions. Unlike the CARLA approach, this method does not aim to generate a huge amount of data, but instead uses real images to generate its clones to help the generalization process of the characteristics from the original dataset objects.

2.3 Bidimensional State-of-the-art Object Detectors

Currently there are multiple DL based 2D detectors with state-of-the-art performances. Some of the more popular are multi-stage systems such as fast-RCNN

[10] and faster-RCNN [11] or one-stage detectors such as Yolo [12][13][14] and EfficientDet [15]. Table 1 shows differences between the latest and heaviest models in Average Precision (AP) on COCO and their inference time.

Table 1. Performance of different state-of-art 2D detectors [11,14,15]

Model	COCO AP 0.5	Latency (ms)
Yolov5L	67.3	2.7
Yolov5X	68.9	4.8
EfficientDet D6	71.5	92.8
EfficientDet D7	74.3	153
Faster RCNN	42.7	200

Faster RCNN obtains much worse results than the rest of the networks, so it is discarded. EfficientDet obtains slightly better results than Yolov5, however, the difference in execution time is very high. Due to the focus on autonomous driving tasks, low inference times are required, so Yolov5 is chosen as detector.

2.4 2D Object Detection Architecture

The detector that is going to be trained is Yolov5 [14] due to its good results and the flexibility it offers thanks to its multiple models. The base architecture is the one shown in Fig. 2, with an image input size of 640x640 and using layers such as CSP Bottleneck with 3 convolutions (C3) or Spatial Pyramid Pooling (SPPF). This architecture scales following the idea presented in EfficientDet [15] to offer various performance and execution times, shown in the Table 2. For more information about Yolov5 please refer to [14].

Table 2. Yolov5 models comparison. Validation on COCO val2017.

Model	mAP^{val} 0.5	Speed V100 (ms)	Params (M)
Yolov5n	46.0	6.3	1.9
Yolov5s	56.0	6.4	7.2
Yolov5m	63.9	8.2	21.2
Yolov5l	67.2	10.1	46.5
Yolov5x	68.9	12.1	86.7

From the data shown in Table 2 it was decided to use the Yolov5L model as it offers similar results to Yolov5X with half the parameters in the network, and the Yolov5S model as it is the lightest model without taking into account Yolov5N which does not offer sufficient results for the task to be performed.

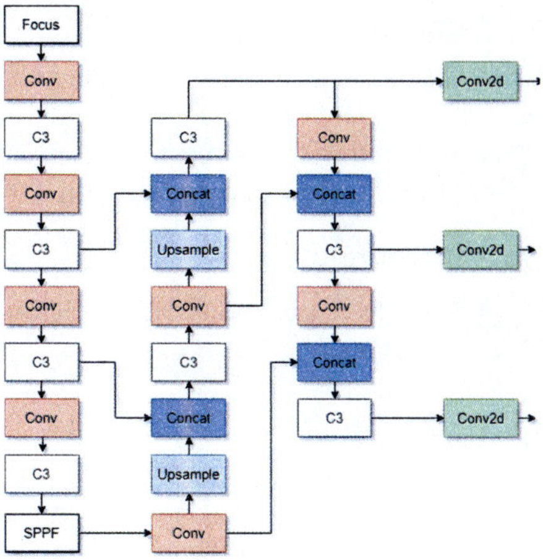

Fig. 2. Yolov5 detector layers

3 CAMERA DATA

3.1 COCO and Waymo Datasets

COCO [16] is one of the largest datasets for general object detection, with 200k images labeled representing 80 object classes. The authors of Yolov5 provide weights with a pre-training of 300 epochs on this dataset which will be used in the training of the paper, particularizing the knowledge only for the class *'Car'*. It should be noted that in order to train with the different datasets, a transformation of the ground truth annotations to the COCO data format has been performed. Waymo [2] is used to evaluate the trained detectors on a real dataset different from the ones used for training. This dataset specializes in autonomous driving tasks, offering data segments recorded by multiple real-world environments from a vehicle with information from multiple sensors compressed into multiple files. Camera information was extracted from the files and transformed into COCO format for evaluation.

3.2 Kitti and Virtual-Kitti Datasets

KITTI dataset [1] is one of the most widely used autonomous driving datasets since its release in 2013. Like Waymo, KITTI provides information from different sensors on a vehicle moving through real environments. In this paper KITTI will be used in practically all experiments as the source of real camera data, using the 4781 images it provides, of which 80% have been used for training, 10% for validation and the rest for testing. The KITTI labeling format provides object information such as class, camera visibility, position, rotation, 3D dimensions and the 2D bonding box.

Virtual-Kitti [4,5] (VKitti) uses a semi-automatic technique to generate virtual environments on Unity's engine based on camera information provided by

KITTI. These *'proxy'* environments are modified to reflect different weather conditions or time of day, allowing the trained DL model to better generalize information about the original dataset, which in this case is KITTI. The first version of the dataset was released in 2015 and provided visual information and the ground truth of the objects, while the second one has been released in 2020 and improves the graphic section as well as providing additional stereo camera and ground truth information such as class segmentation or depth. The object information from both datasets is transformed into COCO format for the training and evaluation process.

3.3 AD PerDevkit Dataset

The synthetic data from the CARLA simulator is obtained through our AD PerDevkit dataset [9]. The synthetic data coming from the CARLA simulator are obtained through our tool. Using the ROS bridge the information from the sensors is exported, both the raw data (images, LiDAR pointcloud and Radar pointcloud) and the labeled objects in a csv with the required information such as each object class or 2D bounding box.

The dataset is composed of 11 challenging scenes recorded in 5 towns included in default CARLA under different weather conditions, each containing between 1400 and 3200 frames with full set of annotations. Only about 10000 daytime synthetic images will be used to maintain the similarity with KITTI, which only has daytime images.

(a) (b)

Fig. 3. a) KITTI and VKITTIv2 image samples. b) CARLA and KITTI 2D ground truth example.

4 Experimental Results

Multiple experiments have been conducted to evaluate the effect of using synthetic data with other techniques such as transfer learning or classical data augmentation to improve the network performance. All training processes are performed on a GTX 1080Ti with 25 epochs and a batch size of 32 for Yolov5L and 64 for Yolov5S models, applying the geometric data augmentation techniques presented in Sect. 2.1 and calculating the precision, recall and mAP metrics. The tables presented below highlight in one color the best results for each case, with black being the best using the 100% of Kitti training data, orange using the 50% and blue using the 25%.

4.1 Training Without Pretrained Weights

The first experiment was to train the detectors from scratch, in other words, without any prior knowledge of COCO. The training was performed on separate datasets and various combinations with the multiple synthetic sources.

The Table 3 shows the results of training on individual databases without prior knowledge. As expected, the best results are obtained by evaluating on the same dataset used for training. With no prior knowledge, the network only has the information of the characteristics of one database, which can lead to overfitting on the training data, deteriorating the metrics on the rest of the datasets.

Table 3. Training results on one database from scratch

Model	Train	Test	P	R	mAP
Yolov5S	Kitti	Kitti	0.887	0.807	0.908
		Waymo	0.520	0.294	0.327
		Carla	0.298	0.303	0.172
	Carla	Kitti	0.152	0.058	0.043
		Waymo	0.016	0.029	0.013
		Carla	0.640	0.890	0.767
	VKitti	Kitti	0.671	0.380	0.425
		Waymo	0.029	0.053	0.016
		VKitti	0.973	0.921	0.980
Yolov5L	Kitti	Kitti	0.924	0.869	**0.943**
		Waymo	0.618	0.350	0.409
		Carla	0.276	0.198	0.154
	Carla	Kitti	0.078	0.077	0.038
		Waymo	0.023	0.023	0.013
		Carla	0.621	0.893	0.780
	VKitti	Kitti	0.681	0.416	0.462
		Waymo	0.048	0.047	0.017
		VKitti	0.982	0.941	0.984

Tables 4 and 5 show the results of training by combining a percentage of KITTI data with each of the synthetic datasets.

Using KITTI and CARLA (Table 4) the detector can generalize the knowledge better than if it is trained only on one database, obtaining the same or better results on these datasets even than in the case of training and evaluating the network with only one of them (Table 3). For the Yolov5L model, KITTI improves from a mAP of 0.943 to 0.949, thus improving performance on the real data base, which is the desired effect.

In the case of using VKitti and KITTI (Table 5) both datasets have more similar features since VKitti is built from KITTI, this is reflected in the results of evaluating on KITTI, which improve even more than training with CARLA, reaching a mAP of 0.961 if we use Yolov5L and train with both complete datasets.

It is worth noting that Waymo also benefits slightly from training with mixed datasets, but its results still leave room for improvement.

4.2 Transfer Learning Experiments with Pretrained Weights

Once the training tests had been performed without the knowledge from COCO, the training processes were repeated starting from the pre-trained weights.

To evaluate the initial level of knowledge provided by the pre-trained weights, an initial evaluation was performed on the real-world datasets used to evaluate the results of the other experiments: KITTI and Waymo.

Table 4. Training results on CARLA and KITTI from scratch

Model	Train	Test	P	R	mAP
Yolov5S	Kitti 100% + Carla	Kitti	0.884	0.853	0.925
		Waymo	0.533	0.315	0.352
		Carla	0.669	0.879	0.776
	Kitti 50% + Carla	Kitti	0.925	0.873	0.901
		Waymo	0.755	0.519	0.605
		Carla	0.699	0.796	0.811
	Kitti 25% + Carla	Kitti	0.831	0.755	0.841
		Waymo	0.476	0.268	0.292
		Carla	0.675	0.804	0.724
Yolov5L	Kitti 100% + Carla	Kitti	0.933	0.874	**0.949**
		Waymo	0.598	0.350	0.356
		Carla	0.687	0.767	0.772
	Kitti 50% + Carla	Kitti	0.910	0.841	**0.930**
		Waymo	0.521	0.277	0.314
		Carla	0.657	0.775	0.753
	Kitti 25% + Carla	Kitti	0.873	0.794	**0.889**
		Waymo	0.506	0.263	0.303
		Carla	0.652	0.769	0.723

Table 5. Training results on VKitti and KITTI from scratch

Model	Train	Test	P	R	mAP
Yolov5S	Kitti 100% + Vkitti	Kitti	0.913	0.850	0.936
		Waymo	0.578	0.348	0.395
		Vkitti	0.968	0.926	0.981
	Kitti 50% + Vkitti	Kitti	0.889	0.803	0.901
		Waymo	0.615	0.347	0.407
		Vkitti	0.968	0.900	0.970
	Kitti 25% + Vkitti	Kitti	0.864	0.758	0.862
		Waymo	0.580	0.353	0.404
		Vkitti	0.958	0.919	0.979
Yolov5L	Kitti 100% + Vkitti	Kitti	0.943	0.900	**0.961**
		Waymo	0.652	0.374	0.433
		Vkitti	0.987	0.931	0.984
	Kitti 50% + Vkitti	Kitti	0.918	0.878	**0.940**
		Waymo	0.644	0.360	0.420
		Vkitti	0.975	0.939	0.982
	Kitti 25% + Vkitti	Kitti	0.887	0.820	**0.908**
		Waymo	0.637	0.325	0.385
		Vkitti	0.978	0.945	0.986

The Table 6 shows that the pre-trained weights lead to better performance results in Waymo compared to those on KITTI. This may be due to the fact that Waymo images are more similar to those used for pre-training in COCO than the ones in KITTI. Despite this, it is observed that COCO provides a high degree of initial knowledge for both datasets, since before any training is performed, mAP values of 0.78 for Waymo and 0.6 for KITTI are already obtained. The results of training on a database starting from the COCO weights are represented in the Table 7, in which each model has been trained on the synthetic datasets and on KITTI and various percentages of its data to test the metrics when having limited real-world data. The evaluation is performed on Waymo and KITTI and a synthetic database depending on the training case. Comparing with the

Table 6. Pretrained weights evaluated on KITTI and Waymo

Model	Test	P	R	mAP
Yolov5S	Kitti	0.753	0.531	0.590
	Waymo	0.815	0.650	0.763
Yolov5L	Kitti	0.743	0.567	0.609
	Waymo	0.778	0.692	0.784

results of training without COCO weights (Table 3) the improvement is pretty evident. Even training with exactly the same data leads to results improvement, for example Yolov5L trained from scratch with KITTI reaches a mAP of 0.943 which rises to 0.969 with pre-training.

Table 7. Training result on one database with transfer learning

Model	Train	Test	P	R	mAP
Yolov5S	Kitti 100%	Kitti	0.903	0.823	0.911
		Waymo	0.763	0.559	0.647
		Carla	0.504	0.541	0.419
	Kitti 50%	Kitti	0.862	0.668	0.807
		Waymo	0.793	0.568	0.674
	Kitti 25%	Kitti	0.783	0.654	0.751
		Waymo	0.789	0.586	0.695
	Carla	Kitti	0.638	0.427	0.459
		Waymo	0.741	0.531	0.617
		Carla	0.691	0.763	0.762
	VKitti	Kitti	0.788	0.586	0.632
		Waymo	0.820	0.594	0.709
		VKitti	0.942	0.913	0.970
Yolov5L	Kitti 100%	Kitti	0.953	0.910	**0.969**
		Waymo	0.776	0.582	0.674
		Carla	0.409	0.559	0.439
	Kitti 50%	Kitti	0.921	0.867	**0.942**
		Waymo	0.786	0.568	0.693
	Kitti 25%	Kitti	0.869	0.781	**0.883**
		Waymo	0.798	0.622	0.716
	Carla	Kitti	0.651	0.466	0.532
		Waymo	0.614	0.509	0.552
		Carla	0.710	0.842	0.793
	VKitti	Kitti	0.797	0.622	0.656
		Waymo	0.838	0.663	0.769
		VKitti	0.966	0.935	0.983

For the case of training KITTI in combination with CARLA (Table 8) there is a slight deterioration of the mAP with respect to training from scratch if not all KITTI data is used. The reason for this is that the pretrained weights are trained on COCO, therefore lowering the amount of KITTI data leads to a worse acquisition of the KITTI knowledge maintaining decent results on Waymo.

Table 8. Training result on CARLA and KITTI with transfer learning

Model	Train	Test	P	R	mAP
Yolov5S	Kitti 100%+ Carla	Kitti	0.886	0.828	0.912
		Waymo	0.767	0.581	0.670
		Carla	0.660	0.734	0.728
	Kitti 50%+ Carla	Kitti	0.760	0.810	0.806
		Waymo	0.784	0.578	0.674
		Carla	0.685	0.749	0.742
	Kitti 25%+ Carla	Kitti	0.767	0.764	0.775
		Waymo	0.794	0.570	0.672
		Carla	0.717	0.730	0.742
Yolov5L	Kitti 100%+ Carla	Kitti	0.950	0.911	**0.969**
		Waymo	0.800	0.579	0.676
		Carla	0.745	0.760	0.829
	Kitti 50%+ Carla	Kitti	0.802	0.871	**0.889**
		Waymo	0.792	0.600	0.694
		Carla	0.610	0.901	0.800
	Kitti 25%+ Carla	Kitti	0.817	0.808	**0.846**
		Waymo	0.799	0.570	0.673
		Carla	0.702	0.731	0.776

Training with a combination of KITTI and VKitti (Table 9) achieves surprising results due to the generalization of information induced by the synthetic dataset. First of all, it manages to improve the results compared to not using transfer learning (Table 5) especially in the cases with lower KITTI data proportion. For example, for Yolov5L with only 25 percent of KITTI increases the mAP from 0.908 to 0.931 without affecting Waymo as much (decreases only to 0.679 instead of the 0.385 it had without transfer learning). On the other hand, this combination of data also achieves better results than training with transfer learning but only on the equivalent single database (Table 7). The difference in results increases again as the KITTI data decreases, resulting in an 5% higher mAP when mixed with the 25% of KITTI. From this data it can be deduced that adding VKitti to KITTI training will lead to improved results if the real world data is limited to a small number of images respect the synthetic data. If the amount of images is similar, adding the synthetic source dos not achieve any improvement resulting in similar mAP.

It should be noted that some tests have also been performed in which synthetic data was used for training and then fine tuning was performed with KITTI data, but they do not provide any improvement in results, so they have not been incorporated. For example training on CARLA and performing fine tune on all KITTI training images results on exactly the same mAP as training only on KITTI.

Table 9. Training result on VKitti and KITTI with transfer learning

Model	Train	Test	P	R	mAP
Yolov5S	Kitti 100%+ Vkitti	Kitti	0.896	0.813	0.904
		Waymo	0.783	0.568	0.664
		Vkitti	0.945	0.910	0.969
	Kitti 50%+ Vkitti	Kitti	0.864	0.772	0.868
		Waymo	0.782	0.567	0.663
		Vkitti	0.955	0.906	0.970
	Kitti 25%+ Vkitti	Kitti	0.815	0.732	0.816
		Waymo	0.804	0.570	0.675
		Vkitti	0.945	0.909	0.970
Yolov5L	Kitti 100%+ Vkitti	Kitti	0.949	0.915	**0.967**
		Waymo	0.770	0.597	0.683
		Vkitti	0.973	0.929	0.983
	Kitti 50%+ Vkitti	Kitti	0.938	0.882	**0.955**
		Waymo	0.779	0.597	0.682
		Vkitti	0.971	0.925	0.982
	Kitti 25%+ Vkitti	Kitti	0.895	0.878	**0.931**
		Waymo	0.781	0.586	0.679
		Vkitti	0.979	0.930	0.984

5 Disscussion and Conclusions

The experiments conducted have made it possible to quantify the impact of various techniques on the results of a training process. First, multiple trainings from scratch have been performed with different databases (Table 3) and mixing real and synthetic data (Tables 4 and 5). Subsequently, these trainings have been repeated but applying the transfer learning technique, that is, starting from pre-trained weights in COCO (Tables 7, 8 and 9). It has been observed how applying a transfer learning technique helps to avoid overfitting problems that appear in training from scratch, obtaining even higher performances. This is why in DL network training processes it is recommended to use this technique whenever possible due to the benefits it provides. On the other hand, the results have also allowed us to verify which approach to obtain synthetic data between CARLA and VKitti provides the most improvement. Training with only one of these two datasets does not provide any improvement over real data (KITTI or Waymo), so it is advisable to mix them with real-world data to improve the results. In the case of performing this combination, it is obtained that in the case of training from scratch, both data sources provide better results than using only the real database. When using pre-trained weights, CARLA does not achieve better results than training with a single real database, but VKitti does achieve an improvement, especially as the percentage of KITTI is reduced. The use of transfer learning for DL training is recommended as long as the original database has similarities to those to be used for the particular case. For example, COCO [16] can be used for general camera detection. The proxy worlds approach used by VKitti [4] provides a more noticeable improvement than using data from a simulator such as CARLA. However, obtaining data by the former method is more expensive and more limited than using CARLA, which is able of generating unlimited labeled data in a simple way.

Acknowledgements. This work has been funded in part from the Spanish MICINN/FEDER through the Techs4AgeCar project (RTI2018-099263-B-C21) and from the RoboCity2030-DIH-CM project (P2018/NMT- 4331), funded by Programas de actividades I+D (CAM), cofunded by EU Structural Funds and Scholarship for Introduction to Research activity by University of Alcalá

References

1. Geiger, A., Lenz, P., Urtasun, R.: Are we sready for autonomous driving? the kitti vision benchmark suite. In: Conference on Computer Vision and Pattern Recognition, CVPR (2012)
2. Sun, P., et al.: Scalability in perception for autonomous driving: Waymo open dataset. In: Proceedings of the IEEE/CVF Conference on Computer Vision and Pattern Recognition, CVPR (2020)
3. Dosovitskiy, A., Ros, G., Codevilla, F., Lopez, A., Koltun, V.: CARLA: An open urban driving simulator. In: Proceedings of the 1st Annual Conference on Robot Learning, pp. 1–16 (2017)
4. Gaidon, A., Wang, Q., Cabon, Y., Vig, E.: Virtual worlds as proxy for multi-object tracking analysis. In: Proceedings of the IEEE conference on Computer Vision and Pattern Recognition, pp. 4340–4349 (2016)
5. Cabon, Y., Murray, N., Humenberger, M.: Virtual kitti 2 (2020)
6. Unity. https://unity.com/
7. Shorten, C., Khoshgoftaar, T.M.: A survey on image data augmentation for deep learning. J. Big Data **6**(1), 1–48 (2019). https://doi.org/10.1186/s40537-019-0197-0
8. Weng, X., et al.: All-In-One Drive: A Large-Scale Comprehensive Perception Dataset with High-Density Long-Range Point Clouds (2021)
9. de la Peña, J., et al.: AD PerDevKit: An autonomous driving perception development kit using CARLA simulator and ROS - Accepted ITSC (2022)
10. Girshick, R.: Fast R-CNN (2015)
11. Ren, S., He, K., Girshick, R., Sun, J.: Faster R-CNN: Towards real-time object detection with region proposal networks (2016)
12. Redmon, J., Farhadi, A.: YOLOv3: An incremental improvement. arXiv:1804.02767 (2018)
13. Bochkovskiy, A., Wang, C.Y., Liao, H.Y.M.: YOLOv4: Optimal speed and accuracy of object detection. arXiv:2004.10934 (2020)
14. Jocher, G., et al.: ultralytics/yolov5: v6.0 - yolov5 models, roboflow integration, tensorflow export, opencv dnn support. (2021). https://doi.org/10.5281/zenodo.5563715

15. Tan, M., Pang, R., Le, Q.V.: EfficientDet: Scalable and efficient object detection. In: Proceedings of the IEEE/CVF Conference on Computer Vision and Pattern Recognition, pp. 10781-10790 (2020)
16. Lin, T.-Y., et al.: Microsoft COCO: common objects in context. In: Fleet, D., Pajdla, T., Schiele, B., Tuytelaars, T. (eds.) ECCV 2014. LNCS, vol. 8693, pp. 740–755. Springer, Cham (2014). https://doi.org/10.1007/978-3-319-10602-1_48

Unsupervised Hyperbolic Action Recognition

John-Alejandro Castro-Vargas(✉), Alberto Garcia-Garcia,
Pablo Martinez-Gonzalez, Sergiu Oprea, and Jose Garcia-Rodriguez

3D Perception Lab, University of Alicante, Alicante, Spain
{jacastro,agarcia,pmartinez,soprea,jgarcia}@dtic.ua.es

Abstract. Methods based on Deep Geometric Learning allow the development of solutions with a geometric approximation in different applications. In particular, the curved feature of hyperbolic space has the ability to describe hierarchical structures in a better manner. In this paper, we aim to define an unsupervised learning model for action recognition. The curved feature space is intended to be used to describe a hierarchical relationship between the clips that compose a complete video sequence. These, in turn, are related to each other by means of a triplet loss function and a VAE (Variational Auto-Encoder) neural architecture, which establishes a similarity relationship between clips to identify actions from a set of unlabelled data.

Keywords: Deep geometric learning · Action recognition · Unsupervised · Hyperbolic

1 Introduction

In the course of a single day, humans perform a large number of actions. These actions can be defined in different ways depending on the temporality we choose. That is, depending on the time window we select in the same sequence of data, we can identify, move a foot, walk or make a cup of coffee.

The detection of these actions in their different temporal ranges is of interest in multiple situations. It can be use as a mechanism of security to detect strange actions or behaviour in shops, and also as a tool for the care of dependent people, as it could allow an early detection of possible accidents. In case of detecting one, it could automatically act accordingly, either by contacting the health services or a caregiver.

In order to address this issue, different types of data have been used to make predictions over the years. These include skeletal traces, optical flow, depth maps, point clouds and images.

Each type of data has its advantages and disadvantages, but images have remained of particular interest. Among other things, this is due to the greater ease of acquiring this type of data and the reduced cost of cameras. In recent years, most approaches using this type of data have been focused on the use of

D. Tardioli et al. (Eds.): ROBOT 2022, LNNS 590, pp. 479–488, 2023.
https://doi.org/10.1007/978-3-031-21062-4_39

deep learning models, especially those based on convolutional networks [22,26] and recurrent networks [19]. However, despite the improvement in the results obtained in action recognition, this type of challenge still faces the need to acquire labelled data to feed the learning models.

Among the possible solutions that we can find to facilitate the acquisition of data for action recognition we can find: recording and labelling actions, segmenting and labelling public videos (e.g. YouTube) or generating them synthetically with automatic labelling. However, the first and second cases require hard work, both in generation and labelling, especially taking into account the temporal characteristic of this type of data, which requires playback for the correct observation of the action being performed.

The latter case involves the development or use of a tool for the synthetic generation of body actions. However, although this approach has raised a lot of interest and different options for such solutions have been proposed, there are usually some drawbacks with such approaches. The main one is the visual difference between the features generated in the synthetic approach compared to the images obtained in real environments. These differences may limit the transference of results obtained from this type of data to real environments.

Another line of research has been around the use of semi-supervised and unsupervised learning models, which can benefit from the use of a larger number of videos and reduce the time needed for data labelling. These models typically attempt to learn features by taking into account the temporality in the image sequences. That is, from the images following a sequence of images, an attempt is made to establish a criterion to describe these sequences.

This paper addresses action recognition by means of unlabelled video sequences and a geometric model. For this purpose, we consider the videos as behaviours, which according to [3] are considered as a level of abstraction composed of several actions. To recognise these actions we proposed an architecture inspired in the advances presented in [9,20]. The authors of these articles use geometric approximations to model the solutions in their respective applications, specifically hyperbolic geometry.

Our goal is taking advantage of the curvature characteristic in hyperbolic space to infer a hierarchical relationship between the clips that compose a behaviour. This structure aims to establish a similarity relationship between similar clips, while hierarchically distancing the most "relevant" clips to describe the behaviour [20]. Furthermore, to train the model without the necessity of pre-labelled data, we rely on a triplet loss function and a VAE (Variational Auto-Encoder) network. The relationship between the video clips is defined by calculating the clusters for the latent space values, which are generated in hyperbolic space [9].

2 Related Works

Clustering approaches aim to relate groups of features to each other. Over the years, different proposals have been tested that address this strategy in action recognition, and among them two fundamental phases can be observed. On the one hand, the extraction of features from the image sequence, and on the other hand, the algorithm used to relate these features to each other and define the clusters [1].

2.1 Feature Detection

With respect to the feature extraction process, we will focus on those models that address this process by means of neural networks and self-supervised or unsupervised approaches. Much of the recent work revolves around these models due to their good results, especially those approaches that use convolutional layers. Most of these approaches aim to make the proposed machine learning models learn to extract descriptive features from the corresponding sequences. The reason for this approach is to take advantage of the large amount of videos available on the Internet while reducing the need to deal with the labelling of new data. To this end, we seek to use the features implicit in the videos, both spatially and temporally, to provide pretext tasks with which to train neural models. Once trained, these networks have the capacity to generate descriptive feature vectors.

The temporal connection that links the information between frames can be used as a pretext task [6,24], where the order of the frames can be used for that purpose. The application of spatial transformations [12], or the use of texture information as target data [14] were also explored as mechanisms with which to self-supervise the network. Similarly, the environment information in the image [5] [13] and the resemblance between regions in the image and the surrounding environment [8] [24] have also been used as a self-training mechanism.

Most of the self-supervised models propose neural architectures made up of convolutional layers, where the last layer is used as a feature vector [12,24]. These models learn between layers sets of patterns from which to extract the most representative information from the clips. Another possible way to obtain a representative vector of the input dataset can be the use of variational auto-encoders [17]. This type of networks, rather than searching for a feature vector, works on finding a continuous latent space. This objective is approached by posing the network as a problem of probability distributions, where the latent space is defined as a prior $P(Z)$. This architecture consists of two stages, an encoder and a decoder. The first aims to find the operations necessary to represent the input data set in the latent space, a process that is usually carried out to reduce the dimensionality of the input data, that is, it computes $P(Z|X)$. The second aims at the opposite operation, starting from the probability distribution of the latent space to find the operations necessary to reconstruct the input data, $P(X|Z)$. This approach in the architecture allows the network to keep close and continuous those input data that are semantically related. This feature can be observed when reconstructing nearby information in the representation space, where we can observe an interpolated evolution of the samples.

2.2 Clustering

The clustering phase consists of relating those data sets that have similar or close features to each other. Its application has been useful in image segmentation [9], customer behaviour analysis [10], behaviour analysis in daily activities [18], among others.

Once the features have been extracted, there are different strategies used to generate the clusters. Among them we can highlight partitional methods such as k-means [4], which by means of the distance between the features it groups them into hypersurfaces; hierarchical methods, which also by means of distance relates features and clusters into hierarchies; spectral clusters, which make use of the spectrum of the similarity matrix to group the features; and methods based on machine learning such as SOM [11] (self-organising maps).

2.3 Geometric Approach

There are different approaches based on geometric applications that have been used in action recognition. Several of them can be seen as spatial transformations that modify the three-dimensional viewpoint in order to improve the response of models to different perspectives [7]. However, one of the most attractive lines of research is that which merges the geometric capacity to better describe certain environments with machine learning methods [2].

Traditionally, most of the operations used to perform estimations are based on the conception of a Euclidean space. This notion is intrinsic to the operations performed by neural models, which is why in the process, situations that can be represented by other types of structures, such as graphs or manifolds, obtain results that are bounded or approximate to the Euclidean plane during the training process [2].

With respect to graph-based networks, they offer the possibility of relating features which have a certain connection relationship with other adjacent features. Several approaches have used this property to learn from skeletons that represent the poses between frames of action. Spatio-Temporal Graph Convolutions [25] have the property to learn from both, the features of connected joints and their connection over time. Furthermore, networks based on manifolds propose other geometries to the Euclidean plane to approximate solutions that are closer to the desired objective. In [21] they take advantage of hyperbolic curvature to infer levels of abstraction in predicting future actions. Depending on the certainty of what action will occur, they estimate a more specific or generic level for that action.

3 Architecture

One of the fundamental points of the model we use is its non-Euclidean characteristic with constant negative curvature. As it is this property that allows us to define hierarchies in a continuous manner, unlike tree-based approaches. In addition, to perform the necessary operations along the hyperbolic pipeline, we use the Poincaré ball as a geometric space. Moreover, because the inputs and outputs work in Euclidean space, we use the exponential function (Eq. 1) and logarithmic maps (Eq. 2) to convert between the different planes [9], where $\oplus$ referers to the Mobius addition [23]. The distance function is defined by the Eq. 3 to obtain the distance between projected points on the Poincaré ball.

$$exp_z(v) = Z \oplus \left(tanh \left(\frac{\lambda_z ||x - y||}{2} \right) \frac{v}{||v||} \right) \tag{1}$$

$$log_z(y) = \frac{2}{\lambda_z} tanh^{-1}(|| -z \oplus y||) \frac{-z \oplus y}{|| -z \oplus y||} \tag{2}$$

$$d_p(x, y) = cosh^{-1} \left(1 + 2 \frac{||x - y||^2}{(1 - ||x||^2)(1 - ||y||^2)} \right) \tag{3}$$

The method used to obtain a mechanism to train the system in an unsupervised way is based on the use of a variational autoencoder network and a triplet loss function.

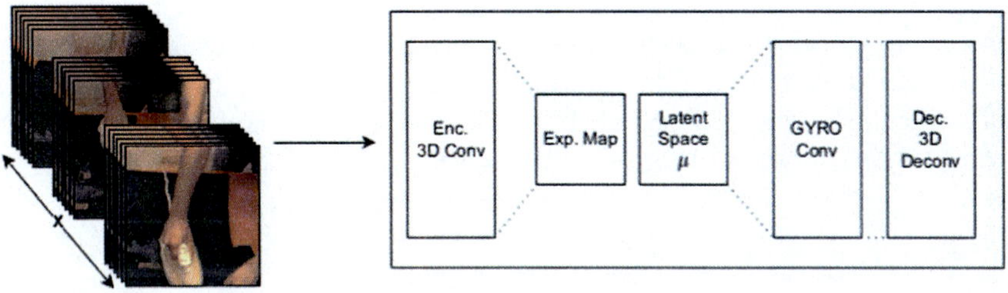

Fig. 1. The system pipeline consists of three phases. Firstly, we define the triplets that relate the video clips to each other, then we train the system using a 3D VAE modified with hyperbolic operations, and finally we generate clusters from the latent values.

The VAE network is used to generate latent values in hyperbolic space from the input clips. This process is defined according to [16], where an exponential map is used at the end of the encoder, together with a gyroplane convolution before the decoder (Fig. 1). These layers are intended to create the network respect the hyperbolic geometry in latent space. The encoder and decoder modules are composed of 3D convolution and deconvolution layers.

$$L_{triplet}(\mu_x, \mu_{pos}, \mu_{neg}) := max(0, d_p(\mu_x, \mu_{pos}) - d_p(\mu_x, \mu_{neg}) + \alpha) \tag{4}$$

The triplet loss function (Eq. 4) is based on relating similar features between different elements. To obtain this capability automatically, we define the triplets by segmenting the video into M clips and relating them to each other by proximity. That is, those clips closer together are expected to be part of the same

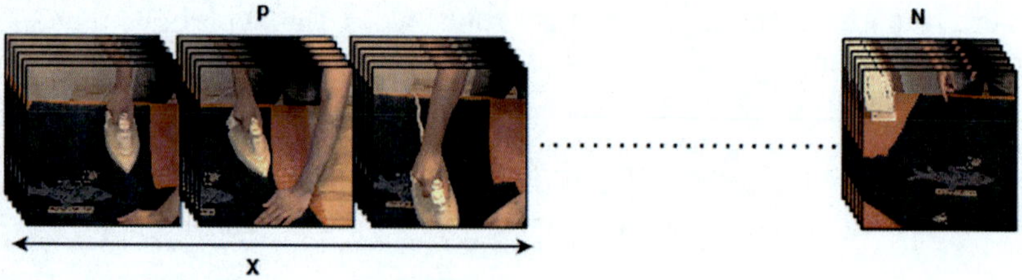

Fig. 2. Triplets are defined by selecting clips based on their adjacency. X is defined selecting three adjacent clips, where the central clip P corresponds to μ_{pos} in latent space. The furthest clip N corresponds to μ_{neg}.

action, while those farther apart are expected to be different. The dissimilarity function used for the triplet elements is the distance in the Poincaré ball d_p.

This process is done by rescaling the clips to a size of $10 \times 10 \times 10$ and converting the videos to grayscale. From the set of clips, we select three adjacent clips to define X. The center clip corresponds to μ_P and the farthest clip to μ_N (Fig. 2).

Finally, after training stage we generate the latent space in the hyperbolic space by means of the encoder. These values are clustered using an analogous version of *k-means* on a Poincaré disk. The centroid and Euclidean distance calculations are computed by their counterparts in Rimannian geometry, the Fréchet mean [15] and the manifold distance.

3.1 Experiments and Results

Table 1. The table shows the results obtained in Precision (P), Recall (R), F1 score (F_1) and Accuracy for the four labels. These values are the result of grouping and labeling the latent values generated in Euclidean and hyperbolic space.

	Euclidean			Hyperbolic		
Action	P	R	F_1	P	R	F_1
Prepare cloth	0.6488	0.625	0.5429	0.8072	0.5104	0.5603
Ironing	0.6703	0.7292	0.6770	0.7188	0.8125	0.7059
Leave Iron	0.8667	0.4688	0.5206	0.6223	0.6875	0.6323
Take Iron	0.7203	0.7708	0.7341	0.8292	0.6563	0.6476
Accuracy	0.6484			0.6667		
Avg. Precision	0.7265			0.7444		
Avg. F_1	0.6186			0.6365		

The experimentation was performed with 19 videos of people ironing clothes published on YouTube. From this task we identified four actions that define the whole activity: "Prepare clot", "Ironing", "Leave Iron" and "Take Iron". Fifteen of these unlabeled videos were used as training sets, while the remaining four were labeled with the actions that compose the activity to evaluate the use of hyperbolic geometry in contrast to Euclidean geometry. In order to compare the Euclidean approximation with the hyperbolic one, we train the VAE network in both its Euclidean and hyperbolic versions. The main difference lies in the exponential map and the gyroplane convolution layer, which allow the generated latent values to be viewed as Euclidean latent values in the case we don't use them, or hyperbolic in case we do.

The training set was constructed by slicing the videos into one-second clips with a frame rate of eight frames per second. We then indexed the clips to define the triplets with, following the method described above (Fig. 2). The parameters used to train the network over fifteen epochs were: a learning rate 5e-10, triplet weight 1e2 and batches of 128.

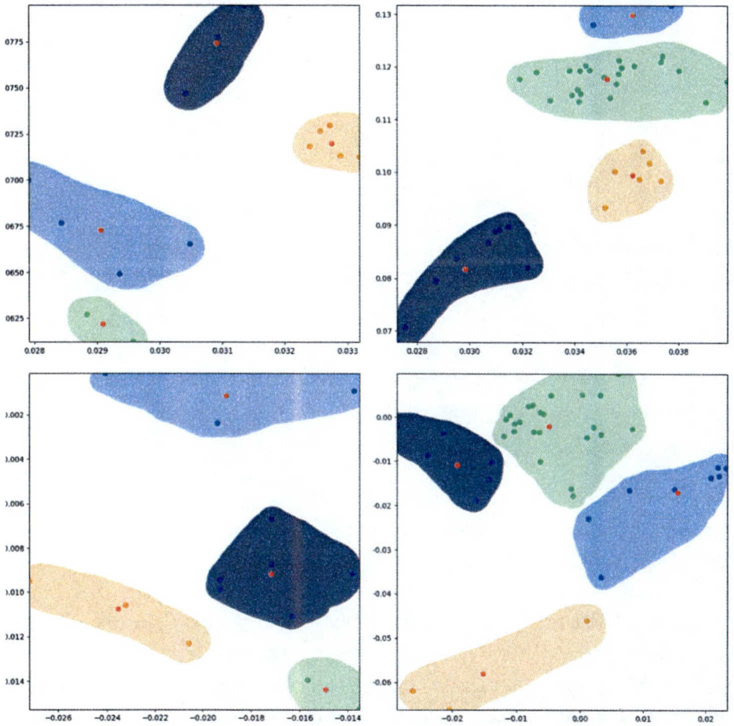

Fig. 3. The top area corresponds to the projections and clusters in hyperbolic space for the clips composing two samples. The bottom one shows the projections for the same samples in Euclidean space. The red points indicate the centroid. (Color figure online)

The test set was labelled by hand, indicating the corresponding labels in those sections where the actions of interest are performed. Similar to the training set, we pre-process the clips with the same number of frames and frame rate, although in the process of generating the latent values, we fill the spaces adjacent to the clip tensor with zeros. That is, the central part of the tensor with size [10, 30] contains the data corresponding to the clip, and the rest, zeros. Furthermore, due to the difference between the execution times of the actions and their occurrence throughout the video, there is an imbalance in the available clips that can alter the calculation of centroids during clustering. To address this situation, we reduce the total number of clips for each action, matching them with the label that occurs the least number of times.

The latent values generated in both spaces are related to each other using *k-means* and its analogous version for the hyperbolic space. This process is performed separately for the clips of each video, which we relate to their corresponding label by analysing a cost matrix weighted with the recall function. That is, we assign the labels to the clusters based on the correct classification of true positives and true negatives.

Table 1 shows the results obtained in the clusters generated for the Euclidean and hyperbolic space. In the first case we can observe an accuracy of 64.84% while in the second case we obtain a slight increase to 66.67%. Locally to the classes, we can observe that in the hyperbolic space we obtain a slightly lower Recall for three of the classes, although in contrast the Precision has a better response in these classes, indicating a lower false positive rate. Analysing the F1 scores we can observe a slightly better performance in the hyperbolic approximation, with the exception of "Take Iron", where due to the higher number of false negatives it shows a lower recall compared to the Euclidean one.

Figure 3 shows the projections of the latent values in Euclidean and hyperbolic space for two samples. In the top row, which corresponds to the hyperbolic space, we can observe the slightly more localised cluster of latent values between them. In the bottom row, the result of the projections in Euclidean space is shown, where slightly more sparse and blended results can be observed.

4 Conclusions

Throughout the paper we have presented the steps taken to process and train unsupervised videos obtained from the Internet. This process has been carried out by leveraging the property of VAE networks to generate probability distributions in their latent space, in addition to using the triplet loss function to relate the clips that compose the video. These latent values have been generated in both Euclidean and hyperbolic space, concretely the Poincaré ball. From these values we have generated clusters and compared the performance in both spaces.

The results obtained and presented in the previous section show similar results, although the hyperbolic approach has shown a slightly better overall performance. In the projections shown in the figures, one can also visually observe more localized regions of clusters of latent values. This more localized dispersion may benefit the process of clustering and localization of their respective centroids.

The main goal of this work lies in analyzing the performance of a geometric approach during the process of grouping actions. These relationships established in a curved space are intended to be exploited to establish a hierarchical relationship between the actions that make up an activity, either in terms of their relevance to describe a given activity or to infer a notion of sequentiality in this hierarchical relationship.

These interests define the line for future work. By preparing a larger data set, with a greater variety of daily activities and the respective actions that compose them, it would be possible to analyze how these actions relate to each other in their respective activities. Based on this, a criterion could be established to recognize in advance the activity being performed or to anticipate a possible future action.

Acknowledgment. We would like to thank "A way of making Europe" European Regional Development Fund (ERDF) and MCIN/AEI/10.13039/501100011033 for supporting this work under the MoDeaAS project (grant PID2019-104818RB-I00). This work has also been supported by two Spanish national grants for PhD studies, FPU17/00166, and UAFPU2019-13 respectively. Furthermore, we would like to thank Nvidia for their generous hardware donation that made these experiments possible.

References

1. Ariza Colpas, P., et al.: Unsupervised human activity recognition using the clustering approach: a review. Sensors **20**(9), 2702 (2020)
2. Bronstein, M.M., Bruna, J., LeCun, Y., Szlam, A., Vandergheynst, P.: Geometric deep learning: going beyond euclidean data. IEEE Signal Process. Mag. **34**(4), 18–42 (2017)
3. Chaaraoui, A.A., Climent-Pérez, P., Flórez-Revuelta, F.: A review on vision techniques applied to human behaviour analysis for ambient-assisted living. Expert Syst. Appl. **39**(12), 10873–10888 (2012)
4. Cook, D.J., Crandall, A.S., Thomas, B.L., Krishnan, N.C.: Casas: a smart home in a box. Computer **46**(7), 62–69 (2012)
5. Doersch, C., Gupta, A., Efros, A.A.: Unsupervised visual representation learning by context prediction. In: Proceedings of the IEEE International Conference on Computer Vision, pp. 1422–1430 (2015)
6. Fernando, B., Bilen, H., Gavves, E., Gould, S.: Self-supervised video representation learning with odd-one-out networks. In: Proceedings of the IEEE Conference on Computer Vision and Pattern Recognition, pp. 3636–3645 (2017)
7. Friji, R., Drira, H., Chaieb, F., Kchok, H., Kurtek, S.: Geometric deep neural network using rigid and non-rigid transformations for human action recognition. In: Proceedings of the IEEE/CVF International Conference on Computer Vision, pp. 12611–12620 (2021)
8. Hoffer, E., Hubara, I., Ailon, N.: Deep unsupervised learning through spatial contrasting. arXiv preprint arXiv:1610.00243 (2016)

9. Hsu, J., Gu, J., Wu, G., Chiu, W., Yeung, S.: Capturing implicit hierarchical structure in 3D biomedical images with self-supervised hyperbolic representations. In: Advances in Neural Information Processing Systems **34**, 5112–5123 (2021)
10. Hu, W.Y., Scott, J.S.: Behavioral obstacles in the annuity market. Financ. Anal. J. **63**(6), 71–82 (2007)
11. Huang, W., Wu, Q.J.: Human action recognition based on self organizing map. In: 2010 IEEE International Conference on Acoustics, Speech and Signal Processing, pp. 2130–2133. IEEE (2010)
12. Jing, L., Yang, X., Liu, J., Tian, Y.: Self-supervised spatiotemporal feature learning via video rotation prediction. arXiv preprint arXiv:1811.11387 (2018)
13. Larsson, G., Maire, M., Shakhnarovich, G.: Colorization as a proxy task for visual understanding. In: Proceedings of the IEEE Conference on Computer Vision and Pattern Recognition, pp. 6874–6883 (2017)
14. Li, Y., Paluri, M., Rehg, J.M., Dollár, P.: Unsupervised learning of edges. In: Proceedings of the IEEE Conference on Computer Vision and Pattern Recognition, pp. 1619–1627 (2016)
15. Lou, A., Katsman, I., Jiang, Q., Belongie, S., Lim, S.N., De Sa, C.: Differentiating through the fréchet mean. In: International Conference on Machine Learning, pp. 6393–6403. PMLR (2020)
16. Mathieu, E., Le Lan, C., Maddison, C.J., Tomioka, R., Teh, Y.W.: Continuous hierarchical representations with poincaré variational auto-encoders. In: Advances in Neural Information Processing Systems, vol. 32 (2019)
17. Pu, Y., et al.: Variational autoencoder for deep learning of images, labels and captions. In: Advances in Neural Information Processing Systems, vol. 29 (2016)
18. Rawassizadeh, R., Dobbins, C., Akbari, M., Pazzani, M.: Indexing multivariate mobile data through spatio-temporal event detection and clustering. Sensors **19**(3), 448 (2019)
19. Sarabu, A., Santra, A.K.: Human action recognition in videos using convolution long short-term memory network with spatio-temporal networks. Emerg. Sci. J. **5**(1), 25–33 (2021)
20. Surís, D., Liu, R., Vondrick, C.: Learning the predictability of the future. In: Proceedings of the IEEE/CVF Conference on Computer Vision and Pattern Recognition, pp. 12607–12617 (2021)
21. Surís, D., Liu, R., Vondrick, C.: Learning the predictability of the future. In: Proceedings of the IEEE/CVF Conference on Computer Vision and Pattern Recognition, pp. 12607–12617 (2021)
22. Tran, D., Wang, H., Torresani, L., Ray, J., LeCun, Y., Paluri, M.: A closer look at spatiotemporal convolutions for action recognition. In: Proceedings of the IEEE Conference on Computer Vision and Pattern Recognition, pp. 6450–6459 (2018)
23. Ungar, A.A.: The möbius gyrovector space. In: Beyond the Einstein Addition Law and its Gyroscopic Thomas Precession, pp. 161–210. Springer (2001). https://doi.org/10.1007/0-306-47134-5_6
24. Wang, X., Gupta, A.: Unsupervised learning of visual representations using videos. In: Proceedings of the IEEE International Conference on Computer Vision, pp. 2794–2802 (2015)
25. Yan, S., Xiong, Y., Lin, D.: Spatial temporal graph convolutional networks for skeleton-based action recognition. In: Thirty-Second AAAI Conference on Artificial Intelligence, pp. 7444–7452 (2018)
26. Yao, G., Lei, T., Zhong, J.: A review of convolutional-neural-network-based action recognition. Pattern Recogn. Lett. **118**, 14–22 (2019)

Multimodal Object Recognition Module for Social Robots

Alejandro Cruces, Alberto Tudela(✉), Adrián Romero-Garcés, and Juan Pedro Bandera

University of Málaga, Málaga, Spain
{acruces,ajtudela,argarces,jpbandera}@uma.es
http://www.uma.es

Abstract. Sensor fusion techniques are able to increase robustness and accuracy over data provided by isolated sensors. Fusion can be performed at a low level, creating shared data representations from multiple sensory inputs, or at a high level, checking consistency and similarity of objects provided by different sources. These last techniques are more prone to discard perceived objects due to overlapping or partial occlusions, but they are usually simpler, and more scalable. Hence, they are more adequate when data gathering is the key requirement, while safety is not compromised, computational resources may be limited and it is important to easily incorporate new sensors (e.g. monitorization in smart environments or object recognition for social robots). This paper proposes a novel perception integrator module that uses low complexity algorithms to implement fusion, tracking and forgetting mechanisms. Its main characteristics are simplicity, adaptability and scalability. The system has been integrated in a social robot and employed to achieve multimodal object and person recognition. Experimental results show the adequacy of the solution in terms of detection and recognition rates, integrability into the constrained resources of a robot, and adaptability to different sensors, detection priorities and scenarios.

1 Introduction

European society faces important changes in the near future. The upcoming *Silver society* [8], the energy and resources recent crisis, the increasing demand for teleworking and reducing the carbon footprint, the new Industry 4.0 and business models focused on data mining and processing, pave the way for a new social model. This model implies a strong presence of technology in daily life settings, as a tool to address many of the challenges produced by these changes, which have been recently augmented (sometimes, drastically) by the COVID-19 pandemic [14]. Smart environments able to monitor, assist and communicate with

This work has been funded by the by the European Union's Horizon 2020 research and innovation programme under grant agreement No 825003 (DIH-HERO SUSTAIN) and by the Spanish National Research Project DEHESA RTI2018-099522-B-C41.

D. Tardioli et al. (Eds.): ROBOT 2022, LNNS 590, pp. 489–501, 2023.
https://doi.org/10.1007/978-3-031-21062-4_40

people are experiencing an exponential development, supported by advances in electronics, Artificial Intelligence, robotics, data mining and ubiquitous computing.

Among the devices found in these technological environments, robots may be one of the most versatile ones. The current advances in robotics point towards using these agents to offer proactivity, ubiquity and adaptability, and they have the potential to act as a natural interface between the human userMesh Bed Compensation and the environment [6]. However, adherence problems are a common issue for these robots working in daily life settings, specially if they cannot adapt to the particular tasks, contexts and needs and preferences of the different users [8]. Hence, it is important to provide these robotic agents, or better, the complete technological ecosystems they belong to, with the capability to robustly detect and recognize objects and people in the environment, and modify their behaviour accordingly.

While isolated sensors can be used to detect and recognize objects and people in an environment, sensor fusion techniques promotes a robustness increase on the perception process, allowing to overcome each sensor's weaknesess using the strenghts of the remainder [7]. Multimodal perception and sensor fusion are no new techniques, but in recent years this research field has received a growing attention due to the expansion of smart environments and sensor networks, such as the ones mounted in autonomous or semi-autonomous vehicles [15]. Among the numerous approaches to multi-sensor data fusion, it is possible to set the following classification: high-level fusion (HLF), low-level fusion (LLF), and mid-level fusion (MLF) [15]. LLF and MLF approaches deal with data fusion among different sensors, allowing for cooperative detection and tracking to construct shared perceptual maps, such as the one proposed in [10] to track objects in 3D space fusing optical flow, scene flow, stereo-depth, and 2D object detections, or the proposal in [9] that relies only in 2D and 3D bounding box detection to produce a more scalable fusion system. Ferreira et al. [7] also rely on low level audiovisual fusion, performed using a Bayesian framework, to drive the attention of a robot, while in [4] detected legs and faces are integrated using an Unscented Kalman Filter. Recognition is addressed in a posterior step through histogram analysis using the Bhattacharyya coefficient.

HLF approaches, on the other hand, carry out object detection or a tracking algorithm independently for each sensor, and subsequently performs fusion. They are usually easier to implement and scale, but they may discard objects with low confidence values in situations such as overlapping [15]. Among the different HLF proposed approaches, there are many based on fuzzy methods such as the one Ban et al. [2] implement to fuse speaker and face recognition. Particle filters are also frequently employed to fuse perceived objects, as in [1] where a multirobot visual tracking particle filter is proposed to track a certain object identified by its color among multiple robots. There are other approaches, finally, that employ computationally lighter approaches. Hence, Reily et al. [13] propose to control the fusion of multiple sensors via a weighted combination of observations. These weights also control the display of observations to a human operator, providing enhanced situational awareness.

This paper presents a new HLF multimodal recognition system, based on a perception integrator module relying on confidence and pose criteria to provide a fused set of objects. The integrator uses low complexity algorithms to implement fusion, tracking and forgetting mechanisms. Its main characteristics are simplicity, adaptability and scalability. While the system can be integrated in any technological ecosystem including any kind of sensors, its HLF nature makes it a better solution for non-critical object detection (as pointed out in [15], more complex LLF or MLF approaches may be better candidates when avoiding false negatives is a must). Hence, we describe and test in the paper the application of the system to a certain use case: object and person recognition for social robots.

2 Software and Hardware Assets

2.1 Cognitive Architecture

This paper proposes a multimodal recognition system that is integrated in the CORTEX cognitive architecture [5]. The main characteristic of CORTEX is the existence of a unified, dynamic working memory that represents information at different abstraction levels, from raw perceptual data to high-level symbols and action plans. This so-called Deep State Representation (DSR) becomes a short-term dynamic representation of the context, both internal (inner state, proprioception) and external (environmental data, objects and people in the environment). Around this DSR, a set of *agents* are deployed that interact through this shared representation to achieve different goals. This structure allows decomposing the functionality of the system into a set of software components, that can be placed anywhere in the reactive-deliberative spectrum, and that are in charge of different functionalities. The CORTEX implementation employed in this paper implements the DSR and the agents using the robocomp framework [11], while processing modules (i.e. software components) connected to these agents may be implemented in different frameworks and middlewares, such as Robot Operating System (ROS) [12].

2.2 Robotic Platform

In this paper, the recognition system considers input data coming from different sources, all of them related to a particular Socially Assistive Robot (SAR), that integrates an instance of the CORTEX architecture. Figure 1(left) shows the robot employed in the tests performed in the paper. The base of this robot is the CLARA robot [3], developed in the CLARC EU Project (ECHORD++ FP7-ICT-601116) to perform Comprehensive Geriatric Assessment (CGA) procedures, and currently deployed in a retirement house to implement different user defined tasks [8]. This robot is already using CORTEX to: (i) navigate autonomously in dynamic daily life environments; (ii) interact with people in constrained scenarios using speech, a tactile screen and other *ad-hoc* devices; (iii) allow healthcare professionals to set an agenda for the different use cases the robot can perform; (iv) detect people in its surroundings using two different mechanisms (LIDAR-based and vision-based) that will be detailed in Sect. 4.

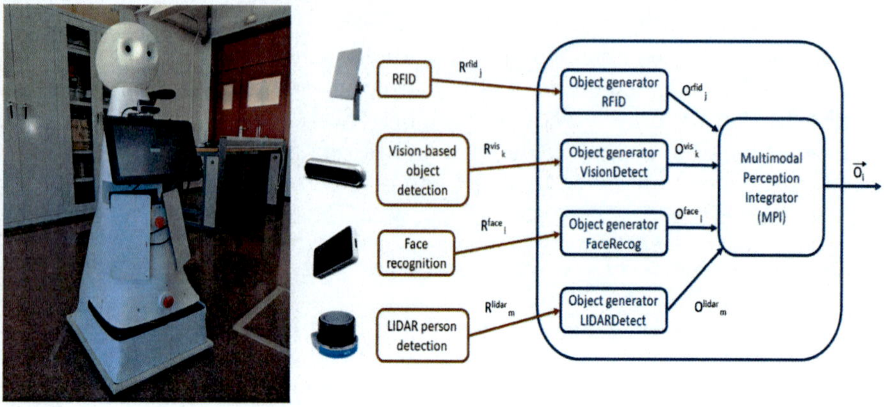

Fig. 1. (Left) Socially Assistive Robot employed in this paper; and (Right) Multimodal recognition system overview

The perceptual capabilities of the robot are extended in the paper including two additional modules devoted to recognition: one of them is an Intel Real Sense F455 camera, specifically designed for face recognition applications, and the other is a Keonn AdvanReader-10 RFID reader, connected to two Advantenna-p12 antennas, allowing the robot to locate RFID tags. These two both systems are also deeply detailed in Sect. 4.

3 Multimodal Recognition System

Figure 1 (right) shows an overview of the proposed multimodal recognition system. All these software components are programmed as ROS nodes. The system is basically composed by a set of object generator modules, that receive data from sensors and generate detected objects ${O^m}_k$, and a Multimodal Perception Integrator (MPI) that, upon receiving any new object ${O^m}_k$ from an object generator module, processes it and produces the output of the system, $\overrightarrow{O_i}$. This output is the set of detected objects, that is sent to a robocomp agent to be inserted into the DSR. Hence, $\overrightarrow{O_i}$ is a vector containing the set of detected objects for iteration i, where a new iteration is performed every time any sensor produces a new detected object. Each of the objects in the vector $\overrightarrow{O_i}$, and each of the objects detected by object generator modules ${O^m}_k$, contain the fields detailed in (1).

$$O = (id, C_{id}, type, C_{type}, \overrightarrow{p}, \overrightarrow{p_{err}}, t) \tag{1}$$

where O stands for a certain detected object, *id* is its identifier (i.e. an unambiguous identifier for that object, such as the passport number of a person), C_{id} is the confidence value associated to *id* in the range [0...1], *type* is the type of object (e.g. "chair" or "person"), C_{type} is the confidence value associated to *type* in the range [0...1], $\overrightarrow{p}$ is the pose (position and orientation) of object O in the format $(x, y, z, yaw, pitch, roll)$, $\overrightarrow{p_{err}}$ codes the error of each pose data, so that the value of each pose coordinate j of the object should be considered as $p(j) \pm p_{err}(j)$, and t is a timestamp recording the time in which that object was detected.

It is interesting to highlight that vector $\overrightarrow{O_i}$ may contain several objects sharing the same object *type* (e.g. several "chair" objects), but it is also possible to have several objects sharing the same *id* (e.g. several detections of a certain person), due to the probabilistic nature of the data provided by the sensors. Another relevant characteristic of the selected notation is that it is possible to have unknown fields for *id* and *type* (the label *unknown* is used in these situations, and the corresponding confidence C_{id} or C_{type} is set to 0.0), and also for the pose data (in that case, the corresponding $\overrightarrow{p_{err}}$ coordinate will be -1).

3.1 Object Generators

Different sensors may provide different data in different formats. In order to keep the Multimodal Perception Integrator (MPI) module independent from the particular data sources and make the system easily scalable, a set of *object generator* modules appears which process the sensor data (R_b^a in Fig. 1 (Right)) and converts them to a standard object representation O_b^a able to be used by the MPI. These object generators check consistency of generated object data (e.g. by assigning a certain *type* to the data provided by a certain sensor), and fill empty fields if required (e.g. by inserting a timestamp in case it is not available in the sensor).

3.2 Multimodal Perception Integrator (MPI)

This module is in charge of producing the set of detected objects to be inserted in the DSR. As the flow diagram in Fig. 2 depicts, upon receiving a new object ${O^m}_k$ from any of the active object generator modules (see Fig. 1 (Right)), the MPI triggers a fusing process that will generate a new output vector $\overrightarrow{O_i}$ by adding or fusing ${O^m}_k$ to the previous instance of the output vector $\overrightarrow{O_{i-1}}$, and deleting (*forgetting*) objects which data can no longer be considered reliable.

The fusing process will try to fuse the input object with the ones already detected, following a certain order that is determined by rearranging $\overrightarrow{O_{i-1}}$ to $\overrightarrow{O'_{i-1}}$. Rearranging is performed using a certain sorting function $f_{sort}(\overrightarrow{O}, R)$ that, given a certain input object R, sort objects in $\overrightarrow{O}$. Then, object ${O^m}_k$ is tested against the objects in $\overrightarrow{O'_{i-1}}$ (see Fig. 2).

The selected sorting function f_{sort} strongly influences the outcome of the system. For the tests conducted in this paper, f_{sort} is implemented as depicted in

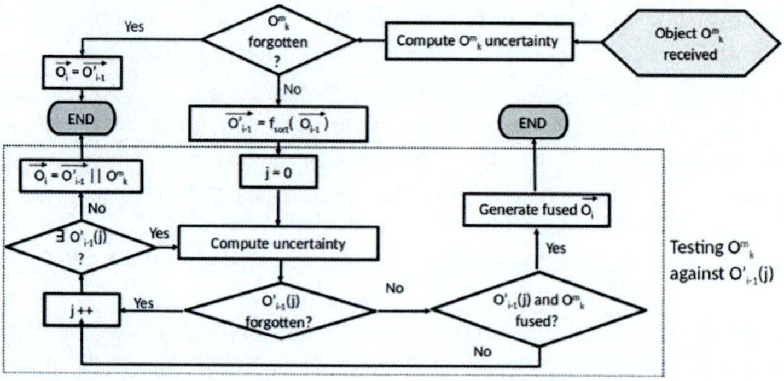

Fig. 2. Flow diagram of the forgetting and fusing process in the MPI

(2), where $\overrightarrow{O(t^X; i^Y)}$ stands for a subset of $\vec{O}$ including all objects with *type* $= X$ and *id* $= Y$. The values that X and Y indexes can have are the following: (i) R when the value is equal to the corresponding value of input object R (e.g. $\overrightarrow{O(t^R; i^R)}$ is the subset of $\vec{O}$ including all objects that have the same *type* and the same *id* than R); (ii) U when the value is *unknown*; and (iii) O when the value is not R nor *unknown* (i.e. the object has a recognized *type* or *id* value, that differs from the one of the input object).

$$f_{sort}(\vec{O}, R) = (\ \overrightarrow{O(t^R; i^R)} || \overrightarrow{O(t^R; i^U)} || \overrightarrow{O(t^R; i^O)} || \\ \overrightarrow{O(t^U; i^R)} || \overrightarrow{O(t^U; i^U)} || \overrightarrow{O(t^U; i^O)} || \\ \overrightarrow{O(t^O; i^R)} || \overrightarrow{O(t^O; i^U)} || \overrightarrow{O(t^O; i^O)}) \tag{2}$$

Equation 2 shows that the proposed implementation firstly looks for a possible fusion in the subset of objects that share the *type* and *id* of R. If no fusion is performed, the function looks in the subset of objects that have the same *type* than R, but have an *unknown id*, and so on. The use of this particular function prioritizes the fusion with *a priori* similar objects, but it is also possible to fuse objects having different *id* or even different types, to deal with noisy or imprecise detectors. It is also interesting to point out that, in the common situation in which setting an *id* for an object always implies setting a *type*, the only effect in the previous function will be that $\overrightarrow{O(t^U; i^R)} = \overrightarrow{O(t^U; i^O)} = \overrightarrow{O(t^O; i^R)} = \emptyset$.

3.3 Forgetting Obsolete Objects

Pose uncertainty can grow over time, depending on the nature of the perceived object. If this uncertainty grows over a certain threshold θ, the object information should be considered obsolete and the object removed from the system. Figure 2 shows the flow diagram followed by the forgetting and fusion algorithm employed in the MPI. As depicted, there are two steps in which object's obsolescence is checked. Checking is based on the time difference between two detected objects,

Δt. This difference is processed by a function f_u that defines how uncertainty grows in time. The implementation of f_u may depend on different factors, such as the quality of the sensor or the object type. In this paper, the algorithm has been set to a worst-case error increment: 2 m per second for position coordinates, and 2π radians per second for orientation coordinates.

The first check is performed to the incoming object ${O^m}_k$, considering a Δt that is the maximum time difference between the timestamp of ${O^m}_k$ and the timestamps of objects in $\overrightarrow{O_{i-1}}$. ${O^m}_k$ will be forgotten if $f_u(\Delta t) > \theta$ and the input object was perceived *before* the object in $\overrightarrow{O_{i-1}}$ which timestamp was used to compute Δt.

Objects in $\overrightarrow{O_{i-1}}$, on the other hand, will be checked for removal before they are compared against ${O^m}_k$. In this case, Δt is computed as the time difference between that particular object and ${O^m}_k$. Again, the object will be discarded if $f_u(\Delta t) > \theta$.

3.4 Fusion Algorithm

Providing none of them has been forgotten, in order to fuse two objects O_1 and O_2, it is firstly necessary to update the pose uncertainty ($\overrightarrow{p_{err}}$) of the first object detected. Being Δt the difference between the timestamps of both objects, the object with an older timestamp will have its $\overrightarrow{p_{err}}$ values modified before the fusion as $p_{err}(k) = f_u(\Delta t) * p_{err}(k)$. Then, the two objects will fuse if and only if all the following criteria are met: (i) their *id* can fuse; (ii) their types can fuse; and (iii) their poses can fuse.

Algorithm 1 shows these steps for the possible *id* fusion of two objects O_1 and O_2. The confidence-based fusion procedure is designed to: (i) fuse objects that are similar (with the same *id*); (ii) fuse *unknown* objects with known objects; and (iii) fuse objects with a low recognition probability with objects more robustly detected. The *type* fusion algorithm follows the same steps as *id* fusion.

In order to understand how the complete MPI works it is important to remember that a fusion will only be performed if the pose fuse criterion, and not only the *id* and *type* fuse criteria, is met. Being $voxel_1$ and $voxel_2$ the 3D volumes in which O_1 and O_2 may be located, pose fusion is achieved as detailed by Algorithm 2.

To summarize, two certain objects will fuse if their possible poses overlap, and: (i) they are *similar* (regarding *id* and/or *type*); (ii) one of them is not classified (i.e. *id* and/or *type* are *unknown*) while the other is; (iii) they are not similar but the confidence in the classification was low for both of them.

4 Experimental Setup: Employed Sensors

While the proposed system may incorporate any different data source, this section details the ones integrated in the robot used in this paper.

Algorithm 1 Confidence fusion algorithm (id)

if $id^1 \neq unknown$ **and** $id^1 = id^2$ **then**
  id^{fused} set to the value of the object with higher C_{id}
  C_{id}^{fused} set to the value of the object with higher C_{id}
end if
if $id_1 = unknown$ **then**
  id^{fused} set to id^2
  C_{id}^{fused} set to C_{id}^2
end if
if $id^1 \neq unknown$ **and** $id^1 \neq id^2$ **then**
  if $(C_{id}^1 + C_{id}^2) \leq 1.0$ **then**
    id^{fused} set to the value of the object with higher C_{id}
    C_{id}^{fused} set to the value of the object with higher C_{id}
  else
    no fusion
  end if
end if

Algorithm 2 Pose fusion algorithm

if $voxel_1$ **overlaps** $voxel_2$ **then**
  for $k = 1, 2, \ldots$ **do**
    if $(p_{err}(k)^1 \geq 0.0)$ **and** $(p_{err}(k)^2 \geq 0.0)$ **then**
      $e = (p_{err}(k)^1 + p_{err}(k)^2)$ **or** e = minimum arch containing overlapping circle sections
      $d = p(k)^2 - p(k)^1$
      $p(k)^{fused} = p(k)^1 + d * p_{err}(k)^1/e$
      $\alpha_1 = (p_{err}(k)^1 - |d * p_{err}(k)^1/e|)$
      $\alpha_2 = (p_{err}(k)^2 - |d * p_{err}(k)^2/e|)$
      $p_{err}(k)^{fused} = min(\alpha_1, \alpha_2)$
    else
      if $(p_{err}(k)^1 \geq 0.0)$ **then**
        $p(k)^{fused} = p(k)^1$
        $p_{err}(k)^{fused} = p_{err}(k)^1$
      else
        if $(p_{err}(k)^2 \geq 0.0)$ **then**
          $p(k)^{fused} = p(k)^2$
          $p_{err}(k)^{fused} = p_{err}(k)^2$
        else
          $p_{err}(k)^{fused} = -1$
        end if
      end if
    end if
  end for
else
  No fusion
end if

RFID Identification. CLARA robot is equipped with an RFID reader and two antennas. The antennas are located at the robot's front with an horizontal angle between them of 60° (30° between each one and the robot's frontal plane), so radiation patterns of both antennas, characterized by a 90° beam width at horizontal plane, overlap. This setup allows to locate RFID tags by comparing the RSSI levels received by each antenna with the estimated attenuation due to path loss and antenna's radiation pattern, modeled as a cardioid.

Vision-Based Person Detection. Our main object generator is a vision based module using an Intel RealSense RGBD camera and an Intel Neural Compute Stick (NCS2). Within this module we load a MobileNet-SSD neural network model inside the NCS2 to compute the bounding box and category of an object from the color camera. Then, depth analysis is performed to extract the bounding box surrounding the object in 3D coordinates.

LIDAR-Based Person Detection. The LIDAR is the sensor that covers a wider area for the robot, with respect to the other sensors mounted. Thus, this module can detect people in ranges where the other object generator modules can not. The LIDAR detector pipeline has three steps. First, the laser scan is splitted in different segments by a Jump Distance Segmentation algorithm which divide two consecutive points in different segments if the euclidean distance between them is larger than a threshold. Secondly, a set of features are extracted from every segment. Thirdly, these features are forwarded to a boosting model previously learned. Once a segment is classified as a person, then, its pose is calculated as the centroid of the segment. Finally, this data is sent to the MPI.

Vision-Based Face Recognition. The last sensor that gets into the mix is the Intel RealSense ID for facial authentication, face camera from go on. This device has the capacity to extract faceprints from the person's face in front of the robot and store it in a secure database. As the bounding box of the face and the faceprint score are the only two outputs we receive from the device, we extract the face pose by matching the face camera image with the RGBD camera image and point cloud described in Sect. 4.

5 Experiments

The system has been tested in a real unconstrained environment, where people move around the robot and different objects are randomly distributed. Detection confidence values for the different sensors were empirically set according to prior evaluation of their isolated performance, and remained fixed for all performed experiments. In order to provide a quantitative evaluation of the solution, an experimental setup was prepared in which a chair is located in a fixed, known position of the map, and a person moves two meters, from one known position to another (see Figs. 3(a) and (b)). Other objects (an additional chair, a table, etc.)

are scattered through the environment. Figures 3(b) and (d) show the associated output of the MPI module during one of the experiments. The box drawn in the virtual map for each object measures its position uncertainty. In this sample, the person (Ale) is correctly identified thanks to the combined use of the RFID detector and the face recognition system. Once labelled, the person is still identified even when the two previous modules are not perceiving him, thanks to the leg and object detection modules keeping an updated pose of the person. Note that the RFID and face recognition modules only can detect the person while he is into the nearest position.

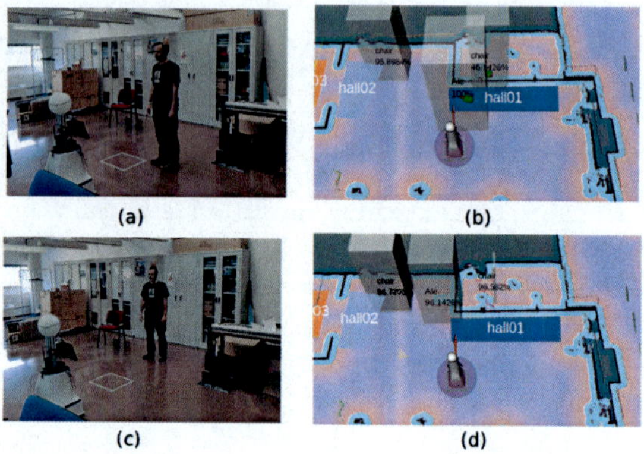

Fig. 3. Experiments. (a): Scene 1 perceived in t; (b) Objects provided by MPI for scene 1; (c) Scene 2 perceived in t+δ; (d) Objects provided by MPI for scene 2.

Tables 1 and 2 show the results offered by the different sensors and the MPI for two objects (the person and the chair on the left) during two tests lasting for one minute each. Each table includes the number of detected *id* and *type* for each object, along with the confidence values average ($\bar{C}$), the position mean error ($\bar{e}$) and its standard deviation (σ_e) with respect to the ground-truth, and average detections per second (fps). Table 1 shows results when the person starts the test close to the robot and moves to a further position, while 2 shows the opposite case (the person moves closer). More than 4000 incoming inputs from the object generator modules where fed to the MPI, coming from different sources, in each test. These tables show how the fusing algorithm takes the best characteristics for each input module (frequency, confidence and accuracy) improving the ones collected by isolated sensors, even when only one sensor is detecting an object, thanks to the fusion mechanism. It can also be seen in Table 1 how recognized objects are tracked and kept on recognized as long as any sensor is still detecting them: in this test, the RFID recognized the person but only while he was close to the robot, at the beginning of the test. However, even when RFID no longer detected the person, the leg and object detection modules kept on doing it. These new detected objects fuse with previous ones and the *id* of the person propagates during this *de facto* tracking process. Moreover, once a sensor is able to provide a high confidence value for an object, even if that sensor no longer detects the object, that confidence value persists thanks to the followed fusion approach (see Algorithm 1). Table 2, on the other hand, shows how the confidence associated to each sensor affects obtained results. Hence, the relatively low *type* confidence included in Table 2 can be explained because initially only LIDAR (with low confidence associated) was detecting the legs of the person.

Table 1. Object detection and recognition results. Person moves away from the robot

		faceRecog	LIDAR	objDetect	RFID	Fused
Person	*id* ($\bar{C}_{id}$)	4 (89.3%)	0	0	132 (100%)	1649 (100%)
	type ($\bar{C}_{type}$)	4 (89.3%)	1304 (18.0%)	209 (99.0%)	132 (100%)	1649 (100%)
	$\bar{e}$	0.12	0.09	0.18	0.44	0.08
	σ_e	0.01	0.04	0.11	0.11	0.05
	fps	0.06	20.17	3.23	2.04	25.51
Chair	*id* ($\bar{C}_{id}$)	0	0	0	0	0
	type ($\bar{C}_{type}$)	0	0	437 (94.2%)	0	437 (95.9%)
	$\bar{e}$			0.54		0.54
	σ_e			0.36		0.46
	fps	0	0	6.76	0	6.76

6 Conclusions and Future Work

The qualitative results show that the proposed fusion technique can overcome the limitations of individual sensors, offering improved results over isolated devices. It can also extend the results of a correct detection or recognition over time thanks to the fusing algorithm acting as a *de facto* tracking system. The solution is lightweight, able to run in autonomous agents, and easily scalable to include new sensors, mobile or fixed. Future work will firstly focus on looking for better matches in the fusing algorithm, as the current proposal just fuses the new detection with the first match. Even if the sorting function is adequate, this criterion is too constrained. More complex settings, including several people and moving devices, will also be addressed. Moreover, fusion with data provided by sensors installed in several robots and/or as part of smart environments will also be evaluated. Ablation studies considering different sensors in the fusion process will be conducted to analyze the influence of each sensor. Context information and learned cues will also be used to fed high-level knowledge to the MPI just like any other sensor (e.g. the daily agenda of a person can help filtering the possible identifications offered by perceptors). Finally, a comparison between the presented HLF approach with other MLF and LLF approaches that use DL techniques will also be addressed.

Table 2. Object detection and recognition results. Person moves closer to the robot

		faceRecog	LIDAR	objDetect	RFID	Fused
Person	*id* ($\bar{C}_{id}$)	2 (88.1%)	0	0	96 (100%)	844 (100%)
	type ($\bar{C}_{type}$)	2 (88.1%)	1257 (16.8%)	152 (97.3%)	96 (100%)	1507 (67.6%)
	$\bar{e}$	0.06	0.10	0.14	0.65	0.10
	σ_e	0.01	0.01	0.04	0.09	0.03
	fps	0.03	18.27	2.21	1.40	21.91
Chair	*id* ($\bar{C}_{id}$)	0	0	0	0	0
	type ($\bar{C}_{type}$)	0	0	497 (94.4%)	0	497 (96.3%)
	$\bar{e}$			0.53		0.53
	σ_e			0.39		0.43
	fps	0	0	7.23	0	7.23

References

1. Amorim, T.G.S., Souto, L.A., Do Nascimento, T.P., Saska, M.: Multi-robot sensor fusion target tracking with observation constraints. IEEE Access **9**, 52557–52568 (2021)
2. Ban, K.D., Kwak, K.C., Yoon, H.S., Chung, Y.: Multimodal user identification in a intelligent robot environment. In: Proceedings of the International Conference on Control, Automation and Systems, 2007 (ICCAS '07), pp. 1172–1177 (11 2007)
3. Bandera, A., et al.: Clarc: a robotic architecture for comprehensive geriatric assessment (2016)
4. Bellotto, N., Hu, H.: Multimodal people tracking and identification for service robots. I. J. Inf. Acquisition **5**(3), 209–221 (09 2008)
5. Bustos, P., Manso, L., Bandera, A., Bandera, J.P., García-Varea, I., Martínez-Gómez, J.: The cortex cognitive robotics architecture: use cases. Cogn. Syst. Res. **55**, 107–123 (2019)
6. Feil-Seifer, D., Mataric, M.: Defining socially assistive robotics. In: 2005 IEEE C9th International Conference on Rehabilitation Robotics, pp. 465–468 (July 2005)
7. Ferreira, J.F., Castelo-Branco, M., Dias, J.: A hierarchical bayesian framework for multimodal active perception. Adapt. Behav. **20**(3), 172–190 (2012)
8. Iglesias, A., et al.: Towards long term acceptance of socially assistive robots in retirement houses: use case definition. In: Proceedings of the International Conference on Autonomous Robot Systems and Competitions (ICARSC), pp. 134–139. IEEE (2020)
9. Kim, A., Ošep, A., Leal-Taixé, L.: Eagermot: 3D multi-object tracking via sensor fusion. In: 2021 IEEE International Conference on Robotics and Automation (ICRA), pp. 11315–11321 (2021)
10. Luiten, J., Fischer, T., Leibe, B.: Track to reconstruct and reconstruct to track. IEEE Robot. Autom. Lett. **5**(2), 1803–1810 (2020)
11. Manso, L.J., Bachiller, P., Bustos, P., Núñez, P., Cintas, R., Calderita, L.V.: Robocomp: a tool-based robotics framework. In: Ando, N., Balakirsky, S., Hemker, T., Reggiani, M., von Stryk, O. (eds.) Simulation, Modeling, and Programming for Autonomous Robots. SIMPAR 2010. Lecture Notes in Artificial Intelligence, vol. 6472, pp. 251–262. Springer Berlin Heidelberg, Darmstadt, Germany (2010). https://doi.org/10.1007/978-3-642-17319-6_25
12. Quigley, M., et al.: Ros: an open-source robot operating system. In: ICRA workshop on open source software, vol. 3, p. 5. Kobe, Japan (2009)

13. Reily, B., Reardon, C., Zhang, H.: Multi-modal sensor fusion and selection for enhanced situational awareness. In: Dennison Jr., M.S., Krum, D.M., Sanders-Reed, J.N., Arthur III, J.J. (eds.) Proceedings of Virtual, Augmented, and Mixed Reality (XR) Technology for Multi-Domain Operations II, vol. 11759, pp. 144–150. SPIE, Florida, USA (april 2021)
14. Vimarlund, V., Borycki, E.M., Kushniruk, A.W., Avenberg, K.: Ambient assisted living: identifying new challenges and needs for digital technologies and service innovation. Yearb. Med. Inform. **30**(1), 141–149 (august 2021)
15. Yeong, D.J., Velasco-Hernandez, G., Barry, J., Walsh, J.: Sensor and sensor fusion technology in autonomous vehicles: a review. Sensors **21**(6), 2140 (2021)

RL-Studio: A Tool for Reinforcement Learning Methods in Robotics

Pedro Fernández de Cabo(✉), Rubén Lucas, Ignacio Arranz, Sergio Paniego, and José M. Cañas

Universidad Rey Juan Carlos, Madrid, Spain
{p.fernandezd.2016,r.lucasz}@alumnos.urjc.es, {sergio.paniego, josemaria.plaza}@urjc.es

Abstract. This paper introduces RL-Studio, an open-source software that eases the implementation of reinforcement learning algorithms to solve a problem. This software enables the integration of any simulator to recreate the problem where a reinforcement learning algorithm can be applied. Some relevant advantages of this tool are the generalization of common software components and the unified architecture. RL-Studio permits to just focus on fine tuning the hyperparameters of the algorithm and redefine the goal and reward function of the previously integrated projects. In case a non-integrated scenario, algorithm or simulator is needed, RL-Studio also provides some already integrated libraries that can be reused to save time. It has been experimentally validated in research projects and some of the canonical problems such as Robot Mesh or Mountain Car.

Keywords: Reinforcement learning · Software tools · Open source · ROS · Gazebo · Robotics · Autonomous vehicles

1 Introduction

The use of autonomous agents has become popular in recent years. It is already common to find robots in extensive areas of society such as logistics automation and warehouse robots, self-driving cars and aerial unmanned vehicles and even home vacuum robots among many examples. It is now possible to travel in driverless vehicles of the Waymo company in San Francisco or acquire home cleaning robots such as Roomba® or Braava®. Traditionally the solution has been built through modular handcrafted pieces of hardware and software, where each one has been designed to solve a minimal case, and altogether has been able to obtain the solution. But autonomous agents are designed to help and even replace human beings in tasks carried out in multidimensional fields. Therefore, the complexity of agent control systems has increased in such a way that it is very difficult to design them with those explicit programming methods. Traditional methods do not scale well nor only in other fields, but even in their application domain. Deep learning (DL) as a function approximator and feature learning has demonstrated its excellence in finding low-dimensional patterns in multidimensional data so has made possible to face real problems in complex environments. For this reason, DL is used

D. Tardioli et al. (Eds.): ROBOT 2022, LNNS 590, pp. 502–513, 2023.
https://doi.org/10.1007/978-3-031-21062-4_41

extensively in visual object detection and perception. Supervised and semi-supervised learning methods, using these DL techniques, process labeled datasets to find similar patterns in the input data. Even with no enough labeled data, data augmentation can help to generate synthetic data that help neural networks achieve good performance. DL has demonstrated robustness in end-to-end solutions based on the direct understanding of the input information for delivering a complete and functional result. There are real business solutions implemented like Tesla Autopilot, in garbage collection to identify containers, in waste recycling or even in garment recognition. However, DL needs tons of labeled data and with these techniques alone it is not easy to guarantee that a neural net understands the real complex environments. It is also necessary to learn to simulate complex scenarios where there could be dangerous events. A dataset which undersamples these important cases might yield a classifier which fails on them. Reinforcement learning (RL) is based on behavioral psychology where the agent learns by interacting in an environment through trial and error seeking to maximize reward and traditionally has been used successfully in control tasks. In this type of problems an agent, a car or robot, is interacting in an environment, such as a road, a city, or a room, at time-step to obtain a reward. The agent has to learn which are the best decisions and actions at each moment to learn the best strategy, the policy, that leads to optimizing the rewards in the long term. The reward is one of the most important distinctions in RL, since the objective sought is the maximization of the expected values of the cumulative sum of a signal. The design of the reward function in an RL algorithm is a mixture of art and science, achieved through testing hard work. Unlike the previous methods, RL algorithms do not need massive training data to learn. The agent receives the information just having contact with the environment in which it operates. But one of the main issues of robotic learning is to collect a large amount of data autonomously and safely. Modern simulation tools and RL methods, such as addressing partial observations, domain-randomization or domain adaptation help reducing the reality gap in sim-to-real transfer scenarios. In RL methods we let the agent find the best solution so that it can be generalized to similar domains without having to program explicitly. But creating a good policy representation is not a trivial problem, and now other kinds of complex issues appear to solve as sample efficiency, credit assignment, exploration vs. exploitation, or representation. To help in domains with high-dimensional states and continuous actions, the use of DL algorithms, as the representation learning part, within RL framework has created the new field of Deep Reinforcement Learning (DRL). Developing RL models involves different steps. One is to select the agent with the appropriate characteristics and sensors to correctly read the information from the environment. Another is to choose the RL algorithm that suits the goal. There are dozens and each of them has its own parameters that must be adjusted. Additionally, each simulator has features that make it suitable for a specific problem, and finally, the work process requires continuous trial and error testing, evaluating the training results, tweaking parameters, or even changing the design of certain components.

The Reinforcement Learning Studio (RL-Studio) tool, the main contribution of this article, is designed with the aim of facilitating the creation of RL models. It is modular, it allows working with different agents, in any environment, and with any RL algorithm. Any type of sensor suitable for collecting information can be designed and allows to integrate any simulation tool. All these features allow research experimental validation. RL-Studio toolkit is open-source and easily downloaded[1].

2 Related Works

In this section we provide a brief perspective of the state-of-art algorithms, methods and tools in RL. Regarding algorithms, there are two main taxonomies in classical RL, *sample-based learning methods* or also denominated *value-based algorithms* and *approximate solution methods* or *policy-based algorithms*. The former is used in actions and states in low-dimensionality problems and inside this category we can find dynamic programming, Monte-Carlo methods, and temporal-difference methods. The latter can work with real scenarios with continuous actions and states space and the solution they provide is only approximate. Policy gradients are the most popular family of methods [7]. DRL algorithms fall in both categories and we discard other methods such as imitation-based algorithms where the agents try to mimic expert actions.

2.1 RL Software Frameworks

There are several frameworks that allow learning RL models without having to program from scratch, although some have different characteristics. Basically, frameworks are modular, cross-platform and ease of running with many pre-implemented RL algorithms allowing easy reproduction of training and results, making it painless to understand and to prototype new research ideas. *OpenAI baselines*, supported by OpenAI company, allows working with different state-of-the-art RL algorithms. It was design specifically for RL research providing built-in simulated environments such as Atari games, board games, 2D and 3D physical simulations, so it allows easily train agents, develop new RL algorithms and compare them. *Keras-rl* based on Keras DL framework and developed by Google, with easy integration with Tensorflow and OpenAI environments. *Ray lib* is an open-source project developed at UC Berkeley that, in addition to having the preceding characteristics, allows building distributed applications in large infrastructures and has been included in the major cloud providers. *ReAgent* by Meta company and thus supports an easy integration with Pytorch. *Acme* developed by DeepMind focus on as a starting block for novel researchers and designed to run single-stream or distributed agents.

2.2 Simulation Tools

Planning and control strategies in real robots are often expensive, time-consuming and sometimes dangerous. So, powerful and realistic dynamic simulation tools have been developed improving Sim2Real practice. The next simulators have been chosen considering their level of maturity, modularity, and popularity among engineers and researchers in RL. All of them allow flexible level of control and can run in a fast and distributed manner in order to provide the iteration time required for experimental research. *Carla* is a domain-specific platform and has been built for the development of algorithms for

[1] https://github.com/JdeRobot/RL-Studio.

autonomous cars. It includes very real scenarios to test results thus it is widely used in the RL research community. *Mujoco* is a free and open-source physics engine designed for the purpose of model-based optimization recently acquired by DeepMind. It has been developed with mind in optimization through contacts. *Unity ML-Agents Toolkit* is a real-time 3D development general platform based in Unity game engine, able to perform sensory, physical and task logic complexity. *Gazebo* is a simulation environment with a built-in physics engine called Ignition. Ignition allows you to simulate the robot with realistic physics. Since Gazebo integrates very well with Robotic Operating System (ROS) which is the main robotics framework, it has been widely adopted in robotics community.

2.3 RL in Robotics

In recent years, RL has achieved human-level control in fields as diverse as video games, Go game, financial industry, traffic and smart cities, education system and health as some examples. In robotics the advantages of RL in online adaptability and self-learning features of complex systems have developed considerable research in learning simple manipulation tasks such as opening doors and stacking Lego blocks, or in to complete object rearrangement tasks such as grasping an apple and putting it on a plate, moving object flexible grasping [8], pushing objects to desired locations [6] visuomotor control [5] object pushing and trajectory following [12], for satisfying relative object positioning tasks [10], and for cloth manipulation such as folding different fabric types [1]. In legged locomotion tasks, RL models are a good promise to design high performance controllers for legged robots [4] and applications in which the robot can learn walking and turning from scratch in the real world [11]. In navigation, RL methods enable mobile robots to self-explore and self-learn by interacting with the dynamic environment. While in traditional programming every task involved could be carefully design, such as environment perception, localization, pose estimations, obstacle avoidance and path planning, with RL techniques have been possible build accurate end-to-end models [13]. In global and local path planning, RL algorithms have achieved improvements in accuracy, stability and robustness [9]. In Trajectory and route tracking, where the robots track in a partially observable nonlinear dynamic environment, RL methods have improved the real time adaptability of obstacle avoidance and navigation by trajectory and route tracking techniques [3]. And finally, in autonomous vehicles RL methods have had remarkable success in end-to-end models ending in commercial proposals such as [2].

3 RL-Studio Tool

RL-Studio is a modular open-source multi-platform command line software application supported in Linux and Mac OS and developed in Python language. It currently works with Gazebo and OpenAi Gym simulators, although Carla will be available shortly. It enables the flexibility of configuring the agents, algorithms, environments and simulation parameters by just modifying the configuration file used at launch time. It enables two main use cases: training and inference mode. In training mode, RL-Studio permits to develop robotic brains that best adapt to a specific domain in stochastic environments. In inference mode, RL-Studio allows to choose any of the trained robotics brains and test them in the same or any other different environments.

3.1 Software Architecture

RL-Studio has been designed with four main logic components in mind. *Agent* contains the definition of the training and inference agents and the necessary software utilities for execution. *RL algorithm* where algorithms are found. *Env* holds the environment features and simulation tools applied in every project. Based on OpenAI Gym, it acts as an information processing engine. And finally in *Simulator* the agent features, sensors and actuators to be launched by the simulator are found. The simulator processes the dynamics physics knowledge of the environment simulating the real world. The information flowing through the nodes is managed by ROS, a standard robotic communication framework, that acts as middleware between agents and environment. To build a new project from scratch, the following classes must be implemented: *Environment class,* located in the *Env* package, interfaces the simulator sensors and actuators to both execute the actions indicated by the agent and to provide the reward and next state using the *step* and *reset* methods. *Algorithm class*, located in the *algorithm* package, is in charge of learning from the experience to finally select the action which maximizes the reward and leads the agent to its objective. *Agent class*, located in the *agents* package as the core class of the project, is in charge of building the simulation according to the models and worlds configurations, loading the trained algorithm in inference mode, calling the environment class methods to execute the selected actions, feeding the algorithm with the resulting rewards, saving the trained algorithm in training mode and tracking the performance making use of the logging and monitoring libraries. In RL-Studio operating cycle, each agent acts as an independent node in ROS, obtaining the data through the onboard sensors, e.g., a camera, and publishing it through the ROS topics. The RL-Studio environment subscribes to the agent topics and receives the data that is used to

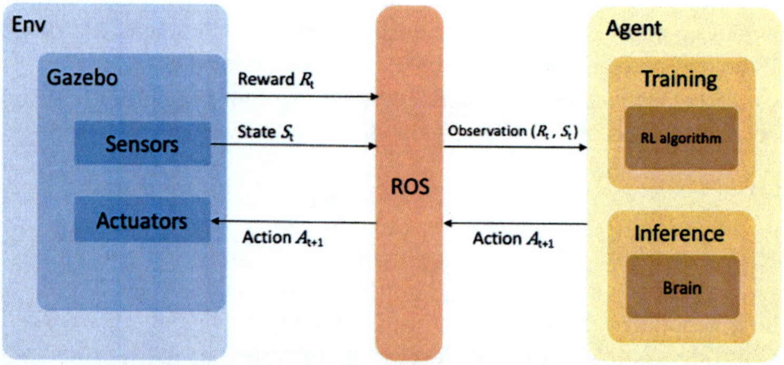

Fig. 1. RL-Studio satisfies the standard RL cycle. ROS handles information between nodes through topics. Highlight on the right, an agent can actuate in training mode where the agent executes the corresponding RL algorithm. Or in inference mode, the already trained brain is executed.

generate the necessary information that it sends back to the agent actuators, also through the ROS topics. Therefore, RL-Studio satisfies the standard RL cycle where the environment generates the state and the reward at each time step and the agent gives back the corresponding action as shown in Fig. 1. RL-Studio currently supports the simulator Gazebo and the algorithms Q-learning and DQN.

3.2 Configuration

The configuration validation has been generalized to just focus on adding, removing or modifying parameters that will affect our agents both in learning or inferencing mode. These configuration parameters have to be provided in a yaml file, where the agent, algorithm, environment, simulation tool and execution mode, training or inference, are setting. RL-Studio builds the scenario and starts its execution, and for that it needs some of the following input parameters: 1. Agent: width, height, weight, sensors, etc. 2. Actions: number of actions and values. 3. Environment: start pose, world specifications, maximum steps per iteration... 4. Algorithm hyper-parameters: learning rate, discount factor, future rewards relevance... 5. Training configuration: save model flag, logs folder, logs level... 6. Inference configuration: inferences file name and actions file name. Depending on the mode input parameter, RL-Studio runs in training or inference mode.

3.3 Training and Inference Mode

Training mode allows to create reinforcement learning models that will be saved for later use. In the training mode execution flow, first RL-Studio loads the training configuration settings and validates the integrity. Then, it initializes the project implementation and starts the training: the simulator engine and the configured training algorithm are launched, the configured environment is built from an existing or newly created implementation and monitoring utilities are called. The environment, as an OpenAI Gym based, plays an important role in this mode, and it is used as a library to interact with the simulated world. It implements the reset and steps execution behavior which retrieves the state the agent can read from sensors and the rewards obtained in every state. The goal of inference mode is to test whether robotic brains can generalize to other environments. Therefore, the execution flow is similar to the training mode, except that the trained brain to be tested must be included as a parameter in the configuration file.

3.4 Logging and Monitoring

Every training carries out with RL-Studio is registered in the desired location through the configuration file. Directories are automatically created, in case they do not exist, to save the trained brains as well as the log files necessary to analyze training or inference execution. Also, some error messages and common helpful information have been generalized to be imported in any integrated project. Some libraries have been implemented to monitor the evolution of training in the RL-Studio projects: *LivePlot* library is based on matplotlib[2] and renders a graph of either episode rewards or episode lengths.

[2] https://matplotlib.org.

Stats_recorder stores in json format the episode lengths, episode rewards, episode types and their timestamps. *Video_recorder* optionally renders a movie of a rollout.

4 Experiments

RL-Studio has been validated building and collaboratively evolving some of the following research projects.

4.1 Robot Mesh

The problem to solve is to make an agent learn the optimal path to the goal as soon as possible inside a mesh. The provided environment permits performing one of four actions, ahead, back, left and right, and each simulation step tried to get closer to the location/state in the board which was configured as goal. This problem has been validated in RL-Studio shown in this video[3] with Q-learning algorithm based on the hyperparameters obtained after several tests, $\gamma = 0.95$, $\alpha = 0.9$, *exploration min* $= 0.00001$ and *exploration decay* $= 0.999$, a sparse reward function with 1 when the goal is reached and 0 otherwise and the input sensory information has been the position of the agent, x and y coordinates, running in Gazebo simulation tool, Fig. 2 left. As we can see in the Fig. 2 right, the agent learned the optimal path to the goal in approximately 100 iterations. In this problem, RL-studio has been especially useful to easily training and validating the inference mode.

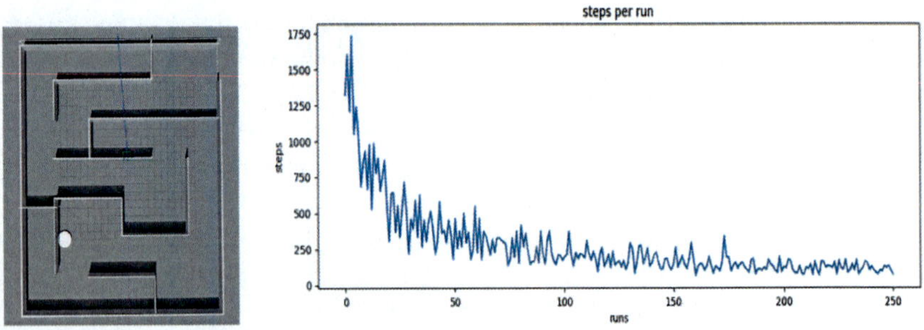

Fig. 2. Left: Robot Mesh in RL-Studio using Gazebo simulator. Right: Training steps per iteration, showing model stabilization after 100 episodes.

4.2 Mountain Car

The objective has been to teach a car how to get out of the valley and reach the peak of a mountain, Fig. 3 left, in less than 300 steps just applying one action, force to right, to left or do nothing, every 0.5 s. It has been needed to get farther from the goal to be able to reach it. Thus, defining the problem, action and states, and the hyper-parameters

[3] https://www.youtube.com/watch?v=36Zko9hSY6k.

configuration has been more complicated so the agent could learn either the optimal or a feasible sequence of steps to accomplish the objective. In this case we have solved the problem using Gazebo, Q-learning algorithm, the state space defined by discretized vehicle coordinates received by the GPS-simulated input sensor, $x = [-10, 10]$, and the discretized linear velocity, $v = [-100, 100]$. The reward function is 1 when the goal is reached and 0 otherwise and the best hyper-parameters configuration tested has been $\gamma = 0.9$, $\alpha = 0.2$, *exploration min* $= 0.01$ and *exploration decay* $= 0.99995$.

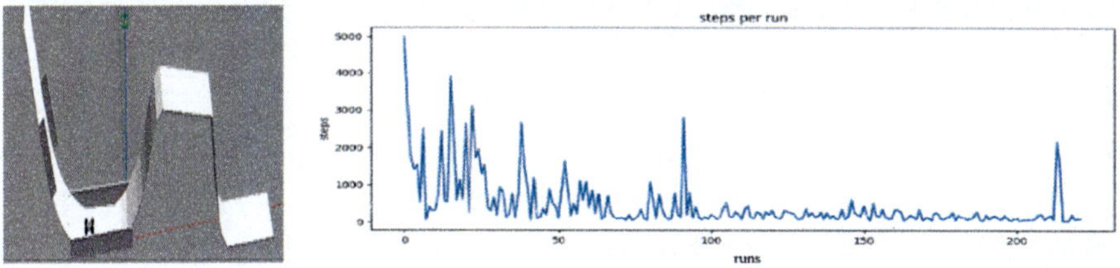

Fig. 3. Left: Mountain-car in RL-Studio using Gazebo simulator. Right: Training steps per iteration, starting model stabilization after 100 episodes.

As we can see in the Fig. 3 right, the agent considerably improved its performance as it learns. In this problem, RL-Studio has been especially useful to build the problem. Thanks to the easy way of configuring the actions to be applied, the agent to be used and the monitoring libraries already integrated, we could perform multiple tests to solve the canonical mountain car problem as it can see in this video[4].

4.3 Formula1 Visual Follow-Line

The agent is based on a real Formula1 (F1) car and it is endowed with a built-in color camera as input sensory data for reading the information retrieved by environment. The environments designed are several circuits with different characteristics of length and curves as can be seen in Fig. 4. They are inspired by real F1 circuits and each one has different light conditions, road textures, size straights and pronounced curves what helps us understand the learning generalization capabilities of the model[5].

Fig. 4. Some of the circuits in RL-Studio. From left to right: Simple circuit, Nürburgring and Montreal.

[4] https://www.youtube.com/watch?v=KZjDe6N-d0k.

[5] https://www.youtube.com/watch?v=3jdxZTjPCss.

The robot goal has been to follow the center red line of the road as closely as possible and without leaving the road as shown in Fig. 5. The environment retrieved a reward that tells the agent how good is the action taken, following the standard RL cycle.

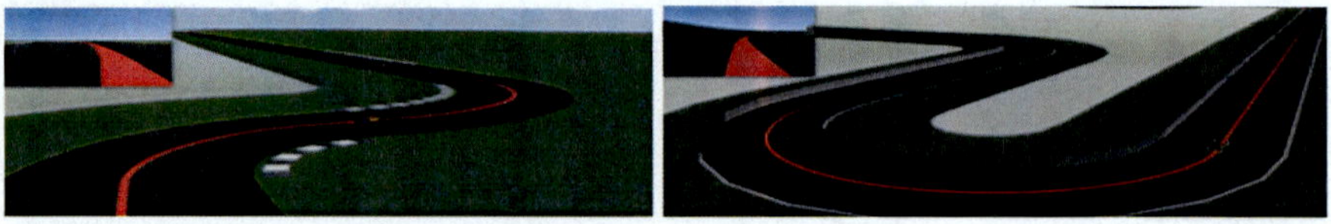

Fig. 5. Left: F1 circulating in the Simple circuit and taking a curve to the left from two perspectives. Right: The same F1 agent in Montreal circuit.

In this project Q-learning has been used as RL algorithm and Gazebo tool as a simulator. Q-learn is a tabular RL algorithm admitting a space of actions and states of low dimensionality. Therefore, the raw input sensory data must have been processed to obtain the appropriate information, simplifying it to a set of positions that indicates how close or far the car is from the center of the red line. Every input frame is separated in 17 vertical columns, from -8 to 8 including 0, each of them representing a position or state. In this project, the F1 has been trained with 1 point of perception, so the state will be between -8 to 8. Knowing only the lower half of the image captured by the camera sensor contains the road information, and after various tests, we take as a perception point a distance of 60 pixels below the middle of the image, Fig. 6. So, the distance from the center of the image to the center red line of the road will give us the state of the car at each time step. In RL-Studio is very easy to configure different points of perception.

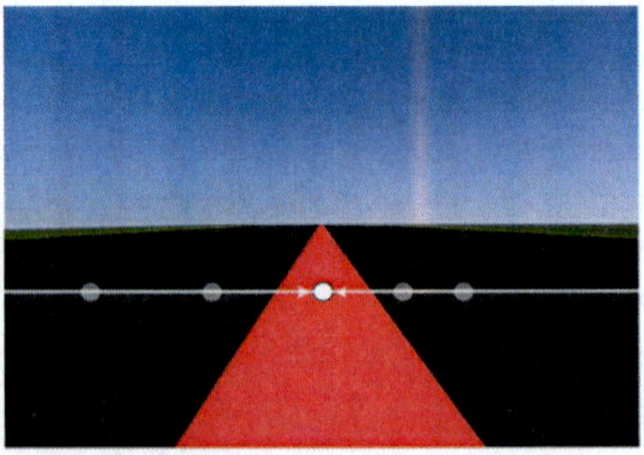

Fig. 6. Simplified perception with one point

Likewise, we have created a set of discrete actions defined as a vector $a = [v, w]$ which contains the two degrees of freedom that a planar vehicle needs, linear and angular velocity in *m/sec* and *rad/sec* respectively. So, we have defined three actions, *action A* = [3,0], *action B* = [1, 2] and *action C* = [2, -1]. Training with low speeds allows us have not taking into account the kinematics and dynamics of the vehicle for the model. The below defined reward function *R*, maximizes the value being close to the center, *c*, of the road line, decreasing as it moves further away. A high penalty has been defined in case the car goes off the track to help make learning faster. Remember that F1 does not have the stop action defined to prevent unwanted behavior.

$$R = \begin{cases} 10\, c < |0.2| \\ 2\, |0.4| > c > |0.2| \\ 1\, |0.9| > c > |0.4| \\ -100\, c > |0.9| \end{cases}$$

Three training have been executed in the Simple circuit for two hours each one and 4000 steps in every episode, green, yellow and blue in Fig. 7 left, with hyperparameters optimized to $\gamma = 0.9$, $\alpha = 0.8$ and *exploration decay* $= 0.998$. In all of them, the agent has been able to complete the circuit and in two of the three trials, green and yellow, converge very quickly in just less than 10 epochs and accumulating high reward. Namely it spends more time in regions nearby the center of the line where the reward returns the highest values. Likewise, the number of steps taken in every training is showing in Fig. 7 right.

Table 1 contains more information about the three experiments. We can see the algorithm has learned by improving in each one of the episodes not only in completed lap time, but in a smaller number of episodes and in less exploration, time given by the variable ε, where an ε nearby to 1 tells the exploration time is low. *Time per lap (min.)* shows time in minutes to complete the first lap. *Total epochs* shows number of episodes carried out in two hours.

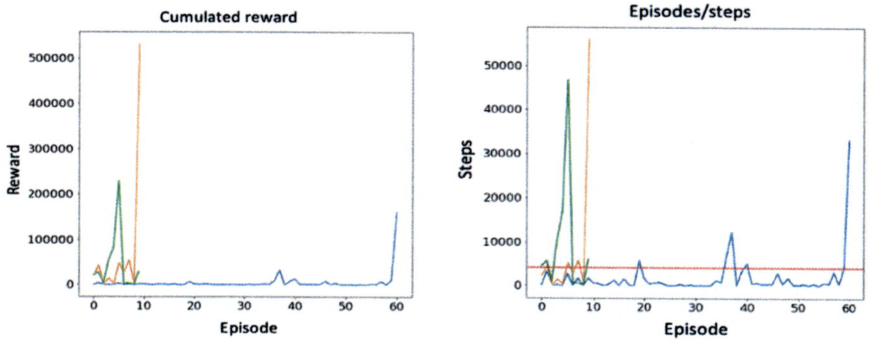

Fig. 7. Results of the same three training sessions in the Simple circuit, with a simplified perception point and three discretized actions. Left: Cumulated reward in three Simple circuit training. Right: Number of steps in every training in Simple circuit training.

We have taken the *time per lap* as a quantitative metric to know how the algorithm has learned and the *percentage of circuit completed*, as a qualitative metric, has allowed to know if the agent has completed the circuit correctly. In this last case as all training converged, the values have been 100%.

Table 1. Three different training in Simple circuit with one point of perception and the three defined actions A, B and C.

Simple circuits			
Training	1	2	3
Time per lap (min.)	20:06	6:53	4:26
Epochs to complete	20	2	1
Total epochs	60	10	10
ε	0.8	0.84	0.85
Circuit completed	100%	100%	100%

5 Conclusions

In this paper we have presented a tool to build and test RL models applied to robotics. The proposed architecture, modular-based, has been validated in three different robotics projects where it has demonstrated its robustness, effectiveness and usability. It allows indirect programming with RL based methods in such a way that defining the space of states and actions, robots can find the policy that maximizes the sought objective. RL-Studio covers the cycle of any standard RL model and is easily configurable for training and inference, allowing to configure all the pieces that make up an RL model. As future work, we plan to integrate other simulators used in the scientific robotics community like Carla, along with more state-of-the-art algorithms in RL such as DDPG and PPO which allow to work with real continuous spaces. We will also take advantage of its modularity along with Docker to create end-to-end solutions with hardly any changes in the configuration.

References

1. Hietala, J., Blanco-Mulero, D., Alcan, G., Kyrki, V.: Learning visual feedback control for dynamic cloth folding. arXiv preprint arXiv:2109.04771 (2021)
2. Kendall, A., et al.: Learning to drive in a day. In: 2019 International Conference on Robotics and Automation (ICRA), pp. 8248–8254. IEEE (2019)
3. Kong, X., Xia, Y., Hu, R., Lin, M., Sun, Z., Dai, L.: Trajectory tracking control for under-actuated hovercraft using differential flatness and reinforcement learning-based active disturbance rejection control. J. Syst. Sci. Complexity **35**(2), 502–521 (2022)
4. Lee, J., Hwangbo, J., Wellhausen, L., Koltun, V., Hutter, M.: Learning quadrupedal locomotion over challenging terrain. Sci. Robotics **5**(47), eabc5986 (2020)
5. Lee, J.W., Kim, K.W., Shin, S.H., Kim, S.W.: Vision-based collision avoidance for mobile robots through sim-to-real transfer. In: 2022 International Conference on Electronics, Information, and Communication (ICEIC), pp. 1–4. IEEE (2022)
6. Serhan, B., Pandya, H., Kucukyilmaz, A., Neumann, G.: Push-to-see: learning non-prehensile manipulation to enhance instance segmentation via deep q-learning. Institute of Electrical and Electronics Engineers (2022)
7. Sutton, R.S., Barto, A.G.: Reinforcement Learning: An Introduction. MIT Press (2018)

8. Tu, Z., Yang, C., Wu, X., Zhu, Y., Wu, W., Jia, N.: Moving object flexible grasping based on deep reinforcement learning. In: 2022 8th International Conference on Control, Automation and Robotics (ICCAR), pp. 34–39. IEEE (2022)
9. Wang, B., Liu, Z., Li, Q., Prorok, A.: Mobile robot path planning in dynamic environments through globally guided reinforcement learning. IEEE Robotics Autom. Lett. **5**(4), 6932–6939 (2020)
10. Xie, A., Singh, A., Levine, S., Finn, C.: Few-shot goal inference for visuomotor learning and planning. In: Conference on Robot Learning, pp. 40–52. PMLR (2018)
11. Yang, Y., Caluwaerts, K., Iscen, A., Zhang, T., Tan, J., Sindhwani, V.: Data efficient reinforcement learning for legged robots. In: Conference on Robot Learning, pp. 1–10. PMLR (2020)
12. Yen-Chen, L., Bauza, M., Isola, P.: Experience-embedded visual foresight. In: Conference on Robot Learning, pp. 1015–1024. PMLR (2020)
13. Zhou, Q., Lyu, L., Liu, H.: Deep reinforcement learning with long-time memory capability for robot mapless navigation. In: 2022 IEEE 25th International Conference on Computer Supported Cooperative Work in Design (CSCWD), pp. 1215–1220. IEEE (2022)

Interactive Social Agents Simulation Tool for Designing Choreographies for Human-Robot-Interaction Research

Olivier Hauterville[1], Camino Fernández[1,2(✉)], Phani Teja Singamaneni[1], Anthony Favier[1,3], Vicente Matellán[1,2], and Rachid Alami[1,3]

[1] RIS Group - LAAS (CNRS), Colonel Roche, 31400 Toulouse, France
{Olivier.Hauterville,phani-teja.singamaneni, anthony.favier,Rachid.Alami}@laas.fr,
[2] Robotics Group, Universidad de León, 24007 León, Spain
{camino.fernandez,vicente.matellan}@unileon.es
[3] ANITI, Univeristé Fédérale Toulouse Midi-Pyrénées, B612, Rue Tarfaya, 31400 Toulouse, France
http://www.laas.fr/public/, https://robotica.unileon.es

Abstract. Social robotics research requires the replicability of the experiments used to validate new technical or scientific claims. However, humans behavior is difficult to replicate. We have designed and implemented a system to simulate interactive agents that can be perceived by a robot in a robotic simulator. These agents can exhibit social behaviour in human group movement, for example, by talking in pairs. These humans are interactive with the robot and can be perceived so that the robot can react to their behavior. The humans can be choreographed to create high-level social behaviours such as waiting for an elevator, getting in and out of it, or standing in front of a store window and moving together to the next one. The tool would provide metrics to evaluate the robot's performance in defined social situations.

Keywords: HRI · Simulation · Social robotics

1 Introduction

One of the basic requirements of science is the replicability of the experiments designed to validate any model, theory, or hypothesis proposed. However, when the research topic involves interaction with people, as in social robotics, this replicability is difficult to achieve. Even detailed descriptions like the one in [9] are really difficult to replicate. Furthermore, small variations in the demeanour of the people have a significant impact on the behaviour of the robot. Software simulators cope with this problem by providing the ability to replicate the exact same situation repeatedly. These simulators should be able to provide "simulated" human behaviour for the development of socially acceptable robotic behaviours, but currently offer only limited capabilities in this area.

D. Tardioli et al. (Eds.): ROBOT 2022, LNNS 590, pp. 514–527, 2023.
https://doi.org/10.1007/978-3-031-21062-4_42

Likewise, developing skills for autonomous robots also requires testing the skills on real robots, but designing and conducting these user studies are costly and time-consuming because they typically involve a large number of people to be statistically significant. Usually, tuning and testing tasks are performed on simulators which, as previously mentioned, should provide simulated humans that can actually be useful when designing skills to interact with people.

Simulations are also used for benchmarking the robotic navigation algorithms. If the same environment and conditions are presented to different algorithms, different metrics can be calculated to assess and compare the performance of the algorithms.

The goal of the tool described in this paper is to provide a means to generate scenarios in which predefined social interactions of groups of reactive humans can be used to test the social performance of a robot behaviour under evaluation. In the rest of the paper, we will refer to it as *tested robot*. The aim is to use the system for benchmarking human-robot interaction behaviours.

The rest of this paper is organized as follows. The state of the art of the simulation of robots in the presence of humans is discussed in Sect. 2. The requirements, analysis and tool design are described in Sect. 3. Section 4 shows the implementation details of the tool and Sect. 5 includes some basic use case study of a specific scenario. Finally, Sect. 6 summarizes the contributions and the envisioned future works.

2 State of the Art

There are several simulators commonly accepted in the robotics research community such as Gazebo [10], Unity [7], Coppelia (V-REP) [17], Morse [11], Unreal, etc. These simulators have proven to be very useful for testing basic service robotics skills, such as obstacle avoidance, localization, global navigation, etc. However, they are not well suited for human simulation. Some of them, like Gazebo, provide basic mechanisms (actors) to simulate humans, but these simulated humans can exhibit only basic behaviors, such as going to a point, and are not responsive and not even aware of the robot's presence; they are not even fully integrated with the physics of the simulation and are closer to "decorations" than to real simulated entities.

The problem of human social navigation is also an active research question in the crowd simulation research community. Crowd simulation has applications in many areas, from transportation to psychology, the design of video games, or digital effects for movies. But usually, they are more concerned about the management of large crowds than about the intelligent behaviours of the individuals. Unfortunately, classical crowd simulators such as Continuum [18], Pedsim [4], or Menge do not simulate robots. Some adaptations have been proposed to include robots in these crowd simulators. The best known is Pedsim-ROS [14], a set of packages developed at Freiburg University for 2D pedestrian simulation integrated with ROS. However, these simulators usually provide simple behaviours for simulated humans. For instance, Pedsim-ROS only provides local collision avoidance based on the Social-Force Model (SFM) [5] (Fig. 1).

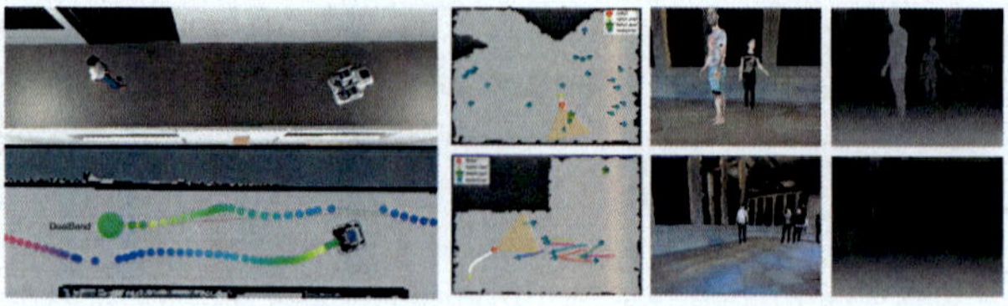

Fig. 1. InHuS and SocNavBench human-robot simulators.

A more recent related work is SOCIALGYM [6], a lightweight 2D simulation environment for benchmarking, training and evaluating social navigation behaviours. Social navigation is modelled as an action selection problem, and the optimal action is selected from a set of four discrete sub-policies: Halt, GoAlone, Follow and Pass. In SOCIALGYM, each simulated human moves back and forth between a starting point and a goal pose by computing a global path of intermediate nodes between them and using a separate local planner (Pedsim-ROS social force model) to handle dynamic obstacles. It also includes a "Navigation Module" responsible for planning the obstacle-free path from the starting point to the goal. Therefore, SOCIALGYM only provides a combined behaviour using the above sub-policies but does not provide a real social interaction.

Regarding benchmarking, currently most proposals to compare social navigation algorithms are based on datasets. Some of them are designed to test perception, such as SRIN (Social Robot Indoor Navigation) [8] which contains images taken by medium size robots. SocNav1 [13] includes 9,280 scenarios containing locations of humans and a robot in an 8×8 area and the subjective level of discomfort generated by the robot's presence in that scenario. Although datasets are essential for machine learning training algorithms, their use as benchmarking tools is not justified, as they are intrinsically static. Therefore, dynamic simulators should be part of the benchmarking process.

SocNavBench [2] is an intermediate testbed that mixes a dataset and some simulation. It comprises a photo-realistic renderer, a set of navigation scenarios based on real-world pedestrian data, and a set of metrics to characterize the performance of robot navigation algorithms. Instead of simulating behaviour, SocNavBench derives real-world pedestrian trajectories from real behaviour datasets. It applies real-world textures on pedestrians and the environment to overcome the inability to obtain sensor inputs for these datasets. Although it is an improvement over mere images, simulated humans have fixed, non-reactive behaviour. They can only exhibit pre-recorded behaviour and only existing behaviors can be reproduced. The creation of new scenes requires the recording new real data.

Liu et al. [12] have also proposed interactive pedestrian simulation in iGibson, sharing some of the ideas with our proposal. They decompose the problem into a "Global Planning" module based on A^* to compute coarse-grained waypoints to guide pedestrians towards their goals and a "Local Planning" module to control pedestrian positions at each time step. The main difference is that they compute these trajectories for each of the simulated humans focusing on learning socially-aware individual navigation behaviours. Our proposal is not oriented towards learning individual social behaviours; we just want to simulate group behaviours that can be used to test the social skills of robots.

The closest tool to the goals of the proposed system is the Intelligent Human Simulator (InHuS) [3], which proposes a generic architecture for simulating intelligent human agents. It provides a complex "Human Behavior Model" module capable of defining goals, managing relations with other agents and building perception from the data received from the simulator. So, it can be considered as a complete cognitive architecture that could be used to control simulated humans or robots. It has been tested for social navigation, but only with one simulated human, which uses CoHAN [16] as the robot controller, but enhanced with two "attitudes": *Harass* and *StopAndLook*.

Finally, there are other approaches for generating the behaviour of simulated entities. For instance, in the classical planning community, there are simulators as PDSim [15], where the behaviour of objects in planning problems can be defined using PDDL. These "animations" may include moving an object onto another object, letting an agent follow a path between two points on a map, etc.

In summary, existing simulation tools that integrate robots and humans do not allow simulating realistic social behaviours of the simulated people. The only available alternative is the use of teleoperated avatars or basic behaviours. The need for simulation tools that provide high-level primitives to define the behaviour of simulated humans has been identified, and some commercial products are starting to appear, for instance, PedSim Pro V1.2 has incorporated the "queuing behaviour".

The goal of IMHuS (Intelligent Multi Human Simulator), the tool described in this paper, is to provide an open-source toolkit for defining high-level reactive simulated humans with the ability to show the behavior of social groups, but using realistic standard robotics simulators that allow researchers to use models of their real robots, both for debugging their algorithms and for benchmarking and repeatability.

3 Design of a Step-Based Social Simulation

The goal of IMHuS is to standardize the validation of the autonomous robot behaviour in the presence of people, allowing researchers to design repeatable human social situations. For example, defining a set of waypoints where different simulated people arrive, meet and move as a group to a different position, or two people facing each other and then moving together to a different location.

Our proposal considers that both individuals and groups have to be included in the simulated environment and that in both cases, each simulated person should exhibit adaptive behaviour. To achieve this goal, we assume that global path planning for the whole set of agents is more suitable for defining fixed social behaviours than the individual path planning approach. This assumption also serves the purpose that humans exhibit a common social behaviour that the robot must be able to detect in order not to cross, for example, through the middle of a social group.

A central process can have perfect knowledge of the simulated environment, accessing the real and accurate positions of all the elements of the simulations (obstacles, robot, etc.). It is not designed to emulate an embedded agent, since the robot simulation process has to obtain the information through the noise-simulated sensor readings provided by the simulator. This process can use the API of the simulator to get all the information directly, without noise and identification problems (it will know which types of elements are in the simulation and their state). For instance, it does not need to identify if something is "a door" or if it is opened. The door state is obtained directly from the simulator.

3.1 Choreography Oriented Simulation

The constraints and assumptions made for IMHuS are:

1. The set of persons $P = \{p_1, p_2, ...p_n\}$ in the environment is defined *a priori*, and each person will be unequivocally identified by its ID (p_i) during each execution.
2. The number of locations $L = \{l_1, l_2..., l_m\}$ for these people is also known *a priori*.
3. The number of locations will be greater than the number of people $|L| \geq |P|$.
4. In a given time-step, only one person can stay at an individual location l_i.
5. Group locations L_i^n will have a maximum number n of individual positions $l_1, l_2, ..., l_n$. For instance, an elevator with four positions will be named L^4.
6. The number of actions is also finite and known.

The goal of the tool is to allow researchers to have a high-level definition of a "choreography" of people moving in a social way. For instance, let's consider an example, first a set of people (p_1 to p_4) is defined. In a first step, p_1, p_2 and p_3 have to move from their initial positions to a "group location" (L^3). This joint navigation is represented in Fig. 2 as a continuous top-opening box grouping the three people at time-step t_0 and a double arrow labelled with the goal destination. In the same way, p_4 will remain in its position but will face a particular direction (30°C), indicated by the circle with an arrow.

At t_1, p_1 and p_2 engage in a pair-move to L^4, while p_3 and p_4 initiate individual moves, indicated by a single arrow. A pair-move is a specific type of group move included in the design because it is usually the most common one and because it can be considered as the smallest group unit. At t_2, the two persons moving as a pair would have reached their destination and p_1 will initiate its movement towards l_{10}, while p_2 remains in its current position with no particular orientation (indicated by a circle wit no arrow), and p_3 and p_4 continue to navigate to their targets.

p_1 at t_2 begins to move individually towards l_{10}, which reaches at t_3 while p_3 and p_4 continue to execute their solo movement. p_3 reaches l_{11} at t_4, when p_1 and p_2 begin to face in the same direction. Finally, at t_5, p_4 reaches its destination (l_{12}) and the choreography ends.

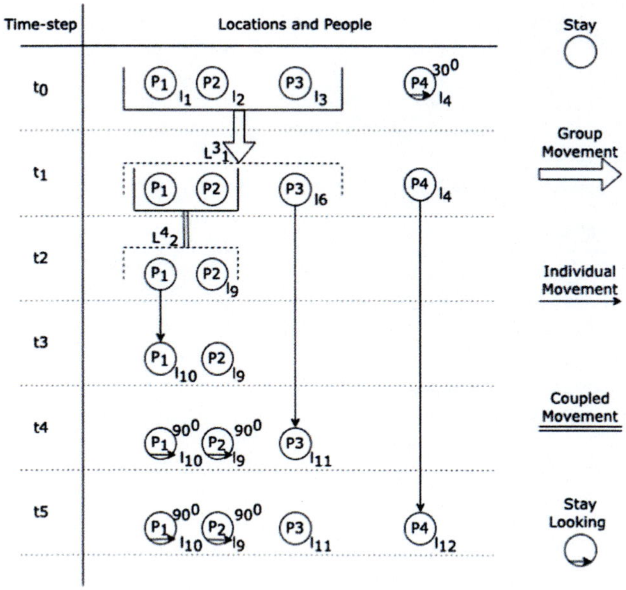

Fig. 2. Schematic representation of the definition of a social navigation choreography

Time-step (t_i) in Fig. 2 means "*choreography steps*", that is, significant moments for the definition of the simulation, it does not refer to a magnitude measured by a clock. We will refer to them as *steps*, which is an increased ordered sequence of discrete points in time at which a given set of events has to occur. For example, $step_0$ usually specifies the initial state of all the elements of the simulation, i.e., the position of the robot, human', and the other elements. A typical step specifies a set of actions that are initiated at that instant, a navigation task for one of the humans, for instance, at $step_1$ (t_1 in Fig. 2).

This is the type of simulation that the tool should be able to generate. Figure 3 shows the general architectural framework. This tool receives the definitions of the social scenarios in a definition file (an XML in its current version). It then manages the simulation, obtaining the simulation data, using existing libraries and other tools (such as *move_base* for calculation of trajectories in ROS), and finally generates the log data for evaluation.

The definition of the choreography shown in the upper left of the Fig. 3 is defined in XML. The design allows five types of *basic actions* that the *agents* (simulated robots or simulated people) can perform:

Navigation actions: Those actions modify the global pose of the agents in the environment. Typical actions in this group are *GoToPose* and *Wait*.

Turning actions: Actions related to the facing of the agents, such as *LookAt* and *Turn* depending on the way the action is specified.

Grouping actions: These actions manage the creation and dismantling of groups.

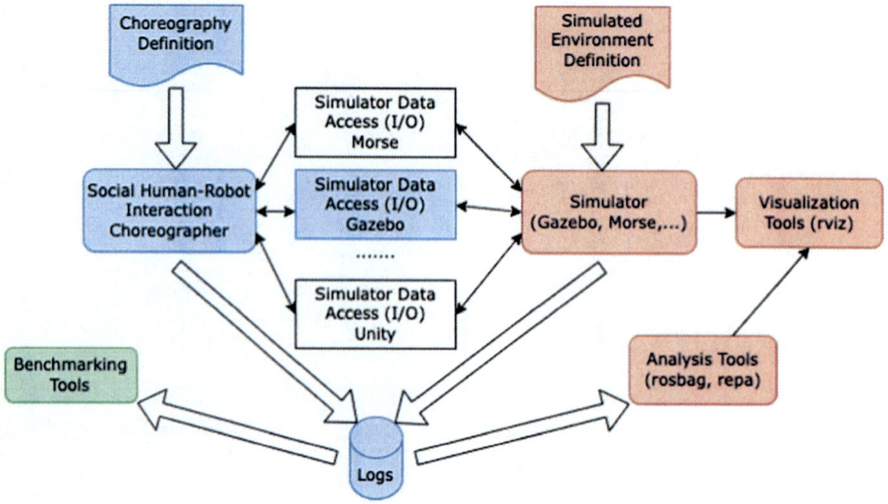

Fig. 3. Global simulation architecture. Red signifies existing tools, blue components are those described in this paper, green signifies on-going development, and uncolored boxes alternative simulators under consideration. (Color figure online)

Attitude actions: Actions modelling the high level behavior of the agent. For instance, a simulated human can be ordered to *harass* the robot.

Synchronous actions: Actions related to the environment to be accomplished by an agent during one specific step of the simulation. They have been standardized as *publish* and *subscribe*.

Asynchronous actions: Actions related to the environment and not associated with a specific step in the timeline of the simulation. They have been standardized as *request* and *respond*.

These actions can be applied to a single agent or a group. Basic actions can be integrated into a *compound_task*. For defining these actions, the following components are used:

- *map*: corresponds to the "world" where the simulation will happen. Inside the map a set of *objects* can be defined specifying their individual ID.
- *poses*: correspond to the "location" used in Fig. 2. They are made up by the x, y position and orientation θ, and a *radius* of tolerance for the motion planner to consider that the goal has been reached.
- *agents*: that can be either *robots* or *humans*, each of them identified by an unique ID. They are given an initial pose where they appear in the world.
- *groups*: they can be created and dismantled during the simulation

The evolution of the simulation is based on steps, as described in the previous section. The regular steps are synchronous, but there is an asynchronous step for interacting with the tested robot:

- A scenario is composed of a set of regular steps and a singular asynchronous step.
- Steps are made up of different actions like navigation, grouping, etc.
- Every action of a step is executed *at the same time*, meaning that the actions are executed in simulated parallelism.
- One step ends when all its actions have finished.
- The asynchronous step runs in parallel to the execution of the synchronous steps.
- The scenario ends when the last regular step ends.

4 Implementation

In the current version choreographies are defined in an XML file that includes four main sections[1]:

- Map = poses and objects. Example: poses definition.
```
<poses> ::= <pose> (<pose>)* ;
    <pose> ::= <poseID> <x> <y> <theta> <radius> ;
```
- Agents = humans/choreographed robots with their initial poses, and groups with their composition. Example: groups definition.
```
<groups> ::= (<group>)* ;
    <group> ::= <groupID> <pair> (<human> (<human>)*) ;
        <pair> ::= <pairID> <human> <human> ;
```
- Tasks = generic actions. Example: compound tasks and look-at action definition.
```
<compound_tasks> ::= <compound_task>* ;
    <compound_task> ::= <compound_taskID> <action> (<action>)* ;
<lookAt_action> ::= <actionID> (<humanID>|<robotID>|<objectID>) ;
```
- Scenarios = step elements of synchronous steps and asynchronous step. Example: scenario definition.
```
<scenario> ::= <name> <step> (<step>)* (async_step);
    <step> ::= <stepID> <stepElement>;
        <stepElement> ::= <agentsID> <pose>|<compound_task>;
        <agentsID> ::= <humanID>|<robotID>|<pairID>|<groupID>;
    <async_step> ::= <respond_event_action>;
```

The prototype has been implemented as a new version of the InHus tool ([1,3]) which connects to the Gazebo simulator to obtain the information about the world. It uses the *move_base* ROS to generate the global plan for each agent and updates the positions of each agent in the next step of the simulation accordingly. The new version is capable of handling the navigation of multiple agents (humans) in parallel while managing conflicts and completing individual tasks

[1] A full version of the syntax can be accessed here (link).

as in the previous version. It also provides an updated graphical user interface for repeatedly selecting and executing scenarios defined in the XML file.

The IMHuS tool shown in Fig. 4 has been structured in three layers: the IMHuS layer (in blue), its configuration in the application layer (in yellow), its communication with the simulator and ROS (in red). The robot whose behavior would be tested in the tool has also been included (in violet).

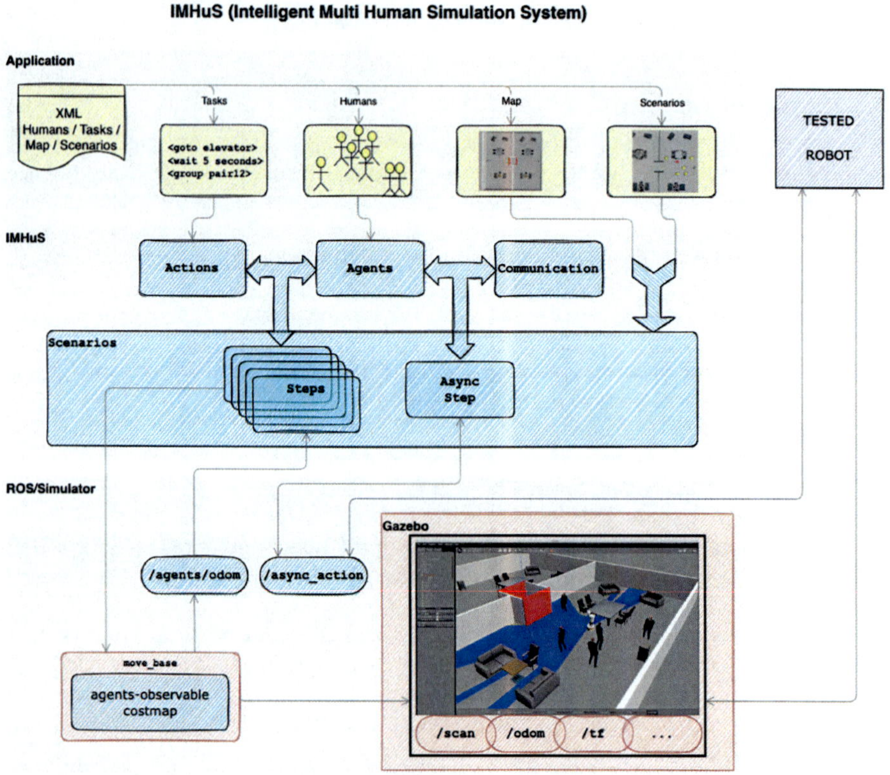

Fig. 4. Structure of the new IMHuS tool.

Application layer This layer includes the XML configuration file needed to use IMHuS by defining its main components: map, agents, tasks and scenarios. The map includes all the locations of the agents to be used during the choreography, as well as those of static objects. These locations are represented as poses and a name is assigned to each of them. The description of the agents includes the name and initial position of all humans and robots choreographed, and also the name and composition of the groups that will appear in any step. Generic tasks are described with no specific subject to perform them, so that they can be reused by several agents. Last, scenarios describe the steps of the choreography. Each step includes a set of tasks assigned to a particular agent or to a groups. They are the step elements.

IMHuS layer The tool interprets the information contained in the XML configuration file to represent the tasks as actions and the humans/robots as agents. When the scenarios are run, a command combines actions with agents for every step element. The commands concerning each agent are executed in a separate thread inside a step. The step finishes when all the threads are done. This is the way simultaneous movement of agents is achieved. This is the behavior of the tool for the choreography, but taking into account that the tested robot will be present, there must be a way for it to communicate with the agents of the simulation, just as it would happen in the real world. The asynchronous step is in charge of this task. At any point in time, the tested robot can ask something, for example, to press the elevator button. This action would be done through the communication module and would be answered by the agents in the simulation in this asynchronous step by means of a respond action. Not every agent can respond to a request from the tested robot. For instance, if the robot asks for someone to press the button and no one is around, the request will be dropped, counting as a failure.

ROS/Simulator layer This layer is used by IMHuS to place the agents in the simulation and to ask for the trajectories to move them around, as well as to communicate with the tested robot. The agents are placed in the world as obstacles so that when `move_base` is asked for a new path, they are avoided. The costmap obstacle layer has been modified to obtain what we call the *agents-observable costmap* that IMHuS needs.

Interface of the tested robot The way to include the software of a robot in the simulator for its social behaviour to be tested would be through the simulator, Gazebo for now. Outside the simulator, the only communication would be through the asynchronous action thanks to which the robot can send a request action to be answered by a respond action of one agent of the simulation.

5 Use Cases

This module is available as Open Source and it has been evaluated in the elevator scenario, as defined for SciRoc competition (IMHuS repository link).

An elevator scenario will be used to show a running example of the behavior of the IMHuS system with a tested robot, and also to explain the communication between the *tested robot* and the agents of IMHuS (video available at this link). The proposed scenario includes five human agents and the *tested robot*, whose goal is to go to the second floor. In order to do that, it has to ask a human agent to push the elevator button. Figure 5 shows the initial situation where the *tested robot* is approaching the elevator, human 2 is walking and the rest of humans are idle.

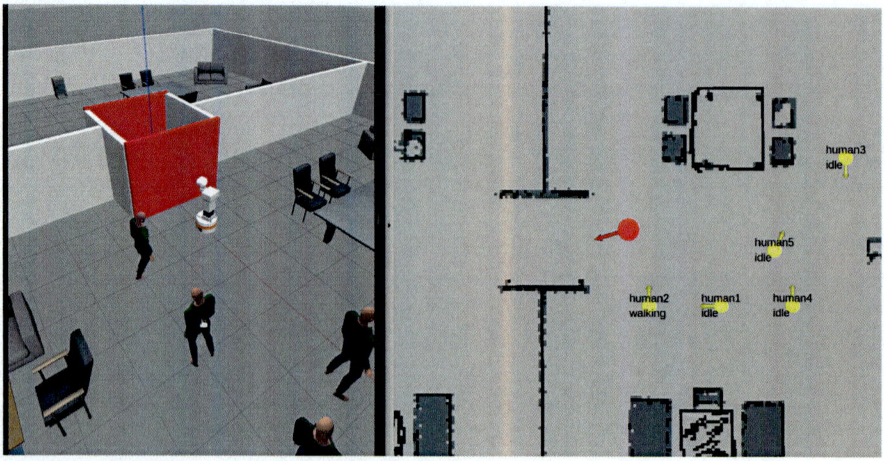

Fig. 5. Initial situation.

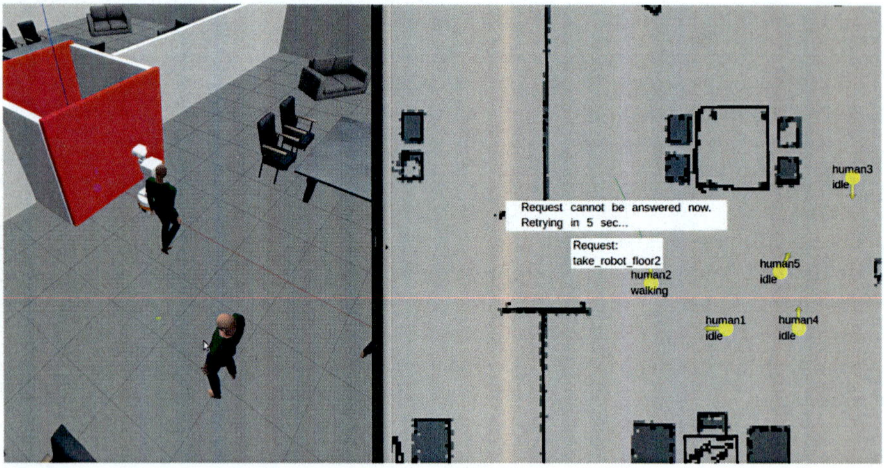

Fig. 6. Request dropped.

Once the *tested robot* gets to the elevator door, it requests an action to be performed by one of the humans. In order for a *tested robot* to trigger an action in IMHuS, it has to publish a specific message type on the `async_action` topic (see Fig. 4). By doing so, it triggers an asynchronous action response from IMHuS. This message should contain the time of emission, the name of the agent requesting, the pose of this agent and the name of the task it is requesting. If the IMHuS' scenario includes in its configuration an asynchronous action response corresponding to the requested action, that action is triggered on the humans' side.

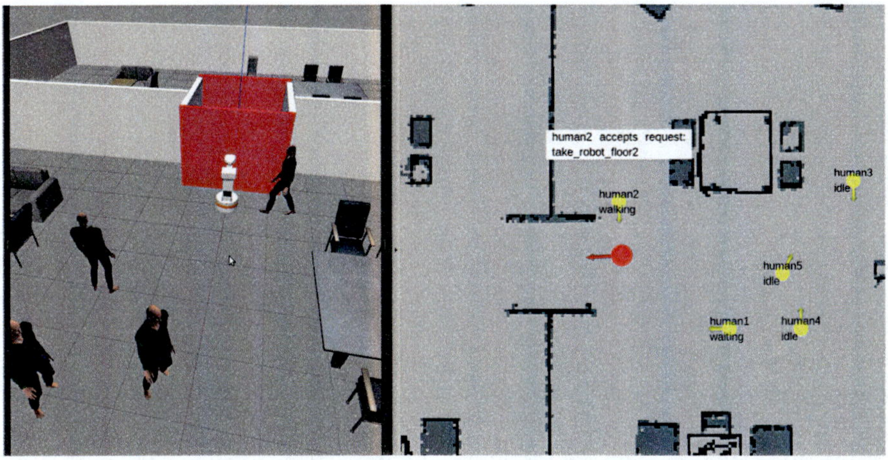

Fig. 7. Request accepted by human2.

Fig. 8. Final situation. *Tested robot* in second floor.

At this point, two different things can happen. If there is no human close enough to the *tested robot*, or there is a human but it is not in an idle state, no one will respond to the request and it will be dropped, as Fig. 6 shows.

The *tested robot* repeats the request every five seconds until human 2 enters an idle state and responds to the request. Figure 7 shows this moment of the simulation. When an asynchronous action request is triggered, IMHuS periodically checks if one human is able to respond. The requirements for a human agent to respond are to be in the idle state and within a three meters radius from the *tested robot* position when it requested the action. If several human agents are able to accept the task, the task will be assigned to the closest human.

Fig. 8 shows the final situation where the *tested robot* has reached the second floor, simulated in the scenario as the room on the other side of the elevator.

6 Conclusions and Future Work

The proposed tool, IMHuS, offers the possibility to create a realistic and challenging simulated environment in which groups of humans can be choreographed to evaluate the behaviour of a *tested robot.* Different scenarios can be easily created using an XML configuration file in which social situations can be defined to measure the behaviour of *tested robots* in a replicable environment. Furthermore, human agents are programmed to respond to interactions related to the particular situation of each scenario and their communication with the *tested robot.*

IMHuS's code is available as Open Source in the IMHuS repository. The current version has been implemented for Gazebo simulator, but the design presented in Sect. 3 can be easily migrated to other simulators, such as MORSE or Unity. The tool could be used for benchmarking of competitions such as SciRock or RoboCup as a previous step for the teams before getting to the physical robot challenges. To create a new scenario with IMHuS, all that is needed is a map and the configuration file to choreograph the human agents. The software has been designed to easily support the addition of both new tasks to be performed by the humans, and new interactions between them and the *tested robot.*

One of the ongoing developments is the automatic generation of simulation metrics such as the distance between the robot and the agents, time-to-collision, etc. Another line of work is the extension of basic actions, especially towards the social behaviour of groups of agents. Last, IMHuS is being tested with CoHAN [16], a human-aware robot navigation planner to challenge CoHAN under several human-robot interaction settings. From this, we plan to identify the areas of improvement for human-aware navigation planners and provide a benchmark for testing these planners.

References

1. Favier, A., Singamaneni, P.T., Alami, R.: Simulating intelligent human agents for intricate social robot navigation. In: RSS 2021- Workshop on Social Robot Navigation, July 2021
2. Biswas, A., Wang, A., Silvera, G., Steinfeld, A., Admoni, H.: SocNavBench: a grounded simulation testing framework for evaluating social navigation. CoRR abs/2103.00047 (2021). https://arxiv.org/abs/2103.00047
3. Favier, A., Singamaneni, P.T., Alami, R.: Challenging human-aware robot navigation with an intelligent human simulation system, June 2022. https://hal.laas.fr/hal-03684245
4. Gloor, C.: Pedsim: Pedestrian crowd simulation. https://pedsim.net
5. Helbing, D., Molnar, P.: Social force model for pedestrian dynamics. Phys. Rev. E **51**(5), 4282 (1995)

6. Holtz, J., Biswas, J.: SOCIALGYM: a framework for benchmarking social robot navigation (2021)
7. Juliani, A., et al.: Unity: a general platform for intelligent agents (2020)
8. Othman, K.M., Rad, A.B., Srin: a new dataset for social robot indoor navigation. Global J. Eng. Sci. **4**(5), 1–6 (2020)
9. Kobayashi, Y., et al.: Robot navigation based on predicting of human interaction and its reproducible evaluation in a densely crowded environment. Int. J. Soc. Robot. **14**(2), 373–387 (2021). https://doi.org/10.1007/s12369-021-00791-9
10. Koenig, N., Howard, A.: Design and use paradigms for gazebo, an open-source multi-robot simulator. In: 2004 IEEE/RSJ International Conference on Intelligent Robots and Systems (IROS) (IEEE Cat. No.04CH37566), vol. 3, pp. 2149–2154 (2004)
11. Lemaignan, S., Echeverria, G., Karg, M., Mainprice, J., Kirsch, A., Alami, R.: Human-robot interaction in the morse simulator. In: Human-Robot Interaction Conference (HRI'2012) (2012)
12. Liu, Y., Li, S., Li, C., Perez-D'Arpino, C., Savarese, S.: Interactive pedestrian simulation in igibson. In: RSS 2021 - Workshop on Social Robot Navigation, July 2021
13. Manso, L.J., Nuñez, P., Calderita, L.V., Faria, D.R., Bachiller, P.: Socnav1: a dataset to benchmark and learn social navigation conventions. Data **5**(1), 1–10 (2020). https://www.mdpi.com/2306-5729/5/1/7
14. Okal, B., Linder, T.: Pedsim: pedestrian crowd simulation (2015). https://github.com/srl-freiburg/pedsim_ros
15. Pellegrin, E.D., Petrick, R.P.: PDSim: planning domain simulation with the unity game engine (2021). https://icaps21.icaps-conference.org/demos/demos/387.pdf
16. Singamaneni, P.T., Favier, A., Alami, R.: Human-aware navigation planner for diverse human-robot interaction contexts. In: IEEE/RSJ International Conference on Intelligent Robots and Systems (IROS), October 2021
17. Rohmer, E., Singh, S.P., Freese, M.: Coppeliasim (formerly v-rep): a versatile and scalable robot simulation framework. In: IEEE/RSJ International Conference on Intelligent Robots and Systems (IROS 2013), pp. 1321–1326 (2013)
18. Treuille, A., Cooper, S., Popović, Z.: Continuum crowds. ACM Trans. Graph. **25**(3), 1160–1168 (2006). https://doi.org/10.1145/1141911.1142008

YASMIN: Yet Another State MachINe

Miguel Á. González-Santamarta(✉), Francisco J. Rodríguez-Lera, Vicente Matellán-Olivera, and Camino Fernández-Llamas

Robotics Group, University of León, León, Spain
mgons@unileon.es

Abstract. State machines are a common mechanism for defining behaviors in robots where each behavior is based on identifiable stages. There are several libraries available for easing the implementation of state machines in ROS 1, however, the community was focused on SMACH or SMACC. Although these tools are still predominant, there are fewer alternatives for ROS 2. Besides, Behavior Trees are spreading fast, but there is a niche for using State Machines. Here, YASMIN is presented as yet another library specifically designed for ROS 2 for easing the design of robotic behaviors using state machines. It is available in C++ and Python, and provides some default states to speed up the development, in addition to a web viewer for monitoring the execution of the system and helping in the debugging.

Keywords: Finite state machines · ROS 2 · Robot behavior · Decision making

1 Introduction

A Decision-making system must effectively drive the behavior generation process by deciding which actions or commands need to be selectively activated or executed in autonomous robots. One widely extended approach is using finite state machines (FSM) for designing these systems.

A state machine is a mathematical model for organizing a process based on transitions among states. State machines have been widely used in robotics to define autonomous robot behaviors understood as transitions from one state, where a set of actions is performed, to another state, where a different set of actions or their parametrization is executed. These actions can be high-level robot tasks, such as navigation, text-to-speech or speech-to-text, or low-level actions such as specific controllers. In the same way, state-based systems can be composed of several state machines that will change from one state to another executing the robot tasks. These systems may become really complex depending on the transitions between the states and the mechanism to select and execute them. For the sake of clarity, and because this research always considers a finite number of states, this paper uses indistinctly state machines and finite-state machines (FSM).

D. Tardioli et al. (Eds.): ROBOT 2022, LNNS 590, pp. 528–539, 2023.
https://doi.org/10.1007/978-3-031-21062-4_43

In ROS 1 there are popular libraries such as SMACH [4] or SMACC [2] for designing robotic behaviors based on FSMs. At the beginning of this research, these libraries were not migrated to ROS 2, so we developed YASMIN (Yet Another State MachINe) as an alternative to SMACH and SMACC, together with a visual debugging tool to monitor their execution. Nowadays, to the best of our knowledge, SMACC and SMACH were migrated. However, the SMACH version still has issues with the visualization tool, which we consider of high importance in debugging processes.

The main contribution of this paper is the development of YASMIN. It is an open-source library for Python and C++ designed for ROS 2. It is intended for designing robot behaviors using state machines. All states and any composition of them will share a blackboard, which is an easy-to-use storage used to share data. In addition, it is necessary to know the transitions and states that compose a state machine.

The rest of this paper is organized as follows. Section 2 overviews the current state-of-the-art solutions for quick development of robot behaviors. Section 3 formally describes some of the elements associated with YASMIN. Section 4 describes the internal features defined in YASMIN. Afterward, Sect. 5 describes the experiment applied for the validation of the tool, at the same time, this experiment was replicated using SMACH. This is followed by a discussion of the results obtained and, finally, conclusions and future work are outlined.

2 Background

As it has been mentioned before, there are libraries to develop behaviors in robots from ROS 1. One of the most famous is SMACH, a contraction derived from "State MACHine". It is a Python library that can be used independently from ROS. It provides states integrated with ROS and hierarchical and concurrently states machines. In addition, a viewer is used, although it has some problems in recent versions of ROS 2.

On the other hand, SMACC is an event-driven, asynchronous, behavioral state machine library for real-time ROS written in C++. It is inspired by SMACH and allows the development of robot behaviors. It uses orthogonal regions to cope with the creation of concurrent state machines. It also has a viewer and is available for ROS 2.

Hierarchical FSM vs Behavior Trees

YASMIN helps to implement robot behaviors using FSM. But there are other mechanisms that can be used to implement behaviors in robotics. For instance, the use of behavior trees to produce modular structures is becoming very popular, and the use of a deliberative system based on symbolic planners has been a traditional approach.

Finite state machines (FSM) are a collection of states, where the execution of the actions in each state depends on the transitions that occur. Hierarchical

FSM (HFSM) allows complex FSM to be generated, using a hierarchy of choices. As a result, transitions can be better organized by optimizing the code used.

In particular, the use of behavior trees (BT) has experienced a huge increase in robotics, [5,10]. However, BTs are similar to HFSM except that nodes are modular and uncoupled. Nodes can be executed from anywhere in the tree, allowing irrelevant nodes to be skipped. BTs can use a blackboard so that all nodes can share data.

FSMs, HFSMs and BTs are very similar but they should be used to address different problems. BTs can produce more complex behavior structures thanks to their modular nature but can take longer to develop. In contrast, YASMIN makes it possible to design FSMs and HFSMs that use a blackboard easily and quickly. Finally, YASMIN is integrated into ROS 2 allowing the creation of HFSMs in Python and C++, the most commonly used languages in robotics.

3 Formalization

This section describes the universal formal definition of a finite state machine (FSM). An FSM is defined as a 7-tuple consisting of a list of states as the conditions that the machine could take over time, a set of input symbols corresponding to the alphabet values that define the input to the machine, a set of output symbols representing the data processed in the system, an initial state from which the machine starts and two functions: one that determines the transition behavior of the states and the other, the processing of the input symbols translated into a sequence of output symbols.

$$FSM = (\sigma, \tau, \delta, \sigma^*, \tau^*, \Delta, \Omega) \tag{1}$$

- σ set of states.
- τ is the input set.
- δ is the output set.
- σ^* is the initial state.
- τ^* is the initial input set.
- Δ is a state transition function, such that: $\Delta : \sigma \times \tau \rightarrow \sigma$, i.e. the function takes as an element of the domain, an ordered pair consisting of a state and an input data, returning as an image, another state. This image represents the state of the system after the processing indicated by the ordered pair (state, input).
- Ω is the output function, $\Omega : \sigma \times \tau \rightarrow \delta$, i.e., like the function Δ, an element of the domain is an ordered pair formed by a state and an input data, which returns as an image, an output symbol. This symbol is the output of the system at that instant of time.

In addition, YASMIN also uses a blackboard to store the data of each state machine. States and nested state machines of inside a state machine use the same blackboard. This means that the input and output symbols are stored in this blackboard. The definition of a blackboard is also presented in Eq. 2. A blackboard is a 5-tuple formed by a set of symbols of the intermediate solutions, a set of knowledge sources and a function that determines which knowledge source to execute next.

$$Blackboard = (\sigma, \tau, \tau^*, \Delta, \Omega) \tag{2}$$

- σ is the set of knowledge sources.
- τ is the set of symbols.
- τ^* is the set of symbols.
- Δ is a scheduler function, such that: $\Delta : \tau \rightarrow \sigma$, i.e. the function takes as an element of the domain, a set of symbols, returning the next knowledge source to execute.
- Ω is an execute function, such that: $\Omega : \sigma \times \tau \rightarrow \sigma$, i.e. the function takes as an element of the domain, an ordered pair consisting of a knowledge source and a set of symbols, returning an output symbol.

Finally, both definitions can be merged to obtain the formal definition of the FSM of YASMIN. The following correspondences are applied:

- States $\rightarrow$ Knowledge sources.
- Input and output set $\rightarrow$ Symbolic set.
- State transition function $\rightarrow$ Scheduler function.
- Output function $\rightarrow$ Execute function.

A FSM of YASMIN is defined as a 4-tuple (Eq. 3) formed by a list of states as the conditions that the machine could take over time, a set of symbols corresponding to the alphabet values that are the intermediate solution of the FSM and two functions: one that determines the transition behavior of the states and the other, the processing of the input symbols translated into a sequence of output symbols.

$$YASMIN\_FSM = (\sigma, \tau, \Delta, \Omega) \tag{3}$$

- σ set of states (knowledge sources).
- τ is the blackboard, i.e., the set of symbols (input and output sets).
- Δ is a state transition function (scheduler function), such that: $\Delta : \sigma \times \tau \rightarrow \sigma$, i.e. the function takes as an element of the domain, an ordered pair consisting of a state and an input data, returning another state.
- Ω is the output function (execute function), $\Omega : \sigma \times \tau \rightarrow \delta$, i.e., the function takes as an element of the domain, an ordered pair consisting of a state and an input data, returning as an image, an output symbol.

4 Architecture

It is possible to say that the FSM could be applied in almost every layer of cognitive architecture. This section presents the implementation of YASMIN (Yet Another State MachINe) and describes all associated software components.

First, it is necessary to provide a high-level overview of where to apply State Machines. To this end, the MERLIN2 cognitive architecture model is proposed, which is an evolution of MERLIN [8]). It is a hybrid architecture consisting of two main components: Deliberative and Behavioral (Fig. 1 illustrate the MERLIN2 architecture).

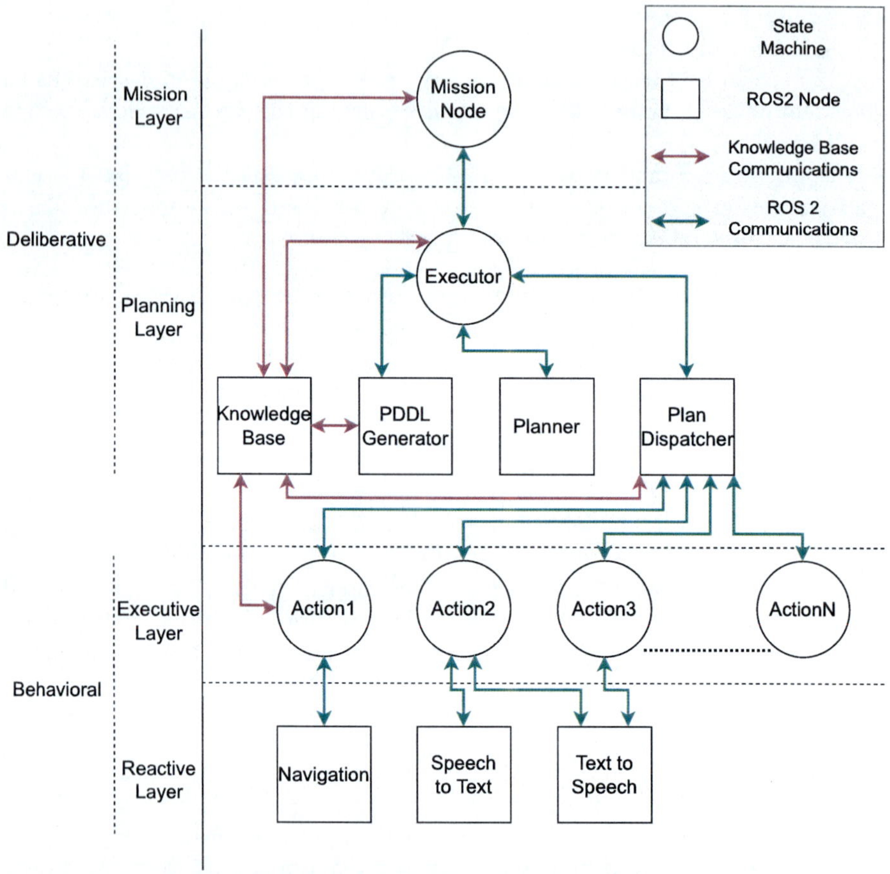

Fig. 1. MERLIN2 architecture.

As can be seen, these components are divided into layers:

- Mission Layer: this is the layer in charge of managing the robot's goals. Its function is to create goals that will be used by the Planning Layer to generate plans. This can be performed by means of a state machine whose states will produce the goals.
- Planning Layer: this layer, which is executed as a state machine at a high level, is in charge of generating and executing plans. To do so, the PDDL [6]

of the current state and domain is generated using KANT [9]. This PDDL is then used in the planner to generate the plan and the sequence of actions. Finally, the sequence of actions is executed by calling each action, which are the components of the Executive Layer.

- Executive Layer: is the layer composed of the actions used to produce plans. In order to improve the management of actions, they have been designed as state machines, whose states can process data or call the systems of the Reactive Layer.
- Reactive Layer: this is the layer formed by the reactive systems. Some reactive systems are navigation, object recognition, speech synthesis and object recognition. All these systems must have a ROS 2 communication interface in order to be used. Again, some of these components could be low-level managed by FSM (although in our approach we are avoiding it).

4.1 YASMIN Description

YASMIN is a library written in Python that allows to develop behaviors by creating state machines. Moreover, YASMIN can be used to implement the State Pattern. This behavioral design pattern allows an object to modify its behavior when its state changes. The YASMIN class diagram is presented in Fig. 2. These classes are organized into the following ROS 2 packages.

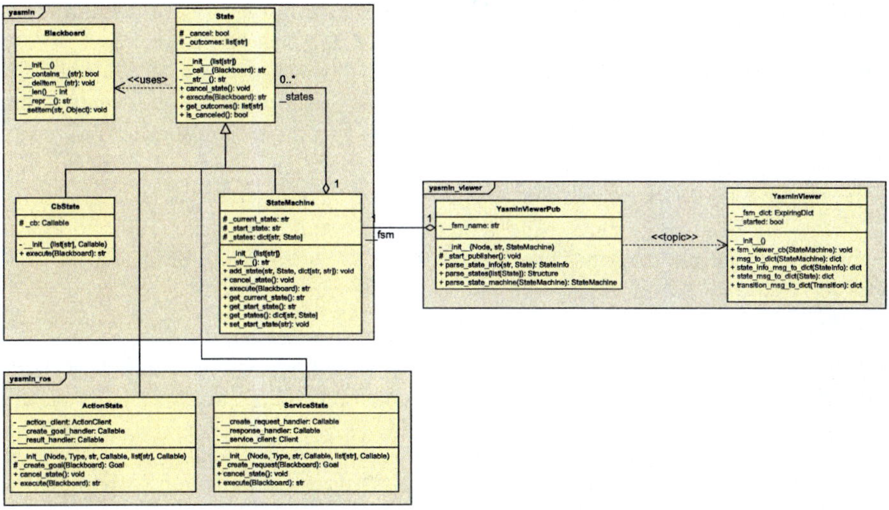

Fig. 2. UML class diagrama of YASMIN.

YASMIN Package. This package is the base of YASMIN. It consists of the following classes:

- **Blackboard**: this class is based on the Blackboard Pattern. This software design pattern includes the use of a global memory structure that contains data from different knowledge sources. Thus, YASMIN states are the knowledge sources that use the Blackboard to store and exchange data. On the other hand, the state machines will be the control components that decide which knowledge sources are executed.
- **State**: this class is the basis for developing new YASMIN states. Thus, to create a new state, it inherits from the State class and implements the execute method. In addition, you can redefine the cancel_state method that allows you to define the way in which the state will be canceled. Finally, it is necessary to define the state outputs (outcomes) that will be strings used later in the definition of state transitions.
- **CbState**: this class represents a type of state that receives a callback function. Thus, a new state can be created by defining a function instead of a new class.
- **StateMachine**: this class represents YASMIN state machines. Using the add_state function, new states can be added. This function receives an instance of State and a dictionary, which represents the transitions of that state. A transition consists of linking one of the state's outputs to another state or one of the state machine's final outputs. Finally, since StateMachine inherits from State, nested state machines can be added.

YASMIN_ROS Package. This package contains state classes to facilitate the integration of YASMIN with ROS 2. The following classes are available:

- **ActionState**: this class represents a state that executes an action client of ROS 2. To use it, just choose the type of the action and define the create_goal function, which instantiates the goal to be sent to the action server. This state is already prepared to cancel the action client if the state is interrupted.
- **Service State**: this class represents a state that executes a ROS 2 service client. To use it, just choose the type of service and define the create_request function, where the request to be sent to the service server is instantiated.

YASMIN_VIEWER Package. This ROS 2 package contains the code needed to view the state of YASMIN state machines. An example of the viewer is presented in Fig. 3. It has the following classes:

- **YasminViewerPub**: this class extracts the information from a state machine and publishes it in a ROS 2 topic.
- **YasminViewerNode**: this class reads the state machine information from the topic and presents it in the state machine viewer. This viewer is based on a web application formed by a backend, created in Python3 with Flask; and a frontend, created with React.js. To display the state machines, Cytoscape.js was used, which is a JavaScript library used to draw graphs. In this way, the state machines can be displayed from any computer using a web browser.

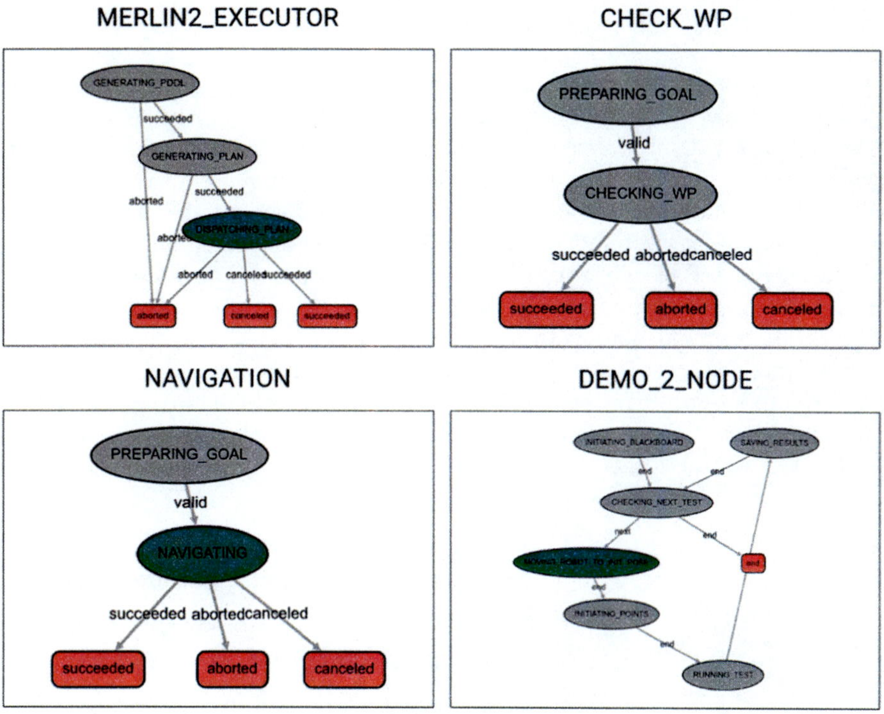

Fig. 3. Example of the YASMIN viewer.

YASMIN_INTERFACES Package. This section presents the ROS 2 interfaces (messages) that have been created in YASMIN. These messages have been used to publish information from the state machines. Thus, we have the following interfaces:

- **Transition.msg**: this message represents the transitions of the states. Each state has a name and, when executed, produces an output. Thus, this message contains two strings, one for the output name and one for the state name.
- **State.msg**: this message represents a state. It contains the state information: its name, a list of transitions (Transition.msg) and its outputs. Moreover, it contains a boolean to show if it is a state machine, which allows sending information about nested state machines. If it is a state machine it also contains the name of its current state. Finally, each state is assigned an ID and the ID of the parent.
- **StateMachine.msg**: this message represents a state machine. It contains a list of states (State.msg). The first element of the list identifies the own state machine. The following elements correspond to the states and nested state machines, referring to the first element as their parent state.

5 Evaluation

To evaluate our proposed solution, a simple task is posed. This task presents a robot that has to patrol some parts of a warehouse. For this purpose, the warehouse is divided into five points the robot has to explore in a specific order. Exploring each point implies rotating the robot 45°, moving forward, rotating 180° and returning to the point. This is repeated eight times to complete a circle.

The moving forward and rotation actions are implemented as a small state machine composed of two states: a state for computing the position to navigate and a state to perform navigation. This way, the exploration task is composed of this state machine along with other states.

The warehouse used is the AWS Robomaker Warehouse world [3]. Figure 4 shows the top view of this warehouse world and the five points to patrol with a table of its exact coordinates and the order of visit. A total of 20 executions were run for each solution due to the fact that robots, computers and the simulator are not perfect systems that can suffer noise.

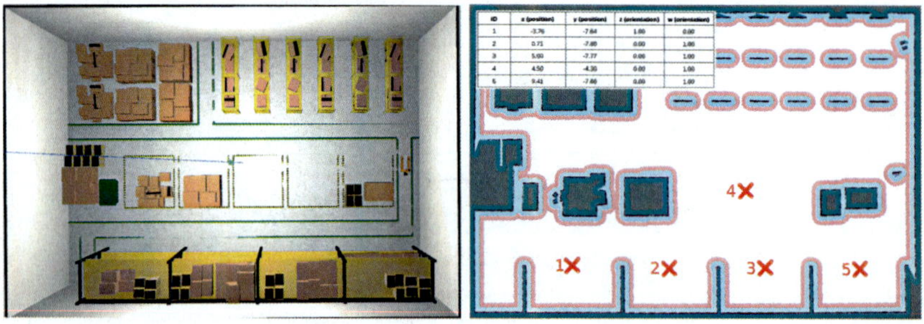

Fig. 4. Top view of the AWS Robomaker Warehouse.

The metrics employed for measuring each FSM, only the performance of the application that runs the FSM, are the following:

- **RAM usage**: this metric allows to measure the impact of each solution in the RAM. As robots are embedded systems, it is important to optimize the use of their resources (CPU, RAM, GPU, etc.).
- **Duration**: this metric allows to measure the time required to reach the five points of each experiment. However, it is important to remark that it is based not only on the time taken to execute states and transition between states, but also depends on the navigation stack used by the robots, i.e. Nav2 [11].
- **Energy consumed**: this metric allows to measure the impact of each solution on the robot's battery. Again, the battery is one of the resources of the autonomous mobile robot that is important to take into account.

5.1 Results and Discussion

A total of 20 executions were run for each solution using a regular desktop computer composed of an Intel(R) Core(TM) i7-6700 CPU @ 3.40GHz, GeForce GTX 1060 6GB and 16 GB DDR4 RAM. There are 10 executions where RAM usage is measured and 10 executions where energy consumption is measured. Both measurements have been performed independently because energy consumption measurement produces an increase in RAM usage.

A set of Python libraries was used for profiling the FSM performance in this experiment: a) *psutil package* [1], is a cross-platform python library for retrieving information and profiling running processes and system utilization in Python. b) *CodeCarbon* application [12]: it is a software package that allows python developers to estimates the amount of carbon dioxide (CO_2) produced by the system used to execute the code.

This task was completed using YASMIN and SMACH under ROS Galactic Version in Ubuntu 20.04. In both cases, the source code employed would be considered similar in terms of the following numbers obtained with Visual Studio Code:

- SMACH: 12 Python files, 1 CMake, 1 XML, 1 YAML. This version presents overall 575 lines: 417 codes, 15 comments, 143 blank lines.
- YASMIN: 12 Python files, 1 CMake, 1 XML, 1 YAML. This version presents overall 565 lines: 407 codes, 14, comments and 144 blank lines.

Table 1 reveals the descriptive statistics of this experiment. These results expose that RAM consumption has similar behavior between the two options. However, the Std. Deviation presents lower values in YASMIN and the range is 1 MB smaller, which means that the decision-making system of YASMIN is more stable. Furthermore, the duration for the same experiment presents a 3.46% reduction in time, which would be considered minor, but represents a reduction of 16 s. Finally, energy consumed presents similar values.

Table 1. Descriptive Statistics of experimental experience

	RAM (Mb)		duration (s)		energy_consumed (Kw)	
	SMACH	YASMIN	SMACH	YASMIN	SMACH	YASMIN
Valid	4653	4602	10	10	10	10
Mean	94.776	93.648	473.718	457.301	0.011	0.010
Std. Deviation	0.983	0.548	30.020	19.794	0.001	$8.073e^{-4}$
Range	5.031	3.941	98.418	58.717	0.004	0.002
Minimum	91.051	91.875	438.435	440.768	0.009	0.009
Maximum	96.082	95.816	536.854	499.485	0.013	0.011
25th percentile	94.480	93.309	455.042	443.652	0.010	0.009
50th percentile	95.117	93.566	462.082	449.634	0.010	0.010
75th percentile	95.426	93.906	489.390	468.610	0.011	0.010

Figure 5 illustrates the statistics of both, SMACH and YASMIN behaviors during the experiment. These plots support the notion that the performance of

YASMIN attending to duration is slightly better. Besides, the interquartile range is reduced compared to SMACH, which means that it appears more stable in this experimental approach, suffers from some limitations due to a single experiment and would be extended to other tasks and used by external researchers. This is made possible by accessing the YASMIN repository [13].

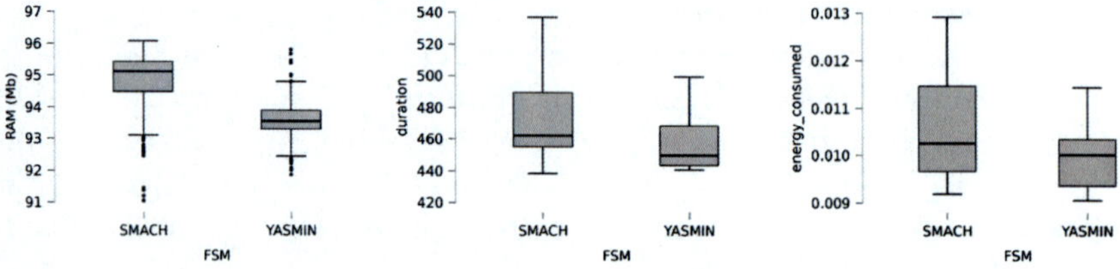

Fig. 5. BoxPlot illustrating RAM consumption, Duration (s) and energy consumed (Kw)

6 Conclusions

YASMIN has been designed and created to update SMACH. Thanks to YASMIN it is possible to develop behaviors by creating state machines. In addition, it is possible to monitor the operation of the different state machines that are running thanks to its viewer. The code can be found in a GitHub repository[1].

The use of YASMIN for the design of robot actions compared to generic State Machine Solutions has several advantages: Fully integrated into ROS 2, it allows the generation of state machines using Python and C++, easily shared information among all the states, and nested state machines thanks to the use of blackboards.

Further work overviews developing performance comparison with the state-of-art behavior trees approach, such as Behavior Tree in ROS 2 and the visualizer and inspection tool Groot [7].

Acknowledgment. This work has been partially funded by the Ministerio de Ciencia e Innovación and the Agencia Estatal de Investigación of the Kingdom of Spain under grant PID2021-126592OB-C21.

References

1. Psutil. https://github.com/giampaolo/psutil
2. Aldrich, B.: State machine asynchronous c++. https://smacc.dev/
3. Aws-Robotics: Aws-robotics/aws-robomaker-small-warehouse-world at foxy-devel. https://github.com/aws-robotics/aws-robomaker-small-warehouse-world/tree/foxy-devel

[1] https://github.com/uleroboticsgroup/yasmin.

4. Bohren, J., Cousins, S.: The SMACH high-level executive [ROS news]. IEEE Robot. Autom. Mag. **17**(4), 18–20 (2010)
5. Colledanchise, M., Natale, L.: On the implementation of behavior trees in robotics. IEEE Robot. Autom. Lett. **6**(3), 5929–5936 (2021)
6. Fox, M., Long, D.: Pddl2.1: an extension to PDDL for expressing temporal planning domains. J. Artif. Intell. Res. (JAIR) **20**, 61–124 (2003)
7. Ghzouli, R., Berger, T., Johnsen, E.B., Dragule, S., Wąsowski, A.: Behavior trees in action: a study of robotics applications. In: Proceedings of the 13th ACM SIGPLAN International Conference on Software Language Engineering, pp. 196–209 (2020)
8. González-Santamarta, M.Á., Rodríguez-Lera, F.J., Álvarez-Aparicio, C., Guerrero-Higueras, Á.M., Fernández-Llamas, C.: MERLIN a cognitive architecture for service robots. Appl. Sci. **10**(17), 5989 (2020). https://doi.org/10.3390/app10175989
9. González-Santamarta, M.Á., Rodríguez-Lera, F.J., Martín, F., Fernández, C., Matellán, V.: Kant: a tool for grounding and knowledge management. In: Ferrández Vicente, J.M., Álvarez-Sánchez, J.R., de la Paz López, F., Adeli, H. (eds.) Bio-inspired Systems and Applications: from Robotics to Ambient Intelligence, vol. 13259, pp. 452–461. Springer, Cham (2022). https://doi.org/10.1007/978-3-031-06527-9_45
10. Iovino, M., Scukins, E., Styrud, J., Ögren, P., Smith, C.: A survey of behavior trees in robotics and AI. Robot. Auton. Syst. **154**, 104096 (2022)
11. Macenski, S., Martin, F., White, R., Ginés Clavero, J.: The marathon 2: a navigation system. In: 2020 IEEE/RSJ International Conference on Intelligent Robots and Systems (IROS) (2020)
12. Schmidt, V., et al.: CodeCarbon: estimate and track carbon emissions from machine learning computing. https://github.com/mlco2/codecarbon, May 2021
13. Uleroboticsgroup: Yasmin (yet another state machine) (2021). https://github.com/uleroboticsgroup/yasmin

Advanced Waypoint Navigation Strategies for Autonomous Mobile Robots in Industrial Settings

María-Teresa Lorente(✉), Lorenzo Montano, and María Teresa Lázaro

Instituto Tecnológico de Aragón (ITAINNOVA), Zaragoza, Spain
{mtlorente,lmontano,mtlazaro}@itainnova.es

Abstract. This work addresses some limitations encountered when developing autonomous robotic systems for resource-constrained industrial applications in Robotic Operating System (ROS). A new version of the map used for local navigation is proposed to augment the visibility horizon of the robot. During robot navigation, new obstacles appearing in the scenario may lead the robot to deadlock situations, getting the robot stuck trying to reach an unreachable waypoint. Several strategies are implemented to deal with these situations and achieve a robust navigation. The solutions proposed are validated in simulation and real experimentation.

Keywords: Robust waypoint navigation · ROS Navigation Stack · Dynamic environments · Deadlock situations

1 Introduction

Advances in the development of autonomous robotic systems have led to a growing interest from companies in the deployment of such systems in industrial environments. These environments are characterized by being very dynamic, where the different elements (pallets, boxes, forklifts, etc.) can change location or new elements can appear in places that were previously empty. In addition, robotic systems in these industrial environments have to share the space with its human co-workers, which introduces two main variables to be taken into account in the robotic system: unpredictability and safety. This is why autonomous, reactive and safe navigation is one of the main technological challenges in these environments.

So far there has been a significant deployment of Automated Guided Vehicles (AGVs) in the industry. Navigation of these robotic systems rely mainly on guiding devices such as magnetic tape, beacons or barcodes to follow predefined and restricted paths. Laser scanners installed on these robots are usually intended to meet safety standards, so that they stop the robot in case an obstacle interferes with its path which remains blocked until the obstacle has been removed. On very rare occasions these robots are able to deviate from their established route. This lack of flexibility and the need to adapt the infrastructure has led to an

D. Tardioli et al. (Eds.): ROBOT 2022, LNNS 590, pp. 540–551, 2023.
https://doi.org/10.1007/978-3-031-21062-4_44

increasing number of companies demanding the implementation of Autonomous Mobile Robot (AMR) systems based on SLAM.

The emergence of ROS[1] in 2007 –and its open source nature– has certainly been one of the significant factors in the rise of development and deployment of AMRs, becoming a de-facto standard programming framework for industry. Its well-known ROS Navigation Stack[2] has been designed to be as general purpose as possible, allowing its application to a wide variety of scenarios and robot typologies. However, such a generic system can present difficulties when applied to some specific problems and may need to be adapted to concrete requirements.

In this paper we analyze a number of limitations encountered when using the ROS Navigation standard in the context of a resource-constrained industrial AMR that must execute predefined routes in a dynamic environment and propose a set of navigation strategies intended to deal with such limitations. The main issues observed and addressed in this paper are the following:

- High CPU and memory consumption which limit the visibility horizon of the local navigation map used for reactive navigation.
- Poor feedback about reachability conditions or success in computing a valid plan which is only given once the goal has been aborted after attempting all recovery behaviours.

The rest of this paper continues as follows: Sect. 2 discusses some works and implementations related to this paper, Sect. 3 describes the framework in which our work is developed, Sects. 4 and 5 present our contributions to improve navigation robustness in dynamic scenarios, Sect. 6 presents our experimental results and, finally, we draw conclusions in Sect. 7.

2 Related Work

Despite the great contribution that ROS and its Navigation Stack have made to the robotics community, its widespread deployment has brought to light its major weaknesses, which are well-known to robotics developers. Based on well-founded research works and algorithm implementations (Occupancy grids [9], A* [5] or Dynamic Window Approach (DWA) [4]), `move_base` –the main ROS package of the Navigation Stack– presents problems of scalability and high CPU consumption. For this reason, different alternative implementations have been proposed in the recent years to overcome this issue. In [6], a lightweight navigation system which achieves similar performance than `move_base` is proposed. In [10], authors propose `move_base_flex`, a backwards-compatible replacement for `move_base` which decouples path planning, motion control and recovery behaviors execution.

Main implementations in the Navigation Stack consider holonomic and non-holonomic robot types with differential drive system. However, there are not too many implementations compatible with the ROS Navigation Stack that contemplate other robots with constrained movements such as Ackermann steering based

[1] https://www.ros.org/.
[2] http://wiki.ros.org/navigation.

robots. The Search-Based Lattice Planner (SBPL) [1] provides a global planner implementation to generate smooth and kinematically feasible paths from the definition of motion primitives. Besides, the Timed Elastic Band (TEB) controller [11] provides time-optimal trajectories that comply with kinodynamic constraints such as maximum velocities and accelerations including those of car-like robots.

With an end-of-life date of May 2025 for the latest ROS distribution –*Noetic,*– the main developers of the Navigation Stack are currently in a process of adapting the current implementations to ROS2 [7], trying to apply a number of lessons learned during this last decade of ROS developments. In this new framework, `Nav2` [8], the new Navigation Stack for ROS2, uses a Behavior Tree (BT) [2] navigator server to coordinate control, planning and recovery tasks and aims at providing secure and reliable navigation with real-time capabilities.

3 Problem Statement

We consider an Ackermann robot R moving in a dynamic environment, where it has to complete a mission. The mission is defined by a set of waypoints previously taught to the robot. Let $wp_i, i = 1..n$ denote the waypoints composing the mission. The robot should move in workspace $WS = \mathbb{R}^2$ following the sequence of waypoints defined in the mission without stopping.

The state of the robot is defined by its location $(x, y, \theta)_t$, its linear velocity v_t and its steering angle ϕ_t, at instant t, $R_t = (x, y, \theta, v, \phi)_t$. In our case, the controls are the linear velocity and steering angle for an Ackermann robot. The motion model for the robot can be expressed by the following equations:

$$\dot{x} = v * cos(\theta); \quad \dot{y} = v * sin(\theta); \quad \dot{\theta} = \frac{v}{L} * \tan(\phi) \tag{1}$$

where L denotes the wheelbase of the robot.

Setup. The ROS Navigation Stack defines `move_base` as its core, which provides an standard interface to implement planning algorithms and their integration into the Navigation Stack. The global planner used to compute a path between the current robot location and the current waypoint of the mission is the algorithm Theta* [3], implemented based on the ROS GlobalPlanner[3] which implements the A* algorithm. The local planner used is TEB. Additionally, a waypoint follower algorithm is implemented to provide the waypoints of the mission in sequence.

Figure 1 shows the map of a simulated scenario and the waypoints of a mission to be accomplished by the robot in RViz[4].

4 Visibility Horizon in the Local Costmap

Deploying an autonomous navigation system requires modelling data from the environment around the robot in a data structure so that a trajectory free of

[3] http://wiki.ros.org/global_planner.
[4] http://wiki.ros.org/rviz.

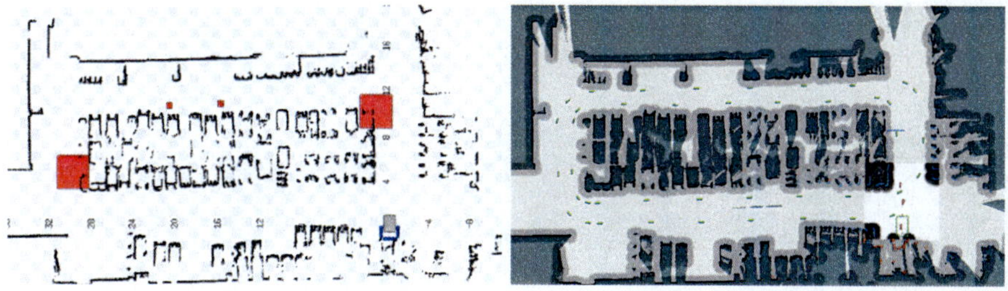

Fig. 1. Stage ROS simulation environment (left) and RViz visualization of the scenario (right), with the arrows representing the waypoints of the mission.

collision can be computed for the robot. These data are perceived by the perception system on board the robot. In ROS, the data structure used is named *costmap*, an occupancy grid with cells containing the cost of traversing a cell according to the presence of obstacles. The *local* costmap is used for reactive robot navigation. It is centered with respect to the current position of the robot, and it updates as the robot moves including data gathered from the sensors in real time. The size of the local costmap establishes the visibility horizon of the robot in order to consider obstacles around, and thus, the capability to react to obstacles and avoid any possible future collision. The size influences the computational cost required to map the information from the sensors into the costmap and to compute a motion for the robot.

Figure 2 shows part of the environment plotted in Fig. 1. The image on the left pictures the local costmap as a white square area centered in the current position of the robot. In applications where the robot should move mainly forwards, having the local costmap centered in the position of the robot is a limitation. Given that the space reserved for the back side of the robot results underused, the robot could benefit from having it available for the front side without increasing the total size of the costmap, and thus, maintaining the computational cost restrictions. Addressing this limitation represents the first contribution of the work: to increase the visibility horizon of the robot in the local costmap.

A new version of the local costmap is developed to adapt the visibility horizon of the robot. A parameter *offset* is defined to represent the displacement to be performed in the costmap. It defines the forward increasing amount (in same units as the costmap) of the costmap with respect to the current position of the robot, obtaining a costmap that rolls as the robot moves and providing the robot with greater visibility to anticipate any possible future collision with any obstacle and compute a smooth trajectory to avoid it. Therefore, the new origin of the costmap updates by following these equations:

$$O_x = x_t - sizeof(costmap_x)/2 + offset * \cos(\theta_t) \tag{2}$$

$$O_y = y_t - sizeof(costmap_y)/2 + offset * \sin(\theta_t) \tag{3}$$

where $sizeof(costmap_x)$ and $sizeof(costmap_y)$ represent the size of the costmap in X and Y axis, respectively; and (O_x, O_y) are the new origin coordinates

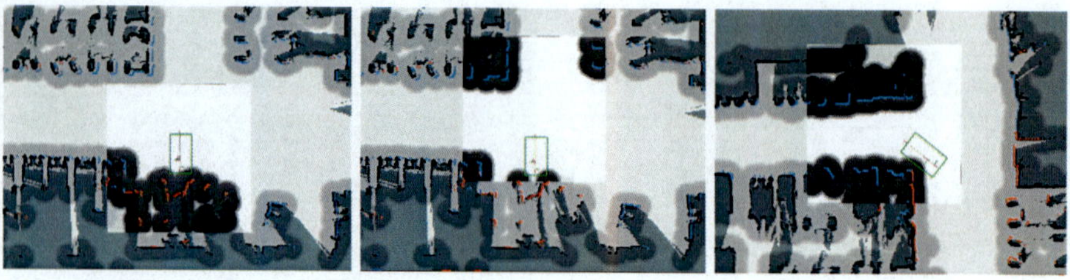

Fig. 2. Local costmap centered in the current position of the robot (left). Defined as a grid of 7 × 7 m and given the size of the AMR, it leaves barely 2 m of visibility horizon for dynamic obstacle avoidance. Local costmap with increased horizon visibility in front of the robot (center and right).

of the costmap with increased *offset* visibility. The center and right images of Fig. 2 show the resultant local costmap of the robot after increasing the visibility horizon. Notice how the space in front of the robot has augmented without increasing the total size of the costmap, creating the effect of the robot pivoting with respect to the center of the costmap depending on its orientation.

Note that the robot also moves backwards, though with very controlled manoeuvres. Also, the value of the parameter offset can be adapted depending on the type of movement the robot desires to perform.

5 Deadlock Situations

In situations where the robot should accomplish a mission while executing a map-based navigation strategy, some new static objects appearing in the scenario that were not present at the time of building the map of the environment can lead to a deadlock situation. The obstacles occupy the space where a waypoint was defined or interfere the path towards it, thus preventing the robot from reaching the waypoint. From our experience during real experimentation, this type of situations represent a limitation when deploying an autonomous navigation system in ROS because the robot finds itself in a deadlock. Despite `move_base` providing recovery behaviours to avoid some deadlock situations, these are ineffective for the situations presented later and for the type of robot movement under consideration (e.g., our robot is unable to perform turnings on itself). In an attempt to replicate these deadlock situations, the red static obstacles in Fig. 1 are placed in purpose after the map construction. An analysis of the data perceived by the robot from this environment leads to identify these situations, in which case, the next waypoint in the mission is sent to the robot.

Figure 3 shows the deadlock situations identified. The first situation is illustrated on the left of Fig. 3, where a waypoint is occupied by a new obstacle in the scenario. In the center of Fig. 3, the second situation is depicted. The robot tries to reach a waypoint very near of an obstacle. However, reaching the waypoint would mean colliding with the obstacle. Finally, on the right of Fig. 3, the last situation is plotted. An obstacle is blocking the path towards the waypoint, causing

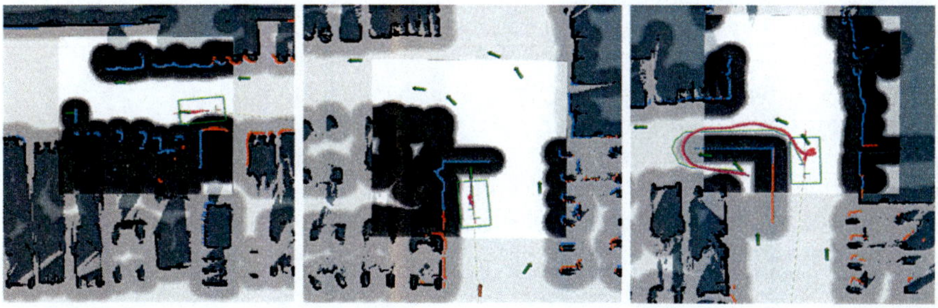

Fig. 3. Deadlock situations caused by new appearing obstacles in the scenario.

the planner to recompute a plan where the robot should manoeuvre and, later, move backwards to reach the waypoint. In scenarios where the robot motion during the mission is forwards, this type of manoeuvres indicates the waypoint is unreachable. In our experience, `move_base` fails to detect these situations and give proper feedback. Therefore, a solution based on reachability check strategies is proposed to evaluate whether a waypoint is reachable. It represents the second contribution of the work.

The implemented strategies are depicted in Algorithms 1–3. A waypoint is reachable if: (i) it does not lie inside an obstacle and (ii) the robot footprint, defined as a polygon, transformed into the waypoint location does not collide with any obstacle and (iii) the global plan transformed with respect to the current waypoint lies behind it. The first one, illustrated in Algorithm 1, checks if the position of the waypoint lies inside an obstacle. The function declared in line 4 obtains the proper costmap depending on the waypoint coordinates in the map; if the waypoint lies outside the local costmap bounds, the global costmap is read to check this condition. Then, the cost of the cell where the waypoint is in the costmap is obtained in line 5 to determine if the waypoint is free of collision. The second strategy verifies if the robot footprint collides with any obstacle at the waypoint location, as shown in Algorithm 2. The local costmap obtained in line 4 contains current data about obstacles around the robot. The robot footprint is transformed with respect to the waypoint and the resultant coordinates of its vertices are converted into map coordinates. Then, the set of cells in the costmap that lie inside this resultant polygon are obtained to check if any of it is occupied by an obstacle (lines 5–15). Finally, the last strategy is shown in Algorithm 3. First, it considers if the angular difference between the current (wp_i) and previous (wp_{i-1}) waypoints is large, to avoid skipping a waypoint placed in opposite direction (lines 5–7). If not, the transform of wp_{i-1} with respect to wp_i is computed to determine if the wp_{i-1} is behind or ahead of wp_i (lines 8–9), and thus, the points in the path transformed with respect to the wp_i should also be behind or ahead of wp_i, respectively, following the desired forward/backward direction of movement (lines 10–15). Otherwise, the robot would be following a deformed and twisted path, which would lead to a non-desirable behaviour in the type of applications under consideration. These strategies are applied according to some conditions such as the enduring of

Algorithm 1. Check if a waypoint lies in an occupied cell in the costmap

```
Require:
wp_i, the current waypoint to be reached in the mission
function IsWpOnObstacle(wp_i)
    mwp = worldToMap(wp_i)        ▷ Convert wp_i coordinates into map coordinates
    costmap = getCostmap(mwp)    ▷ Get the local/global costmap the wp_i lies in
    if getCost(mwp, costmap) != FREESPACE then
        return true
    end if
    return false
end function
```

Algorithm 2. Check if the robot footprint collides with an obstacle at waypoint location

```
Require:
wp_i, the current waypoint to be reached in the mission
fp, the robot footprint
function IsFootprintInCollision(wp_i, fp)
    costmap = getLocalCostmap()                         ▷ Get the local costmap
    fpTf = transformFootprint(fp, wp_i)   ▷ Transform the footprint w.r.t. the wp_i
    for j = 0 to sizeof(fpTf) − 1 do
        mp = worldToMap(fpTf[j])                    ▷ Convert to map coordinates
        polygon.push_back(mp)   ▷ Construct a polygon with the fp in map coord.
    end for
    pCells = convexFillCells(polygon)   ▷ Get the map cells that fill a convex pol.
    for j = 0 to sizeof(pCells) − 1 do
        if getCost(pCells[j], costmap) == LETHALOBSTACLE then
            return true
        end if
    end for
    return false
end function
```

the situation and the distance to the waypoint, in order to consider the case where the new obstacles appearing in the scenario are moving.

6 Experiments

This section presents both simulated and real experiments to show the advantages of the different navigation strategies proposed in this paper. We have tested our methods on an Ackermann drive scrubber machine equipped with two front and rear 2D-LiDAR SICK scanners. Besides, it is provided with two encoders placed on the rear wheels in order to compute the odometry. On a setup step, an operator records the map of the environment while driving manually the machine, and later he/she teaches the trajectory that the machine must execute for its cleaning mission. Then, this trajectory is transformed into a sequence of waypoints that the machine has to follow. Simulated and real experiments

Algorithm 3. Check if the global plan lies behind or ahead of the waypoint

Require:
wp_i, the current waypoint to be reached in the mission
wp_{i-1}, the previous waypoint reached by the robot
path, the global path computed by the global planner
function IsGlobalPlanBehindWp($wp_i, wp_{i-1}, path$)
  if abs(wp_{i-1}.getYaw() - wp_i.getYaw()) $> \pi/2$ **then**
    return true
  end if
  $pwpTf$ = transformPoint(wp_{i-1}, wp_i) ▷ Transform wp_{i-1} w.r.t. wp_i
  $pwpAhead$ = ($pwpTf$.GetX() > 0) ▷ wp_{i-1} is placed ahead of wp_i
  for $j = 0$ to $sizeof(path) - 1$ **do** ▷ Transform the path w.r.t. wp_i
    pTf = transformPoint($path[j], wp_i$)
    if (pTf.getX() > 0 & !$pwpAhead$) || (pTf.getX() < 0 & $pwpAhead$) **then**
      return false
    end if
  end for
  return true
end function

are performed on a PC equipped with Intel(R) Core(TM) i7-1165G7 CPU@2.80 GHz and 16GB RAM running Ubuntu 20.04 and ROS Noetic. Our navigation setup is the one described in Sect. 3.

6.1 Simulation Environment

For the following set of experiments, a simulated scenario has been created in the Stage ROS simulator based on a 2D map captured in a real warehouse, shown in Fig. 1, where we have set up obstacles along the path of a cleaning mission in order to reproduce the different deadlock situations encountered when using the standard ROS Navigation Stack addressed in this paper.

Throughout these experiments, we want to show the improvement in robustness achieved with our approach with respect to the ROS Navigation Stack and compare the performance of both methods in terms of mission completion time, computational cost and the difference between the trajectories computed by both methods. In these experiments, we have tested the behaviour of the robot in two scenarios. First, we have evaluated the performance of the navigation algorithms without using reachability check strategies and, second, using them. In both cases, we also consider three different costmap setups: 7×7 m size with 0m offset, named *setup1*; 12×12 m size with 0 m offset, named *setup2*, in order to cover the frontal distance covered by a 7×7 m costmap with 2.5 m offset; and finally, our method which consists of a 7×7 m size with 2.5 m offset, named *setup3*. The two first of these setups are equivalent to using the costmap standard version of ROS Navigation Stack.

In the first scenario, no reachability strategies are used, so the mission fails regardless the costmap setup. The scrubber finds a waypoint near an obstacle at

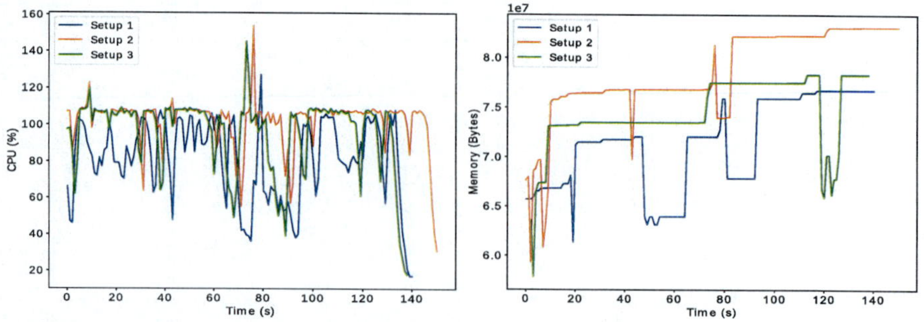

Fig. 4. CPU consumption (left), where the mean of each one are: Setup1 80.96%, Setup2 101.76%, Setup3 95.04%. Memory consumption of the `move_base` process (right) expressed in bytes, in which the tasks related to the navigation are computed.

the beginning of the mission, which leads to a deadlock situation (Fig. 3 center). As reachability check strategies are not applied during this experiment, the trajectory planner is stuck computing a non-feasible trajectory. Even though the scrubber footprint would collide with the obstacle, the plan still remains valid. By applying the reachability check strategies to deal with these deadlock situations, the mission succeeds with every costmap setup. Therefore, from now on, we present evaluation data for this second scenario with the reachability check strategies.

We have analyzed the CPU and Memory consumption of each costmap configuration. In Fig. 4, it is shown that the most favorable case corresponds to the *setup1*, where the frontal visibility horizon range provided by the costmap is 2m. If this distance is needed to be enlarged, the size of the costmap must be increased considerably. However, as shown for *setup2*, this would lead to high CPU and Memory consumption. Therefore, in order to increase the frontal visibility horizon range without increasing the size of the costmap, we have tested the case defined in *setup3*. The CPU and Memory consumption are reduced with respect to *setup2*, whereas they are slightly higher with respect to *setup1*. This might be caused by the local planner requiring to compute a longer motion trajectory due to the increasing visibility horizon introduced by the offset value.

From now on, we will just analyze the behaviour of the scrubber with *setup1* and *setup3* of the costmap. This is due to, in terms of trajectory planning, the behaviour of the *setup2* closely resembles that of *setup3* since the frontal visibility horizon range is the same. Figure 5 shows the angular and linear velocity performed by the scrubber during the mission. Between the seconds 30 and 45, a considerable slow down is observed by using the *setup1*. As the frontal visibility horizon range is smaller than the one used in *setup3* the capability of reactivity is reduced and it leads to sharper movements. On the flip side, the *setup3* allows the scrubber to anticipate the obstacles earlier and thus perform smoother movements. Therefore, the mission completion time is also reduced by using the *setup3* as the smooth movements allow the maximum linear velocity to be maintained for a longer time.

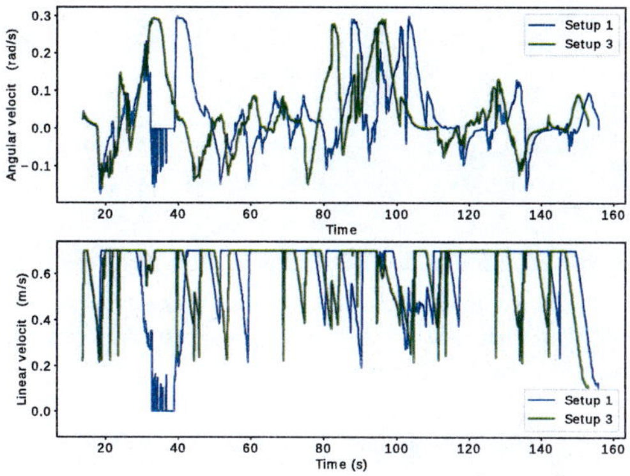

Fig. 5. Angular and linear velocity measured during the mission.

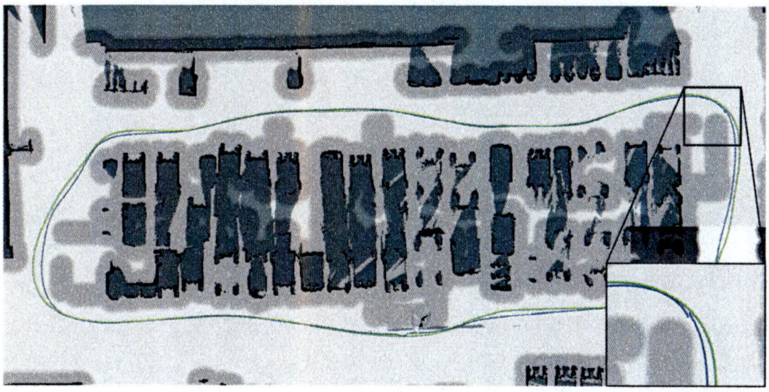

Fig. 6. The blue line represents the trajectory obtained with the *setup1* and the green line is the trajectory obtained with the *setup3*.

Finally, the trajectories performed by using the *setup1* and *setup3* are shown in Fig. 6. The *setup3* allows the scrubber to trace safer trajectories away from obstacles as the visibility horizon range is larger than the one used in *setup1*. The bottom right corner of Fig. 6 zooms in on the part of the trajectory where the scrubber needs to slow down (Fig. 5, seconds 30–45) by using the *setup1*, due to a critical approach to an obstacle. By using the *setup3*, this dangerous situation is avoided to a large extent.

6.2 Warehouse Environment

As mentioned before, we have performed real experiments with an autonomous scrubber machine in a warehouse environment at SERMA S.A. (Fig. 7). As can be seen, this environment is very dynamic, with vans entering and leaving the warehouse throughout the day and a significant amount of forklift traffic, which complicates the normal operation of the scrubbing machine. This type of environment is a challenge to achieve autonomous navigation based on the standard ROS Navigation Stack.

Fig. 7. Warehouse scenario at SERMA S.A. (left, center) and autonomous scrubber machine (right) used in the real experiments.

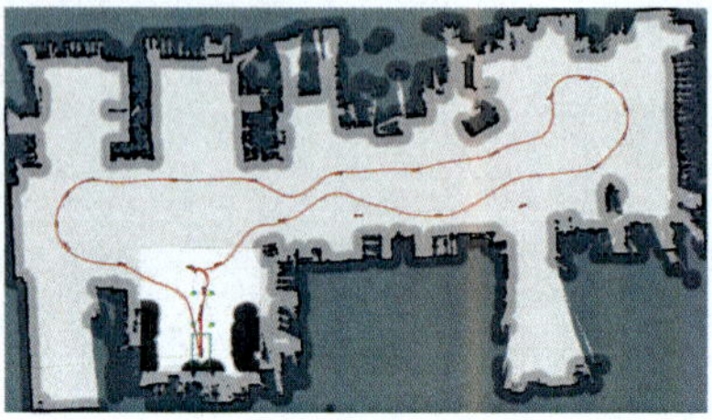

Fig. 8. Trajectory followed by the scrubber.

Figure 8 shows the execution of a cleaning mission in this environment using the new costmap version and the reachability check strategies. It can be shown that the scrubber accomplishes its mission. During navigation, it avoids some obstacles appearing in the scenario without influencing the waypoint following, thanks to the increased visibility horizon. Two waypoints were unable to be reached due to the appearance of two deadlock situations, properly identified with our algorithms. Finally, the scrubber finishes the whole mission without getting stuck, achieving a robust autonomous navigation. Figure 9 illustrates a sequence of instants of a deadlock situation during navigation. The waypoint is located near an obstacle, the scrubber cannot reach the waypoint because its footprint does not fit in the

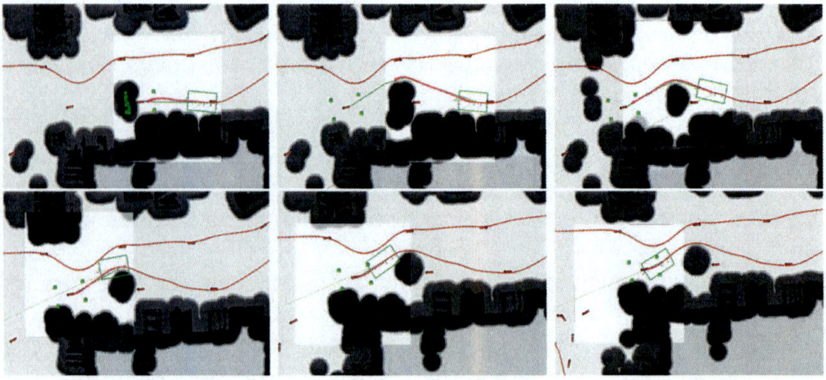

Fig. 9. Sequence of a deadlock situation during execution of a mission.

space between the waypoint and the obstacle. Therefore, the next waypoint of the mission is sent to the scrubber, which is successfully reached.

7 Conclusions

In this paper we have presented a series of strategies aimed at improving the robustness of navigation of AMRs in dynamic environments. Based on our experience in the development of navigation systems in logistics environments, we have identified situations in which the standard ROS Navigation Stack fails and proposed intuitive and effective techniques to deal with those situations.

On future work, we will adapt our implementation for compatibility with `Nav2` for ROS2 and integrate 3D perception on our system. We will also explore techniques for map updating and long-term navigation for dynamic environments in multi-session settings.

Acknowledgments. This work has been supported by the Spanish Project *"Máquina Autónoma Robotizada para Limpieza Industrial"* (IDI-20220063).

References

1. Cohen, B., Chitta, S., Likhachev, M.: Search-based planning for manipulation with motion primitives, pp. 2902 – 2908, June 2010
2. Colledanchise, M., Ogren, P.: Behavior trees in robotics and AI: an introduction, July 2018
3. Daniel, K., Nash, A., Koenig, S., Felner, A.: Theta*: any-angle path planning on grids. J. Artif. Intell. Res. **39**, 533–579 (2010)
4. Fox, D., Burgard, W., Thrun, S.: The dynamic window approach to collision avoidance. IEEE Robot. Autom. Mag. **4**(1), 23–33 (1997)
5. Konolige, K.: A gradient method for realtime robot control. In: Proceedings. 2000 IEEE/RSJ International Conference on Intelligent Robots and Systems (IROS 2000) (Cat. No.00CH37113), vol. 1, 639–646 (2000)
6. Lázaro, M., Grisetti, G., Iocchi, L., Fentanes, J.P., Hanheide, M.: A lightweight navigation system for mobile robots. In: ROBOT 2017: Third Iberian Robotics Conference. Sevilla, Spain, 22–24 November 2017
7. Macenski, S., Foote, T., Gerkey, B., Lalancette, C., Woodall, W.: Robot operating system 2: design, architecture, and uses in the wild. Sci. Robot. **7**(66) (2022)
8. Macenski, S., Martin, F., White, R., Ginés Clavero, J.: The marathon 2: a navigation system. In: 2020 IEEE/RSJ International Conference on Intelligent Robots and Systems (IROS) (2020)
9. Moravec, H., Elfes, A.: High resolution maps from wide angle sonar. In: IEEE International Conference on Robotics and Automation, vol. 2, pp. 116–121 (1985)
10. Pütz, S., Simón, J.S., Hertzberg, J.: Move base flex: a highly flexible navigation framework for mobile robots. In: 2018 IEEE/RSJ International Conference on Intelligent Robots and Systems (IROS), October 2018
11. Rösmann, C., Hoffmann, F., Bertram, T.: Kinodynamic trajectory optimization and control for car-like robots. In: 2017 IEEE/RSJ International Conference on Intelligent Robots and Systems (IROS), pp. 5681–5686 (2017)

Towards LiDAR and RADAR Fusion for Object Detection and Multi-object Tracking in CARLA Simulator

Santiago Montiel-Marín(✉), Carlos Gómez-Huélamo, Javier de la Peña, Miguel Antunes, Elena López-Guillén, and Luis M. Bergasa

Electronics Department, Universidad de Alcalá, Alcalá de Henares, Spain
{santiago.montiel,j.pena,miguel.antunes}@edu.uah.es,
{carlos.gomezh,elena.lopezg,luism.bergasa}@uah.es

Abstract. Detection and Multi-Object Tracking (DAMOT) systems have a critical role to play in scene understanding in the context of autonomous driving. Modern Autonomous Driving Stacks (ADS) require a software processing unit or module that allows them to understand the data in the environment and convert it into vital information for further decision making. In this context, this work develops a DAMOT module based on Machine Learning techniques, such as DBSCAN or BEV-SORT, that receives information from LiDAR and RADAR sensors in CARLA Simulator. This module uses containerisation techniques with Docker and standard robotics communications with ROS. The performance of the method is evaluated in terms of detection in the AD PerDevKit dataset, developed by the authors.

Keywords: LiDAR · RADAR · Object detection · Multi-object tracking · CARLA Simulator

1 Introduction

In recent years, academia and industry have both shown great interest in the development of autonomous vehicles (AV) and self-driving cars as they are seen as the main characters of a booming sector that marks the next transportation revolution. An AV is an intelligent transportation system capable of performing the functions related to the five domains of an AV architecture: perception, localisation, mapping-planning, decision making and control.

Faced with this situation, two types of architectures for solving this problem appear: modular [1] or end-to-end [2]. Moreover, the diversity of these architectures lies not only in the way they deal with the available data, but also in the sources of information they bring into play, the sensors. Typically, LiDAR, RADAR and monocular or stereo cameras are used to capture the instantaneous state of the environment surrounding the ego-vehicle and are accompanied by others such as odometry, Inertial measurement units (IMU), Differential Global

D. Tardioli et al. (Eds.): ROBOT 2022, LNNS 590, pp. 552–563, 2023.
https://doi.org/10.1007/978-3-031-21062-4_45

Navigation Satellite System (D-GNSS) and High-definition Maps (HD Maps), which provide information on the state of the ego-vehicle and its control variables, as well as the environment itself [3].

Fig. 1. Performance of the DAMOT method in a traffic jam use case in AD PerDevKit dataset. Tracked objects are shown as green Bounding Boxes.

This work is framed in the domain of perception in the context of a modular architecture, a layer in charge of understanding the scene and extracting information for decision making in the respective domains. In this work, we propose a work framed in the Object Detection and Multi-Object Tracking (DAMOT) tasks using Machine Learning techniques, achieving satisfactory results in challenging and complex test scenarios in CARLA Simulator [4].

Regarding the sensing of the environment, this work chooses to merge the information coming from a LiDAR and a RADAR synthetically generated in CARLA. Both sensors provide point clouds that describe the space in a three-dimensional way and while LiDAR is much more common in recent literature [5], it shows some shortcomings that RADAR can overcome, such as:

- It can take advantage of the Doppler effect to infer velocity while LiDAR is unable to natively infer it.
- It can excel in long range detections, as it typically performs better given the smaller gradient of information with respect to distance.
- Native velocity inference can also contribute to the Object Tracking task, since more variables are contributed to the model than are normally inferred or predicted by Bayesian models.
- It can also contribute to the categorical identification of an object by means of the Radar Cross Section (RCS) together with the Reflection Intensity of LiDAR. As it is not a feature included in CARLA, this aspect is postponed to future studies.

Therefore, this method will be implemented and validated over the novel AD PerDevKit dataset [6], in which the performance of the Object Detection method will be measured. This tool includes complex test scenarios recorded in CARLA Simulator under diverse weather and environment conditions. Finally, the main contributions of this paper are:

- The development and integration of a DAMOT pipeline based on LiDAR and RADAR into a state-of-the-art autonomous driving architecture [3].
- An evaluation of an Object Detection system in our challenging and open-source AD PerDevKit dataset.

2 Related Works

The following section will discuss different approaches taken by the community in the field of LiDAR and RADAR based Object Detection and Multi-Object Tracking, paying special attention to the methods that have set the guidelines in these fields. Typically these two tasks, which are crucial in the field of autonomous vehicle perception, have been tackled with only one of the two sensors used in this work or by fusion of one with cameras, with LiDAR and RADAR fusion being the least explored in recent literature [7] as can be seen in Fig. 2.

In the field of object detection using LiDAR, [8] set a guideline on which a great diversity of works have been developed. Point-based object detection methods take as input the raw information from point clouds, so the key concept for object detection is the iterative clustering and sampling of groups of points through the concept of voxelisation. The model took as input segmented 3D point clouds on which object classification tasks were performed by partial cloud segmentation. Empowered by a neural network working at point-wise level, transformations were performed using convolutional and max pooling layers.

In [9], it was acknowledged that 3D convolutions were a bottleneck to achieving a robust 3D detection in real time. Therefore, it is proposed a model that uses 2D convolutions with an encoder that learns scene features in pillars (vertical columns). The pillars allow the 3D scene to be converted into a 2D pseudo-image. Moreover, the model does not need hand-tuning to adapt to new sensor conditions (agnosticism), and the authors claim that it is even valid for use with radar point clouds.

With respect to the field of radar object detection, a number of methods to achieve this task are proposed in [10]. Five methods are developed and reviewed, covering a wide spectrum of possibilities with which to tackle this task, from traditional clustering to bird's eye view imaging (BEV):

- DBSCAN [11] clustering for detection and Long Short-term Memory (LSTM) [12] for classification.
- PointNet++ [8] for 3D semantic segmentation and detection with DBSCAN clustering [11].
- Grid-mapping and YOLO-v3 [13] for BEV-image detection.
- Object detection based on PointPillars [9].

– Combined pipeline of PointNet++ [8], DBSCAN [11] and LSTM [12].

In [14], the concept of tracking within the context of LiDAR and RADAR fusion for detection in urban environments was introduced. Applying an Extended Kalman Filter (EKF) [15] to cope with non-linearities coming from RADAR measurements, tracking was made possible.

With respect to similar works in tracking terms, in [16], a pipeline is presented that fuses LiDAR and camera information for object detection and tracking tasks, in addition to a forward prediction of the movement of vehicles surrounding the ego. In this case, BEV-SORT [17] is also used for tracking and the concepts of containerization and standard communications with ROS and Docker are exploited.

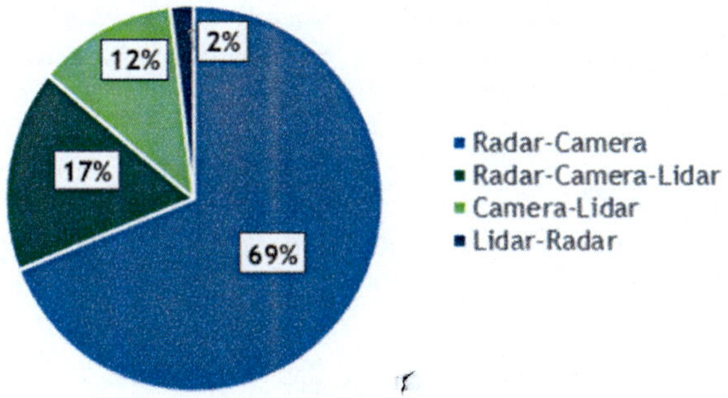

Fig. 2. Status of Multi-sensor Fusion in the context of autonomous driving. Adapted from [7].

3 Methodology

In this work we take as a starting point the premise that the problem of perception in the field of autonomous driving can be solved by the coherent combination of a set of traditional processing techniques, which constitute a modular architecture. This modularisation enables a large problem to be solved by breaking it down into smaller, simpler tasks. Although these systems are weaker and more prone to error propagation between modules, a policy of strict constraints and filtering makes them viable [18]. This type of systems have a greater capacity for explainability and determinism in each of the modules, as opposed to end-to-end systems.

The present architecture is divided into four modules: two Object Detection modules, one each for LiDAR and RADAR, a sensor fusion module and a final Object Tracking module, which are shown in Fig. 3. In addition, our system pipeline exploits the concepts of lightweight Linux containers using Docker [19] to provide the system with isolation, flexibility and portability, and standard

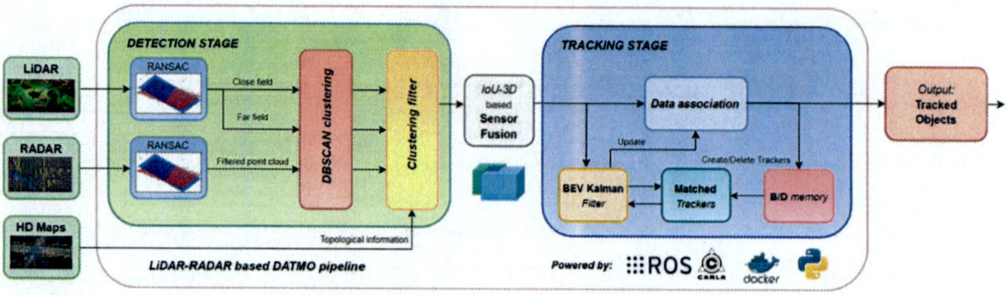

Fig. 3. LiDAR-RADAR based DAMOT method. From left to right all tasks involved in the method can be seen. With the information from the RADAR, LiDAR and HD Maps, the detection module is entered. In this module, the stages of plane filtering, clustering and flitting of tracker candidates are concatenated. Sensory fusion is then carried out, which provides the input for the tracking stage. Finally, the tracking is carried out through successive stages of data association and cycles of the BEV Kalman Filter, which are accompanied by the Birth-Dead memory module for the management of the tracked objects. These objects make up the output of the system.

communication in robotics using the Robot Operating System (ROS) [20]. A use case of this method in action is the Traffic Jam on a highway which can be seen in Fig. 1.

3.1 LiDAR and RADAR Object Detection

The first stage of this modular pipeline is the Object Detection over the inputs of the system: the point clouds provided by a LiDAR mounted on the top and a RADAR mounted on the front of the ego-vehicle in CARLA Simulator. On the one hand, the LiDAR point cloud consists of detections in $P_L = x, y, z, i$ format: Cartesian coordinates and intensity of the reflection. On the other hand, the RADAR point cloud consists of detections in $P_R = x, y, z, r, \theta, \phi, v$ format: Cartesian and spherical spatial coordinates, as well as Doppler velocity. Before proceeding to the object detection, it is convenient to filter the input data, as it will facilitate this task. In addition, it should be noted that the spatial distribution of point clouds is different, so each point cloud will require a specific set of filters.

First, the processing of the LiDAR point cloud will be discussed. Taking into account that this sensor has been configured to imitate a Velodyne HDL-64, it can be stated that the light beams are emitted with a vertical FoV of $[-25^{\circ}, 2^{\circ}]$. This configuration makes it possible to take advantage of the height position of the sensor and to avoid occlusions caused by objects very close to the ego-vehicle. A large number of these light beams hit the ground directly, resulting in high density cluster candidates. To avoid the formation of these, the road plane is segmented using a RANSAC algorithm [21]. It is a robust model fitting algorithm that performs the plane segmentation task in a iterative repeated sequence, generating a random hypothesis and verification. It proposes a random plane for n iterations and it maximizes the number of inliers fitting that plane according to a given threshold and is robust to a strong presence of outliers in the data. In addition, since it was observed that the density of incident points on the objects in the scene decreases considerably with distance, it was decided to divide the cloud into two parts at a distance of 25 m, giving rise to the concepts of close field and far field. This is done in order to be able to apply an adaptive clustering with distance.

Once the point cloud has been filtered and divided according to distance, a DBSCAN-based clustering process is applied for each of the defined zones. This process allows grouping the points belonging to a vehicle for a correct identification or detection. DBSCAN [22] is a clustering model that aims to find regions in a multi-dimensional space whose density is higher than certain user-defined parameters and which are separated by zones of lower density. The concept behind cluster formation is based on two parameters: `minPts` and ϵ. A point can be considered:

- **Core point:** if within a distance defined by a circle or sphere of radius ϵ there is a number of neighbours greater than `minPts`.
- **Border point:** if it is in the neighbourhood of a *core point*, but does not have a number of neighbours greater than `minPts`.
- **Noise point:** if the point do not meet any of the above cases.

Therefore, the resulting clusters will be formed by the *core* and *border points* directly reachable in density. This algorithm allows not having a priori knowledge of the number of vehicles (or clusters) present in the scene. The appearance of objects change with range, and after some distance, very few data points per objects are available to detect an object. This poses some challenges for detection, but this approach allows defining a series of different parameters for the close field and for the far field, in order to adapt to the far field cluster appearance.

On the other hand, RADAR is configured with a horizontal FoV of 160º and a vertical FoV of ±8º, emulating commercial configurations currently on the market or available in datasets [23]. The spatial distribution of the points in this cloud corresponds to a sparse point cloud, whose rays have no tendency to hit the ground. Given the scarcity of points and the importance of obtaining the native Doppler velocity of the detected objects, it is decided not to apply additional filters to those provided by the topological information of the map, as is done in the case of LiDAR. Consequently, another instance of the DBSCAN algorithm is applied to find the clusters belonging to the vehicles in the RADAR field of view. Given the clusters that are candidates for detection a positional filtering is made taking into account the information from HD Maps, remaining only those within the conduct zones. Lastly, a further geometrical filtering step is carried out. Those clusters with much larger dimensions, either in height, width or length or the product of the latter two, are rejected. All the remaining candidates pass to the fusion stage in Bounding Box (BB) format, and are defined by the vector $B = x, y, z, l, w, h, \theta, v$, where $[x, y, z]$ are the position in space of its centroid, $[l, w, h]$ are its dimensions for each axis, θ is the yaw angle and v is Doppler velocity obtained with RADAR.

3.2 Sensor Fusion

The next stage of the pipeline is the sensor fusion of the object candidates clustered in the previous stage. How to fuse objects is an open question in academia and in this work we have chosen to consider as valid those objects detected by both sensors, which leads to redundancy, as well as those that are detected by one of the sensors and are out of the field of view of the other. For the first case, in which both sensors provide a detection, a spatial association mechanism using IoU-3D is introduced, which combines the spatial information of the clusters and assigns the velocity obtained by the RADAR to this candidate. Therefore, it is considered as a detected object and proposed for the tracking stage. It is only consistent to opt for a fusion based on full redundancy of detections when the FoVs of the sensors involved in the fusion are identical, as this implies no detection of occluded objects. Therefore, for the candidates in the second case, the occluded angles for both sensors are examined and objects that are proposed by one sensor, but are out of range of the other, are taken as detected objects. The approach enables the height advantage of LiDAR at close distances to be exploited, although these objects will not have a native velocity inference. Another advantage is the greater ease of clustering at long distances for RADAR. This combination of criteria results in an exploitation of the native strengths of both sensors.

3.3 Multi-object Tracking

The third and last stage of the pipeline is a Multi-Object Tracking (MOT) module based on SORT that performs 3D Tracking from a Bird's Eye View (BEV). Adding the height dimension in solving this problem does not provide a clear improvement in the results and increases the computational cost and the completeness of the module. Since representing the space surrounding the ego-vehicle on a two-dimensional XY axis eliminates the main problem of traditional 2D tracking, occlusion, such an assumption has no negative effect on the result. As the Tracking module is fed by the filtered and refined detections, the concept of Tracking-by-Detection is reached.

BEV-SORT addresses MOT problem as a pipeline resulting from the combination of traditional techniques, mainly Kalman Filter (KF) [24] and Hungarian Algorithm (HA) [25], that are used for state estimation and data association, respectively. The algorithm makes use of a motion model to propagate the identity of a detected object from one timestamp to the next. The identity of each detected object is stored in a vector of 8 variables:

$$B_{trk} = [x, y, l, w, \theta, x', y', \theta']' \tag{1}$$

As a result of the detection and fusion modules, these vectors are fed with 7 of the 8 variables, with only θ' being inferred. When a detection is associated to an object in the scene, a B_{trk} is produced and used to update the state of the object and predict its behaviour at the next timestamp via KF. If there is no detection associated with the object, it is predicted without correction or update

Table 1. Evaluation in the AD PerDevKit dataset for all scenarios tackling object detection with the fusion between LiDAR and RADAR. For evaluation difficulties: E stands for Easy, M for Medium and H for Hard, depending on the evaulation range.

Scene	Town	Weather	Environment	F1 score		
				E (25 m)	M (50 m)	H (75 m)
1	03	Day	Urban	46.84	14.68	10.51
2	03	Night	Urban	43.67	21.09	17.91
3	05	Day	Urban	26.40	14.69	11.99
4	05	Day	Highway	49.05	29.03	21.42
5	06	Day	Highway	**57.45**	29.40	20.07
6	06	Day	Highway	31.47	19.43	14.60
7	06	Rainy	Highway	23.23	14.86	10.97
8	07	Day	Rural	19.86	11.08	8.99
9	07	Rainy	Rural	54.70	**42.86**	**31.66**
10	10HD	Day	Urban	29.26	14.22	10.73
11	10HD	Night	Urban	21.97	9.70	7.28
TOTAL				36.43	20.67	16.31

using the motion model. To associate the detections to the objects stored by the algorithm, a cost matrix is proposed that computes the IoU of all detections with the objects. This provides the necessary input to apply the HA and see which new detection corresponds to the saved object, an assignment problem.

The identities of the objects and their motion models are created as they appear in the scene and are destroyed when they no longer appear in the detections. In addition, the algorithm has two user-customisable parameters that allow defining the behaviour in case of object creation and destruction. As a mechanism to avoid false positives, the user can decide the number of times an object is detected to be considered by the algorithm. In other words, if the same target appears n times in the detections, its identity and its motion model are created so that it counts in the processing. The second parameter is the number of times an object is not detected so that its identity is destroyed and it is not counted.

To conclude the method, it is indicated that the output provided by the approach is a list of tracked and identified objects that will serve as input to the other modules of a modular architecture for the successful completion of autonomous navigation.

4 Validation and Experimental Results

In order to provide qualitative and quantitative metrics that reflect the strengths and weaknesses of the proposed approach, a quantitative validation method is chosen.

4.1 Validation Method

The metrics of the method's performance are obtained by using our AD PerDevKit, a dataset that allows scoring the goodness of Object Detection in use cases of interest and complex scenarios designed in CARLA Simulator. This dataset, in its first version, consists of a series of complex scenarios and use cases recorded on the different CARLA maps, representing conventional autonomous driving scenes in urban, rural and highway environments.

A suite of sensors integrated in the tool is composed of a 360º LiDAR, a 90º frontal RADAR, a frontal monocular camera, and IMU and GPS sensors. Given the nature of our proposal, it is decided to carry out an evaluation on the 11 available scenarios in a 90º frontal field of view to measure the performance of Object Detection. Goodness of the approach is validated with F-beta score. For the case in point, $\beta = 1$.

- **F-beta score (F_β) [Eq. 2]:** Measure of a test accuracy calculated from the precision and the recall: True Positives (TP), False Negative (FP) and False Negative (FN) detections.

$$F_\beta = \frac{(1+\beta^2) \cdot TP}{(1+\beta^2) \cdot TP + \beta^2 \cdot FN + FP} \tag{2}$$

To carry out the validation, an efficient evaluator is developed which takes as input the groundtruth provided by the dataset and the detections made by the presented method and calculates the offline metrics.

4.2 Experimental Results

A series of experiments is decided to evaluate the performance of the method, comparing the fusion performance against simple Object Detection pipelines with LiDAR or RADAR only. Furthermore, thanks to the diversity of AD PerDevKit scenarios, it is possible to compare the performance against different types of driving environments and weather conditions. The evaluation is performed in three difficulties, differentiated by the range of distance to be detected: Easy (25 m), Medium (50 m) and Hard (75 m). The best scores are marked in blue in Tables 1 and 2.

Table 1 shows the performance of the algorithm for each and every AD PerDevKit scenario. On Easy difficulty, a maximum score of 57.45 is obtained for Scenario 5 (day and highway). For Medium and Hard, the maximum score is obtained in Scenario 9 (rainy and rural environment), with 42.86 and 31.66, respectively. Weighing all scenarios, a score of 36.43, 20.67 and 16.31 is obtained for each difficulty. Regarding the driving environments, Fig. 4 (left) shows that the best environment in which the method performs is Highway for Easy and Rural for Medium and Hard. In the case of weather conditions, the method performs best in rainy scenarios, as shown in Fig. 4 (right), presumably thanks to RADAR information.

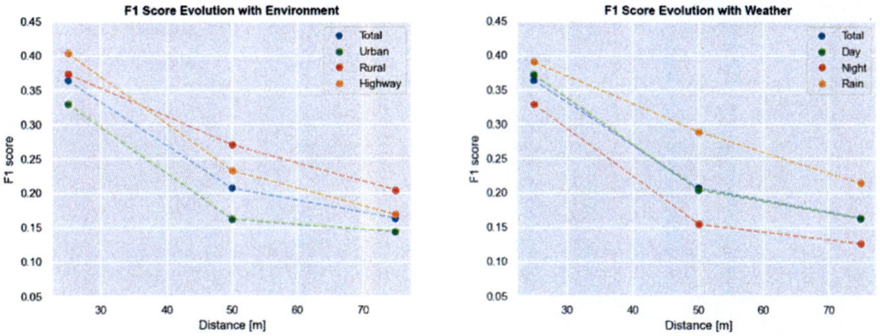

Fig. 4. Evolution of F1 score with respect to the distance evaluated over the different environments (left) and climate (right) scenarios offered by AD PerDevKit.

Table 2 shows an ablation between the Object Detection performance and the full fused module and the sensors alone. It can be seen that the system improves significantly (11.21 points) by keeping the same detection method but including fusion and tracking over the sensed detections.

Table 2. Complete evaluation in the AD PerDevKit dataset.

Mode	F1 score (%)		
	Easy (25 m)	Medium (50 m)	Hard (75 m)
Only LiDAR	25.22	16.14	12.13
Only RADAR	16.23	7.74	5.96
LiDAR+RADAR	**36.43**	**20.47**	**16.31**

5 Conclusions and Future Works

This paper presents the development, implementation and validation of an Object Detection and Multi-Object Tracking pipeline module using LiDAR and RADAR in CARLA Simulator. Using state-of-the-art software tools, such as ROS and Docker, and Machine Learning techniques such as DBSCAN and BEV-SORT. In addition, a quantitative evaluation is performed over AD PerDevKit dataset, which demonstrates that sensor fusion is beneficial over the inclusion of a single sensor, while maintaining the same detection methods. This work opens the way for future works on carrying out an ablation on Tracking and how to improve this fusion with state-of-the-art methods, or with different configurations or arrangements of the same sensors at different mounting positions.

Acknowledgements. This work has been funded in part from the Spanish MICINN/FEDER through the Techs4AgeCar project (RTI2018-099263-B-C21) and from the RoboCity2030-DIH-CM project (P2018/NMT-4331), funded by Programas de actividades I+D (CAM), cofunded by EU Structural Funds and Scholarship for Introduction to Research activity by University of Alcalá.

References

1. Gog, I., Kalra, S., Schafhalter, P., Wright, M.A., Gonzalez, J.E., Stoica, I.: Pylot: a modular platform for exploring latency-accuracy tradeoffs in autonomous vehicles. In: 2021 IEEE International Conference on Robotics and Automation (ICRA), pp. 8806–8813. IEEE (2021)
2. Chekroun, R., Toromanoff, M., Hornauer, S., Moutarde, F.: Gri: general reinforced imitation and its application to vision-based autonomous driving. arXiv preprint. arXiv:2111.08575 (2021)
3. Gómez-Huélamo, C., et al.: How to build and validate a safe and reliable autonomous driving stack? a ros based software modular architecture baseline. In: 2022 IEEE Intelligent Vehicles Symposium (IV). IEEE (2022)
4. Dosovitskiy, A., Ros, G., Codevilla, F., Lopez, A., Koltun, V.: CARLA: an open urban driving simulator. In: Levine, S., Vanhoucke, V., Goldberg, K. (ed.) Proceedings of the 1st Annual Conference on Robot Learning, vol. 78 of Proceedings of Machine Learning Research, 13–15 Nov 2017, pp. 1–16. PMLR
5. Patole, S.M., Torlak, M., Wang, D., Ali, M.: Automotive radars: a review of signal processing techniques. IEEE Sig. Process. Mag. **34**(2), 22–35 (2017)
6. de la Pena, J., Bergasa, L. M., Antunes, M., Arango, F., Gómez-Huélamo, C., López-Guillén, E.: Ad perdevkit: an autonomous driving perception development kit using carla simulator and ros. In: 2022 IEEE International Conference on Intelligent Transportation Systems (ITSC). IEEE (2022)
7. Zhangjing Wang, Y.W., Niu, Q.: Multi-sensor fusion in automated driving: a survey. IEEE Access **8**, 2847–2868 (2019)
8. Qi, C. R., Yi, L., Su, H., Guibas, L.J.: Pointnet++: deep hierarchical feature learning on point sets in a metric space. In: Advances in Neural Information Processing Systems, vol. 30 (2017)
9. Lang, A.H., Vora, S., Caesar, H., Zhou, L., Yang, J., Beijbom, O.: Pointpillars: fast encoders for object detection from point clouds. In: Proceedings of the IEEE/CVF Conference on Computer Vision and Pattern Recognition, pp. 12697–12705 (2019)
10. Scheiner, N., Kraus, F., Appenrodt, N., Dickmann, J., Sick, B.: Object detection for automotive radar point clouds-a comparison. AI Perspect. **3**(1), 1–23 (2021)
11. Ester, M., Kriegel, H.-P., Sander, J., Xiaowei, X., et al.: A density-based algorithm for discovering clusters in large spatial databases with noise. In: KDD, vol. 96, pp. 226–231 (1996)
12. Hochreiter, S., Schmidhuber, J.: Long short-term memory. Neural Comput. **9**(8), 1735–1780 (1997)
13. Redmon, J., Farhadi, A.: Yolov3: an incremental improvement. arXiv preprint. arXiv:1804.02767 (2018)
14. Cho, H., Seo, Y.W., Kumar, B.V., Rajkumar, R.R.: A multi-sensor fusion system for moving object detection and tracking in urban driving environments. In: 2014 IEEE International Conference on Robotics and Automation (ICRA), pp. 1836–1843. IEEE (2014)

15. Julier, S.J., Uhlmann, J.K.: New extension of the kalman filter to nonlinear systems. In: Signal processing, sensor fusion, and target recognition VI, vol. 3068, pp. 182–193. International Society for Optics and Photonics (1997)
16. Gómez-Huélamo, C., Bergasa, L.M., Gutiérrez, R., Arango, J.F., Díaz, A.: Smartmot: exploiting the fusion of hdmaps and multi-object tracking for real-time scene understanding in intelligent vehicles applications. In: 2021 IEEE Intelligent Vehicles Symposium (IV), pp. 710–715. IEEE (2021)
17. Bewley, A., Ge, Z., Ott, L., Ramos, F., Upcroft, B.: Simple online and realtime tracking. In: 2016 IEEE International Conference on Image Processing (ICIP), pp. 3464–3468. IEEE (2016)
18. Yurtsever, E., Lambert, J., Carballo, A., Takeda, K.: A survey of autonomous driving: common practices and emerging technologies. IEEE Access **8**, 58443–58469 (2020)
19. Merkel, D., et al.: Docker: lightweight linux containers for consistent development and deployment. Linux J. **2014**(239), 2 (2014)
20. Quigley, M., et al. Ros: an open-source robot operating system. In: ICRA Workshop on Open Source Software, Kobe, Japan, vol. 3, p. 5 (2009)
21. Fischler, M.A., Bolles, R.C.: Random sample consensus: a paradigm for model fitting with applications to image analysis and automated cartography. In: Communications of the ACM, vol. 24, no. 6, pp. 381–395 (1981)
22. Schubert, E., Sander, J., Ester, M., Kriegel, H.P., Xu, X.: Dbscan revisited, revisited: why and how you should (still) use DBSCAN. ACM Trans. Database Syst. (TODS) **42**(3), 1–21 (2017)
23. Schumann, O., et al.: A real-world radar point cloud data set for automotive applications. In: RadarScenes (2021)
24. Kalman, R.E.: A new approach to linear filtering and prediction problems. J. Basic Eng. **82**(1), 35–45 (1960)
25. Kuhn, H.W.: The hungarian method for the assignment problem. Naval Res. Logistics Q. **2**(1–2), 83–97 (1955)

SAFaD: A System for Automatic Fall Detection on Surveillance Imagery

Borja Perez-Lopez, Francisco Gomez-Donoso(✉), and Miguel Cazorla

Institute for Computer Research, University of Alicante, PO-BOX 99, Alicante, Spain
bpl14@alu.ua.es, {fgomez,miguel.cazorla}@ua.es

Abstract. In this work we introduce SAFaD: A System for Automatic Fall Detection on Surveillance Imagery. Our system heavily relies in an intermediate representation that allows us to accurately and optimally encode the motion of a person. An ensemble of different approaches use this volume as input to predict whether a fall event has happened. We tested out system with a state-of-the-art fall detection dataset reaching a 62.86% accuracy.

Keywords: Fall detection · Machine learning · Image analysis · Ensemble

1 Introduction

Falling is one of the main causes of accidents at home world-wide. In the USA, 3 million older people are treated in emergency departments for fall injuries each year, as a study of the Center for Disease Control and Prevention stated [1]. Furthermore, over 800,000 of them are hospitalized finally totaling more than $50 billion dollars in medical cost, mostly covered by public funds. In addition to that, the same study predicts that there will be around 7 deaths caused by falls every hour by 2030.

As it can be seen, this is a serious problem that affects lots of people, being specially severe for older people. Thus, any effort for controlling and preventing this kind of accidents would be much appreciated by the global community.

Thus, bearing this in mind, we introduce SAFaD: A System for Automatic Fall Detection. Our proposal takes surveillance imagery, such as the video stream of a security or IP camera, and is able to state whether a person has suffered a fall. No additional devices or wearables needed. To do this, we rely in a machine learning pipeline that includes a pixel-wise classification method that provides us with the location of the person in the image. Then, this information is used to build the Semantic-Aware Motion History Image that encodes the temporal context in a single frame. Finally, an ensemble of algorithms are used to analyze it and provide the final classification in either fall or not fall. If a fall is detected, an alarm could be issued to any relative or contact person to take further measures. Ultimately, SAFaD could help to improve the quality of life of our elderly people

D. Tardioli et al. (Eds.): ROBOT 2022, LNNS 590, pp. 564–575, 2023.
https://doi.org/10.1007/978-3-031-21062-4_46

whilst reducing the expenses in healthcare. Some qualitative results of our system are shown in Fig. 1.

Specifically, the main contributions of this work are:

- The Semantic-Aware Motion History Image that encodes the temporal context, namely, the motion of a person in a single image. This allows to efficiently analyze temporal events.
- An ensemble of classifiers that provide accurate predictions which rely in automatic image analysis methods, so no additional or wearable devices are needed.
- Exhaustive experimentation involving state-of-the-art datasets that definitely states the performance of the proposal.

The rest of the document is structured as follows. Firstly, we review the main methods for detecting falls in the Sect. 2. Then, in the Sect. 3, we provide an in-depth explanation of the SAFaD pipeline. As mentioned before, we involved a well-known dataset for fall detection in the experimentation, which is specified in the Sect. 4. Next, the results of the exhaustive experimentation we carried out are shown in the Sect. 5. Finally, in Sect. 6 we draw the final conclusion, advantages and limitations of our proposal, and point the future research lines.

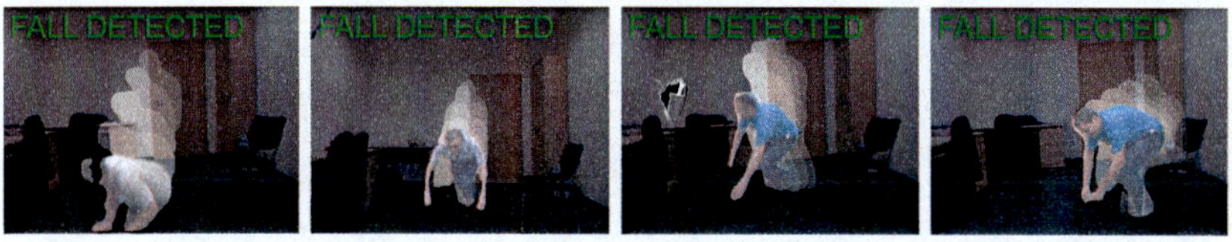

Fig. 1. Some results of detected falls as produced by our proposal SAFaD.

2 Related Works

Automatic fall detection is a hot topic among the scientific community and thus, a huge number or research projects have been arisen lately. Specifically, we can group them in two broad families: those that use inertial devices and those that solely rely in images.

Regarding the fall detection using inertial data, in [2], the authors used a non-linear autoregression neural network to model the fall detection problem. The input to their proposal is inertial data. In [3], the authors followed an ensembling scheme that used threshold-based methods and multiple kernel learning support vector machine to process the inertial measurements of a smartphone. Traditional classifiers as nearest neighbors were also explored in [4]. In this work, a range of different inertial sensors were used.

Regarding those method that rely in image analysis algorithms, we find that deep learning has been intensively used. For instance, in [5] they propose the training a deep spatio-temporal autoencoder to minimize the reconstruction error of activities of daily living video frames. It was hypothesizes that the reconstruction error for unseen falls should be higher. They modelled the problem after this anomaly detection because it is hard to collect a huge dataset of falling events. Some other works, such as [6], rely in intermediate representations. In this case, the authors compute the 3D human poses that are then processed by a fully convolutional architecture with residual connections that takes as input a sequence of 3D poses. Other example is [7] where they use 3D information from different cameras and use colors and movements for tracking and learn visual appearance of the persons. There are also approaches that do not rely in color frames but in the output of different sensors. This is the case of [8], that combines thermal imaging with traditional and deep learning models. This is also the case of [9], that used the emerging millimeter-wave radar sensor to collect the human body's point cloud that is fed to a variational recurrent autoencoder to compute the anomaly level of the body motion based on the acquired point cloud.

In the light of this study, some interesting conclusions can be obtained. First, it is worth mentioning that the reviewed works claim high accuracy but they only provide evaluation on private datasets, hindering comparison and replication. Then, the methods that rely in inertial measurements require a wearable sensor, limiting their advantages. Finally, some methods involve different kinds of input such as color and depth information, color and point clouds or color and pose data. This require special equipment and/or additional computation power. Nonetheless, our proposal only relies in a sequence of color frames that can be obtained with any regular camera. In addition, our proposed Semantic-Aware Motion History Image acting as intermediate representation allows an optimal balance between accuracy and computation requirements. Finally, ours is evaluated with a state-of-the-art dataset, allowing a proper comparison framework.

3 Approach to Automatic Fall Detection

The pipeline of SAFaD is composed of different pieces. First, a sequence of consecutive images are obtained from a regular color camera, webcam, surveillance camera or IP camera. Then, each frame is processed by a pixel-wise classification method. The resting semantic masks are used to build our Semantic-Aware Motion History Image (SAMHI) that optimally encodes the human motion. Finally, the SAMHI is processed by an ensemble of individual classifiers which predictions are combined to state whether the person has suffered a fall or not. Each piece of the pipeline is explained in the following Subsections.

It is worth noting that, as the images provided by the color cameras are actually a continuous stream of frames, we process them by following a sliding window approach. Thus, we accumulate a number of frames in a batch, which are processed together. Then, a number of frames are skipped before considering the next batch of images.

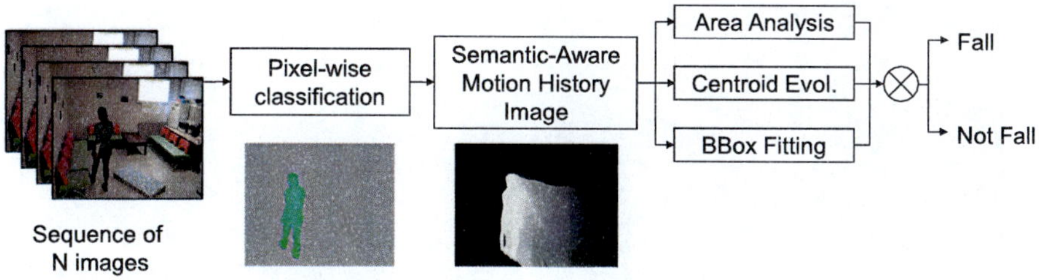

Fig. 2. Diagram of SAFaD pipeline. A sequence of N images are processed by a pixel-wise classification system. Then, the resulting semantic maps are used to build the semantic-aware motion history image which is finally used by the classifiers to produce partial predictions. Finally, these partial predictions are summarized together to return the final classification result.

3.1 Semantic-Aware Motion History Image

As stated before, a Motion history Image (MHI) [10] is a view-based temporal template method which is simple but robust in representing movements. In the MHI, the silhouette sequence is condensed into gray scale images in which the dominant motion information is preserved. Therefore, it can represent a motion sequence in a compact manner. It keeps a history of temporal changes at each pixel location, which then decays over time. The MHI ultimately expresses the motion flow or sequence by using the intensity of every pixel in temporal fashion. Thus, the input for a MHI estimator is a sequence on N consecutive frames, and the output is a greyscale image. MHIs have been widely used for motion and gait analysis, and for detecting human actions [11]. Nonetheless, the pure incarnation of MHI has a severe limitation. As the resulting image encodes the motion, the pixels that do not change during the input sequence are not actually updated. As a result, an inaccurate result could be returned when it comes to human action representation due to some parts of the body remaining still. For instance, if a person is standing and only moves their arm, the corresponding MHI would only depict the motion of the arm, the rest of the body would be not represented. Likewise, movement of background objects would be also depicted Samples of this issue are depicted in Fig. 3. In addition to that, the method is affected by image noise and lighting changes. These events can cause modifications of the pixel values between consecutive frames that would be considered motion and be incorrectly included in the resulting MHI.

To overcome this limitation of the original implementation of the MHI, we introduce the Semantic-Aware Motion History Image. In this version, we firstly detect the pixels of the input images belonging to a person with Mask-RCNN [12], which is a pixel-wise classification algorithm. Mask R-CNN is the current state-of-the-art in terms of image segmentation. This variant of a Convolutional Neural Network (CNN) detects objects in an image and generates a high-quality segmentation mask for each instance. Thus, the input of this architecture is an image and the output is a segmentation mask that encodes the silhouettes of

Fig. 3. Color frames (first row) and corresponding MHI (second row) and SAMHI (third row).

each instance of object that is trained to detect. To do this, Mask-RCNN relies in a CNN pipeline that is stacked on top of a Region Proposal Network (RPN).

Thus, the images in the input sequence of SaFAD are fed to Mask-RCNN so a segmentation mask depicting the shape of a person is produced for each image. Then, the SAMHI is built as follows: it starts with a zero matrix, with the same shape of the input images. The segmentation masks are traversed, from older to newer, one by one. For each mask, the pixels depicting a person adds one unit to the final SAMHI. The values of pixels of the SAMHI depicting a person of previous iterations decay in a $1/n$ factor, being n the number of frames in the input sequence. Then, the resulting image is normalized in the range of $0-255$, thus generating a grey scale image. In the SAMHI, the pixels belonging to a person in the recent events are brighter than those belonging to a person in the older images. Some results provided by SAMHI are depicted in the third row of Fig. 3. As it can be seen, ours provides a more accurate representation of human events during a time frame, leaving no missing body parts and filtering out the background movement. In addition to that, SAMHI is also agnostic to image noise, and shading and lighting changes between frames that would cause undesirable artifacts in the original MHI as they would be considered motion.

3.2 Area Analysis Fall Detection

Once the SAMHI is obtained from the scene, the next step is to feed it to the classifiers. The first classifier is based on the difference between user's area from one frame to the following. The idea is that, when the person is walking or doing housework, the displacement of the silhouette between consecutive frames is so small that it overlaps to the silhouette from the previous frame. Therefore, few data from the older events is kept in the SAHMI. In the other hand, falling is likely to be a quick event which produces large variations between frames, so large traces of previous moments will remain in the SAMHI.

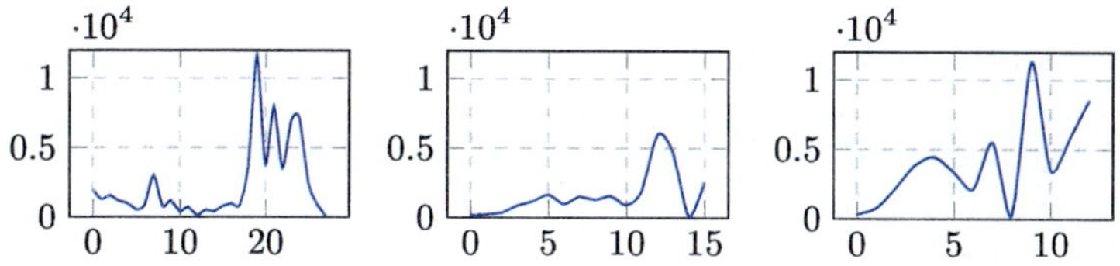

Fig. 4. Evolution of quantity of coloured pixels for area analysis method. Note how the number of coloured pixels increases drastically when the user falls and decreases or stands when walking or moving slowly.

Thus, we can detect the falling event by analyzing the area occupied by the person along the time in the SAMHI. To do this, the classifier goes through each of the pixels of the SAHMI and counts the number of pixels different than zero. Finally, if the total count of person's pixels exceeds a threshold, the system classifies the sequence as falling. To avoid false predictions due to noise or detection failures, it is needed two or more consecutive falling detections to consider it as a definitive fall. As mentioned before, this method heavily relies in the threshold the classifier uses to distinguish between falling and non-falling events.

3.3 Centroid Evolution Fall Detection

This classifier is based on the variation of the center point of the silhouette during the sequence. As mentioned before, falling is a fast movement so the displacement of this center during falling frames is greater than in non-falling frames, being a feature to distinguish between these events. Considering the evolution of the centroid instead of the whole area occupied by a person makes the classifier agnostic to the relative size of the human regarding the image resolution. Centroid variations in multiple cases are shown on Fig. 5.

As falling events mainly occur vertically, the classifier only consider the variation on this axis. We do this to avoid false detections when motion parallel to the camera is registered.

Thus, the classifier calculates the centroid of the shape from the SAMHI to compare the vertical coordinate variation along the sequence. If the variation is above a threshold, it is classified as fall. Similarly to Area Analysis method, this classifier relies in a threshold for being able to differentiate between falling and non-falling and purge some false detections requiring two or more SAHMIs classifieds as falls.

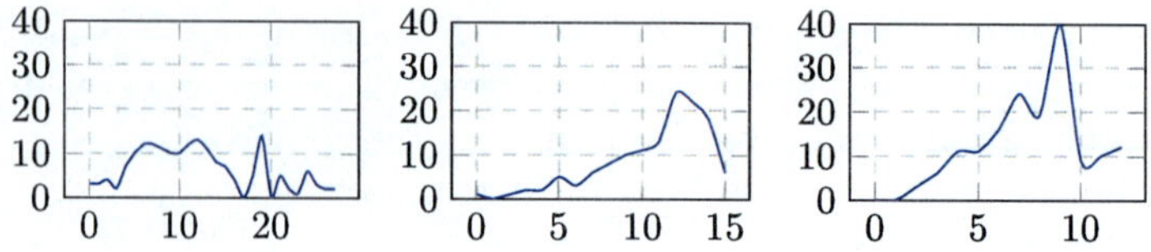

Fig. 5. Evolution of centroid's vertical position for centroid based detection method. Absolute difference of the centroid's vertical value between two consecutive for three different input sequences. It is worth noting how the classifier uses large variation between steps to detect a fall. Accordingly with the image axis, when the user falls, the X value increases.

3.4 Bounding Box Fitting Fall Detection

The last classifier works similar to the centroid one. This is based on height variation of the bounding box calculated from the user. This is due to, when a fall happens, the person lies on the floor where the distribution of the shape in the SAHMI is mainly horizontal, as opposed to when standing. This evolution is shown on Fig. 6. This approach is intended to cover the case in which the subject is already very close to the floor because the subject was previously sitting, or the point of view of the camera hindered the accuracy of the other classifiers. In the same way as the two previous ones, this method require two or more fall detections to consider it as a fall.

Thus, we can classify a sequence as fall when the evolution of bounding box's height decrease above a threshold between SAMHIs. Accordingly, tuning this threshold is critical in this case as well.

3.5 Final Ensemble

Finally, as each method previously explained are strong on some situations and weak on others, we decided to combine them all to achieve a more precise and less false positive prone system. With the objective of getting the best results, the final fall decision of SAFaD is contidioned to being detected as a fall by all the classifiers, thus leaving out those possible false detections from each single classifier.

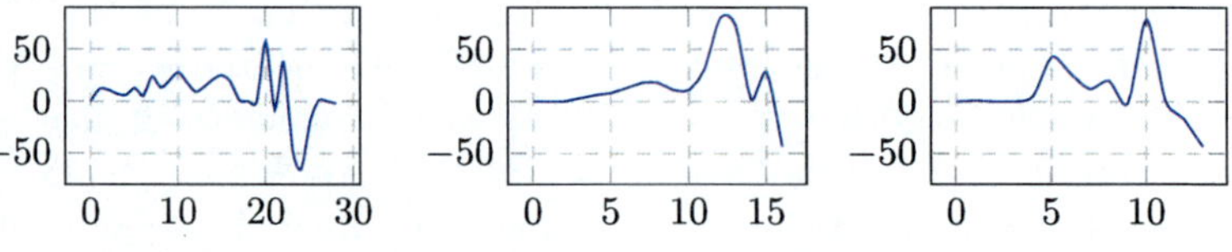

Fig. 6. These are a representation of the bounding box vertical size difference between two consecutive frames for three different input sequences for the Bounding Box fitting detection method. We can observe the slight variation of the bounding box's height when the user is walking and the large variation when it finally falls.

4 Dataset

In order to evaluate SAFaD, we adopted the UR Fall Detection Dataset[1]. This dataset contains 70 sequences, 30 of which are falls and the remaining 40 depict activities of daily living. The dataset provides color images, some of them depicted in Fig. 7, and corresponding depth maps, as the sequences were recorded by a Kinect v1 camera. In addition, the fall events were also recorded by two different inertial sensors. The fall events were also filmed from two points of view: a bird's eye view from the camera attached to the ceiling and pointed to the floor, and a frontal view. The dataset comprises different subjects and environments, and different ways of falling.

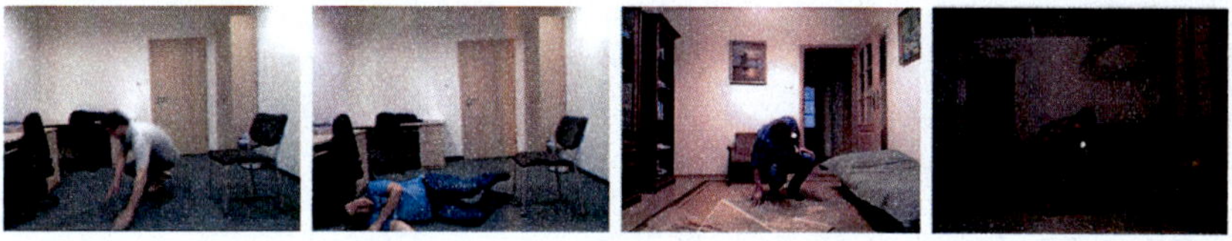

Fig. 7. Random samples of the UR fall dataset.

As our approach is fully visual-based, we only use the video streams from the frontal camera, which are of 640 × 480 pixels resolution. The inertial data and the depth maps are not used by our approach.

5 Experiments and Results

In this Section, we show the experiments and results we carried out with the UR Fall Dataset, which is a well-known fall detection benchmark as we explain in Sect. 4. We focused the experiments in modelling the behaviour of SAFaD in terms of the thresholds, which will be referred as T in the following subsections, used by each classifiers.

5.1 Area Analysis Experimentation

In this way, we tested multiple setups to check which combination of parameters achieve the best results.

Firstly, the main parameter of the area analysis classifier is the threshold, that triggers a fall detection. This value is related to the number of coloured pixels that have changed between frames as explained in Sect. 3.2. It must be tuned to achieve the best accuracy. The results of the different tests are shown in the Fig. 8.

[1] http://fenix.univ.rzeszow.pl/~mkepski/ds/uf.html.

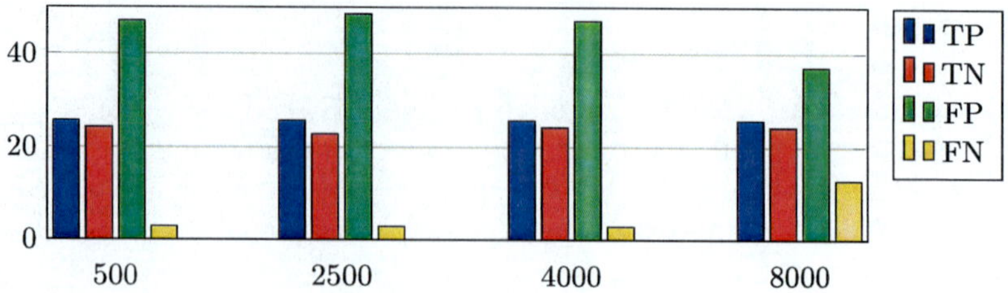

Fig. 8. Number of true positive and false positive reached for each test using the 70 sequences from the UR fall dataset on area analysis' method shown in the Y axis. The horizontal axis is the detection threshold T.

Once these results were obtained, we checked if the detection were made correctly. For this verification, we save in which frame the method have detected the fall. This way, we could manually check if the fall detection was a True Positive or not. Consequently, despite the fact that $T = 8000$ shows better results, it had more missed falls than the $T = 4000$ which detected more falls but at the wrong instant. Although this could be approximated since it is not possible to estimate the perfectly correct fall frame with the naked eye.

5.2 Centroid Evolution Experimentation

On this part, there were multiple experiments to achieve the best threshold for this part. This value, as mentioned above, is focused on vertical variation of the centroid, because when falling, the fast movement is usually vertically. In this manner, we made some tests observing how this centroid varies to adjust which one results in the best setup, shown in the Fig. 9.

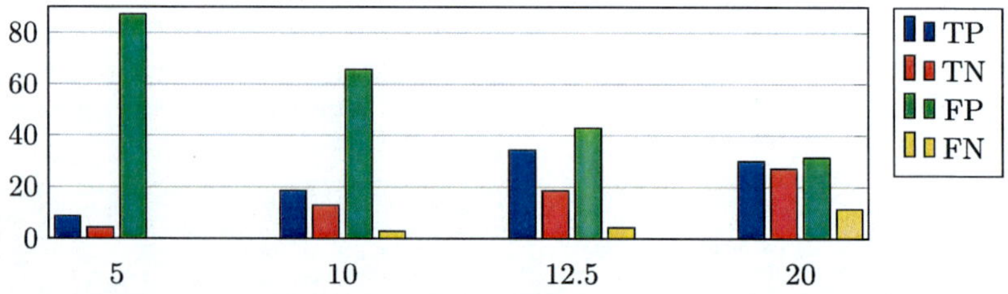

Fig. 9. This plot shows the results from different threshold T value variation between 5, 10, 12.5 and 20 respectively on centroid's evolution experimentation.

These results were verified as well after the process finished. In this manner, we found that, in $T = 5$, the system detected a lot of falls but in the wrong frame resulting in only a few True Positive detections. This way, $T = 20$ shows a better balance between True and False Positive detections achieving an accuracy of 57.14% even though it detects a lower percentage of TP than for $T = 12.5$.

Additionally, we found some issues during the test due to bad detections appearing on the SAMHI. This method relies too much on the correct segmentation of the user silhouette from Mask-RCNN because the calculation of the centroid is about all the area detected as person. This means that, when on the SAMHI the detection did not encompass the user's entire body, the centroid varies drastically.

5.3 Bounding Box Fitting Experimentation

In this method, similarly to Centroid's variation method, we tested which vertical threshold T achieves the best results. As the bounding box is only 2D, we mainly worked with the vertical value. Nevertheless, the horizontal value was also useful because when the person lies on the ground, the bounding box extends primarily horizontal. However, this can lead to errors when the person falls in a direction perpendicular to the camera, as the camera has no depth measurement and the silhouette is still vertical. That's why we focused on vertical movement to detect the fall. The results of different configurations are shown on Fig. 10.

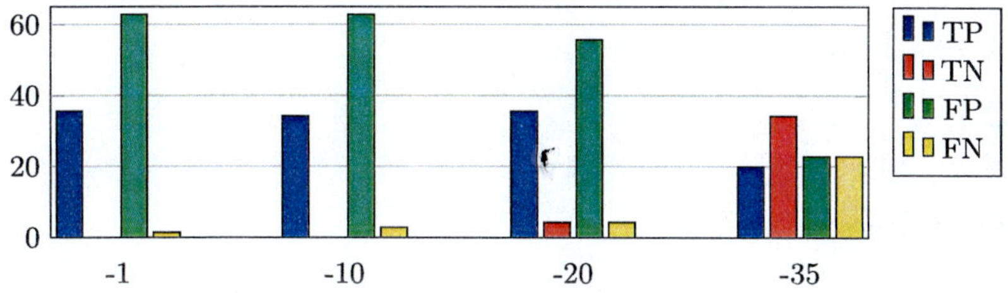

Fig. 10. These results correspond to a threshold T values of -1, -10, -20 and -35 respectively on the bounding box fitting experimentation.

On the basis of these results and their verification, we can see how when the bounding box variation is too high, the system fails to detect the fall because we need a large and fast falling movements to achieve this variation. On the other hand, we can see $T = -20$ achieves a higher number of correct fall and non fall detections while decreasing false detections compared to $T = -1$ and $T = -10$.

It should be noted that, as we work with the sequence of movement, this method requires that the last frames are from the final fall or the ground. That means that, if the falling sequence is too short or the frame rate insufficient, this method could not detect the fall.

5.4 Final Ensemble Experimentation

Finally, we made some tests with all the methods combined, modifying how each method affects to the final result. In this manner, the system only classifies the sequence as fall if all the three methods detect it as a fall. However, there were still some problems due to false detections of the methods or the precision of some methods where two of the methods are successful on detecting the fall, but one fails to do so, so the result is a fall not detected by the system.

The different results between the methods' thresholds are shown on the Fig. 11.

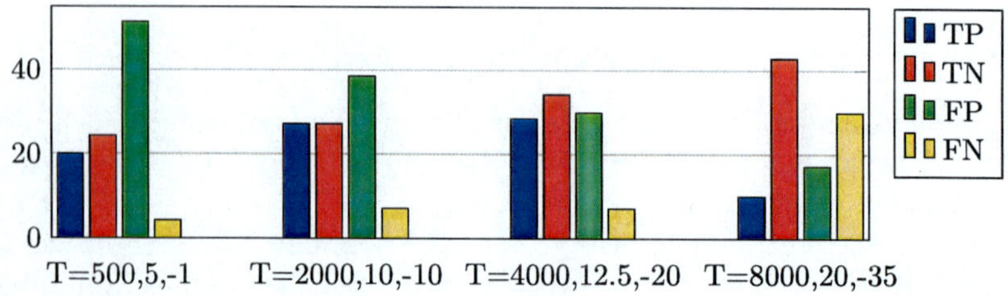

Fig. 11. These are the final results of the system with the corresponding values mentioned on the previous plots.

Note how with T=4000, 12.5, −20 configuration for the method's thresholds, we achieve a large amount of True Results, corresponding to 62.86% of precision discriminating fall and non fall situations. This precision was considered as a success on achieving the objectives proposed for the system.

6 Conclusion and Future Work

In this paper we introduced SAFaD: A System for Automatic Fall Detection on Surveillance Imagery that is composed of an ensemble of soft classifiers that work on different features to state whether a fall has happened in an input sequence of color images. Additionally, the MaskRCNN model is the one provided by the authors, however the subsequent fall detection process is not based on learning so it yields no overfitting to certain persons, backgrounds or situations. On the other hand, there may be situations when the user crouches or sits where the system detects a fall. Nevertheless, if the user subsequently stands up, it would no longer detect fall due to the temporary evolution of the silhouette. Our approach achieves 60% accuracy on the UR Fall Dataset, which is a well-know benchmark for the task. Nonetheless, it has some limitations. For instance, the SAMHI is very dependent of the accuracy of the pixel-wise classifier, as it relies in the provided shapes. It is also dependant on the framerate, as different framerates would require tuning of the thresholds. Finally, it has limitations related to be working in the 2D space as not detecting fall due to its direction respect to the camera. As a future work, we plan to tackle these limitations by involving 3D mesh generation from 2D images using for instance MeshGraphFormer [13].

Acknowledgements. This work has been funded by the Spanish Government PID2019-104818RB-I00 grant, supported with Feder funds.

References

1. Centers for Disease Control and Prevention, Facts About Falls (2021). www.cdc.gov/falls/facts.html. Accessed 06 June 2022
2. Salleh, S.M., Koh, C.Z.: Neural network algorithm-based fall detection modelling. Int. J. Integr. Eng. **12**, 138–150 (2020)
3. Shahzad, A., Kim, K.: Falldroid: an automated smart-phone-based fall detection system using multiple kernel learning. IEEE Trans. Industr. Inf. **15**, 35–44 (2019)
4. Özdemir, A.T., Barshan, B.: Detecting falls with wearable sensors using machine learning techniques. Sensors **14**(6), 10691–10708 (2014)
5. Nogas, J., Khan, S.S., Mihailidis, A.: DeepFall: non-invasive fall detection with deep spatio-temporal convolutional autoencoders. J. Healthc. Inf. Res. **4**(1), 50–70 (2019). https://doi.org/10.1007/s41666-019-00061-4
6. Chen, Z., Wang, Y., Yang, W.: Video based fall detection using human poses. In: Liao, X., et al. (eds.) BigData 2022. CCIS, vol. 1496, pp. 283–296. Springer, Singapore (2022). https://doi.org/10.1007/978-981-16-9709-8_19
7. Cañas, J.M., Marugán, S., Marrón, M., García, J.C.: Intelligent spaces as assistive environments: visual fall detection using an evolutive algorithm. In: Ruano, A.E., Várkonyi-Kóczy, A.R. (eds.) New Advances in Intelligent Signal Processing. SCI, vol. 372, pp. 225–247. Springer, Heidelberg (2011). https://doi.org/10.1007/978-3-642-11739-8_11
8. Mehta, V., Dhall, A., Pal, S., Khan, S.S.: Motion and region aware adversarial learning for fall detection with thermal imaging. In: 2020 25th International Conference on Pattern Recognition (ICPR), Los Alamitos, CA, USA, pp. 6321–6328. IEEE Computer Society (2021)
9. Jin, F., Sengupta, A., Cao, S.: mmfall: fall detection using 4-d mmwave radar and a hybrid variational rnn autoencoder. IEEE Trans. Autom. Sci. Eng. **19**(2), 1245–1257 (2022)
10. Davis, J., Bobick, A.: The representation and recognition of human movement using temporal templates. In: Proceedings of IEEE Computer Society Conference on Computer Vision and Pattern Recognition, pp. 928–934 (1997)
11. Ahad, M.A.R., Tan, J.K., Kim, H., et al.: Motion history image: its variants and applications. Mach. Vis. Appl. **23**, 255–281 (2012). https://doi.org/10.1007/s00138-010-0298-4
12. He, K., Gkioxari, G., Dollár, P., Girshick, R.: Mask r-cnn. In: 2017 IEEE International Conference on Computer Vision (ICCV), pp. 2980–2988 (2017)
13. Lin, K., Wang, L., Liu, Z.: Mesh graphormer. In: ICCV (2021)

Integrating State-Based Multi-Agent Task Allocation and Physical Simulators

Daniel Rivas(✉) and Lluís Ribas-Xirgo

Universitat Autònoma de Barcelona, Cerdanyola del Vallès, Barcelona, Spain
daniel.rivas@uab.cat

Abstract. The multi-robot task allocation (MRTA) systems face the challenge of adapting to dynamic environments where new tasks and communication errors might appear during execution. This paper presents a framework to run agent-based MRTA within a physical simulator to test different algorithms and/or setups. Agents are modeled by a specific type of state machines able to represent deliberative behaviors as well as reactivity. While this adds formality and simplifies implementation, execution of state machines within a physical simulator requires decoupling transitions that imply the passing of time from those occurring instantly. The result framework includes a state machine execution engine that synchronizes with the simulator's engine. Experiments using an auction-based MRTA for an example plant show not only the capability of the framework for modeling a wide range of systems but also that the MRTA method works with on-the-fly task inclusions, varying number of active robots and error occurrences.

Keywords: Computational modeling · Time modeling · Control systems · Mobile robots · State machines · State-based programming

1 Introduction

Using mobile robots to perform logistics tasks is a growing practice nowadays [1–3], and finding the best allocation of tasks to the robots is a critical part of these systems. The task allocation (TA) pursues a specific goal, like minimizing operations time [4]. Obviously, TA's complexity grows exponentially with the number of agents and tasks in the system, which makes it almost impossible for a person alone to solve it for medium to large systems. So, many companies opt for automating the TA or providing their specialists with appropriate decision-support tools.

In static environments, where all the tasks are known beforehand and only the robots of the system are present in the operation's area, centralized MRTA solutions [5,6] based on, e.g. integer linear programming or exploration algorithms, are appropriate, as they can provide pretty optimal allocations. However, they are hard to scale and lack robustness against changes in the system. Furthermore, the quality of the results highly depends on the *a priori* information on

D. Tardioli et al. (Eds.): ROBOT 2022, LNNS 590, pp. 576–587, 2023.
https://doi.org/10.1007/978-3-031-21062-4_47

the tasks, and their performance degrades when dealing with certain randomization levels in the system [7]. These limitations are much worse in dynamic environments, as the TA system must reevaluate its solution every time there is a change in the environment, including the appearance of new tasks.

Distributed solutions adapt better to dynamic environments. However, even these present-day solutions have room for improvements, like including more complex temporal models and more accurate data on the physical system's conditions [4], or optimizing the energy and time consumption of the agents altogether, or embedding robust and intelligent controllers that support hybrid control architectures [8]. Solutions for the latter would use *reactive control* to handle the obstacles and other physical elements in the environment and *deliberative control* for inter-agent communication.

Our approach builds upon a physical simulator which not only captures the details of the physical plant and robots but also includes the MRTA module. This module gets tasks from e.g. a manufacturing execution system and allocates (or re-allocates) them to robots. As the physical simulator's scenario includes the robots and the plant itself, the TA can be done with accurate information, and results can be more exact. Of course, real data could also be obtained from actual plants but, still, the use of a simulator enables testing hazardous situations, plant modifications, *et cetera*, without causing disturbances to them.

We already had a multi-robot scene in the CoppeliaSim [9] simulator with robots that modelled real ones which has been used for educational purposes since 2014. In this scene, each robot is an agent whose controller follows a layered architecture, with the lower levels being embedded both into the real robot and the simulated counterpart, and the upper levels are only virtual. Note that the data going from the upper levels to the lower ones goes to the virtual robots and, if they are connected, to the real ones. In this case, in case of conflict, the data coming from the reality prevails.

In this work we integrate the TA into the multi-robot scene (Sect. 2) so that it can take into account as many physical details as needed, including permanent or spurious disturbances and time, even without the actual plant. Although the TA part can be independent from the robots, in this case, is distributed among them, and constitutes a new layer on top of the existent ones for their control.

As the focus of the work was the integration, we opted to implement a quite straightforward distributed method for task assignment based on simple auctions carried out by agents that represent tasks [10]. However, it is easy to change the TA mechanism for any other, as it is independent of other components of the framework.

For TA operation to synchronize with lower-level controllers, its behavior is described by state machines, just as reactive controllers are. This guarantees that data is used timely and, additionally, provides a means to systematize their development and implementation. As computations linked to deliberations of agents are considered timeless but not so the ones for the reactive parts related to the control of physical elements, transitions of state machines may occur instantly or require that some time passes. The proposed solution includes a state-machine execution module that takes care of execution of the instantaneous transitions at each simulators' time step.

The result framework includes a distributed TA which takes into account data in real-time and adapts to any change in the physical system, including failures of the robots. Thus, it overcomes the previously mentioned limitations of methods shown in literature, but also of our previous TA solutions [10–12].

We have tested the framework with different setups and variations regarding space, number of agents and tasks, as well as with some disturbances at the physical level (communication failures) and measured the system's performance of each scenario (Sect. 3.)

The experiments show that the integration of the MRTA with a physical simulator makes it straightforward to test TA algorithms in dynamic environments and that the transition to controlling an actual plant could be done seamlessly (concluding section.)

2 Proposed Framework

Integrating TA into a physical simulator allows to model every element in a logistics system, introduce disturbances or changes at any level without using actual physical robots, and test tasks assignments taking into account the execution times of physical operations (like traveling or handling goods) and communications.

The model of the proposed framework organizes as a Multi-Agent System (MAS) divided into layers. Each layer deals with a different level of complexity in the allocation solution: one for task allocation, one for route planning, one for route execution, one for traffic management, and one for executing simple movement commands. This segmentation makes it easier to obtain the system answers at different levels, allowing the communication of our framework with many robots or models, regardless of their complexity. Each layer can also be used as an independent module. So, the users may use our framework as a standalone decision-support tool or any of its modules to complement other architectures.

Decentralized control architectures provide systems with adaptability and robustness when facing dynamic environments [7], and allow distributing computation system components. Indeed, many approaches have this type of architecture to implement TA based upon auctions [13].

In our case, the model contains two types of agents: *managers* and *executors*. Managers offer the task they represent for bidding and executors are bidders [10, 11]. The task allocation is made through a series of closed-envelope auctions, where the bidders do not know the bidding values of other participants.

2.1 Agents' Communication

The framework operates similarly to how a taxi company would: the managers are passengers while the executors are taxis. When a task enters the system for its execution, a passenger (manager) is created at the task's initial point and asks for transportation to its destination. Each taxi (executor) replies to the call with the cost to reach the passenger, and the latter selects the closest as the winner. After that, the taxi picks the passenger up, takes it to the destination, and drops it off. Figure 1 shows the interaction between two agents while completing a task. When a task enters the system, its manager sends a *call-for-proposals* message (CFP), with its position (P_I) and destination (P_F), to all executors in the system and collects their answers for a while. Upon receiving the CFP, all available executors plan the shortest route they would travel to perform the task and send the distance as their bidding value in a *proposal* message (PRP, *bid*). Then the manager evaluates all proposals, picks the closest one as the winner, sends an ACCEPT message to it, and waits for its arrival. When an executor receives an ACCEPT, it begins to travel the planned route, and when it arrives at P_I, it stops and sends a READY message to the manager. The loading operations begin and, when completed, the manager sends an ON message to the executor to continue with the travel. Upon arrival to P_F, the executor sends a DONE message to the manager, to start the unloading operations. Once the items are out, the manager sends an OFF message to the executor, thus finishing the task.

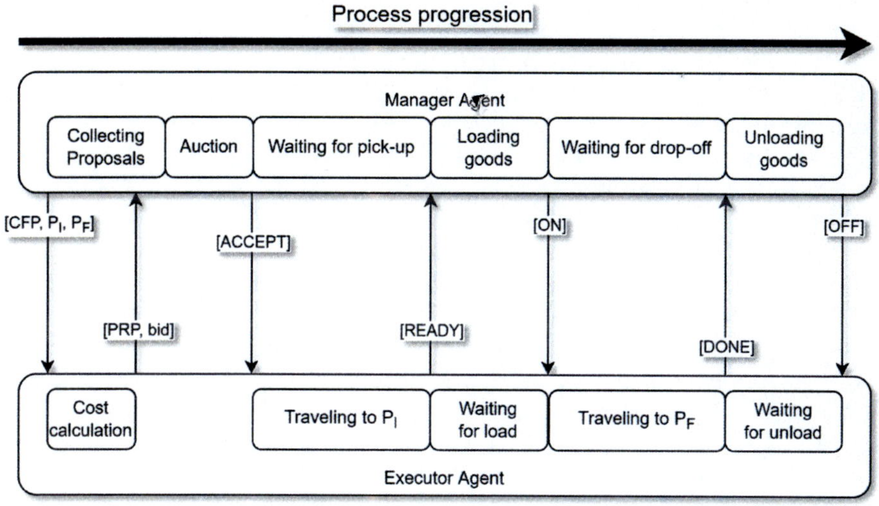

Fig. 1. Agents' communication during the execution of a task

2.2 Controllers' Representation

To perform those interactions, the agents need controllers capable of making, modifying, and aborting plans easily. Since most of our TA algorithm and general framework relies on agent behavior, we used state machines to define all the controllers' protocols.

For the modeling, we used Extended Finite State Stack Machines (EFSSM) [12], a method of our own making that we use extensively to teach controller design in our institution. EFSSM is a super-set of Extended Finite State Machines (EFSM) that include a states' stack where the next state is always at the top. This characteristic allows to easily make, rearrange, or discard plans by pushing and popping states into the states' stack. Thanks to this, EFSSM diagrams are simpler than those from regular EFSM when designing complex controllers.

Like EFSM, EFSSM ensures a formal description of the system's behavior to preserve predictability and verifiability. This type of language eases the design, analysis, and customization of the controllers' operations [14]. Also, state machines maintain synchronicity among the variables' values, which allows for determining the exact situation of the system at any time during its operation. EFSSM uses the same graphical elements as EFSM with a few new symbols, so the code synthesis from EFSSM diagrams is as simple as from EFSM ones. Figure 2 shows a simplified version of the deliberative controller (DLB) from a manager agent using EFSSM. To make the resulting code independent of the instructions' order, most variables present two fields: one for writing the value the variable will take on the next cycle and one for reading the value it has during the current cycle. To refer to the next cycle value, we use a plus sign (+) at the end of the variable name. As in EFSM graphs, the big circles represent states and the arrows the transitions between them. The new elements are the small circles and hexagons at the tip of the arrows. The small circles represent pushing states into the stack, while the hexagons represent popping them out. The small circles specify what state to push, and when empty, to push the current state. (If they appear at the origin of an arrow, they mean to pop the corresponding state or the top state from the stack.) Using these operations, the DLB can plan and adapt as required. For example, at the start, the controller is at INIT state and proceeds to send a CFP to all executors, and push the current state (INIT) and AUC into the stack before going to GET, to collect the *proposals*. With this procedure, the controller can advance, after the collection, to the auction state (AUC) by simply popping that state from the stack. Also, if later there is a need for starting a new auction, it can pop the INIT state from the stack to do so, e.g. if the assigned executor sends a FAIL message, because it is no longer able to perform the task, while DLB is at WAIT.

2.3 Time Model

The diagram in Fig. 2 shows that the whole auction process (from the CFP to announcing the winner) takes several cycles. This means that by executing the controller with a regular physics simulator's execution engine, time will pass during the whole operation. Under these conditions, after the controller finishes its operations for the current cycle will have to wait for another controller to finish a movement before proceeding with its actions. This situation can cause a mismatch between the environmental information a controller is working with and the actual environment, which can cause severe problems in the task assignment.

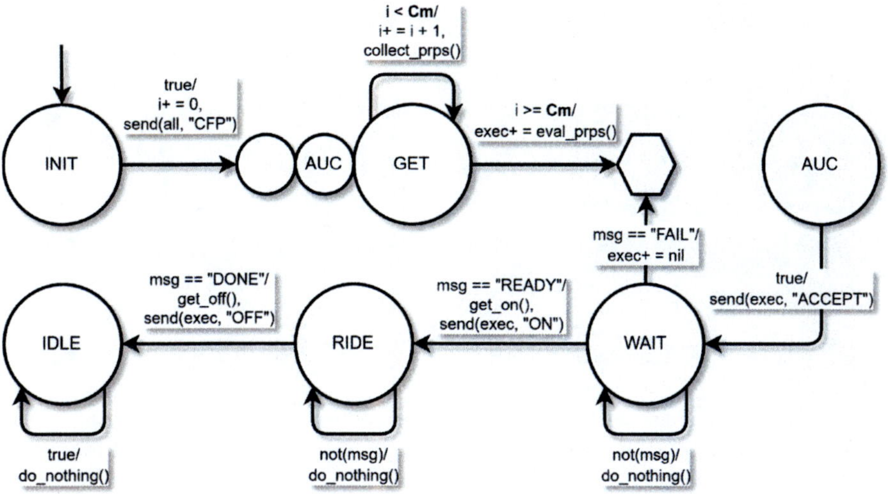

Fig. 2. Simplified EFSSM diagram of the deliberative controller for a manager agent

For example, imagine that three free robots (R1, R2, and R3) are passing, one right after the other, in front of the manager for a transport task, represented by a red dot in Fig. 3 (left), right at its starting time. The less costly solution in this situation would be to assign the robot right in front of the starting point (R2) since it is the closest one. However, since the auction lasts more than one cycle, the robots will keep moving, and R2 will pass the pick-up point, represented with a green dot in Fig. 3 (right), before the winning notification reaches it. At this point, R2 will have to go around the block to pick up the task (since it cannot go backward because the traffic flow does not allow it), making it the worst candidate to pick the task up: R3 is the closest one and R1 is ahead of R2 in the route around the block, represented by a red arrow in Fig. 3 (right). This bad assignment can greatly degrade the system performance in environments with a high concentration of robot agents, where this situation will be more common.

Our approach avoids this problem by executing immediate tasks like planning and communications, which have no time consumption, separated from those that do consume time. So, for the case described previously, there would be no delay between asking for transportation to the executors and informing the winner. In this case, R2 stays in front of the task's starting point from the moment it receives the CFP to the instant when it receives the assignment since no time passes between the two events, and it just needs to stop to pick up the goods.

Algorithm 1 shows the execution of this time model. Each simulation starts by executing *init*() to initialize the model elements: agents, communications, etc. Then the simulation executes the main loop. Instead of a single loop structure that calls all functions cyclically, our model relies on a nested loop inside the main one for operation differentiation. Each execution of the main loop implies

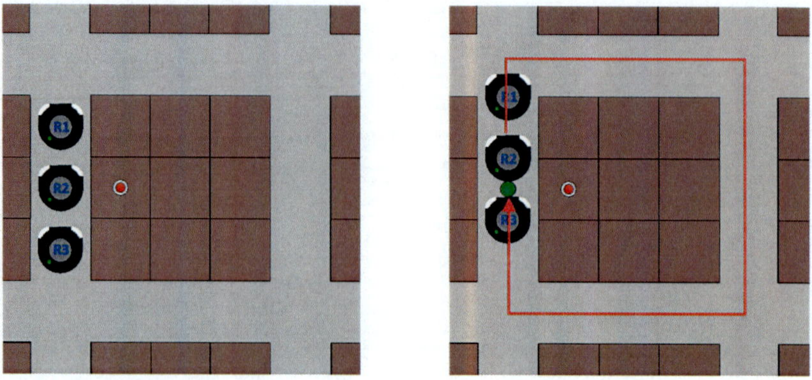

Fig. 3. TA solution degradation due to time consumption during the assignment. Left: Agents' positions when the proposals collection happens. Right: Agents' positions when R2 wins the assignment

the passing of time, denoted by the increment of the simulation time (*simTime*) by a time step (*dt*) at the end of the main cycle. The internal loop, which only contains immediate operations, can execute them many times without time advancement. The internal loop keeps on executing while there are changes inside any controller. So, when there is a change in any internal variable value, input, or output of any of the controllers, the time stops until all internal operations finish. When a controller has no more internal changes, we say it reached a stability point. There are two ways to achieve this condition: the controller is waiting for some external event, like a signal from another controller or a timeout, or the controller must change the environment and waits for time to flow, to do it. When all controllers reach a stability point, the internal loop execution stops, and time starts to run again. Since time does not pass during the internal loop execution, any operation that depends on time, like the interactions between the agents and the environment, happens outside the internal loop. *read_environment*() encapsulates all sensing operations from the environment, while all actuation operations are in *write_environment*().

The internal operations loop only contains functions that do not consume time. Through *read_inputs*(), all controllers register the data sent by other controllers. However, all controllers do not send their signals the same way. Some maintain the signal value for some time, while others send it only for one time step. In this last case, if time stops right at the reception of the signal, the controller may believe that it is a new notification. To avoid this, the controllers must treat the inputs differently when time is flowing than when it is still. Given that the internal loop executes once in a time loop when time flows and more times when time stops, *read_inputs*() receives the parameter *first* to differentiate these situations. The *step*() function determines the values of all variables for the next cycle, according to the current values of the inputs and variables. Also, all controllers send signals and data to other controllers through *write_outputs*().

Finally, *update*() assigns the values determined by *step*() to the corresponding variables. Since all writing operations become valid during *update*(), the variables maintain the same value for all reading operations in *step*(). Such distribution ensures that all variables remain the same during a cycle, which allows generating the code with order independence between the instructions [12]. The return value of *update*() tells if the internal loop execution must go on or not by comparing all variables' current values with their previous ones. If at least one value is different, the loop keeps executing. If all controllers report no changes, the system is at a stability point, and the main loop execution may proceed.

Our time model ensures that all controllers operate with environmental information adjusted to the current situation in the model and also allows a representation of the modeled transportation system closer to reality, where different processes consume different amounts of time. This capability would make it easier to integrate our model with a digital twin of a physical system. The digital twin would provide precise data from the physical system, like travel speeds, waiting times, and task distribution, which would allow a highly accurate representation of the system with our model. This model could later be used to study how different changes in the system (like the number of agents, routes, task distribution, etc.) could affect its performance.

In [15] there is a public repository containing a brief explanation of each controller's functions and operations, the EFSSM diagrams that model their behavior, the code corresponding to those diagrams, and the files for executing our program in CoppeliaSim.

Algorithm 1. Time model execution

```
dt ← 0.05                                   // Simulation time step
simTime ← 0                                 // Simulation time
init()                                      // Initialization
while true do                               // Main simulation loop
    int_ops ← true                          // Internal operations flag
    first ← true                            // First internal cycle indicator
    read_environment(simTime)               // Sense the environment
    while int_ops do                        // Internal operations loop
        read_inputs(first)                  // Receive information
        step()                              // Calculate next values
        write_outputs()                     // Send information
        int_ops ← update()                  // Establish next values
        first ← false                       // Finish the first internal cycle
    end while                               // End of internal operations loop
    write_environment()                     // Modify the environment
    simTime ← simTime + dt                  // Advance the simulation time
end while                                   // End of main simulation loop
```

3 Results

To evaluate the framework's performance when dealing with changes in the plan and errors, we simulated the operation of several mobile robots in a warehouse with a Manhattan distribution. The environment contains four blocks with bidirectional roads on their sides, as shown in Fig. 4. Each blocks has a port on its most external corner (represented by a white dot), where the goods are to be picked-up or dropped-off. Each port has gates on each adjacent road (represented by black dots), where the executors must stand to interact with the port. The tests consisted in 5 mobile agents executing 25 transportation tasks in total. The tasks are totally random, which means that they appear in the system at random times, at a random port and need transportation to another random port (excluding the one they are at.) This randomness helps to assess the TA algorithm's adaptability, since there is no information on the future tasks to perform and the TA planning must be revised every time a new task enters the system. Since the auctions to determine the TA only include the active executors that answer the calls, our system automatically adjusts its solution to the number of robots present at each moment. If the number of participating robots goes under the number of active tasks, the time separation between tasks' completions will grow to the value it would have had if the system started with that amount. To evaluate the systems' resistance to errors, we mimicked communication errors by eliminating a percentage of the messages exchanged during the assignment process. The error probability ranged from 0 to 100%, in steps of 10%. The parameter selected for measuring the system performance was the total time it took the agents to perform the totality of the tasks. These measurements depend on the agents' traveling times. A more traditional approach would estimate these times using the distances to travel and the agents' speed. This simplification is valid if nothing interferes the agents' movements, like other agents or other physical elements. The physical simulator allows our model to represent the agents' movements, as well as their reaction to any interference they may encounter while traveling. Each agent is provided with a distance sensor in the front, to detect any object in its path and avoid collisions.

The executions' results appear in Fig. 5. The graphic shows the total time, in seconds, for the agents to execute all tasks against the communication errors probability, in percentage. We ran 100 executions for each different error probability value to calculate the total completion time average and standard deviation. There is little variation in the total completion time from 0 to 40% error probability. In this range, the completion time average remains between 100 and 115 s, with a standard deviation around ± 6 s. This variability comes from the tasks' randomness, which affects the distance the agents must travel to complete the tasks. On the other side, the small affection from the errors in this group comes from the fact that if more than one executor agent is available when a task manager performs an auction, the error probability for that manager to not receive a *proposal*, and wait a while to perform another auction, is the joint error probability of those executors, which diminishes it significantly. However, from 50% to greater probability errors appears a steeper ascending tendency in the

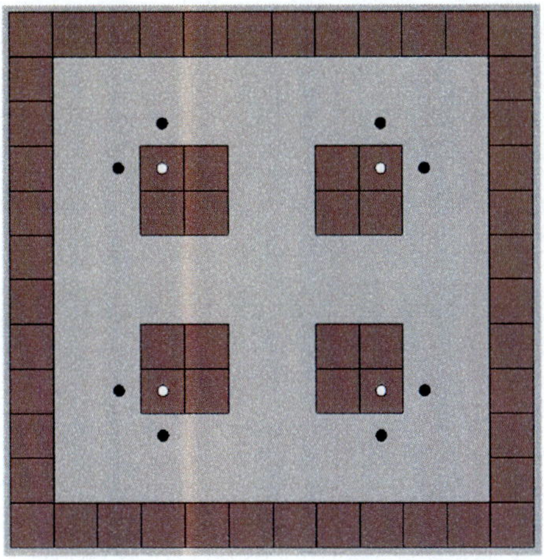

Fig. 4. Plant distribution for the tests

graphic, which shows the impact of those errors in the performance. For a 90% error probability the total completion time increases in almost 70% respect the execution without errors. We did not include the results for 100% error probability in Fig. 5 since the agents could not communicate between them and the tasks did not get attended, as expected for this case.

Our experiments demonstrated that our TA algorithm is both adaptable to changes and robust against errors. The plan changes had a small effect on the end results, producing small variations in the completion times due to the randomness of the distances for the agents to travel. On the other side, the communication errors degrade the system performance, but the system remains capable of finding and executing a TA solution.

4 Conclusion

In order to make MRTA applicable to real cases and, what is more, to test new or potentially hazardous scenarios, it is interesting to have a dynamic MRTA that can be linked to a physical simulator.

We have proposed a MAS-based MRTA integrated into a physics simulator that uses auctions as the base mechanism for TA. The whole stack of controllers for the manager and executor agents in the system have been modeled by EFSSM. These formal, graphical descriptions are directly synthesized into programs by using a template.

The integration of the MRTA with the physical simulator makes it straightforward to simulate new facilities served by mobile robots or errors, either physical or software-related, changes in the traffic network or any other scenario without disturbing the actual plant.

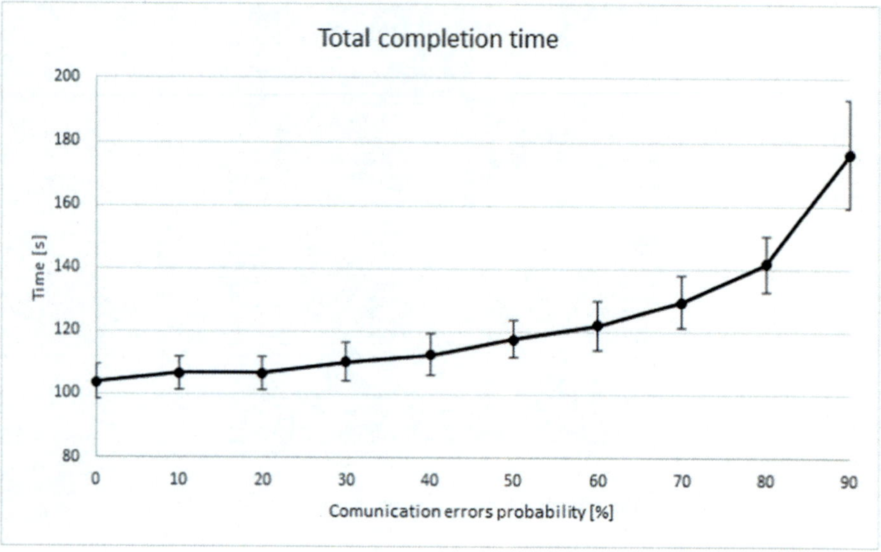

Fig. 5. Tasks' completion time vs. communication errors probability

As the execution of the EFSSM is embedded into the physical simulation engine which runs objects' software repeatedly with time increasing at every iteration, EFSSMs require a specific execution engine that enables them to transition from one state to the next one without consuming time for those computation-only operations that are assumed to occur instantaneously. We have also included this execution engine so that the upper levels of the controllers can communicate with each other without consuming time for operations that do not involve physical time. Additionally, this makes possible for the MRTA block to run independently of the physical simulator.

The experiments show not only that the framework works fine but that the auction-based MRTA is capable of dealing with dynamic cases where new tasks are defined along time and mobile robots might add or quit the system during operation. The latter is simulated by injecting communication errors which cause some random robots temporarily not taking part in auctions. It has been shown that this causes a degradation in the performance of the system but all tasks are carried out.

In the near future, we plan to tune-in the parameters of the EFSSM that model the MRTA so that it minimizes the time to complete the tasks.

References

1. Pol, R.S., Murugan, M.: A review on indoor human aware autonomous mobile robot navigation through a dynamic environment survey of different path planning algorithm and methods. In: International Conference on Industrial Instrumentation and Control (ICIC), pp. 1339-1344 (2015). https://doi.org/10.1109/IIC.2015.7150956

2. Ribas-Xirgo, Ll., Chaile, I.: An agent-based model of autonomous automated-guided vehicles for internal transportation in automated laboratories. In: ICAART 2013 - 5th International Conference on Agents and Artificial Intelligence, vol. 1, pp. 262-268 (2013). https://doi.org/10.5220/0004257702620268
3. Ribas-Xirgo, Ll., Miro-Vicente, A., Chaile, I.F., Velasco-Gonzalez, A.J.: Multi-agent model of a sample transport system for modular in-vitro diagnostics laboratories. In: 2012 IEEE 17th International Conference on Emerging Technologies & Factory Automation (ETFA 2012), pp. 1-8. IEEE (2012). https://doi.org/10.1109/ETFA.2012.6489591
4. Nunes, E., Manner, M., Mitiche, H., Gini, M.: A taxonomy for task allocation problems with temporal and ordering constraints. Robot. Auton. Syst. **90**, 55–70 (2017). https://doi.org/10.1016/j.robot.2016.10.008
5. Sarkar, C., Paul, H.S., Pal, A.: A scalable multi-robot task allocation algorithm. In: IEEE International Conference on Robotics and Automation (ICRA), pp. 5022-5027 (2018). https://doi.org/10.1109/ICRA.2018.8460886
6. Kim, M.-H., Kim, S.-P., Lee, S.: Social-welfare based task allocation for multi-robot systems with resource constraints. Comput. Ind. Eng. **63**(4), 994–1002 (2012). https://doi.org/10.1016/j.cie.2012.06.011
7. Schneider, E., Sklar, E.I., Parsons, S.: Mechanism selection for multi-robot task allocation. In: Gao, Y., Fallah, S., Jin, Y., Lekakou, C. (eds.) TAROS 2017. LNCS (LNAI), vol. 10454, pp. 421–435. Springer, Cham (2017). https://doi.org/10.1007/978-3-319-64107-2_33
8. Zool, I., Nohaidda S.: A survey and analysis of cooperative multi-agent robot systems: challenges and directions. In: Applications of Mobile Robots (2018). https://doi.org/10.5772/intechopen.79337
9. E. Rohmer, S. P. N. Singh, M. Freese. V-REP: a versatile and scalable robot simulation framework. In: 2013 IEEE/RSJ International Conference on Intelligent Robots and Systems, pp. 1321-1326 (2013). https://www.coppeliarobotics.com
10. Rivas, D., Jiménez-Jané, J., Ribas-Xirgo, L.: Auction model for transport order assignment in AGV systems. In: Fuentetaja Pizán, R., García Olaya, Á., Sesmero Lorente, M.P., Iglesias Martínez, J.A., Ledezma Espino, A. (eds.) WAF 2018. AISC, vol. 855, pp. 227–241. Springer, Cham (2019). https://doi.org/10.1007/978-3-319-99885-5_16
11. Rivas, D., Ribas-Xirgo, Ll.: Agent-based model for transport order assignment in AGV systems. In: 24th IEEE International Conference on Emerging Technologies and Factory Automation (ETFA), pp. 947–954 (2019). https://doi.org/10.1109/ETFA.2019.8869302
12. Rivas, D., Das, P., Saiz-Alcaine, J., Ribas-Xirgo, Ll.: Synthesis of controllers from finite state stack machine diagrams. In: 23rd IEEE International Conference on Emerging Technologies and Factory Automation (ETFA), pp. 1179-1182. Torino, Italy (2018). https://doi.org/10.1109/ETFA.2018.8502451
13. Otte, M., Kuhlman, M.J., Donald, S.D.: Auctions for multi-robot task allocation in communication limited environments. Auton. Robot. **44**, 547–584 (2019). https://doi.org/10.1007/S10514-019-09828-5
14. Cohenour, C.: Teaching finite state machines (FSMs) as part of a programmable logic control (PLC) Course. In: 2017 ASEE Annual Conf. & Exposition. Columbus, Ohio (2017). https://doi.org/10.18260/1-2-28917
15. Rivas, D.: Software EFSSM models and code repository (2022). https://github.com/DanielRivas88/MRTA-in-physical-simulator.git

Inner Speech: A Mechanism for Self-coordinating Decision Making Processes in Robotics

Adrian Romero-Garcés[1], Alejandro Hidalgo-Paniagua[2], Pablo Bustos[3], Rebeca Marfil[1], and Antonio Bandera[1](✉)

[1] Universidad de Málaga, Málaga, Spain
{argarces,rebeca,ajbandera}@uma.es
[2] SCALIAN, Labège, France
alejandro.hidalgo@scalian.com
[3] University of Extremadura, Badajoz, Spain
http://www.uma.es, https://www.scalian.com, http://www.uex.es

Abstract. The experience of inner speech is a common one for humans, playing a relevant role in generating spontaneous responses to the context but also in regulating how we think and behave. Intimately tied to our sense of self, the inner speech provides, via a mechanism of internalization, a running monologue of thoughts. This monologue plays a basic role in relevant aspects related, for instance, with our ability for structuring, regulating, and shaping our activities. In this paper, we emphasize this specific aspect of the inner speech and run some examples in the CORTEX software architecture, where semantic tokens (words and sentences) are employed not only for internalizing the world but also for regulating the decision making activities.

Keywords: Inner speech · Cognitive architecture · Inner world

1 Introduction

Different theories have formulated the inner language in people as a form of private speech, which we can summarize as the silent production of words in one's mind. In psychology, there is currently an open discussion about the precise nature of this inner speech [4,7,8]. For authors such as Gregory [8], the inner speech is reactive, i.e. it occurs as a spontaneous and uncontrolled response to the context the person finds herself in. It is then the responsible of the inner generation of more complex, elaborated perceptions. The production of these words or sentences are not under our control, and do not involve any effort. On the other hand, other authors, although agree with this proposal that some inner speech utterances are reactive, argue that not all of them are. Thus, in general, they recognize the importance of inner speech in the self-regulation of cognition and behaviour [7].

D. Tardioli et al. (Eds.): ROBOT 2022, LNNS 590, pp. 588–599, 2023.
https://doi.org/10.1007/978-3-031-21062-4_48

For this last school of thought, we usually engage in inner speech when we are deliberating about what to do in a specific scenario. This inner speech then plays a cognitive and self-regulatory role in the control of one's own behavior. Briefly, as we deliberate, we are aware, and sentences, or at least words, in natural language come to our minds [7]. In this way, egocentric language, whose aim is to internalize and guide thought and behavior, helps us to elaborate a plan with respect to the activity we perform, and helps us to mental orientation and conscious understanding.

With the idea of implementing self-awareness in robots currently gaining relevance [2,15], the aim of mimicking the concept of inner speech is particularly interesting. To achieve this goal, we must not to confuse inner speech with other inner, non-verbal experiences [3]. Any solution that will not be supported by symbols is not an inner speech instance. On the other hand, the topic is inline with the Language of Thought Hypothesis (LOTH) proposed by Fodor [6]. The LOTH proposes that thinking occurs in a mental language. This mental language (Mentalese) shares with spoken language to be organized in meaningful words that can be combined into sentences. The meaning of these sentence depends on the meanings of its component words and how they are combined. That is, simple concepts are combined in organized ways according to certain rules of grammar to create thoughts.

The goal of encoding information related to perceptions and actions as semantic tokens was one of the driving forces behind the design of the CORTEX cognitive robotic architecture [1,13]. The central core of CORTEX is the Deep State Representation (DSR), a graph-based representation where all information coming from the agents in the architecture is annotated. A relevant part of the information in this DSR describes the context, but also the robot's intents and current activities, using semantics tokens. In this paper, we extend the CORTEX architecture to implement some kind of inner speech for self-organizing the robot's behaviour. Through words and sentences, our proposal is able to structure an inner dialogue, that allows to coordinate the activity of several decision makers.

The rest of the paper is organized as follows: Sect. 2 provides some details about related work, putting the focus on the relevance of inner speech as one of the mechanisms involved in self-regulation. Section 3 describes an instantiation of the robotics cognitive architecture CORTEX for the intralogistic robot and how the symbolic information in the DSR can be considered an internal monologue that allows the robot to self-organize its actuation in the outer world. Section 4 briefly introduces the real robot and scenario used for testing. Some key performance indicators obtained from our evaluation are provided. Finally, Sect. 5 draws some conclusions and introduces our future work.

2 Related Work

Inner speech is an area of interest in artificial intelligence and machine consciousness [10]. In the earlier work by Steels [18], the process of reentering generated speech as speech understanding (re-entrant loop) was used for pushing language but also its underlying meaning towards greater complexity. Re-entrancy was then mainly proposed for checking the intelligibility of an utterance in their own reception systems, being linked to the ability for generating complex grammars in natural language. In this work, the relevance of the inner speech is restricted to the area of natural language understanding or generation. Moreover, as Clowes [4] pointed out, having inner speech a conscious component, the construction of grammatical sentences is however usually considered an unconscious activity. On the other hand, Clowes [4] emphasizes the role of inner speech for organizing consciousness, regulating and shaping ongoing activities, and driving attention. This self-regulation model was evaluated in several experiments in which groups of agents had to execute different tasks [5].

Inner imagery and grounded inner speech appear as the relevant items in the 'consciousness test' by Haikonen [9]. Thus, a machine will be conscious if it has these phenomena without being pre-programmed, can describe their contents, and recognize them as its own results.

Chella [3] proposes a cognitive architecture for inner speech implementation in a Pepper robot, based on the Standard Model of Mind proposed by Laird et al. [12]. Considering their relevant ties, this proposal focuses on inner speech as a mechanism for reaching self-awareness in robotics. As in our work, they assume that the robot has linguistics abilities.

Large Language Models (LLMs) have demonstrated to be useful for dealing with problems in computational linguistics (speech recognition, natural language generation, machine translation...). But recently, they have also shown their ability for managing a rich internalized knowledge about the world [14], and even for answering questions that appear to require some degree of reasoning [11]. Considering these new capabilities of LMMs, Huang et al. [10] extend these models for becoming an interactive problem solver and serving as reasoning models combining multiple sources of feedback. Thus, in the Inner Monologue (IM), the actuation of the robot is self-adapted considering the feedback from a language-based scene description and the one provided by a human user that the robot is cooperating with. As in our case, the actions to be executed are chosen from a set of pre-trained robotic skills. If these skills are mapped in the IM to textual descriptions that can be invoked by the language model, in our approach they are mapped to graph structures that can be detected by the agents in CORTEX. If the IM chains together perceptions, robotic skills and human feedback in a shared language prompt [10], our approach chains perceptions, robotic actions and human requests in the DSR.

3 Inner Speech in the CORTEX Architecture

Figure 1 shows an instantiation of the CORTEX architecture for the domain of robotic intra–logistics. This is the scenario that we will use for demonstration in Sect. 4. Briefly, the goal of the robot is to manage roll containers in the store, satisfying the petitions from a team of human pickers.

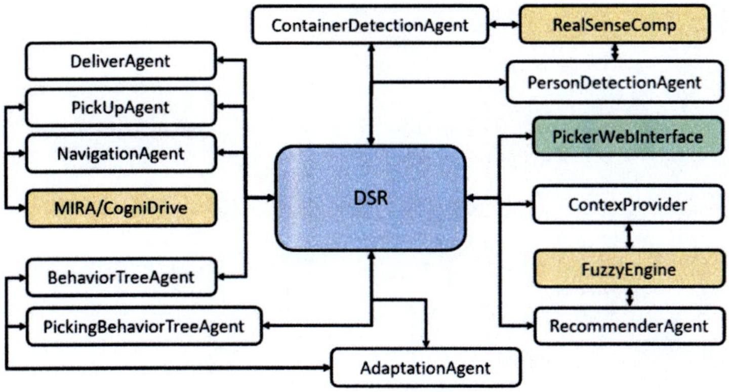

Fig. 1. Instantiation of the CORTEX architecture in the domain of robotic intralogistic.

The central item in the architecture is the DSR. The DSR is a multi-labelled directed graph that holds symbolic and geometric information within the same structure. Figure 2 shows one simplified example. It works as a working memory. The **robot** and the **roll container** are geometrical entities, linked to the **world** node (a specific anchor providing the origin of coordinates) by a rigid transformation. At the same time that we can compute the metric relationship between **robot** and **roll container** ($RT^{-1} \times RT''$), this **roll container** can be located close to the **robot**, and hence, the robot can launch the procedure for picking it up. In parallel, an agent can annotate that the **robot** is_not_detecting **people**. Features as the level of the **battery** are annotated as properties of the specific node linked to the **robot**. In this example, most of the nodes are present when we woke up the robot. Other ones, such as the **roll container**, are added by specific low-level perceptive modules. The verbs, encoded in the arcs (e.g. is_not_detecting) are updated by this same set of modules,

3.1 Perception and Action

In this scenario, the software architecture employs ten agents that are connected to the DSR. Robot localization and navigation is addressed by the MIRA/CogniDrive software from Metralabs. This stack is connected to the NavigationAgent, which is the responsible for updating all the needed information in the DSR. For instance, this agent monitors the battery status or the presence of close obstacles. The PersonDetection and the ContainerDetection agents use an Intel RealSense D435i RGBD camera for detecting and localizing people and roll containers in the environment. Some context items are fuzzified (e.g. this is the case for the **trolleypayload**, encoded into {EMPTY, LOW, MEDIUM, HIGH, FULL}). This enriches the dialogue (looking at the DSR, we can now for instance read that the **trolleypayload** is **FULL**).

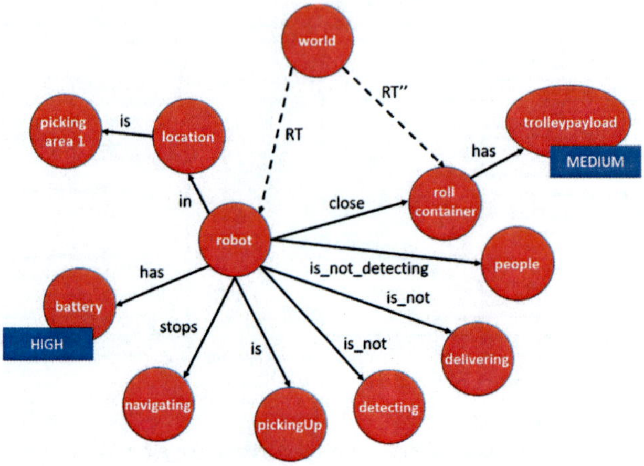

Fig. 2. Example of the DSR. Edges labelled as close or is_not denote logic predicates between nodes. On the other hand, edges starting at **world** and ending at **robot** and **roll container** are geometric and they encode a rigid transformation (RT and RT'' respectively) between them. Geometric transformations can be chained or inverted to compute changes in coordinate systems (see text).

More elaborated concepts related to non-functional properties such as safety are also generated and updated in the DSR at runtime. A fuzzy logic framework was designed for managing these high-level perceptions. The ContextProvider agent takes specific context information from the DSR and provides this to the FuzzyEngine. These context items are fuzzified and employed in a fuzzy inference engine for estimating, in this case, metrics related with safety, mission_completion and power_autonomy (basically, they are the non-functional properties considered in this intralogistic scenario). Figure 3(Top) provides a snapshot of estimation of these perceptions. The robot can then internalize that the **safety** is **MEDIUM - VERY HIGH**, the **mission_completion** is **VERY LOW - LOW**, and the **power_autonomy** is **LOW**. This situation is the unconscious result of a context situation (see Fig. 3(Bottom)). All this corpus of information, related with responses to changes on the context, constitutes in our approach the reactive voice of inner speech [7].

But the novelty here is to use this framework for involving inner speech in the 'episodes of deliberation' [7]. At deliberative level, the nominal course of action is encoded in the robot using two Behaviour Trees, designed using the BehaviourTree.CPP library by Davide Faconti[1]. In Fig. 1, the modules in charge of executing these BTs are the BehaviourTreeAgent and the PickingBehaviourTreeAgent. Following the strategy described in [16], both BTs implement a nominal course of action (i.e. the set of actions to be sequentially executed when all is going fine) as its main branch, but are extended with the variability expressed in alternative branches that we define at design-time. Figure 4 shows

[1] https://github.com/BehaviorTree/BehaviorTree.CPP.

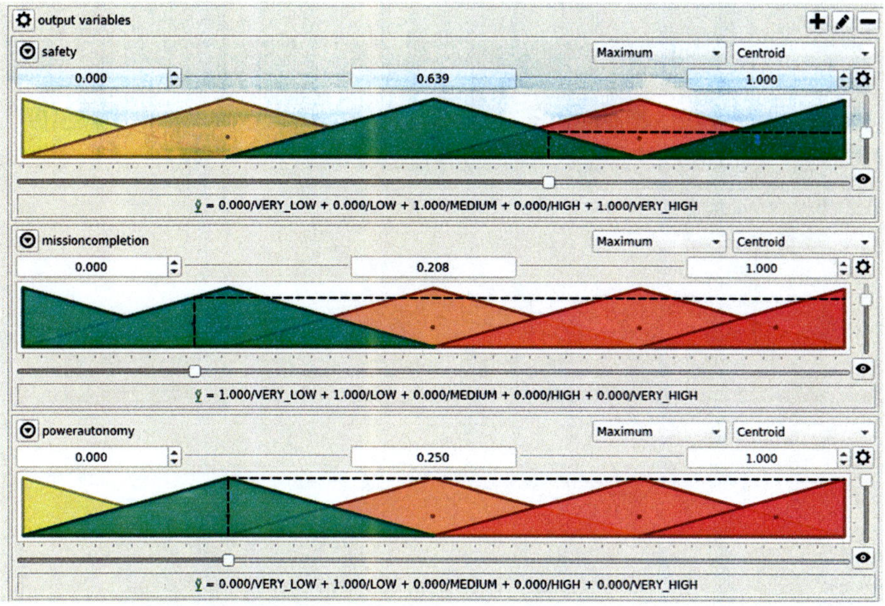


```
if robotbump is PRESSED then safety is VERY_LOW
if peopleinroom is EMPTY then safety is HIGH
if peopleinroom is LOW then safety is HIGH
if peopleinroom is MEDIUM then safety is MEDIUM
if peopleinroom is HIGH then safety is LOW
if peopleinroom is FULL then safety is VERY_LOW
if trolleypayload is FULL then safety is VERY_LOW
if trolleypayload is HIGH then safety is LOW
if trolleypayload is MEDIUM then safety is MEDIUM
if trolleypayload is LOW then safety is HIGH
if trolleypayload is EMPTY then safety is VERY_HIGH
if objectdelivered is DELIVERED then missioncompletion is VERY_HIGH
if objectdelivered is not DELIVERED then missioncompletion is VERY_LOW
if objectdetected is IDENTIFIED then missioncompletion is HIGH
if objectdetected is not IDENTIFIED then missioncompletion is LOW
if batterylevel is VERY_LOW then missioncompletion is VERY_LOW
if batterylevel is LOW then missioncompletion is LOW
if batterylevel is MEDIUM then missioncompletion is MEDIUM
if batterylevel is HIGH then missioncompletion is HIGH
if batterylevel is VERY_HIGH then missioncompletion is HIGH
if objectpicked is PICKED then missioncompletion is VERY_HIGH
if objectpicked is not PICKED then missioncompletion is VERY_LOW
if goal is REACHED then missioncompletion is VERY_HIGH
if goal is not REACHED then missioncompletion is VERY_LOW
if batterylevel is VERY_LOW then powerautonomy is VERY_LOW
if batterylevel is LOW then powerautonomy is LOW
if batterylevel is MEDIUM then powerautonomy is MEDIUM
if batterylevel is HIGH then powerautonomy is VERY_HIGH
if batterylevel is VERY_HIGH then powerautonomy is VERY_HIGH
if trolleypayload is FULL then powerautonomy is VERY_LOW
if trolleypayload is HIGH then powerautonomy is LOW
```

Fig. 3. (Top) Evaluation of high-level perceptions using fuzzy logic: (safety, mission_completion and power_autonomy; and (Bottom) Rules set for determining the current high-level perceptions

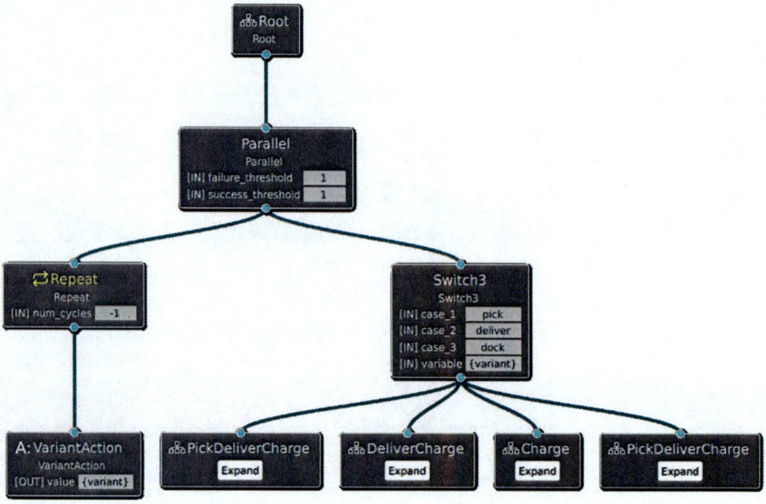

Fig. 4. Behaviour Tree encoding the behaviour of the robot at the higher level (see text)

the BT executed by the BehaviourTreeAgent. In the right side of the BT, under the Switch3 node, we have three possible robot's behaviour. The PickDeliverCharge is the nominal one: the robot moves to a picking area for picking up a roll container, delivering in a second area, and then returns to the charging area. The second branch forces the robot to deliver the roll container in the Delivery area (e.g. because the container is already full), and the third one commands the robot to abort the mission and goes to the charging area because battery level is very low. The branch to be chosen depends on an input command, the variant value, which is captured by the VariantAction node. Changing this value (pick, deliver, dock), we can modify the course of action at runtime.

The PickingBehaviourTreeAgent executes a BT that encodes the algorithm for allowing the robot to approach and pick up a roll container. There is also a nominal behaviour and alternatives, encoded as we have shown for the main BT in Fig. 4. The commands emanated from both BT Executors, which will be simultaneously active, must be coordinated, and also correctly synchronized with the information reflecting the context dynamics. Being all the information about perceptions and actions refreshed in the DSR, this coordination is addressed by the inner dialogue that automatically emerges. We detail an example in Sect. 3.2. In this situation, the robot is deliberating and engage itself in a inner speech. Contrary to the reactive speech, the inner speech involving the BT Executors is performed for a reason (satisfying a mission) and it involves trying (and sometimes failing). It then meets the criteria demanding an event to be qualified as an action [7,8].

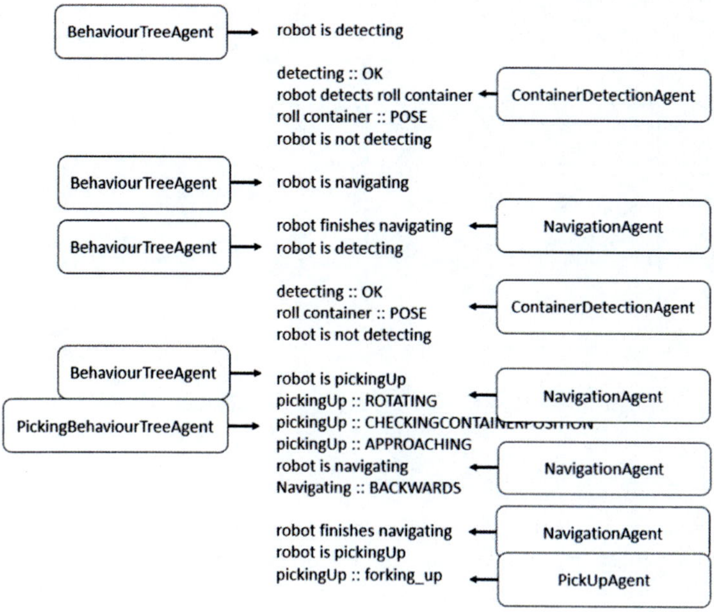

Fig. 5. Inner dialogue in the DSR.

Finally, Fig. 1 also shows two additional agents. The PickUp and the Deliver agents are responsible of lifting up and down the roll container once it has been detected and localized.

3.2 Self-coordinating the Robot's Behaviour

The robot receives the command of collecting a roll container from a specific picking area. The BehaviourTreeAgent drives the robot to this area and then launches the detection procedure. Figure 5 illustrates a simplified view of the inner speech (emphasizing the relevant items for this case) in the DSR and the agents involved in annotating the information. The information coming from the BT Executors (e.g. requiring to detect the container) is complemented with the linguistic (e.g. adding a roll container to the DSR when this is detected) and geometrical (e.g. pose of the container) data provided in a reactive manner by the perception modules . When the BehaviourTreeAgent annotates that the **robot** is **pickingUp**, the PickingBehaviourTreeAgent knows that it must take the control of the robot.

The execution in Fig. 5 is satisfying the requirements of a nominal use case. But it is relevant to note that the agents in the architecture are not asleep, waiting to be awakened by the deliberative modules. On the contrary, they are always running. Figure 6 shows a situation where people is detected close to the roll container. This information is annotated in the DSR by the PersonDetectionAgent. In this situation, The FuzzyEngine returns a low value for the safety, which is updated in the DSR by the RecommenderAgent. Taken into account this

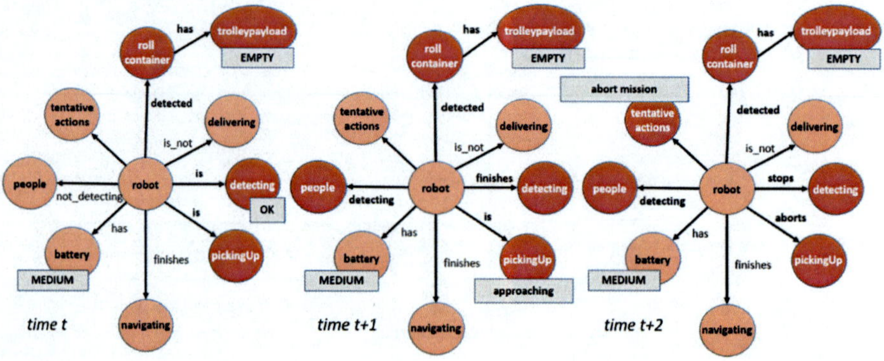

Fig. 6. The nominal execution is aborted because people is detected very close to the roll container. The figure shows a simplified view of the evolution of the DSR.

data, the AdaptationAgent decides that it is needed to change the course of action, and sends an indication (a variant value) for changing the executed branch to the BehaviourTreeAgent. A second fuzzy inference engine in the FuzzyEngine module quantifies what the best action to conduct is. The BehaviourTreeAgent should then stop the current action and executes the proposed new action, updating the DSR for maintaining the coordination among agents. All the inner dialogue narrating this 'episode of deliberation' can be traced by checking the linguistic terms annotated in the DSR. In Sect. 4, we document another example of self-adaptation.

4 Experimental Evaluation

For correctly moving roll containers in the retail scenario, we have employed the CARY robot from MetraLabs GmbH. Each container can be moved by the robot from one picking position to another one if this is required by a human operator, and will be delivered if there is not a new petition or when the roll container is full. For detecting the roll container, the robot uses a camera placed in its frontside. However, for fine approaching, the robot uses a laser range finder placed in its backside (at the front of the fork, very close to the ground). For navigation, it uses the four laser range finders on the front and the lidar and camera placed in its frontside.

The robot was deployed in a real retail store (Eroski) sited in Casarabonela (Malaga, Spain) in February 2022. The shop is approximately 300 m^2 in size, with narrow corridors (close to 2 m). The deployment was easily conducted: we need to capture the map of the shop and then divide up the space into picking areas. In each of these areas, we set a pair of observation poses and a delivering pose. When a human picker asks for a roll container in a picking area, the robot will try to deliver it in the delivering pose. On the other hand, when the operator asks for a container to be removed, the robot will search for it in the picking

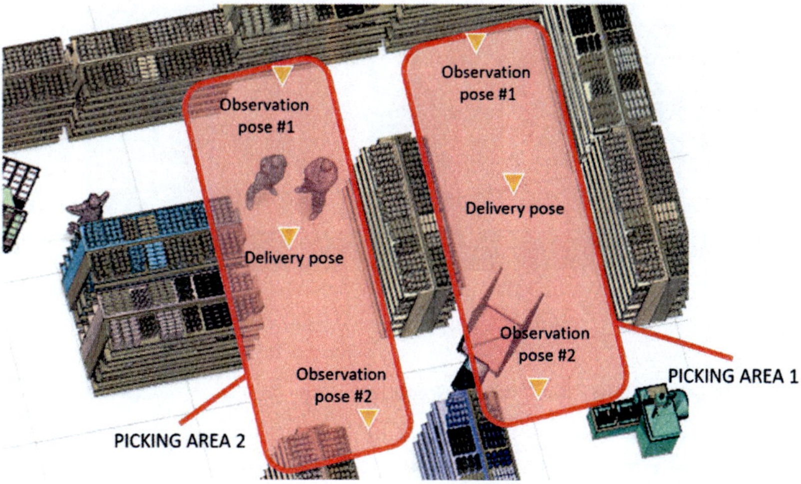

Fig. 7. Layout of the space: picking areas and relevant poses [17].

area from the observation points. If it fails to search from the first point, it will try to search from the second. Figure 7 shows an example of distribution of areas and relevant poses. Figure 8 provides some snapshots of the robot moving in the shop.

Next, we detail one mission. The robot is asked to move to one of the picking areas for picking up a roll container and deliver it to a second area. The robot arrives to the observation pose #1 and starts detecting the roll container. The container is detected and the robot pick it ups. The inner dialogue is the same that the one illustrated in Fig. 5. Now, the BehaviourTreeAgent asks the robot to move the container to a second picking area. However, the robot detects that the roll container is loaded with its maximum payload. The inner dialogue is shown in Fig. 9. It illustrates how the changes in the context are considered and they can modify the course of action. When the payload is FULL, the situation is considered unsafe by the robot, and the RecommenderAgent recommends to execute the default deliver action (move the roll container to the deliver area instead of delivering it to the next requested picking point). The AdaptationAgent sends a deliver value to the BehaviourTreeAgent, which aborts the current execution and annotates in the DSR that the robot should move to the default deliver area.

5 Conclusions and Future Work

The robotic software architecture CORTEX has been widely used in several projects, and has been endowed into robots working in real scenarios. The idea of annotating all the information coming from the agents in the architecture in a central graph-based representation has been a direct step towards real composability and has eased configurability. The information in this representation

Fig. 8. The CARY robot moving a roll container through the Eroski store.

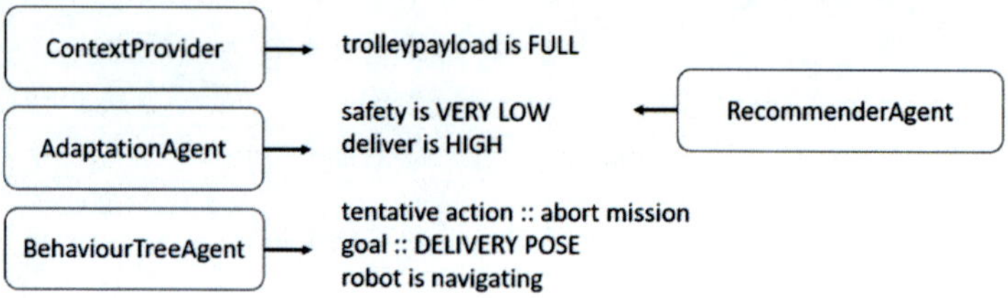

Fig. 9. The inner speech verbalizes that the context has changed and the system takes the decision of moving the roll container directly to the Delivery area.

can be geometric or symbolic. And being this symbolic information the result of our natural manner of encoding a problem-solving procedure, it has been unconsciously structured using natural language. This can be traced back to our previous work in other robotic domains, where there is already a body of topics or ontology that has been maintained and updated over time. A process of internalization thus emerges, in which an instrument of thought has gone from being used by researchers to define how to interact with the environment or solve problems to determining how the robot organises itself internally to deal with these same issues.

This paper describes how to employ this inner speech for synchronizing the execution of deliberative modules. In our examples, the dialogue is simple and we could argue that the same result could be obtained by substituting the annotated words by specific commands or messages. However, all this inner speech is here available to the robot. A stream of sentences such as "the container is FULL" or "the battery level is LOW" can be obtained from the DSR. We are currently working on using this inner stream as a source of input data for a Natural Language Processing framework. This source can complement the external data coming from a speech recognition system, allowing the robot to use all the linguistic information in the DSR for augmenting the interaction with human users. Future work will also focus on testing more complex decision makers based on Automated Planning.

Acknowledgements. This work has been partially developed within SA3IR (funded by EU H2020 ESMERA Project, Grant Agreement 780265), the project RTI2018-099522-B-C4X, funded by the Gobierno de España and FEDER funds, and the B1-2021_26 project, funded by the University of Málaga.

References

1. Bustos, P., Manso, L., Bandera, A., Bandera, J., García-Varea, I., Martínez-Gómez, J.: The CORTEX cognitive robotics architecture: use cases. Cogn. Syst. Res. **55**, 107–123 (2019)
2. Chella, A., Cangelosi, A., Metta, G., Bringsjord, S.: Editorial: consciousness in humanoid robots. Front. Robot. AI **6**, 17 (2019)
3. Chella, A., Pipitone, A., Morin, A., Racy, F.: Developing self-awareness in robots via inner speech. Front. Robot. AI **7**, 16 (2020)
4. Clowes, R.: A self-regulation model of inner speech and its role in the organisation of human conscious experience. J. Conscious. Stud. **14**, 59–71 (2007)
5. Clowes, R., Morse, A.F.: Scaffolding cognition with words. In: Proceedings of the Fifth International Workshop on Epigenetic Robotics: Modeling Cognitive Development in Robotic Systems, pp. 101–105 (2005)
6. Fodor, J.A.: The Language of Thought. Thomas Y. Crowell, New York (1975)
7. Frankfort, T.: Action and reaction: the two voices of inner speech. Teorema **16**(1), 51–69 (2022)
8. Gregory, D.: Are inner speech utterances actions? Teorema **39**(3), 55–78 (2020)
9. Haikonen, P.: Robot Brains: Circuits and Systems for Conscious Machines. Wiley, Hoboken (2007)
10. Huang, W., et al.: Inner monologue: embodied reasoning through planning with language models (2022). https://arxiv.org/abs/2207.05608
11. Kojima, T., Gu, S.S., Reid, M., Matsuo, Y., Iwasawa, Y.: Large language models are zero-shot reasoners (2022). https://arxiv.org/abs/2205.11916
12. Laird, J.E., Lebiere, C., Rosenbloom, P.S.: A standard model of the mind: toward a common computational framework across artificial intelligence, cognitive science, neuroscience, and robotics. AI Mag. **38**(4), 13–26 (2017)
13. Marfil, R., et al.: Perceptions or actions? grounding how agents interact within a software architecture for cognitive robotics. Cogn. Comput. **12**(2), 479–497 (2019). https://doi.org/10.1007/s12559-019-09685-5
14. Petroni, F., et al.: Language models as knowledge bases? CoRR abs/1909.01066 (2019)
15. Reggia, J.A.: The rise of machine consciousness: studying consciousness with computational models. Neural Netw. **44**, 112–131 (2013)
16. Romero-Garcés, A., Freitas, R.S.D., Marfil, R., et al.: QoS metrics-in-the-loop for endowing runtime self-adaptation to robotic software architectures. Multimed. Tools Appl. **81**, 3603–3628 (2022). https://doi.org/10.1007/s11042-021-11603-7
17. Romero-Garcés, A., Hidalgo-Paniagua, A., González-García, M., Bandera, A.: On managing knowledge for MAPE-K loops in self-adaptive robotics using a graph-based runtime model. Appl. Sci. **12**(17), 8583 (2022)
18. Steels, L., et al.: Language re-entrance and the 'inner voice'. J. Conscious. Stud. **10**(4–5), 173–185 (2003)

Approaches to Automatic Assembling of Plastic Toys

Daniel Sanchez-Martinez[1(✉)], Carlos A. Jara[2], and Francisco Gomez-Donoso[3]

[1] AIJU Instituto Tecnológico, Av. de la Industria, 23, 03440 Ibi, Spain
danielsanchez@aiju.es

[2] Human Robotics, University of Alicante, 03690 San Vicente del Raspeig, Spain
carlos.jara@ua.es

[3] Institute for Computer Research, University of Alicante, PO-BOX 99, Alicante, Spain
fgomez@ua.es

Abstract. Nowadays, there are a lot of repetitive and tedious tasks carried out by people who can be replaced by an intelligent robotic system, allowing the operators to perform other kind of dexterous works. In this paper, the tackled task is the automatic assembling of a toy doll in a real environment. In this work, authors present different intelligent systems that are able to perform this kind of task. The main challenge is the soft material the dolls are made of, whose physical behaviour dynamic, as their features change depending of the applied force and temperature. Our proposal is a comparison of different approaches to perform the task, focusing on the handling of such flexible materials. On the one hand, the proposed method acquires the information of the process by the previous demonstration of an expert operator, allowing to record all necessary data (movements, positions, velocities, etc). On the other hand, a perception module is developed employing vision-based algorithms to detect the pieces and to perform the assembly using a robot manipulator.

Keywords: Robotics · Soft manipulation · Computer vision · Intelligent assembly

1 Introduction

In an industrial environment, it is very common to find repetitive tasks performed by operators where the risk increases due to the long time periods. This great amount of times that workers perform these tasks, they are exposed to hazardous products or to make great physical efforts, leading to injuries.

Nowadays, there are many industrial processes that require dexterous actions, as for example the processes of the toy industry. Manufacturing a toy doll is a defined process which always follows the same steps. One of these steps is the doll assembly (insert arms, legs and the head into the body). The main difficulty is the manipulation of this flexible and deformable objects. This work includes

D. Tardioli et al. (Eds.): ROBOT 2022, LNNS 590, pp. 600–611, 2023.
https://doi.org/10.1007/978-3-031-21062-4_49

a comparison between different methods of automatic solutions for this task by using a collaborative robot. One method is by copying the movements of an expert operator during the task performance. The other one is a vision-based algorithm that is able to detect the pieces, grasp by the robot and assemble them.

The main goal of this study is to reduce the effort of the expert operators by removing the part of the task they need to apply manual force. The fact of being able to replicate complex movements makes this development really scalable and easy to move to other kind of tasks.

The rest of the document is structured as follows. First, in Sect. 2 other kind of related contributions are explained. In Sect. 3, the main developments are explained by differentiating the two methods mentioned previously. Then, in Sect. 4, results of the comparison between both methods are shown. Finally, in Sect. 5 some conclusions are described.

2 Related Works

Robot manipulation of soft materials is a field of research which is really on the rise. For instance, in [1], authors present a survey about robot manipulation and sensing with deformable objects in diverse industries (food handling, medicine, etc.). Another similar work can be seen in [2], where the authors provide a tutorial on foundational aspects about dynamics and shape models, where the objective is to reach a desired deformation with this information. In [3], it is presented a framework where the robot uses contact points of the environment to reach the shaping of cables in different configurations. This research is related with the assembly task of the dolls, since the object manipulated has to come into contact with other. A similar approach is explained in [4], where authors present the assembly of a rigid object into a deformable hole. In this case, they have developed a Deep Reinforcement Learning algorithm where the robot learns to perform the task from past actions and observations. In [5], it is explained a really critical point, the error identification during the assembly. They are able to classify success or failure patterns and predict future results.

As explained, the object to manipulate in the task is deformable, therefore its hardness has to be taken into account. In [6], it is explained how to estimate the hardness with a GelSight Tuch sensor. In [7], present a motion planning method to manipulate clothes and arrive to a goal shape from an initial one. They expose that sometimes is not possible to achieve the goal by a simple operation, probably multiple operations might be necessary. In [8], authors present an intuitive method of data collection based in a real time interface using monitors and a head-mounted display to teleoperate the robot to perform a folding task.

One of the points to be developed in this approach is the demonstration based learning of the complex task. As related work, it can be cited [9], where is presented a method for learning force-based manipulation skills from demonstration. Finally, in [10], it is described a vision-based manipulation system to perform a manufacturing task such as soldering flexible PCBs. That work is related to the another main point of our proposal, a vision-based system to perform the assembly task.

3 Approaches to Automatic Assembly

As explained before, two different methods have been defined to perform the automatic assembly task. The first one, which is described in Sect. 3.2, relies in a motion capture system. This setup allows us to acquire all necessary data to adapt the dexterous movements to the robot as explained in Sect. 3.1. The second approach, explained in Sect. 3.3, consists of a vision-based algorithm which is able to detect all pieces and to perform the assembly task.

In order to develop the task with both proposed methods, a robotic cell has been designed and developed (Fig. 1). For the setup, an operation table has been placed in front of the robot (UR 10e). This table is divided in two different parts: on the one hand, a part for the detection and grasping of the pieces; on the other hand, an assembly area has been developed.

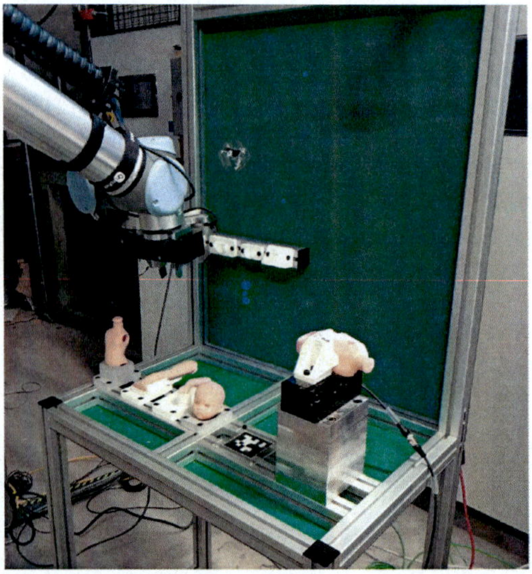

Fig. 1. Robotic cell setup involved in the experiments.

In the grasping part, there is one support for each piece. They are made with the negative shape of the pieces and consequently, they will ensure the fixation of the pieces in the table. However, all pieces are placed by a human, so they will not be exactly in the same position each time.

In the another part of the table, there is an extra gripper to support the robot for the assembly. The robot will put the body in that gripper and then, it will assemble the rest of the pieces there. To ensure the fixation in the gripper, there have been designed some special fingers with the negative shape and mirror pose of the body.

Basically, the process of the assembly task follows next steps: first, the placement of the pieces in the grasping part. Then, the detection of the grasping point of each piece. Once detected, the body is grasped and placed in the extra gripper. Afterwards, the arm and leg are grasped and assembled. Finally, the body is turned, and the rest of the pieces are assembled.

To automatize this kind of manual tasks, the following two different approaches have been defined.

3.1 Motion Capture of the Manual Process

Assembly a doll seems a simple task, but it is not really. There are several of dexterous movements performed by the human hands. A robot manipulator has not the same degrees of freedom as the human hand, so the target has been moved from the operator hand to the tool they use. These tools are pliers.

To record all information during the task performance, some sensors and markers have been installed and coupled to the pliers. These modifications in the original tool allow us to detect the position and the orientation of the tool around the table during its use. The final desired result is a 6D (position and orientation) trajectory which the robot can reproduce (Fig. 2).

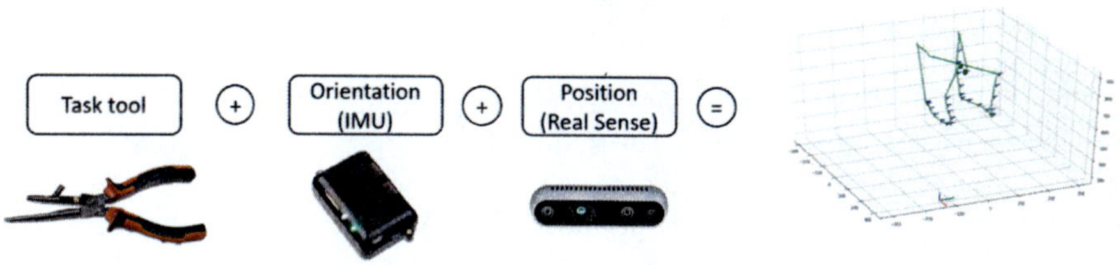

Fig. 2. Elements involved in the trajectory acquisition process and a sample of the trajectory recorded.

As shown in Fig. 2, data acquisition is captured by two different devices, a camera and an IMU (Inertial Measurement Unit). The first one detects the markers placed in the modified tool to know its 3D relative position. The camera is an Intel Real Sense d435i, connected by USB 3.0. This model provides RGB-D information of the environment, which means three color channels and depth data. To ensure a real time logging, computer vision is used to segment the colour markers placed in the tool. Finally, the marker point detected is transformed into a 3D point, knowing the relative distance between the marker and the camera. The IMU provides us information about the rotation. This device is a NGIMU from X-IO technologies, which has a WiFi module to connect the PC and to get the information in real time.

3.2 Automatic Assembly by Copying the Manual Process

One of the challenges is to move the dexterity and experience of an expert operator to the robot, reducing the data acquired to positions, orientations, velocities, and forces.

This method is divided in two different phases. The first one has been explained in the previous section, the motion recording of the expert worker. The second phase is the imitation of those trajectories by the robot. The expected result is the robot performance of the same movements than the expert worker did.

Because of the stored data is only positions and orientations, the robot applies the force depending of the difference between the target and current positions. This method allows to apply the maximum force possible along the trajectory.

This is the first step to achieve a further objective related to demonstration learning, where the robot will learn how to interpret all data recorded to perform the task properly.

3.3 Automatic Assembly by Using Perception

The second implemented method uses the same 3D camera detailed previously to detect all pieces needed using the RGB-D image and point clouds. Then, the robot performs the grasping, the placement of the body, and the assembly of the other parts. The algorithm developed uses 3D information and 2D information to detect the interest point of each piece to perform the grasping. The pipeline developed is adaptable to many other tasks, and it is composed by the steps of the Figs. 3 and 4.

Once the initial 3D data is obtained as point cloud, it is possible to remove sections of it with an estimated threshold. This step provides us basically the grasping area of the robotic cell. To estimate the position of the pieces, RANSAC is used to detect the plane of the supporting rack. Then, due to the pieces are in a fixed position, one bounding box per piece is manually defined and split up from the original point cloud. All 3D process is shown in the Fig. 3.

In this step, the 3D data is transformed to 2D. As the initial point cloud is composed by the initial RGB and depth image, pixels from the 2D image correspond to the points of the point cloud. Following, it is proceeded to remove pixels from the image according to the outlier points of the bounding box for each piece. Once the piece has been segmented, the center is computed with a vision-based algorithm. This 2D processing is shown in the Fig. 4.

3.4 Computer Vision Methods

Once removed useless pixels in the image (outside of the bounding box), an extra processing is needed. In normal conditions of lighting, the RGB component of the piece is known, hence it is easily to segment it. Finally, the whole piece is detected in the image, but only one concrete point is needed. To find this point, different computer vision methods (from OpenCV library) have been employed and compared:

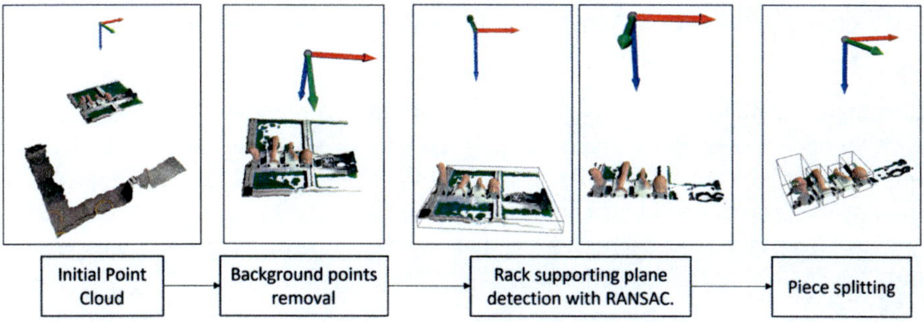

Fig. 3. 3D data processing from the initial point cloud to the final bounding box of each piece.

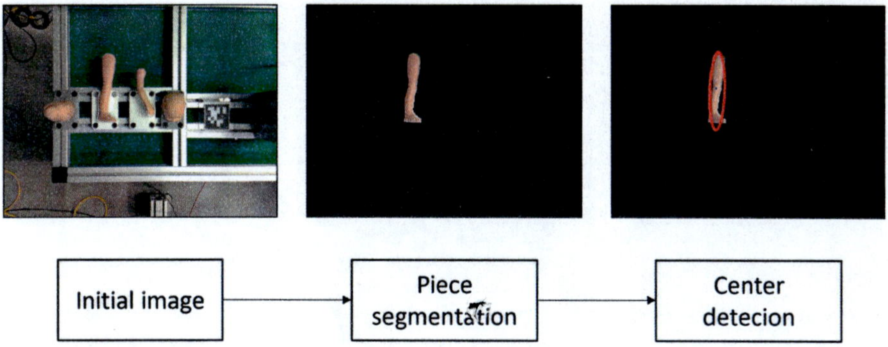

Fig. 4. Initial image and segmented one in the leg case and its grasping point and the best-fit ellipse.

- **Mass center detection:** in this case, the area of the leg in the image works as a simple rigid object with uniform density, so the mass center is located at the centroid. This is a good option due to the fact that pieces are always the same four.
- **Center of the best-fit ellipse:** Open CV uses LIN algorithm [11]. It is a more complex algorithm than the previous one, but provides more information (ellipse, direction, axis, etc.).

The main point in this part is to find a stable and accurate method able to provide the same point each iteration. Both methods need the area of the object in the image and the points that compose it.

4 Experiments and Results

In this section, it is compared the different methods and algorithms developed in order to take relevant conclusions from the results.

4.1 Comparison Between the Trajectory Recorded and the Replicated by the Robot

This experiment is focused in the comparison between the trajectories recorded from an expert operator during the performance of the task, and the trajectory performed by the robot. The result of this experiment will allow to compute the amount of error in the algorithm and if it is affordable for the performance.

Next graphics (Fig. 5) represent the trajectory recorded from the operator and the replicated by the robot.

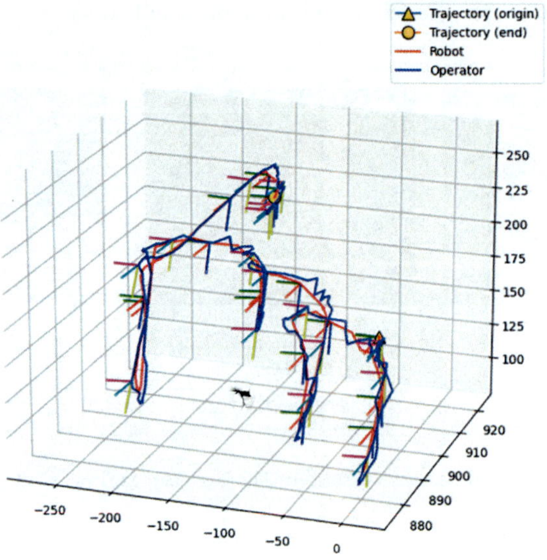

Fig. 5. Comparison between the trajectory of the operator and the trajectory of the robot.

Blue path is the operator's trajectory and the frames marked in cyan, magenta and yellow, represent the orientation. Red path is the robot's replicated trajectory with the orientation frames in red, green and blue. Next figures show the position and orientation error of the robot during the trajectory.

The main conclusions obtained from these graphics are that rotation error (measured in radians) is small to reproduce the trajectory accurate. However, in the position error, it is possible to see the high error peaks in the zones that the robot changes the direction completely. This error is due to the inertia and the accelerations of the own robot during the movements.

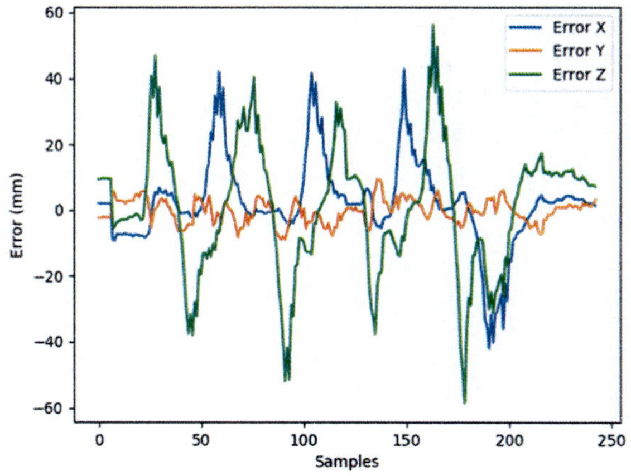

(a) Position.

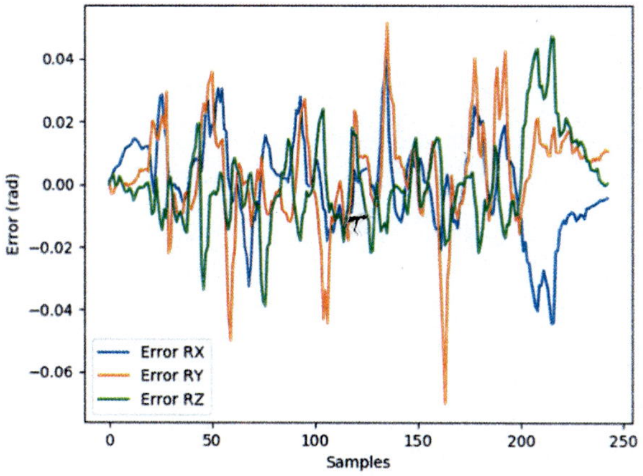

(b) Orientation.

Fig. 6. Position (6a) and orientation (6b) error during the repetition of the robot.

4.2 Comparison Between Grasping Point Detection in Motion and Static

One of the most important steps is the detection of the grasping point of each piece. This point is obtained from the vision-based methods as a distance in "X" and "Y" coordinates. As a high repeatability is needed, this test tries to compare the distances measured along 150 iterations in two different situations. In the first one, the robot is static between iterations; in the second one, the robot moves to another position, and returns to the same position in each iteration. All detections are performed with an eye-in-hand camera system.

Next graphic (Fig. 7) shows all the "Y" coordinate normalized distances measured during the 150 detections. It is possible to see the difference between the detection moving or being at the same position at each detection. Blue dots (static measurements) has to be more stable than red ones (with movement between iterations).

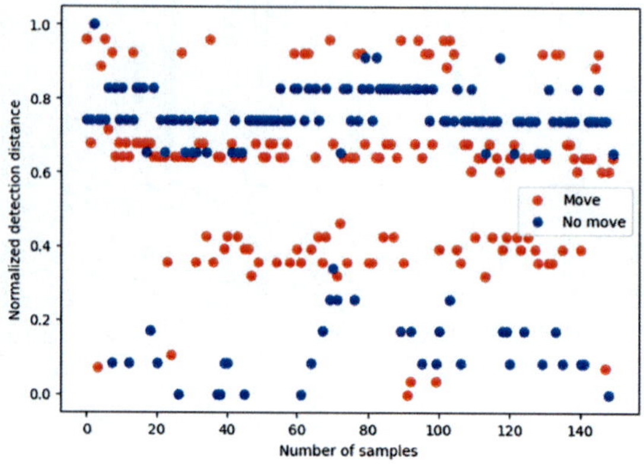

Fig. 7. "Y" coordinate is the distance (normalized) measured in each iteration between detection with the robot at the same position and moving every iteration.

It is possible to see that measurements of the static situation are more stable than the other. However, this situation is not real because in the real application, the robot will perform different movements between detections.

In conclusion, measurements of the static situation are more stable than the another situation. This difference between the measures is due to the repeatability of the robot. When it moves and returns to the detection position, there is an error, and it affects the detection algorithm. However, we have to deal with this fact as it corresponds to the real situation.

4.3 Comparison Between Grasping Point Detection with Mass Center and Fitting Ellipse Methods

Once tested the repeatability, we need to ensure the accuracy of the different computer vision methods explained previously. To test this feature, the robot will perform other 150 detections, and the mass center and fitting ellipse provided point will be compared at each iteration.

In this case, the Fig. 8 shows the normalized "X" coordinates detected during the test of one piece.

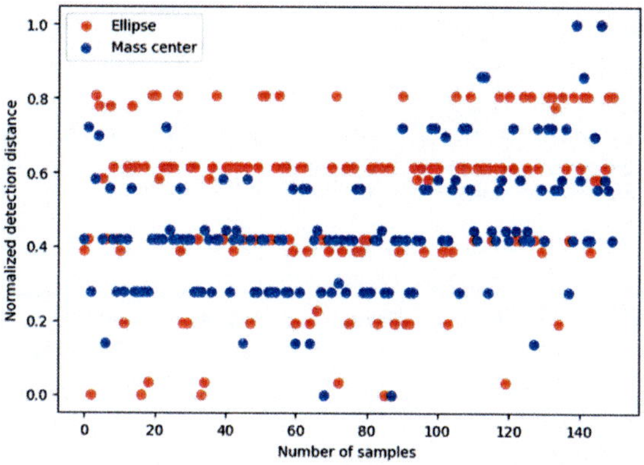

Fig. 8. "Y" coordinate is the distance (normalized) measured in each iteration comparison between mass center and ellipse fitting methods. Test made with the robot moving every iteration.

As we can see, there is no clear evidences of which method is more accurate than the other one. However, blue dots (mass center method) are less distributed than red ones (fitting ellipse method).

4.4 Comparison Between Grasping Point Detection Methods Errors

In the previous experiments explained, it is compared the repeatability and accuracy of one coordinate obtained from one piece. In this one, it is compared the standard deviation of all coordinates of all pieces in the two same situations the first experiment. Comparing the standard deviation, it is possible to see the different between all the measurements. Next figures represent the deviation error of each coordinate detected by the vision methods mentioned previously for all the pieces.

In Fig. 9a, it is shown the standard deviation of the measurements of "X" and "Y" coordinates provided by both computer vision methods. Results shown lead us to the conclusion that mass center measures are more stable and accurate with each other. However, this test does not represent the real application as the robot is static.

In Fig. 9b, it is shown the results with the robot moving at each iteration. In contrast to the previous graphic, this one represents the results of the application in the real situation. The error is higher in this experiment but still being affordable.

In conclusion, according to the results of the test, it is possible to say that the mass center is better than the ellipse fitting regarding to the accuracy and repeatability. The higher error provided by the mass center method in the real case is 0,64 mm, thus it is reached a really impressive repeatability.

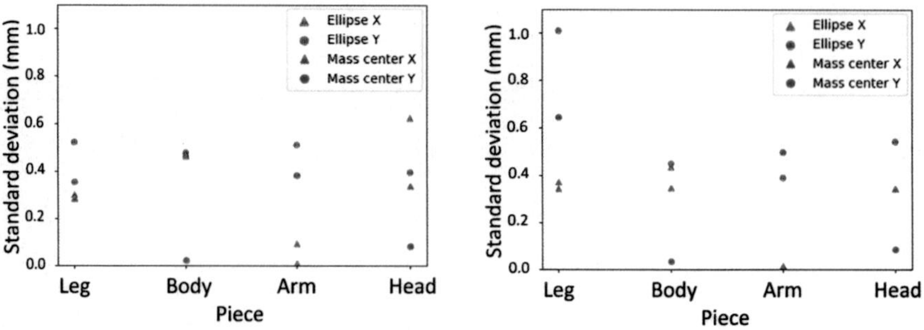

(a) Robot static.

(b) Robot in motion.

Fig. 9. Deviation comparison of every coordinate between ellipse and mass center method for each piece. Experiment performed in two situations: Robot static and robot in motion between detections.

5 Conclusion and Future Work

In this paper, a solution for the automatic assembly of dolls in an industrial environment is proposed. Two different methods are explained and compared. First method is based in the acquisition of the data from an expert operator and then, the robot replicates the task. However, due to the acquisition (processing and replication), an error is accumulated and makes difficult to carry on an accurate task performance. The second method is a vision-based algorithm which needs 3D and 2D data to detect the pieces. Nevertheless, this method needs to present the pieces in fixed positions to define the bounding boxes.

In conclusion, because of the assembly task needs a stable robot grasping of the pieces, the vision-based algorithm produces better results in terms of accuracy and repeatability according with the performed experiments.

As future work, we plan to apply neural networks to use the data from the operator and make the robot learn different ways to perform the tasks by the demonstration learning, including not only position and orientation, but also forces. In addition, it is planed to expand this method to other fields and use different tools to make the robot learn different dexterous tasks.

Acknowledgements. This work has been supported with the scholarship referenced as UAIND21-06B. And it was done in the context of the SOFTMANBOT project, which received funding from the European Union's Horizon 2020 research and innovation programme under grant agreement number 869855.

References

1. Sanchez, J., Corrales, J.-A., Bouzgarrou, B.-C., Mezouar, Y.: Robotic manipulation and sensing of deformable objects in domestic and industrial applications: a survey. Int. J. Robot. Res. **37**(7), 688–716 (2018)

2. Arriola-Rios, V.E., Guler, P., Ficuciello, F., Kragic, D., Siciliano, B., Wyatt, J.L.: Modeling of deformable objects for robotic manipulation: a tutorial and review. In: Front Robot (2020)
3. Zhu, J., Navarro, B., Passama, R., Fraisse, P., Crosnier, A., Cherubini, A.: Robotic manipulation planning for shaping deformable linear objects withenvironmental contacts. IEEE Rob. Autom. Lett. **5**(1), 16–23 (2020)
4. Luo, J., Solowjow, E., Wen, C., Ojea, J.A., Agogino, A.M.: Deep reinforcement learning for robotic assembly of mixed deformable and rigid objects. In: 2018 IEEE/RSJ International Conference on Intelligent Robots and Systems (IROS), pp. 2062–2069 (2018)
5. Hayami, Y., Shi, P., Wan, W., Ramirez-Alpizar, I.G., Harada, K.: Multi-dimensional error identification during robotic snap assembly. In: Uhl, T. (eds.) Advances in Mechanism and Machine Science. IFToMM WC 2019. Mechanisms and Machine Science, vol. 73. Springer, Cham (2019). https://doi.org/10.1007/978-3-030-20131-9_217
6. Yuan, W., Srinivasan, M.A., Adelson, E.H.: Estimating object hardness with a gelsight touch sensor. In: 2016 IEEE/RSJ International Conference on Intelligent Robots and Systems (IROS), pp. 208–215 (2016)
7. Tanaka, D., Arnold, S., Yamazaki, K.: Emd net: an encode-manipulate-decode network for cloth manipulation. IEEE Robot. Autom. Lett. **3**(3), 1771–1778 (2018)
8. Yang, P.-C., Sasaki, K., Suzuki, K., Kase, K., Sugano, S., Ogata, T.: Repeatable folding task by humanoid robot worker using deep learning. IEEE Robot. Autom. Lett. **2**(2), 397–403 (2017)
9. Lee, A.X., Lu, H., Gupta, A., Levine, S., Abbeel, P.: Learning force-based manipulation of deformable objects from multiple demonstrations. In: 2015 IEEE International Conference on Robotics and Automation (ICRA), pp. 177–184 (2015)
10. Li, X., Su, X., Liu, Y.-H.: Vision-based robotic manipulation of flexible PCBS. IEEE/ASME Trans. Mechatron. **23**(6), 2739–2749 (2018)
11. Fitzgibbon, A., Fisher, R.: A buyer's guide to conic fitting (1970)

Initial Results with a Simulation Capable Robotics Cognitive Architecture

Guillermo Trinidad Barnech[1(✉)], Gonzalo Tejera[1], Juan Valle-Lisboa[1], Pedro Núñez[2], Pilar Bachiller[2], and Pablo Bustos[2]

[1] Universidad de la República, Montevideo, Uruguay
gtrinidad@fing.edu.uy
[2] Universidad de Extremadura, Cáceres, Spain

Abstract. In this paper, we present some conceptual and experimental results obtained from the integration of a Robotics Cognitive Architecture (RCA) with an embedded Physics simulator. The RCA used, CORTEX, is based on a highly efficient, distributed working memory (WM) called Deep State Representation (DSR). This WM already provides a basic ontology, state persistency, geometric and logical relationships among elements and tools to read, update and reason about its contents. The hypothesis that we want to explore here is that integrating a physics simulator into the architecture facilitates the enacting of a series of additional functionalities that, otherwise, would require extensive coding and debugging. Also, we characterize these functionalities in broad types according to the kind of problem they tackle, including occlusion, model-based perception, self-calibration, scene's structural stability and human activity interpretation. To show the results of these experiments, we use CoppeliaSim as the embedded simulator, and a Kinova Gen3 robotic arm as the real scenario. The simulator is kept synchronized with the stream of real events and, depending on the current task, several queries are computed, and the results projected to the working memory, where the participating agents can take advantage of them to improve the overall performance.

1 Introduction

According to G. Hesslow [16], the *simulation hypothesis* states that a simulated action can elicit perceptual activity that resembles the activity that would have occurred if the action had actually been performed. Closely related to this line of thought, the field of *Intuitive Physics* has gained relevance in recent years. Following Kubricht [2], "...humans are able to understand their physical environment and interact with objects and substances that undergo dynamic state changes, making at least approximate predictions about how observed events will unfold". Moreover, recent experiments show evidence that humans might have some sort of embedded mental game engine to help them reason about their environment [1,5]. This research is tightly connected to Robotics, and has early precedents in perception, with experiments showing internal visualization

D. Tardioli et al. (Eds.): ROBOT 2022, LNNS 590, pp. 612–623, 2023.
https://doi.org/10.1007/978-3-031-21062-4_50

for navigation [19,20], and in architectures, with the adding to SOAR of an internal simulator [18]. Although some authors have raised objections to the theory, arguing that it is only fruitful when both time and space scales are small [3], continuous advances in simulation technology and computer power are pushing the limits further with new demonstrations [4,12].

In this paper, we present preliminary results on the integration of a PS (Physics Simulator) in a Robotics Cognitive Architecture (RCA) and how it can be used to solve known problems in robot perception. In particular, we aim to embed a simulation-based geometric reasoning pipeline into CORTEX. We have selected three use cases for this preliminary work that show the use of simulation to: (I) correct object poses perceived by the robot according to a physically plausible model, (II) detect pose perception errors and provide an automatic recalibration procedure and (III) reason about the persistence of out of sight objects. The following sections describe in more detail the CORTEX architecture, the use of simulation as an internal tool for complex scene understanding, and a description of the performed experiments to validate our application of this concept.

2 The CORTEX Architecture

The CORTEX cognitive architecture is a very dynamic proposal that has been evolving since its conception in 2016 [9,10]. In its current version, CORTEX defines a distributed architecture organized around a working memory (WM) and a set of agents that have access to it. The working memory is called Deep State Representation (DSR) and is formally a directed graph with vertices holding metric or symbolic data, and whose edges represent geometric or logic predicates. Vertices or nodes are concepts of the ontology, and edges are relationships between them. Being a WM, DSR is intended to represent the current situation involving the robot's body and intentions, and the space and objects proximal to it and relevant to the current task. Agents are responsible for creating and maintaining this representation, implementing the functionality of perception and motor modules, as well as the procedural, declarative and episodic memories of the Standard Model [23]

CORTEX establishes a way of operating within the working memory. The node representing the robot is always connected through an *RT* (Rotation-Translation geometric transformation) edge to one of the existing nodes representing a zone of space, typically a room in indoor scenarios. This edge is modified when the robot changes its location. From the robot, two distinct nodes, *body* and *mind* are connected downwards. The rest of the robot's parts are connected to *body* according to their kinematic relationships. These parts include rigid segments and pieces, joints, sensors and actuators. Raw data from sensors is stored in the node's attributes and made available to all agents. This is possible thanks to an efficient software design and implementation [6,7]. The other branch, *mind*, holds the current intention (or goal) of the robot. This intention is transformed into plans by deliberative agents, which create additional nodes hanging from

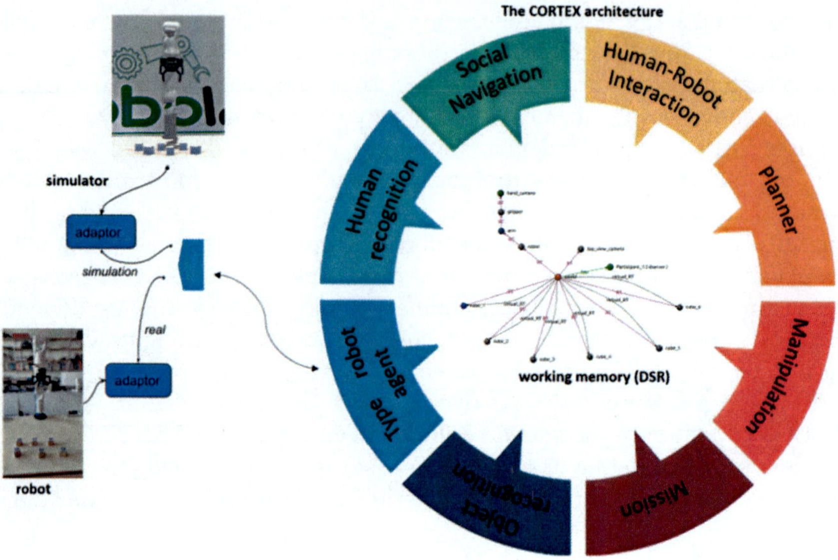

Fig. 1. The figure shows an instance of the CORTEX architecture with a double connection to the robot's body and a simulator.

intention to store and advertise the current list of subgoals or actions. All agents are made aware of the current intention and plans and take the actions of their choice to achieve the goal. Once this happens, the intention node is deleted and the agents go back to their local activities. Intentions are managed through a special agent called *mission-manager* that usually offers a GUI to the roboticist, accepts new missions from an interacting human or uses a scheduler to activate periodic missions. Figure 1 shows an abstract representation of a CORTEX deployment and Fig. 2 a real case scenario for social navigation. More details on the implementation of CORTEX can be found in [11].

3 What Can Be Obtained from an Embedded Simulator?

The working memory in CORTEX provides a persistent, structured state representation that can be used to implement complex cognitive mediated behaviours. However, there are many situations where direct perception and persistence are not enough to solve simple problems. What might be needed is some sort of pre-existing knowledge that could be injected into the WM, and provide a basic kind of common-sense. This knowledge can come from different sources and have various formal representations. A usual source in Cognitive Robotics is an ontology with rich relationships among concepts and an inference engine to answer queries, as in [17]. In this work, we also want to obtain predicates relating objects in the current scene, but we will focus on predicates that can be obtained from a physics simulator synchronized with the working memory. To explore this path

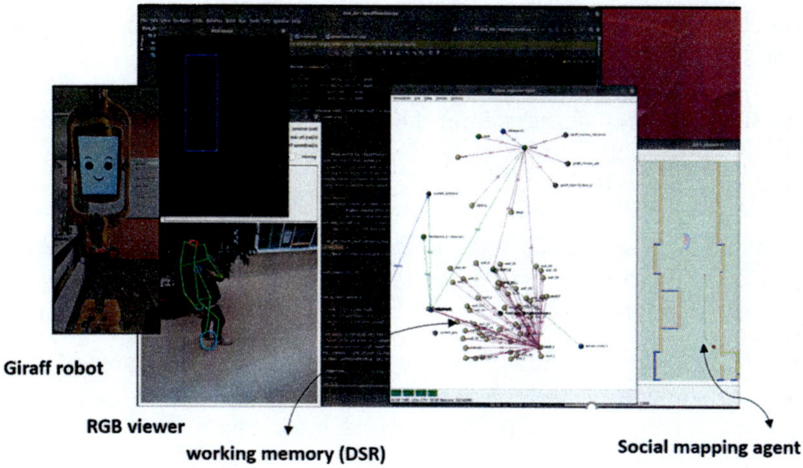

Fig. 2. Indoor example: Giraff robot navigation in a real scenario with people.

in a more systematic way, we have studied some examples and scenarios in which this common-sense knowledge would clearly improve the resolution of the task. The examples have been grouped into a set of informal categories that highlight common patterns. Three of them will be discussed here, as aforementioned in the Introduction, and the rest are briefly commented in the Conclusions section, as current research lines.

4 Experimental Setup

The experimental scenario includes of a Kinova Gen3 robotic arm placed on a table, and a set of 4cmx4cm cubes marked with a distinctive AprilTag [15] on one face. The arm has a RGBD camera (Realsense D415) attached to the wrist. Simulation is performed with the CoppeliaSim [14] using a choice of Physics engines[1]. The working memory in CORTEX is initialized from a file with a subgraph representing the robot and its sensors, and some additional nodes representing the scenario. An example configuration is presented in Fig. 4, where (a) shows the real world state and (c) how it is reproduced inside CoppeliaSim. All simulations run at 20 hertz on an Intel i9 10th generation and a RTX3090 GPU.

Three CORTEX agents have been created and deployed for these experiments.

1. **arm-controller**, this agent reads and inserts the gripper pose, read from the Kinova arm, into the working memory in real-time as a $SE(3)$ spatial transformation from the arm base. It also injects the raw stream of RGBD

[1] Bullet, ODE, Newton or Vortex.

data obtained from the RealSense camera as an attribute of the *hand_camera* node.
2. **scene-estimator**, detects AprilTags and inserts model cubes into the working memory. The cubes hang from the node *camera* through an RT edge with the estimated relative pose.
3. **simulation-handler**, is in charge of synchronizing the working memory with the simulation bidirectionally. It reads cubes poses from the graph to update the simulation, and publishes them back as *Virtual_RT* edges. *Virtual_RT* edges are not part of the RT tree since they would induce loops. Instead, they are treated as symbolic edges representing an *opinion* from the simulator.

In CORTEX, agents run autonomously and can only communicate indirectly through the working memory. An example state of the DSR is given in Fig. 3 with all the information provided by the presented agents.

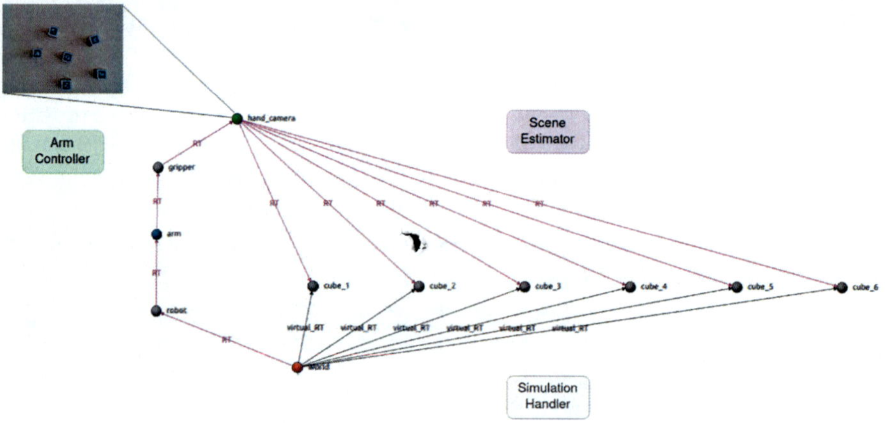

Fig. 3. Example DSR state. Arm Controller writes camera readings as an attribute of the hand_camera node. Position estimations of the cubes are presented as RT edges from *hand_camera*, inserted by the Scene estimator agent. Virtual_RT edges represent geometric transformations from the origin to every cube, and are given by the Simulation Handler agent.

We now describe three experiments, in which the use of the same integrated simulator, facilitates the approach to three well-known problems in Robotics.

4.1 Model-Based Perception

The perception of objects in the world by the robot is always a noisy process, subject to positioning errors. When the working scenario includes objects placed on top of, or leaning over other objects, the perception errors will yield physically unfeasible configurations, with floating and intruding objects. The simulator can apply the internal physics laws to quickly find a feasible configuration, and project it back to the WM. The result being that the robot perceives a configuration of objects that has been corrected by a complex internal model following the laws of Physics (Fig. 4). By embedding this pipeline within the working memory, all agents gain access to this kind of physical reasoning, and a better understanding of the observed scene. Examples of this functionality have been shown in [12,13]. In the following sections, we describe some ways that this can improve the robot's performance and precision.

(a) (b) (c)

Fig. 4. When the robot encounters state (a) using the RGBD information results in estimation (b) which suggest poses where cubes appear to be floating above the table. After applying the simulation physics, scene (c) is generated, correcting the initial erroneous perceptions.

4.2 Self Calibration

As a result of model-based perception, there is a residual error computed between the estimated pose and the model corrected pose. This error can be systematic and derived from an incomplete calibration of the robot's sensors. Most frequently, the cause can be a miss alignment of the sensing device in the kinematic chain, i.e. RGBD camera or LIDAR, since this are usually hard coded by a human. In this case, the computed error can be used to re-calibrate the sensor by computing the corresponding derivatives.

Let c_i^r be the real world pose of cube i, p^r the camera's pose relative to its parent frame and c_i^e the pose for cube i estimated by the system. Using C^e (the set of all c_i^e) the simulation is synchronized, and after its physics are applied, c_i^s is obtained, as the corrected pose for cube i. In the following experiment, we aim to find a value for p^r that minimizes the distance between C^e and C^s, using the simulator's corrections as a way to approximate C^r. The error function used is the average across all detected cubes of the distance described in Eq. 1, and Scipy's implementation of the Powell minimization method is executed.

$$dist(c_i, c_j) = \langle c_i^{rot}, c_j^{rot} \rangle^2 + ||c_i^{trans} - c_j^{trans}|| \quad (1)$$

Recalibration can come as a consequence of two distinct scenarios: (i) A well calibrated system which suffered some change (i.e. the sensor moved unexpectedly) or (ii) The initial sensor position is erroneous, and the correct one has to be found. We will explore the first situation here, but the same process can be applied for (ii).

Every time a new cube is detected, it is placed inside the simulation. In this scenario, the robot re encounters previously placed cubes, but the discrepancy between C^e and C^s (calculated with the error function) is too big. Many events could explain this difference, but a recalibration is triggered nonetheless to evaluate if a change in the camera pose can eliminate the error. If a value for p^r that aligns C^e to C^s is found, it is taken as the new camera position from this point on, since it is more likely to be the case than a coordinated reposition of all cubes in the scene.

To evaluate this functionality, a simple scene is created, similar to the one presented in Fig 4. After the cubes are detected and placed inside the simulation, the kinematic chain to the camera (p^r) is corrupted, in particular, the spatial transformation from the arm's tip to the camera is substituted by a random one. This generates the desired effect, where all cubes remain visible but the difference between detected (C^e) and remembered (C^s) positions is bigger than expected. For a setup of 4 cubes and a top view, 50 trials were conducted. Figure 5 presents how the error evolved during the optimization process. This graph shows that the method is capable of finding low error values for p^r even when starting from highly erroneous guesses. This convergence does not assure that p^r is in fact the real pose, but results show that the variance across trials is low and close to the measured position of the camera (Fig. 6).

4.3 Occlusion

When the robot ceases to see an object that is being tracked, its existence and continuity in the WM has to be updated by internal reasoning processes. As a representative example of occlusion, a recent experiment [12] shows a human holding a ball over two boxes and dropping it inside one of them. Then, the boxes are interchanged, and the system is asked for the location of the ball. For an RCA with an embedded simulator, the answer to the query is rather simple if we let gravity and collision detection algorithms operate on the free ball.

Once the system inserts any object in the simulator, its position is always reported in the WM as a Virtual RT. Unseen interactions are automatically computed, and the position information is available at any time to any agent that needs it.

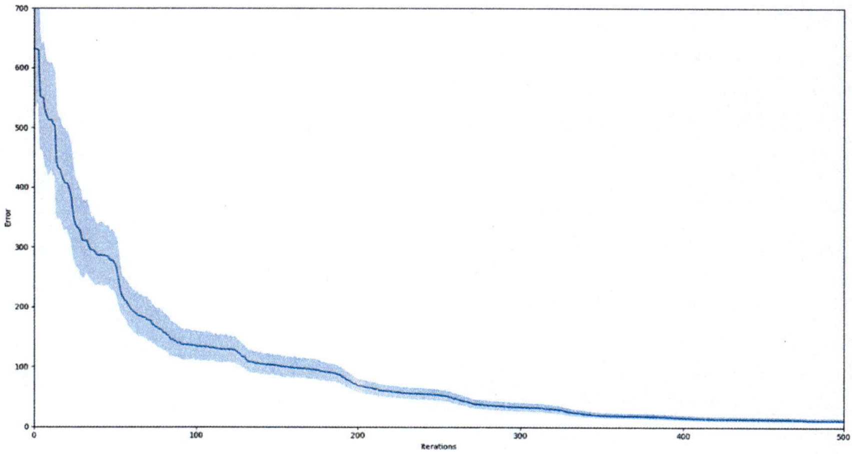

Fig. 5. Results of the recalibration after an undesired change in camera position. The blue line presents the mean across trials with the 95% confidence interval, showing how the system can find a low error solution despite initiating from highly erroneous guesses.

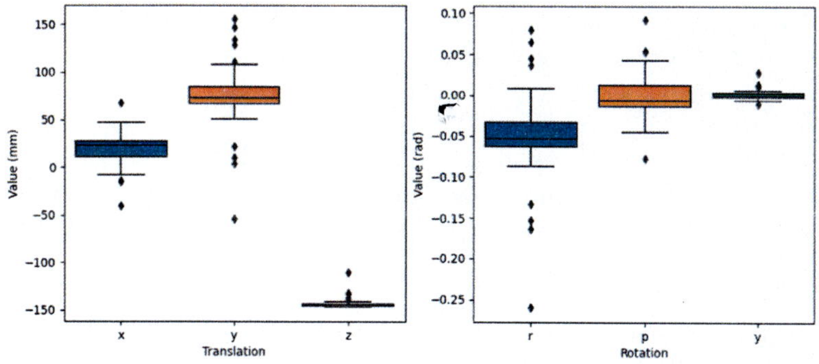

Fig. 6. Values for translation (x, y, z) and rotation (raw, pitch, yaw) from the arm tip to the camera found across trials.

In their experiments, Sallami et al. [12] detect human intervention with a state machine. When an object is presented in a configuration that violates strongly the effect of gravity, they conclude that it is being held by someone. Instead, we take advantage of the geometric reasoning provided by our setup, and use it for grasp detection. A new agent has been developed which detects human hands using the MediaPipe Hands algorithm [8] in combination with depth information, and places finger positions in the working memory. The *simulation-handler* agent then inserts the fingertips into the simulation as simple spheres. While one or more of these spheres is colliding with an object, gravity is not applied to it. Grasping and touching are not differentiated by the algorithm. This is one example of how having a physical duplicate of reality can endow the system with simple ways of detecting complex interactions, such as contact and gravity, that would be hard to describe with the use of rules or ad hoc algorithms.

Using these elements, we performed a similar experiment replicating the famous game of cups and balls (Fig. 7). The robot is presented with a cube and two boxes (Fig. 7a). One box is placed over the cube (Fig. 7b), and then the boxes are interchanged (Fig. 7d). At one point, the box containing the cube is moved over the edge of the table, dropping it to the floor (Fig. 7e). Thanks to the synchronized simulator, the system can detect this situation and continues to correctly track the cube position even though it is out of sight almost all the time. Being able to determine where the cube is without using extensive sets of rules or some case specific heuristics is a simple example of the advantages of our approach. Using internal simulation, this information is readily accessible to the robot.

5 Conclusions and Future Work

In this paper, we have presented some experiments performed with a Robotics Cognitive Architecture embedded with a Physics simulator. The goal has been to explore and test different uses of the new simulation capability, each of one enhancing some desirable aspects of the architecture. In the three examples, the underlying mechanism is the same, namely injecting back virtual RT edges with poses corrected through the filter of the Physics engine. This preliminary work is the first step in a long-term research focused on expanding the physical reasoning capabilities of autonomous robots by adding geometrical common-sense (GCS).

According to the distributed and reactive nature of CORTEX, the virtual RT edges are injected asynchronously by the simulator into the working memory. Instead of requiring explicit cues to access this geometric knowledge, our system is driven by the activity in the WM, and reacts to it by providing alternative RT edges. This line of though will be exploited in the future, when other common-sense agents handling declarative knowledge or episodic memories are developed.

In future work, we will expand the capabilities of the simulation by adopting several simplifying shortcuts that make the simulation tractable in real time for many objects [21]. This is needed to deal with more realistic situations. Besides, simplification allows for the deployment of multiple simulations and the use of a probabilistic approach to learning that can lead to unsupervised training of the robotic arm in cluttered situations. This in turn will ease the combination of probabilistic and neural network learning, an approach that has been successfully applied to other domains, such as program induction [22].

Another promising line of research targets the interpretation of human activities based on a combination of retargeting the perceived body into the robot's geometry, and the recovery of previous sensorimotor episodes in similar situations, i.e. grasping a box. When using a synchronized simulation where the human hand is retargeted into the robot's gripper, the physics engine can reproduce the most similar gripper configuration in which the object is held, and the

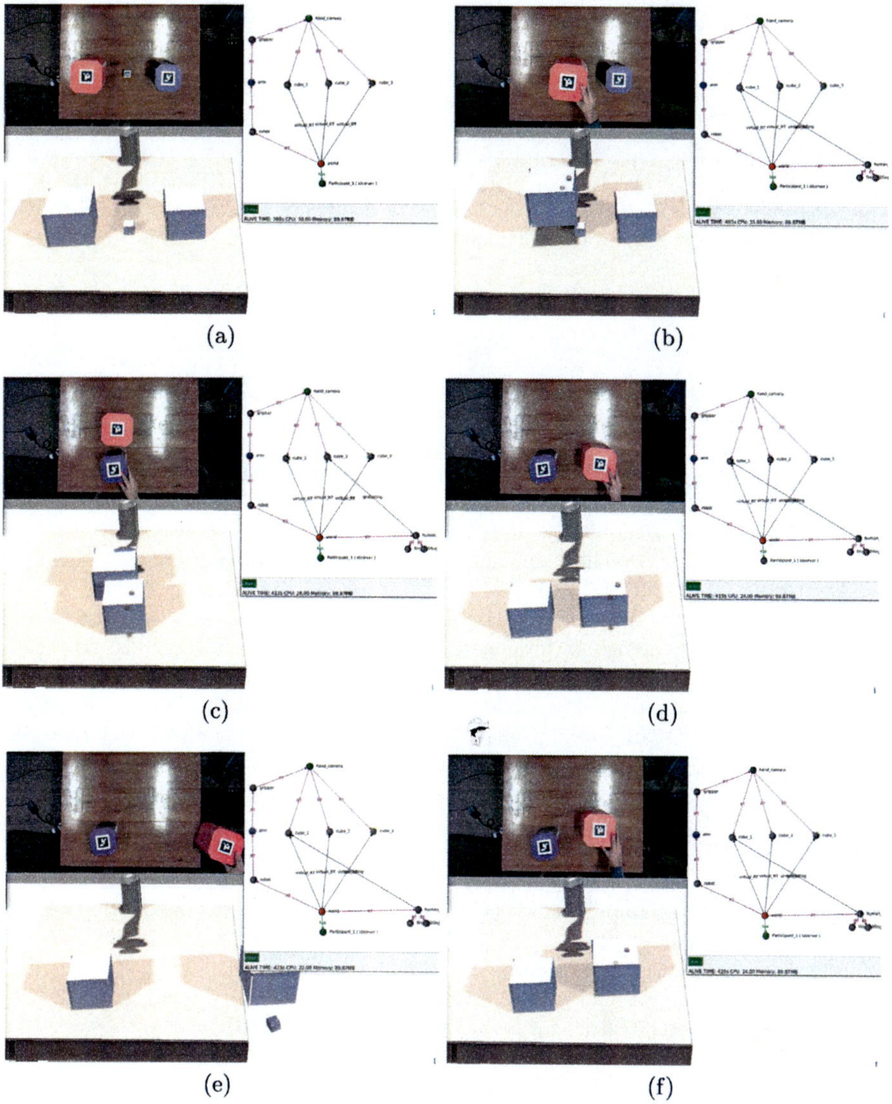

Fig. 7. Cups and balls experiment. (a) Shows the initial configuration. When a human hand is detected for the first time, a new node is inserted and two spheres representing fingertips placed in the simulation (b). Grasp detection can be seen in (c) and (d) where the WM has a *grasping* edge from the hand to the box being manipulated. In (e) the box containing the cube is placed over the edge and the simulator shows the cube falling to the ground.

gripper's force sensors will generate the corresponding response to the suspending effort. On one side, the predicate *hold(gripper, object)* can be evaluated now as a simple function of relative distances and velocities; on the other, we may say that the robot has a grounded interpretation of the current situation, that would enable it to make a description, predict future outcomes or react according to a purpose.

Acknowledgments. This work has been partially supported by the Feder funds and by the Extremaduran Goverment (projects GR21018 and IB18056), the MICINN RTI2018-099522-B-C42, by the Feder project 0770_EuroAGE2_4_E (Interreg V-A Portugal-Spain - POCTEP), and CSIC and CAP from Universidad de la República.

References

1. Ullman, T.D., Spelke, E., Battaglia, P., Tenenbaum, J.B.: Mind games: game engines as an architecture for intuitive physics. Trends Cogn. Sci. **21**(9), 649–665 (2017)
2. Kubricht, J.R., Holyoak, K.J., Lu, H.: Intuitive physics: current research and controversies. Trends Cogn. Sci. **21**(10), 749–759 (2017), ISSN 1364-6613. https://doi.org/10.1016/j.tics.2017.06.002
3. Davis, E., Marcus, G.: The scope and limits of simulation in automated reasoning. Artif. Intell. **233**, 60–72 (2016)
4. Patrick, M., et al.: Imagination-enabled robot perception. In: 2021 IEEE/RSJ International Conference on Intelligent Robots and Systems (IROS). IEEE (2020)
5. Battaglia, P.W., Hamrick, J.B., Tenenbaum, J.B.: Simulation as an engine of physical scene understanding. Proc. Natl. Acad. Sci. **110**(45), 18327–18332 (2013)
6. Garcia, J.C.: G: a low-latency, shared-graph for robotics cognitive architectures. Master Thesis, University of Extremadura (2021)
7. Núñez, P., García, J.C., Bustos, P., Bsahciller, P.: Towards the design of efficient and versatile cognitive robotic architecture based on distributed, low-latency working memory. International Conference in Advanced Robotics and Competitions, Santa Maria da Feira, Portugal (2022)
8. Fan, Z., et al.: Mediapipe hands: On-device real-time hand tracking. arXiv preprint arXiv:2006.10214 (2020)
9. Bustos, P., Luis, M.-A., Bandera, A., Bandera, J.P., García-Varea, I., Martínez-Gómez J.: EUCognition Meeting - Cognitive Robot Architectures, CORTEX: a new Cognitive Architecture for Social Robots, Viena (2016)
10. Bustos, P., Manso, L., Bandera, A., Bandera, J., García-Varea, Martín-Gomez, J.: The cortex cognitive robotics architecture: use cases. Cogn. Syst. Res. **55**, 107–123 (2019)
11. Bustos, P., et al.: DSRD: a proposal for a low-latency, distributed working memory for CORTEX. In: Bergasa, L.M., Ocaña, M., Barea, R., López-Guillén, E., Revenga, P. (eds.) Advances in Physical Agents II. AISC, vol. 1285, pp. 109-122. Springer, Cham (2021). ISBN 978-3-030-62579-5. https://doi.org/10.1007/978-3-030-62579-5_8
12. Sallami, Y., Lemaignan, S., Clodic, A., Alami, R.: Simulation-based physics reasoning for consistent scene estimation in an HRI context. In: IEEE International Conference on Intelligent Robots and Systems, pp. 7834–7841 (2019)
13. Mosenlechner, L., Beetz, M.: Fast temporal projection using accurate physics-based geometric reasoning. In: IEEE International Conference on Robotics and Automation, pp. 1821–1827 (2013)

14. Rohmer, E., Singh, S.P.N., Freese, M.: CoppeliaSim (formerly V-REP): a Versatile and Scalable Robot Simulation Framework. IEEE/RSJ International Conference on Intelligent Robots and Systems (2013). www.coppeliarobotics.com
15. Olson, E.: AprilTag: a robust and flexible visual fiducial system. In: IEEE International Conference on Robotics and Automation (2011). https://doi.org/10.1109/ICRA.2011.5979561
16. Hesslow, G.: Conscious thought as simulation of behaviour and perception. Trends Cogn. Sci. **6**, 242–247 (2002)
17. Michael, B., et al.: Know rob 2.0-a 2nd generation knowledge processing framework for cognition-enabled robotic agents. In: 2018 IEEE International Conference on Robotics and Automation (ICRA). IEEE (2018)
18. Wintermute, S.: Integrating action and reasoning through simulation. In: Proceedings of the 2nd Conference on Artificial General Intelligence, pp. 102–107 (2009). https://doi.org/10.2991/agi.2009.24. (Atlantis Press)
19. Ziemke, T., Jirenhed, D.A., Hesslow, G.: Internal simulation of perception: a minimal neuro-robotic model. Neurocomputing **68**, 85–104 (2005)
20. Jirenhed, D.-A., Hesslow, G., Ziemke, T.: Exploring internal simulation of perception in mobile robots. In: Arras, K., Baerveldt, A.-J., Balkenius, C., Burgard, W., Siegwart, R. (eds.) 2001 Fourth European Workshop on Advanced Mobile Robotics-Proceedings, Lund University Cognitive Studies, vol. 86, pp. 107–113. Lund, Sweden (2001)
21. Bass, I., Smith, K.A., Bonawitz, E., Ullman, T.D.: Partial mental simulation explains fallacies in physical reasoning. Cogn. Neuropsychol. (2022). https://doi.org/10.1080/02643294.2022.2083950
22. Ellis, K., et al.: Tenenbaum, DreamCoder: growing generalizable, interpretable knowledge with wake-sleep Bayesian program learning (2020). arXiv:2006.08381, https://doi.org/10.48550/ARXIV.2006.08381
23. Laird, J.E., Lebiere, C., Rosenbloom, P.S.: A standard model of the mind: toward a common computational framework across artificial intelligence, cognitive science, neuroscience, and robotics. Ai Magazine **38**(4), 13–26 (2017)

Author Index

D. Tardioli et al. (Eds.): ROBOT 2022, LNNS 590, pp. 625–627, 2023.
https://doi.org/10.1007/978-3-031-21062-4